ANALYSIS, MANIFOLDS AND PHYSICS
PART I: BASICS

ANALYSIS, MANIFOLDS AND PHYSICS

Part I: Basics

by

YVONNE CHOQUET-BRUHAT
Membre de l'Académie des Sciences,
Université de Paris VI, Département de Mécanique
Paris, France

CÉCILE DEWITT-MORETTE
University of Texas, Department of Physics
and
Center for Relativity, Austin, Texas, USA

with

MARGARET DILLARD-BLEICK

REVISED EDITION

ELSEVIER
AMSTERDAM · LAUSANNE · NEW YORK · OXFORD · SHANNON · TOKYO

ELSEVIER SCIENCE B.V.
Sara Burgerhartstraat 25
P.O. Box 211, 1000 AE Amsterdam, The Netherlands

First edition 1977
Revised edition April 1982
Reprinted 1982, 1983, 1985 (with minor corrections), 1987, 1989 (with minor corrections), 1991, 1996 (with minor corrections)

Library of Congress Cataloging in Publication Data

Choquet-Bruhat, Yvonne.
Analysis, manifolds, and physics.

Bibliography: p.
Includes index.
1. Mathematical analysis. 2. Manifolds (Mathematics) 3. Mathematical physics. I. De Witt-Morette, Cécile. II. Dillard-Bleick, Margaret. III. Title
QC20.7.A5C48 1981 515 81-11296
ISBN 0-444-86017-7 AACR2

ISBN 0-444-86017-7

This book is printed on acid-free paper

Printed and bound by Antony Rowe Ltd, Eastbourne

INTRODUCTION

All too often in physics familiarity is a substitute for understanding, and the beginner who lacks familiarity wonders which is at fault: physics or himself. Physical mathematics provides well defined concepts and techniques for the study of physical systems. It is more than mathematical techniques used in the solution of problems which have already been formulated; it helps in the very formulation of the laws of physical systems and brings a better understanding of physics. Thus physical mathematics includes mathematics which gives promise of being useful in our analysis of physical phenomena. Attempts to use mathematics for this purpose may fail because the mathematical tool is too crude; physics may then indicate along which lines it should be sharpened. In fact, the analysis of physical systems has spurred many a new mathematical development.

Considerations of relevance to physics underlie the choice of material included here. Any choice is necessarily arbitrary; we included first the topics which we enjoy most but we soon recognized that instant gratification is a short range criterion. We then included material which can be appreciated only after a great deal of intellectual asceticism but which may be farther reaching. Finally, this book gathers the starting points of some great currents of contemporary mathematics. It is intended for an advanced physical mathematics course.

Chapters I and II are two preliminary chapters included here to spare the reader the task of looking up in several specialized books the definitions and the basic theorems used in the subsequent chapters. Chapter I is merely a review of fundamental notions of algebra, topology, integration, and analysis. Chapter II treats the essentials of differential calculus and calculus of variations on Banach spaces. Each of the following chapters introduces a mathematical structure and exploits it until it is sufficiently familiar to become an "instrument de pensée"[1]: Chapter III, differentiable manifolds, tangent bundles and their use in Lie groups; Chapter IV, exterior derivation and the solutions of exterior differential systems; Chapter V, Riemannian structures which, together with the previous structures provide the basic geometric notions needed in physics; Chapter VI, distributions and the Sobolev spaces with recent applications to the theory of partial differential equations. The last

[1]"La Mathématique... instrument de pensée au service de notre effort d'intelligence du monde physique", A. Lichnerowicz.

chapter covers some selected topics in the theory of infinite dimensional manifolds.

At the end of each chapter, several problems are worked out. Most of them show how the concepts and the theorems introduced in the text can be used in physics. They should be of interest both to physicists and mathematicians. A sentence like "The Lagrangian is a function defined on the tangent bundle of the configuration space" helps explain to the physicist what a tangent bundle is and tells a mathematician what a Lagrangian is. A sentence like "The strain tensor is the Lie derivative of the metric with respect to the deformation" helps a physicist to understand the concept of Lie derivatives and defines the strain tensor to a mathematician. To both, they bring an added pleasure.

The pleasure of physical mathematics is well described by Hilbert: learning that some genetic laws of the fruit fly had been derived by the application of a certain set of axioms he exclaimed "So simple and precise and at the same time so miraculous that no daring fantasy could have imagined it"[2].

PREFACE TO THE SECOND EDITION

We are happy that the success of the first edition gave us a chance to prepare a revised edition. We have made numerous changes and added exercises with their solutions to ease the study of several chapters. The major addition is a chapter "Connections on principal fibre bundles" which includes sections on holonomy, characteristic classes, invariant curvature integrals and problems on the geometry of gauge fields, monopoles, instantons, spin structure and spin connections. Other additions include a section on the second fundamental form, a section on almost complex and kählerian manifolds, and a problem on the method of stationary phase. More than 150 entries have been added to the index.

Can this book, now polished by usage, serve as a text for an advanced physical mathematics course? This question raises another question: What is the function of a text book for graduate studies? In our times of rapidly expanding knowledge, a teacher looks for a book which will provide a broader base for future developments than can be covered in one or two semesters of lectures and a student hopes that his purchase will serve him for many years. A reference book which can be used as a text is an answer to their needs. This is what this book is intended to be, and thanks to a publishing company which keeps it moderately priced, this is what we hope it will be.

[2]"Hilbert" Constance Reid, Springer Verlag, 1970.

ACKNOWLEDGEMENTS

It is a pleasure to thank R. Bott, C. and S. Dicrolf, E.E. Fairchild Jr., C. Isham, C. Marle, J.E. Marsden, A.H. Taub, J.W. Vick, E. Weimar, for having read critically parts of the book. Suggestions and criticisms from readers of the first edition have been very useful. We wish to thank K.D. Elworthy, T. Jacobson, T. Piran, G. Sammelmann, B. Simon, W. Thirring and A. Weinstein, and particularly G. Grunberg who has read critically most of the book. We thank D. Goudmand, M.P. Serot Almeras Latour and L. Ferris for a painstaking preparation of the manuscript and the staff of North-Holland for an excellent collaboration. This work has been supported in part by an NSF grant (PHY77-07619 A02) and a NATO Research Grant. Several Centers have provided excellent working conditions, in particular the Institut des Hautes Etudes Scientifiques (Bures sur Yvette), the CNRS Centre de Physique Théorique (Marseilles) and the Institute for Theoretical Physics (Santa Barbara) where much of the revised edition has been written.

We regret that one of us who is now engaged in other pursuits could not contribute to the revised edition. The core of the book remains nevertheless under the trademark of our collaboration.

Y. Choquet-Bruhat
C. DeWitt-Morette

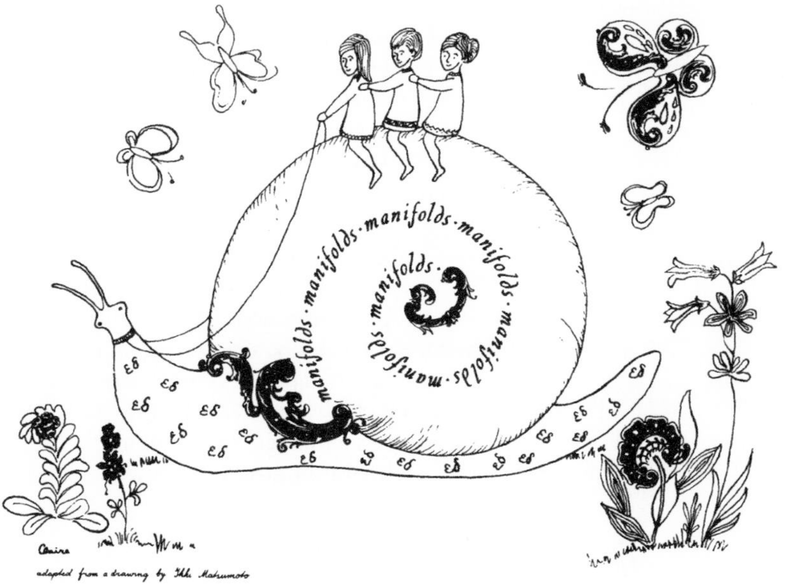

ACKNOWLEDGEMENTS

It is a pleasure to thank R. Bott, C. and S. Darrot, S.F. Fulcher Jr., C. Green, C. Mack, J.L. Marsden, A.H. Taub, J.A. Wolf, P. Wortman for having read only parts of the book. Suggestions and corrections from readers of the first edition have been very useful. We wish to thank K.D. Elworthy, B. Jacobson, F. Pirani, G. Sarzarinis, J. Sanan, W. Tulczyjew and A. Weinstein, and particularly D. Greenberg who has read critically most of the book. We thank D. Cassiannut, M.P. Sorse, Ninette Latour and L. Borns for a painstaking preparation of the manuscript and the staff of North-Holland for an excellent collaboration. This work has been supported in part by an NSF grant (PHY 77-07619 A02) and a NATO Research Grant. Several Centers have provided excellent working conditions, in particular the Institut des Hautes Etudes Scientifiques (Bures-sur-Yvette), the CNRS Centre de Physique Théorique (Marseilles) and the Institute for Theoretical Physics (Santa Barbara) where much of the revised edition has been written.

We regret that one of us who is now engaged in other pursuits could not contribute to the revised edition. The core of the book remains nevertheless under the trademark of our collaboration.

Y. Choquet-Bruhat
C. DeWitt-Morette

CONTENTS

CONTENTS TO PART II: 92 APPLICATIONS

"Entia non multiplicanda praeter necessitatem": Occam

I. REVIEW OF FUNDAMENTAL NOTIONS OF ANALYSIS

A. SET THEORY, DEFINITIONS

1. SETS

We recall below some fundamental notations and axioms of standard set theory[1]. The reader may start the book at Chapter III and come back to this chapter—essentially a dictionary—if he needs it.

Mathematical logic is a complicated subject and will not be discussed here. But we warn the reader that any "collection" is not necessarily a set. Some axioms and properties given for sets are true only for sets. The collection of objects satisfying a property P, for instance the collection of all groups, is not a set. But, if X is a set, the collection of members of X satisfying a property P is a set. The collection of subsets of X is a set.

In the following chapters we will use the words "collection, family, system", loosely, often as synonyms for "set", but not necessarily.

Notation: $x \in X$: x is a member or element of the set X.
$x \notin X$: x is not a member of the set X.

The **null (empty, void) set** is the set with no members, denoted $\emptyset$. — null, empty, void set

If every element of a set X is also an element of a set Y, then X is a **subset** of Y: $X \subset Y$, or $Y \supset X$. $X = Y$ means that $X \subset Y$ and $Y \subset X$. — subset

The subset of X consisting of all $x \in X$ possessing the property P is denoted $\{x; P\}$.

Let A be a subset of a set X. The function — characteristic function

$$\chi_A(x) = 1 \qquad x \in A$$
$$\chi_A(x) = 0 \qquad x \in X \qquad x \notin A$$

is called the **characteristic function** of A.

[1] See for instance [Halmos 1967].

proper — A **proper** subset X of the set Y is a subset which is not equal to Y. We shall denote it $X \subset Y$, $X \neq Y$, or sometimes $X \subsetneq Y$.

union — The **union** of the sets X and Y, denoted $X \cup Y$, is the set of all elements which belong either to X or to Y.

intersection — The **intersection** of X and Y, denoted $X \cap Y$, is the set of all elements which belong to both X and Y.

disjoint — Two sets are **disjoint** if their intersection is void. A class of sets is disjoint if every pair of distinct sets in the class is disjoint.

complement $\complement A$ $X\backslash A$ — Let A be a subset of the set X, the **complement** of A relative to X is the totality of points $x \in X$ which are not contained in A. It is denoted $\complement A$ or $X\backslash A$. Notice that the symbol $\backslash$ is not the quotient symbol $/$.

difference — Let A and B be two subsets of the set X, the **difference** $A - B$ is the totality of points $x \in A$ which are not contained in B: $A - B = A \cap \complement B$.

symmetric difference — The **symmetric difference** of A and B is $A \triangle B = (A - B) \cup (B - A)$.

cartesian product — The **cartesian product** $X \times Y$ is the set of all ordered pairs (x, y) where $x \in X$ and $y \in Y$.

partition — A **partition** of X is a class $\{X_i\}$ of non-empty disjoint subsets of X whose union is X, where i belongs to a family I of indexes.

power set — If X has a finite number n of elements, the set of all subsets of X has 2^n elements; it includes $\emptyset$. Motivated by this fact, one denotes by 2^X the set of all subsets of X and calls it the **power set**.

2. MAPPINGS

mappings — *Notation*: $f: X \to Y$ by $x \mapsto y = f(x)$, also written $y = fx$.

domain, range — The **mapping** f from X into Y associates with every element $x \in X$ a uniquely determined element $y = f(x) \in Y$. The **domain** of f, denoted $D(f)$, is X. Its **range**, denoted $R(f)$, is contained in Y.

function — A mapping of X into Y is also called a **function** on X with values in Y. A function on X with values in $\mathbb{R}$ (real line) or in $\mathbb{C}$ (complex plane) is also

numerical function — called a **numerical function** on X, often abbreviated in this book to "function on X".

composite — The **composite mapping** of the two mappings $f: X \to Y$ and $g: Y \to Z$ is the mapping $g \circ f: X \to Z$ by $x \mapsto g(f(x))$.

image — The symbol $f(M)$ denotes the subset $\{f(x); x \in M\}$ of Y; $f(M)$ is the **image** of M under the mapping f.

inverse image — The symbol $f^{-1}(N)$ denotes the subset $\{x; f(x) \in N\}$ of X; $f^{-1}(N)$ is the **inverse image** of N under the mapping f. It has the properties

$$Y_1 = f(f^{-1}(Y_1)) \quad \text{for all } Y_1 \subset f(X)$$
$$X_1 \subset f^{-1}(f(X_1)) \quad \text{for all } X_1 \subset X.$$

If for every $y \in f(X)$ there is only one $x \in X$ such that $f(x) = y$ then f has an **inverse mapping** f^{-1} and is said to be **one–one (injective)**,

inverse
injective
one–one

$$f^{-1}: f(X) \to X \quad \text{by} \quad y \mapsto x = f^{-1}(y).$$

The mapping f is said to map X **onto** Y (is **surjective**) if $f(X) = Y$. The mapping f is a **bijection** if it is one–one and onto.

surjective
onto
bijection

Example: Consider the following mappings from $\mathbb{R}$ into $\mathbb{R}$:

$x \mapsto x^3$ is a bijection,

$x \mapsto \exp(x)$ is one–one but not onto,

$x \mapsto x^3 + x^2$ is onto but not one–one,

$x \mapsto \sin x$ is neither onto nor one–one.

The mapping f is an **extension** of g (g is a **restriction** of f) if $D(f)$ contains $D(g)$ and $f(x) = g(x)$ for every x in $D(g)$. The restriction of f to a domain U is denoted $f|_U$.

extension
restriction
$f|_U$

A **sequence** (A_n) of subsets of a set X is a mapping from the set of natural numbers $\mathbb{N}$ into 2^X by $n \mapsto A_n$.

sequence

A set X is **countably infinite** if there is a bijection between X and the set of natural numbers $\mathbb{N}$. A set is **countable** if it is either finite or countably infinite.

countable
set

More generally two sets are called **equipotent** (have the same **cardinality**) if there is a bijection between them.

equipotent
cardinality

A **category** is a class[1], whose members are called **objects**, together with sets **Hom(*X*, *Y*)**, one for each ordered pair of objects (X, Y), and mappings

category
objects
Hom

$$\mathrm{Hom}(X, Y) \times \mathrm{Hom}(Y, V) \to \mathrm{Hom}(X, V) \quad \text{by} \quad (f, g) \mapsto h = g \circ f$$

such that

1) If $f \in \mathrm{Hom}(X, Y)$, $g \in \mathrm{Hom}(Y, V)$ and $h \in \mathrm{Hom}(V, U)$ then $h \circ (g \circ f) = (h \circ g) \circ f$.
2) For each object X there is a unique element in $\mathrm{Hom}(X, X)$, denoted Id_X, such that for all objects Y

$$f \circ \mathrm{Id}_X = f, \qquad \forall f \in \mathrm{Hom}(X, Y)$$
$$\mathrm{Id}_X \circ g = g, \qquad \forall g \in \mathrm{Hom}(Y, X).$$

The members of $\mathrm{Hom}(X, Y)$ are called **morphisms** and the operation $\circ$ is called **composition.**

morphism
composition

Example: The **identity mapping** id: $X \to X$ by $x \mapsto x$ is bijective. The

identity
mapping

[1]Not usually a set.

constant mapping

constant mapping $f\colon X \to Y$ defined by $f(x) = c$ with c fixed in Y is not injective [surjective] if X [if Y] has more than one element.

isomorphism

A morphism $f \in \mathrm{Hom}(X, Y)$ is called an **isomorphism** if there exists $g \in \mathrm{Hom}(Y, X)$ such that

$$f \circ g = \mathrm{Id}_Y, \qquad g \circ f = \mathrm{Id}_X.$$

Example: The class of all sets together with the mappings between them is a category: the isomorphisms are the bijections.

covariant functor

Let $\mathscr{A}$ and $\mathscr{B}$ be two categories, $\mathrm{Mor}(\mathscr{A})$ and $\mathrm{Mor}(\mathscr{B})$ the classes of their morphisms. A **covariant functor** from $\mathscr{A}$ to $\mathscr{B}$ is a map $C\colon (\mathscr{A}, \mathrm{Mor}(\mathscr{A})) \to (\mathscr{B}, \mathrm{Mor}(\mathscr{B}))$ such that (cf. exercise p. 122)

1) if f is a morphism $X \to Y$, $C(f)$ is a morphism $C(X) \to C(Y)$,
2) if f and g are morphisms in $\mathscr{A}$ which can be composed then $C(g) \circ C(f) = C(g \circ f)$,
3) $C(\mathrm{id}_X) = \mathrm{id}_{C(X)}$.

f_*

$C(f)$ is then often denoted f_*.
For a covariant functor from $\mathscr{A}$ to $\mathscr{B}$ we have the diagram

$$X \xrightarrow{f} Y \xrightarrow{g} Z, \qquad X, Y, Z \in \mathscr{A}$$

$$C(X) \xrightarrow{C(f)} C(Y) \xrightarrow{C(g)} C(Z), \qquad C(X), C(Y), C(Z) \in \mathscr{B}.$$

contravariant functor

A mapping $C\colon (\mathscr{A}, \mathrm{Mor}(\mathscr{A})) \to (\mathscr{B}, \mathrm{Mor}(\mathscr{B}))$ is called a **contravariant functor** from $\mathscr{A}$ to $\mathscr{B}$ if

$$X \xrightarrow{f} Y \xrightarrow{g} Z, \qquad X, Y, Z \in \mathscr{A}$$

$$C(X) \xleftarrow{C(f)} C(Y) \xleftarrow{C(g)} C(Z), \qquad C(X), C(Y), C(Z) \in \mathscr{B}$$

and

$$C(f) \circ C(g) = C(g \circ f),$$
$$C(\mathrm{id}_X) = \mathrm{id}_{C(X)}.$$

f^*

$C(f)$ is then often denoted f^*.

We will deal later with other categories whose objects are sets with additional structure and whose morphisms are mappings preserving this structure: for instance the class of all groups with their homomorphisms. An isomorphism will then be a bijective morphism whose inverse is also a morphism. However, since we will often have to consider several structures simultaneously we will in general use a more explicit terminology, and reserve the word isomorphism for the most basic category within the scope of this book, namely, the category whose objects are the topological vector spaces, and whose morphisms are the continuous linear maps (see p. 22).

3. RELATIONS

A **relation** between X and Y is a subset R of $X \times Y$. x and y are said to be R-related if $(x, y) \in R$, also written xRy. relations

A relation $R \in X \times X$ is an **equivalence relation** in X if it is equivalence relation

$$\begin{array}{lll} \text{reflexive:} & (x, x) \in R & \forall x \in X, \\ \text{symmetric:} & (x, y) \in R \Rightarrow (y, x) \in R & \forall x, y \in X, \\ \text{transitive:} & (x, y) \in R \text{ and } (y, z) \in R \Rightarrow (x, z) \in R, & \forall x, y, z \in X. \end{array}$$

If x is related to y by an equivalence relation R, we write $x \sim y$ or $x = y \pmod R$.

The **equivalence class** of $x \in X$, denoted $[x]$, is the subset of X $\{y ; y \sim x\}$. X is the disjoint union of equivalence classes. The set of these equivalence classes is denoted X/R. equivalence class

Example: Make an equivalence class with points $x \in \mathbb{R}$ by the equivalence relation: $x \sim x_0$ iff $x - x_0 = \pm 2\pi n$, $n = 0, 1, 2, 3 \ldots$. $[x_0] = \{x : x \pm 2\pi n = x_0\}$. One of these equivalence classes $[x]$ may be considered as a point on a circle.

4. ORDERINGS

A relation $R \subset X \times X$ is a **partial ordering** in X if partial ordering

$$\begin{array}{ll} (x, x) \in R \quad \forall x \in X & \text{(reflexivity)}, \\ (x, y) \in R \text{ and } (y, x) \in R \Rightarrow x = y & \text{(antisymmetry)}, \\ (x, y) \in R \text{ and } (y, z) \in R \Rightarrow (x, z) \in R & \text{(transitivity)}. \end{array}$$

Notation: $x \leq y$ (x precedes y or y follows x): x is related to y by the partial ordering $\leq$.

Let P be a partially ordered set; let $a, b, c \in P$; if $a \leq c$ and $b \leq c$, c is said to be an **upper bound** for a and b. If furthermore $c \leq d$ for all $d \in P$ which are upper bounds for a and b, then c is said to be the **least upper bound**, or the **supremum**, of a and b, $c = \sup(a, b)$. upper bound least upper bound supremum

Similarly, one defines a **lower bound** and the **greatest lower bound**, or the **infimum**, of a and b and denotes it $\inf(a, b)$. lower bound infimum

A partially ordered set P is **directed** if every pair of elements of P has an upper bound. directed set

If $\sup(a, b)$ and $\inf(a, b)$ exist for every pair (a, b), P is a **lattice**. lattice maximal

Let $m, p \in P$; m is **maximal** if $m \leq p$ implies $m = p$, that is no element other than m itself follows m.

linearly ordered

A partially ordered set P is **linearly ordered**, or **totally ordered**, if for any two elements $a, b \in P$ either $a \leq b$ or $b \leq a$.

Zermelo's axiom of choice. Given any non empty set of disjoint non empty sets X_i, $i \in I$, a set can be formed which contains precisely one element x_i taken from each set X_i.

Zorn's lemma. Let P be a non empty partially ordered set such that every linearly ordered subset of P has an upper bound in P. Then for each $x \in P$ there exists a maximal element $y \in P$ such that $x \leq y$. Zorn's lemma is equivalent to Zermelo's axiom of choice.

Example 1: Let P be the set $\mathbb{N}$ of all positive integers, let $m \leq n$ mean that m divides n:
sup (m, n) = least common multiple of m and n,
inf (m, n) = greatest common divisor of m and n,
$\mathbb{N}$ has no maximal element.

Example 2: Let P be the set $\mathbb{R}$ of all real numbers, let $x \leq y$ have its usual meaning:
$\sup(x, y) = \max\{x, y\}$, $\inf(x, y) = \min\{x, y\}$,
$\mathbb{R}$ has no maximal element.

Example 3: Let P be the class of all subsets of some given set U, let $A \leq B$ mean $A \subset B$: $\sup(A, B) = A \cup B$, $\inf(A, B) = A \cap B$. P has a single maximal element, the set U.

Example 4: Let P be the set of all real functions defined on a non-empty set X, let $f \leq g$ mean $f(x) \leq g(x)$ $\forall x \in X$.
$\sup(f, g)$ is the real function defined by
$\sup(f, g)(x) = \max\{f(x), g(x)\}$,
$\inf(f, g)$ is the real function defined by
$\inf(f, g)(x) = \min\{f(x), g(x)\}$,
P has no maximal element.

B. ALGEBRAIC STRUCTURES, DEFINITIONS

internal operation
external operation
operators

An **internal operation** on a set X is a mapping from $X \times X$ into X. Let A and X be sets. A mapping from $A \times X$ into X is called an **external operation** on X. The elements of A are called **operators** on X.

1. GROUPS

A **group** is a set X together with an internal operation $X \times X \to X$ by $(x, y) \mapsto xy$ such that

group

1) the operation is associative

$$(xy)z = x(yz) \text{ for all } x, y, z \in X,$$

2) there is an element $e \in X$ called the identity such that

$$xe = ex = x \text{ for all } x \in X,$$

3) for each $x \in X$ there is an element of X called the inverse of x, written x^{-1}, such that $x^{-1}x = xx^{-1} = e$.

The group operation is often called multiplication.
Groups form a category whose morphisms, usually called **homomorphisms**, are the mappings that preserve the group structure, i.e.

homomorphism

$$f \in \text{Hom}(G_1, G_2) \Leftrightarrow f(g_1h_1) = f(g_1)f(h_1) \qquad \text{for every } g_1, h_1 \in G_1.$$

The group is **abelian (commutative)** if $xy = yx$ for all $x, y \in X$. The operation is then often called addition and written $(x, y) \mapsto x + y$. In this case the inverse of x is written $-x$ and the identity is denoted by 0.

abelian
commutative

Example: Let X be the set of all $n \times n$ matrices; this set together with matrix addition is an abelian group; the subset of invertible matrices together with matrix multiplication is a non-abelian group.

The **center** of a group X is the set of elements which commute with all the elements of the group.

center

A subset A of X is a **subgroup** of X if $xy \in A$ and $x^{-1} \in A$ for each $x \in A$, $y \in A$ (equivalently, $xy^{-1} \in A$).

subgroup

Let A be a subgroup of X. The set $xA = \{xa\,;\, a \in A\}$ is called a **left coset** of A in X. The set $Ax = \{ax\,;\, a \in A\}$ is called a **right coset** of A in X. A left [right] coset can be defined as the equivalence set $[x] = xA$ [the equivalence set $[x] = Ax$] determined by the equivalence relation $y \sim x$ iff $y^{-1}x \in A$ [iff $xy^{-1} \in A$].

coset

A subgroup A of X is **invariant (normal)** if $xax^{-1} \in A$ for every $a \in A$ and every $x \in X$. The right coset Ax and the left coset xA of an invariant subgroup A are identical.

invariant
normal
subgroup

The collection of all the cosets of an invariant subgroup A in X is a group called the **quotient (factor) group** of X by A; it is denoted X/A. The group operation in X/A is

quotient
factor
group

$$(xA)(yA) = (xy)A.$$

Example: The quotient group $\mathbb{T} = \mathbb{R}/\mathbb{Z}$ of the group of real numbers $\mathbb{R}$

by the group of integers $\mathbb{Z}$ is the additive group of real numbers modulo 1, it is called the **1-torus.** It is homeomorphic (p. 18) to the 1-sphere (circle).

torus

2. RINGS

ring

A **ring** is a set X together with two internal operations $(x, y) \mapsto xy$ and $(x, y) \mapsto x + y$, called respectively **multiplication** and **addition**, such that

1) X is an abelian group under addition,
2) multiplication is associative and distributive with respect to addition, that is

$$(xy)z = x(yz)$$
$$x(y+z) = xy + xz$$
$$(y+z)x = yx + zx \qquad \text{for all } x, y, z \in X.$$

ring with identity

If in addition there is an element $e \in X$ such that $ex = xe = x$ for all $x \in X$, then the ring is a **ring with identity (unity)**. If the ring is abelian under multiplication, it is called an abelian ring.

regular invertible

Let X be a ring with identity. If an element $x \in X$ has an inverse, x is said to be **regular (invertible, non singular)**.

field

A ring with identity is called a **field** if all its elements except zero (neutral element of addition) are regular.[1]

$\mathbb{K}$, $\mathbb{R}$, $\mathbb{C}$

We shall denote by $\mathbb{K}$ the fields of complex or of real numbers, $\mathbb{K} = \mathbb{C}$ or $\mathbb{R}$.

ideal

A **left ideal** I of a ring X is a subring (a subset with ring properties) of X such that

$$i \in I \Rightarrow xi \in I \qquad \forall x \in X.$$

A **right ideal** is defined similarly; an **ideal** is one which is both a right and a left ideal.

factor ring

Let I be an ideal in X; the relation $x \sim y$ iff $x - y \in I$ is an equivalence relation. The set X/I of distinct equivalence sets $[x] = \{x + i; i \in I\} = x + I$ form a ring under multiplication $([x], [y]) \to [x][y] = [xy]$ and addition $([x], [y]) \to [x] + [y] = [x + y]$. The ring X/I is called the **quotient (factor) ring** of X by I.

3. MODULES

module

A **module** X over the ring R is an abelian group X together with an

[1]A number of authors call the fields defined here division rings, and reserve the word field for a commutative division ring.

external operation, called scalar multiplication, $R \times X \to X$ by $(\alpha, x) \mapsto \alpha x$ such that

$$\alpha(x+y) = \alpha x + \alpha y$$
$$(\alpha+\beta)x = \alpha x + \beta x$$
$$(\alpha\beta)x = \alpha(\beta x)$$

for all $\alpha, \beta \in R$ and $x, y \in X$. If the ring R has an identity e then

$$ex = x \text{ for all } x \in X.$$

4. ALGEBRAS

An **algebra** A is a module over a ring R with identity together with an internal associative operation, usually called multiplication such that

algebra

1) A is a ring,
2) the external operation $(\alpha, x) \to \alpha x$ is such that

$$\alpha(xy) = (\alpha x)y = x(\alpha y).$$

Most algebras are linear spaces together with an internal operation called multiplication. The general definition given here includes such algebras as the algebra of tensor fields over the ring of functions.

5. LINEAR SPACES

A **linear space (vector space)** is a module for which the ring of operators is a field. Usually, and always in this book, the field is the field $\mathbb{K}$ of the real numbers $\mathbb{R}$ or the complex numbers $\mathbb{C}$. Elements of a linear space will be called **vectors.**

linear (vector) space

vectors

Let L and M be linear subspaces of the linear space X. If each vector $z \in X$ is uniquely expressible in the form $z = x + y$ with $x \in L$ and $y \in M$, then X is said to be the **direct (geometric) sum**, denoted $L \oplus M$, of the subspaces L and M. L and M are then called **complementary**. They are necessarily such that $L \cap M = \{0\}$.

geometric direct sum

complementary

A subset A of a linear space X is called **linearly independent** if for every non-empty finite subset $\{x_i : i = 1, 2, \ldots, n\}$ of A the relation $\Sigma_{i=1}^{n} \lambda_i x_i = 0$, $\lambda_i \in \mathbb{K}$, implies $\lambda_i = 0$ for all $i \leq n$.

linearly independent

A maximal linearly independent subset of X is called a **Hamel basis** of X; such a basis always exists by Zorn's lemma.

Hamel basis

If $\{e_i;\, i$ belonging to an arbitrary indexing set$\}$ is a Hamel basis of X, then there exists, for each $x \in X$, a finite sum $\sum_{i=1}^{n} \lambda_i e_i = x$.

Any set of linearly independent vectors is a subset of a Hamel basis; thus every linear subspace L of X has a complement (not unique!) in X.

dimension

Two Hamel bases have the same cardinality. It defines the **dimension** of the space. If the number of elements in a basis is a finite number n, then any other basis also has n elements and the space is n-dimensional. In the other cases, the dimension is infinite, countable or not; for example, most spaces whose elements are functions are infinite dimensional.

codimension

The **codimension** of a subspace L is the dimension of a complementary subspace.

convex

A subset A of a vector space is **convex** if $\lambda x + (1-\lambda)y \in A$ for all $x, y \in A$ and $0 \leq \lambda \leq 1$; a convex set is a set which contains the line segment joining any pair of its points.

affine

An **affine** subspace (affine hyperplane) of a linear space X is a set of elements of X which can be written

$$x = y + x_0, \qquad y \in L,$$

where L is a linear subspace of X and x_0 is a given point of X. Let X and Y be linear spaces over the field $\mathbb{K}$.

linear mapping

kernel

The mapping f: $X \to Y$ is a **linear mapping** if $f(\alpha x + \beta y) = \alpha f(x) + \beta f(y)$ for all $x, y \in X$ and $\alpha, \beta \in \mathbb{K}$. The **kernel** of a linear mapping f: $X \to Y$ is the set $\ker f = \{x \in X;\, f(x) = 0\}$.

Theorem. A linear mapping f is injective iff $\ker f = \{0\}$ *(the null vector).*

Theorem. The inverse of a bijective linear mapping is a linear mapping.

$L(X, Y)$

The set $L(X, Y)$ of all linear mappings from X to Y becomes a linear space over $\mathbb{K}$ when addition and scalar multiplication are defined by $(f+g)(x) = f(x) + g(x)$ and $(\lambda f)(x) = \lambda f(x)$. The notation $L(X, Y)$ will designate the set together with its natural linear space structure. The set $L(X, \mathbb{K})$ of linear functions on X is called the **algebraic dual** of X and is denoted X^*. The elements of $L(X, \mathbb{K})$ are also called **linear forms**, or **linear functionals** when X is a function space.

$L(X, \mathbb{K})$

algebraic dual

forms

functional

sesquilinear mapping

Let X be a linear space over $\mathbb{C}$ or $\mathbb{R}$. A mapping $X \times X \to \mathbb{C}$ [or $\mathbb{R}$] by $(x, y) \mapsto (x|y)$ is a **sesquilinear mapping** if

$$(x|y) = \overline{(y|x)}$$
$$(\alpha x + \beta y|z) = \alpha(x|z) + \beta(y|z)$$

where the bar means complex conjugate. Note that in physics the usual practice is to take $(\alpha x|z) = \bar{\alpha}(x, z)$.

Let $f: X \to X^*$ by $x \mapsto (x|\cdot)$. The sesquilinear mapping is **nondegenerate** if f is bijective. nondegenerate
If X is finite dimensional, this condition is equivalent to

$$(y|x) = 0 \quad \forall x \in X \Rightarrow y = 0.$$

Example 1: Let $X = \mathbb{R}^2$ and let $(y|x) = x^1 y^1$, where x is the ordered pair (x^1, x^2). This sesquilinear mapping is degenerate: $x^1 y^1 = 0 \quad \forall x$ implies $y^1 = 0$ but does not imply $y^2 = 0$, hence does not imply $y = 0$.

Example 2: Let $X = C^0(U)$, the space of continuous functions (p. 17) defined on a closed bounded interval $U \subset \mathbb{R}$; and let $(y|x) = \int_U y(t)\bar{x}(t)\, dt$. The sesquilinear mapping is degenerate; indeed f is not surjective, all linear forms on X are not of the form $(y|\cdot)$. Nevertheless $(y|x) = 0 \quad \forall x$ implies $y = 0$.

A sesquilinear mapping is **positive** if positive

$$(x|x) \geq 0 \qquad \forall x \in X.$$

A sesquilinear mapping is **strictly positive** if it is positive and if positive strictly

$$(x|x) = 0 \text{ implies } x = 0.$$

A linear space X together with a strictly positive sesquilinear mapping $X \times X \to \mathbb{C}$ (or $\mathbb{R}$) is called a **pre-Hilbert space**. The sesquilinear mapping is called the **inner product (scalar product)**. If a sesquilinear mapping is positive then we have the **Cauchy–Schwarz inequality** (p. 53) pre-Hilbert inner product scalar product Cauchy–Schwarz inequality

$$|(x|y)|^2 \leq (x|x) \cdot (y|y).$$

C. TOPOLOGY

A topological space is a set with a structure allowing for the definition of neighboring points and continuous functions; Hausdorff devised the topological axioms so as to extend the concepts of neighborhood, continuity, compactness, connectedness, to spaces more general than metric spaces which are themselves a generalization of the familiar euclidean spaces.

1. DEFINITIONS

A system $\mathcal{U}$ of subsets of a set X defines a **topology** on X if $\mathcal{U}$ contains topology

1) the null set and the set X itself,
2) the union of every one of its subsystems,
3) the intersection of every one of its finite subsystems.

open

The sets in $\mathcal{U}$ are called the **open sets** of the topological space $(X, \mathcal{U})$, often abbreviated to X.

usual topology on $\mathbb{R}$

Exercise: a) Show that the open sets of $\mathbb{R}$, defined by unions of open intervals $a < x < b$ and the null set, satisfy the above properties. This topology is called the **usual topology** on $\mathbb{R}$.
b) Show that if infinite intersections were allowed in the defining properties of a topology, the above sets would not define a topology on $\mathbb{R}$.

Solution: a) The properties 1 and 2 are obviously satisfied. To verify 3, set

$$A = \bigcup_{i \in I} A_i, \quad B = \bigcup_{j \in J} B_j, \quad A_i \text{ and } B_j \text{ open sets,}$$

$$A \cap B = \bigcup_{\substack{i \in I \\ j \in J}} (A_i \cap B_j)$$

is open since the intersection of two open intervals is either an empty set or an open interval.
b) Any point a of $\mathbb{R}$ is an infinite intersection of open intervals: for instance $(-1/n + a, 1/n + a)$, $n \in N$. On the other hand a point is not the union of open intervals $a < x < b$, and hence is not open. Actually in a topology where points are open sets every subset is open (cf. discrete topology p. 12), a property which is undesirable for many purposes.

trivial

Let X be any non-empty set and let the open sets consist of $\emptyset$ and X; this topology is called **trivial**.

discrete

Let X be any non-empty set and let the open sets consist of all subsets of X, $\emptyset$ and X included. This topology is called the **discrete** topology.

coarser

finer

If $\mathcal{U}_1$ and $\mathcal{U}_2$ are two topologies on X such that $\mathcal{U}_1 \subset \mathcal{U}_2$, $\mathcal{U}_1$ is **coarser** than $\mathcal{U}_2$, and $\mathcal{U}_2$ is **finer** than $\mathcal{U}_1$. The coarsest topology is the trivial topology, the finest is the discrete topology. The words weaker and stronger are sometimes used but their meanings are not quite standardized: in this book a weak topology on X will be coarser than a strong topology.

neighborhood

A **neighborhood** of a point x [of a set A] in X is a set $N(x)$ [a set $N(A)$] containing an open set which contains the point x [the set A].
A family of neighborhoods of x introduces a notion of "nearness to x".

Theorem. A subset $A \subset X$ is open iff it is a neighborhood of each of its points.

A point $x \in X$ is a **limit point (accumulation point)** of $A \subseteq X$ if every neighborhood $N(x)$ of x contains at least one point $a \in A$ different from $x: (N(x) - \{x\}) \cap A \neq \emptyset, \forall N(x)$.
$A \subset X$ is **closed** if $A' = X \backslash A$ is open.

accumulation limit point

closed sets

Theorem. A is closed iff it contains all its limit points.

The **closure** $\bar{A}$ of A in X is the union of A and all its limit points; it is the smallest closed set containing A.

closure

The **support** of a function f, denoted supp f, is the smallest closed set outside which f vanishes identically. Note that supp f is not the set $\{x; f(x) \neq 0\}$ but is the closure of this set. For instance let $f: \mathbb{R} \to \mathbb{R}$ by $x \mapsto x$; supp f is $\mathbb{R}$ and not $\mathbb{R} - \{0\}$.
The **interior** of a set A is the largest open set contained in A. The set A is **dense** in X if $\bar{A} = X$.

support supp

interior dense

Example: The set of rational numbers is dense in the set of real numbers.

A is **nowhere dense** in X if $\bar{A}$ has an empty interior.
A space is **separable** if it has a countable dense subset.

nowhere dense separable

2. SEPARATION

The following separation conditions describe the fineness of a topology so that one knows if the supply of open sets is rich enough for a given purpose.
A topological space is **Hausdorff (separated)** if any two distinct points possess disjoint neighborhoods. In a Hausdorff space the points are closed subsets. The usual topology on $\mathbb{R}$ is Hausdorff. The discrete topology is Hausdorff. The trivial topology is not Hausdorff. There are other separation definitions[1]. For instance, let X be the set of all real numbers $\mathbb{R}$ and let the open sets consist of all the sets $x_m = \{x; -1/m < x < 1/m\}$, where m is any positive real number, together with $\emptyset$ and X. This space satisfies a weaker separation condition than Hausdorff: if x and y are any two points, there is a neighborhood of one point to which the other does not belong.
A space is **normal** if it is Hausdorff and if any pair of disjoint closed sets F_1 and F_2 have disjoint open neighborhoods U_1 and U_2. An open set of $\mathbb{R}^n$ is normal.

Hausdorff separated

normal

[1]See for instance [Simmons].

3. BASE

base

A **base** $\mathscr{B}$ for a topology $\mathscr{U}$ is a subsystem of $\mathscr{U}$ which satisfies either one of the following equivalent conditions.

1) Every element of $\mathscr{U}$ is the union of elements of $\mathscr{B}$.
2) For each $x \in X$ and $U \in \mathscr{U}$, with $x \in U$, there exists $B \in \mathscr{B}$ such that $x \in B$ and $B \subset U$.

$\mathscr{U}$ is then said to be generated by $\mathscr{B}$ and is designated $\mathscr{U}_{\mathscr{B}}$.

Example: A base for the usual topology of R consists of all open intervals $(a, b) = \{x; a < x < b\}$.

Conversely, if a class $\mathscr{B}$ of subsets of X covers X and includes all the finite intersections of its elements and the empty set, then $\mathscr{B}$ is a base for a topology on X; this topology has for open sets all the elements of $\mathscr{B}$ and all unions of elements of $\mathscr{B}$.

base of neighborhoods

A **base of neighborhoods** of a point x is a family $\mathscr{N}$ of neighborhoods of x such that every neighborhood of x contains a member of $\mathscr{N}$.

second countable

A **second countable space** (a space which satisfies the second axiom of countability) is a space whose topology has a countable base.

first countable

A **first countable space** is a space in which each point has a countable base of neighborhoods.

4. CONVERGENCE

filter

A **filter** on a set X is a family $\mathscr{F}$ of subsets of X such that

1) the empty set does not belong to $\mathscr{F}$;
2) the intersection of two subsets of X belonging to $\mathscr{F}$ is an element of $\mathscr{F}$;
3) any subset of X which contains an element of $\mathscr{F}$ belongs to $\mathscr{F}$.

Example: The set of neighborhoods of $x \in X$ is a filter $\mathscr{F}(x)$.

ultrafilter

A filter $\mathscr{F}$ is finer than a filter $\mathscr{F}'$ if every element of $\mathscr{F}'$ is an element of $\mathscr{F}$. An **ultrafilter** $\mathscr{F}$ on X is a filter such that any finer filter is identical to $\mathscr{F}$. Filters and ultrafilters are particularly useful in the study of the properties of continuity and convergence in non metrizable topological spaces. We shall not use them in this book; we shall only use sequences and nets, which are somewhat more concrete.

sequence

A **sequence** (x_n) of points in X is a mapping $\mathbb{N} \to X$ by $n \mapsto x_n$. Originally a sequence (x_n) did not mean the mapping but the set of points x_n. A sequence (x_n) **converges** to $x \in X$ if for every neighborhood $N(x)$ there exists $n_0 \in \mathbb{N}$ such that $N(x)$ contains all points x_n for which $n \geq n_0$. We shall write $(x_n) \rightsquigarrow x$. If X is a first countable space the topology can be described in terms of sequences. convergence

Let D be a directed set (p. 5). A **net** (x_α) in a space X is a mapping $D \to X$ by $\alpha \mapsto x_\alpha$. net

A net converges to $x \in X$ if for any neighborhood $N(x)$ therc exists an $\alpha_0 \in D$ such that $x_\alpha \in N(x)$ for all $\alpha \geq \alpha_0$.

A point $x \in X$ is a **limit point (cluster point)** of the net (x_α), if for every neighborhood $N(x)$ and every $\alpha \in D$ there is an $\alpha' \geq \alpha$ such that $x_{\alpha'} \in N(x)$. limit (cluster) point

Let B and D be directed sets. A **subnet** (y_β) of (x_α) is a mapping $B \to X$ by $\beta \mapsto x_{\phi(\beta)} = y_\beta$ where ϕ is a mapping $B \to D$ such that for each $\alpha \in D$ there is a $\beta \in B$ for which $\beta' \geq \beta \Rightarrow \phi(\beta') \geq \alpha$. subnet

5. COVERING AND COMPACTNESS

Given a topology, it is interesting to extract from all the open sets simpler systems which make a covering of the space, a subcovering and a refinement of a covering. The concepts of compactness and paracompactness are defined in terms of the properties of coverings.

A system $\{U_i\}$ of [open] subsets of X is a [open] **covering** if each element in X belongs to at least one U_i (i.e. $\cup U_i = X$). If the system $\{U_i\}$ has a finite number of elements, the covering is said to be finite. Unless otherwise specified a covering will always be an open covering. coverings

A **subcovering** of the covering $\mathcal{U}$ is a subset of $\mathcal{U}$ which is itself a covering. subcovering

The covering $\mathcal{V} = \{V_i\}$ is a **refinement** of the covering $\mathcal{U} = \{U_i\}$ if for every V_i there exists a U_j such that $V_i \subset U_j$. refinement

A covering $\mathcal{U}$ is **locally finite** if for every point x there exists a neighborhood $N(x)$, which has a non empty intersection with only a finite number of members of $\mathcal{U}$. locally finite

A subset $A \subset X$ is **compact** if it is Hausdorff and if every covering of A has a finite subcovering. compact

Theorem. A compact subspace of a Hausdorff space is necessarily closed. Any closed subspace of a compact space is compact.

Bolzano-Weierstrass theorem. A Hausdorff space is compact iff every net has a convergent subnet.

A useful criterion, valid only for metric spaces is the following:

Theorem. In a metric space (*p.* 23) *a set A is compact if and only if every sequence in A contains a convergent subsequence with limit in A.*

Heine–Borel theorem. The compact subsets of $\mathbb{R}^n$ *are the closed bounded subsets of* $\mathbb{R}^n$.

This property is not necessarily true for an arbitrary topological space: a closed bounded subset with non empty interior of an infinite dimensional normed space is NEVER compact, under the norm topology.

relatively compact

$A \subset X$ is **relatively compact** if the closure $\bar{A}$ is compact.

locally compact

A space is **locally compact** if every point has a compact neighborhood. Euclidean spaces are locally compact but not compact.

para-compact

A Hausdorff space is **paracompact** if every covering has a locally finite refinement.

Theorem. All metric spaces (*p.* 23) *are paracompact.*

A locally compact second countable Hausdorff space is paracompact. Paracompactness is the property which a manifold X (p. 111) must possess so that one can construct a partition of unity (p. 214) on X, a necessary device in the theory of integration on X.

compactification

A **compactification** of a topological space X is a pair (f, Y) where Y is a compact space and f is a homeomorphism from X onto a dense subspace of Y.

Let X be a Hausdorff locally compact space. Let $X^+ = X \cup \{z\}$ where $z \notin X$; X^+ together with the following topology is compact: all the original open sets of X, the complement in X^+ of all the compact subsets of X, and, X^+ itself. It can be proved that X is a topological subspace of X^+. X^+ is called the **one point compactification** of X.

one point compactification

6. CONNECTEDNESS

connected

The concept of connectedness is fairly intuitive.

If a topological space X is such that there exist two disjoint non empty subsets A_1 and A_2 both open in X and such that $A_1 \cup A_2 = X$, then X is said to be **disconnected** (is not connected). Since A_2 is the complement of the open set A_1, it is closed as well as open. Similarly A_1 is closed as well as open. This proposition leads to the following theorem.

Theorem. A topological space is connected iff the only subsets which are both open and closed are the void set and the space X itself.

This theorem provides a characterization of connected spaces which is often used as a definition. X is **locally connected** if any neighborhood of any point x contains a connected neighborhood.

locally connected

A space may be connected without being locally connected. Multiple and simple connectedness are defined in section 8.

7. CONTINUOUS MAPPINGS

A mapping f from a topological space X to a topological space Y is **continuous** at $x \in X$ if given any neighborhood $N \subset Y$ of $f(x)$ there exists a neighborhood M of $x \in X$ such that $f(M) \subset N$. f is continuous on X if it is continuous at all points x of X.

continuous mappings

Theorem. A mapping $f: X \to Y$ is continuous iff for every set U open in Y the set $V = f^{-1}(U) \subset X$ is open in X.

Proof: Let U be open in Y and let $V = f^{-1}(U)$. Let $x \in V$. Then $f(x) \in U$. Now U is a neighborhood of $f(x)$. Thus there exists $N(x)$ such that $f(N(x)) \subset U$ if f is continuous; therefore $N(x) \subset V$. Hence given any $x \in V$ it has a neighborhood contained in V. Thus V is open.
Conversely, let W be a neighborhood of $f(x)$. Then there is an open V in W containing $f(x)$, and $f^{-1}(V)$ is open in X. Now $f^{-1}(W) \supset f^{-1}(V)$. Therefore $f^{-1}(W)$ is a neighborhood of x such that $f(f^{-1}(W)) \subset W$. ■

Theorem. f is continuous iff the net $(f(x_\beta))$ converges to $f(x)$ whenever the net (x_β) converges to x.

If f is continuous, then the sequence $(f(x_n))$ converges to $f(x)$ whenever the sequence (x_n) converges to x; but the converse is not true for non countable spaces.

Theorem. The image by a continuous mapping of a compact space is compact.

Corollary. Every continuous function on a compact space takes on its minimum and maximum values.

proper

A continuous mapping is said to be **proper** if the reciprocal image of every compact set is compact.

Remark: If a mapping $f: X \to Y$ is continuous for topologies $\mathcal{U}$ on X and $\mathcal{V}$ on Y it is also continuous for all finer topologies on X (which have more open sets than $\mathcal{U}$) and coarser topologies on Y (which have less open sets than $\mathcal{V}$).

All mappings $f: X \to Y$ are continuous for the discrete topology on X; all mappings $f: X \to Y$ are continuous for the trivial topology on Y.
One may wish to determine the coarsest topology on X [the finest topology on Y] for which given mappings are continuous. This approach leads to the construction of projective (kernel) topologies and inductive (hull) topologies:
Let X be a set; let $\{X_\alpha; \alpha \in A\}$ be a family of topological spaces. Let $\{f_\alpha; \alpha \in A\}$ be a family of mappings $f_\alpha: X \to X_\alpha$; the **projective topology** on X with respect to $\{(X_\alpha, f_\alpha)\}$ is the coarsest topology on X for which each f_α is continuous.

projective topology

Let $\{g_\alpha; \alpha \in A\}$ be a family of mappings $g_\alpha: X_\alpha \to X$; the **inductive topology** on X with respect to $\{(X_\alpha, g_\alpha)\}$ is the finest topology on X for which each g_α is continuous.

inductive topology

homeomorphism

A **homeomorphism** is a bijection f which is bicontinuous (f and f^{-1} are continuous).
In a homeomorphism the images and inverse images of open sets are themselves open. The existence of a homeomorphism between topological spaces is an equivalence relation.
Notice that it is possible to construct[1] continuous one–one mappings whose inverse is not continuous so that the requirement f^{-1} continuous is necessary in order to have an equivalence relation.

Theorem. A continuous bijection $f: X \to Y$ is a homeomorphism in the following important cases.

1) *X and Y are compact.*
2) ***(Banach theorem) X and Y are Banach spaces** (p. 28) **and f is linear.***

topological invariant

A **topological invariant** is a property of topological spaces which is preserved under a homeomorphism, for example, the separation properties, compactness, connectedness, as well as the following concepts.

[1]See for instance [Patterson, p. 28].

8. MULTIPLE CONNECTEDNESS

A **covering space** of a topological space X is a pair $(\tilde{X}, f)$ where $\tilde{X}$ is a connected and locally connected space, and f a continuous mapping of $\tilde{X}$ onto X, such that, roughly speaking, f^{-1} is a multivalued homeomorphism; more precisely, each point $x \in X$ has a neighborhood $V(x)$ such that the restriction of f to each connected component C_α of $f^{-1}(V(x))$ is a homeomorphism from C_α onto $V(x)$.

Two covering spaces $(\tilde{X}_1, f_1)$ and $(\tilde{X}_2, f_2)$ are isomorphic if $\tilde{X}_1$ is homeomorphic to $\tilde{X}_2$ by φ and $f_2 = f_1 \circ \varphi^{-1}$.
X is **simply connected** if X is connected and locally connected, (p. 16), and every covering space $(\tilde{X}, f)$ is isomorphic to the trivial covering space $(X, \text{Id.})$, where Id is the identity mapping.

simply connected

Examples: $\mathbb{R}$ is simply connected. The torus T^1, quotient of $\mathbb{R}$ by $\mathbb{Z}$ is multiply connected. A covering space of T^1 is $(\mathbb{R}, \pi)$, π the canonical projection.

$(\tilde{X}, f)$ is a **universal covering** space for X if it is a covering space and $\tilde{X}$ is simply connected.
X is **locally simply connected** if every point $x \in X$ has at least one simply connected neighborhood.

universal covering space

locally simply connected

Theorem. A connected and locally connected space has a universal covering space.

Theorem. If a space X admits a universal covering space, it admits only one – up to isomorphisms.

Let X be a space which admits a universal covering space $(\tilde{X}, f)$. The group of homeomorphisms φ of $\tilde{X}$ onto itself such that $f \circ \varphi = f$ is called the **fundamental group**[1].
Since two universal covering spaces of X are isomorphic, so are their fundamental groups. The corresponding abstract group is called the fundamental group of X.
The concept of simple connectedness and the fundamental group becomes more intuitive through the notion of arcwise connectedness[2]. A topological space X is **arcwise connected**, if given any two points a and b in X there exists a continuous path between them; that is a continuous map $q: [0, 1] \to X$, such that $q(0) = a$, and $q(1) = b$.

fundamental group

arcwise connected

[1]See for instance [Chevalley, p. 53].
[2]A detailed exposition may be found in [Singer and Thorpe, ch. III].

locally arcwise connected

X is **locally arcwise connected** if for each point $x \in X$ and each neighborhood $V(x)$ there exists a neighborhood $U(x) \subset V(x)$, which is arcwise connected.
It is not difficult to prove[1] that if a topological space is arcwise connected [locally arcwise connected] then it is connected [locally connected]. The converse of this statement is false.

homotopic

Two paths q_1 and q_2 are said to be **homotopic** if there exists a continuous mapping F: $[0,1] \times [0,1] \to X$ such that $F(t,0) = q_1(t)$ and $F(t,1) = q_2(t)$.

Theorem. If X is arcwise connected and locally arcwise connected, it is simply connected if every closed path q in X is homotopic to a constant mapping (p. 4).

sphere

Example: The **sphere** S^n defined by $\Sigma_{i=0}^n (x^i)^2 = 1$ is simply connected when $n \geq 2$.

9. ASSOCIATED TOPOLOGIES

Corresponding to the concepts of subset A of a set X and Cartesian product of sets $X_1 \times X_2$ one defines the relative topology on A and the product topology on $X_1 \times X_2$ as topologies coherently related respectively to the topology $\mathscr{U}$ on X and the topologies $\mathscr{U}_1$ and $\mathscr{U}_2$ on X_1 and X_2.

relative topology

The **relative topology** on A is $\mathscr{U}_A = \{A \cap U;\ U \in \mathscr{U}\}$.
Let $(A, \mathscr{V})$ and $(X, \mathscr{U})$ be topological spaces where A is subset of X.

topological inclusion $\hookrightarrow$

$A \subset X$ is a **topological inclusion**, sometimes denoted $\hookrightarrow$, if the topology $\mathscr{V}$ is finer than the relative topology $\mathscr{U}_A$ on A induced by X (i.e. $\mathscr{U}_A \subset \mathscr{V}$).

product topology

A base for the **product topology** on $X_1 \times X_2$ is the set of pairs $\{(U_1, U_2);\ U_1 \in \mathscr{U}_1,\ U_2 \in \mathscr{U}_2\}$.

usual topology on $\mathbb{R}^n$

All the open balls on $\mathbb{R}^n$, $\Sigma_{i=1}^n (x^i - a^i)^2 < b^2$ form a base for the **usual topology on** $\mathbb{R}^n$. It can be shown that it is equivalent to the product topology on $\mathbb{R} \times \ldots . \times \mathbb{R}$, n times.

projection mapping

The αth **projection mapping** P_α is the mapping P_α: $X_1 \times X_2 \to X_\alpha$ by $(x_1, x_2) \mapsto x_\alpha$, where $\alpha = 1$ or 2. The product topology is the coarsest topology in which each P_α is continuous.
Let $\{X_\alpha;\ \alpha \in A\}$ be an arbitrary family of topological spaces indexed by the set A. The cartesian product $\times_{\alpha \in A} X_\alpha$ is the set of all mappings x on A

[1]See for instance [Singer and Thorpe, p. 45].

such that $x(\alpha) = x_\alpha \in X_\alpha$. The projection mapping P_α maps $\times_{\beta \in A} X_\beta \to X_\alpha$ by $x \mapsto P_\alpha(x) = x_\alpha$ where $x = (x_\alpha)_{\alpha \in A}$. The product topology on $\times_{\beta \in A} X_\beta$ is the coarsest topology in which each P_α is continuous. A base for this product topology consists of products of open sets, one from each factor X_α, with all, but finitely many of these, being the whole space X_α; that is to say, the sets of the base have the form

$$U_{\alpha_1} \times \ldots\ldots \times U_{\alpha_n} \underset{\beta \in A}{\times} X_\beta$$

$$\beta \neq \alpha_1, \ldots, \alpha_n$$

where U_{α_i} is an open subset of X_{α_i}.

This topology is nice in the sense that the product of compact spaces is compact. This result is known as the *Tychonoff theorem.* It is not true if we do not require all but finitely many U_α to be the whole space. See Exercise 2, p. 68.

Tychonoff theorem

10. TOPOLOGY RELATED TO OTHER STRUCTURES

A topology on a set which possesses another structure is meaningful only if it is compatible with the other structure, i.e. if the mappings which define the other structure are continuous.

A set X together with a group operation and a topology is said to be a **topological group** if the mappings which define its group structure are continuous, i.e. if the mappings

topological group

$$X \times X \to X \text{ by } (x, y) \mapsto x \cdot y = xy$$
$$X \to X \text{ by } x \mapsto x^{-1}$$

are continuous.

A topological space X which is also a vector space on $\mathbb{K}$ is a **topological vector space** if the operations of addition and scalar multiplication are continuous, i.e. if the mappings

topological vector space

$$X \times X \to X \text{ by } (x, y) \mapsto x + y$$
$$\mathbb{K} \times X \to X \text{ by } (\lambda, x) \mapsto \lambda x$$

are continuous.

Example: $\mathbb{R}^n$ together with its usual topology (p. 20) is a topological vector space. It can be shown that the topologies making $\mathbb{R}^n$ into a Hausdorff topological vector space are all equivalent[1].

[1]See for instance [Choquet].

base

A **base** in a topological vector space X is a linearly independent subset A such that

$$X = \overline{\text{sp } A}$$

separable

where sp A is the linear subspace with Hamel basis A (p. 9), and $\overline{\text{sp } A}$ is its closure. A topological vector space is called **separable** if it admits a countable base: it can be shown that it is then separable in the topological sense given previously (p. 13). Note that the base defined here should not be confused with a Hamel basis (p. 9) which is an algebraic concept. For instance $\{\sin nx\,;\, n \in \mathbb{N}\}$ is a (topological) base of $L^2([0, \pi])$ but not a Hamel basis. A Hamel basis for $L^2([0, \pi])$ is not countable.

locally convex

A topological vector space is **locally convex** if it admits a topological base made of convex sets (p. 10).

Theorem. A linear mapping $T: X \to Y$, where X and Y are topological vector spaces, is continuous iff it is continuous at the origin of X.

topological dual

$\langle , \rangle$

Let $\mathscr{L}(X, Y)$ be the set of all linear continuous mappings from X into Y. The **topological dual** X' of X is the space $\mathscr{L}(X, \mathbb{K})$ of all linear continuous functions (linear forms) on X.
Throughout the book we will denote the duality between X and X' by $\langle , \rangle$; i.e. the element x' of X' is the continuous linear mapping from X to $\mathbb{K}$ defined by

$$x' : x \mapsto \langle x', x \rangle.$$

Example 1: $X = \mathbb{R}^n$, any linear form on $\mathbb{R}^n$ is continuous and may be written

$$x = \{x^i\} \to \sum x'_i x^i, \qquad \{x'_i\} \in \mathbb{R}^n.$$

Thus the dual of $\mathbb{R}^n$ is (may be identified with) $\mathbb{R}^n$ and

$$\langle x', x \rangle = \sum x'_i x^i \text{ if } x' = \{x'_i\} \text{ and } x = \{x^i\}.$$

Example 2: $X = L^2(0, 1)$ (p. 53). Any continuous linear form may be written

$$x \mapsto \int_0^1 x'(t)\overline{x(t)}\, dt, \qquad x' \in L^2(0, 1).$$

Thus the dual of $L^2(0, 1)$ is (may be identified with) $L^2(0, 1)$ and $\langle x', x \rangle = (x'|x)$ (scalar product).

A **bounded set** in a topological vector space is a set which can be mapped inside any neighborhood of the origin by a homothetic transformation of sufficiently small ratio centered on the origin ($x \mapsto \epsilon x$). bounded sets

Example: **A subset of a normed vector space (p. 27) is bounded if and only if it is contained in a ball of finite radius (p. 23).**

A net (x_α) in a topological vector space is a **Cauchy net** if $x_\alpha - x_\beta$ converges to zero in the sense of a net, i.e., if D is the indexing set, for every neighborhood N of the origin there is an $\alpha_0 \in D$ such that $x_\alpha - x_\beta \in N$ for all $\alpha, \beta > \alpha_0$. Cauchy net

A **Cauchy sequence** is a sequence which is a Cauchy net. Cauchy sequence

A subset A of a topological vector space is **complete** if each Cauchy net in A converges to some point in A. complete space

A subset of a topological vector space is **sequentially complete** if each Cauchy sequence in the set converges to a point in the set. sequentially complete

11. METRIC SPACES

Normed spaces and metric spaces are the natural generalization of $\mathbb{R}^n$.

A **metric space** is a set X together with a mapping $d: X \times X \to R$ such that metric space

$$d(x, y) \geq 0$$

$$d(x, y) = 0 \qquad \text{iff } x = y$$

$$d(x, y) = d(y, x) \qquad \text{symmetry}$$

$$d(x, z) \leq d(x, y) + d(y, z) \qquad \text{triangle inequality (subadditivity).}$$

$d(x, y)$ is called the **distance** between x and y. distance

An **open ball** $B_a(x_0)$ of radius a about x_0 is the set of points $x \in X$ such that $d(x, x_0) < a$. open ball

Proposition. The class $\mathcal{B}$ of balls $B_a(x)$ for all points $x \in X$ and all radii a is a base for a topology on X. The open sets are unions of balls.

The topology on X generated by $\mathcal{B}$ is called the **topology induced by the metric** d. Unless otherwise specified, the topology on a metric space will be the topology induced by its metric. metric topology

A metric space is first countable; at each point the balls of radii $1/n$ form a local countable base. A metric space is Hausdorff and normal (p. 13).

Example 1: The discrete metric d_0

$$d_0(x, y) = \begin{cases} 0 \text{ if } x = y \\ 1 \text{ otherwise.} \end{cases}$$

The topology induced by this metric is the discrete topology (p. 12).

Example 2: Let X be the plane $\mathbb{R}^2$; let (x_1, y_1) and (x_2, y_2) be the coordinates of the points P_1 and P_2, respectively.
The three following metrics induce the same topology

$$d_1(P_1, P_2) = [(x_2 - x_1)^2 + (y_2 - y_1)^2]^{1/2} \text{ the euclidean metric}$$
$$d_2(P_1, P_2) = \max \{|x_2 - x_1|, |y_2 - y_1|\}$$
$$d_3(P_1, P_2) = |x_2 - x_1| + |y_2 - y_1|.$$

The ball of radius a about the origin relative to each of these metrics is indicated by the shaded areas.

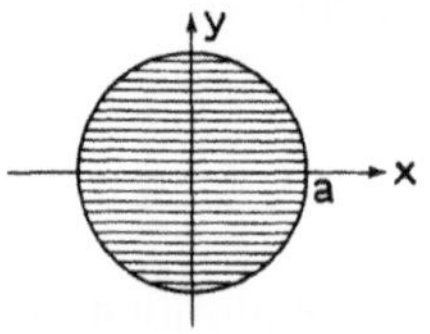

$d_1(P, 0) \leq a$

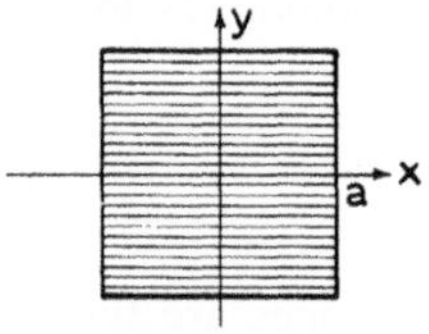

$d_2(P, 0) \leq a$

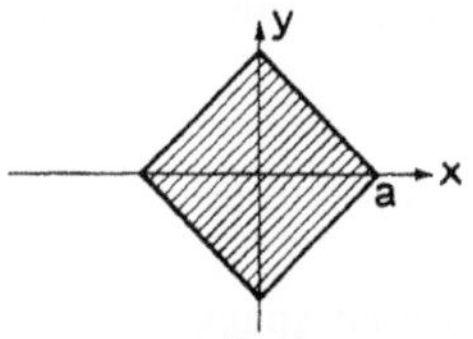

$d_3(P, 0) \leq a$

Example 3: Unless otherwise specified, $\mathbb{R}^n$, as a metric space, will be the set of n-tuples $\{x^1 \ldots . x^n; x^i \in \mathbb{R}\}$ with the euclidean metric: $d(x, y) = [\Sigma (x^i - y^i)^2]^{1/2}$. "The euclidean topology on $\mathbb{R}^n$" means the topology induced by the euclidean metric; it is equivalent to the product topology $\underbrace{\mathbb{R} \times \ldots \times \mathbb{R}}_{n \text{ copies}}$.

Example 4: Let a and b be two points of a metric space X with metric d, let q be a continuous parametrized path from a to b, $q: [0, 1] \to X$ by $t \mapsto q(t) \in X$, such that $q(0) = a$ and $q(1) = b$. A metric is defined on the space of these paths q by

$$d_1(q, q') = \sup_{0 \leq t \leq 1} d(q(t), q'(t)).$$

Exercise: Show that the geometric paths, equivalence classes of parametrized paths under changes of parametrization ([0, 1] → [0, 1], monotone and continuous) also form a metric space, under the metric

$$D(q, q') = \inf_{\mathcal{P}} \sup_{0\le t\le 1} d(q(t), q'(t)),$$

where $\mathcal{P}$ is the set of all possible parametrizations.
In the chapter on riemannian manifolds, another metric is defined for the space of paths $q:[0, 1]\to$ riemannian manifold (Chapter V).

The following theorem makes it possible, in studying metric spaces, to replace the study of continuity of maps by the study of convergence of sequences.

Theorem. Let X be a metric space and let Y be a topological space. A mapping $f: X\to Y$ is continuous at $x\in X$ if and only if the sequence $(y_n = f(x_n))$ converges to $y = f(x)$ in Y whenever (x_n) converges to x in X.

Proof: Let us consider a countable base $V_1, V_2, \ldots V_n \ldots$ of neighborhoods of x_0 and a sequence of points $x_n \in V_n$ converging to x_0 such that, V being a given neighborhood of $y_0 = f(x_0)$, $x_n \not\in f^{-1}(V)$. Such a sequence exists if f is not continuous at x_0; in this case $f(x_n)\not\in V$ and $f(x_n)$ cannot converge to $f(x_0)$. The converse – f continuous implies (y_n) converges to y whenever (x_n) converges to x – is readily proved. ■

A **Cauchy sequence in a metric space** is a sequence (x_n) such that — Cauchy sequence

$$\lim_{n,m\to\infty} d(x_n, x_m) = 0,$$

i.e. there exists N such that, for $n, m > N, d(x_n, x_m) < \epsilon$ for every preassigned $\epsilon > 0$.
A Cauchy sequence is not necessarily convergent; for example, in the metric space $0 < x < 1$, the sequence $x_n = 1/(n+1)$ is a Cauchy sequence but is not convergent since $x = 0$ is not a point of the space. A metric space is said to be **complete** if every Cauchy sequence in the space is convergent. — complete
An incomplete space can be made complete by adding new elements to the space, namely, the limits of Cauchy sequences with an appropriate equivalence relation. Completeness is not a topologically invariant property. Consider, for example, the subspaces $[0, \infty[$ and $[0, 1[$ of $\mathbb{R}$. The first is complete, the second is not. However, the mapping f: $[0, \infty)\to[0, 1)$ by $x\mapsto x/(1+x)$ is a homeomorphism. — completion
A space is a **Baire space** if it is not a countable union of nowhere dense subsets (p. 13). The following theorem is important in functional analysis on metric spaces. — Baire space

Theorem. A complete metric space is a Baire space.

metrizable

A topological vector space is said to be **metrizable** if its topology can be induced by a metric invariant by translation, i.e.

$$d(x, y) = d(x + z, y + z).$$

Two kinds of completeness have already been defined for topological vector spaces: completeness (convergence of Cauchy nets) and sequential completeness (convergence of Cauchy sequences) (p. 23). In a metrizable topological vector space completeness, sequential completeness and metric completeness are equivalent[1].

Fréchet space

A topological vector space that is metrizable and complete is called a **Fréchet space**.

We will return to Fréchet spaces in Chapter VI.

12. BANACH SPACES

norm

The mapping $x \mapsto \|x\|$ of a vector space X on $\mathbb{K}$ into $\mathbb{R}$ is a **norm** if for $x \in X$ and $\lambda \in \mathbb{K}$

$\|x + y\| \leq \|x\| + \|y\|$	triangle inequality (subadditivity)
$\|\lambda x\| = \lvert\lambda\rvert \|x\|$	scale property
$\|x\| = 0$ iff $x = 0$.	

The first two conditions imply that $\|x\| \geq 0$ and $\|x - y\| \geq \lvert \|x\| - \|y\| \rvert$, and moreover that $\|0\| = 0$ which makes the last condition partly redundant.

seminorm

A mapping which satisfies only the first two axioms is a **seminorm** (see Chapter VI).

Exercise: Prove that the first two axioms imply $\|0\| = 0$, $\|x\| \geq 0$ and $\|x - y\| \geq \lvert \|x\| - \|y\| \rvert$.

Answer:

$$\|0\| = \|0x\| = 0\|x\| = 0.$$
$$0 = \|x - x\| \leq \|x\| + \|-x\| = 2\|x\|.$$

$\|x - y\| + \|y\| \geq \|x\|$ hence $\|x - y\| \geq \|x\| - \|y\|$; on the other hand $\|x - y\| = \lvert -1 \rvert \|y - x\| \geq \|y\| - \|x\|$.

[1]See for instance [Kelley].

A norm induces a metric on X which is invariant by translation

$$\|x - y\| = d(x, y)$$

and hence a topology.

Remark: Even if it is invariant by translation, a metric is not always equivalent to the norm $\|x\| = d(0, x)$ because $d(\cdot, \lambda x)$ is not necessarily equal to $|\lambda| d(\cdot, x)$.

Remark: Let $\|\ \|$ and $\|\ \|'$ be two different norms on X such that $\|x\| < \|x\|'$ for all $x \in X$, the corresponding balls $B_\epsilon(x_0)$ and $B'_\epsilon(x_0)$ satisfy $B'_\epsilon(x_0) \subset B_\epsilon(x_0)$ and the corresponding topologies satisfy $\mathscr{U} \subset \mathscr{U}'$ ($\mathscr{U}'$ is finer than $\mathscr{U}$)

$$\|x\|' < \epsilon \ \Rightarrow\ \|x\| < \epsilon \ \Rightarrow\ B'_\epsilon \subset B_\epsilon.$$

Conversely if $\mathscr{U} \subset \mathscr{U}'$ then there exists a constant λ such that $\|x\| < \lambda \|x\|'$ for every $x \in X$. Indeed the open ball $B_\epsilon = \{x\,;\, \|x\| < \epsilon\}$ contains an open ball $B'_\eta = \{x\,;\, \|x\|' < \eta\}$; thus $\eta x / 2\|x\|$, which is in B'_η, by virtue of

$$\left\|\frac{\eta}{2} \cdot \frac{x}{\|x\|'}\right\|' = \frac{\eta}{2}, \quad \text{is in } B_\epsilon \text{ and } \quad \left\|\frac{\eta x}{2\|x\|'}\right\| < \epsilon, \quad \text{i.e. } \|x\| < \frac{2\epsilon}{\eta}\|x\|'.$$

The vector space X together with a norm topology (the **normed vector space** X) is a topological vector space. normed vector space

Exercise: Prove this proposition.

Answer: We have to prove that the mappings $(x, y) \mapsto x + y$ and $(\lambda, x) \mapsto \lambda x$ are continuous (p. 17). The mapping f: $(x, y) \mapsto x + y$ is continuous at (x_0, y_0) if one can find a δ-neighborhood of x_0 and a δ'-neighborhood of y_0 such that the image of these neighborhoods under f is contained in any preassigned ϵ-neighborhood of $x_0 + y_0$, $\|x + y - (x_0 + y_0)\| < \epsilon$. The δ and δ' neighborhood defined respectively by

$$\|x - x_0\| < \frac{\epsilon}{2} \quad \text{and} \quad \|y - y_0\| < \frac{\epsilon}{2}$$

satisfy this requirement, by virtue of the triangle inequality

$$\|x + y - x_0 - y_0\| \leq \|x - x_0\| + \|y - y_0\|.$$

Similarly the mapping f: $(\alpha, x) \mapsto \alpha x$ is continuous at (α_0, x_0) because

$$\|\alpha x - \alpha_0 x_0\| = \|\alpha(x - x_0) + (\alpha - \alpha_0)x_0\| \leq |\alpha|\,\|x - x_0\| + |\alpha - \alpha_0|\,\|x_0\|$$

so that the neighborhoods of α_0 and x_0 defined respectively by

$$|\alpha - \alpha_0| < \epsilon/\|x_0\| \text{ and } \|x - x_0\| < \epsilon/(|\alpha_0| + \epsilon/\|x_0\|)$$

are mapped into the neighborhood $\|\alpha x - \alpha_0 x_0\| < 2\epsilon$. ■

A norm topology on a vector space X makes it a locally convex topological vector space (p. 22). It suffices to prove the convexity of a ball $B_a(x)$ (B-ball) (p. 23) defined by the metric induced by the norm. Indeed the triangle inequality implies the convexity of the B-balls and vice versa.

Exercise: Prove this proposition.

Answer: It is sufficient to prove it for B-balls centered at the origin since a B-ball centered at any other point is obtained by translation from one centered at the origin.
Let $x, y \in B_\epsilon(0)$; the triangle inequality says that

$$\begin{aligned}\|\lambda x + (1-\lambda)y\| &\le \lambda\|x\| + (1-\lambda)\|y\| \\ &\le \lambda\epsilon + (1-\lambda)\epsilon = \epsilon.\end{aligned}$$

Hence $\lambda x + (1-\lambda)y \in B_\epsilon(0)$.
Conversely, assume $x, y, \lambda x + (1-\lambda)y \in B_\epsilon(0)$; if the triangle inequality did not hold we would have

$$\|\lambda x + (1-\lambda)y\| \ge \lambda\|x\| + (1-\lambda)\|y\| \ge \epsilon$$

in contradiction with the hypothesis. ■

The fact that a norm topology on a vector space implies that it is a locally convex topological vector space has been proved by using only the first two axioms of a norm; thus the implication is more precisely: a seminorm topology on a vector space implies that it is a locally convex topological vector space. Conversely, it can be proved that the topology on a locally convex vector space can always be defined by a family of seminorms (p. 423). The relation between these seminorms and an eventual metrizability of the topological vector space will be discussed in Chapter VI.

Banach space

A complete normed vector space is a **Banach space** or **B-space**.

Example 1: The euclidean metric on $\mathbb{R}^n$ induces a norm $\|x\| = d(0, x)$ for which $\mathbb{R}^n$ is complete. It can be proved[1] that all the norms compatible

[1] See for instance [Choquet].

with the vector space structure of $\mathbb{R}^n$ induce the same topology, namely the euclidean topology.

Example 2: The vector space $\mathscr{C}^0(X)$ of all bounded continuous functions on the topological space X, with the norm $\|f\|_{\mathscr{C}^0(X)} = \sup_{x\in X} |f(x)|$. This norm is called the **uniform norm**. This space is complete because if f is the uniform limit of a sequence of bounded continuous functions, then f itself is bounded and continuous. $\mathscr{C}^0(X)$ uniform norm

Example 3: The space $\mathscr{C}^m(\bar{U})$ of functions on an open set U of $\mathbb{R}^n$ which have continuous and uniformly bounded derivatives of order $\leq m$, with the norm $\mathscr{C}^m(\bar{U})$

$$\|f\|_{\mathscr{C}^m(\bar{U})} = \operatorname*{Sup}_{x\in U} \sum_{|j|\leq m} |D^j f(x)|$$

where, j being a multi-index,

$$D^j f(x) = \left(\frac{\partial}{\partial x^1}\right)^{j_1} \cdots \left(\frac{\partial}{\partial x^n}\right)^{j_n} f(x)$$

$$j = (j_1, \ldots, j_n), \qquad |j| = j_1 + j_2 + \cdots j_n.$$

Example 4: The space of continuous, square integrable functions on $U \subset \mathbb{R}^n$, with the norm

$$\|f\|^2_{L^2(U)} = \int_U |f(x)|^2\, \mathrm{d}x, \qquad \mathrm{d}x = \mathrm{d}x^1 \ldots \mathrm{d}x^n$$

is not complete. It is not a Banach space.

We will see other important examples of Banach spaces (L^p spaces) in the next section.

Strong and weak topologies, compactness. The compact subsets of $\mathbb{R}^n$ are the closed and bounded subsets (Heine–Borel theorem, p. 16). It is easy to prove that the compact subsets of a metric space are also closed and bounded, i.e. included in some ball of finite radius. But the converse is not generally true, as shown by the following theorem.

Theorem. A closed bounded subset with non empty interior – for instance the unit ball – of an infinite dimensional normed space is never compact in the norm topology.

This striking difference between finite and infinite dimensional spaces is an indication of the pitfalls into which our intuition, based on the properties of

$\mathbb{R}^n$, can lead us. The norm topology on a normed space is also called the **strong topology**. Thus the sequence (f_n) in X converges strongly to $f \in X$ if

strong topology

$$\lim_{n=\infty} \|f - f_n\| = 0$$

which implies

$$\lim_{n=\infty} \|f_n\| = \|f\|.$$

The norm topology is not the only topology which can be given meaningfully to a normed space. Another topology, called the **weak topology** and defined in terms of the dual space plays an important role. Let X be a topological vector space, not necessarily a normed space, let X' be its topological dual. The sequence (f_n) in X converges weakly to $f \in X$ if for every $g \in X'$

weak topology

$$\langle g, f_n \rangle \rightsquigarrow \langle g, f \rangle \text{ as } n \to \infty.$$

Weak convergence in a normed space does not imply $\|f_n\| \rightsquigarrow \|f\|$, but a weaker statement, namely

$$\|f\| \leq \lim_{n=\infty} \inf \|f_n\|$$

(infimum in $\mathbb{R}$ of the limit points of the sequence $\|f_n\|$). Weak convergence, together with $\lim_{n=\infty} \|f_n\| = \|f\|$ imply strong convergence.

weak star topology

The sequence (f_n) in X' converges to $f \in X'$ in the **weak star topology** if for every $g \in X$ $\langle f_n, g \rangle \rightsquigarrow \langle f, g \rangle$ as $n \rightsquigarrow \infty$.

If the space X is reflective – i.e. if the dual of X' coincides with X (p. 59), then on X' the weak and weak star topologies coincide.

Theorem. All bounded subsets are relatively compact (p. 16) *in the weak topology of a reflexive Banach space.*

13. HILBERT SPACES

A strictly positive sesquilinear mapping on a complex vector space induces a norm

$$\|x\| = (x|x)^{1/2}.$$

Thus a pre-Hilbert space (p. 11) is a normed vector space.

Hilbert space

A complete pre-Hilbert space is a **Hilbert space**.

A Hilbert space is said to be real if the underlying linear space is defined on $\mathbb{R}$.

Examples: See in the next section p. 56 and Chapter VI.

Riesz theorem: There is an isomorphism between a Hilbert space $\mathcal{H}$ and its dual $\mathcal{H}'$ defined by

$$x \in \mathcal{H} \mapsto (x|\cdot) \in \mathcal{H}'.$$

In other words every continuous linear form on $\mathcal{H}$ is defined by a scalar product with an element of $\mathcal{H}$

$$\langle x', x\rangle = (x'|x).$$

Remark: The Riesz theorem expresses the fact that the sesquilinear product $(x, y) \mapsto (x|y)$ is non degenerate.

A subset $\{x_\alpha\}$ $\alpha \in A$ of a Hilbert space $\mathcal{H}$ is called an **orthonormal base** of $\mathcal{H}$ if it is a base (cf. p. 22) such that orthonormal base

$$(x_\alpha|x_\beta) = 0 \qquad \alpha \neq \beta$$
$$(x_\alpha|x_\beta) = 1 \qquad \alpha = \beta.$$

It can be proved that a Hilbert space $\mathcal{H}$ always admits an orthonormal base, that two orthonormal bases have the same cardinality, and that two Hilbert spaces with orthonormal bases of the same cardinality are isomorphic.

Most Hilbert spaces that arise in practice are separable – thus isomorphic to $L^2((0, 1))$ (p. 53). Each element $x \in \mathcal{H}$ may then be written as the sum of a convergent series

$$x = \sum_{n=1}^{\infty} (x|x_n)x_n.$$

D. INTEGRATION

We shall in this section review only the main definitions and some useful theorems of integration theory[1]. Our aim is to introduce the L^p spaces, and to give some of their most important properties that we shall use later.

[1]For a study of integration theory the reader may consult [Halmos 1950], [Bartle], [Royden], [Dunford and Schwartz, Ch. III], [Bourbaki], [Marle], [Descombes].

1. INTRODUCTION

A Riemann integral of a function $f: \mathbb{R} \to \mathbb{R}$ is defined as the limit, when it exists, of the sum of the areas of narrow rectangles approximating more

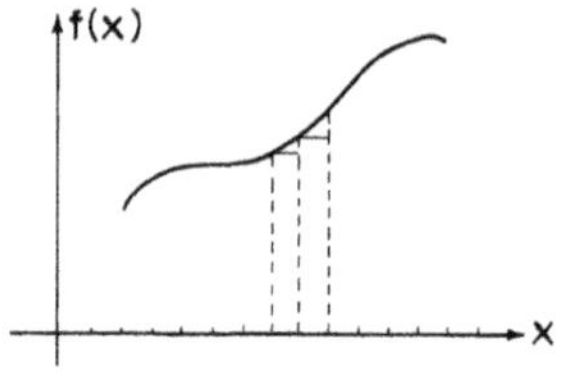

and more closely the surface under the curve $x \mapsto f(x)$. The construction of the rectangles gives more emphasis than necessary to the value of the function at all points and not enough emphasis to the properties of the space on which the function is defined.

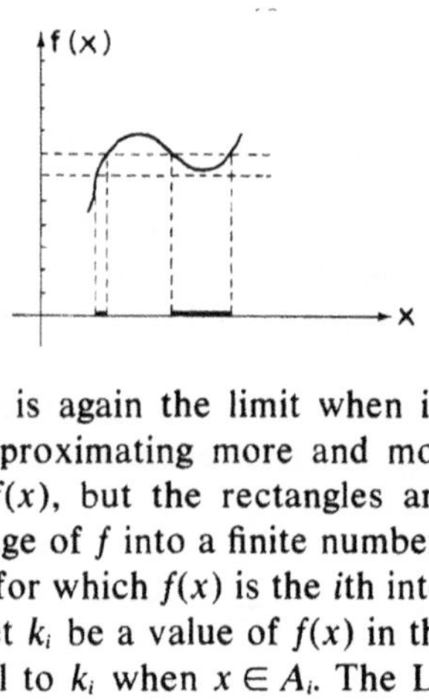

The Lebesgue integral is again the limit when it exists of the sum of areas of rectangles approximating more and more closely the surface under the curve $x \mapsto f(x)$, but the rectangles are now constructed as follows. Divide the range of f into a finite number of small intervals and find the set A_i of all x for which $f(x)$ is the ith interval; assign a measure $m(A_i)$ to the set A_i. Let k_i be a value of $f(x)$ in the ith interval; let f_n be the step function equal to k_i when $x \in A_i$. The Lebesgue integral is the limit when it exists of

$$\sum_i k_i m(A_i)$$

when the sequence of functions (f_n) tends to f in a sense to be made precise later in this chapter. Lebesgue integration generalizes Riemann integration to a larger class of functions on $\mathbb{R}^n$, and gives spaces of integrable functions which are complete (p. 23). By introducing the sets A_i and measure $m(A_i)$ one brings out the role played by the space on which f is defined. It is then possible to define integration for functions on spaces more general than $\mathbb{R}^n$, in particular for functions on locally compact spaces. Integration on non locally compact infinite dimensional spaces raises new problems (Chapter VII D). We shall give first the properties which a family of subsets of a space must possess so that a meaningful measure can be assigned to them, then the properties which a function defined on that space must possess to be integrable.

2. MEASURES

Not any family of subsets of a space X is suitable as a substratum for measures. The useful families are the rings, the fields, the σ-rings and particularly, the σ-fields of subsets, as defined below. A fundamental example will be the Borel subsets of a topological space.

A **ring of subsets** of a set X is a non empty class $\mathcal{R}$ of subsets of X such that ring of subsets

$$A, B \in \mathcal{R} \Rightarrow A \backslash B \in \mathcal{R} \text{ and } A \cup B \in \mathcal{R}.$$

It follows that $\emptyset \in \mathcal{R}$, and that $\mathcal{R}$ is closed under finite unions, and finite intersections.

Exercise[1]: Show that the ring $\mathcal{R}$ is a ring in the algebraic sense if the sum and product are defined by

$$A + B = A \triangle B \quad \text{symmetric difference}$$
$$AB = A \cap B \quad \text{intersection.}$$

The neutral element and the unit element are respectively $\emptyset$ and X.

Example: The power set 2^X of X is a ring of subsets.

The ring $\mathcal{R}$ is called a **field** (or boolean algebra) if $X \in \mathcal{R}$; it is then often denoted $\mathcal{A}$. field

The ring $\mathcal{R}$ [The field $\mathcal{A}$] is called a **σ-ring** [a **σ-field**] if $\mathcal{R}$ [if $\mathcal{A}$] is closed under countable unions σ-ring σ-field

$$A_i \in \mathcal{R}, i \in \mathbb{N} \Rightarrow \bigcup_{i=1}^{\infty} A_i \in \mathcal{R}.$$

Exercise: Show that a σ-field $\mathcal{A}$ can equally well be defined by the following axioms:

$$A \in \mathcal{A} \Rightarrow A' = X \backslash A \in \mathcal{A}$$

$$A_i \in \mathcal{A} \Rightarrow \bigcup_{i=1}^{\infty} A_i \in \mathcal{A}.$$

[1]For detailed proofs, see for instance [Marle].

It follows immediately that $A \cup A' = X \in \mathcal{A}$.

Let $\mathscr{E}$ be a class of subsets of a set X which does not necessarily satisfy the σ-field axioms, let $\mathcal{A}(\mathscr{E})$ be the smallest class of subsets of X containing $\mathscr{E}$ such that $\mathcal{A}(\mathscr{E})$ is a σ-field, it can be shown that this σ-field exists and is unique; it is called the σ-field generated by $\mathscr{E}$.
A pair $(X, \mathcal{A})$ where X is a set, and $\mathcal{A}$ a σ-field of subsets of $\mathcal{A}$ is called a measurable space.

In most cases the set X to be endowed with a measure is already a **topological space**. It is then desirable to have a measure related, in a certain sense, with the topology. This can be achieved by taking as the σ-field on X the Borel σ-field, whose definition follows.

Borel σ-field

Borel sets

The **Borel σ-field** of a topological space X is the σ-field generated by the open sets of X (or, equivalently the closed sets of X). An element of the Borel σ-field is called a **Borel set**.

All countable unions and countable intersections of open and closed sets are Borel sets – but they are not the only ones[1].

Exercise: Show that the Borel σ-field of $\mathbb{R}$ is generated by the open intervals (a, b) or equivalently by the closed intervals $[a, b]$, or the semi closed $[a, b)$ or $(a, b]$ or $(-\infty, a]$ or $(-\infty, a)$ or $[a, +\infty)$ or $(a, +\infty)$.

Answer: Every open set in $\mathbb{R}$ is a countable union of open intervals, therefore the σ-field generated by open intervals is identical with the Borel σ-field of $\mathbb{R}$. To prove the other statements either use the property that if $A \in \mathscr{B}$ then $\mathbb{R}\backslash A \in \mathscr{B}$, or the fact that

$$(a, b) = \bigcup_{n=1}^{\infty} \left[a + \frac{\epsilon}{n}, b\right) \quad \text{and} \quad [a, b) = \bigcup_{n=1}^{\infty} \left[a, b - \frac{\epsilon}{n}\right].$$

3. MEASURE SPACES

positive set function

A **positive set function** on a space X is a mapping from a family $\mathcal{A}$ of subsets of X, containing the empty set $\emptyset$, into the extended positive real numbers

$$m: \mathcal{A} \to \mathbb{R}^{+} \cup \{+\infty\}.$$

[1]A countable intersection of open sets [union of closed sets] is called a G_δ set [a F_σ set]. A countable union of G_δ sets, called a $G_{\delta,\sigma}$, is, if X is not a countable set, neither a F_σ nor a G_δ.

m is **finitely additive** if for every disjoint finite family of subsets, $(A_1, \dots, A_n)$ in $\mathscr{A}$, with union in $\mathscr{A}$, finitely additive

$$m\left(\bigcup_{i=1}^{n} A_i\right) = \sum_{i=1}^{n} m(A_i)$$

and

$$m(\emptyset) = 0.$$

m is **countably additive** if for every disjoint countable family of subsets. $(A_1, \dots, A_n \dots)$ in $\mathscr{A}$, with union in $\mathscr{A}$, countably additive

$$m\left(\bigcup_{i=1}^{\infty} A_i\right) = \sum_{i=1}^{\infty} m(A_i)$$

and

$$m(\emptyset) = 0.$$

A **positive measure** m on a space X is a countably additive positive set function from a σ-field $\mathscr{A}$ of X into $\mathbb{R}_+ \cup \{+\infty\}$. The space X endowed with the measure m is called a **measure space** and denoted $(X, \mathscr{A}, m)$. The elements of $\mathscr{A}$ are the **measurable subsets** of X, $m(A)$ is the measure of $A \in \mathscr{A}$. positive measure measure space measurable sets

A definition of a measure in terms of finite additivity is insufficiently restrictive to build a theory of integration with nice properties: the definition of integrable functions would be more involved[1], the Lebesgue dominated convergence theorem (p. 44) would not hold, etc. . . . On the other hand, to require non-countable additivity would obviously be overly restrictive since any subset is a union of its points.
The following theorem is useful to assess the existence of a measure on a σ-field of subsets, when a countably additive set function is given on a field of subsets.

Hahn extension theorem. Let μ be a positive countably additive set function on a field Σ of subsets of X. There exists a positive measure $\tilde{\mu}$ on the σ-field generated by Σ, whose restriction to Σ is μ. If μ is σ-finite[2] on Σ, then the extension $\tilde{\mu}$ is unique.

We shall later on consider non positive and complex measures, but the results can always be obtained by considering associated real positive measures.

[1]See for instance [Dunford and Schwartz, p. 112].
[2]See the definition below.

σ-finite
finite

m is **σ-finite** if X is expressible as a countable union of sets A_i such that $m(A_i) < \infty$. m is **finite** if $m(X) < \infty$.

probability space

A **probability space** is a measure space $(X, \mathscr{A}, m)$ with $m(X) = 1$. The elements A of $\mathscr{A}$ are called events and $m(A)$ is the probability of A.

m-almost everywhere (a.e.)

A property is said to hold **m-almost everywhere** (or **a.e.**) if it holds for all points of X except possibly for points of a set A of measure $m(A) = 0$. It is often convenient, and always possible[1], to extend the class of measureable sets in such a way that every subset of a set with measure zero is measurable, and has measure zero. The measure thus obtained is said to be **completed** from the original measure.

completed

Borel measure

It is customary to reserve the name **Borel measure** for the positive measures on the Borel sets of a locally compact Hausdorff topological space, that have the property that the measure of every compact set – closed, and hence measurable – is bounded.

The following definition is a link between abstract measure theory and the original definition of measurable sets on $\mathbb{R}$.
A positive finitely additive set function μ defined on a field Σ of subsets of a topological space X is said to be **regular** if for each $A \in \Sigma$ and $\epsilon > 0$ there are sets F and G in Σ, with closure $\bar{F}$ and interior $\mathring{G}$, such that $\bar{F} \subset A \subset \mathring{G}$ and

regular set function

$$\mu(G \backslash F) < \epsilon.$$

The following theorem is useful in proving the countable additivity of regular set functions on locally compact spaces; we give its proof to show where the compactness hypothesis comes in.

Alexandroff theorem. Let μ be a finite regular positive finitely additive set function on a field Σ of subsets of a compact, topological space X. Then μ is countably additive.

Proof: Let $\epsilon > 0$, let $\{A_n\}$ be a countable disjoint family of sets in Σ, with $\cup_{n=1}^{\infty} A_n = A \in \Sigma$. There exists $F \in \Sigma$ such that $\bar{F} \subset A$ and $\mu(A \backslash F) < \epsilon$. For each n there exists $G_n \in \Sigma$ such that $\mathring{G}_n \supset A_n$ and $\mu(G_n \backslash A_n) < \epsilon/2^n$. Therefore $\mu(A_n) \geq \mu(G_n) - \epsilon/2^n$. But since X is compact, there exists an integer m such that

$$\bigcup_{n=1}^{m} \mathring{G}_n \supset A, \text{thus} \bigcup_{n=1}^{m} G_n \supset F,$$

$$\sum_{n=1}^{m} \mu(G_n) \geq \mu\left(\bigcup_{n=1}^{m} G_n\right) \geq \mu(F) > \mu(A) - \epsilon.$$

[1]See for instance [Royden], [Marle].

Therefore

$$\sum_{n=1}^{m} \mu(A_n) \geq \sum_{n=1}^{m} \mu(G_n) - \epsilon > \mu(A) - 2\epsilon.$$

On the other hand, since for any integer m

$$A = \bigcup_{n=1}^{\infty} A_n, \qquad A \supset \bigcup_{n=1}^{m} A_n$$

$$\mu(A) \geq \sum_{n=1}^{m} \mu(A_n), \qquad \forall m;$$

thus the series $\{\mu(A_n)\}$ has a finite sum such that

$$\mu(A) \geq \sum_{n=1}^{m} \mu(A_n).$$

Comparison of the two inequalities obtained for $\mu(A)$ gives

$$\mu(A) = \sum_{n=1}^{\infty} \mu(A_n). \qquad \blacksquare$$

According to the definition of regularity, a Borel measure is regular iff for each Borel set A and $\epsilon > 0$ there is a closed set F and an open set U such that $F \subset A \subset U$ and $\mu(U \backslash F) < \epsilon$.

Let P be a mapping from the set X to the set Y; the measure space $(Y, \mathscr{B}, n)$ is the **image** by P of the measure space $(X, \mathscr{A}, m)$ if the measurable sets $B \in \mathscr{B}$ are the subsets of Y such that $P^{-1}(B) \in \mathscr{A}$ and

image of a measure space

$$m(P^{-1}(B)) = n(B) \text{ for every } B \in \mathscr{B}.$$

measure invariant by P

A measure $(\mathscr{A}, m)$ on X is **invariant by** P: $X \to X$ if $P^{-1}(A) \in \mathscr{A}$ for every $A \in \mathscr{A}$ and $m(P^{-1}(A)) = m(A)$.

product measure

Let $(X, \mathscr{A}, m)$ and $M' = (X', \mathscr{A}', m')$ be two σ finite measure spaces. We define the σ-field $\mathscr{A} \otimes \mathscr{A}'$ of subsets of the direct product $X \times X'$ as the smallest σ-field containing all the sets of the form $A \times A'$, $A \in \mathscr{A}$, $A' \in \mathscr{A}'$. It can be proved that there exists a unique measure $m \otimes m'$ defined on $\mathscr{A} \otimes \mathscr{A}'$ such that for all $A \in \mathscr{A}$, $A' \in \mathscr{A}'$.

$$(m \otimes m')(A \times A') = m(A)m'(A').$$

The measure space $M \times M' = (X \times X', \mathscr{A} \otimes \mathscr{A}', m \otimes m')$ is called the **product measure space** of X and X'.

We shall now define the Lebesgue measure on $\mathbb{R}$; it can be proved that it is the only regular Borel measure on $\mathbb{R}$ (up to multiplication by a constant number) invariant by translation. The existence and uniqueness of the measure defined in the following paragraph can be proved[1] by using the Alexandroff and Hahn theorems, and the results of the exercise given below.

Lebesgue measure on $\mathbb{R}$

We shall designate simply by $\mathbb{R}$ the measure space $(\mathbb{R}, \mathcal{B}, m)$, where $\mathbb{R}$ is the real line with its usual topology, $\mathcal{B}$ its Borel σ-field generated by open intervals (cf. exercise p. 36) and l the additive positive set function defined on the open intervals (a, b), $b > a$, by

$$l((a, b)) = b - a.$$

Exercise: 1) Extend l, defined above on the open intervals, to the field (boolean algebra) generated by these intervals. Show that if $b \geq a$,

$$l([a, b]) = l([a, b)) = l((a, b]) = b - a.$$

2) Show that if $[a, b) = \bigcup_{i=1}^{\infty} [a_i, b_i)$, we can write $l([a, b)) = \Sigma_{i=1}^{\infty} l([a_i, b_i))$ provided that the intervals $[a_i, b_i)$ are pairwise disjoint.

Answer: 1) We have, if $a < b < c$,

$$l((a, c)) = c - a = c - b + b - a = l((a, b)) + l((b, c));$$

but

$$(a, b) \cup (b, c) = (a, c) \backslash \{b\}.$$

Therefore, by the additivity,

$$l(\{b\}) = 0.$$

The other results follow.
2) Apply the same ideas as in the proof of the Alexandroff theorem, using the fact that $[a, b]$ is compact. ■

Note that the set of finite unions of semi-closed intervals $[a, b)$ are a field of subsets of $\mathbb{R}$.

Since a point in $\mathbb{R}$ has Lebesgue measure zero, countable unions of points are also of measure zero, but there are other sets of measure zero.

Example: The Cantor set $C = [0, 1] \backslash S$ – where $S = S_1 \cup S_2 \cup \ldots$, $S_1 = (\frac{1}{3}, \frac{2}{3})$ $S_2 = (\frac{1}{9}, \frac{2}{9}) \cup (\frac{7}{9}, \frac{8}{9})$, $S_3 = (\frac{1}{27}, \frac{2}{27}) \cup \ldots$ – is an uncountable set of measure 0. The set S is of measure 1.

[1]For a detailed proof see for instance [Marle, ch. I § 7].

Some authors define the Lebesgue measure on **R** as the measure completed (p. 36) from the one defined above: a subset A of **R** is then said to be measurable if for every $\epsilon > 0$ there exists an open set U_ϵ and a closed set F_ϵ such that

$$F_\epsilon \subset A \subset U_\epsilon$$
$$m(U_\epsilon \backslash F_\epsilon) \leq \epsilon .$$

The measure of A is then the common limit of $m(U_\epsilon)$ and $m(F_\epsilon)$ as ϵ tends to zero and U_ϵ is a decreasing sequence of open sets and F_ϵ is an increasing sequence of closed sets.

The **Lebesgue measure on** $\mathbf{R}^n$ is the product measure $l \otimes l \otimes \ldots \otimes l$. It can be proved that it is, up to multiplication by a constant, the only regular Borel measure on $\mathbf{R}^n$ invariant by translations and rotations.

Lebesgue measure on $\mathbf{R}^n$

Moreover it can be proved that if a measure on $\mathbf{R}^3$ is required to be invariant by translation and rotation, then not every subset can be given a measure since by cutting a unit ball in wild pieces and reassembling it, one can obtain two unit balls (Banach–Tarski paradox). The existence and uniqueness of the Lebesgue measure on $\mathbf{R}^n$ is a particular case of the theorem stated in the following paragraph.

When the locally compact space X is a topological group G, a left invariant regular Borel measure on G (i.e. invariant under the mappings $G \to G$ by $x \mapsto gx$) is called a **Haar measure** (p. 180).

Haar measure

Theorem[1]. *There exists strictly one such measure modulo a constant factor.*

Image of a regular Borel measure. We have given previously (p. 37) the conditions under which the measure space $(Y, \mathscr{B}, n)$ is the image of a measure space $(X, \mathscr{A}, m)$ by a mapping $P: X \to Y$. We consider now the case where X and Y are both locally compact and countable unions of compact sets, and where m is a regular Borel measure on X. It can be shown that the mapping $P: X \mapsto Y$ endows Y with a regular Borel measure $n = P(m)$ by the formula

$$n(B) = m(P^{-1}(B))$$

if and only if

1) the reciprocal image of an open [closed] set of Y is measurable in X
2) the m-measure of the reciprocal image $P^{-1}(K)$ of every compact set $K \subset Y$ is finite.

[1]See for instance [Halmos 1950, ch. XI].

The conditions are obviously necessary. When P satisfies these two conditions it is said to be ***m*-proper**.

m-proper

Every proper (p. 17) continuous mapping is obviously m-proper. Thus every homeomorphism is m-proper.
We will use in Chapter VII the fact that every measurable mapping is m-proper if m is finite (bounded).

4. MEASURABLE FUNCTIONS

measurable mapping

A mapping u from the measurable space $(X, \mathscr{A})$ to the measurable space $(Y, \mathscr{B})$ is called **measurable** if $u^{-1}(B) \in \mathscr{A}$ for every $B \in \mathscr{B}$.

Example 1: If $(Y, \mathscr{B}, n)$ is the image of $(X, \mathscr{A}, m)$ by the mapping P (p. 37) then P is measurable.

Example 2: If X and Y are topological spaces and if $\mathscr{A}$ and $\mathscr{B}$ are the Borel sets of X and Y respectively then every continuous mapping from X into Y is measurable.

If f and g are measurable mappings respectively from $(X, \mathscr{A})$ to $(Y, \mathscr{B})$ and from $(Y, \mathscr{B})$ to $(Z, \mathscr{C})$ then $f \circ g$ is a measurable mapping from $(X, \mathscr{A})$ to $(Z, \mathscr{C})$. The measurable spaces are the members of a category whose morphisms are the measurable mappings.

measurable function

A real function on the measure space $(X, \mathscr{A}, m)$ is **measurable** if it is a measurable map from X into $\mathbb{R}$. An equivalent condition is

$$\{x; a < f(x) < b\} \in \mathscr{A}, \quad \forall a, b \in \mathbb{R}.$$

A complex valued function $f + ig$ is measurable if the real valued functions f and g are measurable.
Let f and g be measurable functions on X, and let $\lambda \in \mathbb{K}$, then the functions $\lambda f, f + g, fg, |f|$ are measurable. These properties are not always true for mappings into arbitrary measure spaces (cf. p. 50).

Exercise: Show that if $(X, \mathscr{A}, m)$ is a complete measure space and the two functions f and g are equal a.e. on X, then g is measurable if f is measurable.

Answer: Let E be the subset, of measure zero, where $f(x) \neq g(x)$. $\{x; a < g(x) < b\}$ is measurable since it may be constructed from four measurable sets – three of them of measure zero

$$\{x; a < g(x) < b\} = \{x; a < f(x) < b\} \cup \{x \in E, a < g(x) < b\}$$
$$\backslash \{x \in E, g(x) \geq b\} \cup \{x \in E, g(x) \leq a\}.$$

The following theorem gives a feeling for what a measurable function is on a locally compact space.

Lusin theorem. Let X be a locally compact space, which is a countable union of compact sets. Let m be a Borel measure on X. A function f on X is measurable if and only if for every compact set K and $\epsilon > 0$ there exists a compact $K_\epsilon \subset K$ and a continuous function g defined on K_ϵ such that $m(K - K_\epsilon) < \epsilon$ and f is equal to g on K_ϵ. This function g may be extended to a continuous function with compact support in X, in such a way that

$$\sup_{x \in X} |g(x)| = \sup_{x \in X} |f(x)|.$$

convergence in measure

Let (f_n) be a sequence of measurable functions on $(X, \mathcal{A}, m)$ that are finite a.e. The sequence is said to **converge in measure** $(f_n) \overset{m}{\rightsquigarrow} f$, to the measurable function f, if

$$\lim_{n=\infty} m(\{x: |f_n(x) - f(x)| > \epsilon\}) = 0 \qquad \text{for every } \epsilon > 0.$$

Convergence in measure does not imply point-wise convergence a.e., but if $m(X) < \infty$, point-wise convergence a.e. implies convergence in measure through a stronger property, uniform convergence a.e. More precisely:

Egorov theorem. If a sequence (f_n) converges point-wise a.e. in a finite measure m on X to a finite measurable function f, then for every $\epsilon > 0$ there exists a subset A of X such that $m(X - A) \leq \epsilon$ and the convergence of (f_n) to f is uniform on A.

5. INTEGRABLE FUNCTIONS

simple functions

The following propositions apply to both real valued and complex valued functions. A function on $(X, \mathcal{A}, m)$ is said to be **simple**[1] **(finitely valued)** if it is zero except on a finite number n of disjoint sets $A_i \in \mathcal{A}$ of finite measure $m(A_i)$ where the function is equal to a finite constant k_i.

step function

[1]If $X = \mathbb{R}$ and if the A_i's are intervals, a simple function is often called a **step function**.

The integral of a simple function is, by definition,

$$\int_X f \, dm = \sum_{i=1}^{n} k_i m(A_i).$$

It can be proved[1] that if f is an extended real valued (with values in $\bar{\mathbb{R}} = \{-\infty\} \cup \mathbb{R} \cup \{+\infty\}$) measurable function defined on X then there exists a sequence (f_n) of simple functions converging point-wise to f. If moreover f is positive, then the f_n may be chosen positive and the sequence (f_n) increasing

$$f_n \geq 0, f_n \leq f_{n+1}.$$

integral

This property paves the way for the following definition of the **integral** of a positive function.

Let f be a positive extended real valued measurable function on the measure space $(X, \mathcal{A}, m)$. Then $\int_X f \, dm$ is the supremum of the integrals $\int_X \rho \, dm$ as ρ ranges over all simple functions with $0 \leq \rho \leq f$.

integrable function

If $\int_X f \, dm$ is finite f is said to be **integrable**, its integral is $\int_X f \, dm$. Properties of the integral are more easily obtained through the following theorem.

Monotone convergence theorem. If an increasing sequence of positive integrable functions ($f_n \geq 0$) converges a.e. to f and if $\lim \int_X f_n \, dm$ *exists, then f is integrable and*

$$\int_X f \, dm = \lim \int_X f_n \, dm.$$

This theorem is a consequence of the following lemma.

Fatou lemma. Let (f_n) be a sequence of positive measurable functions which converge almost everywhere to f. Then

$$\int_X f \, dm \leq \liminf \int_X f_n \, dm.$$

Recall that

$$\liminf a_n = \lim_{n=\infty} \{\inf a_k, k > n\}.$$

An arbitrary real valued function f can always be written $f = f^+ - f^-$ with f^+ and f^- positive functions. It is said to be integrable if both f^+ and f^-

[1] See a simple construction in [Marle, 2.1.15].

are integrable. Its integral is then

$$\int_X f\,dm = \int_X f^+\,dm - \int_X f^-\,dm.$$

It is easy to see that f is integrable iff it is measurable and $\int_X |f|\,dm$ is finite, hence the following theorem.

Theorem. A measurable function f is integrable if and only if $|f|$ is integrable.

Integration on a subset $A \in \mathscr{A}$ of X. Let χ_A be the characteristic function of A. By definition

$$\int_A f\,dm = \int_X f\chi_A\,dm.$$

If $A = \cup_{i=1}^{\infty} A_i$ with $\{A_i\}$ disjoint, then

$$\int_A f\,dm = \sum_{i=1}^{\infty} \int_{A_i} f\,dm.$$

Properties of integrals.
1) An integral is a linear function on the space of integrable functions on A,

$$\int_A (\lambda f + \mu g)\,dm = \lambda \int_A f\,dm + \mu \int_A g\,dm.$$

2) If $|f| \leq |g|$ a.e. and if g is integrable and f measurable, then f is integrable.
3) If $f \leq g$ a.e. and if f and g are integrable, then

$$\int_A f\,dm \leq \int_A g\,dm.$$

Exercise: Use these properties to deduce the following.
1) If f is measurable and $|f|$ bounded on a measurable set A of finite measure, then f is integrable on A and

$$\left|\int_A f\,dm\right| \leq M\,m(A), \qquad \text{when } |f| \leq M \text{ on } A.$$

2) If $f \geq 0$ and $A \subset B$ then $\int_A f\,dm \leq \int_B f\,dm$.
Lebesgue dominated convergence theorem. If a sequence of integrable functions (f_n) converges almost everywhere to a function f, and if $|f_n| \leq g$

with g integrable, then f is integrable and

$$\lim_{n=\infty} \int_X f_n \, dm = \int_X \lim_{n=\infty} f_n \, dm = \int_X f \, dm.$$

Example: In the following example the sequence (f_n) of integrable functions is bounded from below by an integrable function g, BUT IS NOT bounded from above. One then finds that $\lim_{n=\infty} \int_X f_n \, dx$ IS NOT equal to $\int_X (\lim_{n=\infty} f_n) \, dx$.

Let $f_n(x) = n^2 x \, e^{-nx}$ with $x \in [0, 1]$.

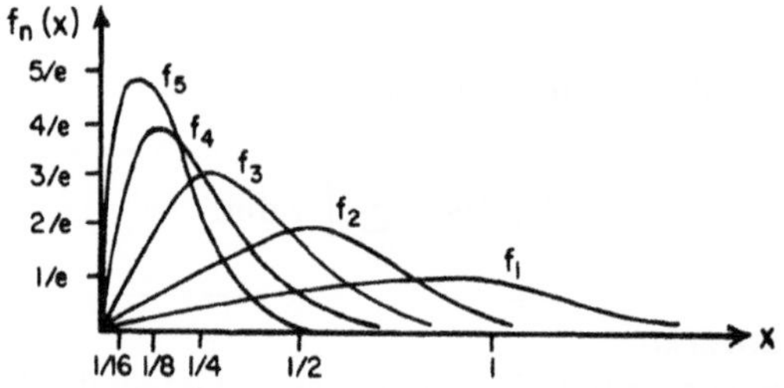

The maximum of f_n is n/e at $x = 1/n$.
The sequence (f_n) is bounded from below by $g(x) = 0$, it is not bounded from above; for n sufficiently large the peak of f_n can be made higher than any preassigned value.
The sequence (f_n) converges not uniformly but pointwise to the function $f = 0$: for each $x \in [0, 1]$, $\lim_{n=\infty} f_n(x) = 0$, and $\int_0^1 \lim_{n=\infty} f_n \, dx = 0$.
On the other hand

$$\lim_{n=\infty} \int_0^1 f_n \, dx = \lim_{n=\infty} \int_0^1 n^2 x \, e^{-nx} \, dx = \lim_{n=\infty} \int_0^n u \, e^{-u} \, du = \int_0^\infty u \, e^{-u} \, du = 1.$$

Hence

$$\lim_{n=\infty} \int_0^1 f_n \, dx > \int_0^1 \left(\lim_{n=\infty} f_n\right) dx. \qquad \blacksquare$$

If the measure space $(X, \mathscr{A}, m)$ is a product $(M \times M', \mathscr{A} \otimes \mathscr{A}', m \otimes m')$

(cf. p. 37) the integral I of a function integrable on $M \times M'$ is denoted

$$I = \int_{M\times M'} f\,\mathrm{d}m\,\mathrm{d}m'.$$

One can prove the following theorem, important in many applications,

Fubini theorem. A measurable function f on $M \times M'$ is integrable if and only if one of the integrals

$$\int_M \left\{\int_{M'} |f|\,\mathrm{d}m'\right\}\mathrm{d}m \quad\text{and}\quad \int_{M'}\left\{\int_M |f|\,\mathrm{d}m\right\}\mathrm{d}m'$$

exists (is finite).

If f is integrable, then

$$\int_M\int_{M'} f\,\mathrm{d}m\,\mathrm{d}m' = \int_M\left\{\int_{M'} f\,\mathrm{d}m'\right\}\mathrm{d}m = \int_{M'}\left\{\int_M f\,\mathrm{d}m\right\}\mathrm{d}m'.$$

Example: The measurable function on $\mathbb{R}^2$: $f(x, y) = (x^2 - y^2)/(x^2 + y^2)^2$ is not integrable on the rectangle

$$B = \{(x, y); 0 \le x \le 1, 0 \le y \le 1\} \subset \mathbb{R}^2.$$

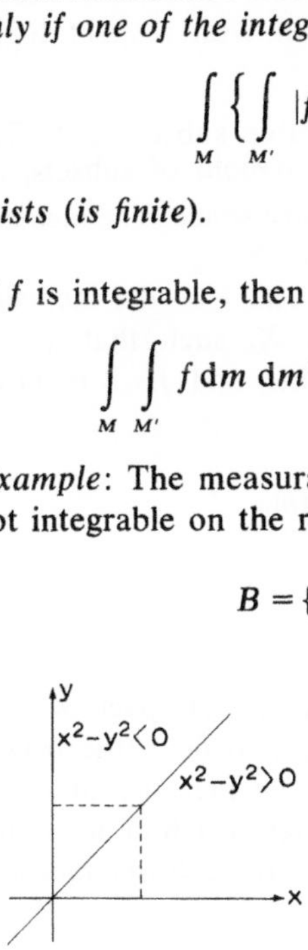

$$\int_0^1 \mathrm{d}y \int_0^1 \mathrm{d}x\, f(x, y) = -\frac{\pi}{4}$$

$$\int_0^1 \mathrm{d}x \int_0^1 \mathrm{d}y\, f(x, y) = \frac{\pi}{4}$$

It follows from $\displaystyle\int_y^1 f(x, y)\,\mathrm{d}x = -\frac{1}{y^2+1} + \frac{1}{2y}$ and

$\displaystyle\int_0^y (-f(x, y))\,\mathrm{d}x = \frac{1}{2y}$ that $\displaystyle\int_0^1 \mathrm{d}y \int_0^1 \mathrm{d}x |f(x, y)|$ is infinite.

Image measure and integral. Let $(Y, \mathcal{B}, n)$ be a measure space, image under the mapping u of the measure space $(X, \mathcal{A}, m)$. It can be proved that a measurable function on Y is integrable on Y if and only if $f \circ u$ is integrable on X. Then

$$\int_Y f \, \mathrm{d}n = \int_X f \circ u \, \mathrm{d}m.$$

Remark: If f is measurable, it is clear that $f \circ u$ is measurable. A particular case of the above formula is the change of variables in the Lebesgue integral on $\mathbb{R}^n$ (see below). Another useful formula, not to be confused with the preceding one, is the integral with respect to an induced measure, which we shall give presently.

induced measure

Let $(X, \mathscr{A}, m)$ be a measure space, Y a measurable subset of X. The subsets $\bar{A} = Y \cap A$, $A \in \mathscr{A}$ are in $\mathscr{A}$ and are a σ-field of subsets, $\bar{\mathscr{A}}$. Therefore $(Y, \bar{\mathscr{A}}, \bar{m})$, with $\bar{m}(\bar{A}) = m(\bar{A})$ is a measure space. The measure $(\bar{\mathscr{A}}, \bar{m})$ is said to be the **measure induced** by m on Y.
Let χ_Y be the characteristic function of $Y \subset X$, and i be the inclusion map $Y \to X$. If f is a measurable function on X, such that $\chi_Y f$ is measurable and if, moreover, $\chi_Y f$ is m-integrable and $f \circ i$ is $\bar{m}$-integrable then the integrals are equal

$$\int_Y f \circ i \, \mathrm{d}\bar{m} = \int_X \chi_Y f \, \mathrm{d}m = \int_Y f \, \mathrm{d}m.$$

6. INTEGRATION ON LOCALLY COMPACT SPACES

We have defined a Borel measure on a locally compact space X as a measure (countably additive mapping) on the Borel σ-field (the σ-field generated by the open sets) which is such that the measure of every compact set is finite. An easy consequence of such a definition is that every continuous function with compact support on a Borel measure space is integrable.[1]

Lebesgue integral. A case of special importance for applications is the Lebesgue integral on $\mathbb{R}^n$.

Lebesgue integral

The integral with respect to the Lebesgue measure on $\mathbb{R}^n$ (p. 38) is called the **Lebesgue integral** and denoted

$$\int_{\mathbb{R}^n} f \, \mathrm{d}x.$$

All the previously given properties of the integral apply to that case. The formula for the integral with respect to an image measure under a

[1]We shall see (p. 53) that these functions are dense in the space of integrable functions.

given diffeomorphism is called the "change of variable formula" and takes the following explicit form, as in the theory of Riemann integration.

Change of variable theorem. Let f be a Lebesgue integrable function on $V \subset \mathbb{R}^n$. Let φ: $U \to V$ by $u \mapsto v = \varphi(u)$ be a diffeomorphism of $U \subset \mathbb{R}^n$ onto V with Jacobian determinant at u, $\det \varphi'(u) = \det[\partial\varphi^i(u)/\partial u^j] = \det[\partial v^i/\partial u^j] \equiv D(v^i)/D(u^j)$. *Then*

$$\int_U f(\varphi(u))|\det \varphi'(u)|\, du^1 \cdots du^n = \int_{V=\varphi(U)} f(v)\, dv^1 \cdots dv^n.$$

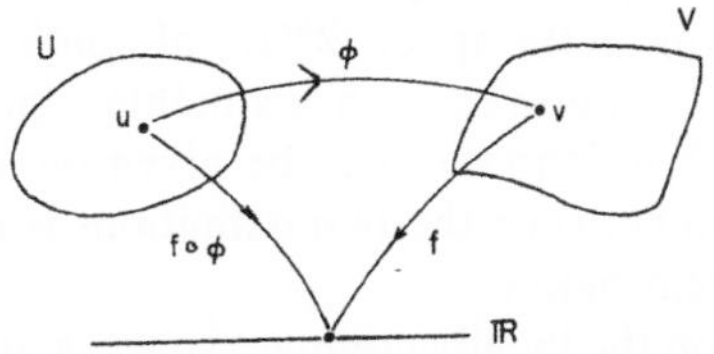

Remark: Remember that, by a previous theorem, a measurable function f is Lebesgue integrable if and only if $|f|$ is integrable. However, there are interesting functions, important for physical applications, which are not Lebesgue integrable on $\mathbb{R}$ and for which there exist well defined – often so-called "improper" – Riemann integrals.

Example 1: A Lebesgue integrable function which is not Riemann integrable. Let

$$f: [0, 1] \to \mathbb{R} \text{ by } f(x) = 1 \quad \text{if } x \text{ is irrational}$$
$$f(x) = 0 \quad \text{if } x \text{ is rational.}$$

The Riemann integral of f is not defined; its Lebesgue integral is equal to 1.

Example 2: Important functions which are not Lebesgue integrable. The functions $\sin x/x$, $\cos x^2$, $\sin x^2$, etc... are not Lebesgue integrable although they have finite improper Riemann integrals. An **improper Riemann integral on R** is the limit of $\int_a^b f\, dx$ when $a \rightsquigarrow -\infty$ and $b \rightsquigarrow +\infty$. There is no reasonable definition of an improper Riemann integral on $\mathbb{R}^n$ because there are quite reasonable functions on $\mathbb{R}^n$ for which two different quite reasonable limiting processes give different results. For example $I = \int\int_D \sin(x^2 + y^2)\, dx\, dy$ where D is the first quadrant in R^2. We can define I either by

improper Riemann integral

$$I = \lim_{n=\infty} \int_0^n dx \int_0^n dy \sin(x^2 + y^2) = \frac{\pi}{4}$$

or by

$$I = \lim_{n=\infty} \int_0^{\pi/2} d\theta \int_0^n r\, dr \sin(r^2) = \lim_{n=\infty} \frac{\pi}{4}(1 - \cos(n^2))$$

which does not exist.

If a function is Lebesgue integrable and Riemann integrable, the two integrals are equal.

Radon measure. We shall see in Chapter VI (distributions) another definition of a measure, on an open set U of R^n, as a distribution of order zero (linear form on the space $\mathscr{D}^0(U)$ of continuous functions with compact support in U endowed with a suitable topology), also called a Radon measure. Such a definition can be given on any locally compact space. The connexion between the two definitions is made precise by the Riesz–Markov theorem below.

Radon measure

$\mathscr{D}^0(X)$

A **Radon measure** μ on the locally compact space X is a continuous linear form on the space $\mathscr{D}^0(X)$, where $\mathscr{D}^0(X)$ is the space of continuous functions on X with compact support, endowed with the inductive limit topology of the topologies of uniform convergence on compact sets (cf. Chapter VI). A linear form μ on $\mathscr{D}^0(X)$ is continuous iff for every compact set $K \subset X$ there is a positive number $C(K)$ such that for every continuous f with support in K,

$$|\mu(f)| \leq C(K) \sup_{x \in K} |f(x)|.$$

positive Radon measure

A Radon measure μ is **positive** if $\mu(f) \geq 0$ for all $f \geq 0$.
If m is a Borel measure on the locally compact space X, we can associate with it a positive Radon measure μ on X, by setting

$$\mu(f) = \int_X f\, dm, \qquad f \in \mathscr{D}^0(X).$$

The above relation is bijective under the conditions stated in the following theorem.

Riesz–Markov theorem. If X is a locally compact space, that is also a countable union of compact sets, then for every positive Radon measure μ, there exists a unique regular Borel measure m such that

$$\mu(f) = \int_X f\, dm.$$

7. SIGNED AND COMPLEX MEASURES

We shall show in this paragraph that signed and complex measures are essentially defined in terms of positive measures.

signed measures

A **signed measure space** $(X, \mathcal{A}, m)$ has the same definition as a measure space, except that the mapping m is now from the σ-field into the extended real line, $\mathbb{R} \cup \{+\infty\}$. Note that $\{-\infty\}$ is still excluded due to the impossibility of defining $\{-\infty\}+\{+\infty\}$.

Jordan decomposition theorem. A signed measure $(\mathcal{A}, m)$ is in a unique way the difference of two positive disjoint measures

$$m = m^+ - m^-$$

with m^+ and m^- positive measures such that there exists $A \in \mathcal{A}$ with

$$m^-(A) = 0, \qquad m^+(X \backslash A) = 0.$$

total variation

The positive measure $(\mathcal{A}, |m|)$ with $|m| = m^+ + m^-$ is called the **total variation** of $(\mathcal{A}, m)$.

f is integrable with respect to m iff it is integrable with respect to $|m|$. Its integral is

$$\int_X f \, \mathrm{d}m = \int_X f \, \mathrm{d}m^+ - \int_X f \, \mathrm{d}m^-.$$

It satisfies the inequality

$$\left| \int_X f \, \mathrm{d}m \right| \leq \int_X |f| \, \mathrm{d}|m|.$$

complex measures

The definition of a complex measure is analogous, but the mapping m now takes its values in the extended complex numbers, $\mathbb{R} \cup \{+\infty\} + \mathrm{i}(\mathbb{R} \cup \{+\infty\})$; thus $m = m_1 + \mathrm{i}m_2$, m_1 and m_2 signed measures. The total variation $|m|$ is defined by

$$|m|(A) = \sup \sum_{i=1}^{\infty} |m(A_i)|$$

where the sup is taken for all countable families $\{A_i; i \in N\}$ of pair wise disjoint elements of $\mathcal{A}$ included in A. Obviously, m being countably additive,

$$|m(A)| \leq |m|(A).$$

It can be shown that the definition given previously of the total variation of a signed measure in terms of its Jordan decomposition agrees with this one.

It can be proved that $(\mathcal{A}, |m|)$ is a measure on X.

Let f be a complex valued function defined a.e. on X, $f = f_1 + \mathrm{i} f_2$, f_1 and f_2 real. f is said to be integrable with respect to the complex measure $m = m_1 + \mathrm{i} m_2$ if f_1 and f_2 are integrable with respect to m_1 and m_2, its integral being given by

$$\int_X f \, \mathrm{d}m = \int_X f_1 \, \mathrm{d}m_1 - \int_X f_2 \, \mathrm{d}m_2 + \mathrm{i} \int_X f_1 \, \mathrm{d}m_2 + \mathrm{i} \int_X f_2 \, \mathrm{d}m_1.$$

It can be proved that f is integrable with respect to m iff it is measurable and $|f|$ is integrable with respect to $|m|$. Then

$$\left| \int_X f \, \mathrm{d}m \right| \leq \int_X |f| \, \mathrm{d}m.$$

8. INTEGRATION OF VECTOR VALUED FUNCTIONS

We shall in this paragraph, give some results about the integration of a map $f: X \to E$ from a measure space $(X, \mathcal{A}, m)$ into a Banach space E. Though the theory given previously for numerical functions carries essentially through, there are a few technical difficulties. The first difficulty is that if f and g are measurable mappings from $(X, \mathcal{A}, m)$ into $(E, \mathcal{B})$ – Banach space E with its Borel σ-field – the sum $f + g$ is not necessarily a measurable mapping. The second difficulty is that if f is a measurable mapping, it is possible that there does not exist a sequence (f_n) of simple functions[1] converging point wise to f. However, the technical difficulties do not arise if we restrict[2] the theory to the case where $(X, \mathcal{A}, m)$ is a σ-finite measure space and E a separable Banach space.

simple mapping

A mapping $f: X \to E$ is called a **simple mapping** if it has value zero except on a finite number A_i, $i = 1, \ldots n$ of subsets $A_i \in \mathcal{A}$ of X, with finite measure $m(A_i) < \infty$, where it has a constant value. A simple mapping f can be written

$$f = \sum_{i=1}^{n} a_i \chi_{A_i}, \qquad a_i \in E$$

with χ_{A_i} the characteristic function of A_i.

[1]See below the definition which generalizes the definition of a simple function, p. 41.
[2]For the general case, see for instance [Marle, ch. II].

It can be shown that, if $(X, \mathscr{A}, m)$ is a σ-finite (cf. p. 36) measure space and E a separable Banach space then every measurable map $f: X \to E$, E endowed with its Borel σ-field, is the limit, point wise almost everywhere, of a sequence (f_n) of simple maps.
The integral of a step map is

$$\int_X f \, \mathrm{d}m = \sum_{i=1}^{n} a_i m(A_i) \in E.$$

It is obviously linear on the space of simple maps.
A **Cauchy sequence of simple maps** is a sequence (f_n) such that, for every $\epsilon > 0$ there exists N such that

Cauchy sequence of simple maps

$$\int_X \|f_n - f_k\| \, \mathrm{d}m < \epsilon \qquad \text{for } n, k > N,$$

where $\| \quad \|$ is the norm in E. The map $x \mapsto \|f(x)\|$ is a real valued function. It is not difficult to show that for a simple map

$$\left\| \int_X f_n \, \mathrm{d}m \right\| \leq \int_X \|f_n\| \, \mathrm{d}m.$$

Therefore the sequence $(\int_X f_n \, \mathrm{d}m)$ is a Cauchy sequence in E if (f_n) is a Cauchy sequence of simple maps in the sense defined above. Since E is a Banach space the sequence $(\int_X f_n \, \mathrm{d}m)$ has therefore a limit in E. We are then led to the following definition.

The mapping $f: X \to E$ is m-**integrable** if there exists a Cauchy sequence (f_n) of simple maps, converging a.e. to f. The integral of f is the limit

integrable mapping

$$\int_X f \, \mathrm{d}m = \lim_{n=\infty} \int_X f_n \, \mathrm{d}m.$$

It can be shown that the limit is independant of the choice of the sequence (f_n).
It can also be shown that, in the case where f is a numerical function, the definition just given is equivalent to the preceding one (cf. p. 42). A theorem analogous to the one about the equivalence of integrability and absolute integrability is still valid.

Theorem. A measurable mapping f from X into the Banach space E is integrable if and only if the mapping $\|f\|: X \to \mathbb{R}^+$ is integrable, and then

$$\left\| \int_X f \, \mathrm{d}m \right\| \leq \int_X \|f\| \, \mathrm{d}m.$$

9. L^1 SPACE

Let $(X, \mathscr{A}, m)$, here abbreviated to X, be a measure space with a positive measure m. The space $\mathscr{L}^1(X)$ of integrable functions over X is the space of measurable functions such that the following integral is finite

$$\int_X |f| \, dm < \infty.$$

Clearly the function $f \mapsto \int_X f \, dm$ is linear on $\mathscr{L}^1(X)$ and the mapping

$$f \mapsto \int_X |f| \, dm$$

has the properties of subadditivity and positive homogeneity required of a norm (cf. p. 26). It is not true that $\int_X |f| \, dm = 0$ implies $f = 0$, but it can be proved that

$$\int_X |f| \, dm = 0 \text{ implies } f = 0 \text{ a.e.}$$

Exercise: Show the above statement.

Answer: Assume $\int_X |f| \, dm = 0$. Let $A_n = \{x; |f(x)| \geq 1/n\}$. Then $|f| \geq (1/n)\chi_{A_n}$, therefore $\int_X |f| \, dm \geq (1/n)m(A_n)$ which implies $m(A_n) = 0$. Since the set where $|f| > 0$ is the union of the set A_n, it has measure zero.

Thus if we identify functions which are equal almost everywhere – strictly speaking if we consider equivalence classes of functions under the relation

$$f \sim g \text{ iff } f = g \text{ almost everywhere}$$

then the mapping $f \mapsto \int_X |f| \, dm$ has the properties required of a norm. ■

$L^1(X)$ The space of (classes of) functions defined a.e. and integrable on X, with respect to the measure m, together with the norm

$$f \mapsto \|f\|_{L^1(X)} = \int_X |f| \, dm$$

is called $L^1(X)$.

A fundamental property of $L^1(X)$ is the following

Fischer–Riesz theorem. $L^1(X)$ is a Banach space (complete normed space).

Another interesting property when X is a locally compact topological space, and m a Borel measure on X is the following.

Theorem. The space $\mathscr{D}^0(X)$ of continuous functions with compact support is dense in $L^1(X)$.

All the properties listed in this paragraph hold when f takes its values in a Banach space[1].

10. L^p SPACE

Let $(X, \mathscr{A}, m)$, here abbreviated to X, be a measure space; let $L^p(X)$ denote the space of (classes of) measurable functions defined a.e. on X, such that $|f|^p$ is integrable. Set

$$\|f\|_{L^p(X)} \equiv \|f\|_p = \left\{\int_X |f|^p \,\mathrm{d}m\right\}^{1/p}.$$

$L^p(X)$ $p \geq 1$

Minkovski inequality

The fact that for $p \geq 1$ $L^p(X)$ is a vector space, and $\| \quad \|_p$ a norm on this space is a consequence of the so-called **Minkovski inequality**

$$\|f + g\|_p \leq \|f\|_p + \|g\|_p$$

Hölder inequality

which is itself a consequence of the so-called **Hölder inequality**, valid for $p \geq 1$, $q \geq 1$, $1/p + 1/q = 1$

$$\|fg\|_1 \leq \|f\|_p \|g\|_q.$$

When $p = 2$, $q = 2$, this is the Cauchy–Schwarz inequality (p. 11).

$L^2(X)$

The norm in $L^2(X)$ can be deduced from the following scalar product $(f|g) = \int_X f\bar{g}\,\mathrm{d}x$, namely $\|f\|_2^2 = (f|f)$.

Exercise: Prove the Hölder and Minkovski inequalities.

Answer: Set $|f|^p = F$, $|g|^q = G$, $1/p = \alpha$, $1/q = \beta = 1 - \alpha$.
From the inequality $t^\alpha \leq \alpha t + 1 - \alpha = \alpha t + \beta$ – expressing that the tangent to the curve $y = t^\alpha$ at $t = 1$ is above the curve – we obtain by substitution of t/v for t

$$t^\alpha v^\beta \leq \alpha t + \beta v.$$

[1]See for instance [Marle].

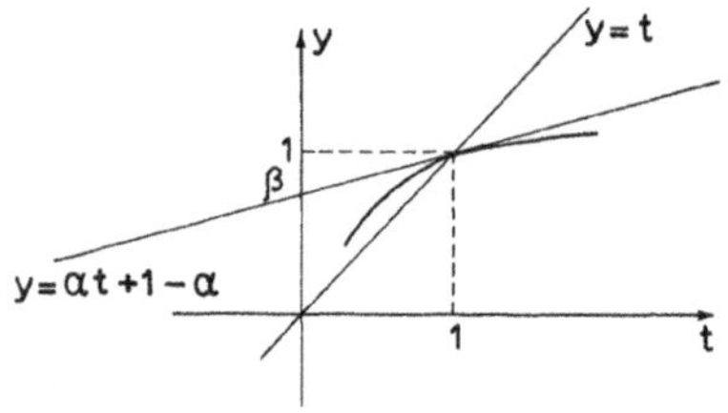

Setting $F(x)/\int_X F(x)\,\mathrm{d}m = t$ and $G(x)/\int_X G(x)\,\mathrm{d}m = v$ in this inequality and integrating, one finds

$$\int_X F^\alpha G^\beta\,\mathrm{d}m \Big/ \Big(\int_X F\,\mathrm{d}m\Big)^\alpha \Big(\int_X G\,\mathrm{d}m\Big)^\beta \leq \alpha + \beta = 1.$$ ■

mean convergence

In the space $L^p(X)$ strong convergence is also called **convergence in the mean** of order p.

The spaces $L^p(X)$, $1 \leq p \leq \infty$, are locally convex. For the sake of illustration, consider the space $L^p(\{1,2\})$ of functions $x = (x^1, x^2)$ on the measure space $\{1,2\}$ where each point has measure 1. The ϵ-open balls defined by the norm

$$\|x\|_p = (|x^1|^p + |x^2|^p)^{1/p}, \qquad p \geq 1$$

are shown in the diagram for $p = 1, 2, \ldots \infty$. When p increases from 1 to ∞, the ball swells continuously from the inner square to the outer square.
The $\lim_{p\to\infty} \|x\|_p$ is equal to $\|x\|_\infty \equiv \max\{|x^1|, |x^2|\}$.

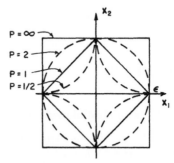

Indeed if $|x^1| = |x^2|$, then

$$\lim \|x\|_p = \lim 2^{1/p}\,|x^1| = |x^1|;$$

if $|x^1| < |x^2|$, then

$$\lim \|x\|_p = \lim (|x^1/x^2|^p + 1)^{1/p} |x^2|.$$ ■

The mapping $x \mapsto (|x_1|^p + |x_2|^p)^{1/p}$ for $p < 1$ is not a norm because it violates the triangle inequality; the set $\{x; (|x_1|^p + |x_2|^p)^{1/p} < 1\}$ with $p < 1$ is not convex.

The spaces $L^p(X)$ with $p \geq 1$ are Banach spaces; the space $L^2(X)$ is a Hilbert space, as a consequence of the following theorem.

Riesz theorem. The spaces $L^p(X)$ with $p \geq 1$ are complete, that is, there is an $f \in L^p(X)$ such that

$$\lim_{n=\infty} \|f_n - f\|_p = 0 \quad \text{if} \quad \lim_{m,n=\infty} \|f_m - f_n\|_p = 0.$$

Exercise: Show that the subspace of $L^2(a, b)$ which consists of continuous functions defined on the interval $[a, b]$ is not complete.

Answer: Consider for example the sequence (f_n) of continuous functions in $L^2(-1, 1)$ defined by

$$f_n(x) = \begin{cases} 0 & \text{if } -1 \leq x \leq -1/n \\ (nx+1)/2 & -1/n \leq x \leq 1/n \\ 1 & 1/n \leq x \leq 1 \end{cases} \qquad n = 1, 2, 3, \ldots$$

then $\lim \|f_m - f_n\| = 0$ as $m, n \to \infty$, but

$$f_n(x) \rightsquigarrow f(x) = \begin{cases} 0 & \text{if } -1 \leq x < 0 \\ 1 & \text{if } \ 0 < x \leq 1. \end{cases}$$

Hence f is the limit of f_n in $L^2(-1, 1)$ but is not in the subspace of continuous functions. ■

Inclusion theorem. If X is of finite measure – for instance if X is a compact subset of $\mathbb{R}^n$ together with the Lebesgue measure (cf. p. 39)– then

$$L^p \subset L^{p'} \qquad \text{for} \qquad 1 \leq p' < p.$$

Proof: Using Hölder's inequality with $g = 1$ we obtain

$$\|f\|_1 < \|f\|_p (m(X))^{1/q} \text{ with } 1/p + 1/q = 1 \text{ i.e. } L^p \subset L^1 \text{ for } p \geq 1.$$

On the other hand

$$\|f\|_{p'} = \{\||f|^{p'}\|_1\}^{1/p'} \leq \{\||f|^{p'}\|_r (m(X))^{1/r'}\}^{1/p'} \text{ for } 1/r + 1/r' = 1.$$

Using $(\||f|^{p'}\|_r)^{1/p'} = \|f\|_{rp'}$, setting $rp' = p$, and using $p \geq 1$ we obtain the desired result.

Example: In a finite interval $(a, b) \subset \mathbb{R}$ if $f \in L^2(a, b)$ then $f \in L^1(a, b)$ but the converse is not true

$$\frac{1}{\sqrt{x}} \in L^1(0,1) \text{ and } \frac{1}{\sqrt{x}} \notin L^2(0,1).$$

If the interval is infinite, there is no inclusion property:

$$\frac{1}{x} \text{ is in } L^2(1,\infty) \text{ but not in } L^1(1,\infty);$$

$$\frac{1}{\sqrt{x}}\frac{1}{x+1} \text{ is in } L^1(0,+\infty) \text{ but not in } L^2(0,\infty).$$

$L^\infty(X)$ Another example of a Banach space is the space $L^\infty(X)$ of (classes of a.e. defined) measurable functions bounded almost everywhere on X which has been given the norm

$$\|f\|_\infty = \operatorname*{ess\,sup}_X |f(x)|.$$

"ess" stands for "essentially", more precisely the norm is the smallest value M for which $|f(x)| \leq M$ almost everywhere.

Duality properties. It is clear from the Hölder inequality that any function $g \in L^q(X)$, $1 \leq q \leq \infty$, defines a continuous linear form on $L^p(x)$, $1/p + 1/q = 1$, by

$$L^p \to \mathbb{C} : f \mapsto \int_X fg \, dx.$$

Through this identification of $g \in L^q$ with an element of $(L^p)'$, dual of L^p, we can write

$$L^q \subset (L^p)', \qquad 1 \leq p \leq \infty, \qquad 1 \leq q \leq \infty, \qquad 1/p + 1/q = 1.$$

The following theorem states that this identification is a bijective mapping when $1 < p < \infty$.

Riesz representation theorem. The dual of $L^p(X)$, $1 < p < \infty$ is (isomorphic to) $L^q(X)$, $1/q = 1 - 1/p$.

The case $p = 2$ is a special case of a theorem already stated for Hilbert spaces (p. 31).

For most spaces the identification mapping given above is still bijective for $p = 1$. Namely, if X is of σ-finite measure, the dual of $L^1(X)$ is (isomorphic to) $L^\infty(X)$. Conversely it is in general *not true* that the dual of $L^\infty(X)$ is $L^1(X)$.

Example: Let X be $\mathbb{R}$ with its usual Lebesgue measure. Consider the linear form defined on continuous bounded functions f by

$$l: f \mapsto f(x_0)$$

x_0 being a given point of $\mathbb{R}$.
This linear form is continuous in the $L^\infty(\mathbb{R})$ norm since

$$|f(x_0)| \leq \|f\|_{L^\infty(\mathbb{R})}.$$

Therefore it can be extended to a continuous linear form on $L^\infty(\mathbb{R})$ (cf. Hahn–Banach theorem, Ch. VI). This linear form cannot be given by

$$l(f) = \int_{\mathbb{R}} fg \, \mathrm{d}x, \qquad g \in L^1(X).$$

See, for example, the proof (p. 438) that the Dirac measure cannot be represented by a locally integrable function.

E. KEY THEOREMS IN LINEAR FUNCTIONAL ANALYSIS

Some of the classical properties of linear operators on $\mathbb{R}^n$ (matrices) extend to linear operators on normed and metrizable vector spaces – particularly to compact operators. The problems and the results are, however, much more rich and varied in the infinite dimensional case. The theory of linear operators on Banach spaces, and particularly on Hilbert spaces, is treated in many excellent texts, old and new.[1] In this section we shall give only the definitions and the fundamental results.
A mapping from X into Y is also called an operator on X, but the name "operator on X" is also used for a mapping T from a subset of X (called the domain $D(T)$ of T) onto a subset of Y (called the range $R(T)$ of T).

1. BOUNDED LINEAR OPERATORS

A continuous linear operator T on a vector space X can always be extended to a continuous linear operator with domain X, since by the definition of continuity, $T(x)$ is defined for all x in a neighborhood of a point $a \in D(T)$, and therefore on all of X by linearity.
We have already seen (p. 22) that a linear operator is continuous if and only if it is continuous at the origin.

[1]See in particular, in relation with modern developments in theoretical physics, [Reed and Simon].

The space of continuous linear mappings from X into Y, denoted $\mathscr{L}(X, Y)$, can naturally be endowed with a vector space structure, and with topologies depending on the topologies of X and Y. The following theorems show that $\mathscr{L}(X, Y)$ has a natural structure as a normed space if X and Y are normed spaces, and as a Banach space if Y is a Banach space.

bounded

A linear operator $T: X \to Y$ between normed spaces X and Y is said to be **bounded** if there is a real number $K \geq 0$, called the bound of T, such that

$$\|T(x)\| \leq K\|x\|, \qquad \text{for every } x \in X.$$

Theorem. A linear operator $T: X \to Y$, X and Y being normed spaces, is continuous if and only if it is bounded.

Proof: That bounded implies continuous is obvious, let us prove the converse. Assume there does not exist a K. For each n, one can find an x_n such that $\|T(x_n)\| > n\|x_n\|$. Hence $\|T(x_n/n\|x_n\|)\|$ is larger than 1 and does not tend to zero when n tends to infinity, even though $\|x_n/n\|x_n\|\| = 1/n$ tends to zero. ■

norm

The **norm** of a continuous operator between normed spaces is given by

$$\|T\| = \inf\{K; \|T(x)\| \leq K\|x\| \quad \forall x \in X\}.$$

One can see that, by the linearity of T, it is equivalent to

$$\|T\| = \sup\{\|T(x)\|; \|x\| = 1\}.$$

Example: An unbounded operator. Let X be the space of C^∞ functions f on the interval $[0, 1]$ with the uniform norm $\|f\| = \sup|f(x)|$. The differential operator $\mathrm{d}/\mathrm{d}x$ is a linear operator from X into X which is not bounded. For example let $f = \sin kt$; then $\|(\mathrm{d}/\mathrm{d}t)f\| = k\|f\|$. Since k can take arbitrarily large values, there is no K such that

$$\left\|\frac{\mathrm{d}}{\mathrm{d}t} f\right\| \leq K\|f\|.$$

On a finite dimensional normed vector space, a linear operator is always continuous and hence bounded.

Theorem. Let Y be a Banach space, let X be a normed space not necessarily complete, then the space $\mathscr{L}(X, Y)$ of linear continuous mappings from X into Y, endowed with the norm $\|T\|$, is a Banach space.

It follows from this theorem that the dual $X' = \mathscr{L}(X, \mathbb{K})$ of a normed space X is a Banach space. The **bidual** $X'' = \mathscr{L}(X', \mathbb{K})$ is therefore also a Banach space. One says that X is a subspace of X'' in the following sense.

bidual

Theorem. There exists an isomorphism J of X onto a subspace of X''.

Proof: Take $J: X \to X''$ *by* $x \mapsto \langle \cdot, x\rangle$, where $\langle \cdot, x\rangle$ is the continuous linear form on X' given by $x' \mapsto \langle x', x\rangle$. It is easy to show that J is linear, injective and an isometry in the sense that

$$\|J(x)\|_{X''} = \|x\|_X. \qquad \blacksquare$$

X is said to be **reflexive** if $X'' = X$ (if the mapping J is an isomorphism of X onto X'').

reflexive

Example: $L^p(X)$ (p. 53) is reflexive for $1 < p < \infty$.

A **Banach algebra (B-algebra)** is a Banach space together with an associative internal operation – usually called multiplication. If X is a Banach space, $\mathscr{L}(X, X) \equiv \mathscr{B}(X)$ together with the multiplication defined by

Banach algebra

$$(T_1 T_2)x = T_1(T_2 x)$$

is a Banach algebra. It follows from the definition that

$$\|T_1 T_2\| \le \|T_1\| \, \|T_2\|.$$

An **involutive B-algebra** is a B-algebra with a norm preserving involution. An involution in a Banach algebra $\mathscr{B}$ is a mapping $\mathscr{B} \to \mathscr{B}$ by $T \mapsto T^*$ such that[1]

involutive B-algebra involution

$(T+S)^* = T^* + S^*$	The involution preserves the linear structure
$(\alpha T)^* = \bar{\alpha} T^*$	of the space of operators.
$(ST)^* = T^* S^*$	It preserves the algebraic structure.
$T^{**} = T$	It is an isomorphism.
$\|T^*\| = \|T\|$	It is continuous.

T^* is called the **adjoint** of T.

adjoint

A **C*-algebra** is an involutive Banach algebra which in addition satisfies the condition

C* algebra

$$\|T^* T\| = \|T\|^2.$$

T is said to be **self-adjoint** if $T = T^*$. A subalgebra of an involutive

self-adjoint

[1]An involutive algebra satisfies only the first four axioms.

algebra is said to be self-adjoint if it contains the adjoint of each of its elements.

normal

T is said to be **normal** if $T^*T = TT^*$.

Example 1: In $\mathbb{C}$, the mapping $z \mapsto \bar{z}$ is an involution.

Example 2: Let X be a Hilbert space; a [canonical] involution on $\mathcal{B}(X)$ is defined by $(Tf|g) = (f|T^*g)$.

Families of bounded operators on Hilbert spaces such that $\|T^*T\| = \|T\|^2$ can be considered as representations of a C*-algebra. Such representations of C*-algebras play an important role in physics.

transposed operator

If T is a linear operator on a Banach space, the **transposed operator**, denoted T',[1] or more often in this book $\tilde{T}$, is a linear operator on the dual space X' defined by

$$\langle \tilde{T}x, y\rangle = \langle x, Ty\rangle, \qquad x \in X', \quad y \in X$$

where $\langle\ ,\ \rangle$ denotes the duality between X' and X.

Remark: If X is a Hilbert space it may be identified with its dual, $\tilde{T}$ is then an operator on X and corresponds to the canonical adjoint T^*.

spectrum

point spectrum

eigenvalues

eigenvectors

continuous spectrum

residual spectrum

Let T be a continuous operator on a Banach space X. A complex number λ is said to be in the **spectrum** $\sigma(T)$ of T if $T - \lambda$ Id is not bijective.
1) If $T - \lambda$ Id is not injective (if $Tx - \lambda x = 0$ has a solution $x \neq 0$) λ is called an **eigenvalue** of T, and is said to be **in the point spectrum** of T. The corresponding solutions are called **eigenvectors**.
2) If $T - \lambda$ Id is injective but not surjective
   a) λ is said to be in the **continuous spectrum** of T if the range of $T - \lambda$ Id is dense in X,
   b) otherwise λ is said to be in the **residual spectrum** of T.

If X is finite dimensional, T has only a point spectrum.

Many properties of linear operators on Banach spaces, and of Banach algebras, can be obtained from the spectrum. We list only a very few of the most important ones in practice.

Theorem. The spectrum of a continuous operator T on a Banach space X is not empty, and is contained in the closed disc of the complex plane of radius $\|T\|$.

[1]Also called T^* in some books. Here, T^* is the adjoint.

Theorem. If X is a Hilbert space and T is self-adjoint then
1) *T has no residual spectrum,*
2) *the spectrum is real, and $\|T\| = \operatorname{Sup}_{\lambda \in \sigma(T)} |\lambda|$,*
3) *eigenvectors corresponding to distinct eigenvalues are orthogonal.*

We will use and prove later the following properties of Banach algebras.

Theorem. 1) *Every element $T \in \mathcal{B}$ such that $\|T - \mathrm{Id}\| < 1$ is regular (invertible). Its inverse is*

$$T^{-1} = \mathrm{Id} + \sum_{n=1}^{\infty} (\mathrm{Id} - T)^n.$$

2) *The set of regular elements in $\mathcal{B}$ is open, hence the set of singular elements is closed.*
3) *The mapping $T \mapsto T^{-1}$ on the set of regular elements is a homeomorphism.*

2. COMPACT OPERATORS

Due to their fundamental importance in many problems of mathematical physics we will review here some properties of linear compact operators. We return to non linear compact operators in Chapter VII.

Let X and Y be two metric spaces, a mapping $f\colon X \to Y$ is **compact** if a) it is continuous and b) it maps every bounded subset of X into a **relatively compact** (p. 16) subset of Y. compact mapping

When X and Y are Hausdorff topological vector spaces and f linear, the continuity of f is a consequence of the second property. The proof is trivial when X and Y are normed spaces.

A frequently used criterion for compactness on function spaces is given by the Ascoli–Arzela theorem, which concerns equicontinuous families. A family $\mathcal{F}$ of functions defined on a metric space X with metric d is called **equicontinuous** if for each $\epsilon > 0$ there is a $\delta > 0$ such that equicontinuous

$$d(x, x') < \delta \Rightarrow |f(x) - f(x')| < \epsilon, \qquad \forall x, x' \in X, \quad f \in \mathcal{F}.$$

Ascoli–Arzela theorem[1]. *Let X be a compact metric space. A bounded and equicontinuous subset K of the space $\mathscr{C}(X)$ consisting of continuous functions on X with the uniform norm (p. 29) is compact.*

The theorem is still valid if the functions of the family $\mathcal{F}$ are replaced by

[1]The proof can be found in textbooks on functional analysis.

continuous maps from the compact metric space X into a complete metric space Y. In the definition of equicontinuity, $|f(x)-f(x')|$ has to be replaced by $D(f(x), f(x'))$, where D is the distance in Y. In the theorem, the hypothesis that K is bounded must be replaced by the hypothesis that for each $x \in X$ the set $\{f(x), f \in K\}$ is a relativity compact set in Y.

Exercise: Show that if $Y = \mathbb{R}$ the second formulation is equivalent to the first.

Example 1: Let X and Y be compact spaces; for instance, closed bounded intervals of $\mathbb{R}$. Let m be a regular Borel measure on Y (p. 36). Let $K: (x, y) \mapsto K(x, y)$ be a continuous function on $X \times Y$. Define an operator $T: \mathscr{C}(Y) \to \mathscr{C}(X)$ by

$$T: f \in \mathscr{C}(Y) \mapsto g \in \mathscr{C}(X)$$

$$g(x) = \int_Y K(x, y) f(y) \, dm(y)$$

where $\mathscr{C}(Y)$ and $\mathscr{C}(X)$ are the spaces of continuous functions on Y and X with the uniform norm.
Using the Ascoli theorem, it is straightforward to show that T is a compact linear operator.

Example 2: Let X and Y be measure spaces (p. 34) and define a linear operator T on $L^2(Y)$ by

$$g(x) = \int_Y K(x, y) f(y) \, dm(y)$$

where K is a function on $L^2(X \times Y)$. g is defined a.e. and belongs to $L^2(X)$ by the Fubini theorem. The operator T can be proved to be compact, as a limit of operators with range in a finite dimensional space (p. 564).
Hilbert–Schmidt It is moreover a Hilbert–Schmidt operator. Namely if $\{e_i\}$ is a base of L^2 then

$$\sum_{i=1}^{\infty} \|Te_i\|^2 < \infty.$$

Fredholm alternative. Let T be a compact linear operator on a Banach space X and let $\lambda \neq 0$ be a complex number. One and only one of the two following statements is true.

1) *$Tf - \lambda f = g$ has one solution f for each $g \in X$, (i.e. $T - \lambda$ Id is an isomorphism (cf. p. 4).*

2) *$Tf - \lambda f = 0$ has non zero solutions (i.e. λ is an eigenvalue of T). For each λ except possibly $\lambda = 0$ the solutions span a finite dimensional subspace of B.*

Riesz–Schauder theorem. The spectrum of a compact operator has no accumulation point other than, possibly, zero.

Adjoint theorem. $Tf - \lambda f = 0$ has a non zero solution iff $\tilde{T}f - \lambda f = 0$ has a non zero solution.
Let λ_α be an eigenvalue of T, $Tf - \lambda_\alpha f = g$ has a solution iff g is such that $\langle \tilde{f}_\alpha, g\rangle = 0$ for all $\tilde{f}_\alpha$ such that $\tilde{T}\tilde{f}_\alpha - \lambda_\alpha \tilde{f}_\alpha = 0$.

Hilbert–Schmidt theorem. Let T be a self adjoint compact operator on a Hilbert space H, then there exists an orthonormal base of H made of eigenvectors of T.

Example: The inverse of the Laplace operator defined on a C^∞ compact manifold is a compact operator on the appropriate Sobolev space (p. 486).

3. OPEN MAPPING AND CLOSED GRAPH THEOREMS

The three following key theorems of linear functional analysis in Banach spaces are also valid in Fréchet spaces (p. 26).

Uniform boundedness theorem. Let $T_n : X \to Y$ be a sequence of continuous linear maps between Fréchet spaces. If $\{\|T_n x\|_Y\}$ is uniformly bounded (i.e. independent of n) for each $x \in X$, then $\{\|T_n\|_{\mathscr{L}(X,Y)}\}$ is also uniformly bounded.

The proof (p. 26) rests on the fact that a Fréchet space is a Baire space (p. 25).

Exercise: Show that the above theorem implies that if $\lim_{n=\infty} T_n x$ exists for each $x \in X$ then the limit $Tx = \lim_{n=\infty} T_n x$ is linear and continuous.

Open mapping theorem. A linear continuous surjective mapping of a Fréchet space X onto a Fréchet space Y is an open mapping, that is the image of each open set X is open in Y.

The following corollary is known as the Banach theorem in the case of Banach spaces.

Corollary. A linear continuous bijective mapping of a Fréchet space X onto a Fréchet space Y has a continuous linear inverse (i.e. is an isomorphism).

Exercise: Given the theorem, prove the corollary.

closed map

A map $T: X \to Y$ between metric spaces is called **closed** if for each sequence $\{x_n\}$ in the domain of T converging to $x \in X$, and such that $\{Tx_n\}$ converges to a point $y \in Y$, x is also in the domain of T and $Tx = y$.

graph

The **graph** of an operator T with domain $D(T)$ is the subset of $X \times Y$ with elements

$$\{x, Tx\}, \qquad x \in D(T).$$

Exercise: Show that, if X and Y are metric spaces, the above definition is equivalent to "the graph of T is closed in $X \times Y$".

Closed graph theorem. A closed linear map T defined on the entire Fréchet space X, and with values in a Fréchet space Y, is continuous.

Exercise: Prove this statement using the open mapping theorem and the fact that the graph of T is a closed linear subspace of the Fréchet space $X \times Y$, hence a Fréchet space Z.

PROBLEMS AND EXERCISES

PROBLEM 1. CLIFFORD ALGEBRA; SPIN(4)

These results will be needed for Problems III2 and V bis 4.

Introduction. *Let $V^n_{(s)}$, $s \in \mathbb{Z}^+$, $s \leq n$ be an n dimensional vector space over the real numbers with inner product $(\boldsymbol{v}|\boldsymbol{w})$ and basis $(\boldsymbol{e}_i)$ such that*

$$\begin{aligned} (e_i|e_j) &= 0 && i \neq j \\ (e_i|e_j) &= 1 && i = j = 1, \ldots, s \\ (e_i|e_j) &= -1 && i = j = s+1, \ldots, n. \end{aligned}$$

Introduce a product $\boldsymbol{vw}$ of vectors in $V^n_{(s)}$ which is associative and distributive with respect to addition and which satisfies the condition

$$\boldsymbol{vw} + \boldsymbol{wv} = 2(\boldsymbol{v}|\boldsymbol{w}).$$

The resulting algebra of all possible sums and products is called the

Clifford algebra *$C(V^n_{(s)})$ of $V^n_{(s)}$. Note in particular that* Clifford algebra

$$e_ie_j + e_je_i = \pm 2\delta_{ij}$$
$$(e_i)^2 = \pm 1$$
$$v^2 = (v|v)$$
$$e_ie_j = -e_je_i, \quad i \neq j.$$

The Clifford algebra is itself a linear space of dimension $\Sigma_{p=0}^{n}\binom{n}{p} = 2^n$ with basis

$$(1, e_I, e_{I_1}e_{I_2}, \ldots, e_1e_2\ldots e_n).$$

where capital letters label ordered natural numbers: $I_j < I_{j+1}$.

a) *Show that $C(V^1_{(0)})$ is $\mathbb{C}$ and that $C(V^2_{(0)})$ is the algebra of quaternions.*

Answer: A basis for $V^1_{(0)}$ is e, $e^2 = -1$ and any element of $C(V^1_{(0)})$ is of the form $a + be$, $a, b \in \mathbb{R}$.
A basis for $V^2_{(0)}$ is (e_1, e_2) with $(e_i|e_j) = -\delta_{ij}$. Any element of $C(V^2_{(0)})$ is of the form

$$a + \alpha e_1 + \beta e_2 + \gamma e_1e_2.$$

Let

$$i = e_1, \qquad j = e_2, \qquad k = e_1e_2.$$

Then

$$ij = -ji = k, \qquad jk = -kj = i, \qquad ki = -ik = j$$
$$i^2 = j^2 = k^2 = -1.$$

b) *The linear subspace of C spanned by the $\binom{n}{p}$ products $(e_{I_1}e_{I_2}\ldots e_{I_p})$ is denoted C_p. The linear subspaces* C_p

$$C_+ = \bigoplus_{p \text{ even}} C_p \qquad \text{and} \qquad C_- = \bigoplus_{p \text{ odd}} C_p$$

are called the even and odd subspaces of C. C_+ is also a subalgebra of C. The dimension of both C_+ and C_- is 2^{n-1}.
The algebra C_+ is isomorphic to the Clifford algebra $C(V^{n-1}_{(s)})$ for certain values of s.
Show that the even subalgebra of the **Dirac algebra** *$C(V^4_{(1)})$ is the* **Pauli algebra** *$C(V^3_{(3)})$ and continue the sequence until the real numbers are reached.* Dirac algebra Pauli algebra

Answer: A basis for the even subalgebra of the Dirac algebra D is[1]

$$1,\ e_0e_1,\ e_0e_2,\ e_0e_3,\ e_1e_2,\ e_1e_3,\ e_2e_3,\ e_0e_1e_2e_3.$$

Let $P = C(V^3_{(3)})$ be the Pauli algebra and let (f_i) be an orthonormal basis for $V^3_{(3)}$. The identification

$$f_1 \mapsto e_0e_1, \qquad f_2 \mapsto e_0e_2, \qquad f_3 \mapsto e_0e_3$$

defines an isomorphism between D_+ and P. Similarly the even subalgebra of the Pauli algebra is isomorphic to the **quaternions**. The even subalgebra of the quaternions is isomorphic to the complex numbers. The even subalgebra of the complex numbers is $\mathbb{R}$.

quaternions

c) *Show that the center Z of the algebra $C(V^n_{(s)})$ is C_0 when n is even and $C_0 + C_n$ when n is odd where C_0, C_n are the linear subspaces defined on p. 65.*

Answer: First note that if $e_B = e_{I_1}e_{I_2}\ldots e_{I_p}$, then

$$e_je_B = (-1)^q e_Be_j$$

where $q = p$ when $j \neq I_i$, $\forall i$ and $q = p - 1$ when $j = I_i$ for some i.

It follows that e_j can always be chosen to anticommute with e_B except for the case n odd and $p = n$.
1) n even. It is clear that C_0 is in the center. Suppose that $c = c^A e_A \in Z$ and $c^B \neq 0$. Choose e_j so that $e_je_B = -e_Be_j$. Then

$$c = e_jce_j^{-1} = c^A e_je_Ae_j^{-1} = c^A(\pm e_A)$$

where in particular the minus sign holds when $A = B$. Therefore c can only be a scalar.
2) n odd. The proof is almost identical.

isometry

d) *An* **isometry** *of the vector space $V^n_{(s)}$ is a linear transformation $A = [A^i{}_j]$ by $e_j \mapsto A^i{}_je_i$ such that*

$$(Av|Aw) = (v|w),\ \forall v, w \in V^n_{(s)}.$$

For $s = n$ these are the orthogonal transformations; for $s = 1$ they are the **Lorentz transformations.** *It is clear that any transformation of the form $v \to \pm \Lambda v \Lambda^{-1}$, $\Lambda \in C(V^n_{(s)})$ which is such that $\Lambda v \Lambda^{-1} \in C_1$, $\forall v \in C_1$ is an isometry. Use the fact* (Riesz, p. 74) *that any isometry can be decomposed into a product of at most n reflections to study the structure of Λ.*

Lorentz transformation

[1]Note that the basis vectors (e_0, e_1, e_2, e_3) for $V^4_{(1)}$ have been labelled by a subscript running from 0 to 3, instead of from 1 to 4; this is the usual physics notation.

Answer: Any non-isotropic (p. 286) vector $\boldsymbol{u}$ determines a reflection

$$\boldsymbol{v} \mapsto \boldsymbol{v} - \frac{2\boldsymbol{u}(\boldsymbol{u}|\boldsymbol{v})}{(\boldsymbol{u}|\boldsymbol{u})} \equiv \boldsymbol{w}.$$

The vector $\boldsymbol{w}$ is the mirror image of $\boldsymbol{v}$ with respect to the hyperplane orthogonal to $\boldsymbol{u}$.

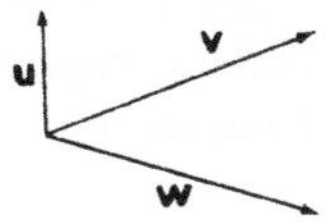

Since $2(\boldsymbol{u}|\boldsymbol{v}) = \boldsymbol{uv} + \boldsymbol{vu}$ and $\boldsymbol{u}^{-1} = \boldsymbol{u}/(\boldsymbol{u}|\boldsymbol{u})$ this can be rewritten

$$\boldsymbol{w} = -\boldsymbol{uvu}^{-1}.$$

Given any non-isotropic vectors $\boldsymbol{u}_1, \boldsymbol{u}_2, \dots \boldsymbol{u}_k$ the transformation

$$\boldsymbol{v} \mapsto (-1)^k \boldsymbol{u}_k \dots \boldsymbol{u}_1 \boldsymbol{v} \boldsymbol{u}_1^{-1} \dots \boldsymbol{u}_k^{-1}$$

is an isometry A. For k odd $\det[A^i{}_j] = -1$, for k even $\det[A^i{}_j] = +1$. We now restrict our attention to the real vector space $V^4{}_{(1)}$. First note that the minus sign can be removed from the transformation formula

$$\boldsymbol{v} \mapsto -\boldsymbol{u}_1 \dots \boldsymbol{u}_k \boldsymbol{v} \boldsymbol{u}_k^{-1} \dots \boldsymbol{u}_1^{-1}, \; k \text{ odd}.$$

Let $\boldsymbol{e} = \boldsymbol{e}_0\boldsymbol{e}_1\boldsymbol{e}_2\boldsymbol{e}_3$; then $\boldsymbol{e}\boldsymbol{e}_\alpha = -\boldsymbol{e}_\alpha\boldsymbol{e}$. Therefore $\boldsymbol{e}$ anticommutes with any vector

$$\boldsymbol{ev} = -\boldsymbol{ve} \text{ or } \boldsymbol{eve}^{-1} = -\boldsymbol{v}.$$

So we have for k odd

$$\boldsymbol{v} \mapsto \boldsymbol{u}_1 \dots \boldsymbol{u}_k \boldsymbol{eve}^{-1} \boldsymbol{u}_k^{-1} \dots \boldsymbol{u}_1^{-1}.$$

It follows that any Lorentz transformation has the form

$$\boldsymbol{v} \mapsto \boldsymbol{\Lambda v \Lambda}^{-1}$$

where $\boldsymbol{\Lambda}$ is a product of non-isotropic vectors.

e) *Use the above decomposition to define the* **reversion** $\tilde{\boldsymbol{\Lambda}}$ *of* $\boldsymbol{\Lambda}$ reversion

$$\boldsymbol{\Lambda} = \boldsymbol{u}_1 \dots \boldsymbol{u}_k, \; \tilde{\boldsymbol{\Lambda}} = \boldsymbol{u}_k \dots \boldsymbol{u}_2 \boldsymbol{u}_1.$$

The group **Spin(4)** *is defined to be the group of all elements* $\boldsymbol{\Lambda} \in C(V^4{}_{(1)})$ Spin (4)
such that

$$\boldsymbol{\Lambda v \Lambda}^{-1} \in C_1, \qquad \forall \boldsymbol{v} \in C_1$$
$$\boldsymbol{\Lambda}\tilde{\boldsymbol{\Lambda}} = \pm 1$$

where C_1 *is the subspace of* C *spanned by* $(\boldsymbol{e}_i)$.

Verify that this is a group and that the second condition implies that Λ can always be written

$$\Lambda = u_1 u_2 \dots u_k, \quad u_i^2 = \pm 1.$$

Prove that the mapping $\mathcal{H}$: Spin(4)$\to L(4)$ *by* $\Lambda \mapsto [a^\alpha{}_\beta]$, *where* $\Lambda e_\beta \Lambda^{-1} = a^\alpha{}_\beta e_\alpha$, *is a 2–1 homomorphism. The change has been made to Greek indices running from 0 to 3 in order to agree with the convention of Chapter V.*

Answer: The mapping $\mathcal{H}$ is a homomorphism because

$$\Lambda' \Lambda e_\alpha \Lambda^{-1} \Lambda'^{-1} = a'^\lambda{}_\alpha a^\beta{}_\lambda e_\beta.$$

It is onto because any element of $L(4)$ can be obtained as a product of reflections.
To find the kernel of $\mathcal{H}$ suppose

$$\Lambda e_\alpha \Lambda^{-1} = e_\alpha \qquad \forall \alpha$$

Then

$$\Lambda \mathbf{e}_\alpha = e_\alpha \Lambda \qquad \forall \alpha$$

It follows that Λ is an element of the center of $C(V^4{}_{(1)})$. However the only scalar elements of Spin(4) are ± 1, e.g. $e_0 e_0 = +1$, $e_1 e_1 = -1$.

f) *Show that elements of* Spin(4) *which correspond respectively to spatial reflection and time reflection are* $\pm e_0$ *and* $\pm e_1 e_2 e_3$. *Show that*

$$\mathcal{H}\left(\cos\frac{\varphi}{2} + \sin\frac{\varphi}{2}\, e_2 e_3\right) = \begin{bmatrix} 1 & 0 & 0 & 0 \\ 0 & 1 & 0 & 0 \\ 0 & 0 & \cos\varphi & -\sin\varphi \\ 0 & 0 & \sin\varphi & \cos\varphi \end{bmatrix}.$$

$$\mathcal{H}\left(\cosh\frac{\varphi}{2} + \sinh\frac{\varphi}{2}\, e_1 e_0\right) = \begin{bmatrix} \cosh\varphi & \sinh\varphi & 0 & 0 \\ \sinh\varphi & \cosh\varphi & 0 & 0 \\ 0 & 0 & 1 & 0 \\ 0 & 0 & 0 & 1 \end{bmatrix}.$$

References: M. Riesz, Clifford numbers and spinors, University of Maryland, Institute for Fluid Mechanics and Applied Mathematics, Lecture series, no. 38. P. K. Raševskii, The theory of spinors, Am. Math. Soc. Transl. 6 (1957). D. Shale and Stinespring, States of the Clifford Algebra, Ann. Math. 80, 365–381 (1964).

EXERCISE 2. PRODUCT TOPOLOGY

Let $\mathcal{U}_\alpha$ *be a topology on* X_α *where the indexing set is the set* $\mathbb{N}$ *of natural numbers. Let* X_α *be the closed unit interval* $I \subset \mathbb{R}$ *for every* α. *Let* U_α *be* *an open subset of I. Show that the topology generated on $\times_{\alpha\in\mathbb{N}} X_\alpha$ by the sets $\times_{\alpha\in\mathbb{N}} U_\alpha$ is not compact.*

Answer: We shall construct a covering of the product space $\times_{\alpha\in\mathbb{N}} X_\alpha$ which has no countable subcovering, let alone a finite one. Let $A \subset \mathbb{N}$ and $V_A = \times_{\alpha\in\mathbb{N}} U_\alpha$ be an element of the topology on the product space defined as follows:

$$U_\alpha = \begin{cases} [0, 2/3) & \text{if } \alpha \in A \\ (1/3, 1] & \text{if } \alpha \notin A \end{cases}.$$

Then $\{V_A; A \subset \mathbb{N}\}$ is an open covering of the product space; it has no countable subcovering. Indeed let $\{V_{A_i}\}_i$ be any countable subset of $\{V_A; A \subset \mathbb{N}\}$. One can always find a point $x = \{x_1, x_2, \ldots\}$ in the product space such that $x \notin \{V_{A_i}\}_i$. For example, let

$$x_i = \begin{cases} 1/6 & \text{if } i \notin A_i \\ 5/6 & \text{if } i \in A_i \end{cases} \quad \text{then } x \notin \{V_{A_i}\}_i.$$

Contributed by J. Labelle.

PROBLEM 3. STRONG AND WEAK TOPOLOGIES IN L^2

a) *Show that the sequence $\{f_n = \sin n(\cdot)\}$ does not tend to a limit in $L^2[0, \pi]$ (strong topology); show that it tends to $f = 0$ in the weak topology.*
b) *Show that the set $\{f_n = \sin n(\cdot)\}$ in $L^2[0, \pi]$ is closed and bounded but is not compact.*

Answers: Let $\| \quad \|$ be the L^2 norm; then $\|f_n\|^2 = \pi/2$.
a) $\|\sin nx - \sin mx\|^2 = \|\sin nx\|^2 + \|\sin mx\|^2 = \pi$ for $m \neq n$. Hence $\|f_n - f_m\|$ cannot be smaller than ϵ for n and m larger than N. On the other hand, for $\varphi \in C_0^\infty(0, \pi)$, the space of C^∞ functions on $(0, \pi)$ with compact support,

$$\int_0^\pi \varphi(x) \sin nx \, dx = \int_0^\pi \varphi'(x) \frac{1}{n} \cos nx \, dx$$

tends to zero when n tends to infinity.
The space $C_0^\infty(0, \pi)$ is dense in $L^2(0, \pi)$ (cf. p. 488), therefore the sequence $\{f_n\}$ tends to $f = 0$ in the weak topology. One checks easily that $\|f\| \leq \lim_{n\to\infty} \inf \|f_n\|$; indeed $0 \leq \sqrt{\pi/2}$.

b) $\{f_n\}$ is discrete (has no accumulation point) because $\|f_n - f_m\| = \sqrt{\pi}$ for $n \neq m$; it follows that it is discrete. It is bounded since $\|f_n\|$ is in the ball of finite radius $\|f\| \leq \sqrt{\pi}/2$.
However it is not compact because the open covering by balls of radius $a < \sqrt{\pi}$

$$\{f \in L^2[0, \pi]; \|f - \sin n(\cdot)\| < a\}$$

has no finite subcovering; indeed, we cannot select a finite number of these balls $n = n_1, \ldots n_p$ from this open covering such that their union covers the set $\{\sin n(\cdot)\}$.

EXERCISE 4. HÖLDER SPACES

$C^{k,\alpha}(\bar{\Omega})$ is the subspace of $C^k(\bar{\Omega})$ of functions which, together with their derivatives of order $\leq k$, satisfy a Hölder condition of order α, $0 < \alpha < 1$. That is

$$\sup_{\substack{x, y \in \bar{\Omega} \\ x \neq y}} \frac{|f(x) - f(y)|}{|x - y|^\alpha} < \infty.$$

a) *Show that the mapping $f \mapsto \|f\|_{C^{0,\alpha}\bar{\Omega}}$ with*

$$\|f\|_{C^{0,\alpha}(\bar{\Omega})} = \|f\|_{C^0(\bar{\Omega})} + \sup_{\substack{x, y \in \bar{\Omega} \\ x \neq y}} \frac{|f(x) - f(y)|}{|x - y|^\alpha}$$

is a norm on $C^{0,\alpha}(\bar{\Omega})$. Analogous question for $C^{k,\alpha}(\bar{\Omega})$, with

$$\|f\|_{C^{k,\alpha}(\bar{\Omega})} = \sum_{|j| \leq k} \|D^j f\|_{C^{0,\alpha}(\bar{\Omega})}.$$

b) *Show that the normed spaces $C^{k,\alpha}(\bar{\Omega})$ are Banach spaces.*
c) *Show that $C^{k,\alpha}(\bar{\Omega})$ is a Banach algebra.*

Answer: The proofs are straightforward applications of the definitions and of the Banach property of $C^k(\bar{\Omega})$.

II. DIFFERENTIAL CALCULUS ON BANACH SPACES

A. FOUNDATIONS

The reader acquainted with differential calculus on $\mathbb{R}^n$ may wish to postpone reading this chapter until the material is needed for Chapter VII.

This section treats the essentials of differential calculus and calculus of variations on Banach spaces[1]. The generalization from differential calculus on $\mathbb{R}^n$ to differential calculus on Banach spaces is remarkably smooth, so smooth that many theorems are similarly phrased and a cursory glimpse at this section might give the impression that it is only a summary of a beginning course in calculus. The generalization to arbitrary topological vector spaces is an entirely different matter because on such spaces the implicit function theorem is not valid without further assumptions. Considerations of topological vector spaces more general than Banach spaces will be limited here to the definition of differentials on locally convex topological vector spaces.

1. DEFINITIONS. TAYLOR EXPANSION

Let X and Y be two Banach spaces, U an open set of X and f a mapping $f: U \to Y$; f is said to be **differentiable at** $x_0 \in U$ if there exists a continuous linear mapping $\mathrm{D}f|_{x_0}$ of X into Y such that

differentiable at a point

$$f(x_0+h)-f(x_0)=\mathrm{D}f|_{x_0}h+R(h),\ h\in X,$$
$$x_0,\ x_0+h\in U,$$
$$f(x_0),\ f(x_0+h),\ R(h)\in Y,$$

with $\|R(h)\| = o(\|h\|)$, (i.e. $\lim_{\|h\|=0} \|R(h)\|/\|h\| = 0$).

Fréchet differential

$\mathrm{D}f|_{x_0}$ is called the **differential of** f at x_0, or sometimes the **Fréchet differential**. It is also denoted $\mathrm{D}f(x_0)$, f'_{x_0}, $f'(x_0)$ according to typographical convenience[2]. The differential of a mapping is also called a **derivative**; the differential

differential at a point

derivative

[1] A basic reference is [Cartan 1967(b)], see also [J.T. Schwartz].

[2] We shall avoid the notation f'_{x_0} and use preferably $f'(x_0)$ when f'_{x_0} could easily be confused with $f'_{x^i} \equiv \partial f/\partial x^i$, a convenient notation often used for partial derivatives.

functional derivative

of a mapping from a function space into $\mathbb{K}$ (i.e. a functional) is called a **functional derivative**.

$Df|_{x_0}$ is an element of the space $\mathscr{L}(X, Y)$ of continuous linear mappings of X into Y; it acts on $h \in X$ to give an element $Df|_{x_0}h \in Y$.

Example 1: Let $f: U \subset \mathbb{R} \to \mathbb{R}$, it is differentiable if it has a derivative in the usual sense. $Df|_{x_0} = df/dx|_{x_0}$ is a number.

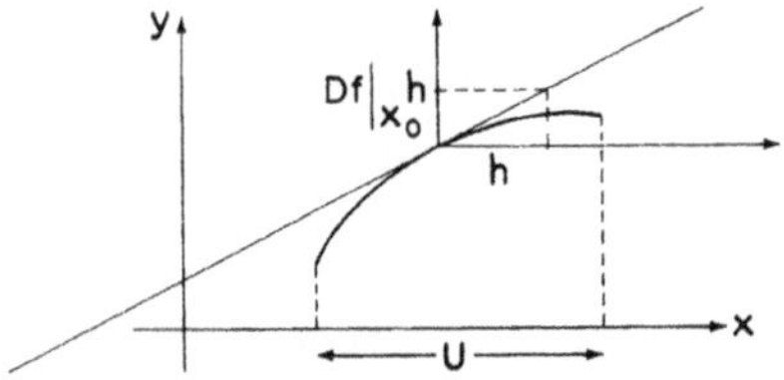

Example 2: The differential of a linear mapping $l: X \to Y$ is the mapping l; the differential of the identity mapping Id: $X \to X$ by $x \mapsto x$ is the identity mapping: $Dl|_{x_0} = l$, $D\mathrm{Id}|_{x_0} = \mathrm{Id}$, $\forall x_0$.

Example 3: The differential of a constant mapping $C: X \to Y$ by $x \mapsto c$ is zero; i.e. $C'_x h = 0$, $\forall h \in X$.

Example 4: Let $f: \mathbf{R} \to X$ be a function of $t \in \mathbf{R}$ with values in X. Its differential $Df|_t = f'(t) \in \mathscr{L}(\mathbf{R}, X)$. If $h \in \mathbf{R}$, $f'(t)h \in X$, thus $f'(t)$ may be identified with an element of X. Hence f' is also a function of $t \in \mathbf{R}$ with values in X.

Exercise: Show that, if $f: \mathbf{R}^n \to \mathbf{R}^p$ by the p functions of n variables

$$y^\alpha = f^\alpha(x^i), \qquad \alpha = 1, \ldots, p, \qquad i = 1, \ldots, n,$$

then, if the f^α admit continuous partial derivatives in the classical sense of elementary calculus, f is differentiable and

$$Df|_{x_0}h = \left\{ \sum_i \frac{\partial f^\alpha}{\partial x^i}\bigg|_{x_0} h^i \right\}, \text{ is also denoted } \{\partial_i f^\alpha|_{x_0} h^i\}.$$

jacobian matrix rank critical jacobian

$Df|_{x_0}$ is then called the **jacobian matrix** of f at x_0. Its rank is called the **rank** of f. A point where the rank is not maximal is called **critical**. If $n = p$ the determinant of Df is called the **jacobian**.

Exercise: Show that, if f is differentiable on the Banach space X, its restriction to a subspace $Y \subset X$ is differentiable if the subspace Y of X is endowed with a norm such that $\| \ \|_Y \geq \| \ \|_X$.

If X and Y are nonbanachisable, but locally convex topological vector spaces, the differential of a mapping $X \to Y$ can be defined similarly, the condition $\|R(h)\| = o(\|h\|)$ being replaced by the condition that R is "tangent to zero". Let $o(t)$ be a real function of the real variable t for $|t| < 1$ such that $\lim_{t=0} o(t)/t = 0$. Then R is **tangent to zero** if given any neighborhood N of zero in Y there exists a neighborhood U of zero in X such that

tangent to zero

$$R(tU) \subset o(t)N.$$

R maps $tU \subset X$ into $o(t)N$ which "shrinks" faster than tN as t goes to 0.

A mapping $f\colon U \subset X \to Y$ is **differentiable in U** if it is differentiable at each point of U. The differential Df is a mapping $U \to \mathcal{L}(X, Y)$ by $x \mapsto \mathrm{D}f|_x$; if the mapping Df is continuous, f is said to be **continuously differentiable** or of **class C^1**.

differentiable in U

continuously differentiable

class C^1

2. THEOREMS

Theorem (composite mapping). Let X, Y, Z be three Banach spaces, U an open set in X, V an open set in Y, $f\colon U \to Y$, $g\colon V \to Z$ and $x_0 \in U$ such that $f(x_0) \in V$; if f is differentiable at x_0 and g differentiable at $y_0 = f(x_0)$ then $h = g \circ f$ is differentiable at x_0 and

$$h'_{x_0} = g'_{y_0} \circ f'_{x_0}.$$

The proof follows from the definition of a differential.

Theorem (inverse mapping). If $f\colon U \subset X \to V \subset Y$ is invertible and differentiable at $x_0 \in U$ and if $f'(x_0) \equiv f'_{x_0}$ is an isomorphism $X \to Y$, then f^{-1} is differentiable at the point $y_0 = f(x_0) \in V$ and

$$(f^{-1\prime}_{y_0}) = (f'_{x_0})^{-1}.$$

Proof: $f(x_0 + h) - f(x_0) - f'_{x_0} \cdot h = R(h)$ may be written, if $f'(x_0)$ is invertible,

$$(f'_{x_0})^{-1}k - (x_0 + h - x_0) = (f'_{x_0})^{-1} \cdot R(h)$$

if one sets $y_0 = f(x_0)$, $y_0 + k = f(x_0 + h)$. The differentiability of f^{-1} at x_0, and the value of its derivative, follow if

$$\lim_{\|k\|=0} \frac{\|(f'_{x_0})^{-1}R(h)\|}{\|k\|} = 0.$$

We have

$$\left\|(f'_{x_0})^{-1}k\right\| = \left\|h + (f'_{x_0})^{-1}R(h)\right\| \geq \left\|h\right\| + \left\|(f'_{x_0})^{-1}R(h)\right\| \geq \left\|h + (f'_{x_0})\right\| \left\|R(h)\right.$$

Since f^{-1} is continuous at y_0, $\|h\|$ tends to zero with $\|k\|$ and if $\|h\|$ is small enough, $\|k\| > \lambda\|h\|$; thus

$$\lim_{\|k\|=0} \frac{\|(f'_{x_0})^{-1}R(h)\|}{\|k\|} = \lim_{\|h\|=0} \frac{\|(f'_{x_0})^{-1}R(h)\|}{\|h\|} \cdot \frac{\|h\|}{\|k\|} = 0. \qquad \blacksquare$$

3. DIFFEOMORPHISMS

$f: U \to V$ is a diffeomorphism if f is a bijection with f and f^{-1} continuously differentiable (of class C^1).

Remark: A homeomorphism of class C^1 is not necessarily a diffeomorphism. For example $f: \mathbb{R} \to \mathbb{R}$ by $x \mapsto y = x^3$ is a homeomorphism of class C^1; the inverse f^{-1}: $y \mapsto x^{1/3}$ is not differentiable at the origin. This remark is reminiscent of a similar remark made for homeomorphisms (p. 18).

Theorem. A homeomorphism f: $U \to V$ of class C^1 is a diffeomorphism if and only if $f'(x)$ is an isomorphism for every $x \in U$.

Proof: If f is a diffeomorphism $f^{-1} \circ f = \text{Id}$. Thus by the composite mapping theorem $f'(x)$ is an isomorphism, with inverse $(f^{-1})'(f(x))$. Conversely let f be a homeomorphism of class C^1 and $f'(x)$ an isomorphism. By the inverse mapping theorem, f^{-1} is differentiable at each point $y \in V$. Moreover $(f^{-1})'(y)$ depends continuously on y (i.e. f^{-1} is C^1) since $y \mapsto (f^{-1})'(y)$ is the composite of the three continuous mappings $y \mapsto x = f^{-1}(y)$, $x \mapsto f'(x)$ and $f'(x) \mapsto (f^{-1})'(y)$. ■

partial derivative at a point

Let f be a mapping from $U \subset X$ into a Banach space Y where X is a product of Banach spaces $X = X_1 \times \ldots \times X_n$. For each point $a = (a^1, \ldots, a^n) \in U$ let us define a "partially constant"[1] mapping e_i: $X_i \to X$ by

$$e_i: x^i \mapsto (a^1, \ldots, a^{i-1}, x^i, a^{i+1} \ldots a^n).$$

e_i is an affine and thus differentiable mapping. Its derivative e'_i maps X_i into X by the "partial identity" mapping,

$$e'_i: h^i \mapsto (0, \ldots, h^i, \ldots, 0).$$

[1]Cf. examples 2 and 3, p. 72.

The partial derivative $\partial f/\partial x^i|_a$, also denoted $f'_{x^i}(a)$, of $f\colon U \to Y$ at a, if it exists, is the derivative of the composite mapping $f \circ e_i$ at a_i. It always exists if f is differentiable and then

$$f'_{x^i}(a) = (f \circ e_i)'_a = f'(a) \circ e'_i(a^i).$$

It follows that

$$\sum_i f'_{x^i}(a)h^i = \sum_i f'(a)[e'_i(a^i)h^i] = f'(a)h.$$

Remark: The existence of the differential f' implies the existence of the partial derivatives, but not conversely. For example the function $f\colon \mathbb{R}^2 \to \mathbb{R}$ by

$$f(x, y) = \frac{xy}{\sqrt{(x^2+y^2)}}, \qquad f(0,0) = 0$$

has partial derivatives at $x = y = 0$, namely

$$f'_x(0,0) \equiv \frac{\partial}{\partial x} f(x,0)|_{x=0} = 0, \qquad f'_y(0,0) = 0.$$

Nevertheless, f is not differentiable at the origin because

$$f(h_x, h_y) = \frac{h_x h_y}{\sqrt{((h_x)^2 + (h_y)^2)}} \neq f(0,0) + h_x f'_x(0,0) + h_y f'_y(0,0) + R(h)$$

with $R(h)$ such that $\lim_{\|h\|=0} \|R(h)\|/\|h\| = 0$.
Note that $f'_x(0,0) \neq \lim_{y=0} f'_x(0,y)$; the discontinuity of f'_x at the origin rules out a priori the continuous differentiability of f at the origin but not the differentiability of f.

f'_{x^i} maps $U \to \mathcal{L}(X_i, Y)$ by $x \mapsto f'_{x^i}(x)$. It can be shown that a necessary and sufficient condition for f to be of class C^1 is that the mappings f'_{x^i} exist and be continuous. That is

$$\text{existence of } f' \underset{\not\Leftarrow}{\Rightarrow} \text{existence of partial derivatives}$$

whereas

$$\text{existence and continuity of } f' \Leftrightarrow \text{existence and continuity of partial derivatives.}$$

Example 1: Finite dimensional spaces (see example, p. 72). Let $f\colon \mathbb{R}^n \to \mathbb{R}^p$ by $(x^i) \mapsto f^\alpha(x^i)$. A partial derivative of f is

$$f'_{x^i}(x_0) \in \mathbb{R}^p, \text{ its coordinates are } \{\partial f^\alpha/\partial x^i|_{x_0}\}.$$

Example 2: Let $P\colon u \to P(D^m u)$ be a non linear partial differential operator of order m on $\mathbb{R}^n$, mapping $U \subset C^m(\bar{\mathbb{R}}^n)$ into $V \subset C^0(\bar{\mathbb{R}}^n)$; we denote by $D^m u$ the set of derivatives of u of order $\leq m$, and by P a function C^1 in all its arguments. Then P is differentiable at $u_0 \in U$ and its differential is the following linear partial differential operator of order m, called the linearization of P

$$h \mapsto \sum_{|j| \leq m} \frac{\partial P}{\partial(D^j u)} (D^m u_0) \cdot D^j h$$

(cf. Problem 1, p. 98).

4. THE EULER EQUATION

Let $[a, b]$ be a closed interval in $\mathbb{R}$ and let L be a continuous function $[a, b] \times \mathbb{R} \times \mathbb{R} \subset \mathbb{R}^3 \to \mathbb{R}$ by $(x, y, z) \mapsto L(x, y, z)$ which is continuously differentiable on $\mathbb{R} \times \mathbb{R}$ with respect to its two last variables y, z.[1]
Let $S: C^1([a, b]) \to \mathbb{R}$ by

$$S(q) = \int_a^b L(x, q(x), q'(x))\, dx, \qquad q \in C^1([a, b]), \qquad q'(x) = dq(x)/dx.$$

We shall compute the differential of S on the Banach space $C^1([a, b])$ with the norm

$$\|q\| = \sup_{x \in [a, b]} \{|q(x)| + |q'(x)|\}.$$

Now

$$S(q + h) = \int_a^b L(x, q(x) + h(x), q'(x) + h'(x))\, dx, \qquad h \in C^1([a, b])$$

and using the usual Taylor's formula for L, we obtain

$$S(q + h) = S(q) + \int_a^b \{h(x) L'_y(x, q(x), q'(x)) + h'(x) L'_z(x, q(x), q'(x))\, dx\}$$

$$+ \int_a^b \alpha(|h(x)| + |h'(x)|)\, dx$$

where α goes to zero with $|h(x)| + |h'(x)|$, hence with $\|h\|$. It follows that S is differentiable on $C^1([a, b])$ and that its differential is

$$S'(q)h = \int_a^b \{h(x) L'_y(x, q(x), q'(x)) + h'(x) L'_z(x, q(x), q'(x))\}\, dx.$$

[1]Note for further use that L'_y is the derivative with respect to the second argument, etc.

Assuming that it is possible to integrate by parts, for instance if L and q are C^2, then

$$\int_a^b h'(x)L'_z(x, q(x), q'(x))\,\mathrm{d}x$$

$$= [h(x)L'_z(x, q(x), q'(x))]_a^b - \int_a^b h(x)\frac{\mathrm{d}}{\mathrm{d}x}[L'_z(x, q(x), q'(x))]\,\mathrm{d}x.$$

In particular, the differential of S on the subspace of $C^2([a, b])$ characterized by $q(a) = q(b) = 0$ is given by

$$S'(q)h = \int_a^b \mathscr{E}(L)h\,\mathrm{d}x,$$

where

$$\mathscr{E}(L) = L'_y - \frac{\mathrm{d}}{\mathrm{d}x}L'_z.$$

$\mathscr{E}(L) = 0$ is a second order differential equation for $q(x)$ called **Euler's equation** for the function q. Euler equation

Remark: The differential of S is defined by the same operator on the affine subspace of $C^2([a, b])$ characterized by $q(a) = \alpha$, $q(b) = \beta$.

Euler's equation for several variables and several unknown functions. Let $q: \bar{\Omega} \to \mathbb{R}^p$ be a C^1 mapping on the closure of a bounded open set $\Omega \subset \mathbb{R}^n$, $q = (q^\alpha; \alpha = 1, \ldots, n)$ where $q^\alpha: \bar{\Omega} \to R$, $\mathrm{D}q = (\partial q^\alpha/\partial x^i)$.
Let $L: \mathbb{R}^{np+p+1} \to \mathbb{R}$ be a C^1 function.
Let S be the function on the product space $C^1(\bar{\Omega}) \times \ldots \times C^1(\bar{\Omega})$ (p times) defined by

$$S(q) = \int_{\bar{\Omega}} L(x, q(x), \mathrm{D}q(x))\,\mathrm{d}x.$$

S is differentiable and its differential is

$$S'(q)\cdot h = \sum_{\alpha=1}^{p}\sum_{j=1}^{n}\int_{\bar{\Omega}}\left(h^\alpha L'_{q^\alpha} + \frac{\partial h^\alpha}{\partial x^j}L'_{\partial_j q^\alpha}\right)\mathrm{d}x.$$

Assuming L of class C^2 and restricting f to mappings q on $\bar{\Omega}$ of class C^2

which vanish on $\partial\bar{\Omega}$, or take given values on $\partial\bar{\Omega}$ – see remark above – we can write

$$S'(q)h = \sum_{\alpha=1}^{p} \int_{\bar{\Omega}} h^\alpha \mathscr{E}_\alpha(L)\, dx$$

where the Euler operators $\mathscr{E}_\alpha(L)$ for q are

$$\mathscr{E}_\alpha(L) = L'_{q^\alpha} - \sum_j \frac{\partial}{\partial x^j} L'_{\partial_j q^\alpha}.$$

5. THE MEAN VALUE THEOREM

Theorem. Let $f: U \to Y$ be a C^1 mapping from U, a convex open set of X, into Y, where X and Y are Banach spaces. If $x \in U$ and $x + h \in U$, then

$$f(x+h) - f(x) = \int_0^1 f'(x + th) \cdot h\, dt.$$

Proof: Let u: $[0, 1] \to Y$ with $u = f \circ g$, where g is the affine mapping $t \mapsto (x + th)$. Thus u is C^1 on $(0, 1)$ and

$$u'(t) = f'(x+th)g'(t) = f'(x+th) \cdot h.$$

The mean value theorem follows from the definition of the integral of vector valued functions on $\mathbb{R}$. ■

When f: $U \subset \mathbb{R} \to \mathbb{R}$ this theorem takes its elementary familiar form.

Lemma. If $f: U \to Y$ is a continuous mapping from a topological space U into a Hausdorf topological space Y, f locally constant and U connected, then f is constant on U.

Proof: Let $a \in U$, $b = f(a)$. Consider the subset $f^{-1}(b)$ of U. This subset is closed because $\{b\}$ is closed when Y is Hausdorf, and because f is continuous. On the other hand $f^{-1}(b)$ is open since f is locally constant. Therefore, $f^{-1}(b) = U$ if U is connected. ■

Corollary. If the differential of a mapping $f: U \to Y$ is zero in the connected open set U, then f is a constant mapping in U.

Proof: By the mean value theorem f is locally constant, i.e. f is constant in a neighbourhood (an open ball, convex set) of each of its points. ■

The following estimate is a trivial consequence of the mean value theorem

$$\|f(x+h)-f(x)\|_Y \leq \sup_{t\in[0,1]} \|f'(x+th)\|_{\mathscr{L}(X,Y)}\|h\|_X.$$

6. HIGHER ORDER DIFFERENTIALS

Let $f\colon U \subset X \to Y$ be a C^1 mapping.
If its differential $Df\colon U \to \mathscr{L}(X, Y)$ is a differentiable mapping, one can define the differential of Df at x. It is called the **second differential** of f at x (differential of order 2) and is denoted $D^2f|_x$ or $D^2f(x)$ or f''_x or $f''(x)$ according to typographical convenience. second differential
It follows from its definition that $f''(x)$ is an element of $\mathscr{L}(X, \mathscr{L}(X, Y))$. The second differential f'' on U is a mapping $U \to \mathscr{L}(X, \mathscr{L}(X, Y))$ by $x \mapsto f''(x)$. It is also called the **second variation** of f. second variation

If f'' is a differentiable mapping one can in turn define its differential and, by recurrence, proceed to define differentials of order p.
A mapping is said to be of **class** C^p on U if D^pf exists on U and is continuous. C^p
A diffeomorphism f is said to be of class C^p on U if f and f^{-1} are of class C^p. It can be shown easily that a diffeomorphism (p. 74) is of class C^p if f is of class C^p.

Remark: Let X, Y, Z be Banach spaces, there is a natural (canonical) isomorphism between $\mathscr{L}(X, \mathscr{L}(Z, Y))$ and $\mathscr{L}(X \times Z, Y)$ the space of bilinear continuous mappings $g\colon X \times Z \to Y$, together with the following norm: for $g\colon X \times Z \to Y$, the norm $\|g\|$ is the greatest lower bound of the numbers K such that $\|g(x, z)\| \leq K\,\|x\|\,\|z\|$.

Proof: Let $f \in \mathscr{L}(X, \mathscr{L}(Z, Y))$: $x \mapsto f(x) \in \mathscr{L}(Z, Y)$.
For every $x \in X$ and $z \in Z$,

$$f(x)z = y \in Y.$$

Let $g\colon X \times Z \to Y$ be the bilinear mapping defined by $(x, z) \mapsto f(x)z = y$. Then $\|g\| \leq \|f\|$ because

$$\|g(x,z)\| = \|f(x)z\| \leq \|f(x)\| \cdot \|z\| \leq \|f\| \cdot \|x\| \cdot \|z\|.$$

We have thus defined a morphism $f \mapsto g$ of $\mathscr{L}(X, \mathscr{L}(Z, Y))$ into

$\mathscr{L}(X \times Z, Y)$ of norm ≤ 1. Consider now the morphism from $\mathscr{L}(X \times Z, Y)$ into $\mathscr{L}(X, \mathscr{L}(Z, Y))$ defined by associating

$$g \in \mathscr{L}(X \times Z, Y): (x, z) \mapsto y = g(x, z)$$

with

$$f \in \mathscr{L}(X, \mathscr{L}(Z, Y)): x \mapsto \text{the mapping defined by } z \mapsto g(x, z).$$

This morphism is the inverse of the morphism $f \to g$ and is also of norm ≤ 1. Hence both morphisms are of norm 1, and we have defined a natural isomorphism between $\mathscr{L}(X, \mathscr{L}(Z, Y))$ and $\mathscr{L}(X \times Z, Y)$. ∎

It follows that the second differential $f''(x_0)$ at a given point x_0 is an element of $\mathscr{L}(X \times X, Y)$

$$f''(x_0): (h, k) \mapsto (f''(x_0)h)k \in Y.$$

Theorem. If $f: U \to Y$ is twice differentiable at x_0, $f''(x_0)$ is a bilinear, symmetric mapping

$$(f''(x_0)h)k = (f''(x_0)k)h \qquad \forall\, k, h \in X.$$

The proof is easy in the finite dimensional case, it is straightforward but longer in the infinite dimensional case.[1]

quadratic form

A bilinear symmetric mapping of X into $\mathbb{K}$ is called a **quadratic form** on X.

hessian

The quadratic form $f''(x)$ is called the **hessian** of f at x.

Example 1: Finite dimensional spaces. Let

$$f: \mathbb{R}^h \to \mathbb{R} \text{ by } (x^i) \mapsto f(x^i),$$

then the hessian of f is

$$f''_{x_0}(h, k) = \frac{\partial^2 f}{\partial x^i \partial x^j} h^i k^j.$$

Let $f: \mathbb{R}^n \to \mathbb{R}^p$ by $(x^i) \mapsto f^\alpha(x^i)$. Then

$$f''_{x_0}(h, k) = \left\{ \frac{\partial^2 f^\alpha}{\partial x^i \partial x^j} h^i k^j \right\} \in \mathbb{R}^p.$$

Example 2: Banach spaces. Let

$$S: C^1([a, b]) \to \mathbb{R} \text{ by } S(q) = \int_a^b L(x, q(x), q'(x))\, dx.$$

[1]See for instance [H. Cartan 1967(b), p. 65].

The second differential of f is the differential of the mapping

$$C^1([a,b]) \to \mathscr{L}(C^1([a,b]), \mathbb{R}) \text{ by } q \mapsto S'(q),$$

thus it is a mapping of $C^1([a,b]) \times C^1([a,b]) \to \mathbb{R}$.
If L is $C^2(\mathbb{R}\times\mathbb{R})$ for its last two variables, this mapping is differentiable and

$$S''(q)(h,k) = \int_a^b \{hkL''_{yy}(x, q(x), q'(x)) + (hk' + h'k)L''_{yz}(x, q(x), q'(x))$$
$$+ h'k'L''_{zz}(x, q(x), q'(x))\}\, dx \qquad h, k \in C^1([a,b]).$$

We shall see examples of $S''(q)$, called the second variation of S, in later chapters and in several problems, for instance Problems 2, p. 100 and 3, p. 105.

Taylor's expansion – integral remainder.

Theorem. Let f: $U \subset X \to Y$ be of class C^n on $U \supset [x_0, x_0 + h]$. Then

$$f(x_0+h) = f(x_0) + f'(x_0)h + \tfrac{1}{2}f''(x_0)(h,h) + \cdots + \frac{1}{(n-1)!} f^{(n-1)}(x_0)(h)^{(n-1)} + R$$

with

$$R = \frac{1}{(n-1)!}\int_0^1 (1-t)^{n-1} f^{(n)}(x_0+th)(h)^n\, dt.$$

Proof: Let $u: [0,1] \to Y$ by $t \mapsto f(x_0+th)$, as defined above. u is now C^n on $[0,1]$ and by recurrence we obtain

$$u^{(n)}(t) = f^{(n)}(x_0+th)\cdot(h)^n.$$

Taylor's formula is then proved by integrating over t from 0 to 1 the following identity valid for functions u of the real variable t with values in a Banach space

$$\frac{d}{dt}\left(u(t) + (1-t)u'(t) + \cdots + \frac{1}{(n-1)!}(1-t)^{n-1}u^{(n-1)}(t)\right)$$
$$= \frac{1}{(n-1)!}(1-t)^{n-1}u^{(n)}(t). \qquad \blacksquare$$

Corollary. Under the hypotheses made for f, $f^{(n)}(x_0+th)$ is bounded for $t\in[0,1]$, i.e., there exists M such that

$$f^{(n)}(x_0+th)\cdot(h)^n \le M\|h\|^n \text{ for } 0 \le t \le 1,$$

hence we have Taylor's formula with Lagrange remainder

$$\|f(x_0+h)-f(x_0)-f'(x_0)h-\cdots-\frac{1}{(n-1)!}f^{(n-1)}(x_0)(h)^{n-1}\|\le M\frac{\|h\|^n}{n!}.$$

Taylor's expression – asymptotic remainder. It can be shown[1] that if f is $n-1$ differentiable in U and n differentiable at x_0 then

$$\|f(x_0+h)-f(x_0)-f'(x_0)h-\cdots-\frac{1}{n!}f^{(n)}(x_0)(h)^n\|=o(\|h\|^n).$$

B. CALCULUS OF VARIATION

In this paragraph we consider continuous functions f on an open set U of a Banach space $X, f: U\subset X\to\mathbb{R}$. We will ask ourselves the necessary conditions – and eventually the sufficient conditions – for the existence of maxima and minima.

relative minimum

f has a **relative minimum** at $a\in U$ if there exists a neighborhood $V(a)\subset U$ such that

$$f(x)\ge f(a) \qquad \forall x\in V(a).$$

strict relative minimum

f is said to have a **strict relative minimum** at a if there exists a neighborhood $V(a)\subset U$ such

$$f(x)>f(a) \qquad \forall x\in V(a), \quad x\ne a.$$

Remark: The definitions of relative maximum and strict relative maximum are analogous. If f has a [strict] relative maximum, it is clear that f has a [strict] relative minimum. We shall consider only minima.

1. NECESSARY CONDITIONS FOR MINIMA

Theorem. Let f be differentiable at a, a necessary condition for f to have a relative minimum at a is

$$f'(a)=0.$$

Proof: This theorem is elementary for a mapping $g:\mathbb{R}\to\mathbb{R}$, and we shall extend it to mappings $f: U\subset X\to\mathbb{R}$ by using it for the function g

[1] See for instance [H. Cartan 1967(b), p. 78].

defined by

$$g(t) = f(a + th)$$

for $|t| < t_0$ sufficiently small and $h \in X$. We have $g'(t) = f'(a + th)$.
If f has a relative minimum for $x = a$, then g has a relative minimum for $t = 0$ and $g'(0) = 0$; now

$$g'(0) = f'(a)h,$$

thus $g'(0)$ vanishing for every $h \in X$ implies $f'(a) = 0$. ■

Theorem. Let f be twice differentiable at a, a necessary condition for f to have a relative minimum at a is

$$f''(a) \geq 0.$$

Remark: The second differential f'' of f is a bilinear symmetric form on X. The condition $f''(a) \geq 0$ expresses that this form is positive, namely for every $h \in X$ we have $f''(a) \cdot (h, h) \geq 0$.

Proof: Taylor's formula with $f'(a) = 0$ reads

$$f(a + h) - f(a) = \tfrac{1}{2} f''(a) \cdot hh + \epsilon(h)\|h\|^2$$

where $\epsilon(h)$ goes to zero with h. f has a relative minimum at a if, for $\|h\|$ sufficiently small

$$f''(a)hh + 2\epsilon(h)\|h\|^2 \geq 0.$$

Fix h and choose a real number λ; for $\|\lambda|$ sufficiently small

$$f''(a)\lambda h \lambda h + 2\epsilon(\lambda h)\lambda^2\|h\|^2 \geq 0 \quad \text{with } \epsilon(\lambda h) \rightsquigarrow 0 \text{ as } \lambda \rightsquigarrow 0$$
$$f''(a) \cdot hh + 2\epsilon(\lambda h)\|h\|^2 \geq 0.$$

This condition remains satisfied when λ tends to 0 only if

$$f''(a) \cdot hh \geq 0 \qquad h \in X.$$ ■

2. SUFFICIENT CONDITIONS

To state a sufficient condition for the existence of a minimum, we shall define nondegenerate second differentials.
The second differential $f''(a)$ of f is said to be **nondegenerate** if the mapping $X \to \mathscr{L}(X, R) = X'$ by $h \mapsto f''(a)h$ is an isomorphism of Banach spaces. non-degenerate

Remark: This condition implies that $f''(a)h = 0$ if and only if $h = 0$; the converse – namely $f''(a)h = 0$ if and only if $h = 0$ implies $f''(a)$ non-degenerate – is true only for finite dimensional spaces X. In this case, one can say that $f''(a)$ is nondegenerate if and only if the determinant $|f''_{x^i x^j}|$ of the hessian matrix is nonvanishing.
However, due to the fact that we have included here in the definition of "non degenerate" the continuity of the mapping $h \mapsto f''(a)h$, and therefore also of the inverse mapping (Banach theorem) we can now show that if $f''(a)$ is non degenerate and positive, then it is positive definite, and even bounded away from zero.

Theorem. If $f''(a)$ is positive and nondegenerate, there exists $\lambda > 0$ such that $f''(a)hh \geq \lambda \|h\|^2$.

Proof: Since the mapping $h \mapsto f''(a)h$ is an isomorphism of X onto X' (linear continuous invertible mapping) there exists $\mu > 0$, such that

$$\|h\|_X \geq \mu \|f''(a)h\|_{X'} = \mu \sup_{\|k\|_X = 1} |f''(a)hk|.$$

Hence there exists $k \in X$ such that $\|k\| = 1$ and

$$\|h\|_X \leq \mu |f''(a) \cdot hk|.$$

Therefore, by the Cauchy–Schwarz inequality, if $f''(a)$ is a positive quadratic form, then

$$\|h\|^2{}_X \leq \mu^2 (f''(a)hh)(f''(a)kk).$$

Since $\|k\| = 1$ and $f''(a)$ is continuous, $f''(a)kk$ is bounded (p. 23); thus there exists $M > 0$ such that

$$\|h\|_X^2 \leq M(f''(a)hh). \qquad \blacksquare$$

We are now in a position to state a sufficient condition for f to have a strict relative miminum. Let $f\colon U \to R$ be a C^2 function at $a \in U$.

Sufficient condition theorem. A sufficient condition for f to have a strict relative minimum is

$$\begin{cases} f'(a) = 0 \\ f''(a) \text{ positive and nondegenerate.} \end{cases}$$

Proof: With $f'(a) = 0$, Taylor's formula reads

$$f(a+h) = f(a) + \tfrac{1}{2} f''(a)hh + \epsilon(h)\|h\|^2 \text{ with } \epsilon(h) \rightsquigarrow 0 \text{ as } h \rightsquigarrow 0.$$

The proof is concluded by taking h such that $|\epsilon(h)| < \lambda/2$ where λ is a positive number such that $f''(a)hh \geq \lambda\|h\|^2$. ■

Other sufficient conditions are expressed in terms of the second derivative in a neighborhood of a.

A sufficient condition theorem. Let $f: U \to Y$ be a C^2 mapping where U is a convex open set in X. A sufficient condition for f to have a minimum at $a \in U$ is that

$$\begin{cases} f'(a) = 0 \\ f''(x) \geq 0 \qquad x \in U. \end{cases}$$

There is a similar theorem for a relative minimum, the condition being $f'' \geq 0$ in a neighborhood of a.

Proof:

$$f(x) = f(a) + \frac{1}{2}\int_0^1 (1-t)f''(a + t(x-a))(x-a)^2 \, dt$$
$$\geq f(a).$$ ■

Example: Let Ω be an open bounded set of $\mathbf{R}^n$, let $S: C^1(\bar{\Omega}) \to \mathbf{R}$ be defined by

$$S(q) = \int_{\bar{\Omega}} \sum_{i=1}^{n} \left(\frac{\partial q}{\partial x^i}\right)^2 dx, \qquad q \in C^1(\bar{\Omega}).$$

S is differentiable and its differential is defined by

$$S'(q)h = \sum_i \int_{\bar{\Omega}} 2 \frac{\partial q}{\partial x^i} \frac{\partial h}{\partial x^i} \, dx.$$

If S is restricted to $C^2(\bar{\Omega})$, then

$$S'(q)h = \sum_i \left\{ -2 \int_{\bar{\Omega}} \frac{\partial^2 q}{(\partial x^i)^2} h \, dx + 2 \int_{\bar{\Omega}} \frac{\partial}{\partial x^i}\left(\frac{\partial q}{\partial x^i} h\right) dx \right\}.$$

If the boundary $\partial\bar{\Omega}$ of $\bar{\Omega}$ is regular, then

$$S'(q)h = \sum_i \left\{ -2 \int_{\bar{\Omega}} \frac{\partial^2 q}{(\partial x^i)^2} h \, dx + 2 \sum_i \int_{\bar{\Omega}} \frac{\partial q}{\partial x^i} n^i \omega \right\}$$

with (p. 449) $n^i\omega = (-1)^{i-1} dx^1 \wedge \ldots \wedge dx^{i-1} \wedge dx^{i+1} \ldots \wedge dx^n$.

If q takes on given values[1] on $\bar{\Omega}$, then $h = 0$ on $\bar{\Omega}$ and $S'(q)h$ reduces to

$$S'(q)h = -2 \int_\Omega \sum_i \frac{\partial^2 q}{(\partial x^i)^2} h \, \mathrm{d}x.$$

The conditions for S to have a minimum at q_0 are

$$\begin{cases} \Delta q_0 \equiv \sum_i \dfrac{\partial^2 q_0}{(\partial x^i)^2} = 0 & \text{Euler's equation} \\ S''(q_0)hh \geq 0 & \text{second order condition.} \end{cases}$$

The second order differential is obtained readily by remarking that the mapping $C^1(\bar{\Omega}) \to \mathscr{L}(C^1(\bar{\Omega}), \mathbb{R})$ by $q \mapsto S'(q)$ is linear. Hence its differential is identical to it. That is, $S''(q)$ is the mapping, independent of q, defined by

$$h \mapsto S'(h),$$

and the quadratic form $S''(q)$ is

$$S''(q)hk = S'(h)k = 2 \int_\Omega \sum_i \frac{\partial h}{\partial x^i} \frac{\partial k}{\partial x^i} \, \mathrm{d}x.$$

It follows that the second order condition is always satisfied. Moreover q_0 is an absolute minimum.

This example is a particular case of the Lagrangian problems which are fundamental in physics and which have played and are still playing an important role in the development of the calculus of variations and stability problems. We shall introduce them in the following paragraph and return to them in Chapter VI, after we have studied generalized derivatives and Sobolev spaces, as these Banach spaces permit a simple formulation of necessary and of sufficient conditions for an extremum to be a minimum or a maximum.

3. LAGRANGIAN PROBLEMS (see also p. 80)

Let X be the subspace of functions $q \in C^1([a, b])$ (cf. p. 29) which vanish on the boundary: $q(a) = q(b) = 0$. Let S be the function on X defined by

[1] The set of q's which take given non-zero values on $\partial\bar{\Omega}$ is not a vector space: the property is not additive. If $q|_{\partial\Omega}$ is a given function one should, strictly speaking, make a change of variable $q \mapsto \bar{q}$ such that $\bar{q}|_{\partial\Omega} = 0$ to derive this result.

$$S(q) = \int_a^b L(x, q(x), q'(x))\,dx.$$

A necessary condition for S to have a relative minimum at $q_0 \in X$ is that q satisfies both Euler's equation[1]

$$\mathscr{E}(L) = \left(L'_{q(x)} - \frac{d}{dx} L'_{q(x)}\right)_{q=q_0} = 0$$

and the second order condition

$$S''(q_0)hh \geq 0 \qquad h \in X.$$

It is easy to show that $S''(q)$ can be written

$$S''(q)hk = \int_a^b (A(x)h'k' + B(x)hk)\,dx \qquad \text{where } h, k \in X$$

with

$$A(x) \equiv L''_{zz}(x, q(x), q'(x))$$

$$B(x) = L''_{yy}(x, q(x), q'(x)) - \frac{d}{dx} L''_{yz}(x, q(x), q'(x)).$$

Equivalently

$$S''(q)hk = \int_a^b J_q(h) \cdot k\,dx$$

where $J_q(h)$ is the following second order differential operator on h

$$J_q(h) = -\frac{d}{dx}(A(x)h') + B(x)h.$$

The equation

$$J_q(h) = 0$$

is called[2] the **Jacobi equation**. A **Jacobi field** along q is a $C^2([a, b])$ solution of the Jacobi equation. If there exists along q_0 a non zero Jacobi field belonging to X (that is such that $h(a) = h(b) = 0$), then the points $x = a$ and $x = b$ of the path q_0 are said to be **conjugate**. It is clear that if $x = a$ and $x = b$ are conjugate along q_0, the quadratic form $S''(q_0)$ is degenerate: the mapping $h \in X \to S''(q_0)h \in \mathscr{L}(X, R)$ is not injective.

Jacobi equation

Jacobi field

conjugate points

[1]One can show (see [Gelfand and Fomin]), that $S'(q_0)h = 0$ for every $h \in C^1([a, b])$ implies that $L'_{q(x)}$ is differentiable.

[2]It has been introduced in different contexts under several different names, e.g. if S is the energy function (p. 344) the Jacobi equation is called the equation of geodetic deviation (p. 345).

In Chapter VI (p. 533) we shall prove the following theorems.

Legendre condition. A necessary condition for $S''(q_0)$ to be positive – therefore for q_0 to be a relative minimum of S – is that

$$A(x) \equiv L''_{zz}(x, q_0(x), q'_0(x)) \geq 0, \qquad a \leq x \leq b.$$

Jacobi condition. A necessary condition for q_0 to be a relative minimum of S is that $x = a$ has no conjugate point $x = c$ along q_0, $a < c < b$.

C. IMPLICIT FUNCTION THEOREM. INVERSE FUNCTION THEOREM

The implicit function theorem is the generalization to Banach spaces of the following.

Classical theorem. Let f be a function of two real variables $f: (x, y) \mapsto f(x, y)$ such that

$$\begin{cases} f(x_0, y_0) = 0 \\ f \text{ differentiable in a neighborhood of } (x_0, y_0) \\ f'_y(x_0, y_0) \neq 0. \end{cases}$$

Then the equation $f(x, y) = 0$ has exactly one continuous solution *$y = \varphi(x)$ for x in a neighborhood of x_0 such that $\varphi(x_0) = y_0$.*

The theorems which are used in the proof of the implicit function theorem are important in their own right.

1. CONTRACTING MAPPING THEOREMS

contracting mapping

Let F be a closed set on a complete metric space X, a **contracting mapping** is a mapping $f: F \to F$ such that

$$d(f(x), f(y)) \leq kd(x, y) \qquad 0 \leq k < 1 \qquad d \text{ distance in } X.$$

lipschitzian

One also says that "f is **lipschitzian** of order $k < 1$".

Contracting mapping theorem. A contracting mapping f has strictly one fixed point; i.e. there is one and only one point a such that $a = f(a)$.

Proof: The proof is by iteration. Let $x_0 \in F$, then $f(x_0) \in F, \ldots f^n(x_0) = f(f^{n-1}(x_0)) \in F$,

$$d(f^n(x_0), f^{n-1}(x_0)) \leq kd(f^{n-1}(x_0), f^{n-2}(x_0)) \leq \ldots \leq k^{n-1}d(f(x_0), x_0).$$

Since $k < 1$ the sequence $f_n(x_0)$ is a Cauchy sequence (see the proof for geometric series in elementary calculus) and tends to a limit $a \in F$ when n tends to infinity

$$a = \lim_{n=\infty} f^n(x_0) = \lim_{n=\infty} f(f^{n-1}(x_0) = f(a).$$

The uniqueness of a results from the defining property of contracting mappings: assume that there is another point b such that $b = f(b)$, then on the one hand $d(f(b), f(a)) = d(b, a)$ and on the other hand $d(f(b), f(a)) \leq kd(b, a)$, $k < 1$; hence $d(b, a) = 0$ and $b = a$. ■

This second form of the contracting mapping theorem is valid for Banach spaces and will be used in the proof of the inverse function theorem.

Contracting mapping theorem. Let $B(a, R)$ be an open ball $\|x - a\| < R$ in a Banach space X; let f: $B(a, R) \to X$ be a contracting mapping of ratio k. Then $\varphi = \mathrm{Id} - f$ is a homeomorphism of an open set $V \in B(a, R)$ onto $B(a - b, (1 - k)R)$ with $b = f(a)$. Moreover the inverse mapping φ^{-1} is $1/(1 - k)$ lipschitzian.

Proof:

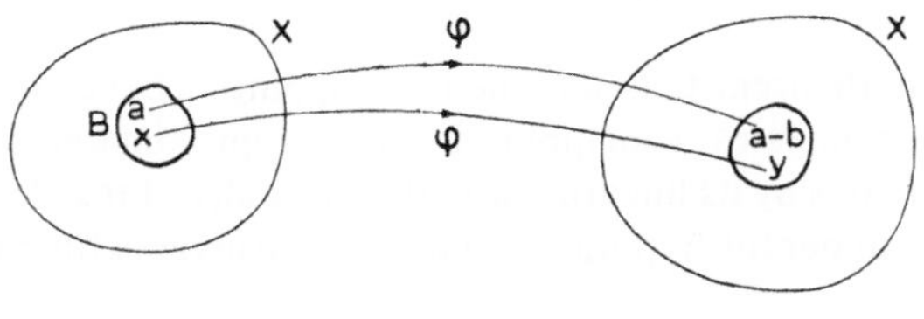

1) Given $y \in B(a - b, (1 - k)R)$, the point x such that $\varphi(x) = y$ is unique if it exists:

$$\|\varphi(x) - \varphi(x')\| \geq \|x - x'\| - \|f(x) - f(x')\| \geq (1 - k)\|x - x'\|.$$

2) x can be constructed as the limit of a sequence (x_n) defined by

$$x_0 = a, \qquad x_1 = f(x_0) + y, \ldots, x_{n+1} = f(x_n) + y, \ldots.$$

First we check that $x_n \in B(a, R) \Rightarrow x_{n+1} \in B(a, R)$ by recurrence as follows

$$\|x_{n+1}-x_n\| \leq k\|x_n - x_{n-1}\| \ldots \leq k^n\|x_1 - x_0\| = k^n\|y-(a-b)\|,$$

moreover

$$\begin{aligned}\|x_{n+1}-a\| &\leq \|x_n - a\| + \|x_{n+1}-x_n\| \\ &\leq \|x_1 - x_0\| + \ldots \|x_{n+1}-x_n\| \\ &\leq (1+k+\ldots k^n)\|y-(a-b)\| \leq \frac{1}{1-k}\|y-(a-b)\|.\end{aligned}$$

Thus

$$\|x_{n+1}-a\| < R \text{ if } \|y-(a-b)\| < (1-k)R.$$

This condition on y implies $\|x_{n+1}-x_n\| \leq k^n(1-k)R$; the sequence (x_n) is a Cauchy sequence, its limit x satisfies the equation

$$x - f(x) = y.$$

Hence φ is invertible on $B(a-b,(1-k)R)$. The inverse mapping φ^{-1} satisfies the inequality (equivalent to that in 1, above)

$$\frac{1}{1-k}\|x-x'\| \geq \|\varphi^{-1}(x) - \varphi^{-1}(x')\|$$

and it is $1/(1-k)$ lipschitzian. ■

2. INVERSE FUNCTION THEOREM

Basically this theorem tells us when a mapping behaves locally like its differential. In physics, a mapping is often approximated locally by its differential, that is by its linearization: this is valid, as far as local inversion is concerned, under the hypothesis stated by the inverse function theorem.

Lemma. The linear group $\mathrm{GL}(X, Y) = \mathrm{Isom}\,(X, Y)$ *is open in* $\mathscr{L}(X, Y)$.

Proof: This lemma is trivial in the finite dimensional case. Let X and Y be Banach spaces, let $g_0 \in \mathrm{GL}(X, Y)$, and g be such that

$$\|g_0^{-1} \circ g - \mathrm{Id}\|_{\mathscr{L}(X,X)} < 1.$$

Then $h = g_0^{-1} \circ g$ belongs to $\mathrm{GL}(X, X)$, its inverse being defined by the following convergent series in $\mathscr{L}(X, X)$

$$h^{-1} = \mathrm{Id} + k + \cdots + (\circ\, k)^n + \cdots, \text{ with } k = \mathrm{Id} - h.$$

Thus $g = g_0 \circ h \in \mathrm{GL}(X, Y)$. The above inequality on g is satisfied if g is in the neighborhood of g_0

$$\|g - g_0\|_{\mathscr{L}(X,Y)} < \frac{1}{\|g_0^{-1}\|_{\mathscr{L}(X,Y)}}. \qquad \blacksquare$$

Inverse function theorem. Let X and Y be two Banach spaces, U an open neighborhood of a in X and V an open neighborhood in Y containing $b = f(a)$; let $f: U \to V$ be a C^1 mapping such that $f'(a)$ is an isomorphism $X \to Y$. Then there exists an open neighborhood $u(a) \subset U$ and an open neighborhood $v(b) \subset V$ such that f^{-1} is a C^1 diffeomorphism of $v(b)$ onto $u(a)$.

Remark: If X and Y are finite dimensional, $f'(a)$ is the jacobian matrix $\partial f^\alpha / \partial x^i$; it is an isomorphism if and only if it is a square matrix with a nonvanishing determinant. This determinant is called the **jacobian** of f and often is denoted $J_{f'}$. jacobian

Proof: Consider $\varphi: U \to X$ by $\varphi = \mathrm{Id} - (f'(a))^{-1} \circ f$. The mapping φ is C^1 in U and its differential at a is

$$\varphi'(a) = \mathrm{Id} - (f'(a))^{-1} \cdot f'(a) = \mathrm{Id} - \mathrm{Id} = 0.$$

Hence for R sufficiently small, there exists $k < 1$ such that

$$\|\varphi'(a + tR)\| < k < 1 \text{ for } 0 < t < 1.$$

φ is a contracting mapping in the ball $B(a, R)$, hence $(f'(a))^{-1} \circ f$ is a homeomorphism of $U' \subset B(a, R)$ onto $B(b, (1-k)R)$. Since $(f'(a))^{-1}$ is an isomorphism of X onto Y, it follows that f is an homeomorphism of U' onto $V' = f'(a)\, B(b, (1-k)R)$. In particular, f is invertible on V', the equation $f(x) = y$ has strictly one solution $x \in U'$ for $y \in V'$.

Differentiability of f^{-1} at b follows from previous results (cf. p. 73), the differential of f^{-1} at b is equal to $(f'(a))^{-1}$. Then by the above lemma, $f'(x)$ is an isomorphism for $x \in u(a) \subset U'$, thus f^{-1} is C^1 on $v(b)$ (cf. theorem p. 73).

3. IMPLICIT FUNCTION THEOREM

Implicit function theorem. Hypothesis: *X, Y, Z are Banach spaces; U is an open set of $X \times Y$; $f: U \to Z$ is a C^1 mapping, $(a, b) \in U$ and $f(a, b) =$*

0. *The partial derivative* $f''_y(a, b)$ *is an isomorphism of* Y *onto* Z.
Conclusion: *There exists an open set* $W \subset X$, $a \in W$, *an open set* $V \subset U \subset X \times Y$, $(a, b) \in V$ *and a* C^1 *mapping* $g: W \to Y$ *such that*

$$(x, y) \in V, f(x, y) = 0 \Leftrightarrow x \in W, y = g(x).$$

In other words, the implicit equation $f(x, y) = 0$ has for $x \in W$ a solution $y = g(x)$ of class C^1 such that $(x, y) \in V$. This solution is unique in an open set $W' \subset W$.

Proof: Set $F: U \to X \times Z$ by $F(x, y) = (x, f(x, y))$, then

$$F' = \begin{pmatrix} 1 & 0 \\ f'_x & f'_y \end{pmatrix}.$$

$F'(a, b)$ is an isomorphism of $X \times Y$ onto $X \times Z$. The inverse function theorem applied to $F(x, y) = (x, f(x, y)) = (x, 0)$ yields the implicit function theorem. ■

4. GLOBAL THEOREMS

The inverse function theorem is a local theorem of existence and unicity. The essential task of nonlinear analysis is to find global theorems.

The conditions that $f: X \to Y$ is a C^1 mapping and $f'(x)$ is an isomorphism for each $x \in X$ are not sufficient to insure that f is a diffeomorphism, even if $X \equiv \mathbb{R}^n$.

Example:

$$f: \mathbb{R}^2 \to \mathbb{R}^2 \text{ by } (x, y) \mapsto (e^x \cos y, e^x \sin y).$$

Then the derivative $f'(x, y)$ is the linear mapping $\mathbb{R}^2 \to \mathbb{R}^2$ with matrix

$$f'(x, y) = \begin{pmatrix} e^x \cos y & -e^x \sin y \\ e^x \sin y & e^x \cos y \end{pmatrix}.$$

Its determinant is $e^{2x} \neq 0$ for every $(x, y) \in \mathbb{R}^2$, but f is not injective since

$$f(x, y) = f(x, y + 2k\pi), \qquad k \in \mathbb{Z}.$$

We give here two theorems which are easy to prove but whose hypotheses are either (theorem 2) more restrictive than the conditions satisfied in many problems or (theorem 1, condition a) difficult to check in a given problem.

Theorem 1. *Let U be an open set in the Banach space X, let f be a C^1 mapping from U into Y such that*
a) *f is injective,*
b) *$f'(x)$ is an isomorphism from X into Y for every $x \in U$;*
then f is a diffeomorphism from U into $f(U) \subset Y$.

Proof: According to hypothesis a, f is a bijection of U onto $f(U)$. The local inversion theorem shows that $f(U)$ is open in Y. Indeed let $b \in f(U)$ and a be such that $f(a) = b$, then there exist open neighborhoods $u(a)$ of $a \in U$ and $v(b)$ of $b \in Y$ such that the restriction of f to $u(a)$ is a diffeomorphism onto $v(b)$

$$v(b) = f(u(a)) \subset f(U).$$

This inclusion shows that $f(U)$ is open in Y. Similarly the image by f of any open set in U is open, hence the mapping $f^{-1}: f(U) \to U$ is continuous[1]. f is a C^1 homeomorphism of U onto $f(U)$ and, according to the theorem (p. 73) is also a diffeomorphism. ■

Theorem 2. *Let X be a Banach space; let f be a C^1 mapping from X to Y such that, for every $x \in X, f'(x)$ is an isomorphism and $(f'(x))^{-1}$ is uniformly bounded*

$$\|f'(x)^{-1}\| \leq M < \infty, \qquad x \in X.$$

Then f is a diffeomorphism of X onto Y.

Proof: We need only to prove that f is a bijection.
1) f is surjective. Let $a \in X$, $b = f(a)$; let y_0 be an arbitrary point of Y. To show the existence of x_0 such that $f(x_0) = y_0$, we consider the segment from b to y_0 in Y

$$y(t) = b(1-t) + ty_0, \qquad 0 \leq t \leq 1.$$

According to the implicit function theorem, there exists a neighborhood $u(a) \subset X$ such that f is a diffeomorphism of $u(a)$ onto a neighborhood $v(b) \subset Y$. Set

$$x(t) = f^{-1}(y(t)) \qquad \text{for } y(t) \in v(b).$$

We shall show that there exists a maximum value t_0 of $t, 0 \leq t_0 \leq 1$ such that $x(t)$ is defined for every t in the closed interval $[0, t_0]$. Let us assume that $x(t)$ is defined on the semiclosed interval $[0, t_0)$, then

$$x'(t) = (f'(x(t)))^{-1} y'(t) \qquad \text{with } 0 \leq t < t_0 \leq 1.$$

[1]Since the reciprocal image $f(v)$ of every open set $v \subset U$ is open.

Hence according to the hypothesis and the definition of $y(t)$

$$\|x'(t)\|_X \leq M\|y_0 - b\|_Y,$$

and according to the mean value theorem

$$\|x(t) - x(\tau)\| < M\|y_0 - b\|_Y |t - \tau| \quad \text{for every pair } (t, \tau)$$

such that

$$0 \leq t < t_0, \qquad 0 \leq \tau < t_0.$$

Hence the existence of the limit in X for $t = t_0$ with $t < t_0$, $\lim x(t) = x(t_0)$, which together with the continuity of f implies $f(x(t_0)) = y(t_0)$. Moreover this maximum value of t_0 cannot be strictly less than 1: the local inverse function theorem applied to the neighborhoods of $x(t_0)$ and $y(t_0)$ makes it possible to determine $x(t) = f^{-1}(y(t))$ in a neighborhood of $g(t_0)$ and $y(t_0)$, hence for values $t > t_0$.

2) f is injective. The proof is an ab absurdo proof which rests on the construction of the following family of closed curves. According to arguments similar to the arguments used in the previous paragraph, if there exist two distinct points x_1 and x_2 such that $f(x_1) = f(x_2) = y$, then one can construct a family of closed curves through y, images by f of a family of curves joining x_1 to x_2 and homotopic (continuously deformable) to the curve consisting only of the point y. The existence of the curve whose image is y contradicts the local inverse function theorem. ■

D. DIFFERENTIAL EQUATIONS

We give here only the fundamental results which will be necessary in the following chapters. In this section X is a real Banach space, φ and its derivative φ' are functions of a real variable t with values in X (see example 4, p. 72).

1. FIRST ORDER DIFFERENTIAL EQUATION

$$\frac{dx}{dt} = f(t, x), \qquad x \in X, \qquad f: U \subset \mathbb{R} \times X \to X \text{ continuously.}$$

A solution of this differential equation is a C^1 function $\varphi: I \subset \mathbb{R} \to X$ such that

$$\begin{cases} (t, \varphi(t)) \in U & t \in I \\ \varphi'(t) = f(t, \varphi(t)). \end{cases}$$

Example: Let $X = \mathbb{R}^n$, $x = (x^1, \ldots, x^n)$, $f = (f^1, \ldots, f^n)$; a differential equation on X is equivalent to a system of n scalar differential equations

$$\frac{dx^i}{dt} = f^i(t, x^1, \ldots, x^n), \qquad i = 1, \ldots, n.$$

Remark: An equation of order n on X which reads, with obvious notations

$$\frac{d^n x}{dt^n} = f\left(t, x, \frac{dx}{dt}, \ldots, \frac{d^{n-1}x}{dt^{n-1}}\right)$$

is equivalent to a first order equation on $(\times X)^n$, that is to n equations of first order on X

$$\frac{dx}{dt} = x^1$$

$$\frac{dx^1}{dt} = x^2$$

$$\vdots$$

$$\frac{dx^{n-1}}{dt} = f(t, x, x^1, \ldots, x^{n-1})$$

where $(x, x^1, \ldots, x^{n-1}) \in (\times X)^n$

2. EXISTENCE AND UNIQUENESS THEOREMS FOR THE LIPSCHITZIAN CASE

A mapping f between two Banach spaces $f: X \to Y$ is said to be **k lipschitzian** in $U \subset X$ if *k* lipschitzian

$$\|f(x) - f(x_0)\| \leq k\|x - x_0\|, \qquad x, x_0 \in U.$$

$f: I \times U \subset \mathbb{R} \times X \to X$ is **locally lipschitzian** if for every point (t_0, x_0) there exists a neighborhood $N(t_0, x_0) \subset I \times U$ and a number $k > 0$ such that locally lipschitzian

$$\|f(t, x_1) - f(t, x_2)\| \leq k\|x_1 - x_2\|, \qquad (t, x_1), (t, x_2) \in N.$$

Theorem. Local existence. If $f: I \times U \to X$ is continuous and locally lipschitzian, and if $(t_0, x_0) \in I \times U$, then there exists $a > 0$ such that the differential equation

$$dx/dt = f(t, x)$$

has a solution $\varphi: [t_0 - a, t_0 + a] \to X$ with $\varphi(t_0) = x_0$.

Picard's method

Proof: Iteration method, called **Picard's method**. Let B be a closed ball of center x_0, radius ϵ; let l be a closed interval of center t_0 (with $(t_0, x_0) \in I \times U, l \times B \subset N(t_0, x_0)$). Set

$$\sup_{\substack{t \in l \\ x \in B}} \|f(t, x)\| \leq M.$$

Note the inequality

$$\|f(t, x)\| \leq \|f(t, x_0)\| + k\|x - x_0\|.$$

The equation

$$\mathrm{d}x_1/\mathrm{d}t = f(t, x_0)$$

has a unique solution which satisfies $x_1(t_0) = x_0$, namely (cf. p. 81)

$$x_1(t) = \int_{t_0}^{t} f(\tau, x_0)\,\mathrm{d}\tau + x_0, \qquad t \in I.$$

Now

$$\|x_1(t) - x_0\| \leq |t - t_0| \sup_{t_0 < \tau < t} \|f(\tau, x_0)\| \leq M|t - t_0|,$$

hence there exists positive $a \in l$ such that $x_1(t) \in B$ for every $t \in [t_0 - a, t_0 + a]$, namely any a such that $a < \epsilon/M$.
By induction, one sets

$$x_n(t) = \int_{t_0}^{t} f(\tau, x_{n-1}(\tau))\,\mathrm{d}\tau + x_0$$

and one has

$$x_n(t) \in B \text{ for } t \in [t_0 - a, t_0 + a],$$

and

$$\|x_n(t) - x_{n-1}(t)\| \leq \int_{t_0}^{t} \|f(\tau, x_{n-1}(\tau)) - f(\tau, x_{n-2}(\tau))\|\,\mathrm{d}\tau$$

$$\leq |t - t_0| k \int_{0}^{t} \|x_{n-1}(\tau) - x_{n-2}(\tau)\|\,\mathrm{d}\tau.$$

Hence, by induction

$$\|x_n(t) - x_{n-1}(t)\| \leq \frac{M^n k^{n-1} |t - t_0|^n}{n!}.$$

The sequence $(x_n(t))$ is thus a Cauchy sequence in B for every $t \in$

$[t_0 - a, t_0 + a]$, it tends to a limit $x(t) \in B$ which satisfies the equation

$$x(t) = \int_{t_0}^{t} f(\tau, x(\tau))\, \mathrm{d}\tau + x_0$$

and hence the given differential equation. ■

Remark: If f is linear and continuous in x, on X, then it is globally liphitzian on X:

$$\|f(t, x_1) - f(t, x_2)\| \leq k\|x_1 - x_2\|, \qquad \forall\, x_1, x_2 \in X.$$

In the proof of the existence theorem we can take $B = X$, and obtain the result that the solution exists globally, and is unique, on X.

Remark: If X is finite dimensional the existence – but not uniqueness – can be proved without the locally lipschitzian assumption, for f continuous, through a compactness argument. This result does not hold if X is infinite-dimensional.

Global uniqueness theorem. If f is continuous and locally lipschitzian and if $(t_0, x_0) \in I \times U$ – the same hypotheses as for the local existence theorem –, then there is a maximum interval $J \ni t_0$ for which there exists a solution $\psi: J \to X$ of the equation $\mathrm{d}x/\mathrm{d}t = f(t, x)$ which satisfies the initial value condition $\psi(t_0) = x_0$. This solution is unique. It is called the maximal solution for the initial value (t_0, x_0).

Proof: 1) Local uniqueness. Let $\varphi_1(t)$ and $\varphi_2(t)$ be two solutions with $\varphi_1(t_0) = \varphi_2(t_0) = x_0$, then, if $(t, \varphi_1(t))$ and $(t, \varphi_2(t))$ are in a convex neighborhood $N(t_0, x_0)$, the following inequality is satisfied

$$\|\varphi_1(t) - \varphi_2(t)\| \leq k|t - t_0| \sup_{t_0 \leq \tau \leq t} \|\varphi_1(\tau) - \varphi_2(\tau)\|.$$

Hence if $k|t - t_0| < 1$

$$\|\varphi_1(t) - \varphi_2(t)\| = 0, \qquad \varphi_1(t) = \varphi_2(t).$$

2) Global uniqueness.
a) We shall show that if φ_1 and φ_2 are solutions in an interval $j \ni t_0$ and if $\varphi_1(t_0) = \varphi_2(t_0)$ then $\varphi_1(t) = \varphi_2(t)$ in j. The set of t such that $\varphi_1(t) = \varphi_2(t)$ is a set closed in j because φ_1 and φ_2 are continuous, it is also open in j because of the local uniqueness theorem, hence it is j.

b) Consider the set of pairs (j, φ) where φ is a solution on $j \ni t_0$ such that $\varphi(t_0) = x_0$. If (j_1, φ_1) and (j_2, φ_2) are two such pairs, $\varphi_1 = \varphi_2$ on $j_1 \cap j_2$ by the preceding argument. The maximal solution ψ is the function defined on the union J of all such j whose restriction to each j of a pair (j, φ) is the function φ. ∎

Example 1: A solution which does not exist in the whole of $\mathbb{R}$ although f is continuous and locally lipschitzian on $\mathbb{R} \times X$. Consider the equation

$$\mathrm{d}x/\mathrm{d}t = x^2 \qquad X = \mathbb{R}.$$

The maximal solution equal to x_0 at $t = t_0 = 0$ is, if $x_0 > 0$

$$\varphi(t) = x_0/(1 - tx_0) \quad \text{defined on} \quad -\infty < t < 1/x_0.$$

Under the hypotheses made in the two previous theorems it can easily be shown that the solution is a continuous function of the initial data x_0, locally lipschitzian at x_0.

Example 2: A differential equation depending on a parameter $\lambda \in T \subset Y$ where Y is a topological space

$$\mathrm{d}x/\mathrm{d}t = f(t, x; \lambda).$$

If

$$\|f(t, x; \lambda)\| \le M \text{ on } I \times B \times T$$

and

$$\|f(t, x_1; \lambda) - f(t, x_2; \lambda)\| \le k\|x_1 - x_2\| \text{ for } t \in I, \quad x_1, x_2 \in B, \quad \lambda \in T$$

– in other words for f k-lipschitzian in x with k independent of t and λ – then $\varphi(t, \lambda)$ is a continuous function of $\lambda \in T$ for $t \in [t_0 - a, t_0 + a]$.

PROBLEMS AND EXERCISES

PROBLEM 1. BANACH SPACES, FIRST VARIATION, LINEARIZED EQUATION

a) *Consider the quasi-linear second order operator on real valued functions on Ω, an open set of $\mathbb{R}^n$,*

$$P: U \to V, \quad U \subset C^2(\bar{\Omega}), \quad V \subset C^0(\bar{\Omega})$$

by

$$u \mapsto Pu \equiv a^{ij}\left(x^l, u, \frac{\partial u}{\partial x^l}\right) \frac{\partial^2 u}{\partial x^i \partial x^j} + b^i\left(x^l, u, \frac{\partial u}{\partial x^l}\right) \frac{\partial u}{\partial x^i}.$$

Assume that a^{ij} and b^i are C^1 functions on $\bar{\Omega} \times \bar{I} \times \bar{I}_1 \times \cdots \times \bar{I}_n$ where $\bar{I}, \bar{I}_1, \bar{I}_n$, are intervals in $\mathbb{R}$ such that if $u \in U$, then $u(x) \in I$, $\partial u/\partial x^i \in I_i$

for all $x \in \bar{\Omega}$. Show that if $u_0 \in U$, the mapping P is differentiable at u_0 and that its differential (called "first variation", or "linearized operator") at u_0 is the following linear operator

$$P'(u_0): C^2(\bar{\Omega}) \to C^0(\bar{\Omega})$$

$$h \mapsto a^{ij}\left(x^l, u_0, \frac{\partial u_0}{\partial x^l}\right) \frac{\partial^2 h}{\partial x^i \partial x^j} + \frac{\partial a^{ij}}{\partial u}\left(x^l, u_0, \frac{\partial u_0}{\partial x^l}\right) \frac{\partial^2 u_0}{\partial x^i \partial x^j} h$$

$$+ \frac{\partial a^{ij}}{\partial(\partial u/\partial x^l)}\left(x, u_0, \frac{\partial u_0}{\partial x^l}\right) \frac{\partial^2 u_0}{\partial x^i \partial x^j} \frac{\partial h}{\partial x^l} + \frac{\partial b^i}{\partial u}\left(x^l, u_0, \frac{\partial u_0}{\partial x^l}\right) h$$

$$+ \frac{\partial b^i}{\partial(\partial u/\partial x^l)}\left(x^l, u_0, \frac{\partial u_0}{\partial x^l}\right) \frac{\partial h}{\partial x^l}.$$

b) *Show that when a^{ij} and b^i are $C^{1,\alpha} \times C^2 \times \cdots \times C^2$ functions on $\bar{\Omega} \times \bar{I} \cdots \times \bar{I}_n$, $P'(u_0)$ is still the differential of $u \mapsto Pu$, considered as a mapping from $C^{2,\alpha}(\bar{\Omega})$ into $C^{0,\alpha}(\bar{\Omega})$ (cf. p. 70).*

Answer: a) Let $u_0, h, u = u_0 + h \in U$

$$P(u) - P(u_0) \equiv \left(a^{ij}\left(x^l, u, \frac{\partial u}{\partial x^l}\right) - a^{ij}\left(x^l, u_0, \frac{\partial u_0}{\partial x^l}\right)\right) \frac{\partial^2 u}{\partial x^i \partial x^j}$$

$$+ a^{ij}\left(x^l, u_0, \frac{\partial u_0}{\partial x^l}\right) \frac{\partial^2 h}{\partial x^i \partial x^j} + b^i\left(x^l, u, \frac{\partial u}{\partial x^i}\right) - b^i\left(x^l, u_0, \frac{\partial u_0}{\partial x^l}\right).$$

If $P'(u_0)$ is the given linear operator, the indicated result is obtained by using the mean value theorem for the functions a^{ij} and b^i in order to bound

$$\|P(u) - P(u_0) - P'(u_0)h\|_{C^0(\bar{\Omega})}.$$

b) It has been proved (cf. Problem I4, p. 70) that $C^{k,\alpha}(\bar{\Omega})$ is an algebra. Moreover if $f: \bar{\Omega} \to I \subset \mathbb{R}$ belongs to $C^{k,\alpha}(\bar{\Omega})$ and $g: \bar{I} \times \mathbb{R} \to \mathbb{R}$ belongs to $C^{k+1}(\bar{I})$, then the composite mapping $g \circ f$ belongs to $C^{k,\alpha}(\bar{\Omega})$ (use the mean value theorem for the proof). Thus under the given hypothesis P maps $C^{2,\alpha}(\bar{\Omega})$ into $C^{0,\alpha}(\bar{\Omega})$.
To show that $P'(u_0)$ is still the differential of the mapping P considered as a mapping from $C^{2,\alpha}(\bar{\Omega})$ into $C^{0,\alpha}(\bar{\Omega})$ we have to show that

$$\|P_u - P_{u_0} - P'(u_0)h\|_{C^{0,\alpha}(\bar{\Omega})}, \quad u_0, u_0 + h = u \in C^{2,\alpha}(\bar{\Omega})$$

tends to zero faster than $\|h\|_{C^{2,\alpha}(\bar{\Omega})}$. The proof is straightforward, using for instance the fact that (p. 81) (we take a^{ij} depending only on u to simplify the writing):

$$a^{ij}(u(x)) - a^{ij}(u_0(x)) - \frac{\partial a^{ij}}{\partial u}(u_0(x))h(x) = \int_0^1 \frac{\partial^2 a^{ij}}{\partial u^2}(u_0(x) + th(x))h^2(x)\,dt.$$

PROBLEM 2. TAYLOR EXPANSION OF THE ACTION; JACOBI FIELDS; THE FEYNMAN–GREEN FUNCTION (THE GREEN FUNCTION OF THE SMALL DISTURBANCE EQUATION); THE VAN VLECK MATRIX (THE HESSIAN OF THE ACTION FUNCTION); CONJUGATE POINTS; CAUSTICS.

Let $\mathscr{C}^1(\bar{\Omega})$ *be the space of* C^1 *mappings* q *on* $\Omega = (a, b) \subset \mathbb{R}$ *into* $\mathbb{R}^n$, *together with the uniform norm. Let* S *be the action of a system* S

$$S(q) = \int_{\Omega} L(q(x), \dot{q}(x), x)\, dx, \quad \dot{q}(x) = dq(x)/dx.$$

Let q_0 *be a critical (stationary) point of* S, $S'(q_0) = 0$; *if it minimizes* S, q_0 *is a classical path of the system* S; *otherwise we shall call it a stationary path. Note that when* q *is a mapping of* Ω *into a manifold* M *other than* $\mathbb{R}^n$, *one cannot write* $q(x) = q_0(x) + h(x)$. *In order to have results which can easily be generalized to the case* $q: \Omega \to M$, *and also to simplify the present case, we introduce a one parameter family of functions* $\bar{\alpha}(u) \in \mathscr{C}^1(\bar{\Omega})$ *where* $u \in U \subset \mathbb{R}$, $U = [0, 1]$, *such that* $\bar{\alpha}(0) = q_0$, $\bar{\alpha}(1) = q$. $\{\bar{\alpha}(u)\}$ *is often called a variation of* q_0. *Set* $\bar{\alpha}(u)(x) = \alpha(u, x)$.

a) *Compute* $d^2(S \circ \bar{\alpha})/du^2$.

b) *Construct the Jacobi fields (solutions of the small disturbance equation).*

c) *Construct the Feynman–Green function.*

d) *Let* $\bar{S}$ *be the two point action function defined for a stationary path* q_0 *by*

$$\bar{S}(A, B) = S(q_0) \text{ where } B = q_0(b), A = q_0(a).$$

Compute the hessian of $\bar{S}$.

e) *Define and characterize conjugate points along* q_0.

Answer:

a) $$S: \mathscr{C}^1(\bar{\Omega}) \to \mathbb{R} \text{ by } q \mapsto S(q),$$

$$S \circ \bar{\alpha}: U \subset \mathbb{R} \to \mathbb{R} \text{ by } u \mapsto S(\bar{\alpha}(u)),$$

$$(S \circ \bar{\alpha})(u) = (S \circ \bar{\alpha})(0) + \sum_{n=1}^{\infty} \frac{u^n}{n!} \frac{d^n(S \circ \bar{\alpha})}{du^n}(0),$$

where $\dfrac{d(S \circ \bar{\alpha})}{du}(u) = S'(\bar{\alpha}(u))\bar{\alpha}'(u)$ by the composite mapping theorem,

$$\bar{\alpha}'(u) \text{ is a mapping } [a, b] \to \mathbb{R}^n \text{ by } x \mapsto \bar{\alpha}'(u)(x) = \frac{\partial \alpha(u, x)}{\partial u},$$

$$\frac{d^2(S \circ \bar{\alpha})}{du^2}(u) = S''(\bar{\alpha}(u))\bar{\alpha}'(u)\bar{\alpha}(u) + S'(\bar{\alpha}(u))\bar{\alpha}''(u),$$

$$\frac{d^2(S \circ \bar{\alpha})}{du^2}(0) = S''(q_0)\bar{\alpha}'(0)\bar{\alpha}'(0).$$

Set $\bar{\alpha}'(0) = h$, $h(x)$ is a vector at $q_0(x)$, tangent to the curve $u \mapsto \alpha(u, x)$ at $u = 0$. We say that $\bar{\alpha}'(0)$ is a vector field along q_0.

Two kinds of variations of q_0 are particularly useful.
1) Variations keeping the end points fixed: $\alpha(u, a) = A$, $\alpha(u, b) = B$. It follows that $\partial\alpha(u, a)/\partial u = \partial\alpha(u, b)/\partial u = 0$, $h(a) = h(b) = 0$;
2) Variations through stationary paths not necessarily keeping the end points fixed. We shall call such variations $\bar{\beta}(u)$: $\bar{\beta}(u)$ is a stationary path from A_u to B_u.

Let $\bar{\alpha}(u)$ be an arbitrary variation of q_0; let L_1 and L_2 be the derivatives of L with respect to its first and second argument respectively, let L_{11} and L_{12} be the derivatives of L_1 with respect to its first and second argument respectively, L_{21} and L_{22} the derivatives of L_2 with respect to its first and second argument respectively.

$$(S \circ \bar{\alpha})'(u) = L_2 \frac{\partial\alpha}{\partial u}(u, x)\Big|_{x=a}^{b} + \int_\Omega \left(L_1 - \frac{dL_2}{dx}\right)\frac{\partial\alpha}{\partial u}(u, x)\,dx,$$

$$(S \circ \bar{\alpha})''(u) = \int_\Omega \left(L_{11}\frac{\partial\alpha}{\partial u}\frac{\partial\alpha}{\partial u} + L_{12}\frac{\partial\alpha}{\partial u}\frac{\partial^2\alpha}{\partial u\partial x} + L_{21}\frac{\partial^2\alpha}{\partial u\partial x}\frac{\partial\alpha}{\partial u}\right.$$

$$\left. + L_{22}\frac{\partial^2\alpha}{\partial u\partial x}\frac{\partial^2\alpha}{\partial u\partial x}\right)dx$$

$$= \left(L_{21}\frac{\partial\alpha}{\partial u}\frac{\partial\alpha}{\partial u} + L_{22}\frac{\partial\alpha}{\partial u}\frac{\partial^2\alpha}{\partial u\partial x}\right)\Big|_{x=a}^{b} + \int_\Omega \left(\left(L_{11} - \frac{d}{dx}L_{21}\right)\frac{\partial\alpha}{\partial u}\frac{\partial\alpha}{\partial u}\right.$$

$$\left. + \left(L_{12} - L_{21} - \frac{d}{dx}L_{22}\right)\frac{\partial\alpha}{\partial u}\frac{\partial^2\alpha}{\partial x\partial u} - L_{22}\frac{\partial\alpha}{\partial u}\frac{\partial^3\alpha}{\partial x^2\partial u}\right)dx.$$

If $\bar{\alpha}$ is a variation keeping the end points fixed, then

$$(S \circ \bar{\alpha})'(0) = S'(q_0)\bar{\alpha}'(0) = \int_\Omega L_{q_0}(x)h(x)\,dx,$$

$$(S \circ \bar{\alpha})''(0) = S''(q_0)\bar{\alpha}'(0)\bar{\alpha}'(0) = \int_\Omega (J_{q_0}h(x))h(x)\,dx,$$

where $L_{q_0}(x) = L_1 - dL_2/dx$

and $\quad J_{q_0}h(x) = -L_{22}\ddot{h}(x) + \left(L_{12} - L_{21} - \frac{d}{dx}L_{22}\right)\dot{h}(x) + \left(L_{11} - \frac{d}{dx}L_{21}\right)h(x),$

the derivatives of L being taken at $\alpha(0, x) = q_0(x)$.

$J_{q_0}h_0(x) = 0$ is the Jacobi equation (small disturbance equation)

Legendre matrix

$(L_{22})q_0$ is sometimes called the Legendre matrix.

b) We shall show that, if $\bar{\beta}(u)$ is a variation through stationary paths, $\bar{\beta}'(0)$ is a solution h_0 of the Jacobi equation and use this property of variations through stationary paths to compute the complete set of Jacobi fields.
Since $\bar{\beta}(u)$ is a stationary path for every $u \in U$,

$$L_{\bar{\beta}(u)}(x) = L_1 - \frac{\mathrm{d}L_2}{\mathrm{d}x} = 0,$$

$$0 = \frac{\mathrm{d}}{\mathrm{d}u}(L_{\bar{\beta}(u)}(x))$$

$$= L_{11}\frac{\partial\beta}{\partial u} + L_{12}\frac{\partial^2\beta}{\partial u\partial x} - \frac{\mathrm{d}}{\mathrm{d}x}\left(L_{21}\frac{\partial\beta}{\partial u} + L_{22}\frac{\partial^2\beta}{\partial u\partial x}\right)$$

$$= J_{\bar{\beta}(u)}\frac{\partial\beta}{\partial u}(u, x).$$

The Jacobi equation $J_{q_0}h_0(x) = 0$ is a system of n second order differential equations. To obtain the complete set of Jacobi fields, one introduces a $2n$ parameter variation $\bar{\beta}(u_i, v_i)$ with $i = 1, \ldots n$, of the stationary path q_0.
The $2n$ vector fields $(\partial\beta/\partial u_i)(0)$, $(\partial\bar{\beta}/\partial v_i)(0)$ are $2n$ solutions of $J_{q_0}h_0(x) = 0$, usually linearly independent.
If the stationary solutions are known, this method yields readily the solutions of the small disturbance equation, which are often very difficult to compute directly. See for instance the next problem.
In general a Jacobi field h_0 can be defined by its values at the end points $h_0(a)$ and $h_0(b)$. If, however there exists a non zero Jacobi field vanishing at a and b, the Jacobi fields cannot be determined by their Dirichlet data, they are instead characterized by their Cauchy data $h_0(a)$ and $\dot{h}_0(a)$.
To express a Jacobi field in terms of its Dirichlet data or in terms of its Cauchy data, it is convenient to introduce the following mappings

$$J(x, a)\colon T_{q_0(a)}\mathbb{R}^n \to T_{q_0(x)}\mathbb{R}^n \quad \text{and} \quad K(x, a)\colon T_{q_0(a)}\mathbb{R}^n \to T_{q_0(x)}\mathbb{R}^n,$$

both J and K, as functions of x, satisfy the Jacobi equation. They are defined by the following boundary conditions

$$J(a, a) = 0, \qquad \frac{\partial J}{\partial x}(x = a, a) = A^{-1}(a),$$

$$K(a, a) = A^{-1}(a), \qquad \frac{\partial K}{\partial x}(x = a, a) = 0$$

where $A(x) \overset{\text{def}}{=} \partial^2 L/\partial\dot{q}_0(x)\partial\dot{q}_0(x)$ is the Legendre matrix.

J is known as the commutator function. Let $M(x, y)$ be the negative inverse matrix of $J(y, x)$, that is $M_{\alpha\beta}(x, y)J^{\beta\gamma}(y, x) = \delta_\alpha^\gamma$.
Then

$$h_0(x) = -J(x, a)M(a, b)h_0(b) - J(x, b)M(b, a)h_0(a) \quad \text{(Dirichlet data)}$$

or

$$h_0(x) = J(x, a)A(a)h_0(a) + K(x, a)A(a)h_0(a) \qquad \text{(Cauchy data).}$$

$J(x, a)$ can be obtained by a variation $\bar{\beta}$ of q_0 through stationary paths keeping the initial point $q_0(a)$ fixed, indeed

$$\frac{\partial\beta}{\partial u}(0, x) = J(x, a)\frac{\partial^2\beta}{\partial x\partial u}(0, a).$$

Similarly $J(x, b)$ can be obtained by a variation $\bar{\beta}$ of q_0 through stationary paths keeping the final point $q_0(b)$ fixed.

c) The Green function of the small disturbance operator vanishing on the boundary is called the Feynman–Green function; it is equal to

$$G(x, y) = Y(x-y)J(x, b)M(b, a)J(z, y) - Y(y-x)J(x, a)M(a, b)J(b, y),$$

where Y is the stepfunction equal to 1 for positive argument and 0 otherwise.

Proof: One checks readily that G vanishes for x or y equal to a or b. Off the diagonal G is a two point Jacobi field. On the diagonal, G is continuous. The discontinuity of its derivative across the diagonal

$$\frac{\partial G}{\partial x}(x = y^+, y) - \frac{\partial G}{\partial x}(x = y^-, y)$$

$$= \frac{\partial J}{\partial y}(y, b)M(b, a)J(a, y) + \frac{\partial J}{\partial y}(y, a)M(a, b)J(b, y).$$

On the other hand

$$J(x, a)M(a, b)J(b, y) + J(x, b)M(b, a)J(a, y) = -J(x, y).$$

Indeed, as a function of x, the left hand side is a linear combination of solutions of the Jacobi equation. Similarly as a function of y. It suffices to check the equation for x equal to a or b and for y equal to a or b. Finally take the derivative of both sides of the equation with respect to x and set $x = y$; it becomes

$$\frac{\partial G}{\partial x}(x = y^+, y) - \frac{\partial G}{\partial x}(x = y^-, y) = -A^{-1}(y)$$

and

$$J_{q_0x}G(x, y) = \delta_y.$$

d) Expand both sides of the equation

$$\bar{S}(\beta(u, a), \beta(u, b)) = S \circ \bar{\beta}(u)$$

with respect to u. Since $\bar{\beta}(u)$ is a stationary path for every $u \in U$,

$$(S \circ \bar{\beta})'(u) = L_2 \frac{\partial \beta}{\partial u}(u, x)\Big|_{x=a}^{b},$$

$$(S \circ \bar{\beta})''(u) = \left(L_{21} \frac{\partial \beta}{\partial u}\frac{\partial \beta}{\partial u} + L_{22} \frac{\partial \beta}{\partial u}\frac{\partial^2 \beta}{\partial u \partial x}\right)(u, x)\Big|_{x=a}^{b}.$$

Using $\partial\beta/\partial u(0, x) = h_0(x) = -J(x, a)M(a, b)h_0(b) - J(x, b)M(b, a)h_0(a)$ we obtain

$$\frac{\partial \bar{S}}{\partial B} = p(b), \qquad \frac{\partial \bar{S}}{\partial A} = -p(a),$$

where $p(x) = L_2(q_0(x), \dot{q}_0(x), x)$ and, abbreviating $L(q_0(a), \dot{q}_0(a), a)$ to $L(a)$,

$$\begin{bmatrix} \dfrac{\partial^2 \bar{S}}{\partial A \partial A} & \dfrac{\partial^2 \bar{S}}{\partial A \partial B} \\ \dfrac{\partial^2 \bar{S}}{\partial B \partial A} & \dfrac{\partial^2 \bar{S}}{\partial B \partial B} \end{bmatrix} = \begin{bmatrix} -L_{21}(a) + L_{22}(a)\dfrac{\partial J}{\partial a}(a, b)M(b, a) & M(a, b) \\ -M(b, a) & L_{21}(b) - L_{22}(b)\dfrac{\partial J}{\partial b}(b, a)M(a, b) \end{bmatrix}.$$

It follows that the Van Vleck matrix is the inverse of the commutator function

$$\frac{\partial^2 \bar{S}}{\partial B \partial A} = J^{-1}(b, a).$$

This result plays an important role in many problems of physics. Note $J^{\alpha\beta}(b, a) = -J^{\beta\alpha}(a, b)$.

e) Two points $A = q_0(a)$ and $B = q_0(b)$ are said to be conjugate along a stationary path q_0 if there exists a non zero Jacobi field h_0 vanishing at A and B. If there exist m linearly independent such Jacobi fields, the conjugate points are said to have multiplicity m. If there is a non zero Jacobi field vanishing on the boundary the bilinear form $S''(q_0)$ is degenerate. The dimension of its null space is called its nullity. The nullity is equal to the multiplicity of the conjugate points.
We shall give two characterizations of conjugate points.
1) Consider a one parameter family of stationary paths $\{\bar{\beta}(u)\}$ going through A: $\beta(u, a) = A$ for every $u \in U$. If this family has an envelope,

the contact point $B_u = \beta(u, b)$ of a stationary path $\bar{\beta}(u)$ with its envelope is conjugate to A along $\bar{\beta}(u)$.

Proof: Let $q_0 = \bar{\beta}(0)$ be tangent to its envelope at $q_0(b)$, then $\beta(u, b) - \beta(0, b)$ is at least of order u^2, i.e.

$$\left.\frac{\partial \beta(u, b)}{\partial u}\right|_{u=0} = 0.$$

There exists then a non zero Jacobi field $h = \bar{\beta}'(0)$ vanishing at a and b, $q_0(b)$ is conjugate to $q_0(a)$. The converse is proved by the reverse argument. ■

2) A and B are conjugate along q_0 if and only if

$$\left(\det \frac{\partial^2 \bar{S}(B, A)}{\partial B \partial A}\right)^{-1} = 0.$$

Proof: Let h_0 be a Jacobi field vanishing at A, $h_0(a) = 0$, then

$$h_0(b) = J(b, a) A^{-1}(a) \dot{h}(a),$$

$$h_0(b) = 0 \text{ if and only if } \det J(b, a) = \left(\det \frac{\partial^2 \bar{S}(B, A)}{\partial B \partial A}\right)^{-1} = 0. \qquad ■$$

The next problem shows the role of conjugate points in determining whether or not a stationary path is a classical path.

PROBLEM 3. THE SOAP BUBBLE PROBLEM; EULER–LAGRANGE EQUATION; THE SMALL DISTURBANCE EQUATION; JACOBI FIELDS; CONJUGATE POINTS.

Dip two wire circles of radius r and R in a soap solution, then remove them. Assume that the two circles, originally concentric, are gradually pulled apart so that their planes remain perpendicular to the axis joining their centers. The soap film forms a surface of revolution of minimum area and eventually breaks into two circular disks bounded by the wire circles. Solve the Euler–Lagrange equation for this problem, show that some of its solutions are not extremal.

Answer: The surface of revolution generated by the curve $y = q(x)$ is

$$S(q) = 2\pi \int_0^l q(x)(1 + (\dot{q}(x))^2)^{1/2} \, dx.$$

The Euler–Lagrange equation $S'(q_0) = 0$ gives $-q_0(x)\ddot{q}_0(x) + \dot{q}_0^2(x) + 1 = 0$, its solutions are

$$q_0(x) = C \cosh \frac{1}{C}(x - D),\ q_0 \text{ is a catenary.}$$

The two constants of integration C and D are the coordinates of the lowest point of the curve q_0. If instead of characterizing a solution of the Euler–Lagrange equation by the values of C and D, we wish to characterize it by the values of l and R, we remark that for a given pair (l, R) we may have 2 solutions, 1 solution, 0 solution; indeed

$$R = C \cosh \frac{l-D}{C}, \qquad r = C \cosh \frac{D}{C},$$

or $R = C \cosh \left(\frac{l}{C} - \left| \operatorname{arc\,cosh} \frac{r}{C} \right|\right)$. For a given l, R is a function of C such that

C	0		$C_0(l)$		r
R	$+\infty$	$\searrow$	$R_0(l)$	$\nearrow$	$+\infty$

where C_0 is the solution of

$$\frac{dR}{dC} = \sinh \frac{l-D}{C}\left(\coth \frac{l-D}{C} - \frac{l-D}{C} + \coth \frac{D}{C} - \frac{D}{C}\right) = 0$$

provided $\sinh(D/C) \neq 0$, the case $D = 0$ is discussed at the end.
For a given l, the equation $R - C \cosh(l/C - |\operatorname{arc\,cosh}(r/C)|) = 0$ has

2 solutions C_1 and C_2	for $R > R_0(l)$,
1 solution C_0	for $R = R_0(l)$,
0 solution	for $R < R_0(l)$.

Are all the solutions of the Euler–Lagrange equation

$$\left\{q_0^i(x) = C_i \cosh \frac{x - D_i}{C_i}, \qquad r = C_i \cosh \frac{D_i}{C_i}, \qquad i = 0, 1, 2\right\}$$

extrema of S?
A solution q_0 is a relative minimum [maximum] of S if $S''(q_0)$ is positive [negative] in the space $C_0^2([a, b])$ of C^2 functions on (a, b) vanishing on the boundary. Using the results of the previous problem (§a) we compute easily

$$S''(q_0)hh = 2\pi \int_0^l h(x)(1 + \dot{q}_0(x))^{-3/2}(-q_0(x)\ddot{h}(x) + 2\dot{q}_0(x)\dot{h}(x) - \ddot{q}_0(x)h(x))\,dx.$$

If there exists a solution h_0 of the small disturbance equation $J_{q_0}h_0 = 0$ vanishing on the boundary, then $S''(q_0)$ is not positive [negative] definite in $C_0^2([a, b])$, q_0 is possibly not an extremum.

The Jacobi fields h_0 can easily be computed from the solutions $q_0(x) = C\cosh\{(x-D)/C\}$ of the Euler–Lagrange equation (see previous problem §b): $\mathrm{d}q_0(x)/\mathrm{d}C$ and $\mathrm{d}q_0(x)/\mathrm{d}D$ are two linearly independent Jacobi fields, and the general solution is path q

$$h_0(x) = c\sinh\frac{x-D}{C} + d\left(\cosh\frac{x-D}{C} - \frac{x-D}{C}\sinh\frac{x-D}{C}\right).$$

There exists a non zero Jacobi field $h_0(x)$ vanishing on the boundary if and only if there is a non zero solution (c, d) to the set of equations $h_0(0) = 0$, $h_0(l) = 0$, i.e. iff

$$\begin{cases} r = C\cosh\dfrac{D}{C} \\ \sinh\dfrac{D}{C}\sinh\dfrac{l-D}{C}\left(\coth\dfrac{l-D}{C} - \dfrac{l-D}{C} + \coth\dfrac{D}{C} - \dfrac{D}{C}\right) = 0. \end{cases}$$

For a given l, this equation is solved for $C = C_0$, i.e. $R = R_0(l)$. It follows that q_0^0 is possibly not an extremum of S.
$q_0^0(l)$ is conjugate to $q_0^0(0)$ (cf. previous problem §e) hence $q_0^0(l)$ is on the envelope of catenaries going through $(0, r)$. The envelope is the curve R_0 defined by

$$\begin{cases} R_0(x) = C\cosh\dfrac{x-D}{C}, \quad r = C\cosh\dfrac{D}{C} \\ \sinh\dfrac{D}{C}\sinh\dfrac{x-D}{C}\left(\coth\dfrac{x-D}{C} - \dfrac{x-D}{C} + \coth\dfrac{D}{C} - \dfrac{D}{C}\right) = 0. \end{cases}$$

In conclusion when $R = R_0(l)$, there is only one solution q_0^0 which may not be an extremum.

We shall now examine the two solutions path q_0^2

$$q_0^i(x) = C_i\cosh\frac{x-D_i}{C_i}, \quad r = C_i\cosh\frac{D_i}{C_i}, \quad i = 1, 2$$

which exists when $R > R_0(l)$.

Let $C_2 < C_0 < C_1$; the solution q_0^1 has no contact with the envelope $R_0(x)$; the solution q_0^2 has a contact with the envelope, set $(a, q_0^2(a))$ this contact point.
Let $h \in C_0^2([0, l])$ be defined by

$$h = \begin{cases} \text{the non zero Jacobi field vanishing at } 0 \text{ and } a \text{ for } 0 < x < a \\ 0 \text{ for } a < x < l, \end{cases}$$

$S''(q_0^2)hh = 0$, q_0^2 is possibly not an extremum of S.

Indeed we can construct a family of non stationary paths which generates surfaces of revolution having the same area as q_0^2.

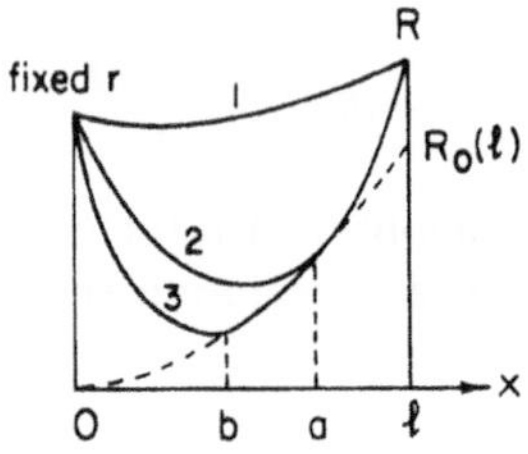

Let $g(b)$ be the function on $(0, a)$ defined as follows

$$g(b) = \begin{cases} \text{the catenary from } r \text{ to } R_0(b) \text{ for } 0 \leq x \leq b \\ R_0(x) \text{ for } b \leq x \leq a. \end{cases}$$

Let $q_0^2|[0, a]$ be the restriction of q_0^2 to the interval $(0, a)$. It can be proved that (Bliss)

$$S(g(b)) = S(q_0^2|[0, a]).$$

The family of paths defined by

$$\begin{cases} g(b) & \text{for } 0 \leq x \leq a \\ q_0^2|[a, l] & \text{for } a \leq x \leq l \end{cases}$$

is a family of non stationary paths which generates surfaces having the same area as q_0^2. Hence q_0^2 is not an extremum.

path q_0^1 There are no conjugate points on the path q_0^1; moreover, the Legendre condition is satisfied:

$$\partial^2 L/\partial \dot{q}_0^2 = q_0/(1 + \dot{q}_0^2)^{3/2} = C \cosh^{-2} \frac{x - D}{C} > 0.$$

It follows, by the argument given (p. 533) that

$$S''(q_0)hh = 2\pi C \int_0^l \cosh^{-2}\left(\frac{x - D}{C}\right)\left(\dot{h}^2(x) - \frac{1}{C} h^2(x)\right) dx \geq 0.$$

Although we expect q_0^1 to be a relative minimum, it is actually difficult to prove it in the present framework. Even if we can prove that $S''(q_0)hh = 0$ implies $h = 0$, we still would have to prove that $S''(q_0)hh \geq \lambda \|h\|^2$, a condition used in the sufficiency criterion in order that the second variation dominates the remaining terms in the Taylor expansion. It will be easy to prove that q_0^1 is a relative minimum in the context of the Sobolev space H^1 (see Chapter VI). In the particular example treated here it can be proved directly (see for instance Bliss) that q_0^1 is a relative minimum.

D is the value which minimizes q_0. D can be zero only if $l = 0$. In that case the solutions $q_0^1(0)$, $q_0^2(0)$ are respectively the interval $[r, R]$ and the interval $[r, 0] + [r, R]$ along the ordinate axis. case $D = 0$

We have now all the elements necessary to describe when and how the soap film breaks. We have explained how to make a model showing the variation of the soap film with l, R and r being fixed.

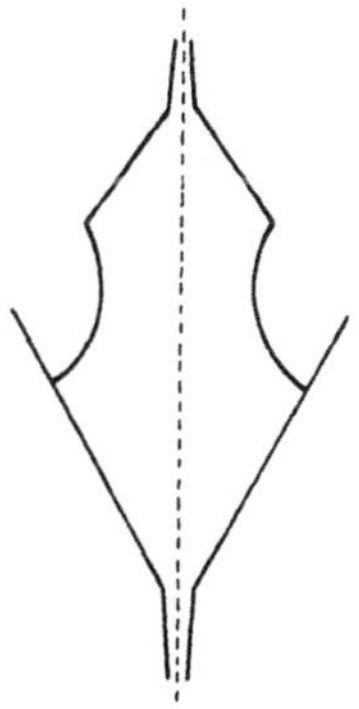

If we want to vary l and R, r being fixed, we can set an inverted funnel in a larger funnel moistened with soap solution. When we pull them apart, a soap film is formed between the rim of the small funnel and a circle sliding up or down the large funnel. The bubble breaks when R is conjugate to r.
The film collapses to the discs bounded by the r-circle and the R-circle, its surface being then $\pi r^2 + \pi R^2 < S(q_0^0)$.
If $R > R_0(l)$, the area of the film for given r, R and l is

$$S(q_0^1) = \pi C_1^2\left(\sinh\frac{l-D_1}{C_1}\cosh\frac{l-D_1}{C_1} + \frac{l-D_1}{C_1} + \sinh\frac{D_1}{C_1}\cosh\frac{D_1}{C_1} + \frac{D_1}{C_1}\right),$$

where $r = C_1 \cosh(D_1/C_1)$ and $R = C_1 \cosh((l - D_1)/C_1)$.
If $R < R_0(l)$, the area is $\pi R^2 + \pi r^2$.

This problem has been prepared in collaboration with P. Tschumi.
Reference: See for instance G. A. Bliss, *Calculus of Variations* (Open Court, 1925).

III. DIFFERENTIABLE MANIFOLDS

A. DEFINITIONS

1. DIFFERENTIABLE MANIFOLDS

Introduction. The concept of a manifold generalizes the concept of a surface or a curve in $\mathbb{R}^3$. However the definition will be given without reference to an embedding in $\mathbb{R}^n$. Rather it will generalize the idea of a "parametric representation" of a surface, that is a homeomorphic map from an open piece of the surface into the plane $\mathbb{R}^2$. Such a parametric representation is called a chart, or coordinate system. The surface is then covered by the domains of the charts (open pieces of the surface) in such a way that, in the overlap of two domains, the two sets of parameters (coordinates) are continuously related. Charts are used to define on manifolds objects and attributes originally defined on $\mathbb{R}^n$.
The concept of a *differentiable* manifold generalizes the concept of a differentiable surface in $\mathbb{R}^3$, that is of a surface with a tangent plane at each point. The tangent vectors to a manifold will be defined intrinsically, that is without reference to a particular coordinate system.

We now give the precise definition. A **(topological) manifold** is a Hausdorff topological space such that every point has a neighborhood homeomorphic to $\mathbb{R}^n$. topological manifold

A **chart** (U, φ) of a manifold X is an open set U of X, called the domain of the chart, together with a homeomorphism $\varphi : U \to V$ of U onto an open set V in $\mathbb{R}^n$. chart
The coordinates $(x^1, \ldots, x^n)$ of the image $\varphi(x) \in \mathbb{R}^n$ of the point $x \in U \subset X$ are called the **coordinates** of x in the chart (U, φ) (local coordinates of x). A chart (U, φ) is also called a **local coordinate system**. local coordinate system

An **atlas** of class C^k on a manifold X is a set $\{(U_\alpha, \varphi_\alpha)\}$ of charts of X such that the domains $\{U_\alpha\}$ cover X and the homeomorphisms satisfy the following compatibility condition. atlas
The **maps** $\varphi_\beta \circ \varphi_\alpha^{-1}$: $\varphi_\alpha(U_\alpha \cap U_\beta) \to \varphi_\beta(U_\alpha \cap U_\beta)$ are maps of open sets of $\mathbb{R}^n$ into $\mathbb{R}^n$ of class C^k. In other words, when (x^i) and (y^i) are respectively the coordinates of x in the charts $(U_\alpha, \varphi_\alpha)$ and (U_β, φ_β), the mapping $\varphi_\beta \circ \varphi_\alpha^{-1}$ is given in $\varphi_\alpha(U_\alpha \cap U_\beta)$ by n real valued C^k functions of n variables maps

$$(x^i) \mapsto y^j = f^j(x^i).$$

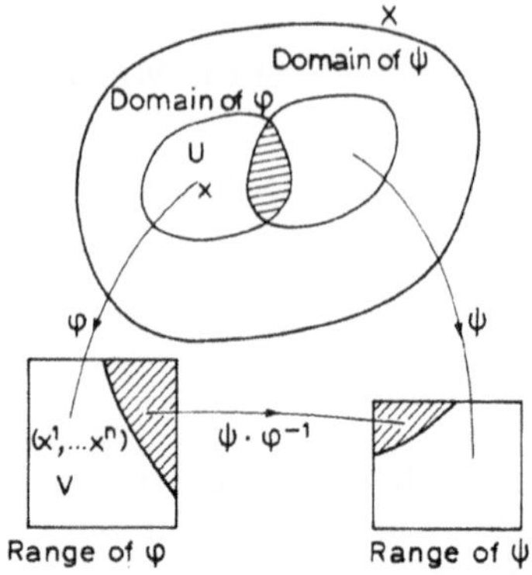

Two charts on a manifold.

equivalent atlases

Two C^k atlases $(U_\alpha, \varphi_\alpha)$ and $(U_{\alpha'}, \varphi_{\alpha'})$ are **equivalent** if and only if their union – domains $(U_\alpha, U_{\alpha'})$, homeomorphisms $(\varphi_\alpha, \varphi_{\alpha'})$ – is again a C^k atlas.

C^k manifold

A topological manifold X together with an equivalence class of C^k atlases is a C^k structure on X; we say that X is a $\boldsymbol{C^k}$ **manifold**.

differentiable manifold

smooth manifold

Strictly speaking a **differentiable manifold** is a manifold such that the maps $\varphi_\beta \circ \varphi_\alpha^{-1}$ of open sets of $\mathbb{R}^n$ into $\mathbb{R}^n$ are differentiable (p. 71), but not necessarily continuously differentiable (p. 73); but the expressions differentiable manifold, **smooth manifold**, are often used to mean a C^k manifold where k is large enough for the given context, eventually $k = \infty$.

real analytic manifold

A **real analytic** (C^ω) **manifold** is defined similarly: the maps $\varphi_\beta \circ \varphi_\alpha^{-1}$ have to be real analytic, i.e. given by n functions of n variables which can be expanded in a convergent Taylor series in a neighborhood of each point.

complex analytic manifold

A **complex analytic manifold** is defined similarly with $\mathbb{C}^n$ replacing $\mathbb{R}^n$ and the mappings $\varphi_\beta \circ \varphi_\alpha^{-1}$ being holomorphic (complex analytic).

Banach manifolds

In Chapter VII we shall define manifolds modelled on infinite dimensional vector spaces, in particular Banach spaces.

Exercise: Is a cone a manifold? More precisely is the set of points in $\mathbb{R}^3$

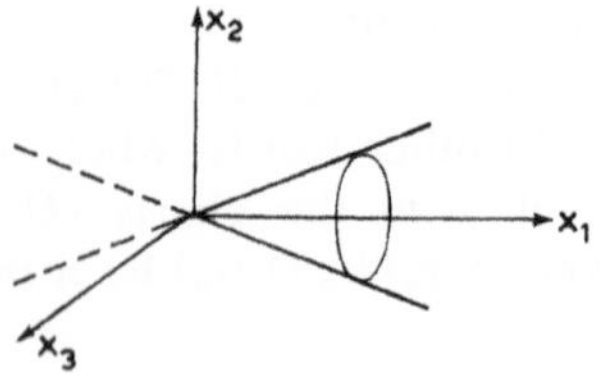

defined by $(x^1)^2-(x^2)^2-(x^3)^2=0$ (the "double cone"), together with the topology induced by the usual topology on $\mathbf{R}^3$, a manifold?

Answer: It is not a manifold because the origin does not have a neighborhood homeomorphic to $\mathbf{R}^2$. The set $(x^1)^2-(x^2)^2-(x^3)^2=0$ with $x^1 \geq 0$ (the "half-cone") is a C^0 manifold homeomorphic to $\mathbf{R}^2$.

Example: Problems 5, p. 181 and 6, p. 190.

Remark 1: Sometimes a manifold is defined only as a locally euclidean topological space X, together with appropriate requirements for the charts which map an open set of X into $\mathbf{R}^n$. But then X can have the undesirable feature of being non Hausdorff as will be shown in the following example. Consider the space which consists of two closed half lines $(-\infty, 0]$ and one open half line $(0, +\infty)$ together with a basis of open

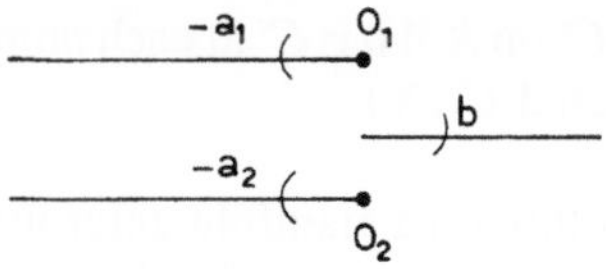

neighborhoods O_i, $i=1,2$ of the form $(-a_i, 0_i] \cup (0, b)$. This space is locally euclidean; it is not Hausdorff because any two open sets containing O_1 and O_2 will contain a segment $(0, b)$.

Remark 2: We have not included here in the definition of a differentiable manifold an axiom of countability of the domains of the charts. We shall require such a property in later chapters.

Remark 3: Submanifolds are discussed on pp. 239 ff and 290 ff.

In Chapter II we have defined the differentiability of a function on a vector space. Charts make it possible to define the differentiability of a function on a manifold. Consider a function f on the manifold X, $f: X \to \mathbf{R}$ by $x \mapsto f(x)$.

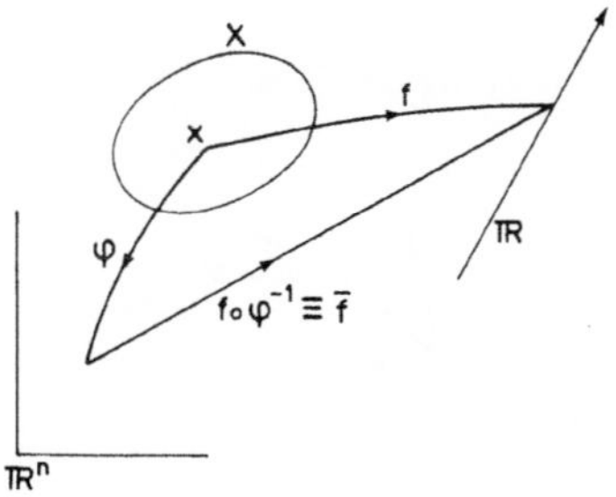

$f \circ \varphi^{-1}$

Let (U, φ) be a chart at x (i.e. $x \in U$), then, $f \circ \varphi^{-1}$ is a mapping from $V = \varphi(U)$, an open set of $\mathbb{R}^n$, into $\mathbb{R}$. Just as the coordinates of $\varphi(x)$ represent x in the local chart (U, φ), the mapping $f \circ \varphi^{-1}$ represents f in the local chart.

differentiable function

Definitions: The function f is **differentiable** at x on a differentiable manifold if, in a chart at x, $f \circ \varphi^{-1}$ is differentiable at $\varphi(x)$.
The definition does not depend on the chart: if $f \circ \varphi^{-1}$ is differentiable at $\varphi(x)$ for a chart (U, φ) at x, then $f \circ \tilde{\varphi}^{-1}$ is differentiable at $\tilde{\varphi}(x)$ for every chart $(\tilde{U}, \tilde{\varphi})$ at x, because

$$f \circ \tilde{\varphi}^{-1} = (f \circ \varphi^{-1}) \circ (\varphi \circ \tilde{\varphi}^{-1});$$

C^r at x

the proposition results from the differentiability of a composite mapping. **The function f is C^r at x on a C^k manifold ($k \geq r$) if, in a chart at x, $f \circ \varphi^{-1}$ is** C^r at $\varphi(x)$. The mappings $\varphi \circ \tilde{\varphi}^{-1}$ are of class C^k on a C^k manifold: the definition is chart independent if $r \leq k$.

$C^r(X)$

A function $f: X \to \mathbb{R}$ is C^r on X if it is C^r at each point $x \in X$. The space of C^r functions on X is denoted $C^r(X)$.

The differentiable structure on a manifold determines the maximum order of differentiability of the functions defined on the manifold.

coordinate functions

It is sometimes desirable to write down some expressions more explicitly as follows.
Let a^i be the **coordinate functions** on $\mathbb{R}^n$ which map a point $u = (u^1 \dots u^n) \in \mathbb{R}^n$ into $\mathbb{R}$ by

$$a^i(u^1 \dots u^n) = u^i.$$

$\varphi^i \equiv a^i \circ \varphi$

$a^i \circ \varphi(x)$ is what we have abbreviated to x^i.
The composite mapping $a^i \circ \varphi \equiv \varphi^i$ maps X into $\mathbb{R}$ by $x \mapsto x^i$. Its local chart representation $\varphi^i \circ \varphi^{-1} = a^i$ is the coordinate function

$$\bar{\varphi}^i(x^1 \dots x^n) = x^i.$$

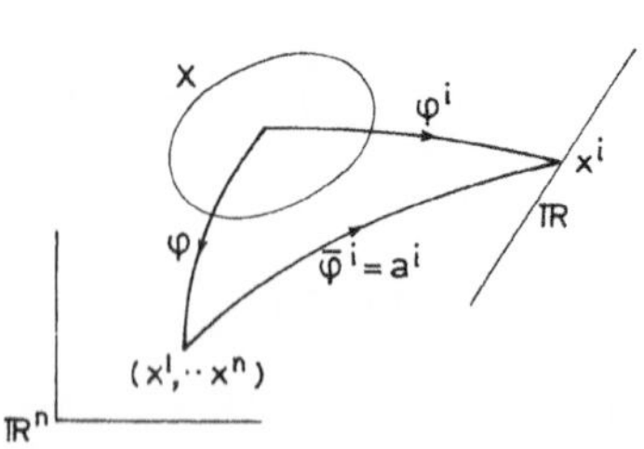

Direct product of manifolds. The direct product of two manifolds with,

underlying sets X^n, Y^p, topologies $\mathcal{U}_X$, $\mathcal{U}_Y$ and differentiable structures (U_X, φ_X), (U_Y, φ_Y) are defined in a natural way to be the manifold Z^{n+p} with underlying set $X^n \times Y^p$, the product topology, and the product atlas made up of charts (U_Z, φ_Z), where $U_Z = U_X \times U_Y$ and

$$\varphi_Z(x, y) = (\varphi_X(x), \varphi_Y(y)),\ x \in U_X,\ y \in U_Y,\ (x, y) \in U_Z.$$

2. DIFFEOMORPHISMS

Let X^n and Y^p be two differentiable (C^k) manifolds of dimension n and p, respectively; let $f: X^n \to Y^p$.

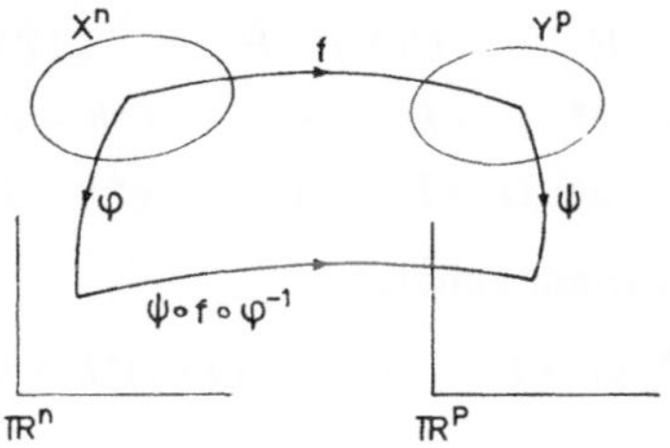

The function $\psi \circ f \circ \varphi^{-1}$ represents f in the local charts (U, φ), (W, ψ) of X^n and Y^p. The differentiability of $f: X^n \to \mathbb{R}$ was simply a particular case of the situation now considered.

f is **C^r differentiable** at x for $r \leq k$ if $\psi \circ f \circ \varphi^{-1}$ is C^r differentiable at $\varphi(x)$. *C^r differentiable at x*
In other words, f is differentiable $[C^r]$ at x if the coordinates $(y^\alpha = f^\alpha(x^i))$ of y are differentiable $[C^r]$ functions of the coordinates (x^i) of x.
f is a **C^r mapping** from X^n to Y^p if f is C^r at every point $x \in X^n$. *C^r mapping*
Other attributes of differentiable mappings, such as rank, which have been studied in Chapter II, carry over to f – first locally in a given chart, then globally if they are defined in all charts of an atlas of X.
In particular f is a (C^r) **diffeomorphism** if f is a bijection and f and f^{-1} are continuously differentiable (C^r). Diffeomorphisms are to differentiable manifolds what homeomorphisms are to topological spaces, and what isomorphisms are to vector spaces. *diffeomorphism*

Let $f: X \to Y$ and let g be a function on Y.

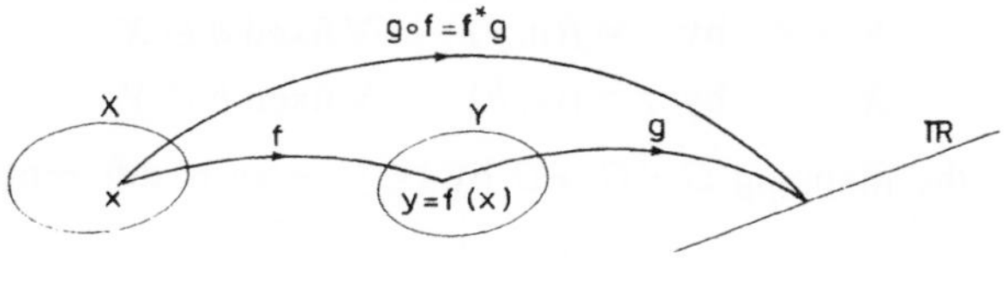

reciprocal image pull back

The **reciprocal image (pull back)** under f of the function g is the function $g \circ f : x \mapsto (g \circ f)(x) = g(f(x))$; it is denoted

$$g \circ f = f^*(g).$$

The same definition applies for the pull back (reciprocal image) f^*g of a mapping $g : Y \to Z$ under $f : X \to Y$

$$f^*g = g \circ f.$$

We have here an example of a contravariant functor (p. 4). Indeed let X, Y, Z be differentiable manifolds; let f: $X \to Y$ and g: $Y \to Z$ be differentiable mappings; let E, F, and G be vector spaces of functions defined respectively on X, Y and Z; and let

$$C(X) = E \qquad C(Y) = F \qquad C(Z) = G$$

$$C(f) = f^*: F \to E \quad \text{by} \quad f^*h = h \circ f$$

$$C(g) = g^*: G \to F \quad \text{by} \quad g^*k = k \circ g.$$

Then, C is a contravariant functor:

$$(g \circ f)^*: G \to E \qquad \text{by} \qquad (g \circ f)^*k = k \circ g \circ f.$$

On the other hand

$f^* \circ g^*$: $G \to E$ in two steps: $(f^* \circ g^*)k = f^*(g^*k) = f^*(k \circ g) = k \circ g \circ f$
and
$$\mathrm{id}_X^* = \mathrm{id}_E.$$

3. LIE GROUPS

Lie group

A **Lie group** G is a group that is also a differentiable manifold such that the differentiable structure is compatible with the group structure, i.e. such that the operation $G \times G \to G$ by $(x, y) \mapsto xy^{-1}$ is a differentiable mapping.

isomorphic

Two Lie groups are **isomorphic** if there exists a diffeomorphism between them which preserves the group structure (which is a group homomorphism).

Exercise: Let G be a Lie group; show that the mapping $G \to G$ by $x \mapsto x^{-1}$ is differentiable; show in general that, given the differentiable mapping $X \times Y \to Z$ by $z = f(x, y)$, the following mappings are differentiable:

$$Y \to Z \quad \text{by} \quad z = f(a, y) \qquad \forall \text{ fixed } a \in X$$

$$X \to Z \quad \text{by} \quad z = f(x, b) \qquad \forall \text{ fixed } b \in Y.$$

Show that the mapping $G \times G \to G$ by $(x, y) \mapsto xy$ is differentiable.

Example 1: $\mathbb{R}^n$ together with the operation of vector addition is a Lie group when endowed with its usual differentiable structure.

Example 2: The space of linear bijective mappings of $\mathbb{R}^n$ onto $\mathbb{R}^n$ together with the operation of map composition is a Lie group $GL(n, \mathbb{R})$.

Example 3: Problem 2, p. 172.

A one dimensional Lie group is usually called a **one parameter group**.

one parameter group

Proposition: *A connected one parameter group G is isomorphic to* $\mathbb{R}$ *or to the* 1-*torus* $\mathbb{T} = \mathbb{R}/\mathbb{Z}$.
This property is not a consequence of the differentiable structure of Lie groups; it can be proved[1] that a connected topological group which possesses a neighborhood of the origin homeomorphic to $\mathbb{R}$ is isomorphic to $\mathbb{R}$ or to $\mathbb{T}$.

A **local Lie group** is a neighborhood of the origin (identity element) of the group. A local one parameter group is isomorphic to an interval $I \subset \mathbb{R}$ containing the origin.

local Lie group

B. VECTOR FIELDS; TENSOR FIELDS

1. TANGENT VECTOR SPACE AT A POINT

The tangent vector space $T_x(X)$ of a manifold X at a point $x \in X$ is used to define differential properties of objects in a neighborhood of x independently of local coordinates. The tangent vector space $T_x(X)$ "models" the manifold at x. Most approximations in physics consist in replacing locally a given manifold by its tangent vector space at a point. Such an approximation can be called a local linearization. $T_x(X)$ is isomorphic to $\mathbb{R}^n$ if X is a manifold of dimension n.
We shall give three equivalent definitions of tangent vectors at a point.

Definition: a) The most direct definition is the following.
A **tangent vector** $\boldsymbol{v}_x$ to a differentiable manifold X at a point x is a linear function from the space of functions defined and differentiable on some

tangent vector $\boldsymbol{v}_x$

[1] See for instance [Bourbaki 1975] or [Milnor 1965].

neighborhood of $x \in X$ into $\mathbb{R}$, which satisfies the Leibniz rule:

$v_x(\alpha f + \beta g) = \alpha v_x(f) + \beta v_x(g)$ $\alpha, \beta \in \mathbb{R}$, f, g functions on X differentiable at x;

$v_x(fg) = f(x)v_x(g) + g(x)v_x(f)$ Leibniz rule.

These two properties together imply $v_x(\alpha) = 0$. *Hint*: let $f = g = 1$.

derivation

directional derivative

A linear mapping which satisfies the Leibniz rule is called a **derivation**. It follows from this definition (see p. 119) that $v_x(f)$ is the **directional derivative** of f along v_x.

tangent vector space $T_x(X)$

The space $T_x(X)$ of tangent vectors to X at x together with addition and scalar multiplication defined by

$$(\alpha u_x + \beta v_x)(f) = \alpha u_x(f) + \beta v_x(f)$$

is a vector space, called the **tangent vector space**.

germ

This definition of a tangent vector can be formulated more precisely by identifying functions which coincide on a neighborhood of x: two functions f_1 and f_2 on X differentiable at x have the same **germ** at x if there exists a neighborhood of x where they coincide. The equivalence class of differentiable functions at x which have the same germ as a function f is called a **germ** of f. The disjoint equivalence classes are called the germs of differentiable functions at x.

Germs form an algebra: we already know that they form a vector space; they form an algebra because the pointwise multiplication of two germs gives another germ.[1] Thus we arrive at the first definition.

tangent vector

A **tangent vector** is a derivation on the algebra of germs of differentiable functions at x.

A tangent vector defined as a mapping on functions on the manifold, does not require the embedding of the manifold in a euclidean space.

components v_x^i

In the chart (U, φ) the local coordinates (**components**) of a tangent vector are the set of numbers

$v_x(\varphi^i) \equiv v^i$ where $\varphi^i = a^i \circ \varphi$, a^i are the coordinate functions (p. 114) $\varphi^i : x \mapsto x^i$.

Let f be a C^1 function on a neighborhood of $x_0 \in X$. In the local chart (U, φ), the mean value theorem reads, with $\bar{f} = f \circ \varphi^{-1}$,

[1]Whereas the set of germs and also the set of differentiable functions on a manifold are algebras, the set of functions differentiable on some neighborhood of a point is not quite an algebra: there is no unique zero element, the two zero functions $f + (-f)$ and $g + (-g)$ are identical only if the domain of f is identical with the domain of g.

$$f(x) = \bar{f}(\varphi(x_0)) + (\varphi^i(x) - \varphi^i(x_0)) \frac{\partial \bar{f}}{\partial x^i}\bigg|_{\varphi(x_0)+(\varphi(x)-\varphi(x_0))s} \quad \text{for some } 0 < s < 1,$$

where summation over repeated indices is assumed.
At x_0 the derivative of f along the vector v_x is

$$v_{x_0}(f) = v^i \partial \bar{f}/\partial x^i|_{\varphi(x_0)} \quad \text{where } v^i = v_{x_0}(\varphi^i).$$

From the coordinate expression for $v_x(f)$ we obtain the expression for the **representative** of v_x in these coordinates, which is written in a somewhat abbreviated form as

representative of v_x

$$v^i \partial/\partial x^i.$$

The vectors of T_x that are represented by $(\partial/\partial x^i \ldots \partial/\partial x^n)$ form a basis for the tangent vector space which is called the **natural basis**. It follows that the dimension of T_x is the same as the dimension of the manifold; a chart (U, φ) at x thus induces an isomorphism of T_x onto $\mathbb{R}^n$.

natural basis

If $\partial \bar{f}/\partial x^i|_x = 0$ for all i, i.e. if the function f has a **critical point** (is **stationary**) at x, then $v_x(f) = 0$ and conversely. The condition for a function to have a critical point stated in terms of a local chart has therefore an intrinsic meaning: if the partial derivatives at x vanish in one system of coordinates, they will do so in any other coordinate system (cf. below).

critical point stationary

Definition: b) The time honored definition of a tangent vector by its transformation law under a change of coordinate system is as follows. Consider two charts (U, φ) and (U, φ').

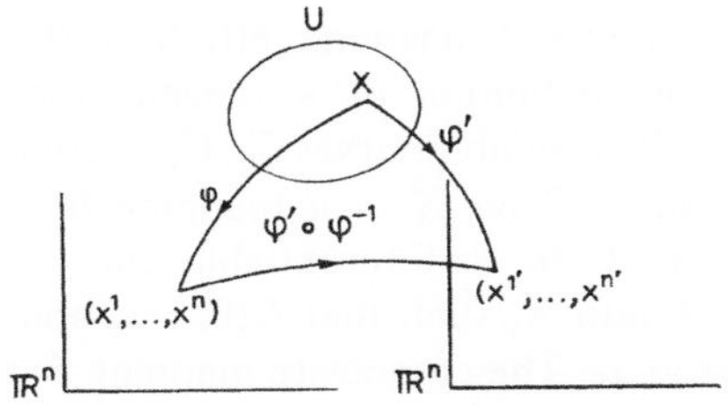

The expressions for f in the two charts are respectively

$$\bar{f} = f \circ \varphi^{-1} \quad \text{and} \quad \bar{f}' = f \circ \varphi'^{-1}, \text{ thus } \bar{f} = \bar{f}' \circ \varphi' \circ \varphi^{-1}$$

and

$$\frac{\partial \bar{f}(x^1 \ldots x^n)}{\partial x^i} = \frac{\partial \bar{f}'(x^{1'} \ldots x^{n'})}{\partial x^{j'}} \frac{\partial (a^{j} \circ \varphi' \circ \varphi^{-1})(x^1 \ldots x^n)}{\partial x^i}$$

which may be abbreviated, incorrectly but conveniently, to

$$\frac{\partial \bar{f}}{\partial x^i} = \frac{\partial \bar{f}'}{\partial x^{j'}} \frac{\partial x^{j'}}{\partial x^i}.$$

We wish to express the components of a vector in one chart in terms of its components in another chart. This gives

$$\boldsymbol{v}_x(f) = v^i \frac{\partial \bar{f}}{\partial x^i}\bigg|_x = v^i \frac{\partial \bar{f}'}{\partial x^{j'}} \frac{\partial x^{j'}}{\partial x^i}\bigg|_x;$$

hence the local coordinates of $\boldsymbol{v}_x$ in the primed system are

$$v^{j'} = v^i \frac{\partial x^{j'}}{\partial x^i}\bigg|_x.$$

The set (v^i) determines a vector $\boldsymbol{V} \in \mathbb{R}^n$ and the set $(v^{i'})$ determines another vector $\boldsymbol{V}' \in \mathbb{R}^n$ such that

$$\boldsymbol{V}' = D(\varphi' \circ \varphi^{-1})|_{\varphi(x)} \boldsymbol{V}$$

where $D(\varphi' \circ \varphi^{-1})|_{\varphi(x)}$ is the jacobian matrix of $\varphi' \circ \varphi^{-1}$ at $\varphi(x)$.
Both $\boldsymbol{V}$ and $\boldsymbol{V}'$ represent the same vector $\boldsymbol{v}_x$ in their respective local charts and we arrive at the second definition of a tangent vector in terms of its local representation $\boldsymbol{V}$.

tangent vector

A **tangent vector** $\boldsymbol{v}_x$ is a triple $(x, \varphi, \boldsymbol{V})$ such that $(x, \varphi, \boldsymbol{V})$ and $(x, \varphi', \boldsymbol{V}')$ define the same vector if $\boldsymbol{V}' = D(\varphi' \circ \varphi^{-1})|_x \boldsymbol{V}$. This is the old definition of a vector used when the taxonomy of tensors is done by means of their transformation properties: "An object is a vector if it transforms like a vector".

Definition: c) More intuitive than the definition of a tangent vector as a derivation but still expressed independently from the choice of a system of coordinates is the definition of a tangent vector in terms of an equivalence class of differentiable curves $C_1, C_2, \ldots$ on X passing through x.

parametrized curve

A **(parametrized) curve** C on X is a mapping from $I \subset \mathbb{R}$ into X by $t \in I \mapsto C(t) \in X$. Let C be a differentiable curve, i.e. a differentiable mapping from $I \subset \mathbb{R}$ into X, such that $C(0) = x_0$ and let f be a function on X differentiable at x_0. The composite mapping $f \circ C$ is a function on I differentiable at $t = 0$. The tangent vector to the curve C at x_0 is a mapping $f \mapsto \boldsymbol{v}^C_{x_0}(f)$ from the space of germs of C^1 functions at x_0 into $\mathbb{R}$:

$$\boldsymbol{v}^C_{x_0}(f) = \frac{\mathrm{d}}{\mathrm{d}t}(f \circ C)(t)|_{t=0}.$$

$\boldsymbol{v}^C_{x_0}(f)$ is called the derivative of f along C at x_0. $\boldsymbol{v}^C_{x_0}(f)$ is sometimes designated $C(f)$. If two curves C_1 and C_2 are such that, for every f,

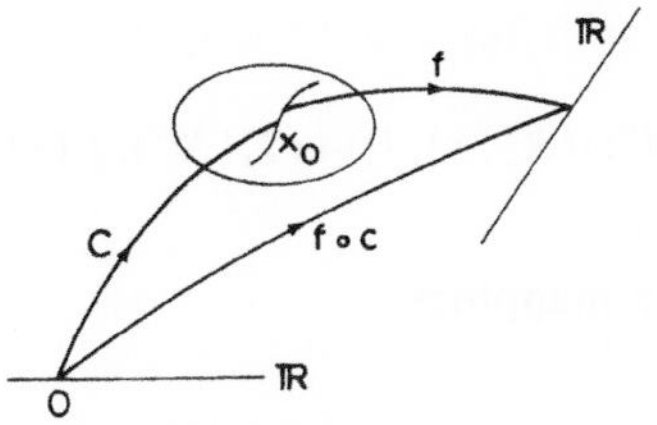

$C_1(f)=C_2(f)$, they are said to be tangent at x_0. We also say they have the same tangent vector at x_0. Conversely given a tangent vector $\boldsymbol{v}_{x_0}$ it is always possible to find a differentiable curve through x_0 to which $\boldsymbol{v}_{x_0}$ is tangent. Thus it is possible to identify a tangent vector at x_0 with an equivalence class of curves tangent at x_0 and we arrive at the third definition.

An equivalence class of curves tangent at x is called a **tangent vector** $\boldsymbol{v}_x$ at x.

tangent vector

The components of the tangent vector $\boldsymbol{v}_{x_0}$ to the curve C at x_0 are

$$(v_{x_0}^C)^i=\boldsymbol{v}_{x_0}^C(\varphi^i)=\left.\frac{\mathrm{d}(\varphi^i\circ C)}{\mathrm{d}t}\right|_{t=0}=\left.\frac{\mathrm{d}C^i(t)}{\mathrm{d}t}\right|_{t=0}.$$

This equation generalizes to curves on X^n the classical result for curves in $\mathbb{R}^n$.

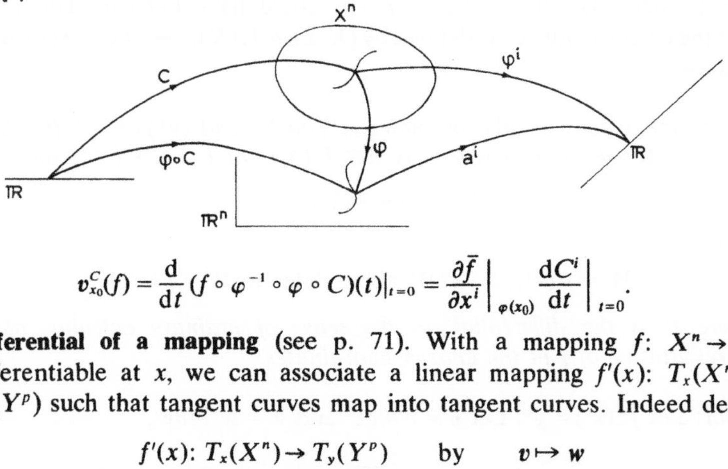

$$\boldsymbol{v}_{x_0}^C(f)=\frac{\mathrm{d}}{\mathrm{d}t}(f\circ\varphi^{-1}\circ\varphi\circ C)(t)|_{t=0}=\left.\frac{\partial\bar{f}}{\partial x^i}\right|_{\varphi(x_0)}\left.\frac{\mathrm{d}C^i}{\mathrm{d}t}\right|_{t=0}.$$

Differential of a mapping (see p. 71). With a mapping $f\colon X^n\to Y^p$, differentiable at x, we can associate a linear mapping $f'(x)\colon T_x(X^n)\to T_y(Y^p)$ such that tangent curves map into tangent curves. Indeed define

$$f'(x)\colon T_x(X^n)\to T_y(Y^p)\qquad \text{by}\qquad \boldsymbol{v}\mapsto\boldsymbol{w}$$

is such that for every differentiable function h at $y=f(x)\in Y^p$

$$\boldsymbol{w}(h)=\boldsymbol{v}(h\circ f).$$

$\boldsymbol{w}$ is called the **image** of $\boldsymbol{v}$ under f. The mapping $f'(x)$ is called the **differential of** f **at** x. It is also denoted $Df(x)$ or $Tf(x)$ or $f_*(x)$.

image
differential

Exercise: Show that

$$C: ((X^n, x), f) \mapsto (T_x(X^n), f'(x))$$

is a covariant functor.
Answer: Consider the mappings

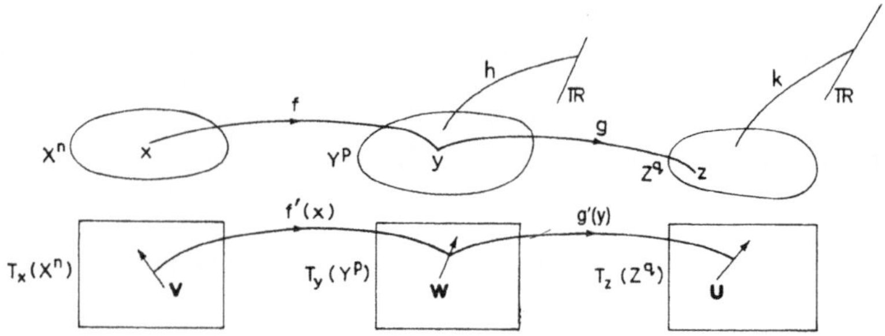

$$[(g \circ f)' \boldsymbol{v}](k) = \boldsymbol{v}(k \circ g \circ f) = (f' \boldsymbol{v})(k \circ g) = [g'(f' \boldsymbol{v})](k) = [(g' \circ f')\boldsymbol{v}](k).$$

If f is the identity mapping $f'(x)$ is also the identity mapping. ■

The property $(\overline{f \circ g})' = (\bar{f} \circ \bar{g})' = \bar{f}' \circ \bar{g}'$ could have been used to prove that the correspondence C defined by $(X, x) \mapsto T_x(X), f \mapsto f'(x)$ is a covariant functor.

Proposition. Choose local coordinates (x^i) on X^n and (y^α) on Y^p. If $\mathbf{V} \in \mathbb{R}^n$ and $\mathbf{W} \in \mathbb{R}^p$ represent respectively $\boldsymbol{v} \in T_x(X^n)$ and $\boldsymbol{w} \in T_y(Y^p)$ such that

$$\boldsymbol{w} = f'(x)\boldsymbol{v}$$

then

$$\mathbf{W} = \bar{f}'(x^1, \ldots, x^n)\mathbf{V}, \text{ abbreviated to } \mathbf{W} = f'(x^i)\mathbf{V}$$

where $\bar{f}'$ is the differential, in the sense of ordinary calculus, of the representative of f in the chosen coordinates.

Proof: Let $\bar{f}^\alpha(x^i) = y^\alpha$. Let $\boldsymbol{v} = v^i\partial/\partial x^i$ and $\boldsymbol{w} = w^\alpha\partial/\partial y^\alpha$,

$$\boldsymbol{w}(h) = \boldsymbol{v}(h \circ f) = v^i \frac{\partial(\overline{h \circ f})}{\partial x^i} = v^i \frac{\partial \bar{h}}{\partial y^\alpha} \frac{\partial \bar{f}^\alpha}{\partial x^i} = w^\alpha \frac{\partial \bar{h}}{\partial y^\alpha}$$

hence $w^\alpha = v^i \partial \bar{f}^\alpha / \partial x^i$

$$\mathbf{W} = \bar{f}'(x^i)\mathbf{V}.$$ ■

Remark: The transformation law of the components of a vector under a change of coordinate system (p. 120) can be interpreted by this proposition.

Exercise: Given a curve $C: \mathbb{R} \to X$, a vector v_x^C tangent to C at $x = C(0)$, and a mapping $f: X \to Y$, show that $f' v_x^C$ is tangent to the image of C under f at $y = f(x)$.

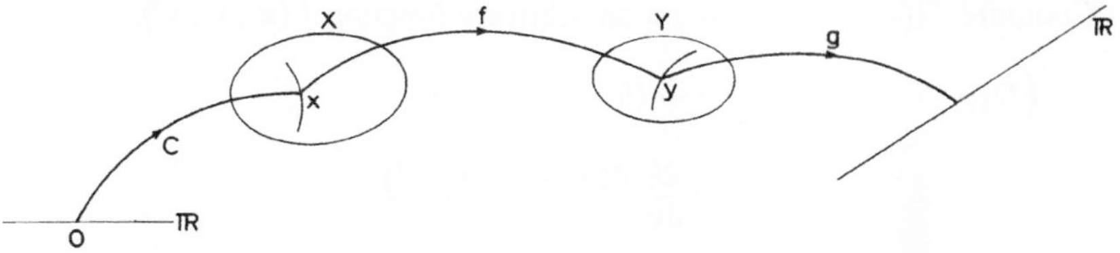

Answer: The image of v_x^C under f is defined by its value for an arbitrary function g

$$(f' v_x^C)(g) = v_x^C(g \circ f).$$

On the other hand v_x^C being tangent to C at x

$$v_x^C(g \circ f) = \frac{\mathrm{d}}{\mathrm{d}t}((g \circ f) \circ C(t))|_{t=0}$$

$$= \frac{\mathrm{d}}{\mathrm{d}t}(g \circ (f \circ C(t)))|_{t=0} \qquad \text{by associativity}$$

which is the value of the tangent vector to the curve $f \circ C$ at y for the function g. ■

Exercise: Show that $f'(x) v_x^C = \frac{\mathrm{d}}{\mathrm{d}t}(f(C(t)))|_{t=0}$.

Exercise: In spherical coordinates (r, θ, φ) in $\mathbb{R}^3$, $\partial/\partial r$ is the unit tangent

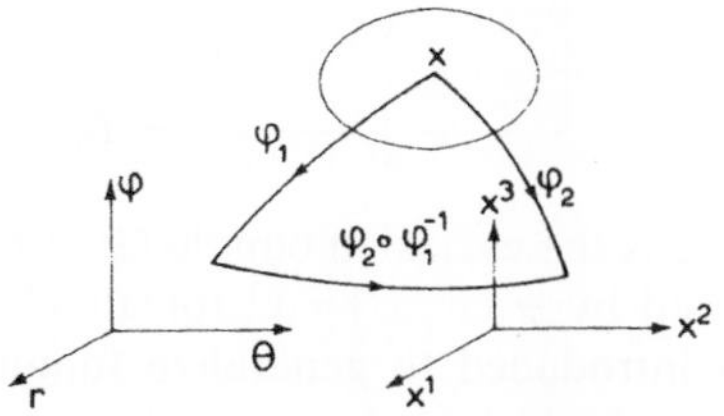

vector to the curve $\theta = \text{constant}$, $\varphi = \text{constant}$. What are the components of $\partial/\partial r$ in cartesian coordinates (x^1, x^2, x^3)?

$$\varphi_1(x) = (r, \theta, \varphi) \qquad \varphi_2(x) = (x^1, x^2, x^3) \qquad a^i \circ \varphi_2(x) = x^i$$

$\varphi_2 \circ \varphi_1^{-1}$: $(r, \theta, \varphi) \mapsto (x^1, x^2, x^3)$

$$\text{by} \begin{cases} x^1 = a^1 \circ \varphi_2 \circ \varphi_1^{-1}(r, \theta, \varphi) = r \sin\theta \cos\varphi \\ x^2 = a^2 \circ \varphi_2 \circ \varphi_1^{-1}(r, \theta, \varphi) = r \sin\theta \sin\varphi \\ x^3 = a^3 \circ \varphi_2 \circ \varphi_1^{-1}(r, \theta, \varphi) = r \cos\theta. \end{cases}$$

Compute $D(\varphi_2 \circ \varphi_1^{-1})\, \partial/\partial r$ on an arbitrary function $h(x^1, x^2, x^3)$,

$$\left(D(\varphi_2 \circ \varphi_1^{-1})\frac{\partial}{\partial r}\right)(h) = \frac{\partial}{\partial r}(h \circ \varphi_2 \circ \varphi_1^{-1})$$
$$= \frac{\partial h}{\partial x^i}\frac{\partial(a^i \circ \varphi_2 \circ \varphi_1^{-1})}{\partial r},$$

$$D(\varphi_2 \circ \varphi_1^{-1})\frac{\partial}{\partial r} = \sin\theta\cos\varphi\frac{\partial}{\partial x^1} + \sin\theta\sin\varphi\frac{\partial}{\partial x^2} + \cos\theta\frac{\partial}{\partial x^3}. \quad \blacksquare$$

Now that a tangent vector space at a point $x \in X$ is a reasonably familiar object, we wish to consider the set of all tangent vectors $\bigcup_{x \in X} T_x(X)$. This set together with the manifold X is a particular case of a structure called a bundle which we shall proceed to investigate. Afterwards we shall pursue the understanding of $T_x(X)$ by the study of its very interesting dual.

2. FIBRE BUNDLES

bundle
base

A **bundle** is a triple (E, B, π) consisting of two topological spaces E and B and a continuous surjective mapping π: $E \to B$. B is called the **base.**

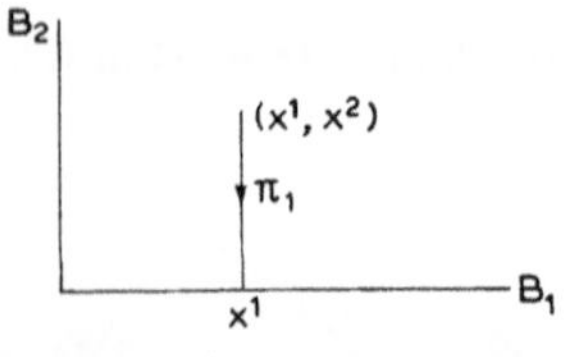

The simplest example is the cartesian bundle $(B_1 \times B_2, B_1, \pi_1)$ with π_1 the first projection defined by $\pi_1(x^1, x^2) = x^1$ for all $x^1 \in B_1$ and $x^2 \in B_2$.
Bundles have been introduced to generalize topological products. The need to generalize topological products can be seen already in a simple example: a cylinder obtained by glueing a strip of paper is the cartesian product of a one-sphere S^1 (circle) with a line segment I, and $(S^1 \times I, S^1, \pi_1)$ is a product bundle; but a Möbius band obtained by twisting and

then glueing a strip of paper cannot be described globally as a topological product. It can be done locally: for $U \subsetneq S^1$, the topological product $U \times I$ describes a segment of the Möbius band; but we need some mechanism to say that twisting occurs somewhere.

We shall restrict ourselves to situations in which the topological spaces $\pi^{-1}(x)$ for all $x \in B$ are homeomorphic to a space F.

F_x
fibre at x
typical fibre
vector bundle

$\pi^{-1}(x)$ is then called the **fibre at** x, denoted F_x, and F is called the **typical fibre.** If the bundle also has certain additional structure involving a group of homeomorphisms of F and a covering of B by open sets, it is called a fibre bundle. If F is a vector space and the group is the linear group, the fibre bundle is called a **vector bundle.**

fibre bundle

A **fibre bundle** (E, B, π, G) is a bundle (E, B, π) together with a typical fibre F, a topological group G of homeomorphisms of F onto itself, and a covering of B by a family of open sets $\{U_j;\ j \in J\}$, such that:

trivial bundle

a) Locally the bundle is a **trivial bundle**, i.e. it is homeomorphic to a product bundle; more precisely $\pi^{-1}(U_j)$ is homeomorphic to the topological product $U_j \times F$ for all $j \in J$. The homeomorphism φ_j: $\pi^{-1}(U_j) \to U_j \times F$ has the form $\varphi_j(p) = (\pi(p), \overset{\Delta}{\varphi}_j(p))$ and the following diagram is commutative.

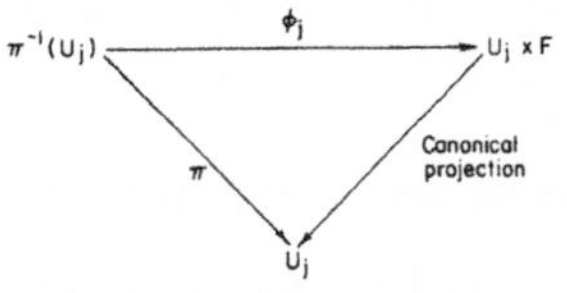

trivialization
$\overset{\Delta}{\varphi}_{j,x}$

The sets $\{U_i, \varphi_i\}$ are called a family of **local trivializations** of the bundle. Let $x \in U_j$, we note that $\overset{\Delta}{\varphi}_j|_{F_x}$, or $\overset{\Delta}{\varphi}_{j,x}$ to simplify notation, is a homeomorphism from F_x onto F.

b) There is a correlation of the trivial subbundles defined on the open sets U_j covering the base. Let $x \in U_j \cap U_k$.

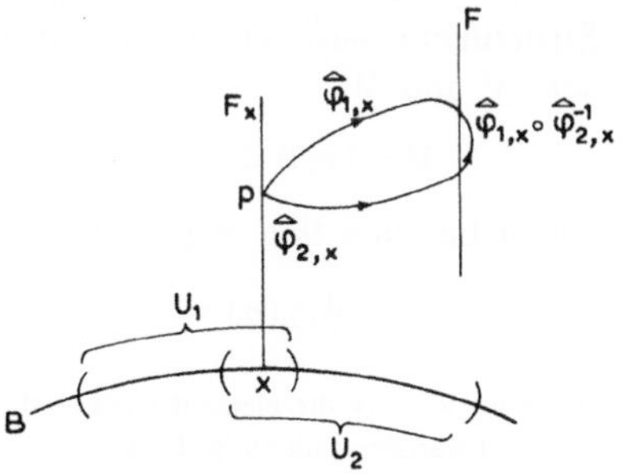

The relationship between the mappings $\overset{\Delta}{\varphi}_{j,x}$ and $\overset{\Delta}{\varphi}_{k,x}$ gives the structure of the fibre bundle; for instance, it gives the twisting of the Möbius band.

structural group

The homeomorphism $\overset{\Delta}{\varphi}_{k,x} \circ \overset{\Delta}{\varphi}{}^{-1}_{j,x}$: $F \to F$ is an element[1] of the **structural group** G for all $x \in U_j \cap U_k$ and all $j, k \in J$. If G has only one element the bundle is trivializable.

transition function

c) The induced mappings $g_{jk}: U_j \cap U_k \to G$ by $x \mapsto g_{jk}(x) = \overset{\Delta}{\varphi}_{j,x} \circ \overset{\Delta}{\varphi}{}^{-1}_{k,x}$ are continuous. They are called **transition functions**. They satisfy the relation

$$g_{jk}(x) g_{ki}(x) = g_{ji}(x).$$

equivalent fibre bundles

This definition must be completed by the notion of **equivalent fibre bundle structures**, analogous to the notion of equivalent atlases on a manifold (p. 112).

Example: Möbius band.

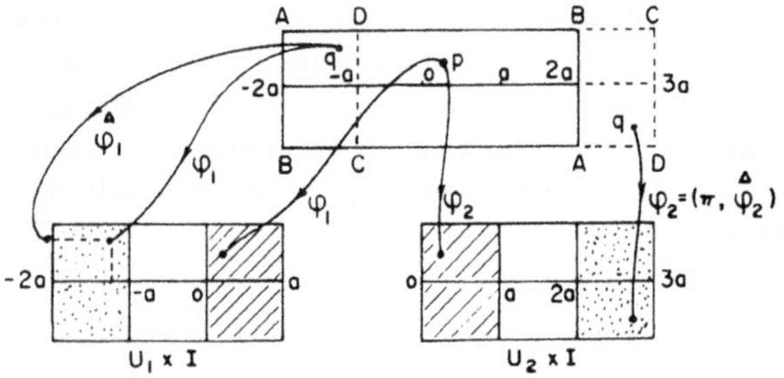

Points with the same label A, B, C, D are identified.

Base: S^1 of length $4a$ covered by two open sets U_1 and U_2

$$U_1 = \{x; -2a < x < a\} \qquad U_2 = \{x; 0 < x < 3a\}$$

where x is the coordinate of a point of S^1, hence $x \pm 4na = x$.
Projection: $\pi(p) \in S^1$; identifying a point of S^1 with its coordinate we shall write $\pi(p) = x_p$.
Typical fibre: $I \subset \mathbb{R}$ $\quad \varphi_1: \pi^{-1}(U_1) \to U_1 \times I$ by $\varphi_1(p) = (\pi(p), \overset{\Delta}{\varphi}_1(p))$.
Let i_p be the coordinate of $\overset{\Delta}{\varphi}_1(p) \in I$, we shall write $\overset{\Delta}{\varphi}_1(p) = i_p$.
Structural group: The intersection of U_1 and U_2 consists of two disjoint sets V and W

$$V = \{x; 0 < x < a\} \qquad \text{and} \qquad W = \{x; -2a < x < -a\}.$$

Let p be such that $\pi(p) \in V$,

$$\overset{\Delta}{\varphi}_{1,x_p}(p) = i_p \qquad \overset{\Delta}{\varphi}_{2,x_p}(p) = i_p \qquad \overset{\Delta}{\varphi}_{1,x_p} \circ \overset{\Delta}{\varphi}{}^{-1}_{2,x_p} = e.$$

[1]Here we identify the element $g \in G$ with the transformation σ_g: $F \to F$, an element of the group of transformations (p. 152).

Let q be such that $\pi(q) \in W$

$$\overset{\Delta}{\varphi}_{1,x_q}(q) = i_q, \qquad \overset{\Delta}{\varphi}_{2,x_q}(q) = -i_q, \qquad \overset{\Delta}{\varphi}_{1,x_q} \circ \overset{\Delta}{\varphi}{}^{-1}_{2,x_q} = g,$$

where e and g are the elements of the symmetry group of order 2. This group with the discrete topology is a topological group.

Consider 2 bundles (E_1, B_1, π_1) and (E_2, B_2, π_2); the mappings of interest between two bundles are the mappings which preserve the fibre structure (the local product bundle structure), i.e. the mappings which map fibres into fibres. They are called **bundle morphisms**.

bundle morphisms

A bundle morphism is a pair of maps (F, f) such that

$$F: E_1 \to E_2, \qquad f: B_1 \to B_2$$

and such that the following diagram is commutative

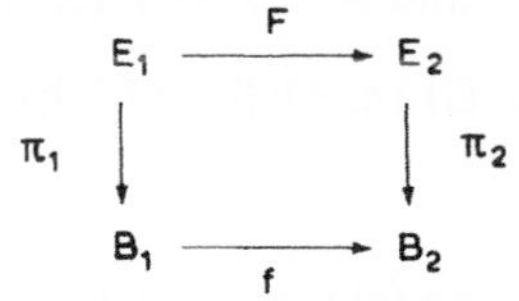

Given the map F, the map f, if it exists, is unique.

bundle category

A **bundle category** consists of bundles and bundle morphisms.

differentiable fibre bundle

A bundle (E, B, π, G) is said to be a **differentiable** $[C^k]$ **fibre bundle** if E, B and the typical fibre are differentiable $[C^k]$ manifolds, π is a differentiable $[C^k]$ mapping, G is a Lie group, and the covering of B being the domains of an admissible atlas (see p. 543), the mappings g_{jk} of condition c) are differentiable $[C^k]$.

Coordinates on a fibre bundle. All systems of coordinates appropriate to E as a manifold are not appropriate to E as a fibre bundle.

fibre coordinates

Consider a differentiable fibre bundle (E, B, π) where E and B are manifolds of dimension $n+p$ and n. A chart (U, φ) on E defines **fibre coordinates** on E if the mapping $\varphi: U \to \mathbb{R}^{n+p}$ is a bundle morphism, with $\mathbb{R}^{n+p}$ having the natural bundle (cartesian product) structure $\mathbb{R}^{n+p} = \mathbb{R}^n \times \mathbb{R}^p$.

Tangent bundle. Let $T(X^n)$ be the space of pairs $(x, \boldsymbol{v}_x)$ for all x in the differentiable manifold X^n and all $\boldsymbol{v}_x \in T_x(X^n)$. It can be given a fibre bundle structure $(T(X^n), X^n, \pi, GL(n, \mathbb{R}))$ as follows.

Fibre at x: $T_x(X^n)$. **Typical fibre F:** $\mathbb{R}^n$. **Projection** π: $(x, \boldsymbol{v}_x) \mapsto x$.

Covering of X^n: $\{U_j;\ \{U_j, \psi_j\}$ is an atlas of $X^n\}$.

Homeomorphism φ_j: φ_j is the pair $(\pi, \psi_j' \circ \pi_2)$ where $\pi_2(x, \boldsymbol{v}_x) = \boldsymbol{v}_x$, and

$\psi_j'(v_x)$ is the representative of v_x in the chart $\{U_j, \psi_j\}$

$$(\pi, \psi_j' \circ \pi_2)\colon \pi^{-1}(U_j) \to U_j \times \mathbb{R}^n \text{ by } (x, v_x) \mapsto (x, \psi_j'(v_x)).$$

The fibre coordinates on $T(X^n)$ are given by the mappings

$$(\psi_j, \text{identity}) \circ (\pi, \psi_j' \circ \pi_2)\colon \pi^{-1}(U_j) \to \mathbb{R}^n \times \mathbb{R}^n.$$

The coordinates of a point $p = (x, v_x) \in \pi^{-1}(U_j) \subset T(X^n)$ are thus

$$(x^1, \dots x^n, v_x^1, \dots v_x^n).$$

A change of fibre coordinates on $T(X^n)$ is entirely determined by a change of charts on X^n; it is a bundle morphism of the form $((f, f'), f)$ where $f\colon X^n \to X^n$ is a differentiable (C^k) mapping.[1]

Structural group G: GL(n, $\mathbb{R}$) the group of linear automorphisms of $\mathbb{R}^n$ (isomorphisms of $\mathbb{R}^n$ onto itself) whose matrix representation is the set of $n \times n$ matrices with non-vanishing determinant. Namely, if $\psi_{1,x}'\colon v_x \in T_x(X^n) \mapsto V \in \mathbb{R}^n$ and $\psi_{2,x}'\colon v_x \in T_x(X^n) \mapsto V' \in \mathbb{R}^n$, then

$$\psi_{1,x}' \circ \psi_{2,x}'^{-1} \in \mathrm{GL}(n, \mathbb{R})\colon \mathbb{R}^n \to \mathbb{R}^n \text{ by } V' \mapsto V;$$

GL(n, $\mathbb{R}$) is a Lie group.

differentiable tangent bundle $T(X^n)$

If X^n is a differentiable manifold of class C^k, $T(X^n)$ is a differentiable manifold of class C^{k-1}; $T(X^n)$ is then called a **differentiable** [C^{k-1}] **tangent bundle**.

There is no canonical isomorphism between a fibre at a point and the typical fibre, and hence no canonical isomorphism between fibres at different points, unless the fibre bundle is given an additional structure; for instance parallel displacement (Ch. Vbis).

frame

Frame bundle. Let ρ_x be an arbitrary **frame** in $T_x(X^n)$; ρ_x is a set of n linearly independent vectors $(e_1, \dots e_n)$ which can be expressed as a linear combination of the elements of a particular basis $(\bar{e}_1, \dots, \bar{e}_n)$ of $T_x(X^n)$.

$$e_i = a_i^j \bar{e}_j \qquad (a_i^j) = a \in \mathrm{GL}(n, \mathbb{R}).$$

frame bundle $F(X^n)$

Thus there is a bijection between the set of all frames in $T_x(X^n)$ and GL(n, $\mathbb{R}$). Let $F(X^n)$ be the space of pairs (x, ρ_x) for all $x \in X^n$; it can be given a differentiable fibre structure $(F(X^n), X^n, \pi, \mathrm{GL}(n, \mathbb{R}))$:

Typical fibre: GL(n, $\mathbb{R}$), which has a natural manifold structure as an open set of $\mathbb{R}^{n^2}$, cf. Problem 2, p. 172.

Structural group of diffeomorphisms of the typical fibre onto itself: GL(n, $\mathbb{R}$). (See for example Problem 6, p. 190.)

[1]For example, the lagrangian is a function on the tangent bundle of the configuration space of the system. A change of coordinates $(q, \dot{q}) \mapsto (\bar{q}, \dot{\bar{q}})$ is a bundle morphism. See Problem 1, p. 169.

A frame ρ_x can also be thought of as a nonsingular linear mapping ρ_x: $\mathbf{R}^n \to T_x(X^n)$ by $\rho_x(v_x^1, \ldots, v_x^n) = \boldsymbol{v}_x$.

Exercise: Spell out the topology and an atlas on $F(X^n)$ when an atlas is given on X^n. Notice that with a maximal atlas the typical fibre is the full group $\mathrm{GL}(n, \mathbb{R})$; with a simpler atlas (one which contains all others) one can restrict oneself to a subgroup of $\mathrm{GL}(n, \mathbb{R})$.

A fibre bundle in which the typical fibre F and the structural group G are isomorphic and in which G acts on F by left translation (p. 154) is called a **principal fibre bundle**. The bundle of frames is a principal fibre bundle. We shall return later to principal fibre bundles (p. 357), and to the frame bundle in particular (p. 376). **principal fibre bundle**

We shall need the following definition of the **right action** of G on the principal fibre bundle (E, X, π, G). Let $\{U_i\}$ be the covering of X used to define the fibre bundle structure. We first define the mapping $\tilde{R}_g$ on $\pi^{-1}(U_i)$ and then show that it can be defined coherently on the whole of the bundle E. **right action $\tilde{R}_g$**

Let $p \in F_x$, $x \in U_i$ and define g_i by

$$g_i = \hat{\varphi}_{i,x}(p)$$

where $\hat{\varphi}_{i,x}$ is the homeomorphism from F_x onto G (p. 125). By definition

$$(\tilde{R}_g p)_i = \hat{\varphi}_{i,x}^{-1}(R_g g_i) = \hat{\varphi}_{i,x}^{-1}(g_i g), \qquad p \in \pi^{-1}(U_i).$$

Clearly $\tilde{R}_{g_1}\tilde{R}_{g_2}p = \tilde{R}_{g_2 g_1}p$, that is $\{\tilde{R}_g, g \in G\}$ is a group (anti) isomorphic to G, which acts on the right on $\pi^{-1}(U_i)$. Note that p and $\tilde{R}_g p$ belong to the same fibre and that the group $\{\tilde{R}_g; g \in G\}$ acts transitively on each fibre.

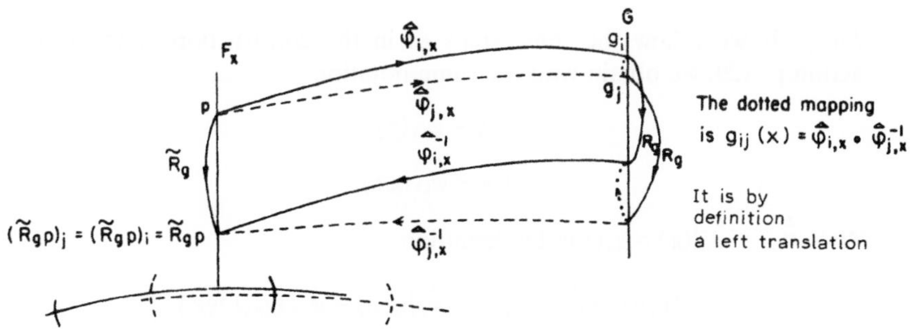

Theorem. For $p \in \pi^{-1}(U_i \cap U_j)$

$$(\tilde{R}_g p)_i = (\tilde{R}_g p)_j.$$

Proof: For $p \in F_x$, $x \in U_i \cap U_j$ we have

$$\overset{\Delta}{\varphi}_{j,x}(p) = g_j \quad \text{and} \quad \overset{\Delta}{\varphi}_{i,x}(p) = g_i \quad \text{so that} \quad g_i = \overset{\Delta}{\varphi}_{i,x} \circ \overset{\Delta}{\varphi}{}^{-1}_{j,x}(g_j).$$

By definition G acts on $F = G$ by left translation; for $x \in U_i \cap U_j$, let $g_{ij}(x)$ be the element of G such that

$$\overset{\Delta}{\varphi}_{i,x} \circ \overset{\Delta}{\varphi}{}^{-1}_{j,x}(g) = g_{ij}(x)g \quad \forall g \in G.$$

Then

$$g_i = g_{ij}(x)g_j$$

and

$$\begin{aligned}(\tilde{R}_g p)_j &= \overset{\Delta}{\varphi}{}^{-1}_{j,x}(g_j g)\\ &= \overset{\Delta}{\varphi}{}^{-1}_{i,x} \circ \overset{\Delta}{\varphi}_{i,x} \circ \overset{\Delta}{\varphi}{}^{-1}_{j,x}(g_j g)\\ &= \overset{\Delta}{\varphi}{}^{-1}_{i,x}(g_{ij}(x)g_j g) = \overset{\Delta}{\varphi}{}^{-1}_{i,x}(g_i g)\\ &= (\tilde{R}_g p)_i.\end{aligned}$$

■

Since the mapping $\tilde{R}_g$ is independent of the choice of the open set U_i containing $\pi(p)$, it is well defined over all of E, and we can write

$$\tilde{R}_g(p) = \overset{\Delta}{\varphi}{}^{-1}_{i,x} \circ R_g \circ \overset{\Delta}{\varphi}_{i,x}(p), \qquad x = \pi(p).$$

One also uses the simplified notation pg instead of $\tilde{R}_g p$.

$\tilde{R}_g$ plays an important role in the definition of a connection (p. 358). The differential mapping $R'_g(p)$ defines a canonical isomorphism from the Lie algebra of the structural group into the tangent space $T_p(F_x)$ (p. 359).

Exercise 1: Show that, in general, the left action of G on the principal fibre bundle (E, X, π, G) does not define a fibre preserving global action.

Proof: If we follow the same steps as in the construction of the right action p. 129, we obtain with the same notation

$$g_i = \overset{\Delta}{\varphi}_{i,x}(p)$$

$$(\tilde{L}_g p)_i = \overset{\Delta}{\varphi}{}^{-1}_{i,x}(g g_i).$$

Since $\overset{\Delta}{\varphi}_{i,x} \circ \overset{\Delta}{\varphi}{}^{-1}_{j,x}(g) = g_{ij}(x)g$ by definition,

$$(\tilde{L}_g p)_j = \overset{\Delta}{\varphi}{}^{-1}_{i,x} \circ \overset{\Delta}{\varphi}_{i,x} \circ \overset{\Delta}{\varphi}{}^{-1}_{j,x}(g g_j) = \overset{\Delta}{\varphi}{}^{-1}_{i,x}(g_{ij}(x) g g_j).$$

Here the argument $g_{ij}(x)gg_j$ cannot be simplified to gg_i and $(\tilde{L}_g p)_i$ is not equal to $(\tilde{L}_g p)_j$. ■

Exercise 2: Write down the right action of $G = \mathrm{GL}(n, \mathbb{R})$ on the frame bundle.

Answer: An element g of $\mathrm{GL}(n, \mathbb{R})$ is a matrix. The fibre F_x at x is the set of all frames at x. Let $p \in \pi^{-1}(U_i)$, and let $\overset{\Delta}{\varphi}_{i,x}(p) = g_i = a_{(i)\lambda}^{\ \mu}$. The action of G on the typical fibre is

$$A^\mu_\alpha a_{(j)\lambda}^{\ \alpha} = a_{(i)\lambda}^{\ \mu}, \qquad \text{i.e.} \qquad \overset{\Delta}{\varphi}_{i,x} \circ \overset{\Delta}{\varphi}{}^{-1}_{j,x} = A^\mu_\alpha$$

and the right action of G on the frame bundle is

$$(\tilde{R}_g p)_i = \overset{\Delta}{\varphi}{}^{-1}_{i,x}(a_{(i)\alpha}^{\ \mu} G^\alpha_\lambda) = \overset{\Delta}{\varphi}{}^{-1}_{j,x}(a_{(j)\alpha}^{\ \mu} G^\alpha_\lambda) \quad \text{with} \quad g = G^\alpha_\beta.$$

The last equality follows from

$$\overset{\Delta}{\varphi}{}^{-1}_{i,x} \circ \overset{\Delta}{\varphi}_{i,x} \circ \overset{\Delta}{\varphi}{}^{-1}_{j,x}(a_{(j)\alpha}^{\ \mu} G^\alpha_\lambda) = \overset{\Delta}{\varphi}{}^{-1}_{i,x}(A^\mu_\alpha a_{(j)\beta}^{\ \alpha} G^\beta_\lambda) = \overset{\Delta}{\varphi}{}^{-1}_{i,x}(a_{(i)\beta}^{\ \mu} G^\beta_\lambda). \qquad ■$$

Reduction. See also Chapter Vbis, p. 380. Given a fibre bundle (E, B, π, F, G) it is interesting to know if it admits an equivalent structure defined with a subgroup G_1 of its structure group G_2. If so, it admits a family of local trivializations with transition functions g_{jk} taking their value in a subgroup G_1 of G; the structure group G is said to be **reducible** to G_1. reducible

Example: The structure group $\mathrm{GL}(n, \mathbb{R})$ of the tangent bundle of the differentiable manifold $\mathbb{R}^n$ is reducible to the identity.
More generally it can be proved that *every fibre bundle with base* $\mathbb{R}^n$ *is reducible to a trivial bundle*: the proof[1] rests on the fact that $\mathbb{R}^n$ is a contractible space, i.e. homotopic to a point.

In the case of a principal fibre bundle (E, X, π, G) it is said to be **reducible** to the principal fibre bundle (E_1, X, π_1, G_1) if G_1 is a subgroup of G and E_1 a subspace of E, such that the injection $f: E_1 \to E$ is a bundle morphism which commutes with the action of G_1: reducible principal bundle

$$\pi f(p) = \pi_1 p, \qquad \forall p \in E_1$$
$$f(\tilde{R}_g p) = \tilde{R}_g f(p), \qquad \forall p \in E_1, \qquad g \in G_1.$$

[1]See for instance [Steenrod, p. 53] or [Osborn II 4].

Example: It will be shown that the frame bundle on a differentiable manifold is reducible to the bundle of orthogonal frames or Lorentz frames.
The relation between the two definitions of reducibility is given by the following theorem which we shall prove later (p. 381).

Theorem: *A differentiable principal fibre bundle* (E, X, π, G) *is reducible to* (E_1, X, π_1, G_1), *with* G_1 *a Lie subgroup of the Lie group* G *(p. 242), if and only if it admits a family of local trivializations whose transition functions take their value in* G_1.

For examples of fibre bundles used in physics, see for instance [Trautman 1970], and Chapter Vbis and the references therein.

3. VECTOR FIELDS

cross-section

A **cross-section** of the bundle (E, B, π) is a mapping

$$f: B \to E \text{ such that } \pi \circ f = \text{identity}.$$

vector field

A **vector field** $\boldsymbol{v}$ on X^n is a cross-section of the tangent bundle $T(X^n)$; a vector field associates to each point $x \in X^n$ a tangent vector $\boldsymbol{v}_x \in T_x(X^n)$ by the mapping $\boldsymbol{v}: x \mapsto (x, \boldsymbol{v}_x)$, often abbreviated to $x \mapsto \boldsymbol{v}_x$.

In local coordinates, a vector $\boldsymbol{v}_x$ is defined by its n components; hence a vector field $\boldsymbol{v}$ is defined in the domain of a chart by n functions v^i of n real variables.

differentiable vector field

By definition a vector field $\boldsymbol{v}$ on a C^k manifold is **differentiable** (C^r) if the mapping $\boldsymbol{v}: X^n \to T(X^n)$ is differentiable (C^r). On a C^k manifold there can only exist C^r vector fields with $r \leqslant k-1$. A vector field $\boldsymbol{v}$ is C^r on X^n if and only if in every chart of an admissible atlas on X^n the n functions v^i are of class C^r.

Corresponding to the definition of a tangent vector $\boldsymbol{v}_x$ as a derivation on the algebra of germs of differentiable functions at x, we can define a vector field as a derivation on the algebra $C^k(X^n)$ of functions of class C^k on X^n:

$$\boldsymbol{v}: C^k(X^n) \to C^{k-1}(X^n) \qquad \text{by} \qquad \boldsymbol{v}(f) = \boldsymbol{v}f.$$

Hence $\boldsymbol{v}_x(f) = (\boldsymbol{v}f)(x)$.

In local coordinates

$$(\overline{vf})(x^j) = v_x^i \partial \bar{f}(x^j)/\partial x^i.$$

Theorem. A principal fibre bundle (E, X, π, G) is trivial if and only if it has a continuous cross-section.

Proof:
a) Cross-section $\Rightarrow$ triviality. Assume that E has a cross-section

$$f: X \to E \qquad \pi \circ f(x) = x.$$

It defines a trivialization as follows.

Given $p \in F_x$ there is a unique $g_0 \in G$ such that $p = \tilde{R}_{g_0} f(x)$. Then

$$\varphi_f: E \to X \times G \qquad \text{by} \qquad p \mapsto (x, g_0)$$

is a homeomorphism which preserves the group structure of the fibres

$$\varphi_f(\tilde{R}_{g'} p) = R_{g'} \varphi_f(p) \qquad \forall g' \in G, \quad \forall p \in E.$$

Note that if $p = f(x)$ then $\varphi_f(p) = (x, e)$, where $e \in G$ is the identity element.
b) Triviality $\Rightarrow$ cross-section. Clearly the trivial bundle $X \times G$ has a cross-section, namely

$$f: X \to X \times G \qquad \text{by} \qquad x \mapsto (x, k(x))$$

where $k: X \to G$ is some continuous mapping. ∎

Remark: A local section defines a local trivialization, see Chapter Vbis.

The Lie algebra $\mathscr{X}(X^n)$. The set $\mathscr{X}(X^n)$ of all C^∞ vector fields on a C^∞ manifold X^n can be given the structure of a Lie algebra as follows. $\mathscr{X}(X^n)$
Let $\boldsymbol{v}, \boldsymbol{w} \in \mathscr{X}(X^n)$ and let $f, g \in C^\infty(X^n)$. Addition in $\mathscr{X}(X^n)$ is defined by

$$(\boldsymbol{v} + \boldsymbol{w})f = \boldsymbol{v}f + \boldsymbol{w}f.$$

Multiplication of $\boldsymbol{v} \in \mathscr{X}(X^n)$ by $g \in C^\infty(X^n)$ is defined by

$$(g\boldsymbol{v})f = g(\boldsymbol{v}f).$$

$\mathscr{X}(X^n)$ is a module on the ring $C^\infty(X^n)$.
$\mathscr{X}(X^n)$ is not closed under multiplication defined by $(\boldsymbol{vw})f = \boldsymbol{v}(\boldsymbol{w}f)$ because $\boldsymbol{vw}$ thus defined is not a vector field: it does not satisfy Leibniz rule[1]; one can also check that it is not a first, but a second order differential operator

$$\overline{(\boldsymbol{vw})f} = v^i \partial_i (w^j \partial_j \bar{f}) = v^i w^j \partial_{ij}^2 \bar{f} + v^i \partial_i w^j \partial_j \bar{f}.$$

[1] $(\boldsymbol{vw})(fg) \overset{\text{def}}{=} \boldsymbol{v}(\boldsymbol{w}fg) = \boldsymbol{v}(f\boldsymbol{w}g + g\boldsymbol{w}f) = (\boldsymbol{v}f)(\boldsymbol{w}g) + f\boldsymbol{v}(\boldsymbol{w}g) + (\boldsymbol{v}g)(\boldsymbol{w}f) + g\boldsymbol{v}(\boldsymbol{w}f)$, whereas $g(\boldsymbol{vw})f + f(\boldsymbol{vw})g = g\boldsymbol{v}(\boldsymbol{w}f) + f\boldsymbol{v}(\boldsymbol{w}g)$.

Lie bracket

On the other hand, the **Lie bracket** defined by

$$[v, w] = vw - wv$$

$$\overline{[v, w]f} = (v^i \partial_i w^j - w^i \partial_i v^j)\partial_j \bar{f}$$

is a vector field.

Jacobi identity

The multiplication defined by the Lie bracket is distributive with respect to addition and anticommutative; it is not associative but it satisfies instead the **Jacobi identity**

$$[v_1, [v_2, v_3]] + [v_2, [v_3, v_1]] + [v_3, [v_1, v_2]] = 0.$$

Lie algebra

A module together with an internal operation which satisfies these multiplicative properties is called a **Lie algebra**.

Lie derivative

The Lie bracket $[v, w]$ is the **Lie derivative** of w in the direction of v: $\mathscr{L}_v w = [v, w]$. The general definition of a Lie derivative applicable to any tensor is given later (p. 147).

moving frame
vierbein
tetrad

A set of n linearly independent differentiable vector fields (e_i) which form a basis for the module $\mathscr{X}(U)$, $U \subset X^n$, is called a **moving frame** (if $n = 4$, **vierbein**, **tetrad**). Such a set may not exist globally on X^n.

Image of a vector field under a diffeomorphism. The image of a vector at a point $x \in X^n$ under a differentiable mapping $f: X^n \to Y^p$ has been defined by the numerical equality (p. 121) $(f'v)_y(g) = v_x(g \circ f)$, which can also be written if f is invertible as

$$[(f'v)(g)](y) = [v(g \circ f)](x) = [v(g \circ f)](f^{-1}(y))$$

in which case the image of a vector field under f is defined by the function equality

$$(f'v)(g) = v(g \circ f) \circ f^{-1}.$$

image of vector field

If v is a differentiable C^r vector field on X^n, and if $f: X^n \to Y^n$ is a C^{r+1} diffeomorphism then $f'v$ is a C^r vector field on Y^n. But if f is not invertible, the image of v under $f: X^n \to Y^p$ is not a vector field on Y^p. If f^{-1} exists but is not differentiable the image is not differentiable. However the images of some differentiable vector fields under some differentiable mappings are differentiable vector fields; when this is the case, the field is said to be **projectable**[1] by f. (v and $f'v$ are also said to be f-**related**.[2])

projectable vector field

Theorem. If $f: X^n \to Y^n$ is a C^∞ *diffeomorphism then* f' *is an isomorphism of the Lie algebra* $\mathscr{X}(X^n) \to \mathscr{X}(Y^n)$; *that is, the image of a Lie bracket is the Lie bracket of the image*:

$$f'[v, w] = [f'v, f'w].$$

Proof:

$$f'[v, w](g) = [v, w](g \circ f) \circ f^{-1} = v(w(g \circ f)) \circ f^{-1} - (v \leftrightarrow w).$$

On the other hand

$$[f'v, f'w](g) = f'v(f'w(g)) - f'w(f'v(g))$$

and

$$\begin{aligned} f'v(f'w(g)) = v(f'w(g) \circ f) \circ f^{-1} &= v(w(g \circ f) \circ f^{-1} \circ f) \circ f^{-1} \\ &= v(w(g \circ f)) \circ f^{-1}. \end{aligned}$$ ■

Invariant vector field. A vector field v on X is said to be **invariant** under the diffeomorphism $f: X \to X$ if

invariant vector field

$$f'(x)v_x = v_{f(x)} \text{ for every } x \in X, \text{ often abbreviated to } f'v = v.$$

The invariance of v under f can be expressed

$$f^* \circ v = v \circ f^*.$$

Indeed by the definition of f^* (p. 116)

$$f^*(f'v(g)) = v(f^*(g)),$$

and in terms of operators

$$f^* \circ f'v = v \circ f^*.$$

If v is invariant under f it follows that $f^* \circ v = v \circ f^*$. ■

This relation is satisfied by all differential operators invariant under f.

4. COVARIANT VECTORS; COTANGENT BUNDLES

The dual $T_x^*(X^n)$ to the tangent vector space $T_x(X^n)$ is the space of

[1][Lichnerowicz 1958, p. 19].
[2][Trautman].

cotangent vector space $T_x^*(X^n)$ covariant vectors contravariant vectors

linear forms on $T_x(X^n)$; it is a vector space of dimension n called the **cotangent vector space** to X^n at x. The elements of $T_x^*(X^n)$ are called cotangent vectors, or **covariant vectors**, or covectors, or differential forms (see p. 137); the elements of $T_x(X^n)$ will henceforth be called contravariant tangent vectors or **contravariant vectors**, as well as just tangent vectors, or vectors.

For $\boldsymbol{\omega}_x \in T_x^*(X^n)$ and $\boldsymbol{v}_x \in T_x(X^n)$

$$\boldsymbol{\omega}_x(\boldsymbol{v}_x) \in \mathbb{R}.$$

Because[1] $T_x^{**}(X^n) = T_x(X^n)$ we have $\boldsymbol{v}_x(\boldsymbol{\omega}_x) = \boldsymbol{\omega}_x(\boldsymbol{v}_x)$.

There is no natural (canonical) isomorphism between a space and its dual. However given a base $(\boldsymbol{e}_1 \ldots \boldsymbol{e}_n)$ in T_x we can construct its dual $(\boldsymbol{\theta}^1 \ldots \boldsymbol{\theta}^n)$ in T_x^* as follows.[2]

basis in $T_x^*(X^n)$

Let (v_x^i) be the components of a vector $\boldsymbol{v}_x$ in the basis $(\boldsymbol{e}_i)$, these components constitute n linear forms defined on $\boldsymbol{v}_x$. It is natural to define the form $\boldsymbol{\theta}^i$ by

$$\boldsymbol{\theta}^i(\boldsymbol{v}_x) = v_x^i,$$

then

$$\boldsymbol{\theta}^i(v^j\boldsymbol{e}_j) = v_x^i \Rightarrow \boldsymbol{\theta}^i(\boldsymbol{e}_j) \equiv \langle \boldsymbol{\theta}^i, \boldsymbol{e}_j \rangle = \delta_j^i.$$

Let us designate by $(\mathbf{d}\boldsymbol{x}^i)$ the dual to the natural basis $(\partial/\partial \boldsymbol{x}^i)$

$$\langle \mathbf{d}\boldsymbol{x}^i, \partial/\partial \boldsymbol{x}^j \rangle = \delta_j^i.$$

natural cobasis

Let ω_{xi} be the components of $\boldsymbol{\omega}_x$ in the **natural cobasis** $(\mathbf{d}\boldsymbol{x}^1 \ldots \mathbf{d}\boldsymbol{x}^n)$

$$\langle \boldsymbol{\omega}_x, \boldsymbol{v}_x \rangle = \langle \omega_{xi}\mathbf{d}\boldsymbol{x}^i, v^j\partial/\partial \boldsymbol{x}^j \rangle = \omega_{xi}v_x^i.$$

The covariant vector components are labelled by subscripts, the contravariant vector components are labelled by superscripts. Notice that $\boldsymbol{\theta}^i$ is not the ith component of a cotangent vector but a cotangent vector itself, the set $(\boldsymbol{\theta}^1, \ldots \boldsymbol{\theta}^n)$ being a set of n linearly independent cotangent vectors.

Change of basis in T_x^* induced by a change of basis in T_x. To a change of basis $(e_i) \mapsto (e_{i'})$ in T_x defined by a matrix a, corresponds a change of basis $(\boldsymbol{\theta}^i) \mapsto (\boldsymbol{\theta}^{i'})$ in T_x^* defined by the inverse matrix.

Proof: $\boldsymbol{e}_i = a_i^{k'}\boldsymbol{e}_{k'}$ and in virtue of $\boldsymbol{v}_x = v^i\boldsymbol{e}_i = v^{k'}\boldsymbol{e}_{k'}$ we get

$$v^{k'} = v^i a_i^{k'} \quad \text{or} \quad V' = aV.$$

Also

$$\boldsymbol{\theta}^{k'}(\boldsymbol{v}_x) = v^{k'} = v^i a_i^{k'} = \boldsymbol{\theta}^i(\boldsymbol{v}_x)a_i^{k'} \Rightarrow \boldsymbol{\theta}^{k'} = \boldsymbol{\theta}^i a_i^{k'}$$

[1] See for instance [Halmos].

[2] Henceforth we abbreviate $T_x(X^n)$ and $T_x^*(X^n)$ as T_x and T_x^*.

and since $\boldsymbol{\omega}_x = \omega_i \boldsymbol{\theta}^i = \omega_{k'} \boldsymbol{\theta}^{k'}$ we get[1]

$$\omega_i = a_i^{k'} \omega_{k'} \quad \text{or} \quad W' = Wa^{-1}. \qquad \blacksquare$$

In particular, for a change of natural basis induced by a change of coordinates, the matrix a is the jacobian matrix of the transformation $(x^i) \mapsto (x'^i)$

$$a_i^{k'} = \frac{\partial x'^k}{\partial x^i} \quad \text{and} \quad \mathbf{d}x'^k = \frac{\partial x'^k}{\partial x^i} \mathbf{d}x^i.$$

The notation chosen for the natural basis in the space T_x^* anticipated the fact that $\mathbf{d}x^i$ is indeed the differential of the coordinate function (p. 114). Although there was no reason to label the coordinates with a superscript rather than a subscript, the label of x^i was put in an upper position in anticipation of the transformation law of the basis $(\mathbf{d}x^i)$.

A covariant vector at $x \in X^n$ can also be defined as an equivalence class of triples (U, φ, W), where (U, φ) is a chart of X^n at x and W a vector of $\mathbb{R}^n$, the equivalence relation being defined by the transformation law given above. For example, the set of n numbers $(\partial f/\partial x^i|_{x_0})$ transforms under a change of coordinates as follows

$$\left.\frac{\partial f}{\partial x^i}\right|_{x_0} = \left.\frac{\partial f}{\partial x'^k}\frac{\partial x'^k}{\partial x^i}\right|_{x_0}.$$

This set is the set of natural components of the covariant vector $\mathbf{d}f|_{x_0}$, differential of f at x_0, defined by

$$\mathbf{d}f|_{x_0}(\boldsymbol{v}_{x_0}) \equiv \langle \mathbf{d}f|_{x_0}, \boldsymbol{v}_{x_0}\rangle \stackrel{\text{def}}{=} \boldsymbol{v}_{x_0}(f) = v^i_{x_0} \partial f(x)/\partial x^i|_{x_0}.$$

Hence $\mathbf{d}f(x) = \partial f/\partial x^i \, \mathbf{d}x^i$ in the basis $(\mathbf{d}x^i)$ dual to $(\partial/\partial x^i)$.

At a point x, the cotangent space $T_x^*(X^n)$ is the **space of differentials** of the germs of differentiable functions at x. space of differentials
This does not say that all covariant vector fields (see next paragraph) are exact differentials; the set of *functions* (ω_i) cannot always be expressed as $(\partial \bar{f}/\partial x^i)$, but the set of *numbers* (ω_{xi}) can always be considered as the set of numbers $\partial \bar{f}/\partial x^i|_x$.

Let us compute the value of $\mathbf{d}f|_x \in T_x^*$ at $\boldsymbol{v}_x \in T_x$ in an arbitrary basis. Let $(\boldsymbol{e}_k)$ and $(\boldsymbol{\theta}^k)$ be dual bases in T_x and T_x^* respectively

[1]With the usual convention a_c^r means the element of the r-row c-column of the matrix a; in the equation $V' = aV$, V and V' are "column vectors"; in the equation $W' = Wa^{-1}$, W and W' are "row vectors".

$$\mathbf{d}f|_x = \boldsymbol{e}_k(f)\boldsymbol{\theta}^k|_x \qquad \boldsymbol{v}_x = v_x^i \boldsymbol{e}_i,$$

Pfaff derivative

$\boldsymbol{e}_k(f)$ is called the **Pfaff derivative**[1] of f, and denoted $\partial_k f$.

$T^*(X^n)$
cotangent bundle

The bundle $(T^*(X^n), X^n, \pi, \mathrm{GL}(n, \mathbb{R}))$, where $T^*(X^n)$ is the space of pairs $(x, \boldsymbol{\omega}_x)$ for all $x \in X^n$ and all $\boldsymbol{\omega}_x \in T_x^*(X^n)$, is called the **cotangent bundle**; its study parallels the study of the tangent bundle.

covariant vector field
one-form

A C^r **covariant vector field** is a C^r cross-section of the cotangent bundle. It is often called a **one-form** (see p. 142).

Reciprocal image (pull-back) of a 1-form under a differentiable mapping.

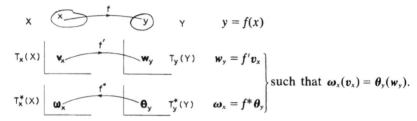

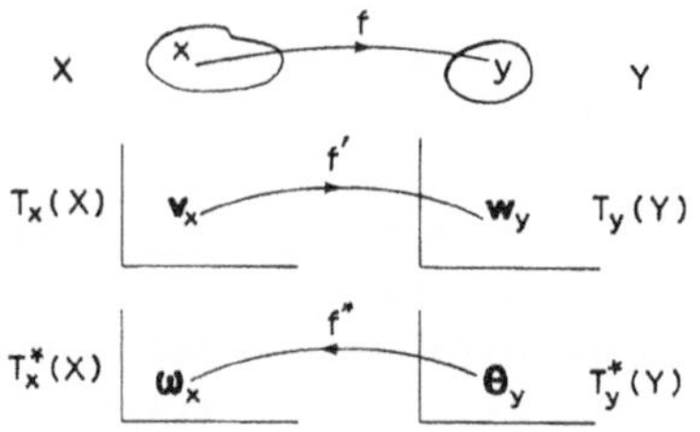

$$y = f(x)$$

$$\left.\begin{array}{l} \boldsymbol{w}_y = f'\boldsymbol{v}_x \\ \boldsymbol{\omega}_x = f^*\boldsymbol{\theta}_y \end{array}\right\} \text{such that } \boldsymbol{\omega}_x(\boldsymbol{v}_x) = \boldsymbol{\theta}_y(\boldsymbol{w}_y).$$

reciprocal image

The **reciprocal image of a covariant vector** $\boldsymbol{\theta}_y$ under a differentiable mapping f is defined by the numerical equality,

$$(f^*\boldsymbol{\theta})_x \boldsymbol{v}_x = \boldsymbol{\theta}_y(f'\boldsymbol{v})_y.$$

The **reciprocal image of a 1-form** $\boldsymbol{\theta}$ under a differentiable mapping f is defined by the function equality

$$(f^*\boldsymbol{\theta})\boldsymbol{v} = \boldsymbol{\theta}(f'\boldsymbol{v}) \circ f.$$

Whereas the expression for the image of a vector field involves f^{-1}, the expression for the reciprocal image of a 1-form (covariant vector field) does not involve f^{-1}; in this respect a covariant vector field is more interesting than a vector field; $f^*\boldsymbol{\theta}$ is always a differentiable covariant vector field if the mapping f and the covariant vector field $\boldsymbol{\theta}$ are differentiable.

5. TENSORS AT A POINT

This section is a straightforward summary of tensor calculus. Tensors at a point of a manifold X^n are defined on the tangent space to the

[1]Note that $\partial^2/\partial x^i \partial x^j = \partial^2/\partial x^j \partial x^i$ but in general $\partial_{ij} f \neq \partial_{ji} f$; indeed let $\boldsymbol{\theta}^k = a_i^k(x)\,\mathbf{d}x^i$, $\partial_k f = (\partial f/\partial x^i)(a^{-1})_k^i(x)$.

manifold at this point, hence on a vector space, and are built exactly like tensors on $\mathbb{R}^n$. For simplicity we shall work on specific cases.

Theorem. The tensor product $\otimes^p T_x^* \otimes^q T_x$ *of p cotangent spaces at x and q tangent spaces at x (the space of p-covariant q-contravariant tensors* $t_{(p)}{}^{(q)}$*), is the vector space of all multilinear forms on the cartesian product*

$$T_x \times \cdots \times T_x \times T_x^* \times \cdots \times T_x^* \qquad (T_x \ p \ \textit{times}, \ T_x^* \ q \ \textit{times}).$$

The space $\otimes^p T_x^* \otimes^q T_x$ *is of dimension* n^{p+q}. Note that $T_x^* \otimes T_x$ is different from $T_x \otimes T_x^*$. Hence when one speaks of a tensor of **order** $(q+p)$, or of **type** (q, p), the nature of the tensor is not in general characterised uniquely. order type

Proof: Consider, for example, the tensor product $T_x^* \otimes T_x^*$ of two cotangent spaces at x, i.e. the space of bilinear forms $\boldsymbol{\omega}_x$: $T_x \times T_x \to \mathbb{R}$, where $\boldsymbol{\omega}_x(\boldsymbol{v}_x, \boldsymbol{w}_x)$ is a linear function of each of its variables $\boldsymbol{v}_x, \boldsymbol{w}_x \in T_x$. We shall prove that $\otimes^2 T_x^*$ is a vector space of dimension n^2. We shall omit the subscript x attached to each tensor, the paragraph being entitled "tensors at a point". The space $\otimes^2 T_x^*$ is a vector space with addition and scalar multiplication defined by

$$(\boldsymbol{\omega} + \boldsymbol{\nu})(\boldsymbol{v}, \boldsymbol{w}) = \boldsymbol{\omega}(\boldsymbol{v}, \boldsymbol{w}) + \boldsymbol{\nu}(\boldsymbol{v}, \boldsymbol{w})$$

$$\alpha\boldsymbol{\omega}(\boldsymbol{v}, \boldsymbol{w}) = \alpha(\boldsymbol{\omega}(\boldsymbol{v}, \boldsymbol{w})).$$

Let us construct a basis in $\otimes^2 T_x^*$ and thereby determine the dimension of the space. Let $(\boldsymbol{e}_i)$ be a basis in T_x

$$\boldsymbol{v} = v^i \boldsymbol{e}_i, \qquad \boldsymbol{w} = w^j \boldsymbol{e}_j.$$

In virtue of the linearity of $\boldsymbol{\omega}$

$$\boldsymbol{\omega}(\boldsymbol{v}, \boldsymbol{w}) = v^i w^j \boldsymbol{\omega}(\boldsymbol{e}_i, \boldsymbol{e}_j).$$

Let $(\boldsymbol{\theta}^i)$ be the basis of T_x^* dual to $(\boldsymbol{e}_i)$ and define $\boldsymbol{\theta}^i \otimes \boldsymbol{\theta}^j$ by

$$\boldsymbol{\theta}^i \otimes \boldsymbol{\theta}^j(\boldsymbol{v}, \boldsymbol{w}) = v^i w^j.$$

Then $\boldsymbol{\omega}$ is clearly a linear combination of the n^2 2-covariant tensors $\boldsymbol{\theta}^i \otimes \boldsymbol{\theta}^j$

$$\boldsymbol{\omega} = \omega_{ij} \boldsymbol{\theta}^i \otimes \boldsymbol{\theta}^j$$

where $\omega_{ij} = \boldsymbol{\omega}(\boldsymbol{e}_i, \boldsymbol{e}_j)$. The set $(\boldsymbol{\theta}^i \otimes \boldsymbol{\theta}^j)$ is linearly independent because $\boldsymbol{\nu} = \nu_{ij} \boldsymbol{\theta}^i \otimes \boldsymbol{\theta}^j = 0$ implies $\nu_{ij} = 0$: Indeed if the vector $\boldsymbol{\nu} = \nu_{ij} \boldsymbol{\theta}^i \otimes \boldsymbol{\theta}^j$ is the null vector of $T_x^* \otimes T_x^*$, $\boldsymbol{\nu}(\boldsymbol{v}, \boldsymbol{w}) = 0$ for all $\boldsymbol{v}, \boldsymbol{w} \in T_x$ in particular $\nu_{ij} = \boldsymbol{\nu}(\boldsymbol{e}_i, \boldsymbol{e}_j) = 0$.

The set $(\boldsymbol{\theta}^i \otimes \boldsymbol{\theta}^j)$ is thus a basis in $\otimes^2 T_x^*$, and the dimension of $\otimes^2 T_x^*$ is n^2. ∎

If one makes a **change of basis** in T_x,

$$e_i = a_i^{k'} e_{k'}, \qquad v^i a_i^{k'} = v^{k'},$$

it follows from

$$\omega(v, w) = v^i w^j \omega_{ij} = v^{k'} w^{l'} \omega_{k'l'} = v^i a_i^{k'} w^j a_j^{l'} \omega_{k'l'}$$

that

$$\omega_{ij} = a_i^{k'} a_j^{l'} \omega_{k'l'}.$$

The generalization of these results to an arbitrary product is obvious.

$\otimes^p T_x$ may also be defined as the dual of $\otimes^p T_x^*$. Its basis will be denoted

$$(e_{i_1} \otimes \cdots \otimes e_{i_p}),$$

and its elements

$$v = v^{i_1 \ldots i_p} e_{i_1} \otimes \cdots \otimes e_{i_p}.$$

Its components obey the transformation law

$$v^{k'_1 k'_2 \ldots k'_p} = a_{i_1}^{k'_1} \ldots a_{i_p}^{k'_p} v^{i_1 \ldots i_p}.$$

A mixed tensor, say once covariant and contravariant, can be either the tensor $t \in T_x^* \otimes T_x$ or the tensor $u \in T_x \otimes T_x^*$ such that

$$t(v, \omega) = v^i \omega_j t(e_i, \theta^j) = t_i{}^j \theta^i \otimes e_j(v, \omega)$$

or

$$u(\omega, v) = \omega_i v^j u(\theta^i, e_j) = u^i{}_j e_i \otimes \theta^j(\omega, v).$$

The components of t transform by

$$t_i{}^j = a_i^{k'} (a^{-1})_{l'}^j t_{k'}{}^{l'};$$

the components of u transform by

$$u^i{}_j = (a^{-1})_{k'}^i a_j^{l'} u^{k'}{}_{l'}.$$

The space of tensors of a given type, for instance $\otimes^p T_x^* \otimes^q T_x$, is a vector space. Let t and u be two of its elements and $\alpha \in \mathbb{R}$: the components of the sum $(t + u)$ are the sum of the corresponding components of t and u, the components of the product (αt) are obtained by multiplying each component of t by α.

addition

scalar multiplication

multiplication

Tensor algebra. We shall now define a **multiplication** between tensors of artitrary order. For example let $u \in \otimes^q T_x$, $\theta \in \otimes^p T_x^*$, the **tensor product**

tensor product

$$t = u \otimes \theta$$

is, by definition, an element of $\otimes^q T_x \otimes^p T_x^*$ such that

$$t(\boldsymbol{\omega}_1 \ldots \boldsymbol{\omega}_q, \boldsymbol{v}_1 \ldots \boldsymbol{v}_p) = \boldsymbol{u}(\boldsymbol{\omega}_1, \ldots \boldsymbol{\omega}_q)\boldsymbol{\theta}(\boldsymbol{v}_1, \ldots \boldsymbol{v}_p)$$

$$\forall\, \boldsymbol{\omega}_1, \ldots \boldsymbol{\omega}_q \in T_x^* \quad \text{and} \quad \forall\, \boldsymbol{v}_1, \ldots \boldsymbol{v}_p \in T_x.$$

In local coordinates

$$t^{i_1 \ldots i_q}_{i_{q+1} \ldots i_{q+p}} = u^{i_1 \ldots i_q}\theta_{i_{q+1} \ldots i_{q+p}}.$$

The space $\otimes^q T_x \otimes^p T_x^*$ is the space of all such products. Tensor multiplication is associative and distributive with respect to addition, *it is not commutative.*

contraction

Contraction. The operation of **contraction** on a given contravariant and a given covariant index is an intrinsic operation, but is most conveniently written in local coordinates. It is the linear mapping of tensors of type (q, p) into tensors of type $(q-1, p-1)$ defined by

$$t^{\cdot j \cdot}_{\cdot \cdot i} \mapsto \sum_{i=1}^{n} t^{\cdot i \cdot}_{\cdot \cdot i}.$$

contracted multiplication

Contracted multiplication is a multiplication followed by the contraction of any pair of indices, one covariant, one contravariant.
One can use totally contracted multiplication of an unknown object by known objects to determine the tensoriality of the unknown object; for instance, if a set $(t_{i_1 \ldots i_p})$ of n^p numbers given for each frame in T_x, is such that for any system of p vectors $v_1, \ldots v_p$, the quantity $t_{i_1 \ldots i_p} v_1^{i_1} \ldots v_p^{i_p}$ is a scalar (a number independent of the choice of frame in T_x), then the set $(t_{i_1 \ldots i_p})$ is, for each frame, the set of components of the same p covariant vector.

Exercise 1: Check the order of a contracted tensor; check that contracted multiplication can be used to determine the tensoriality of an unknown object.

Exercise 2: Let E and F be two vector spaces, show that a linear mapping of E into F is an element of $E^* \otimes F$.

symmetry properties

The **symmetry properties** of a tensor are the symmetry properties of the linear form which defines it. More precisely, let S_p be the group of permutations of the p integers $1, \ldots p$ and let $\pi \in S_p$; by definition

$$(\pi\boldsymbol{\omega})(\boldsymbol{v}_1, \ldots \boldsymbol{v}_p) = \boldsymbol{\omega}(\boldsymbol{v}_{\pi(1)} \ldots \boldsymbol{v}_{\pi(p)})$$

so that ω has the symmetry [antisymmetry] defined by π if

$$\pi\omega = \omega \qquad [\pi\omega = (\text{signature } \pi)\omega]$$

with signature $\pi = \pm 1$ according to whether the permutation of $1 \ldots p$ is even or odd.
In local coordinates

$$(\pi\omega)_{i_1 \ldots i_p} = \omega_{\pi(i_1) \ldots \pi(i_p)}.$$

The [anti] symmetry of q-contravariant tensors is defined similarly. The [anti] symmetry properties of mixed tensors are defined only for indices of the same nature because π permutes only the labels of objects of the same nature.

A symmetrization operator S and an antisymmetrization operator A can be defined as follows.

$$St = \frac{1}{p!}\sum_{S_p} \pi t \qquad \text{is a completely symmetric tensor,}$$

$$At = \frac{1}{p!}\sum_{S_p} (\text{signature } \pi)\pi t \text{ is a completely antisymmetric tensor.}$$

Kronecker tensor

The components of completely antisymmetric tensors can be expressed in terms of the **Kronecker tensor**

$$\epsilon^{k_1 \ldots k_p}_{i_1 \ldots i_p} = \begin{cases} 0 & \text{if } (k_1 \ldots k_p) \text{ is not a permutation of } (i_1 \ldots i_p) \\ +1 & \text{if } (k_1 \ldots k_p) \text{ is an even permutation of } (i_1 \ldots i_p) \\ -1 & \text{if } (k_1 \ldots k_p) \text{ is an odd permutation of } (i_1 \ldots i_p). \end{cases}$$

For example

$$(At)_{i_1 \ldots i_p} = \frac{1}{p!} \epsilon^{k_1 \ldots k_p}_{i_1 \ldots i_p} t_{k_1 \ldots k_p}.$$

p-forms

The next chapter is the study of totally antisymmetric covariant tensor fields which, unexpected as it appears now, deserve, as we shall see, a whole chapter for themselves. Because of their importance, totally antisymmetric covariant tensor fields of order p are given a name: ***p*-forms** (forms of degree p).

6. TENSOR BUNDLES; TENSOR FIELDS

The manifold X together with the set of vector spaces

$\otimes^p T^*_x(X) \otimes^q T_x(X)$ for all $x \in X$ can be given a vector bundle structure; it is then called a **tensor bundle** of **order** $(p+q)$. A **tensor field** of a given **order** on a C^k manifold X is a C^r cross-section $(r \leq k-1)$ of the tensor bundle of the same order. Operations defined for tensors at a point, namely tensor bundle

$$(t+u)^{j_1 \ldots j_q}_{i_1 \ldots i_p}(x) = t^{j_1 \ldots j_q}_{i_1 \ldots i_p}(x) + u^{j_1 \ldots j_q}_{i_1 \ldots i_p}(x)$$

$$(ft)^{j_1 \ldots j_q}_{i_1 \ldots i_p}(x) = f(x) t^{j_1 \ldots j_q}_{i_1 \ldots i_p}(x), \qquad f \in C^r(X^n)$$

are carried over fibrewise to define the similar operations on tensor fields. The set of all C^r tensor fields of the same order, together with these operations is a module on the ring $C^r(X)$.

Exercise: Generalize to tensor fields the results derived for covariant and contravariant vector fields. 1) Let $f: X^n \to Y^p$ by $x \mapsto y$; define the induced mappings $\otimes^q f'$ and $\otimes^q f^*$, often abbreviated to f' and f^*.

$$f' : \otimes^q T_x(X^n) \to \otimes^q T_y(Y^p)$$

$$f^* : \otimes^q T^*_y(Y^p) \to \otimes^q T^*_x(X^n).$$

2) Show that the reciprocal image of a differentiable (C^r) covariant tensor field is differentiable (C^r) if f is differentiable (C^{r+1}) but that, in general, the image of a C^r contravariant tensor field is of class C^r only if f is a C^{r+1} diffeomorphism.
Given $(\boldsymbol{\theta} \otimes \boldsymbol{u}) \in T^*_y(Y) \otimes T_y(Y)$, compute its reciprocal image under the diffeomorphism f

$$[(f^* \otimes f^{-1'})(\boldsymbol{\theta} \otimes \boldsymbol{u})](\boldsymbol{v}, \boldsymbol{\omega}) = (\boldsymbol{\theta} \otimes \boldsymbol{u})(f'\boldsymbol{v}, f^{*-1}\boldsymbol{\omega}) = \boldsymbol{\theta}(f'\boldsymbol{v})\boldsymbol{u}(f^{*-1}\boldsymbol{\omega})$$

$$= f^*\boldsymbol{\theta}(\boldsymbol{v}) f^{-1'}\boldsymbol{u}(\boldsymbol{\omega}) = (f^*\boldsymbol{\theta} \otimes f^{-1'}\boldsymbol{u})(\boldsymbol{v}, \boldsymbol{\omega}). \quad \blacksquare$$

C. GROUPS OF TRANSFORMATIONS

In all this section smooth means "differentiable of class C^k", with k large enough for the statements to be meaningful.

1. VECTOR FIELDS AS GENERATORS OF GROUPS OF TRANSFORMATIONS

Let X be a smooth manifold of dimension n with points $x, y \ldots \in X$, let $\boldsymbol{v}$ be a vector field on X, and let $\sigma: I \subset \mathbb{R} \to X$ be a curve such that the tangent to the curve σ at $x = \sigma(t)$ is the vector[1] $\boldsymbol{v}(x)$. The curve σ satisfies

[1] We shall write $\boldsymbol{v}_x$ or $\boldsymbol{v}(x)$ as needed for typographical convenience.

the differential equation

$$d\sigma(t)/dt = v(\sigma(t)) \qquad t \in I \qquad \text{field differential equation.}$$

integral curve

trajectory

σ is called an **integral curve** of the vector field v.
Dynamical systems are often governed by this type of equation. An integral curve is then called a **trajectory** of the system.

Given a vector field v, can one always find an integral curve through a point x_0? Is this curve unique? The answer – "Yes, locally, if v is C^1" – is given by the following theorem.

Theorem. Suppose v is a C^r vector field on the manifold X, then for every $x \in X$, there exists an integral curve of $v, t \mapsto \sigma(t, x)$, such that

1) *$\sigma(t, x)$ is defined for t belonging to some interval $I(x) \subset \mathbb{R}$, containing $t = 0$, and is of class C^{r+1} there.*
2) *$\sigma(0, x) = x$ for every $x \in X$.*
3) *This curve is unique: Given $x \in X$ there is no C^1 integral curve of v defined on an interval strictly greater than $I(x)$, and passing through x (i.e. such that $\sigma(0, x) = x$).*

The proof is easily obtained from the proof given for ordinary differential equations (p. 97). The following theorem is a consequence of the uniqueness property 3).

Theorem. If t, s, $t + s \in I(x)$ then $\sigma(t, \sigma(s, x)) = \sigma(t + s, x)$.

Proof: The curves $t \mapsto \sigma(t, \sigma(s, x))$ and $t \mapsto \sigma(t + s, x)$ satisfy the same differential equation, and the same initial condition $\sigma(0, \sigma(s, x)) = \sigma(s, x)$. ■

The set of pairs (x, t), $x \in X$, $t \in I(x)$ is an open subset of $X \times \mathbb{R}$, hence a smooth manifold Σ_v of dimension $n + 1$.

flow

The mapping $\sigma: \Sigma_v \to X$ by $(x, t) \mapsto \sigma(t, x)$ is called the **flow** of the C^1 vector field v. If X and v are C^∞, the flow is C^∞.

If X is not a compact manifold, Σ_v is not in general a product $X \times I$. However for every $x_0 \in X$ there exists a neighborhood $N(x_0) \subset X$, and an interval $I(x_0) \subset \mathbb{R}$ such that σ is defined on $N(x_0) \times I(x_0)$, and smooth if X and v are smooth. The proof is the same as for ordinary differential equations (cf. Ch II).

local transformations

The mapping $\sigma(t,.) \equiv \sigma_t$: $x \mapsto \sigma(t, x)$ is defined on $N(x_0) \subset X$ for $t \in I(x_0)$. It is a transformation of $N(x_0)$ – local transformation of X – generated by the vector field v. Under this mapping, a point $x \in N(x_0)$

goes to a point $\sigma_t(x) \in X$ along the integral curve of $\boldsymbol{v}$ at x – the location of $\sigma_t(x)$ along the curve being determined by t.

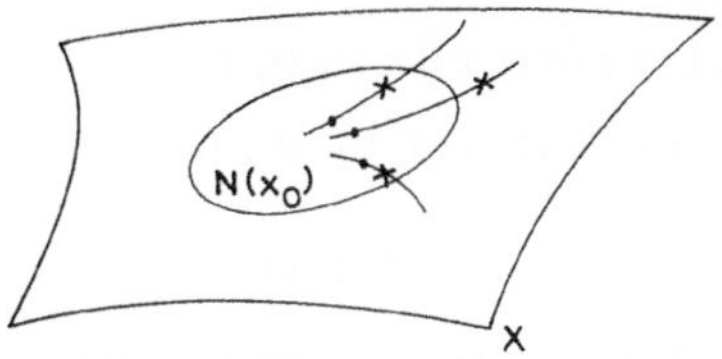

σ_t maps each dot into the cross on the same integral curve.

Usually the interval $I(x_0) \subset \mathbb{R}$ depends on $N(x_0)$. The intersection I of all the intervals $I(x_0)$ corresponding to a set of neighborhoods $\{N(x_0)\}$ covering X may or may not be empty. I is never empty when X is compact, since it is then given by a finite intersection. When I is not empty, then σ_t with $t \in I$ defines a global transformation of X. Moreover we can define σ_t for all $t \in \mathbb{R}$ through the relation

global transformations

$$\sigma_{t+s} = \sigma_t \circ \sigma_s,$$

and each σ_t has an inverse, σ_{-t}. We have thus proved the following.

Theorem. A smooth vector field on a compact manifold X generates a one parameter group of diffeomorphisms of X.

We shall now prove a corollary that we shall use in Chapter VII.

Corollary. A smooth vector field on a manifold X, which vanishes outside a compact set $K \subset X$, generates a one parameter group of diffeomorphisms of X.

Proof: If $x \in X \backslash K$, σ_t is the identity mapping, defined for all t. Therefore we know, by covering K with a finite number of open neighborhoods $N(x_0)$, that there is a non empty interval $I \subset \mathbb{R}$ such that the mapping σ_t is defined on the whole of X for $t \in I$, thus for all $t \in \mathbb{R}$. ∎

Since (p. 144) $\sigma(t, \sigma(s, x)) = \sigma(t+s, x)$ if $t, s, t+s \in I(x)$, we have the composition law

$$\sigma_t \circ \sigma_s = \sigma_{t+s}.$$

The set of mappings σ_t is called a one parameter **local pseudo group**. It is a one parameter local group if σ_t, $t \in I$ defines a global transformation of X; it is a one parameter group if $I = \mathbb{R}$.

local pseudo group

Conversely, any one parameter local pseudo group $\{\sigma_t\}$ of transformations $\sigma_t : x \mapsto \sigma_t(x) = \sigma(t, x)$ can be generated by a vector field $\boldsymbol{v}$ defined uniquely by the equation $\boldsymbol{v}(x) = \mathrm{d}\sigma(t, x)/\mathrm{d}t|_{t=0}$, i.e.

$$\boldsymbol{v}(x) = \mathrm{d}\sigma(t, x)/\mathrm{d}t|_{t=0} \text{ implies } \boldsymbol{v}(\sigma(t, x)) = \mathrm{d}\sigma(t, x)/\mathrm{d}t.$$

Proof: Consider the curve C_x generated from x by σ_t; because of the group property

$$\sigma_{t+s}(x) = \sigma_s \circ \sigma_t(x), \qquad t, s, t+s \in I(x).$$

Take the derivative of both sides of this equation with respect to s

$$\frac{\mathrm{d}}{\mathrm{d}s}\sigma(t+s, x) = \frac{\mathrm{d}}{\mathrm{d}t}\sigma(t+s, x) = \frac{\mathrm{d}}{\mathrm{d}s}\sigma(s, \sigma(t, x))$$

which gives at $s = 0$

$$\frac{\mathrm{d}}{\mathrm{d}t}\sigma(t, x) = \frac{\mathrm{d}}{\mathrm{d}s}\sigma(s, \sigma(t, x))|_{s=0}$$

$$= \boldsymbol{v}(\sigma(t, x)) \qquad \text{by the definition of } \boldsymbol{v}.$$

The vector field $\boldsymbol{v}$ is the generator of the pseudo group of transformations $\{\sigma_t\}$. ■

Physicists often call this vector an "infinitesimal" generator and write the previous equation

$$\delta\sigma = \boldsymbol{v}\delta t$$

Then, since $\sigma(0, x_0) = x_0$,

$$\sigma(\delta t) = x_0 + \boldsymbol{v}\delta t.$$

Theorem. Given a vector field $\boldsymbol{v}$ generating the local pseudo group σ_t, the image $f'\boldsymbol{v}$ of $\boldsymbol{v}$ under the diffeomorphism f of X onto X generates the one parameter local pseudogroup $\{f \circ \sigma_t \circ f^{-1}\}$.

Proof: We shall prove that $\boldsymbol{w}(y)$ defined by

$$\boldsymbol{w}(y) = \frac{\mathrm{d}}{\mathrm{d}t}(f \circ \sigma_t \circ f^{-1}(y))|_{t=0}$$

is equal to $(f'\boldsymbol{v})(y)$. Indeed

$$\frac{\mathrm{d}}{\mathrm{d}t}(f^i \circ \sigma_t(x))|_{t=0} = \frac{\partial f^i}{\partial \sigma_t^j(x)} \frac{\mathrm{d}\sigma_t^j(x)}{\mathrm{d}t}\bigg|_{t=0} = (f'(\boldsymbol{v}(x)))^i.$$

Alternatively we could remark that by its definition, $\boldsymbol{w}(y)$ is the tangent vector to the curve $f \circ \sigma(\cdot, x) \circ f^{-1}$ at y:

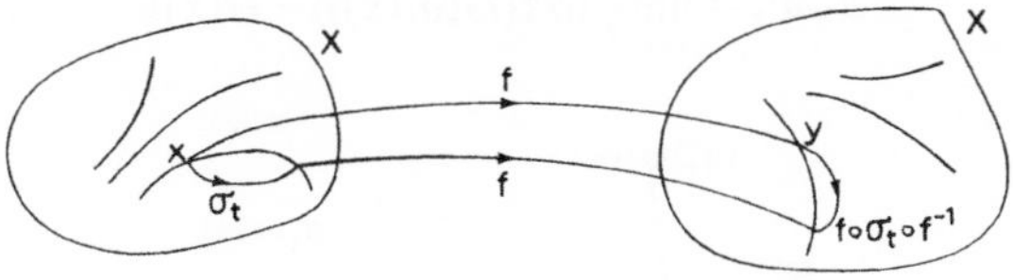

■

Corollary. The vector field $\boldsymbol{v}$ *is invariant under* f *iff* f *commutes with* σ_t.

Proof: $\boldsymbol{v}$ is invariant under f if and only if

$$f'\boldsymbol{v} = \boldsymbol{v}$$

i.e. iff $\frac{\mathrm{d}}{\mathrm{d}t}(f \circ \sigma_t \circ f^{-1}(y))|_{t=0} = \frac{\mathrm{d}}{\mathrm{d}t}\sigma_t(y)|_{t=0}$.

$$f \circ \sigma_t \circ f^{-1} = \sigma_t \quad \Rightarrow \quad f \circ \sigma_t = \sigma_t \circ f.$$

■

2. LIE DERIVATIVES

The flow of a vector field $\boldsymbol{v}$ determines local transformations σ_t of a manifold X, mappings of neighborhoods $N(x) \subset X$ into X.
These mappings induce mappings $\otimes^p \sigma_t'$ of p-contravariant tensors, mappings $\otimes^p \sigma_t^*$ of p-covariant tensors, and mappings of mixed tensors. The Lie derivative of a tensor field characterizes the difference between the tensor field and its [inverse] image under σ_t.
The **Lie derivative $\mathscr{L}_v w$ of the contravariant vector field** w is the contravariant vector field defined by $\mathscr{L}_v w$

$$\mathscr{L}_v \boldsymbol{w}|_x = \lim_{t=0} \frac{1}{t}[\sigma_t^{-1'}(\boldsymbol{w}(\sigma_t(x))) - \boldsymbol{w}(x)]$$

where we need to introduce the inverse mapping $\sigma_t^{-1'}$ to compute the Lie derivative at x.

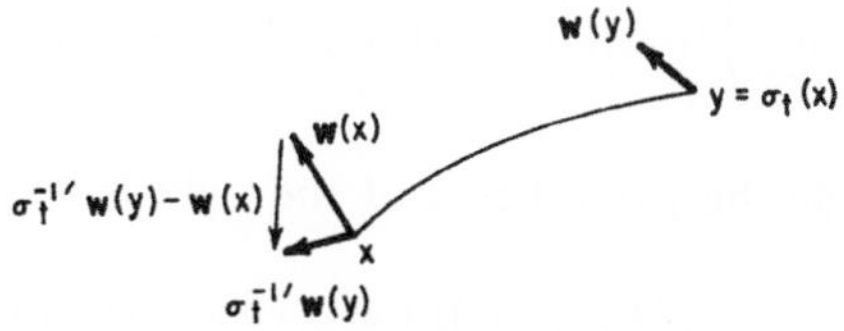

The **Lie derivative $\mathscr{L}_v \omega$ of the covariant vector field ω** is the covariant $\mathscr{L}_v \omega$

vector field defined by

$$\mathcal{L}_v\boldsymbol{\omega}|_x = \lim_{t=0}\frac{1}{t}\,[\sigma_t^*(\boldsymbol{\omega}(\sigma_t(x))) - \boldsymbol{\omega}(x)].$$

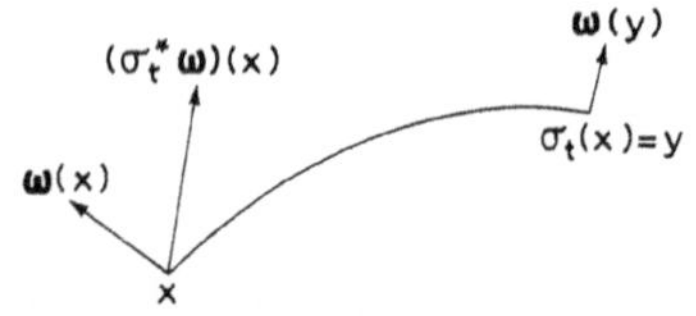

Exercise: Compute the Lie derivative of a function f.

Answer: $$\mathcal{L}_v f|_x = \lim_{t=0}\frac{1}{t}\,[f(\sigma_t(x)) - f(x)] = \boldsymbol{v}(f)|_x.$$

Proposition. The Lie derivative is a local operator:

$$\text{if } \boldsymbol{u} = \boldsymbol{t} \text{ on } N(x) \subset X, \qquad \mathcal{L}_v\boldsymbol{u} = \mathcal{L}_v\boldsymbol{t} \text{ on } N(x);$$

$$\text{if } \boldsymbol{v}_1 = \boldsymbol{v}_2 \text{ on } N(x), \qquad \mathcal{L}_{v_1}\boldsymbol{u} = \mathcal{L}_{v_2}\boldsymbol{u} \text{ on } N(x).$$

Proposition. The Lie derivative is a derivation on the algebra of (germs of) differentiable tensor fields, i.e.

$\mathcal{L}_v$ is an additive operator $\qquad \mathcal{L}_v(\boldsymbol{u}+\boldsymbol{t}) = \mathcal{L}_v(\boldsymbol{u}) + \mathcal{L}_v(\boldsymbol{t});$

$\mathcal{L}_v$ satisfies Leibniz rule $\qquad \mathcal{L}_v(\boldsymbol{u}\otimes\boldsymbol{t}) = \mathcal{L}_v\boldsymbol{u}\otimes\boldsymbol{t} + \boldsymbol{u}\otimes\mathcal{L}_v\boldsymbol{t}.$

Proof: The additive property can be verified readily. The Leibniz property rests on the fact that the transform of a tensor product is the tensor product of the transforms of its factors. Typically

$$(\sigma_t^*\otimes\sigma_t^{-1'})(\boldsymbol{\omega}\otimes\boldsymbol{w}) = \sigma_t^*\boldsymbol{\omega}\otimes\sigma_t^{-1'}\boldsymbol{w}.$$

Thus

$$\mathcal{L}_v(\boldsymbol{\omega}\otimes\boldsymbol{w})|_x = \lim_{t=0}\frac{1}{t}\,[(\sigma_t^*(\boldsymbol{\omega}(\sigma_t(x)))\otimes\sigma_t^{-1'}(\boldsymbol{w}(\sigma_t(x))) - (\boldsymbol{\omega}\otimes\boldsymbol{w})(x)].$$

Let us insert inside the parentheses of the right hand side

$$-\boldsymbol{\omega}(x)\otimes\sigma_t^{-1'}(\boldsymbol{w}(\sigma_t(x))) + \boldsymbol{\omega}(x)\otimes\sigma_t^{-1'}(\boldsymbol{w}(\sigma_t(x)));$$

the Leibniz property of the Lie derivatives follows readily. ∎

Local coordinate expressions of Lie derivatives. If, for instance, the local coordinate expression of an arbitrary tensor is of the form

$$\boldsymbol{t} = t^i_{jk}(x)\frac{\partial}{\partial \boldsymbol{x}^i} \otimes \mathbf{d}\boldsymbol{x}^j \otimes \mathbf{d}\boldsymbol{x}^k$$

then

$$\mathscr{L}_v \boldsymbol{t}|_x = \mathscr{L}_v(t^i_{jk}(x))\frac{\partial}{\partial \boldsymbol{x}^i} \otimes \mathbf{d}\boldsymbol{x}^j \otimes \mathbf{d}\boldsymbol{x}^k + t^i_{jk}(x)\mathscr{L}_v\left(\frac{\partial}{\partial \boldsymbol{x}^i}\right) \otimes \mathbf{d}\boldsymbol{x}^j \otimes \mathbf{d}\boldsymbol{x}^k$$

$$+ t^i_{jk}(x)\frac{\partial}{\partial \boldsymbol{x}^i} \otimes \mathscr{L}_v(\mathbf{d}\boldsymbol{x}^j) \otimes \mathbf{d}\boldsymbol{x}^k + t^i_{jk}(x)\frac{\partial}{\partial \boldsymbol{x}^i} \otimes \mathbf{d}\boldsymbol{x}^i \otimes \mathscr{L}_v\, \mathbf{d}\boldsymbol{x}^j.$$

We have already computed the Lie derivative of a function, it remains to compute the Lie derivatives of the natural basis of vector fields, and of 1-forms in our coordinate system.

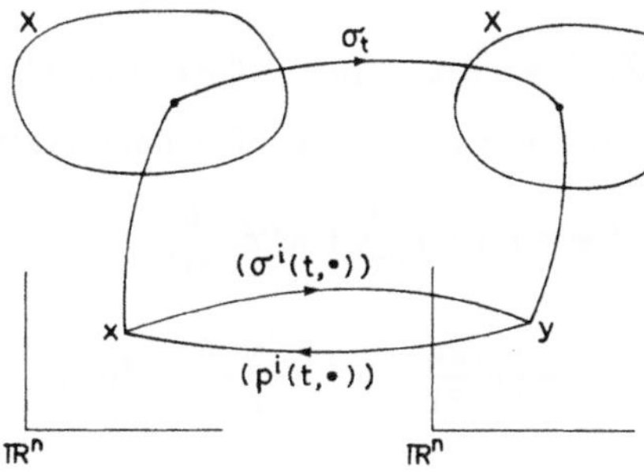

Let us simplify the writing of the coordinate expressions by calling x, $y \in \mathbb{R}^n$ the coordinates of the corresponding points of X, $\sigma^i(t, \cdot)$ the coordinate expression of σ_t, $p^i(t, \cdot)$ the coordinate expression of σ_t^{-1}, $\sigma^i(0, x) = x^i$, $p^i(0, y) = y^i$, $i = 1, \ldots, n$.
By definition

$$\left(\mathscr{L}_v \frac{\partial}{\partial \boldsymbol{x}^i}\right)\bigg|_x = \lim \frac{1}{t}\left[\sigma_t^{-1'}\left(\frac{\partial}{\partial \boldsymbol{x}^i}\bigg|_y\right) - \frac{\partial}{\partial \boldsymbol{x}^i}\bigg|_x\right].$$

The components of $\partial/\partial x^i$ in the natural coordinate system are δ^j_i; the components of its transform $\sigma_t^{-1'}(\partial/\partial x^i)$ are

$$\frac{\partial p^j(t, y)}{\partial y^h}\delta^h_i = \frac{\partial p^j(t, y)}{\partial y^i}.$$

Thus in local coordinates

$$\left(\mathcal{L}_v \frac{\partial}{\partial \boldsymbol{x}^i}\right)\bigg|_x = \lim \frac{1}{t}\left[\frac{\partial p^j(t,y)}{\partial y^i} - \delta_i^j\right]\left(\frac{\partial}{\partial \boldsymbol{x}^j}\bigg|_x\right)$$

$$= \frac{\mathrm{d}}{\mathrm{d}t}\frac{\partial p^j}{\partial y^i}\bigg|_{t=0}\frac{\partial}{\partial \boldsymbol{x}^j}\bigg|_x \qquad \text{using } \delta_i^j = \frac{\partial p^j(0,y)}{\partial y^i}$$

$$= -\frac{\mathrm{d}}{\mathrm{d}t}\frac{\partial \sigma^j}{\partial x^i}\bigg|_{t=0}\frac{\partial}{\partial \boldsymbol{x}^j}\bigg|_x \qquad \text{using } \frac{\mathrm{d}}{\mathrm{d}t}\left(\frac{\partial p^j}{\partial y^h}\frac{\partial \sigma^h}{\partial x^i}\right) = \frac{\mathrm{d}}{\mathrm{d}t}\delta_i^j = 0.$$

Hence

$$\mathcal{L}_v \frac{\partial}{\partial \boldsymbol{x}^i} = -\frac{\partial v^j}{\partial x^i}\frac{\partial}{\partial \boldsymbol{x}^j} \qquad \text{using } \mathrm{d}\sigma(t)/\mathrm{d}t = \boldsymbol{v}(\sigma(t)).$$

A similar calculation gives

$$\mathcal{L}_v\, \mathbf{d}\boldsymbol{x}^i = \frac{\partial v^i}{\partial x^j}\, \mathbf{d}\boldsymbol{x}^j.$$

In summary

$$(\mathcal{L}_v \boldsymbol{t})^i_{jk} = v^l \partial_l t^i_{jk} - t^l_{jk}\partial_l v^i + t^i_{lk}\partial_j v^l + t^i_{jl}\partial_k v^l.$$

Exercise: Compute the Lie derivative of a contravariant vector field.

Answer:

$$\mathcal{L}_v \boldsymbol{w} = \mathcal{L}_v(w^i)\frac{\partial}{\partial \boldsymbol{x}^i} + w^i \mathcal{L}_v \frac{\partial}{\partial \boldsymbol{x}^i}$$

$$= \left(v^j\frac{\partial w^i}{\partial x^j} - w^j\frac{\partial v^i}{\partial x^j}\right)\frac{\partial}{\partial \boldsymbol{x}^i} = [\boldsymbol{v}, \boldsymbol{w}].$$

Tensor fields invariant under a group of transformations.

Theorem. A tensor field is invariant under a (local, pseudo) group $\{\sigma_t\}$ of transformations generated by $\boldsymbol{v}$ if and only if its Lie derivative with respect to $\boldsymbol{v}$ is zero.

Proof: By the definition of the Lie derivative, the invariance of the tensor implies the vanishing of the Lie derivative; typically:

$$\sigma'_t \boldsymbol{w}(x) = \boldsymbol{w}(\sigma_t(x)) \;\Rightarrow\; \mathcal{L}_v \boldsymbol{w} = 0$$

$$\sigma^*_t \boldsymbol{\omega}(\sigma_t(x)) = \boldsymbol{\omega}(x) \;\Rightarrow\; \mathcal{L}_v \boldsymbol{\omega} = 0.$$

The proof of the converse, namely $\mathcal{L}_v \boldsymbol{t} = 0 \Rightarrow$ "$\boldsymbol{t}$ invariant under the group of transformations generated by $\boldsymbol{v}$", will be carried out in a neighborhood of a point x_0 where $\boldsymbol{v} \neq 0$. The general case is obtained by a limiting procedure when v is not identically zero[1] in a neighborhood of x_0. We shall construct a coordinate system where

$$v^n = 1, \qquad v^\alpha = 0, \qquad \alpha = 1, \ldots n-1, \qquad i = 1, \ldots n.$$

Let (x'^i) be a coordinate system in the neighborhood of x_0 such that $x_0'^n = 0$ and $v'^n(x_0) \neq 0$.

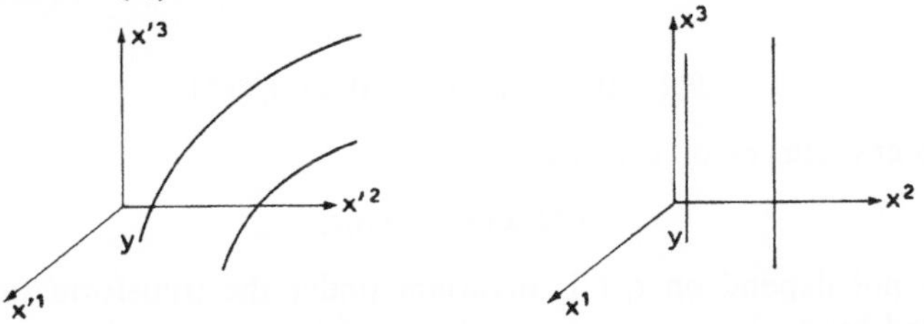

Let (x^i) be the desired coordinate system; in this system the integral curve of v which goes through x at $t = 0$ is

$$\sigma^i(t, x) = x^i + \delta^i_n t.$$

Since a change of coordinates maps integral curves into integral curves, the functions $x'^i(x^n, x^\alpha)$ which characterize the change of coordinates satisfy the equations

$$\mathrm{d}x'^i/\mathrm{d}x^n = v'^i(x'^j(x^n, x^\alpha)),$$

where the variable x^n has been chosen as parameter along the curve. Locally, this system has exactly one solution $x'^i(x^n, y)$ which goes through y for $x^n = 0$

$$x'^i(0, y) = y'^i.$$

For convenience, choose points y such that $0 = y'^n = x'^n(0, y)$ and choose the x-system such that y has the same coordinates in both systems $y'^i = y^i = x^i$; it follows that

$$\left.\frac{\partial x'^i}{\partial x^\alpha}\right|_{x^n=0} = \frac{\partial x'^i(0, y)}{\partial x^\alpha} \qquad \text{using the continuity of } x'^i \text{ in } x^n$$

$$= \delta^i_\alpha \qquad \text{using the initial condition;}$$

thus, the jacobian of the transformation

$$\det\left[\left.\frac{\partial x'^i}{\partial x^j}\right|_{x^n=0}\right] = \left.\frac{\partial x'^n}{\partial x^n}\right|_{x^n=0} = v'^n(x^0) \neq 0.$$

The unique solution $x'^i(x^n, x^\alpha)$ characterizes an admissible change of coordinates such that, in the x-system, the coordinates of v have the

[1]When v is zero in a neighborhood of x_0, it generates only identity transformations of this neighborhood.

desired values. Moreover, in this coordinate system $\partial v^i/\partial x^i = 0$ and the components of the Lie derivative of $\boldsymbol{t}$ reduce to the partial derivative of its components with respect to x^n.
Typically

$$(\mathcal{L}_v \boldsymbol{t})_i^{jk} = \partial t_i^{jk}/\partial x^n.$$

Thus

$$\mathcal{L}_v \boldsymbol{t} = 0 \Rightarrow \partial t_i^{jk}/\partial x^n = 0 \Rightarrow t_i^{jk}(x^\alpha).$$

The integral curves of $\boldsymbol{v}$ being

$$\sigma^i(t, x) = x^i + \delta_n^i t,$$

t_i^{jk} does not depend on t, $\boldsymbol{t}$ is invariant under the transformations $\{\sigma_t\}$ generated by $\boldsymbol{v}$. ■

metric isometry

A **metric** $\boldsymbol{g}$ is a 2-covariant nondegenerate symmetric tensor on X (p. 142). An **isometry** is a diffeomorphism of X which leaves $\boldsymbol{g}$ invariant. The isometries of X form a group of transformations of X. A one parameter group of transformations generated by a vector field $\boldsymbol{v}$ is a group of isometries if and only if

$$0 = (\mathcal{L}_v \boldsymbol{g})_{hk} = v^i(\partial g_{hk}/\partial x^i) + g_{hi}(\partial v^i/\partial x^k) + g_{ik}(\partial v^i/\partial x^h).$$

Set $v_h = g_{hi} v^i$, this equation becomes

$$\frac{\partial v_h}{\partial x^k} + \frac{\partial v_k}{\partial x^h} + v^i\left(\frac{\partial g_{hk}}{\partial x^i} - \frac{\partial g_{hi}}{\partial x^k} - \frac{\partial g_{ik}}{\partial x^h}\right) = 0.$$

It will be shown later (pp. 301 and 308) that this equation can be written

$$\nabla_k v_h + \nabla_h v_k = 0 \qquad \text{where } \nabla_k \text{ are the covariant derivatives.}$$

The tensor $\mathcal{L}_v \boldsymbol{g}$ is the strain tensor generated by the vector field $\boldsymbol{v}$ (cf. Problem 3, p. 177).

D. LIE GROUPS

1. DEFINITIONS; NOTATIONS

Lie group of transformations

So far we have considered the one-dimensional group of transformations (diffeomorphisms) $\{\sigma_t\}$ of X onto X. We shall now consider the finite dimensional group of transformations $\{\sigma_g$: g is an element of a Lie group G of dimension $p\}$. The set $\{\sigma_g\}$ is a **Lie group of transformations** if the mapping

$$\sigma: G \times X \to X \quad \text{by} \quad (g, x) \mapsto \sigma(g, x)$$

is differentiable and if the set of transformations $\{\sigma_g: X \to X; \sigma_g(x) = \sigma(g, x)\}$, together with the composition mapping, follow the group property:

$$\begin{cases} \sigma_g \circ \sigma_h = \sigma_{gh} & \text{(left action of } G \text{ on } X) \\ \text{or } \sigma_g \circ \sigma_h = \sigma_{hg} & \text{(right action of } G \text{ on } X) \\ \sigma_e \text{ is the identity transformation.} \end{cases}$$

It follows that

$$\sigma_{g^{-1}} = \sigma_g^{-1}.$$

G operates **effectively** on X if $\sigma_g(x) = x$ for every $x \in X$ implies $g = e$. effectively
G operates **freely** on X if $\sigma_g(x) \neq x$ unless $g = e$. freely
G operates **transitively** on X if for every $x \in X$ and $y \in X$ there exists $g \in G$ such that $\sigma_g(x) = y$. transitively

The properties of the one-dimensional group of transformations $\{\sigma_t\}$ are largely stated in terms of the vector field v which generates it. Properties similar to the properties of $\{\sigma_t\}$ can be derived for the groups of transformations $\{\sigma_{g(t)}; t \in \mathbb{R}\}$ where $\{g(t); t \in \mathbb{R}\}$ is a one parameter subgroup of G.
A **one parameter subgroup** of a Lie group G is a differentiable curve one parameter subgroup

$$\mathbb{R} \to G \quad \text{by} \quad t \mapsto g(t)$$

such that

$$\begin{cases} g(t)g(s) = g(t+s) \\ g(0) = e. \end{cases}$$

The concepts developed for the group $\{\sigma_t\}$ apply to the group $\{\sigma_{g(t)}\}$:

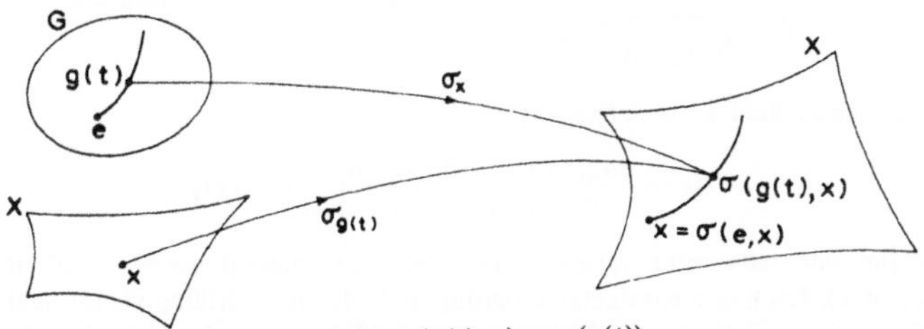

$$\sigma_{g(t)}(x) = \sigma(g(t), x) = \sigma_x(g(t))$$

The curve generated by the transformations $\{\sigma_{g(t)}; t \in \mathbb{R}\}$ operating on x is the image of the one parameter subgroup $\{g(t); t \in \mathbb{R}\}$ by σ_x where $\sigma_x(g(t)) = \sigma(g(t), x) = \sigma_{g(t)}(x)$.

Killing vector fields

The vector field which generates the group of transformation $\{\sigma_{g(t)};\ t \in \mathbb{R}\}$ is called a **Killing vector field on** X relative to the action of the group G.[1]

The integral curve going through x of the Killing vector field $\boldsymbol{v}$ satisfies the equations

$$\begin{cases} \mathrm{d}\sigma_x(g(t))/\mathrm{d}t = \boldsymbol{v}(\sigma_x(g(t))) \\ \sigma_x(e) = x. \end{cases}$$

We shall see (pp. 158–9) that a one parameter subgroup is defined by its tangent vector $\boldsymbol{\gamma}$ at e

$$\boldsymbol{\gamma} = \mathrm{d}g(t)/\mathrm{d}t|_{t=0}.$$

Hence we can label the Killing vector field which generates $\{\sigma_{g(t)};\ \mathrm{d}g(t)/\mathrm{d}t|_{t=0} = \boldsymbol{\gamma},\ t \in \mathbb{R}\}$ by $\boldsymbol{\gamma}$,

$$\boldsymbol{v}^{\gamma}(x) = \frac{\mathrm{d}\sigma_{g(t)}(x)}{\mathrm{d}t}\bigg|_{t=0} = \frac{\mathrm{d}\sigma_x(g(t))}{\mathrm{d}t}\bigg|_{t=0} = \sigma_x'(e)\boldsymbol{\gamma}.$$

The following transformations of G onto itself play a central role in the theory of Lie groups.

left and right translations

$L_g\colon G \to G$ by $L_g(h) = gh$ called **left translation** and
$R_g\colon G \to G$ by $R_g(h) = hg$ called **right translation**
Obviously $R_g(h) = L_h(g)$.

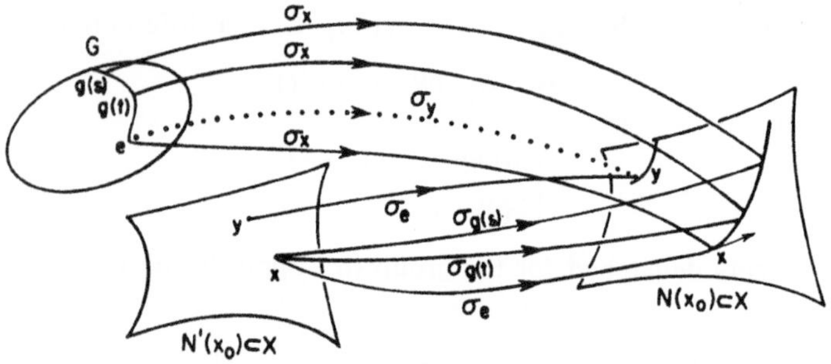

The Killing vector field $\boldsymbol{v}^{\gamma}$ is defined by

$$\boldsymbol{v}^{\gamma}(x) = \frac{d\sigma_{g(t)}(x)}{dt}\bigg|_{t=0} = \frac{d\sigma_x(g(t))}{dt}\bigg|_{t=0} = \sigma_x'(e)\gamma$$

σ_x maps the one parameter subgroup $g(t)$ into the integral curve of $\boldsymbol{v}^{\gamma}$ through x (trajectory of x). Each one parameter subgroup of G defines a Killing vector field on X. If G acts effectively on X, the space of Killing vector fields is isomorphic to the Lie algebra $\mathscr{G}$ of G.

[1]When X is endowed with a metric $\boldsymbol{g}$ – for instance in general relativity – the group G is usually the group which defines isometries.

Notation: We summarize here the notations used in the study of groups of transformations of manifolds, when local coordinates are used. We abbreviate $f(x^1, \dots, x^n)$ to $f(x^i)$.
$x \in X$ with coordinates $(x^i) \in \mathbf{R}^n$ and $g \in G$ with coordinates $(g^\alpha) \in \mathbf{R}^p$.
$\sigma: G \times X \to X$ by $(g, x) \mapsto \sigma(g, x)$ with coordinates $(\sigma^i(g^\alpha, x^j)) \in \mathbf{R}^n$
$\sigma_g: X \to X$ by $\sigma_g(x) = \sigma(g, x)$
$\sigma_x: G \to X$ by $\sigma_x(g) = \sigma(g, x)$
$\sigma_g'(x)$: $T_x(X) \to T_{\sigma_g(x)}(X)$ represented by $(\sigma_g'(x))^i_j = \partial\sigma^i(g^\alpha, x^k)/\partial x^j$
$\sigma_x'(g)$: $T_g(G) \to T_{\sigma_x(g)}(X)$ represented by $(\sigma_x'(g))^i_\alpha = \partial\sigma^i(g^\beta, x^j)/\partial g^\alpha$.

If the group of transformations operates on the group itself then
L: $G \times G \to G$ by $(g, h) \mapsto gh$ with coordinates $(gh)^\alpha = L^\alpha(g^\beta, h^\gamma)$
R: $G \times G \to G$ by $(g, h) \mapsto hg$ with coordinates $(hg)^\alpha = R^\alpha(g^\beta, h^\gamma)$
L_g: $G \to G$ by $L_g(h) = gh$
R_g: $G \to G$ by $R_g(h) = hg$
$L_g'(h)$: $T_h(G) \to T_{gh}(G)$ represented by $(L_g'(h))^\alpha_\kappa = \partial L^\alpha(g^\beta, h^\gamma)/\partial h^\kappa$
$R_g'(h)$: $T_h(G) \to T_{hg}(G)$ represented by $(R_g'(h))^\alpha_\kappa = \partial R^\alpha(g^\beta, h^\gamma)/\partial h^\kappa$.

2. LEFT AND RIGHT TRANSLATIONS OF G; LIE ALGEBRA $\mathscr{G}$ OF G; STRUCTURE CONSTANTS

The two groups of transformations of a Lie group G into G defined by the left and right translations act effectively, transitively and freely on G.

The set of vector fields invariant under the left [right] translations are called the **left [right] invariant vector fields on** G. We shall show that each of these two sets forms a vector space of same dimension as G and that, together with the Lie bracket operation, each is a Lie algebra. left, right invariant vector field

A vector field v on G is left invariant if

$$L_g'(v(h)) = v(L_g h) = v(gh), \quad \forall g, h \in G;$$

it follows that

$$L_g'\gamma = v(g), \quad \text{with } \gamma = v(e).$$

In local coordinates the equation reads

$$v^\alpha(g) = \left.\frac{\partial (gh)^\alpha}{\partial h^\beta}\right|_{h=e} \gamma^\beta.$$

Conversely $L_g'v(e) = v(g)$ implies that v is left invariant:

$$v(L_g h) = v(gh) = L_{gh}'v(e) = (L_g \circ L_h)'v(e) = L_g' \circ L_h'v(e) = L_g'v(h).$$

We thus have proved this theorem.

Theorem. There is a bijective correspondence between the set of left invariant vector fields and the set of vector tangents to G at e, namely the tangent vector space $T_e(G)$.

We shall now prove the following.

Theorem. The set of left [right] invariant vector fields is closed under the Lie bracket operation.

Proof: Let v and w be two left invariant vector fields; one can prove that their Lie bracket is left invariant (p. 134)

$$L'_g[v, w] = [L'_g v, L'_g w] = [v, w]$$

either by using local coordinates or by using the relationship between the Lie derivative of a vector field and the corresponding Lie bracket. The same arguments apply to right invariant vector fields. ■

Lie algebra $\mathscr{G}$

The vector space of left invariant vector fields on G together with Lie bracket multiplication is called the **Lie algebra** $\mathscr{G}$ of the group G.

structure constants

Let $\{v_\alpha;\ \alpha = 1, \ldots p\}$ be a basis of the Lie algebra; there exist numbers $c^\gamma_{\alpha\beta}$ such that:

$$[v_\alpha, v_\beta] = c^\gamma_{\alpha\beta} v_\gamma.$$

The coefficients $c^\gamma_{\alpha\beta}$ are called the **structure constants of the Lie group** G; their properties follow from the properties of the Lie bracket:
antisymmetry of the Lie bracket $\Rightarrow c^\gamma_{\alpha\beta} = -c^\gamma_{\beta\alpha}$
Jacobi identity for the Lie bracket $\Rightarrow c^\gamma_{\alpha\beta} c^\kappa_{\gamma\sigma} + c^\gamma_{\beta\sigma} c^\kappa_{\gamma\alpha} + c^\gamma_{\sigma\alpha} c^\kappa_{\gamma\beta} = 0$
bilinearity of the Lie bracket $\Rightarrow (c^\gamma_{\alpha\beta})$ transforms by a change of the basis on $\mathscr{G}$ as the components of a third order tensor.

Exercise 1: Given a group G by its multiplication table $\{L^\alpha(g^\beta, h^\gamma)\}$, compute the structure constants in the $(v_\alpha(e) = \partial/\partial g^\alpha|_{g=e})$ basis of T_eG.

Answer:

$$c^\gamma_{\alpha\beta} = \left(\frac{\partial^2 L^\gamma}{\partial g^\alpha \partial h^\beta} - \frac{\partial^2 L^\gamma}{\partial g^\beta \partial h^\alpha}\right)_{g=h=e}.$$

Proof:

$$[v_\alpha, v_\beta](g) = c^\gamma_{\alpha\beta} v_\gamma(g).$$

Therefore

$$v^{\kappa}_{\alpha}(g)\,\frac{\partial v^{\lambda}_{\beta}(g)}{\partial g^{\kappa}} - v^{\kappa}_{\beta}(g)\,\frac{\partial v^{\lambda}_{\alpha}(g)}{\partial g^{\kappa}} = c^{\gamma}_{\alpha\beta}v^{\lambda}_{\gamma}(g).$$

Using

$$v^{\lambda}_{\beta}(g) = \left.\frac{\partial L^{\lambda}(g,h)}{\partial h^{\mu}}\right|_{h=e}\gamma^{\mu}_{\beta},$$

where $\gamma^{\mu}_{\beta} = v^{\mu}_{\beta}(e)$, and setting $g = e$ we obtain

$$\gamma^{\mu}_{\beta}\gamma^{\kappa}_{\alpha}\left.\frac{\partial^2 L^{\lambda}}{\partial g^{\kappa}\partial h^{\mu}}\right|_{g=h=e} - \gamma^{\mu}_{\alpha}\gamma^{\kappa}_{\beta}\left.\frac{\partial^2 L^{\lambda}}{\partial g^{\kappa}\partial h^{\mu}}\right|_{g=h=e} = c^{\gamma}_{\alpha\beta}\gamma^{\lambda}_{\gamma}.$$

Now use the fact that $\gamma^{\lambda}_{\gamma} = \delta^{\lambda}_{\gamma}$ in the chosen basis. ■

Conversely it can be proved that given some constants which satisfy the listed properties of the structure constants, there exists a local group G having them for structure constants. If G is simply connected, it is unique modulo an isomorphism.

Exercise 2: Compute the structure constants of the group $\mathrm{GL}(n,\mathbb{R})$ of non singular $n \times n$ real matrices (p. 172).

Answer: Let (g^a_b) be the coordinates of $g \in \mathrm{GL}(n,\mathbb{R})$. Here g^a_b is the coordinate labelled g^{α} in the previous exercise,

$$L^{\alpha}(g^{\beta},h^{\gamma}) = (gh)^a_b = g^a_c h^c_b.$$

The n^2 vector fields v^a_b (with $v^a_b(e) = \partial/\partial g^b_a|_{g=e}$ as in Exercise 1) whose components at g are

$$(v^a_b(g))^l_m = \left.\frac{\partial (gh)^l_m}{\partial h^c_d}\right|_{h=e}(v^a_b(e))^c_d = g^l_b\delta^a_m$$

form a basis for the Lie algebra $\mathscr{GL}(n,\mathbb{R})$. The structure constants can be read off the equation

$$[v^b_a(g), v^d_c(g)] = \delta^b_c v^d_a(g) - \delta^d_a v^b_c(g).$$ ■

Exercise 3: The **affine group of automorphisms** of $\mathbb{R}^n$ (isomorphisms of $\mathbb{R}^n$ onto itself) is the (n^2+n)-dimensional group $\{\sigma_g : \mathbb{R}^n \to \mathbb{R}^n\}$ defined by

affine group of automorphisms

$$(\sigma_g(x))^a = g^a_b x^b + g^a, \qquad g = (g^a_b, g^a) \in G, \qquad (g^a_b) \text{ non singular.}$$

The multiplication table is obtained by computing $\sigma_g \sigma_h = \sigma_{gh}$:

$$gh = ((gh)^a_b, (gh)^a) \qquad \text{with} \qquad (gh)^a_b = g^a_c h^c_b, \quad (gh)^a = g^a_b h^b + g^a.$$

The (n^2+n) vector fields v^b_a, v_a whose components at g are $(v^b_a(g))^l_m = g^l_a \delta^b_m$ as in Exercise 2 and $(v_a(g))^b = g^b_a$ form a basis for the Lie algebra $\mathscr{G}$ of G. The structure constants can be read off from their Lie brackets,

$$[v^b_a(g), v^d_c(g)] = \delta^b_c v^d_a(g) - \delta^d_a v^b_c(g), \qquad [v^b_a(g), v_c(g)] = \delta^b_c v_a(g),$$

$$[v_a(g), v_b(g)] = 0.$$

Alternatively in both Exercises 2 and 3 we can obtain the structure constants from the mixed derivatives of (gh). ■

The structure constants are a local characteristic and usually do not determine G globally. We shall prove presently that $\mathbb{R}^n$ and the n-torus $\mathbb{T}^n = \mathbb{R}^n/\mathbb{Z}^n$ both have vanishing structure constants.

Theorem. A Lie group G has vanishing structure constants if and only if it is locally isomorphic to $\mathbf{R}^n$*, i.e. if and only if it is abelian.*

Proof:

1) $G \overset{\text{locally}}{\simeq} \mathbf{R}^n \Rightarrow c^\gamma_{\alpha\beta} = 0$.

Indeed $G \overset{\text{locally}}{\simeq} \mathbf{R}^n$ implies the existence of a local coordinate system such that

$$(L_g h)^\alpha = (gh)^\alpha \equiv L^\alpha = g^\alpha + h^\alpha$$

$$(L'_g(e))^\alpha_\beta = \partial L^\alpha / \partial h^\beta |_{h=e} = \delta^\alpha_\beta.$$

$(L'_g(e))^\alpha_\beta$ does not depend on g; hence the left invariant vector fields are constant, their Lie brackets vanish and the structure constants vanish.

2) The converse $c^\gamma_{\alpha\beta} = 0 \Rightarrow G \overset{\text{locally}}{\simeq} \mathbf{R}^n$ will be proved in the next chapter (p. 209). ■

3. ONE-PARAMETER SUBGROUPS

Theorem. The one parameter subgroups of G are the integral curves, going through the origin e, of the left [equivalently right] invariant vector fields.

The integral curves of right and left invariant vector fields which do not go through the origin coincide only when the group is abelian.

Proof: Let $g: \mathbb{R} \to G$ be a one parameter subgroup,

$$L_{g(t)}g(s) = g(t+s);$$

hence

$$L'_{g(t)}\frac{dg(s)}{ds} = \frac{dg(t+s)}{ds} = \frac{dg(t+s)}{dt}.$$

Setting $s = 0$

$$L'_{g(t)}\gamma = dg(t)/dt$$

where γ is the tangent vector at $g(0) = e$ to the curve $t \mapsto g(t)$.
This equation says that $g(t)$ is an integral curve of the left invariant vector field equal to γ at e. Similarly it could be shown that $g(t)$ is an integral curve of the right invariant vector field equal to γ at e.
Conversely, for each $\gamma \in T_e(G)$ there is a unique solution $g(t)$, going through the origin, of the differential equation for each left or right invariant vector field. This solution obeys the group property $g(t)g(s) = g(t+s)$ (see p. 144). ■

In contradistinction to the solution $\sigma(t)$ obtained earlier which was defined only for neighborhoods in X and for $t \in I \subset \mathbb{R}$, $g(t)$ is a global transformation of G and is defined for all values of $t \in \mathbb{R}$. In other words $\{\sigma_t\}$ is a local pseudogroup, whereas $\{g(t)\}$ is a group.

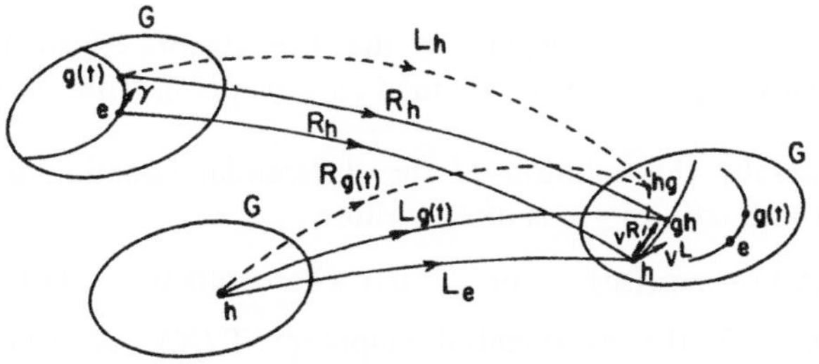

$$v^L(h) = \left.\frac{dL_{g(t)}(h)}{dt}\right|_{t=0} \qquad v^L(h) = R'_h\gamma$$

An element $\gamma \in T_e(G)$ defines the generator v^L [the generator v^R] of a one parameter subgroup of transformations $\{L_{g(t)}\}$ [of transformations $\{R_{g(t)}\}$]; v^L is also the right invariant vector field generated by γ. Similarly the integral curve through h of the left invariant vector field v^R generated by γ is $R_{g(t)}h$. The integral curves through e of the right and left invariant vector fields v^L and v^R generated by the same element of the Lie algebra $\mathscr{G}$ coincide. The isomorphisms of $\mathscr{G}$ with the spaces of right and left invariant vector fields are defined by $\gamma \leftrightarrow v^L$, $\gamma \leftrightarrow v^R$.

Theorem. Let v^L be the generator of the one-parameter group of transformations $\{L_{g(t)};\ t \in \mathbb{R}\}$ of G onto G; then v^L is a right invariant vector field. Similarly the generator v^R of $\{R_{g(t)}\}$ is a left invariant vector field.

Proof:
$$v^L(h) = \left.\frac{\mathrm{d}L_{g(t)}(h)}{\mathrm{d}t}\right|_{t=0} = \left.\frac{\mathrm{d}R_h(g(t))}{\mathrm{d}t}\right|_{t=0} = R'_h\gamma. \qquad \blacksquare$$

Remark: $v^L[v^R]$ is a Killing vector field on G relative to the action $L[R]$ of the group G.

4. EXPONENTIAL MAPPING; TAYLOR EXPANSION; CANONICAL COORDINATES

exponential mapping

The **exponential mapping** maps the line $t\gamma$, in the tangent space $T_e(G)$ at the origin e of a Lie group G, onto the one parameter subgroup $g_\gamma(t)$ of G tangent to γ at the origin, according to the following definition:

$$\exp: T_e(G) \to G \quad \text{by} \quad \gamma \mapsto \exp \gamma = g_\gamma(1).$$

Since $g_{t\gamma}(1) = g_\gamma(t)$, then $\exp(t\gamma) = g_\gamma(t)$. It follows that

$$\exp(t\gamma)\exp(s\gamma) = g_\gamma(t)g_\gamma(s) = g_\gamma(t+s) = \exp(t+s)\gamma$$

which justifies the name "exponential mapping".

Remark: $T_e(G)$ is isomorphic to the Lie algebra $\mathscr{G}$ of G and the exponential mapping is often defined as a mapping from $\mathscr{G}$ into G.

More generally the solution of the differential equation $\mathrm{d}\sigma(t, x)/\mathrm{d}t = v(\sigma(t, x))$ with $\sigma(0, x) = x$ is often written

$$\sigma(t, x) = \exp_x(tv) \qquad \text{or} \qquad \sigma(t, x) = \exp(tv)x, \qquad (v \in T_x).$$

See on p. 325, the exponential mapping: $T_x(X) \to X$, where X is a riemannian manifold, which maps an interval $|t| < \epsilon$ of a line tw into the geodesic curve through x with tangent vector w at x.

Taylor expansion

Proposition. Let f be an analytic function on G. Then f admits the following expansion

$$f(h \exp \gamma) = \sum_{n=0}^{\infty} \frac{1}{n!}(v^n f)(h), \quad h \in G, \gamma \in T_e(G)$$

where v is the left invariant vector field generated by γ.

Proof:

$$(\boldsymbol{v}f)(h)=\boldsymbol{\gamma}(f\circ L_h)=\frac{\mathrm{d}}{\mathrm{d}t}(f\circ L_h\circ g(t))|_{t=0}=\frac{\mathrm{d}}{\mathrm{d}t}f(h\exp t\boldsymbol{\gamma})|_{t=0}.$$

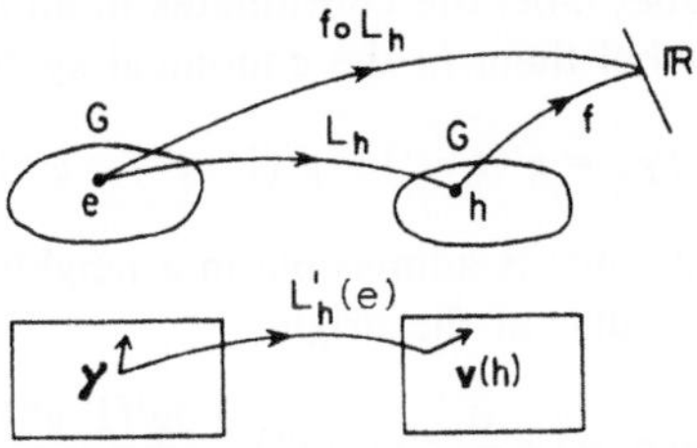

Hence

$$(\boldsymbol{v}f)(h\exp s\boldsymbol{\gamma})=\frac{\mathrm{d}}{\mathrm{d}t}f(h\exp(s\boldsymbol{\gamma})\exp(t\boldsymbol{\gamma}))|_{t=0}=\frac{\mathrm{d}}{\mathrm{d}s}f(h\exp s\boldsymbol{\gamma})$$

and by induction

$$(\boldsymbol{v}^n f)(h\exp s\boldsymbol{\gamma})=\frac{\mathrm{d}^n}{\mathrm{d}s^n}(f(h\exp s\boldsymbol{\gamma})).$$

If f is analytic $\quad f(h\exp s\boldsymbol{\gamma})=\sum_{n=0}^{\infty}\frac{s^n}{n!}\frac{\mathrm{d}^n}{\mathrm{d}s^n}f(h\exp s\boldsymbol{\gamma})|_{s=0}$

and the expression of $f(h\exp\boldsymbol{\gamma})$ follows. ∎

It can be shown, by Taylor expanding both sides, that

$$\exp(t\boldsymbol{\alpha})\exp(t\boldsymbol{\beta})=\exp\{t(\boldsymbol{\alpha}+\boldsymbol{\beta})+\tfrac{1}{2}t^2[\boldsymbol{\alpha},\boldsymbol{\beta}]+\tfrac{1}{12}t^3[\boldsymbol{\alpha},[\boldsymbol{\alpha},\boldsymbol{\beta}]]+\tfrac{1}{12}t^3[[\boldsymbol{\alpha},\boldsymbol{\beta}],\boldsymbol{\beta}]+\mathrm{O}(t^4)\}.$$

In the neighborhood of the identity of G, it is possible to introduce a coordinate system such that the coordinates of the one parameter subgroups take a simple form. Let $g_{\gamma}(t)\equiv g(t,\boldsymbol{\gamma})$ be the one parameter subgroup generated by $\boldsymbol{\gamma}$. Let $\{e_\alpha\}$ be a basis in $T_e(G)$: $\boldsymbol{\gamma}=\gamma^\alpha e_\alpha$. The **canonical (normal) coordinates** of $g(t,\boldsymbol{\gamma})=\exp(t\gamma^\alpha e_\alpha)$ are, by definition

canonical (normal) coordinates

$$g^\alpha(t,\boldsymbol{\gamma})=t\gamma^\alpha.$$

The canonical coordinates are defined with respect to a basis in $T_e(G)$.

We shall prove that, given an arbitrary system of coordinates in G, we can find a canonical system of coordinates, which is unique modulo a change of basis in $T_e(G)$. The proof rests on the following property

$$g(t, a\gamma) = g(at, \gamma)$$

which comes from $\exp(ta\gamma) = \exp(at\gamma)$.
Let the roman alphabet label the coordinates in an arbitrary system and the greek alphabet label them in the canonical system

$$(\exp t\gamma)^i = g^i(t, \gamma^\alpha) = g^i(1, t\gamma^\alpha) = g^i(1, g^\alpha)$$

The change of coordinates is admissible in a neighborhood of the origin if Dg^i/Dg^α is non singular at the origin,

$$v^i(\exp t\gamma) = \frac{\mathrm{d}}{\mathrm{d}t} g^i(1, t\gamma^\alpha) = \frac{\partial g^i(1, g^\alpha)}{\partial g^\beta}\gamma^\beta.$$

Set $t = 0$ $\qquad \gamma^i = \frac{\partial g^i}{\partial g^\alpha}(1, 0)\gamma^\alpha, \qquad \frac{\partial g^i}{\partial g^\alpha}(1, 0) = \delta^i_\alpha.$

The domain of validity of the transformation $g^i(1, g^\beta)$ determined by the range $(I \subset \mathbb{R}) \times (N(0) \subset T_e(G))$ on which the function $g(t, \gamma)$ is defined and differentiable, can actually be characterized simply by $N(0)$ without reference to I, because $g(1, \gamma) = g(a, a^{-1}\gamma)$ and with a proper choice of $N(0)$, a is always small enough to be in any given I. Thus the one parameter subgroups fill in a neighborhood of the identity if the group manifold is finite dimensional. This result can be extended to an infinite dimensional group manifold provided it is modeled on a Banach space (see p. 112). If the group manifold is modeled on an arbitrary topological vector space, this result is not necessarily true[1]; this is unfortunately the case for the group of coordinate transformations on a C^k manifold.

5. LIE GROUPS OF TRANSFORMATIONS; REALIZATION

We now resume the general study of a group $\{\sigma_g; g \in G\}$ of transformations (diffeomorphisms) of a manifold X.
Consider the mapping from G onto the group $\{\sigma_g;\ g \in G\}$ defined by $g \mapsto \sigma_g$; it is a homomorphism since

$$\sigma_{gh} = \sigma_g \circ \sigma_h \qquad (\text{or } \sigma_h \circ \sigma_g).$$

realization The mapping $g \mapsto \sigma_g$ is called a **realization** of G. A realization defines an action of G on X. The composition of the transformations $\{\sigma_g\}$ usually

[1][Freifeld, p. 538].

satisfies the law $\sigma_g \circ \sigma_h = \sigma_{gh}$ and the action of G on X is called a **left action**. If the composition law is $\sigma_g \circ \sigma_h = \sigma_{hg}$, the action of G on X is said to be a **right action**.

left action
right action

Examples of left actions:

1. Left translations, $L_g(h) = gh$.
2. "**Inverse**" **right translations**, $R_g^{-1}(h) = hg^{-1}$.

Indeed let $\bar{\sigma}$: $G \times G \to G$ by $\bar{\sigma}(g, h) = hg^{-1}$, then $\bar{\sigma}(g, \sigma(h, k)) = \bar{\sigma}(gh, k)$.

inverse right translation

3. The inner automorphisms, $L_g R_g^{-1}(h) = ghg^{-1}$.

If the mapping $g \mapsto \sigma_g$ is injective, the realization is said to be **faithful**.

faithful

When the transformations σ_g: $X \to X$ are linear transformations of a vector space X, the homomorphism: $G \to \{\sigma_g\}$ is called a **representation** of G.

representation

An element $\gamma \in T_e(G)$ defines the generator $\boldsymbol{v}^L$ [resp. the generator $\boldsymbol{v}^R$] of a one parameter subgroup of transformations $\{L_{g(t)}\}$ [resp. of transformations $\{R_{g(t)}\}$] of G,

$$\boldsymbol{v}^L(h) = R'_h(e)\gamma \text{ where } R_h g = gh.$$

γ also defines the Killing vector field $\boldsymbol{v}^K$ on X which generates the group of transformations $\{\sigma_{g(t)}\}$ of X, as follows:

$$\boldsymbol{v}^K(x) = \sigma'_x(e)\gamma.$$

The dimension of the space $\{\boldsymbol{v}^K\}$ of the Killing vector fields is equal to the rank r of the mapping $\sigma'_x(e)$; r is equal to or smaller than the dimension p of $T_e(G)$.

Theorem. The four following statements are equivalent:

1) $r = p$.
2) $\boldsymbol{v}^K = 0$ *iff* $\gamma = 0$.
3) *G acts effectively on X.*
4) *The space $\{\boldsymbol{v}^K\}$ of Killing vector fields on X is isomorphic to $T_e(G)$.*

We shall only prove that if G acts effectively on X, then $r = p$, the proof of the other statements being straightforward. The proof is *ab absurdo*: If $r < p$, then there exists a Killing vector field $\boldsymbol{v}^K$ on X which is identically zero, say the Killing vector field generated by $\{\sigma_{g(t)}\}$.
Let $x(t)$ be the curve on X generated by $\{\sigma_{g(t)}(x_0)\}$,

$$\boldsymbol{v}^K(x(t)) = \mathrm{d}\sigma_{g(t)}(x_0)/\mathrm{d}t$$

$$\boldsymbol{v}^K \equiv 0 \quad \Leftrightarrow \quad \sigma_{g(t)}(x_0) = \sigma_{g(0)}(x_0) = x_0, \qquad \forall x_0 \in X.$$

$v^K \equiv 0$ implies that G does not act effectively on X, contrary to the hypothesis. ∎

Exercise: Determine the structure constants of G from the Lie bracket in the space of Killing vector fields $\{v^K\}$ of a faithful realization of G on X.

Answer: Let $(v_{(\alpha)}(e))$ be a basis in T_eG. Let $v_{(\alpha)}(e) = \mathrm{d}g_{(\alpha)}(t)/\mathrm{d}t|_{t=0}$ define $v^K_{(\alpha)}$ as well as the generators $v^L_{(\alpha)}$ and $v^R_{(\alpha)}$ of the one parameter subgroups $\{L_{g(t)}\}$ and $\{R_{g(t)}\}$ of transformations of G. We shall prove that if the structure constants are defined in $\mathscr{G}$ by

$$[v^R_{(\alpha)}, v^R_{(\beta)}] = c^\gamma_{\alpha\beta} v^R_{(\gamma)}$$

then

$$[v^L_{(\alpha)}, v^L_{(\beta)}] = -c^\gamma_{\alpha\beta} v^L_{(\gamma)} \qquad \text{and} \qquad [v^K_{(\alpha)}, v^K_{(\beta)}] = -c^\gamma_{\alpha\beta} v^K_{(\gamma)}. \tag{1}$$

We shall also prove that, $v^{R^{-1}}_{(\alpha)}$ being the generator of $\{R^{-1}_{g(\alpha)(t)}\}$,

$$[v^{R^{-1}}_{(\alpha)}, v^{R^{-1}}_{(\beta)}] = -c^\gamma_{\alpha\beta} v^{R^{-1}}_{(\gamma)}.$$

Since we are interested in relating $\{v^K\}$ and $\{v^L\}$, rather than $\{v^K\}$ and T_eG, we shall obtain $v^K(x)$ by differentiating

$$\sigma(g, \sigma(h, x)) = \sigma(gh, x) \tag{2}$$

with respect to g, rather than differentiating $\sigma(g(t), x)$ with respect to $g(t)$. Thus

$$\partial\sigma^i(g, \sigma(h, x))/\partial g^\alpha = (\partial\sigma^i(gh, x)/\partial(gh)^\beta)\partial(gh)^\beta/\partial g^\alpha. \tag{3}$$

In the basis of the space of right invariant vector fields defined by the basis of T_eG where $v^\beta_{(\alpha)}(e) = \delta^\beta_\alpha$, the components of $v^L_{(\alpha)}$ are

$$v^{L\beta}_{(\alpha)}(h) = (R'_h(e)v_{(\alpha)}(e))^\beta = \partial(g_{(\alpha)}h)^\beta/\partial g^\alpha|_{g=e}.$$

Setting $g = e$ in equation (3) we obtain

$$v^{Ki}_{(\alpha)}(\sigma(h, x)) = (\partial\sigma^i(h, x)/\partial h^\beta)v^{L\beta}_{(\alpha)}(h). \tag{4}$$

To get the Lie bracket $[v^K_{(\gamma)}, v^K_{(\alpha)}]$, we differentiate (4) with respect to h^γ, set $h = e$, and antisymmetrize. We obtain

$$[v^K_{(\gamma)}, v^K_{(\alpha)}]^i(x) = v^{Ki}_{(\beta)}(x)[v^L_{(\gamma)}, v^L_{(\alpha)}]^\beta(e). \tag{5}$$

Consider now the generators of the "inverse" right translations, i.e., the generators of $\{R_g^{-1}, g \in G\}$ where $R_g^{-1}(h) = R_{g^{-1}}(h) = hg^{-1}$.
Let $\bar{\sigma}$: $G \times G \to G$ by $(g, h) \to hg^{-1}$, then $\bar{\sigma}(g, h) = R_g^{-1}(h)$.

We have again

$$\bar{\sigma}(g, \sigma(h, k)) = \bar{\sigma}(gh, k); \tag{6}$$

and the same calculation as before gives

$$[v^{R^{-1}}_{(\gamma)}, v^{R^{-1}}_{(\alpha)}]^{\delta}(k) = [v^{R^{-1}}_{(\beta)}]^{\delta}_{\beta}(k)[v^{L}_{(\gamma)}, v^{L}_{(\alpha)}]^{\beta}(e). \tag{7}$$

Finally $v^{R^{-1}}_{(\alpha)}(k) = \mathrm{d}R^{-1}_{g(t)}(k)/\mathrm{d}t|_{t=0} = \mathrm{d}\, L_k(g^{-1}(t))/\mathrm{d}t|_{t=0} = -v^{R}_{(\alpha)}(k)$, hence

$$[v^{R^{-1}}_{(\gamma)}, v^{R^{-1}}_{(\alpha)}] = -c^{\beta}_{\gamma\alpha} v^{R^{-1}}_{(\beta)}$$

and $[v^{L}_{(\gamma)}, v^{L}_{(\alpha)}]^{\beta}(e) = -c^{\beta}_{\gamma\alpha}$ which inserted in (4) gives the Killing vector field Lie bracket. ■

Lie algebra isomorphism. Let G act effectively on X. As a vector space $\{v^K\}$ is isomorphic to both the space $\{v^L\}$ of right invariant vector fields and the space $\{v^R\}$ of left invariant vector fields (Lie algebra $\mathscr{G}$ of G). The space $\{v^K\}$ together with the Lie bracket is a Lie algebra. Given two Lie algebras $\{v\}$ and $\{w\}$ such that $w = \phi v$, where ϕ is a vector space isomorphism, they are said to be **isomorphic** if $\phi[v_{(\alpha)}, v_{(\beta)}] = [\phi v_{(\alpha)}, \phi v_{(\beta)}] = [w_{(\alpha)}, w_{(\beta)}]$, for all $v_{(\alpha)} \in \{v\}$. isomorphic Lie algebras

Theorem. The Killing vector field Lie algebra on X defined by a left [right] effective action of G on X is isomorphic to the Lie algebra of the generators of left [right] actions of G on G.

Proof: See previous exercise. At $h = e$ equation (3) reads

$$v^{Ki}_{(\alpha)}(x) = \partial\sigma^i(h, x)/\partial h^{\beta}|_{h=e} v^{L\beta}_{(\alpha)}(e)$$

and equation (4) reads

$$[v^{K}_{(\gamma)}, v^{K}_{(\alpha)}]^i(x) = \partial\sigma^i(h, x)/\partial h^{\beta}|_{h=e}[v^{L}_{(\gamma)}, v^{L}_{(\alpha)}]^{\beta}(e) \tag{8}$$

which proves the isomorphism of $\{v^K\}$ and $\{v^L\}$. It is clear that this proof uses only the nature of the composition law (1). In particular the Lie algebra $\{v^K\}$ is isomorphic to the Lie algebra $\{v^L\}$ of the right invariant vector fields, and to the Lie algebra $\{v^{R^{-1}}\}$ of the left invariant vector fields generated by inverse right translations. ■

Remark: The isomorphism $\{v^K\} \to \{v^L\}$ can be called **canonical** in the sense defined by equation (8). There is also an isomorphism $\{v^K\} \to \{v^R\}$ but it is not canonical. See, for instance, the Lie brackets computed on p. 158 and p. 353. canonical isomorphism

Exercise: Given a matrix representation of G on $\mathbf{R}^n$,

$$\sigma_g = [D^i_j(g)]: \mathbf{R}^n \to \mathbf{R}^n \qquad \text{by} \qquad x \mapsto y = D(g)x,$$

show that, in the same basis as in the exercise on p. 164,

$$[v^K_{(\alpha)}, v^K_{(\beta)}]^j(x) = (D^j_{i,\beta}(e)D^i_{k,\alpha}(e) - D^j_{i,\alpha}(e)D^i_{k,\beta}(e))x^k$$
$$= -[D_{,\alpha}, D_{,\beta}]^j_k(e)x^k$$

where $D_{,\alpha}(e) = \partial D(g)/\partial g^\alpha|_{g=e} \in T_eG$ and where the Lie bracket in T_eG is defined by matrix multiplication, antisymmetrized.

Invariance under a group of transformations. We shall show that the necessary and sufficient condition for a tensor field on X to be invariant under a connected group G of transformation $\{\sigma_g\}$ is that it is invariant under its one parameter subgroups $\{\sigma_{g(t)}\}$: there exists a neighborhood $N(e) \subset G$ such that every point $g \in N(e)$ is on exactly one curve of a one parameter subgroup $g(t)$, because in the finite dimensional case the one parameter subgroups fill in a neighborhood of the identity (p. 162). Moreover, every element of G can be constructed by multiplication of elements in an arbitrary neighborhood $N(e)$. Hence a tensor is invariant under $\{\sigma_g; g \in G\}$ if it is invariant under $\{\sigma_g; g \in N(e)\}$, i.e. invariant under $\{\sigma_{g(t)}; g(t) \in N(e)\}$, therefore:

Theorem. The necessary and sufficient condition for a tensor field on a smooth manifold X to be invariant under a group of transformations is that its Lie derivatives with respect to the corresponding Killing vector fields vanish.

Exercise: Show that the Lie derivative $[v^R, v^L] = 0$ where v^R $[v^L]$ is the generator of the right [left] translations. (Hint: v^R is a left invariant vector field.)

6. ADJOINT REPRESENTATION. See Problem 4 (p. 178).

The adjoint representation is a representation of a group G on its Lie algebra $\mathcal{G}$. Since the Lie algebra $\mathcal{G}$ is isomorphic to $T_e(G)$ (p. 156) a mapping from $\mathcal{G}$ into itself can be defined by a mapping from $T_e(G)$ into itself. The mapping

$$L_g R_g^{-1}: h \mapsto ghg^{-1}$$

is an inner automorphism of G that defines a linear mapping $(L_g R_g^{-1})'(e)$ from $T_e(G)$ into itself. The mapping

$$\mathrm{Ad}(g): \mathcal{G} \to \mathcal{G}$$

is by definition the mapping corresponding to $(L_g R_g^{-1})'(e)$ in the isomorphism between $T_e(G)$ and $\mathscr{G}$: Let $\boldsymbol{\alpha}_e$, $\boldsymbol{\gamma}_e \in T_e(G)$; let $\boldsymbol{\alpha}$, $\boldsymbol{\gamma} \in \mathscr{G}$ be the left invariant vector fields generated by $\boldsymbol{\alpha}_e$ and $\boldsymbol{\gamma}_e$ respectively:

$$\boldsymbol{\alpha}_e = (L_g R_g^{-1})'(e)\boldsymbol{\gamma}_e \quad \Leftrightarrow \quad \boldsymbol{\alpha} = \mathrm{Ad}(g)\boldsymbol{\gamma}.$$

We shall often identify $T_e(G)$ and $\mathscr{G}$ and write

$$\mathrm{Ad}(g) = (L_g R_g^{-1})'(e) = L_g'(g^{-1}) R_g^{-1\prime}(e).$$

The mapping from G into the space $L(\mathscr{G}, \mathscr{G})$ of linear mappings from $\mathscr{G}$ into itself

$$\mathrm{Ad}\colon G \mapsto L(\mathscr{G}, \mathscr{G}) \quad \text{by} \quad \mathrm{Ad}\colon g \mapsto \mathrm{Ad}(g)$$

is called the **adjoint representation** of G on $\mathscr{G}$. adjoint representation of G on $\mathscr{G}$

The following abbreviated notation is also used

$$\mathrm{Ad}(g)\boldsymbol{\gamma} = g\boldsymbol{\gamma}g^{-1}, \qquad \boldsymbol{\gamma} \in \mathscr{G}.$$

This equation is literally correct if G is the linear group $\mathrm{GL}(n)$ or one of its subgroups (p. 172). Indeed L_g and R_g are then linear mappings and the derivative of a linear mapping can be identified with the mapping itself.

Exercise: Prove that the mapping Ad is a representation.

Remark: The center (p. 7) of G is the kernel of the homomorphism $g \mapsto L_g R_g^{-1}$, and is also the kernel of the adjoint representation provided G is connected. In a certain sense it measures the non-commutativity of G.

Remark: The adjoint representation is used in the definition of connections on principal fibre bundles (p. 361) where the action of the structural group on the typical fibre is a left translation.

The adjoint representation $\mathscr{A}d$ of $\mathscr{G}$ on $\mathscr{G}$ is the differential of the adjoint representation of G on $\mathscr{G}$ at the origin, modulo the isomorphism of $T_e(G)$ onto $\mathscr{G}$ adjoint representation of $\mathscr{G}$ on $\mathscr{G}$

$$\mathscr{A}d = \mathrm{Ad}'(e).$$

We shall show that

$$\mathscr{A}d\colon \mathscr{G} \to L(\mathscr{G}, \mathscr{G}) \quad \text{by} \quad \mathscr{A}d(\boldsymbol{\delta})(\cdot) = [\boldsymbol{\delta}, \cdot].$$

Indeed in the abbreviated notation $\mathrm{Ad}(g(t))\boldsymbol{\gamma} = g(t)\boldsymbol{\gamma}g(t)^{-1}$,

$$\frac{\mathrm{d}}{\mathrm{d}t}\mathrm{Ad}(g(t))(\boldsymbol{\gamma})\big|_{t=0} = \mathrm{Ad}'(e)\boldsymbol{\delta}(\boldsymbol{\gamma}) = \mathscr{A}d(\boldsymbol{\delta})(\boldsymbol{\gamma})$$

where $\boldsymbol{\delta} = \mathrm{d}g(t)/\mathrm{d}t|_{t=0}$ modulo the isomorphism of $T_e(G)$ onto $\mathscr{G}$. On the other hand,

$$\frac{\mathrm{d}}{\mathrm{d}t} g(t)\gamma g(t)^{-1}|_{t=0} = [\boldsymbol{\delta}, \boldsymbol{\gamma}]. \qquad \blacksquare$$

adjoint map

The mapping $\mathscr{A}d(\boldsymbol{\delta})$: $\mathscr{G} \to \mathscr{G}$ is called an **adjoint map**.

Let $\{\boldsymbol{\iota}_\alpha\}$ be a basis of $\mathscr{G}$, $c^\alpha_{\gamma\beta}$ the structure constants in this basis, and $\boldsymbol{\gamma} = \gamma^\alpha \boldsymbol{\iota}_\alpha$, $\boldsymbol{\delta} = \delta^\alpha \boldsymbol{\iota}_\alpha$. The coordinate expression of $\mathscr{A}d(\boldsymbol{\delta})(\boldsymbol{\gamma}) = [\boldsymbol{\delta}, \boldsymbol{\gamma}]$ gives

$$\mathscr{A}d(\boldsymbol{\delta})^\alpha_\beta = \delta^\gamma c^\alpha_{\gamma\beta}, \qquad \text{also written} \qquad \mathscr{A}d(\boldsymbol{\delta}) = \delta \cdot c.$$

Here two of the three indices of the structure constants play the role of matrix indices, but not any two indices.

Since Ad is an homomorphism from G into $L(\mathscr{G}, \mathscr{G})$, it maps a one parameter subgroup $\{g(t)\}$ of G into a one parameter subgroup $\{\mathrm{Ad}(g(t))\}$ of $L(\mathscr{G}, \mathscr{G})$. Hence there exists a linear transformation A: $\mathscr{G} \to \mathscr{G}$ such that

$$\mathrm{Ad}(g(t)) = \exp tA.$$

A is readily obtained from $\mathrm{d}(\mathrm{Ad}(g(t)))/\mathrm{d}t|_{t=0}$ and found equal to $\mathscr{A}d(\boldsymbol{\delta})$ where $\boldsymbol{\delta}$ is the generator of $\{g(t)\}$. The equation

$$\mathrm{Ad}(\exp \boldsymbol{\delta}) = \exp(\mathscr{A}d(\boldsymbol{\delta}))$$

is also written with the precaution given above

$$\mathrm{Ad}(\exp \boldsymbol{\delta}) = \exp(\delta \cdot c).$$

Killing form

The **Killing form** on $\mathscr{G}$ is the bilinear form $B(\boldsymbol{\gamma}, \boldsymbol{\delta}) = \mathrm{tr}\, \mathscr{A}d(\boldsymbol{\gamma})\mathscr{A}d(\boldsymbol{\delta})$.

7. CANONICAL FORM, MAURER-CARTAN FORM (see Problem Vbis 3)

canonical form
Maurer-Cartan form

The **canonical** or **Maurer-Cartan form** $\boldsymbol{\omega}$ on a Lie group G is a one-form with values in the Lie algebra $\mathscr{G}$ of G defined through the relation

$$\boldsymbol{\omega}(\boldsymbol{v}_g) = \boldsymbol{\gamma} \text{ where } \boldsymbol{\gamma} = L_g^{-1\prime} \boldsymbol{v}_g \in T_e(G).$$

The Maurer-Cartan form is often defined by the equation $\boldsymbol{\omega}(\boldsymbol{v}_g) = g^{-1}\boldsymbol{v}_g$.

This abbreviated notation is literally correct if G is the linear group $GL(n)$ or one of its subgroups since the derivative of a linear mapping can be identified with the mapping itself.

Theorem. The canonical form $\boldsymbol{\omega}$ *is left invariant; its reciprocal image under a right translation satisfies the equation*

$$R_g^*\boldsymbol{\omega} = \mathrm{Ad}(g^{-1}) \circ \boldsymbol{\omega}.$$

Proof:

$$\begin{aligned}(L_h^*\boldsymbol{\omega})(\boldsymbol{v}_{h^{-1}g}) &= \boldsymbol{\omega}(L'_h\boldsymbol{v}_{h^{-1}g}) \quad \text{by the definition of } L_h^* \\ &= L_g^{-1\prime}L'_h\boldsymbol{v}_{h^{-1}g} \quad \text{by the definition of } \boldsymbol{\omega} \\ &= L_{h^{-1}g}^{-1\prime}\boldsymbol{v}_{h^{-1}g} = \boldsymbol{\omega}(\boldsymbol{v}_{h^{-1}g}).\end{aligned}$$

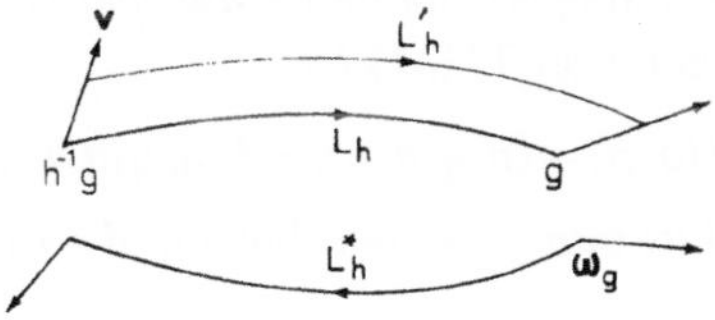

Similarly

$$\begin{aligned}(R_h^*\boldsymbol{\omega})(\boldsymbol{v}_{gh^{-1}}) &= \boldsymbol{\omega}(R'_h\boldsymbol{v}_{gh^{-1}}) \quad \text{by the definition of pull-backs} \\ &= L_g^{-1\prime}R'_h\boldsymbol{v}_{gh^{-1}} \quad \text{by the definition of } \boldsymbol{\omega} \\ &= L_h^{-1\prime}L_{gh^{-1}}^{-1\prime}R'_h\boldsymbol{v}_{gh^{-1}} \\ &= L_h^{-1\prime}R'_hL_{gh^{-1}}^{-1\prime}\,\boldsymbol{v}_{gh^{-1}} \quad \text{since } R' \text{ and } L' \text{ always commute} \\ &= L_h^{-1\prime}R'_h\boldsymbol{\omega}(\boldsymbol{v}_{gh^{-1}}) \quad \text{by the definition of } \boldsymbol{\omega}.\end{aligned}$$

■

PROBLEMS AND EXERCISES

PROBLEM 1. CHANGE OF COORDINATES ON A FIBRE BUNDLE, CONFIGURATION SPACE, PHASE SPACE, LAGRANGIAN, CONJUGATE VARIABLES, HAMILTONIAN.

Let M be the configuration manifold of a dynamical system S; i.e. each point $q(t) \in M$ represents one and only one state of the system at time t. The dimension n of M is equal to the number of degrees of freedom of S.

The lagrangian L of S is a function on $T(M)$, where $T(M)$ is the tangent bundle of M.[1] *The notations mostly used in classical texts do not make a distinction between q as a point of M and q as a trajectory of the dynamical system, mapping an interval of $\mathbb{R}$ into M by $t \mapsto q(t)$. One then writes indifferently $L(q^i, \dot{q}^i)$ for the value of L at a point with natural coordinates $(q^i, \dot{q}^i)$ in $T(M)$, and also for the value at $t \in \mathbb{R}$ of the function $L \circ \tilde{q}$, where $\tilde{q}$ is the mapping from $\mathbb{R}$ into $T(M)$ associated to a smooth trajectory $t \mapsto q(t)$ by, in natural coordinates on $T(M)$,*

$$\tilde{q}\colon t \mapsto \left(q^i(t), \frac{\mathrm{d}q^i(t)}{\mathrm{d}t}\right).$$

a) *Prove that, $p_i \equiv \partial L/\partial \dot{q}^i$ and $\partial L/\partial q^i - (\mathrm{d}/\mathrm{d}t)(\partial L/\partial \dot{q}^i)$ transform under a change of natural coordinates on $T(M)$ as the components of a covariant vector on M at $q(t)$.*

b) *Let $T^*(M)$ be the cotangent bundle (phase space); when can one define the hamiltonian function*[1] *on $T^*(M)$ by*

$$H(q, p) = \langle p, \dot{q}(q, p)\rangle - L(q, \dot{q}(q, p))?$$

The nature of H and its transformation law are discussed in Problem IV 4, p. 272.

Answer: a) To simplify the notation without introducing a possible confusion between q and $q(t)$, we shall denote by (x^i, v_x^i) the natural coordinates in $T(M)$ corresponding to a chart in M, and set $X = (X^\alpha) = (x^i, v_x^i)$. Let $Y = (Y^\alpha) = (y^i, u_y^i)$ be the natural coordinates in $T(M)$ corresponding to another chart in M. The change of fiber coordinates

$$F\colon \mathbb{R}_X^{2n} \to \mathbb{R}_Y^{2n} \quad \text{by} \quad X \mapsto Y$$

is given by the bundle morphism

$$\begin{array}{ccc} (x^i, v_x^i) & \xrightarrow{F} & (y^i, u_y^i) \\ \pi \big\downarrow & & \big\downarrow \pi \\ (x^i) & \xrightarrow{f} & (y^i) \end{array} \quad \text{with} \quad F = (f, f'), \text{ i.e. } u_y^i = \frac{\partial y^i}{\partial x^j} v_x^j.$$

Thus the derivative mapping F' is given by the matrix

$$\left[\frac{\partial Y^\alpha}{\partial X^\beta}\right] = \begin{bmatrix} a_j^i & 0 \\ b_j^i & a_j^i \end{bmatrix} \quad \text{with} \quad a_j^i = \frac{\partial y^i}{\partial x^j} \quad \text{and} \quad b_j^i = \frac{\partial^2 y^i}{\partial x^j \partial x^k} v_x^k.$$

[1]More generally, one may allow L to be a map $T(M) \times \mathbf{R} \to \mathbf{R}$, in which case the lagrangian is said to be "explicitly time dependent"; slight modifications in the discussion would then be necessary, e.g. $\tilde{q}\colon t \mapsto (q^i(t), \mathrm{d}q^i(t)/\mathrm{d}t, t)$. The hamiltonian would become explicitly time dependent also, $H\colon T^*(M) \times \mathbf{R} \to \mathbf{R}$.

Denote by $\bar{L}: Y \mapsto \bar{L}(Y)$ the expression of L in the coordinates (Y^α); the expression $\tilde{L}$ of L in the coordinates (X^α) is

$$\tilde{L} = \bar{L} \circ F.$$

The equation $\partial\tilde{L}/\partial X^\alpha = (\partial\bar{L}/\partial Y^\beta)(\partial Y^\beta/\partial X^\alpha)$ gives the required transformation laws:

$$\frac{\partial\tilde{L}}{\partial x^i} = \frac{\partial\bar{L}}{\partial y^j}\frac{\partial y^j}{\partial x^i} + \frac{\partial\bar{L}}{\partial u_y^j}\frac{\partial^2 y^j}{\partial x^i \partial x^k} v_x^k;$$

$(\partial\tilde{L}/\partial x^i)$ do not obey the law of transformation of the components of a covariant vector on M, but

$$\frac{\partial\tilde{L}}{\partial v_x^i} = \frac{\partial\bar{L}}{\partial u_y^j}\frac{\partial y^j}{\partial x^i};$$

$(\partial\tilde{L}/\partial v_x^i)$ transform as the components of a covariant vector on M. They are called the conjugate variables to x^i.

Finally
$$\frac{\mathrm{d}}{\mathrm{d}t}\frac{\partial\tilde{L}}{\partial v_x^i} = \frac{\mathrm{d}}{\mathrm{d}t}\left(\frac{\partial\bar{L}}{\partial u_y^j}\right)\frac{\partial y^j}{\partial x^i} + \frac{\partial\bar{L}}{\partial u_y^j}\frac{\partial^2 y^j}{\partial x^i \partial x^k} v_x^k;$$

hence $\left(\frac{\partial\tilde{L}}{\partial x^i} - \frac{\mathrm{d}}{\mathrm{d}t}\frac{\partial\tilde{L}}{\partial v_x^i}\right)$ transform as the components of a covariant vector on M.

Remark: The Euler-Lagrange equation

$$\frac{\partial L}{\partial x^i} - \frac{\mathrm{d}}{\mathrm{d}t}\frac{\partial L}{\partial v_x^i} = 0$$

is invariant under a change of *natural* fiber coordinates. It is not invariant under a general change of fiber coordinates:

$$(x^i) \mapsto (y^i(x))$$
$$(v_x^i) \mapsto (u_y^i) \text{ where } u_y^i = a_k^i(x) v_x^i.$$

b) Domain of H:
Consider the mapping $P: T(M) \to T^*(M)$, given in natural coordinates by

$$(x^i, v_x^i) \mapsto (x^i, \partial\tilde{L}/\partial v_x^i).$$

If the mapping has an inverse P^{-1} on a domain $P(D) \subset T^*(M)$ then H is defined on $P(D)$. A necessary condition for P to be invertible from

$P(D)$ onto D is that the derivative mapping be everywhere of rank $2n$ on D. The derivative mapping is

$$\begin{pmatrix} \delta^i_j & 0 \\ \dfrac{\partial^2 \tilde{L}}{\partial x^j \partial v^i_x} & \dfrac{\partial^2 \tilde{L}}{\partial v^j_x \partial v^i_x} \end{pmatrix}.$$

If the rank of $\partial^2\tilde{L}/\partial v^i_x \partial v^j_x$ (in classical notations $\partial^2 L/\partial \dot{q}^j \partial \dot{q}^i$) is n, the rank of the derivative mapping is $2n$.
Sufficient conditions for the existence of H can be deduced from global inverse function theorems (cf. p. 92).

PROBLEM 2. LIE ALGEBRAS $\mathscr{G}$ OF SUBGROUPS OF THE LINEAR GROUP

The space of linear bijective mappings of $\mathbb{R}^n$ *onto* $\mathbb{R}^n$ *is given a group structure by the composite mapping. This group is isomorphic to the group* GL(n, $\mathbb{R}$) *of non singular* (det $M \neq 0$) $n \times n$ *real matrices* M. *There is a bijection between* GL(n, $\mathbb{R}$) *and an open set of* $\mathbb{R}^{n^2}$: $\mathbb{R}^{n^2}$ – set of points where det $M = 0$. *Hence* GL(n, $\mathbb{R}$) *can be given a differentiable structure. Compatibility of the group structure and the differentiable structure makes* GL(n, $\mathbb{R}$) *a Lie group. Its subgroups are also Lie groups. Similarly the space of regular mappings of* $\mathbb{C}^n$ *into* $\mathbb{C}^n$ *is given a group structure by the composite mapping. This group is isomorphic to the Lie group* GL(n, $\mathbb{C}$) *of non singular* $n \times n$ *complex matrices.*
We list in the table the definitions of some subgroups of GL(n, $\mathbb{C}$) [*resp.* GL(n, $\mathbb{R}$)]. *Prove the properties of their Lie algebras listed in the last column.*

Answer: Notation:
$\bar{A}$: complex conjugate; A^{T}: transpose; A^{+}: complex conjugate transpose.

Any matrix M in the neighborhood of the unit element 1 of a group G can be obtained by exponentiating a matrix A in the Lie algebra $\mathscr{G}$ of G. We can obtain the properties of A from the one parameter subgroup $\{M(t)\}$ defined by

$$M(t) = \exp tA = 1 + tA + \frac{t^2}{2!}A^2 + \cdots$$

then

$$A = \mathrm{d}M(t)/\mathrm{d}t|_{t=0}.$$

Symbol	Name of the group	Properties of the matrices	Properties of the mappings	Dimensions of group space	The Lie algebra is the algebra of the following matrices
$GL(n, \mathbb{C})$	Full linear (automorphisms of $\mathbb{C}^n$ onto $\mathbb{C}^n$)	non-singular	regular linear	$2n^2$	any matrix
$GL(n, \mathbb{R})$	Real linear (automorphisms of $\mathbb{R}^n$ onto $\mathbb{R}^n$)	non-singular, real	same as $GL(n, \mathbb{C})$	n^2	real
$SL(n, \mathbb{C})$	unimodular	unit determinant	volume and orientation preserving	$2n^2 - 2$	traceless
$U(n)$	unitary	$UU^+ = 1$	leaves scalar product invariant	n^2	antihermitian
$SL(n, \mathbb{R})$	unimodular	unit determinant	same as $SL(n, \mathbb{C})$	$n^2 - 1$	real, traceless
$SU(n)$	unitary, unimodular	$UU^+ = 1$, det $U = 1$	Same as $U(n)$ and orientation preserving	$n^2 - 1$	antihermitian, traceless
$O(n)$	orthogonal	O real, $OO^T = 1$	Same as $U(n)$	$\frac{1}{2}n(n-1)$	real, antisymmetric
$SO(n)$	rotation (proper orthogonal)	O real, $OO^T = 1$, det $O = 1$	Same as $SU(n)$	$\frac{1}{2}n(n-1)$	real, antisymmetric
$L(4)$	Lorentz	see p. 290	Leaves $c^2t^2 - x^2 - y^2 - z^2$ invariant	6	real matrices $(V^\alpha{}_\beta)$ such that $(\eta_{\alpha\sigma}V^\sigma{}_\beta)$ is antisymmetric
Spin$(2n)$	Spin group	see p. 67		$n(2n-1)$	see answer p. 177

We shall first prove the very useful formula[1]

$$\det \exp A = \exp \operatorname{tr} A.$$

Indeed, the trace and the determinant being invariant under similarity transformations, we can prove the formula for a matrix A which has been made triangular by a similarity transformation; then

$$\exp \operatorname{tr} A = \prod_{i=1}^{n} \exp A_{ii} = \prod_{i=1}^{n} (\exp A)_{ii} = \det \exp A. \quad \blacksquare$$

We shall give another proof valid only when A is non-singular but using properties needed elsewhere:

$$\delta_{ik} \det A = A_{ij} \times (\text{cofactor of } A_{kj}) \text{ and } \delta_{ik} = A_{ij}(A^{-1})_{jk},$$

then

$$\mathrm{d} \det A = \frac{\partial \det A}{\partial A_{ij}} \mathrm{d}A_{ij} = \det A (A^{-1})_{ji}\, \mathrm{d}A_{ij} = \det A \operatorname{tr}(A^{-1}\, \mathrm{d}A),$$

$$\mathrm{d} \log \det A = \operatorname{tr}(\mathrm{d} \log A) = \mathrm{d} \operatorname{tr} \log A,$$

$$\det A = \exp(\operatorname{tr} \log A). \quad \blacksquare$$

The following expansion is also needed elsewhere

$$\det(\mathbb{1} + K) = 1 + \operatorname{tr} K - \tfrac{1}{2} \operatorname{tr}(K^2) + \tfrac{1}{2}(\operatorname{tr} K)^2 + \cdots.$$

1) Lie algebra of $\mathrm{GL}(n, \mathbb{C})$

In the neighborhood of the unit element, any non singular matrix M is of the form $\mathbf{1} + tA$ where A is an arbitrary matrix. Hence the Lie algebra of the group of **automorphisms** on $\mathbb{C}^n$ (isomorphisms of $\mathbb{C}^n$ onto itself) is the algebra of **endomorphisms** on $\mathbb{C}^n$ (homomorphisms of $\mathbb{C}^n$ into itself).

automorphism
endomorphism

2) Lie algebra of $\mathrm{SL}(n, \mathbb{C})$.

$$1 = \det M = \det \exp A = \exp \operatorname{tr} A \Leftrightarrow \operatorname{tr} A = 0.$$

3) Lie algebra of $U(n)$.

$$\exp(tA) \exp(tA)^{+} = 1 \Leftrightarrow A = -A^{+}.$$

An antihermitian matrix A is often written $A = \mathrm{i}H$ where H is a hermitian matrix.

Note that $UU^{+} = 1 \Leftrightarrow \det U \det U^{*} = 1$ but not necessarily $\det U = 1$.

[1]It is sometimes possible to define both sides of this equation when A is an operator on a Banach space, but they are not necessarily equal. See [Nelson, Sheeks] for a case where they are not equal.

Example: The group of rotations in $\mathbb{R}^2$ is isomorphic to the unitary group $U(1)$:
Let $z = x + iy$, let $z' = \exp(i\theta)z$, then

$$x' = x \cos\theta - y \sin\theta$$

$$y' = x \sin\theta + y \cos\theta.$$

Note that

$$\begin{pmatrix} \cos\theta & -\sin\theta \\ \sin\theta & \cos\theta \end{pmatrix}\begin{pmatrix} x \\ y \end{pmatrix} = \begin{pmatrix} x' \\ y' \end{pmatrix} \text{ and } (x, y)\begin{pmatrix} \cos\theta & \sin\theta \\ -\sin\theta & \cos\theta \end{pmatrix} = (x', y').$$

One may choose to represent transformations in $\mathbb{R}^n$ by matrices operating on the left or on the right.
The orientation of rotations will be chosen to be the usual orientation: a rotation in $\mathbb{R}^2$ of angle $+\theta$ corresponds to $z \mapsto \exp(i\theta)z$.

4) Lie algebra of O(n). See Problem III 5, p. 181.

Example: The Lie algebra of O(2) is the algebra of matrices $\begin{pmatrix} 0 & -\theta \\ \theta & 0 \end{pmatrix}$.

Check that $\quad \exp\begin{pmatrix} 0 & -\theta \\ \theta & 0 \end{pmatrix} = \begin{pmatrix} \cos\theta & -\sin\theta \\ \sin\theta & \cos\theta \end{pmatrix}$.

5) Lie algebra of $L(4)$.
With the notation of Chapter V, p. 290, $L(4)$ is the group of all linear transformations L: $(e_\beta) \mapsto (a^\alpha{}_\beta e_\alpha)$ which leave $\eta_{\alpha\beta}v^\alpha v^\beta$ invariant, that is

$$a^\alpha{}_\beta \eta_{\alpha\mu} a^\mu{}_\nu = \eta_{\beta\nu}.$$

Consider a curve on $L(4)$, through the identity, mapping $I \subset \mathbb{R}$ into $L(4)$ by $t \mapsto (a^\alpha{}_\beta(t))$, with $a^\alpha{}_\beta(0) = \delta^\alpha{}_\beta$.

Differentiation of the relation

$$a^\mu{}_\beta(t)\eta_{\mu\lambda}a^\lambda{}_\sigma(t) = \eta_{\beta\sigma}$$

gives upon evaluation at $t = 0$, where $a^\mu{}_\beta(0) = \delta^\mu{}_\beta$,

$$\left.\frac{da^\mu{}_\beta}{dt}\right|_{t=0} \eta_{\mu\sigma} + \eta_{\beta\lambda} \left.\frac{da^\lambda{}_\sigma}{dt}\right|_{t=0} = 0.$$

The Lie algebra of $L(4)$ is the algebra of real matrices $(V^\alpha{}_\beta)$ such that $(\eta_{\sigma\alpha} V^\alpha{}_\beta)$ is antisymmetric.

6) Lie algebra of Spin(4). See Problem I 1, p. 64.
Let $V^4_{(1)}$ be a 4 dimensional vector space over $\mathbb{R}$ with inner product $(v|w)$ and basis (e_α), $\alpha = 0, 1, 2, 3$,

$$(e_\alpha|e_\beta) = 0 \quad \text{for } \alpha \neq \beta$$

$$(e_i|e_i) = -1 \quad \text{for } i = 1, 2, 3; \quad (e_0|e_0) = 1.$$

Consider a product vw such that

$$vw + wv = 2(v|w).$$

Let $C(V^4_{(1)})$ be the corresponding Clifford algebra. A possible basis for $C(V^4_{(1)})$ is the basis generated by the Dirac matrices

$$\gamma_0 = \begin{bmatrix} 1 & 0 & 0 & 0 \\ 0 & 1 & 0 & 0 \\ 0 & 0 & -1 & 0 \\ 0 & 0 & 0 & -1 \end{bmatrix}, \qquad \gamma_1 = \begin{bmatrix} 0 & 0 & 0 & 1 \\ 0 & 0 & 1 & 0 \\ 0 & -1 & 0 & 0 \\ -1 & 0 & 0 & 0 \end{bmatrix},$$

$$\gamma_2 = \begin{bmatrix} 0 & 0 & 0 & -\mathrm{i} \\ 0 & 0 & \mathrm{i} & 0 \\ 0 & \mathrm{i} & 0 & 0 \\ -\mathrm{i} & 0 & 0 & 0 \end{bmatrix}, \qquad \gamma_3 = \begin{bmatrix} 0 & 0 & 1 & 0 \\ 0 & 0 & 0 & -1 \\ -1 & 0 & 0 & 0 \\ 0 & 1 & 0 & 0 \end{bmatrix},$$

i.e. the basis $(1, \gamma_\alpha, \gamma_\alpha\gamma_\beta(\alpha < \beta), \ldots, \gamma_0\gamma_1\gamma_2\gamma_3)$.

Let C_1 be the linear subspace of $C(V^4_{(1)})$ spanned by $\{\gamma_\alpha\}$. The group Spin (4) is defined to be the group of all elements $\Lambda \in C(V^4_{(1)})$ such that

$$\Lambda v \Lambda^{-1} \in C_1 \quad \text{for every } v \in C_1$$

$$\Lambda\tilde{\Lambda} = \pm 1.$$

We have shown in Problem I 1 that the mapping $\mathscr{H}$: Spin(4)$\to L(4)$ by $\Lambda \to [a^\beta{}_\alpha]$, where $\Lambda e_\alpha \Lambda^{-1} = a^\beta{}_\alpha e_\beta$, is a 2–1 homomorphism. This mapping induces a mapping $\mathscr{H}'$: $Spin(4) \to \mathscr{L}(4)$ between the respective Lie algebras which we shall use to determine the Lie algebra of Spin (4). Let $\Lambda(t)$ be a curve on Spin(4) such that $\Lambda(0) = 1$. The image under $\mathscr{H}$ of this curve is a curve $t \mapsto [a^\beta{}_\alpha(t)]$ on $L(4)$ such that $a^\beta{}_\alpha(0) = \delta^\beta{}_\alpha$ and

$$\Lambda(t) e_\alpha \Lambda^{-1}(t) = a^\beta{}_\alpha(t) e_\beta.$$

Now since $\Lambda(t)\Lambda^{-1}(t) = 1$

$$\frac{\mathrm{d}\Lambda(t)}{\mathrm{d}t}\Lambda^{-1}(t) = -\Lambda(t)\frac{\mathrm{d}\Lambda^{-1}(t)}{\mathrm{d}t}.$$

Therefore

$$\frac{\mathrm{d}\Lambda(t)}{\mathrm{d}t}\Lambda^{-1}(t)a^{\beta}{}_{\alpha}(t)\boldsymbol{e}_{\beta} - a^{\beta}{}_{\alpha}(t)\boldsymbol{e}_{\beta}\frac{\mathrm{d}\Lambda(t)}{\mathrm{d}t}\Lambda^{-1}(t) = \frac{\mathrm{d}a^{\beta}{}_{\alpha}(t)}{\mathrm{d}t}\boldsymbol{e}_{\beta}.$$

For $t = 0$ this becomes

$$\left.\frac{\mathrm{d}\Lambda}{\mathrm{d}t}\right|_{t=0}\boldsymbol{e}_{\alpha} - \boldsymbol{e}_{\alpha}\left.\frac{\mathrm{d}\Lambda}{\mathrm{d}t}\right|_{t=0} = \left.\frac{\mathrm{d}a^{\beta}{}_{\alpha}}{\mathrm{d}t}\right|_{t=0}\boldsymbol{e}_{\beta}.$$

We have thus shown that for $\boldsymbol{\lambda} \in Spin(4)$, $\mathscr{H}'(\boldsymbol{\lambda}) \equiv [V^{\beta}{}_{\alpha}]$ is determined by

$$\boldsymbol{\lambda}\boldsymbol{e}_{\alpha} - \boldsymbol{e}_{\alpha}\boldsymbol{\lambda} = V^{\beta}{}_{\alpha}\boldsymbol{e}_{\beta}.$$

A particular solution of this equation is

$$\boldsymbol{\lambda} = -\tfrac{1}{4}V^{\alpha\beta}\boldsymbol{e}_{\alpha}\boldsymbol{e}_{\beta}, \qquad V^{\alpha\beta} = V^{\alpha}{}_{\sigma}\eta^{\sigma\beta}.$$

This can be shown by repeated application of the relation

$$\boldsymbol{e}_{\alpha}\boldsymbol{e}_{\beta} + \boldsymbol{e}_{\beta}\boldsymbol{e}_{\alpha} = 2\eta_{\alpha\beta}.$$

The general solution is $\boldsymbol{\lambda} = -\frac{1}{4}V^{\alpha\beta}\boldsymbol{e}_{\alpha}\boldsymbol{e}_{\beta} + b$, where $b \in \mathbb{R}$. However $b = 0$ due to the requirement $(\exp\boldsymbol{\lambda})(\widetilde{\exp\boldsymbol{\lambda}}) = \pm 1$.
The Lie algebra of Spin(4) is thus generated by the six linearly independent elements $\boldsymbol{e}_{\alpha}\boldsymbol{e}_{\beta} \in C(V^{4}_{(1)})$ with $\alpha < \beta$.
Moreover $\mathscr{H}'$ is an isomorphism of $Spin(4)$ onto $\mathscr{L}(4)$.

PROBLEM 3. THE STRAIN TENSOR; LIE DERIVATIVE

Consider a deformable body S in $\mathbb{R}^3$; $S \subset \mathbb{R}^3$. A displacement with or without deformation of S is a diffeomorphism of $\mathbb{R}^3$ defined in a neighborhood of S. All such diffeomorphisms form a local group generated by the so called displacement vector field $\boldsymbol{v}$. The strain tensor is, by definition, the Lie derivative with respect to $\boldsymbol{v}$ of the metric tensor $\boldsymbol{g}$ in $\mathbb{R}^3$. The metric tensor gives the distance between two points:

$$\boldsymbol{g} = g_{ik}\,\mathbf{d}x^i\,\mathbf{d}x^k = \mathrm{d}s^2.$$

The strain tensor measures the variation of the distance between two points under a displacement generated by $\boldsymbol{v}$. If the distance between two points does not change during the displacement, the displacement is a rigid body motion.

$$\mathscr{L}_{\boldsymbol{v}}\boldsymbol{g} = 0 \quad \textit{for rigid body motion.}$$

Check that the following displacement defined in the euclidean space $\mathbb{R}^3$

is a rigid body motion:

$$x^i \mapsto x^i + \epsilon^i + \epsilon^i_j x^j, \qquad \epsilon^i \text{ and } \epsilon^i_j \text{ do not depend on } x,\ \epsilon_{ij} = -\epsilon_{ji}.$$

Answer: The coordinates of the displacement vector field v are: $v_i = \epsilon_i + \epsilon_{ij}x^j$; in the euclidean space $\mathbb{R}^3$ the metric $g_{ik} = \delta_{ik}$, hence

$$(\mathcal{L}_v g)_{ik} = \frac{\partial v_i}{\partial x^k} + \frac{\partial v_k}{\partial x^i} = \epsilon_{ik} + \epsilon_{ki} = 0.$$

In fluid dynamics, the rate of strain tensor is the Lie derivative of the metric with respect to the velocity vector field of the fluid.

Reference: Cf. for instance L.D. Landau and E.M. Lifshitz. *Theory of Elasticity*. In this book the strain tensor is defined as $\frac{1}{2}\mathcal{L}_v g$.

PROBLEM 4. EXPONENTIAL MAP; TAYLOR EXPANSION; ADJOINT MAP; LEFT AND RIGHT DIFFERENTIALS; HAAR MEASURE

a) *Let G be a Lie group with elements e, g, $h(t)\dots$ and $\mathcal{G}$ be its Lie algebra with elements $\boldsymbol{\alpha}$, $\boldsymbol{\beta}$, $\boldsymbol{\gamma}\dots$. Given* $\exp \boldsymbol{\gamma} = g$ *and* $\exp(\boldsymbol{\gamma} + \boldsymbol{\alpha}t) = gh(t)$ *find the generator $\boldsymbol{\delta}$ of the one parameter subgroup* $\{h(t)\}$. Note that the answer is not trivial because one cannot readily compute

$$\boldsymbol{\delta} = \left.\frac{\mathrm{d}h(t)}{\mathrm{d}t}\right|_{t=0} = \frac{\mathrm{d}}{\mathrm{d}t} g^{-1} \exp(\boldsymbol{\gamma} + \boldsymbol{\alpha}t)|_{t=0}$$

since $\boldsymbol{\alpha}$ and $\boldsymbol{\gamma}$ do not commute in general.

Answer: Let f be a C^∞ function on G. We shall compute $\boldsymbol{\delta}$ such that

$$\boldsymbol{\delta}(f \circ L_g) = \frac{\mathrm{d}}{\mathrm{d}t} f(\exp(\boldsymbol{\gamma} + \boldsymbol{\alpha}t))|_{t=0}.$$

The Taylor expansion (p. 160) of $f(\exp(\boldsymbol{\gamma} + \boldsymbol{\alpha}t))$ gives

$$f(\exp(\boldsymbol{\gamma} + \boldsymbol{\alpha}t)) = \sum_{n=0}^{\infty} \frac{1}{n!}(v_\gamma + tv_\alpha)^n f(e)$$

where v_γ and v_α are the left invariant vector fields generated by $\boldsymbol{\gamma}$ and $\boldsymbol{\alpha}$, respectively. After some careful rearrangements[1] we get

$$\boldsymbol{\delta}(f \circ L_g) = \sum_{k=0}^{\infty} \frac{1}{(k+1)!}(\mathcal{A}d(-\boldsymbol{\gamma}))^k \boldsymbol{\alpha}(f \circ L_g)$$

[1] S. Helgason, Lie Groups and Symmetric Spaces, in *Battelle Rencontres 1967 Lectures in Mathematics and Physics*, eds C. DeWitt and J.A. Wheeler (W.A. Benjamin Inc., 1968) p. 20.

which when formally summed gives

$$\boldsymbol{\delta} = \frac{\mathbb{1} - \exp(-\mathcal{A}d(\boldsymbol{\gamma}))}{\mathcal{A}d(\boldsymbol{\gamma})}\boldsymbol{\alpha}.$$

If $\boldsymbol{\alpha}$ and $\boldsymbol{\gamma}$ commute we can readily check that $\boldsymbol{\delta} = \boldsymbol{\alpha}$.

b) *Show that* $\boldsymbol{\delta} = L'_g(e)^{-1}\exp'(\boldsymbol{\gamma})\boldsymbol{\alpha}$, *with* $\boldsymbol{\alpha}$, $\boldsymbol{\gamma}$, $\boldsymbol{\delta}$ *defined as in question* a.

Answer: Modulo the isomorphism of T_eG onto $\mathcal{G}$ (p. 156)

$$\exp\colon T_eG \to G, \qquad \exp(\boldsymbol{\gamma} + \boldsymbol{\alpha}t) = gh(t)$$

$$\exp'(\boldsymbol{\gamma})\colon T_\gamma T_eG \to T_gG, \qquad \exp'(\boldsymbol{\gamma})\boldsymbol{\alpha} = \frac{\mathrm{d}}{\mathrm{d}t}gh(t)|_{t=0} = L'_g(e)\boldsymbol{\delta}$$

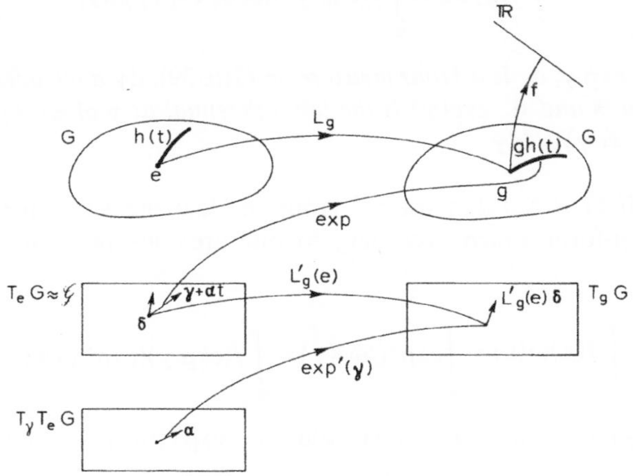

$$\exp'(\boldsymbol{\gamma})\boldsymbol{\alpha} = L'_g(e)\boldsymbol{\delta} = L'_g(e)\frac{\mathbb{1} - \exp(-\mathcal{A}d\boldsymbol{\gamma})}{\mathcal{A}d\boldsymbol{\gamma}}\boldsymbol{\alpha}.$$

c) Let f be a C^1 mapping from a differentiable manifold X into a Lie group G. It is convenient to introduce, beside the differential map $f'(x)$: $T_xX \to T_{f(x)}G$, the **left differential** map $\mathrm{d}_xf\colon T_xX \to T_eG$ defined by left differential

$$\mathrm{d}_xf(x)\colon \boldsymbol{v}_x \mapsto L'_{f(x)}(e)^{-1}(f'(x)\boldsymbol{v}_x) = f(x)^{-1}(f'(x)\boldsymbol{v}_x)$$

in the abbreviated notation p. 167.

One can also introduce a **right differential** as the linear mapping from T_xX into T_eG which maps $\boldsymbol{v}_x \mapsto (f'(x)\boldsymbol{v}_x)f(x)^{-1}$. right differential

With $\boldsymbol{\alpha}$, $\boldsymbol{\gamma}$, $\boldsymbol{\delta}$ *defined as in questions* a *and* b *and* d_γ *the left differential at*

γ, show that

$$\mathrm{d}_\gamma(\exp \gamma) = \sum_{p=0}^{\infty} \frac{1}{(p+1)!} (\mathcal{A}d(-\gamma))^p.$$

Answer:

$$\delta = \sum_{p=0}^{\infty} \frac{1}{(p+1)!} (\mathcal{A}d(-\gamma))^p \alpha = L'_{\exp\gamma}(e)^{-1}(\exp' \gamma)\alpha = \mathrm{d}_\gamma \exp \alpha.$$

For a direct proof see Dieudonné[1].

d) *Prove*

$$\int_G f(g)\,\mathrm{d}g = \int_{\mathcal{G}} f(\exp \gamma)|\det \mathrm{d}_\gamma \exp(\gamma)|\mathrm{d}\gamma,$$

where $g = \exp \gamma$, $\mathrm{d}g$ *is a Haar measure on* G*(p. 39)*, $\mathrm{d}\gamma$ *a euclidean volume element on* $\mathcal{G}$ *and* $\mathrm{d}_\gamma \exp(\gamma)$ *is the left differential at* γ *of* $\exp(\gamma)$ *equal to* $(\mathbb{1} - \exp(-\mathcal{A}d\gamma))/\mathcal{A}d\gamma$.

Answer: If G is a p-dimensional group, a Haar measure on G is a left invariant p-form which according to our previous notations we write $\omega(g)$,

$$\int_G f(g)\omega(g) = \int_{\mathcal{G}} \exp^*(f\omega)(\gamma) = \int_{\mathcal{G}} f(\exp \gamma)(\exp^* \omega)(\gamma)$$

$$(\exp^* \omega)(\gamma)(v_1, \ldots v_p) = \omega(g)(\exp'(\gamma)v_1, \ldots, \exp'(\gamma)v_p) \quad \text{by definition}$$

$$= \omega(e)(\mathrm{d}_\gamma \exp(\gamma)v_1, \ldots, \mathrm{d}_\gamma \exp(\gamma)v_p) \quad \text{since } \omega \text{ is left invariant.}$$

See Problem III 6, p. 190.

e) *Compute* $L'_g(e)$ *in canonical coordinates.*

Answer: Let $\{v_\alpha\}$ be a basis for the Lie algebra $\mathcal{G}$ of G. Let (α^α), $(\gamma^\alpha)\ldots$ be the coordinates of α, γ, ..., respectively, in this basis.
The canonical coordinates of $\exp'(\gamma)$ are

$$(\exp'(\gamma))^\alpha_\beta = \delta^\alpha_\beta.$$

[1][Dieudonné, p. 201].

Indeed if $gh(t) = \exp(\boldsymbol{\gamma} + \boldsymbol{\alpha} t)$ the canonical coordinates of $gh(t)$ are $\gamma^\alpha + \alpha^\alpha t$ and the equation $\exp'(\boldsymbol{\gamma})\boldsymbol{\alpha} = \mathrm{d}(gh(t))/\mathrm{d}t|_{t=0}$ reads in canonical coordinates $\exp'(\boldsymbol{\gamma})^\alpha_\beta \alpha^\beta = \alpha^\alpha$.
The equation $\exp'(\boldsymbol{\gamma}) = L'_g(e)(\mathbb{1} - \exp(-\mathcal{A}d\boldsymbol{\gamma}))/\mathcal{A}d\boldsymbol{\gamma}$ reads in canonical coordinates

$$\delta^\alpha_\gamma = L'_g(e)^\alpha_\beta((\mathbb{1} - \exp(-\boldsymbol{\gamma} \cdot c))/\boldsymbol{\gamma} \cdot c)^\beta_\gamma, \qquad g = \exp \boldsymbol{\gamma}$$

with $(\boldsymbol{\gamma} \cdot c)^\nu_\rho = \gamma^\mu c^\nu_{\mu\rho}$ where $c^\nu_{\mu\rho}$ are the structure constants (p. 168). $L'_g(e)^\alpha_\beta$ is readily obtained from this equation.
For another derivation see [B.S. DeWitt 1965, p. 80]. For explicit values of $L'_g(e)$ in the case of SO(3) and SU(2) see p. 189.

PROBLEM 5. THE GROUP MANIFOLDS OF SO(3) AND SU(2)

Use the Euler angles to show that SO(3), *the group of rotations of* $\mathbb{R}^3$, *is a Lie group. Show that* SU(2) *is a Lie group. Study their adjoint representations. Construct a system of canonical coordinates. Introduce a riemannian structure on* SO(3) *and* SU(2).

Answer: 1) *Group of rotations of* $\mathbb{R}^3$
A **rotation** $R(g)$ of $\mathbb{R}^3$ is a mapping $R(g)$: $\mathbb{R}^3 \to \mathbb{R}^3$ which preserves the euclidean norm of $\mathbb{R}^3$ and its orientation. We shall use the first part of the alphabet to characterize the rotations, e being the zero rotation $R(e)\boldsymbol{V} = \boldsymbol{V}$, $\boldsymbol{V} \in \mathbb{R}^3$. For example g can be the set of **Euler angles** (φ, θ, ψ) defined as follows: $R(\varphi, \theta, \psi)$ is the rotation of the right handed orthogonal frame $Oxyz$ into the right handed frame $Ox'y'z'$ such that

rotation

Euler angles

$$R(\varphi, \theta, \psi) = r_3 \circ r_2 \circ r_1,$$

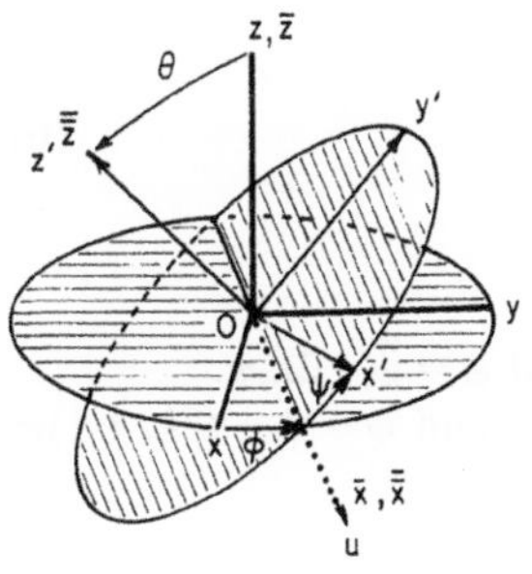

The first rotation φ about Oz sends $Oxyz$ into $O\bar{x}\bar{y}\bar{z}$. The second rotation θ about $Ou = O\bar{x}$ sends $O\bar{x}\bar{y}\bar{z}$ into $O\bar{\bar{x}}\bar{\bar{y}}\bar{\bar{z}}$.
The third rotation ψ about $Oz' = O\bar{\bar{z}}$ sends $O\bar{\bar{x}}\bar{\bar{y}}\bar{\bar{z}}$ into $Ox'y'z'$.
The Oz' axis makes angle θ with Oz; the projection of Oz' onto the xy plane makes angle $\varphi - \pi/2$ with Ox. In quantum mechanics the convention is to take the Ou axis to be along the r_1 rotated Oy axis, rather than the r_1 rotated Ox axis. For a discussion of the various choices of Euler angles see [Goldstein, 2nd ed.].

where r_1 is the rotation of angle φ around Oz that transforms a vector along Ox into a vector along Ou (precession). r_2 is the rotation of angle θ around Ou that transforms a vector along Oz into a vector along Oz' (nutation). r_3 is the rotation of angle ψ around Oz' (proper rotation). Two sets (φ, θ, ψ), $(\bar{\varphi}, \bar{\theta}, \bar{\psi})$ are called equivalent if

$$R(\varphi, \theta, \psi) = R(\bar{\varphi}, \bar{\theta}, \bar{\psi}).$$

group of rotations $\mathcal{R}$

Let E be the equivalence relation thus defined. The quotient space $\mathbb{R}^3/E$ of triples $\{(\varphi, \theta, \psi)\}$ by E is the **group of rotations** $\mathcal{R}$. This equivalence relation reflects the periodicity and the symmetry of rotations: the rotation $(\omega, \boldsymbol{n})$ of angle ω around the unit vector $\boldsymbol{n}$ is the same as the rotation $(\omega + 2\pi, \boldsymbol{n})$ and the rotation $(2\pi - \omega, -\boldsymbol{n})$. To state E explicitly it is convenient to introduce the matrices.

$$\tilde{r}_z(\varphi) = \begin{pmatrix} \cos\varphi & -\sin\varphi & 0 \\ \sin\varphi & \cos\varphi & 0 \\ 0 & 0 & 1 \end{pmatrix}, \qquad \tilde{r}_x(\theta) = \begin{pmatrix} 1 & 0 & 0 \\ 0 & \cos\theta & -\sin\theta \\ 0 & \sin\theta & \cos\theta \end{pmatrix}.$$

Then $r_1 = \tilde{r}_z(\varphi)$, $r_2 = \tilde{r}_z(\varphi)\tilde{r}_x(\theta)\tilde{r}_z(\varphi)^{-1}$, $r_3 = \tilde{r}_z(\varphi)\tilde{r}_x(\theta)\tilde{r}_z(\psi)\tilde{r}_x(\theta)^{-1}\tilde{r}_z(\varphi)^{-1}$ so $\tilde{R}(\varphi, \theta, \psi) = r_3 \circ r_2 \circ r_1 = \tilde{r}_z(\varphi)\tilde{r}_x(\theta)\tilde{r}_z(\psi)$

$$= \begin{pmatrix} \cos\varphi\cos\psi - \sin\varphi\cos\theta\sin\psi & -\cos\varphi\sin\psi - \sin\varphi\cos\theta\cos\psi & \sin\varphi\sin\theta \\ \sin\varphi\cos\psi + \cos\varphi\cos\theta\sin\psi & -\sin\varphi\sin\psi + \cos\varphi\cos\theta\cos\psi & -\cos\varphi\sin\theta \\ \sin\theta\sin\psi & \sin\theta\cos\psi & \cos\theta \end{pmatrix}.$$

(φ, θ, ψ) and $(\bar{\varphi}, \bar{\theta}, \bar{\psi})$ are E related if and only if

$$\tilde{R}(\varphi, \theta, \psi) = \tilde{R}(\bar{\varphi}, \bar{\theta}, \bar{\psi}). \tag{1}$$

Proof: Necessary conditions. Equation (1) implies

$$\cos\theta = \cos\bar{\theta} \quad \text{i.e.} \begin{cases} \text{either} & \theta = \bar{\theta} + 2k_1\pi & \text{(a)} \\ \text{or} & \theta = -\bar{\theta} + 2k_2\pi & \text{(b)} \end{cases}$$

where k_1, k_2, etc. are integers. Two cases

a) $\sin\theta \neq 0$.

Equations (1) and (a) imply $\varphi = \bar{\varphi} + 2k_3\pi$ and $\psi = \bar{\psi} + 2k_4\pi$.

Equations (1) and (b) imply $\varphi = \bar{\varphi} + (2k_5 + 1)\pi$ and $\psi = \bar{\psi} + (2k_6 + 1)\pi$.

b) $\sin\theta = 0$, $\cos\theta = \epsilon$ with $\epsilon = \pm 1$.

Equation (1) implies $\varphi + \epsilon\psi = \bar{\varphi} + \epsilon\bar{\psi} + 2k_7\pi$.

One checks easily that these conditions are also sufficient. ■

The group $\{\tilde{R}(\varphi, \theta, \psi)\}$ is the group of orthogonal matrices with unit determinant called SO(3).

2) SU(2). See p. 173.

a) All matrices in SU(2) are of the form $M = \begin{pmatrix} \alpha & \beta \\ -\bar{\beta} & \bar{\alpha} \end{pmatrix}$ with $\alpha\bar{\alpha} + \beta\bar{\beta} = 1$.

It is profitable to consider SU(2) as a subgroup of the matrix group $Q = \{M\}$ where α and β do not vanish simultaneously but are not otherwise restricted. $Q \cup \{0\}$ is a real four dimensional vector space admitting the basis

$$e = \begin{pmatrix} 1 & 0 \\ 0 & 1 \end{pmatrix}, \rho_1 = \begin{pmatrix} \mathrm{i} & 0 \\ 0 & -\mathrm{i} \end{pmatrix}, \rho_2 = \begin{pmatrix} 0 & 1 \\ -1 & 0 \end{pmatrix}, \rho_3 = \begin{pmatrix} 0 & \mathrm{i} \\ \mathrm{i} & 0 \end{pmatrix}.$$

Set $\alpha = a + ib$ and $\beta = c + id$ then

$$M = ae + b\rho_1 + c\rho_2 + d\rho_3$$

$$M \in \mathrm{SU}(2) \Leftrightarrow a^2 + b^2 + c^2 + d^2 = 1.$$

The mapping f: $\mathrm{SU}(2) \to \mathbb{R}^4$ by $M \mapsto (a, b, c, d)$ is a bijection of SU(2) onto the 3-sphere $S^3 \subset \mathbb{R}^4$.

Remark: $Q \cup \{0\}$ is the algebra of quaternions defined in Problem I.1.

b) The Lie algebra $\mathcal{SU}(2)$ of SU(2) is the algebra of traceless anti-hermitian matrices iA (p. 173)

$$A = \begin{pmatrix} z & x + \mathrm{i}y \\ x - \mathrm{i}y & -z \end{pmatrix}, \qquad (x, y, z) \in \mathbb{R}^3.$$

Let $(\boldsymbol{\sigma}_\alpha)$ be the **Pauli matrices**

$$\boldsymbol{\sigma}_1 = \begin{pmatrix} 0 & 1 \\ 1 & 0 \end{pmatrix}, \quad \boldsymbol{\sigma}_2 = \begin{pmatrix} 0 & \mathrm{i} \\ -\mathrm{i} & 0 \end{pmatrix}, \quad \boldsymbol{\sigma}_3 = \begin{pmatrix} 1 & 0 \\ 0 & -1 \end{pmatrix};$$

$(\mathrm{i}\boldsymbol{\sigma}_\alpha)$ form a basis for the Lie algebra of SU(2)

$$\mathrm{i}A = \mathrm{i}(x\boldsymbol{\sigma}_1 + y\boldsymbol{\sigma}_2 + z\boldsymbol{\sigma}_3).$$

It is convenient to choose the basis $(\mathrm{i}\boldsymbol{\sigma}_\alpha/2)$ which gives the standard normalization to the structure constants.[1]

The Lie algebra is a vector space over $\mathbb{R}$, i.e. $x, y, z \in \mathbb{R}$. In a neighborhood of unity, any element of SU(2) can be obtained by exponentiation of an element of its Lie algebra. Let $n \in \mathbb{R}^3$ with $\|n\| = 1$ and $(n^\alpha\boldsymbol{\sigma}_\alpha)^2 = \mathbf{1}$.

$$M = \exp(\mathrm{i}A) = \exp(\mathrm{i}\omega n^\alpha \boldsymbol{\sigma}_\alpha/2) = \mathbf{1}\cos\tfrac{1}{2}\omega + \mathrm{i}n \cdot \boldsymbol{\sigma} \sin\tfrac{1}{2}\omega.$$

[1]In quantum mechanics the convention is to take $\boldsymbol{\sigma}_2 = \begin{pmatrix} 0 & -\mathrm{i} \\ \mathrm{i} & 0 \end{pmatrix}$ and to expand $\mathcal{SU}(2)$ elements $-\mathrm{i}A$ in terms of the basis $(-\mathrm{i}\boldsymbol{\sigma}_\alpha/2)$.

Note that the mapping ($\omega \mapsto 2\pi - \omega$, $\boldsymbol{n} \mapsto -\boldsymbol{n}$) maps $M \mapsto -M$.

Structure constants: $[\mathrm{i}\boldsymbol{\sigma}_\alpha/2, \mathrm{i}\boldsymbol{\sigma}_\beta/2] = \epsilon_{\alpha\beta\gamma}\mathrm{i}\boldsymbol{\sigma}_\gamma/2$, hence $c^{\gamma}_{\alpha\beta} = \epsilon_{\gamma\alpha\beta}$

where $\epsilon_{\alpha\beta\gamma}$ is the permutation symbol equal to 1 if (α, β, γ) is an even permutation of $(1, 2, 3)$, equal to -1 if it is odd, and 0 otherwise. Although $\epsilon_{\gamma\alpha\beta} = \epsilon_{\alpha\beta\gamma}$, we do not write $c^{\gamma}_{\alpha\beta} = \epsilon_{\alpha\beta\gamma}$ to keep track of the first and last indices of the structure constants which play a special role.

3) *The adjoint representation of* SU(2).
The adjoint representation of SU(2) on its Lie algebra is

$$\mathrm{Ad}: M \mapsto \mathrm{Ad}(M) \text{ where } \mathrm{Ad}(M)\mathrm{i}A = \mathrm{i}MAM^{-1}.$$

Note that Ad(M) is, by definition, a mapping on $\mathscr{SU}(2)$, not on the space of traceless hermitian matrices A. One checks easily that for every $M \in$ SU(2), Ad(M) maps a traceless antihermitian matrix into a traceless antihermitian matrix, and that Ad is group preserving.

4) *Relationship between* SU(2), $\mathscr{R}$, *and* SO(3)
Let h map $\mathbb{R}^3$ into the space of traceless hermitian matrices (*not* $\mathscr{SU}(2)$)

$$\text{by } h(\boldsymbol{V}) = A = \begin{pmatrix} z & x+\mathrm{i}y \\ x-\mathrm{i}y & -z \end{pmatrix}.$$

Let $R(g)$: $\mathbb{R}^3 \to \mathbb{R}^3$ be such that

$$\boldsymbol{V}' = R(g)\boldsymbol{V} \quad \Leftrightarrow \quad h(\boldsymbol{V}') = Mh(\boldsymbol{V})M^{-1}.$$

Prove that the mapping $R(g)$ is a rotation.

Answer: For every $M \in$ SU(2), $Mh(\boldsymbol{V})M^{-1}$ is a traceless hermitian matrix, hence of the form $h(\boldsymbol{V}')$ for some $\boldsymbol{V}' \in \mathbb{R}^3$. It follows that $R(g)$ is defined. $R(g)$ is a rotation; indeed:

$$-\|\boldsymbol{V}\|^2 = \det h(\boldsymbol{V}) = \det h(\boldsymbol{V}') = -\|\boldsymbol{V}'\|^2$$

$$\mathrm{D}(x, y, z)/\mathrm{D}(x', y', z') > 0.$$

In conclusion, $M \in$ SU(2) defines a rotation $R(g) \in \mathscr{R}$. ■

Prove that the mapping f: SU(2)$\to\mathscr{R}$ defined by $M \mapsto R(g)$ is a two-to-one homomorphism with non vanishing jacobian.

Answer: The fact that f is a homomorphism follows readily from its definition. To prove that it is two-to-one, we shall show that $f^{-1}(\mathrm{Id}) = \{\mathbb{1}, -\mathbb{1}\}$. Indeed let $M \in f^{-1}$ (Id), then $h(\boldsymbol{V}) = Mh(\boldsymbol{V})M^{-1}$. It follows that M

commutes with all traceless hermitian matrices. M commutes also with all scalar matrices $\lambda\mathbf{1}$, hence with all real matrices, hence with all pure imaginary matrices, hence with all matrices. In conclusion, M is a scalar matrix $\lambda\mathbf{1}$. Since $M \in \mathrm{SU}(2)$, $\lambda = \bar{\lambda}$ and $\lambda\bar{\lambda} = 1$, hence $M = \pm\mathbf{1}$.
To prove that the jacobian of f is non vanishing we shall prove the following: Let $f(M) = R(g)$, the equation $f(M') = R(g')$ has a unique solution M' in a neighborhood of M. We know that $f(M') = R(g')$ has two solutions M' and $-M'$, but only one is in a neighborhood of M, since α and β cannot be simultaneously small by virtue of $\alpha\bar{\alpha} + \beta\bar{\beta} = 1$. In conclusion, SU(2) is a double covering of $\mathcal{R}$. Since SU(2) is simply connected (p. 19) it is the universal covering of $\mathcal{R}$. $\mathcal{R}$ is doubly connected, i.e. the paths on $\mathcal{R}$ belong to two different homotopy classes. ■

Similarly the mapping $f\colon \mathrm{SU}(2) \to \mathrm{SO}(3)$ defined by $M \mapsto \tilde{R}(g)$ is also a two-to-one homomorphism with non vanishing jacobian.

We shall solve the equation $R(g) = f(M)$ when M is in the neighborhood of $\mathbf{1}$.

$$M = (1 + \mathrm{i}\omega n^\alpha \boldsymbol{\sigma}_\alpha/2) + \mathrm{O}(\omega^2)$$

$$h(V') = (1 + \mathrm{i}\omega n^\alpha \boldsymbol{\sigma}_\alpha/2) h(V) (1 - \mathrm{i}\omega n^\alpha \boldsymbol{\sigma}_\alpha/2)$$

$$= h(V) + \mathrm{i}\frac{\omega}{2}[n^\alpha \boldsymbol{\sigma}_\alpha, h(V)].$$

Assume $n_3 = 1$, $n_1 = n_2 = 0$, then $z' = z$, $x' = x - \omega y$, $y' = y + \omega x$; this is an infinitesimal rotation of angle ω around the z axis. Similarly, $\boldsymbol{\sigma}_2$ is the generator of rotations around the y axis and $\boldsymbol{\sigma}_1$ the generator of rotations around the x axis.
Hence the rotation $R(g)$, solution of $R(g) = f(M)$, is a rotation of angle ω around the axis $\boldsymbol{n}$. $R(g)$ is also a solution of $R(g) = f(-M)$.

5) *Group manifold of* SO(3) *and* $\mathcal{R}$
a) *Lie algebra of* SO(3). We already know (Problem 2, p. 172) that the Lie algebra of SO(3) is the algebra of 3×3 real antisymmetric matrices and we can readily write down a basis for $\mathcal{SO}(3)$. We can also use the mapping $f\colon \mathrm{SU}(2) \to \mathrm{SO}(3)$ to construct the isomorphism $f'\colon \mathcal{SU}(2) \to \mathcal{SO}(3)$. Recall that in the neighborhood of $\mathbf{1}$, f is one-one. The solution of the equation $f(M) = R(g)$ in the neighborhood of $\mathbf{1}$ gives

$$f'(1)(\mathrm{i}\boldsymbol{\sigma}_1/2) = \boldsymbol{\gamma}_1 = \begin{pmatrix} 0 & 0 & 0 \\ 0 & 0 & -1 \\ 0 & 1 & 0 \end{pmatrix}, \quad f'(1)(\mathrm{i}\boldsymbol{\sigma}_2/2) = \boldsymbol{\gamma}_2 = \begin{pmatrix} 0 & 0 & 1 \\ 0 & 0 & 0 \\ -1 & 0 & 0 \end{pmatrix},$$

$$f'(1)(i\sigma_3/2) = \gamma_3 = \begin{pmatrix} 0 & -1 & 0 \\ 1 & 0 & 0 \\ 0 & 0 & 0 \end{pmatrix}.$$

One checks that $[\gamma_\alpha, \gamma_\beta] = \epsilon_{\gamma\alpha\beta}\gamma_\gamma$.

Remark: The solution M of $f(M) = \tilde{R}(\varphi, \theta, \psi)$, where φ, θ, ψ are the Euler angles, is

$$M = \exp(i\varphi\sigma_3/2)\exp(i\theta\sigma_1/2)\exp(i\psi\sigma_3/2).$$

The infinitesimal Euler rotation is $V' = (1 + A)V$ where

$$A = \begin{pmatrix} 0 & -(\psi+\varphi) & 0 \\ \psi+\varphi & 0 & -\theta \\ 0 & \theta & 0 \end{pmatrix}.$$

The parametrization (φ, θ, ψ) is degenerate near the origin. This degeneracy is sometimes removed by introducing the new variables

$$u = \psi + \varphi \quad \text{and} \quad v = \psi - \varphi.$$

b) *Atlas for* $\mathcal{R}$ *and* SO(3). By virtue of the equivalence relation E, the Euler angles in the range $0 \le \theta < \pi$, $0 \le \psi < 2\pi$, $0 \le \varphi < 2\pi$, parametrize the group of rotations $\mathcal{R}$. $\mathcal{R}$ can be given the structure of a C^∞ differentiable manifold with a finite atlas $\{(D_i, \xi_i)\}$: Let $\mathcal{Z}$ and $\mathcal{Y}$ be the set of rotations around the z-axis and the y-axis respectively.
The following charts cover $\mathcal{R} - \mathcal{Z}$; i.e. all rotations such that $\theta \neq 0$.

D_1 defined by	$0<\theta<\pi$,	$0<\psi<2\pi$ and	$0<\varphi<2\pi$
D_2	$0<\theta<\pi$,	$0<\psi<2\pi$ and	$\pi<\varphi<3\pi$
D_3	$0<\theta<\pi$,	$\pi<\psi<3\pi$ and	$0<\varphi<2\pi$
D_4	$0<\theta<\pi$,	$\pi<\psi<3\pi$ and	$\pi<\varphi<3\pi$.

Similarly one can cover $\mathcal{R}-\mathcal{Y}$ with 4 charts. One checks that on each intersection $D_i \cap D_j$, the mapping $\xi_i \circ \xi_j^{-1}$ is C^∞.
Because the parametrization φ, θ, ψ is degenerate at the origin, we need to choose another system of coordinates in the neighborhood of the identity. We can choose (θ, u, v) defined above. We can also choose the three elements of $\tilde{R}(\varphi, \theta, \psi)$ above or below the diagonal. Let D be the domain defined by $\varphi, \theta, \psi \in (-\pi/6, \pi/6)$, let $\Phi: D \to \mathbb{R}^3$ by

$$\xi = \sin\theta\cos\psi$$

$$\eta = \sin\theta\sin\psi$$

$$\zeta = \sin\varphi\cos\psi + \cos\varphi\cos\theta\sin\psi.$$

In the intersection $D_1 \cap D$, the change of coordinates $(\varphi, \theta, \psi) \to (\xi, \eta, \zeta)$ is a C^∞ diffeomorphism and the chart (D, Φ) is compatible with the atlas previously defined.

6) *Canonical coordinates of* SO(3)
Let S be a solid rotating around a fixed point $O \in S$. Let $\boldsymbol{O}$ be a fixed orthogonal frame at O. Let $\boldsymbol{\alpha} = \boldsymbol{\omega}\, \mathrm{d}t$ be the vector rotation characterizing the infinitesimal rotation of an arbitrary point $P \in S$ given by

$$\mathrm{d}\,\vec{\mathrm{OP}}/\mathrm{d}t = \boldsymbol{\omega} \times \vec{\mathrm{OP}}.$$

Let $(\boldsymbol{\gamma}_\alpha)$ be the basis of the Lie algebra of SO(3) *such that $\boldsymbol{\gamma}_\alpha$ is the generator of the rotations around the α-axis of the* **O**-*frame*; *i.e.* $(\mathbb{1} + \epsilon\boldsymbol{\gamma}_\alpha)\boldsymbol{v} = \boldsymbol{w}$ *where $\boldsymbol{w}$ is obtained from $\boldsymbol{v} \in \mathbb{R}^3$ by an infinitesimal rotation of angle ϵ around the α-axis. This basis is given explicitly* *p.* 185. *Let (x^α) be the canonical coordinates of $M \in$* SO(3) *with respect to the $(\boldsymbol{\gamma}_\alpha)$-basis. Show that (x^α) are the coordinates of the vector rotation, in the* **O**-*frame, of the rotation corresponding to M.*

Answer: $M = \exp(x^\alpha \boldsymbol{\gamma}_\alpha)$. Consider first an infinitesimal rotation $M = \mathbb{1} + \epsilon x^\alpha \boldsymbol{\gamma}_\alpha$

$$M\boldsymbol{v} = (\mathbb{1} + \epsilon x^\alpha \boldsymbol{\gamma}_\alpha)\boldsymbol{v} = \boldsymbol{v} + \epsilon \boldsymbol{x} \times \boldsymbol{v}$$

and $\epsilon\boldsymbol{x}$ is the infinitesimal vector rotation. This result remains true for finite values of ϵx^α because rotations around a fixed axis form a one parameter subgroup of SO(3).

Alternative answer: The adjoint representation of SO(3) on its Lie algebra is a representation of SO(3) on a 3-dimensional linear space, i.e. on $\mathbb{R}^3$. In canonical coordinates, $\mathrm{Ad}(M) = \exp x^\alpha c_\alpha$ where c_α is the matrix $(c_\alpha)^\gamma_\beta = c^\gamma_{\alpha\beta} = \epsilon_{\gamma\alpha\beta}$. For infinitesimal (ϵx^α)

$$\mathrm{Ad}(M)\colon\ \boldsymbol{v} \mapsto (\mathbb{1} + \epsilon x^\alpha c_\alpha)\boldsymbol{v} = \boldsymbol{v} + \epsilon\boldsymbol{x} \times \boldsymbol{v},$$

since the 3 matrices c_α whose elements are $(c_\alpha)^\gamma_\beta = \epsilon_{\gamma\alpha\beta}$ are identical with the 3 matrices $\boldsymbol{\gamma}_\alpha$ given p. 185.

7) *Group manifold of* SU(2)
Let $\bar{E}$ be the equivalence relation defined by $(\omega, \boldsymbol{n}) \sim (\omega + 4\pi, \boldsymbol{n})$. The quotient space $\mathbb{R}^3/\bar{E}$ is a double covering of $\mathbb{R}^3/E$. To determine the range of the Euler angles necessary to parametrize SU(2), we state $\bar{E}$ in terms of φ, θ, ψ: $(\bar{\varphi}, \bar{\theta}, \bar{\psi})$ and (φ, θ, ψ) are $\bar{E}$ related in the following cases:

1) $\sin\theta \neq 0$, $\theta = \bar{\theta} + 2k_1\pi$, $\varphi = \bar{\varphi} + 2k_2\pi$, $\psi = \bar{\psi} + 2k_3\pi$.
2) $\sin\theta = 0$, $\cos\theta = \epsilon$ with $\epsilon = \pm 1$, $\varphi + \epsilon\psi = \bar{\varphi} + \epsilon\bar{\psi} + 2k_4\pi$.

The Euler angles in the range $0 \le \theta < \pi$, $0 \le \varphi < 2\pi$, $0 \le \psi < 4\pi$ parametrize the 3-sphere. To compute M as a function $M(\varphi, \theta, \psi)$ of the Euler angles we set

$$M = \begin{pmatrix} a+ib & c+id \\ -c+id & a-ib \end{pmatrix}$$

and solve the set of equations

$$\boldsymbol{V}' = R(\varphi, \theta, \psi)\boldsymbol{V}$$

$$h(\boldsymbol{V}') = Mh(\boldsymbol{V})M^{-1}.$$

A straightforward calculation gives

$$a = \cos\tfrac{1}{2}\theta \cos\tfrac{1}{2}u, \qquad b = \cos\tfrac{1}{2}\theta \sin\tfrac{1}{2}u \qquad \text{with } u = \psi + \varphi$$

$$c = \sin\tfrac{1}{2}\theta \sin\tfrac{1}{2}v, \qquad d = \sin\tfrac{1}{2}\theta \cos\tfrac{1}{2}v \qquad \text{with } v = \psi - \varphi.$$

SU(2) can be given the structure of a C^∞ differentiable manifold by an atlas constructed like the atlas of SO(3). One checks that the point $(\varphi', \theta', \psi')$ solution of $M(\varphi', \theta', \psi') = -M(\varphi, \theta, \psi)$ cannot be in the same chart as (φ, θ, ψ).

In conclusion, SU(2) is diffeomorphic to the 3-sphere. $\mathcal{R}$ and SO(3) are diffeomorphic to the quotient space of the 3-sphere by the antipodal equivalence, hence diffeomorphic to the 3-dimensional **projective space**.

projective space

8) *Riemannian structures of* SU(2) *and* SO(3), *see p.* 285

simple
semisimple

We recall that a **simple** Lie group has no invariant subgroup (p. 7) and a **semisimple** Lie group has no invariant abelian subgroup. Cartan has shown that a necessary and sufficient condition for a Lie group to be semisimple is that the matrix

$$k_{\alpha\beta} = c^{\gamma}_{\alpha\delta} c^{\delta}_{\beta\gamma}$$

Cartan–Killing metric

be non singular. $k_{\alpha\beta}$ defines a metric called **Cartan–Killing**.

Use $k_{\alpha\beta}$ to construct a left invariant metric $\boldsymbol{g}$ *on the group manifold G.*
Answer: A metric is a 2-covariant tensor field $\boldsymbol{g}$. It is left invariant if

$$L_g^* \boldsymbol{g}(L_g h) = \boldsymbol{g}(h), \qquad g, h \in G$$

i.e.

$$L_g^* \boldsymbol{g}(g) = \boldsymbol{g}(e).$$

The metric $\boldsymbol{g}$ whose components are

$$g_{\mu\nu}(g) = L_g^*(g)_\mu^{-1\alpha} L_g^*(g)_\nu^{-1\beta} k_{\alpha\beta}$$

is left invariant. Note that $L_g^*(g)^{-1} = (\partial L_g(h)/\partial h|_{h=g})^{-1}$ (p. 155). The volume element of g is a Haar measure.

Compute this metric for SO(3) *and* SU(2). *Answer*: In the case of SO(3) and SU(2)

$$k_{\alpha\beta} = c^{\gamma}_{\alpha\delta} c^{\delta}_{\beta\gamma} = -2\delta_{\alpha\beta}.$$

Moreover we have shown (p. 181) that for $g = \exp \boldsymbol{\gamma}$

$$L'_g(e)^{-1} = (\mathbb{1} - \exp(-\mathcal{A}d\boldsymbol{\gamma}))/\mathcal{A}d\boldsymbol{\gamma}.$$

Let $(\boldsymbol{\gamma}_\alpha)$ be a basis for the Lie algebra of SO(3), let (x^α) be the canonical coordinates of $g \in \mathrm{SO}(3)$ with respect to $(\boldsymbol{\gamma}_\alpha)$,

$$\boldsymbol{\gamma} = x^\alpha \boldsymbol{\gamma}_\alpha \quad \text{and} \quad (\mathcal{A}d\boldsymbol{\gamma})^\alpha_\beta = x^\gamma c^\alpha_{\gamma\beta} \qquad \text{(p. 168)}.$$

Let $(\mathrm{i}\boldsymbol{\sigma}_\alpha/2)$ be a basis for the Lie algebra of SU(2), let (x^α) be the canonical coordinates of $g \in \mathrm{SU}(2)$ with respect to $(\mathrm{i}\boldsymbol{\sigma}_\alpha/2)$

$$g = \exp \mathrm{i}x^\alpha \boldsymbol{\sigma}_\alpha/2 \quad \text{and} \quad (\mathcal{A}d\ \mathrm{i}x^\alpha\boldsymbol{\sigma}_\alpha/2) = x^\delta c^\alpha_{\delta\gamma}.$$

Hence the metric g is the same in both cases.

If we choose for $(\boldsymbol{\gamma}_\alpha)$ the three matrices given explicitly p. 185, and for $(\boldsymbol{\sigma}_\alpha)$ the three Pauli matrices p. 183, the canonical coordinates (x^α) of $g \in \mathrm{SO}(3)$ and $g \in \mathrm{SU}(2)$ with respect to these bases respectively are the coordinates of the corresponding vector rotation (p. 188).

Set $x \cdot c$ the matrix of elements $x^\gamma c^\alpha_{\gamma\beta}$ and $x = (\delta_{\alpha\beta} x^\alpha x^\beta)^{1/2}$. A straightforward calculation gives

$$((\mathbb{1} - \exp(-x \cdot c))/x \cdot c)^\alpha_\beta = \delta^\alpha_\beta \frac{\sin x}{x} + \frac{x^\alpha x_\beta}{x^2}\left(1 - \frac{\sin x}{x}\right) - \frac{x^\gamma c^\alpha_{\beta\gamma}}{x^2}(1 - \cos x),$$

where $c^\alpha_{\beta\gamma} = \epsilon_{\alpha\beta\gamma}$, and

$$g_{\mu\nu}(g) = -4\delta_{\mu\nu}\frac{1 - \cos x}{x^2} - 2\frac{x_\mu x_\nu}{x^2}\left(1 - 2\frac{1 - \cos x}{x^2}\right).$$

This metric can be obtained from the euclidean metric $\delta_{\mu\nu}$ by the following change of variables: Let $(y^i;\ i = 1, \ldots, 4)$ be the euclidean coordinates of S^3 embedded in $\mathbb{R}^4$ and set

$$y^\alpha = \frac{x^\alpha}{x}\sin\tfrac{1}{2}x, \qquad y^4 = \cos\tfrac{1}{2}x.$$

A straightforward calculation gives

$$\sum (\mathrm{d}y^i)^2 = g_{\mu\nu}\, \mathbf{d}x^\mu\, \mathbf{d}x^\nu.$$

The natural metric on SU(2) and SO(3) has a very simple expression in the coordinate system defined by the Euler angles. Using the results of 7) giving a, b, c, d in terms of the Euler angles, we can compute the metric on S^3 induced by the euclidean metric on $\mathbb{R}^4$

$$\begin{aligned} \mathrm{d}s^2 &= \mathbf{d}a^2 + \mathbf{d}b^2 + \mathbf{d}c^2 + \mathbf{d}d^2 \\ &= \tfrac{1}{4}(\mathbf{d}\theta^2 + \mathbf{d}\varphi^2 + \mathbf{d}\psi^2 + 2\cos\theta\, \mathbf{d}\varphi\, \mathbf{d}\psi). \end{aligned}$$

What is the volume element $\boldsymbol{\tau}$ *(p. 294) defined by the metric* g? *Answer*:

$$\boldsymbol{\tau} = |\det g_{\mu\nu}(g)|^{1/2}\, \mathbf{d}x^1 \wedge \mathbf{d}x^2 \wedge \mathbf{d}x^3 \qquad \text{(in canonical coordinates).}$$

Note that $\boldsymbol{\tau} = |\det(\mathbb{1} - \exp(-\mathcal{A}d\gamma))/\mathcal{A}d\gamma\| \det k_{\alpha\beta}|^{1/2}\, \mathbf{d}x^1 \wedge \mathbf{d}x^2 \wedge \mathbf{d}x^3$. Up to the constant factor $|\det k_{\alpha\beta}|^{1/2}$, the volume element $\boldsymbol{\tau}$ is equal to the volume element obtained p. 180 as could be expected.

References: see for instance S. Helgason, Lie Groups and Symmetric Spaces, in: *Battelle Rencontres 1967 Lectures in Mathematics and Physics*, eds C. DeWitt and J.A. Wheeler (1968, W.A. Benjamin Inc.). For applications see B.S. DeWitt, *Dynamical Theory of Groups and Fields* (1965, Gordon and Breach), L.S. Schulman, Phys. Rev. 176 (1968) 1558.

PROBLEM 6. ATLAS ON THE 2-SPHERE, ORIENTATION OF THE 2-SPHERE MANIFOLD; INTEGRATION ON THE 2-SPHERE; VECTOR FIELDS ON THE 2-SPHERE; FRAME BUNDLE ON THE 2-SPHERE; THE 2-SPHERE IS NOT A GROUP MANIFOLD.

1) *An atlas on the 2 sphere*

sphere

Let the 2 **sphere** S be the C^∞ submanifold of $\mathbb{R}^3$ defined by

$$\sum (x^i)^2 = 1, \qquad i = 1, 2, 3. \tag{1}$$

Since the 1×3 jacobian matrix $[2x^1\, 2x^2\, 2x^3]$ is of rank 1 on S, we can construct a C^∞ atlas A on S consisting of the 6 domains

$$D_{i\epsilon} = \{(x^1, x^2, x^3) \in S;\ \epsilon x^i > 0\} \quad \text{with } \epsilon = \pm 1$$

and the 6 mappings $\phi_{i\epsilon}(x^1, x^2, x^3) = (x^j, x^k)$ with i, j, k being a circular permutation of 1, 2, 3 when $\epsilon = 1$, and a circular permutation of 1, 3, 2 when $\epsilon = -1$.

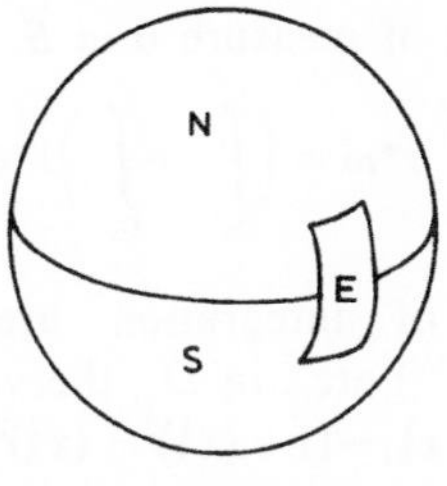

Let $(i, \epsilon) = (3, +) = N$, $(i, \epsilon) = (3, -) = S$, $(i, \epsilon) = (2, +) = E$; the mapping $\phi_N \circ \phi_E^{-1}$ is defined by $x_N^1 = x_E^2$ and $x_N^2 = (1 - (x_E^1)^2 - (x_E^2)^2)^{1/2}$. It is C^∞.

2) *Orientation of the 2-sphere* (p. 212)

Does the atlas A define an orientation on S? Yes, because all the determinants $D(x_N^1, x_N^2)/D(x_E^1, x_E^2)$, $D(x_S^1, x_S^2)/D(x_E^1, x_E^2)$, etc. are positive. We note that on the southern hemisphere the coordinates of a point (x^1, x^2, x^3) are (x^2, x^1), whereas on the northern hemisphere they are (x^1, x^2). If we had chosen the coordinates on the southern hemisphere to be (x^1, x^2) the 2-sphere would not have been oriented, as we can see even without computing the determinants: let us parallel transport (p. 302) a frame (v_1, v_2) along a longitude from the north pole to the

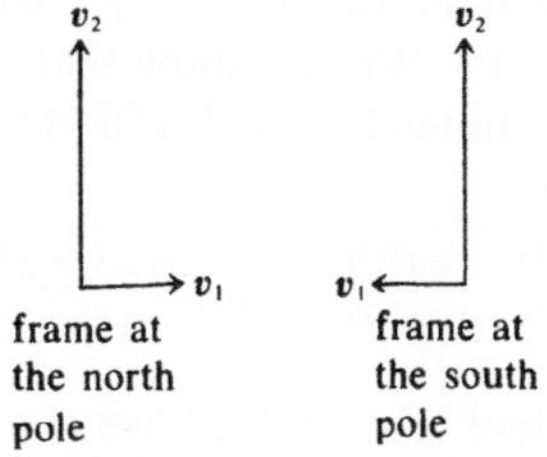

south pole; the projections on the equatorial plane of the frame at the north pole, and at the south pole, do not have the same orientation, unless we interchange the order of the coordinates.

3) *Integration on the 2-sphere* (*see p.* 216 *and example p.* 205)

Let ω be a two form defined on $\mathbb{R}^3$

$$\omega = P(x^i)\, dx^2 \wedge dx^3 + Q(x^i)\, dx^3 \wedge dx^1 + R(x^i)\, dx^1 \wedge dx^2;$$

let $i^*\omega$ be the two form on S induced from ω by the inclusion mapping $i: S \to \mathbb{R}^3$ by

$$(x_N^1, x_N^2) \mapsto (x_N^1, x_N^2, (1 - (x_N^1)^2 - (x_N^2)^2)^{1/2})$$

$$(x_S^1, x_S^2) \mapsto (x_S^2, x_S^1, -(1 - (x_S^1)^2 - (x_S^2)^2)^{1/2}).$$

Since the equator is a set of measure 0 in S

$$\int_S i^*\omega = \left(\int_{D_N} + \int_{D_S}\right) i^*\omega.$$

In D_N, the variables of integration are (x_N^1, x_N^2) and $P(x^i) = P(x_N^1, x_N^2, (1-(x_N^1)^2-(x_N^2)^2)^{1/2})$ etc.; in D_S, the variables of integration are (x_S^1, x_S^2) and $P(x^i) = P(x_S^2, x_S^1, -(1-(x_S^1)^2-(x_S^2)^2)^{1/2})$ etc.

4) *Another atlas on the 2-sphere*

We can construct an atlas B equivalent to A in terms of the coordinates of the **stereographic projections**: let P and Q be the north and south poles respectively, let

stereographic projection

$U = S - \{P\}$ and $V = S - \{Q\}$, let g and h be the stereographic projections of poles P and Q on the plane $x^3 = 0$

$$g\colon U \to \mathbb{R}^2 \qquad \text{by } y^1 = x^1/(1-x^3) \text{ and } y^2 = x^2/(1-x^3)$$

$$h\colon V \to \mathbb{R}^2 \qquad \text{by } z^1 = x^1/(1+x^3) \text{ and } z^2 = x^2/(1+x^3).$$

One can check that B is an atlas:

a) $S = U \cup V$, the mappings g and h are homeomorphisms.

b) B is a differentiable atlas. Let $d = h \circ g^{-1}$ where g^{-1} is understood to be restricted to $g(U \cap V)$. We can show that d is a diffeomorphism of $\mathbb{R}^2 - \{0\}$ into $\mathbb{R}^2 - \{0\}$. Indeed $d\colon (y^1, y^2) \mapsto (z^1, z^2)$ and $d^{-1}\colon (z^1, z^2) \mapsto (y^1, y^2)$ respectively by

$$\begin{cases} z^1 = y^1/((y^1)^2 + (y^2)^2) \\ z^2 = y^2/((y^1)^2 + (y^2)^2) \end{cases} \quad \text{and} \quad \begin{cases} y^1 = z^1/((z^1)^2 + (z^2)^2) \\ y^2 = z^2/((z^1)^2 + (z^2)^2) \end{cases}.$$

c) B is an atlas equivalent to A; indeed, we can show that $A \cup B$ is a C^∞ differentiable atlas by studying all possible intersections $D_{i\epsilon} \cap U$ and $D_{i\epsilon} \cap V$ and the corresponding change of coordinates on these intersections.

5) *Vector fields on the 2-sphere*

Let $\boldsymbol{X}$ be a continuous vector field on S given on U and V respectively by $\boldsymbol{X} = a^\alpha(y^\beta)(\partial/\partial y^\alpha)$ and $\boldsymbol{X} = b^\alpha(z^\beta)(\partial/\partial z^\alpha)$ $(\alpha, \beta = 1, 2)$. Then

$$\begin{pmatrix} a^1 \\ a^2 \end{pmatrix} = \left((z^1)^2 + (z^2)^2\right)^{-2} \begin{pmatrix} (z^2)^2 - (z^1)^2 & -2z^1z^2 \\ -2z^1z^2 & (z^1)^2 - (z^2)^2 \end{pmatrix} \begin{pmatrix} b^1 \\ b^2 \end{pmatrix}$$

We note that if a^1 and a^2 are constant, (b^1, b^2) tends to $(0, 0)$ when (z^1, z^2) tends to 0. Since $\boldsymbol{X}$ is continuous, $b^1(0) = b^2(0) = 0$ and $\boldsymbol{X}$ vanishes at the pole Q; one says that $\boldsymbol{X}$ is not regular at Q. One can show, in general, that it is not possible to define a continuous, regular vector field on S. One possible proof consists in finding the properties that $a^\alpha(y^1, y^2)$ must satisfy for X to be regular at Q. One finds that these properties imply that there exists a point where a^1 and a^2 vanish simultaneously. Hence X is not regular on S. This proof is fairly involved and we shall, instead, prove that the frame bundle $F(S)$ on S is not trivial, thereby proving that it has no cross-section (p. 133): one cannot define a continuous mapping on $F(S)$ which associates a frame $v = (v_1, v_2)$ at each point of S. Let, in the following, X be a continuous field given by $X = X^i v_i$ wherever v is defined.

6) *Frame bundle* $F(S)$
Base: S. Typical fibre: GL(2, ℝ). Structural group: GL(2, ℝ).
Let $\mathcal{B} = \{(\mathcal{U}, \phi_u), \ (\mathcal{V}, \phi_v)\}$ be an atlas on $F(S)$ such that $\phi_u(M, v) = (y^1, y^2, v_1, v_2)$ and $\phi_v(M, v) = (z^1, z^2, w_1, w_2)$. $F(S)$ is trivial if and only if we can find a mapping ϕ_v such that (restricted to the same fibre) $\overset{\Delta}{\phi}_u \circ \overset{\Delta}{\phi}{}_v^{-1} = 1$, i.e. if we can find ϕ_v such that $w_1 = v_1$ and $w_2 = v_2$. We have noted in 2) that if the projection on the equatorial plane of a frame $v = (v_1, v_2)$ at P (north pole) is right handed, its projection after it has been parallel transported to Q (south pole) is left handed. No continuous mapping can map a right handed frame into a left handed frame.
We can find ϕ_v such that

$$\overset{\Delta}{\phi}_u \circ \overset{\Delta}{\phi}{}_v^{-1} = \begin{pmatrix} 0 & 1 \\ 1 & 0 \end{pmatrix} \text{ but not such that } \overset{\Delta}{\phi}_u \circ \overset{\Delta}{\phi}{}_v^{-1} = \mathbb{1}.$$

7) *Tangent bundle* $T(S)$. *The complex 2-sphere*
Base: S. Typical fibre: $\mathbb{R}^2$. Structural group GL(2, ℝ).
We shall use the atlas defined by the domains $\mathcal{U}$ and $\mathcal{V}$ and the local coordinates (y^α, a^α) on $\mathcal{U}$ and (z^α, b^α) on $\mathcal{V}$; i.e. let $\mathcal{G}$ and $\mathcal{H}$ be the homeomorphisms mapping $\mathcal{U}$ and $\mathcal{V}$ respectively on $\mathbb{R}^4$ by $\mathcal{G}$: $(M \in \mathcal{U}, X \in T_M(S)) \mapsto (y^\alpha, a^\alpha)$ with $X = a^\alpha \partial/\partial y^\alpha$, and a similar definition for $\mathcal{H}$. We shall show that the complex 2-sphere is homeomorphic to the tangent bundle of the real 2-sphere. This property is easily generalized to n-spheres.
Let the coordinates of $Z \in \mathbb{C}^3$ be $Z^k = X^k + \mathrm{i}Y^k$ with $k = 1, 2, 3$. The complex 2-sphere is by definition the set of points Z such that $\Sigma(Z^k)^2 = 1$; i.e. such that $\Sigma((X^k)^2 - (Y^k)^2) = 1$ and $\Sigma X^k Y^k = 0$. The mapping of the 2-sphere into $T(S) \subset T(\mathbb{R}^3)$ defined by $(X^k + \mathrm{i}Y^k) \mapsto (X^k/(1 + \Sigma(Y^j)^2)^{1/2}, \ Y^k)$ is a homeomorphism. (Y^k) in $T(\mathbb{R}^3)$ such that $\Sigma X^k Y^k = 0$ is also in $T(S)$.

8) *The 2-sphere is not a group manifold*
We have seen in Problem 5, p. 181 that the 3-sphere is the group manifold of SU(2). It is easy to see that the 1-sphere (the circle) is the group manifold

of the group $\mathcal{R}^1$ of one dimensional rotations of $\mathbf{R}^2$:

$$\text{let } O(\Theta) = \begin{pmatrix} \cos\Theta & -\sin\Theta \\ \sin\Theta & \cos\Theta \end{pmatrix};$$

$O(\Theta)$ is a two-dimensional representation of $\mathcal{R}^1$; indeed

$$O(\Theta'') = O(\Theta')O(\Theta) = O(\Theta' + \Theta)$$
$$O(\Theta + 2\pi) = O(\Theta). \qquad \blacksquare$$

The 2-sphere is not a group manifold. If it were one could construct the left invariant vector fields and the right invariant vector fields generated by the elements of its Lie algebra, i.e. continuous regular vector fields on S. Nonetheless the 2 sphere has "something to do" with the three dimensional rotation group.

Let $\sigma: \mathrm{SO}(3) \times S \to S$ by $(g, x) \mapsto \sigma(g, x)$ and $\sigma_x: \mathrm{SO}(3) \to S$ by $\sigma_x(g) = \sigma(g, x)$. By definition the **auxiliary function** $\sigma^i_\alpha(x)$ is $\partial\sigma^i(g, x)/\partial g^\alpha$ at $g = e$.

auxiliary function

Let $g(t)$ be a one parameter subgroup of SO(3) generated by $\gamma = g^\alpha \gamma_\alpha$

$$\sigma^i_\alpha(x) g^\alpha = v^i_\gamma(x)$$

where v_γ is the vector tangent to the curve $\sigma_x(g(t))$ at $t = 0$.

Under the infinitesimal rotation $1 + \gamma \mathrm{d}t$, the point x is mapped into $x + \mathrm{d}x$ with $\mathrm{d}x^i = \sigma^i_\alpha(x) g^\alpha \,\mathrm{d}t$. On the other hand $\mathrm{d}x^i = g^\alpha \,\mathrm{d}t\, c^i_{\alpha k} x^k$ (see p. 187), where $c^i_{\alpha k} = \epsilon_{i\alpha k}$.

We compute the components of the auxiliary functions in polar coordinates θ, φ defined by $x^1 = \sin\theta \cos\varphi$, $x^2 = \sin\theta \sin\varphi$, $x^3 = \cos\theta$.

A straightforward calculation gives

$$\mathrm{d}\theta = -\sin\varphi g^1 \,\mathrm{d}t + \cos\varphi g^2 \,\mathrm{d}t$$

$$\mathrm{d}\varphi = -\frac{\cos\theta}{\sin\theta} (\cos\varphi g^1 + \sin\varphi g^2) \,\mathrm{d}t + g^3 \,\mathrm{d}t$$

from which we can read off $\sigma^\theta_\alpha(\varphi, \theta)$ and $\sigma^\varphi_\alpha(\varphi, \theta)$ and obtain

$$g^{\theta\theta} = \sigma^\theta_\alpha \sigma^\theta_\beta \gamma^{\alpha\beta} = 1, \quad g^{\theta\varphi} = 0, \quad g^{\varphi\varphi} = \sin^{-2}\theta.$$

The natural metric on S is the familiar spherical metric on S induced by the euclidean metric on $\mathbf{R}^3$. It follows that the spherical metric on S is invariant under rotations of S.

The 2-sphere is the manifold of a transitive realization of the three dimensional rotation group. It can be shown in general that the manifold of a transitive realization of a group is a coset space of the group.

IV. INTEGRATION ON MANIFOLDS

Introduction. In part B we extend to a manifold X^n the theory of integration on $\mathbb{R}^n$ – either ordinary Riemann integrals or Lebesgue integrals (cf. Chapter I.D). We are immediately confronted with the problem of invariance of an integral under a change of coordinates. We know that (p. 47) under the change of coordinates $(y^i) \mapsto (x^i)$ the integral

$$I = \int_{\mathbb{R}^n} f(x^i)\,\mathrm{d}x^1 \ldots \mathrm{d}x^n$$

becomes

$$I = \int_{\mathbb{R}^n} f(x^i(y^j)) \left|\frac{\mathrm{D}(x^i)}{\mathrm{D}(y^j)}\right| \mathrm{d}y^1 \ldots \mathrm{d}y^n,$$

an expression which does not result from straightforward use of the relation $\mathbf{d}x^i = (\partial x^i/\partial y^j)\,\mathbf{d}y^j$. We also know from elementary theory that changing coordinates in, for instance, a surface integral in ordinary 3-space is a rather complicated process. The reason for such awkwardness is that the true nature of the integrand has been left obscure: $f(x^i)$ is in fact the component of an n-form at x defined on $\mathbb{R}^n$ (component at x of a totally antisymmetric covariant tensor field of order n). This explains the change $f(x^i) \mapsto f(x^i(y^j))|\mathrm{D}(x^i)/\mathrm{D}(y^j)|$ under a change of coordinates. More generally, the integrand of a p-integral on X^n is a p-form on X^n; forms are studied in part A.

Part C corresponds to the second meaning of the word integration: integration in the sense of construction of solutions of systems of partial differential equations. In both parts, B and C, the theory that we shall give rests on the theory of exterior differential forms.

A. EXTERIOR DIFFERENTIAL FORMS

1. EXTERIOR ALGEBRA

A totally antisymmetric covariant p-tensor field is called a **(exterior differential) p-form (form of degree p).** *p-form degree*
The space of p-forms of class C^k on a smooth manifold[1] X is a submodule

[1]Smooth means of class C^r, with r large enough for the statement to be meaningful. Here $r \geq k+1$. *smooth*

$\Lambda^p(X)$

$\Lambda^p(X)$ of the module, over the ring of C^k functions, of all covariant C^k tensor fields on X: the sum of two p-forms is a p-form, the product of a p-form and a function f is a p-form. $\Lambda_x^p(X)$ is the space of p-forms at x. A form of degree p superior to the dimension n of the manifold on which it is defined is identically zero because the only non zero components of a totally antisymmetric p-tensor field are those in which all indices are different, a situation which can never exist if $p > n$.

exterior product

The **exterior product** $\wedge$ (**wedge product, Grassman product**) of a p-form and a q-form is a mapping

$$\wedge: (\Lambda^p(X), \Lambda^q(X)) \to \Lambda^{p+q}(X)$$

by $(\boldsymbol{\alpha}, \boldsymbol{\beta}) \mapsto \boldsymbol{\alpha} \wedge \boldsymbol{\beta}$ with $\boldsymbol{\alpha} \wedge \boldsymbol{\beta}$ defined by[1]

$$(\alpha \wedge \beta)(v_1, \ldots v_{p+q}) = \frac{1}{p!q!} \sum_{\pi} (\text{sign } \pi) \pi[\boldsymbol{\alpha}(\boldsymbol{v}_1, \ldots \boldsymbol{v}_p) \boldsymbol{\beta}(\boldsymbol{v}_{p+1}, \ldots \boldsymbol{v}_{p+q})]$$

where $\boldsymbol{v}_i \in T(X)$ and π is a permutation of $(1, 2, \ldots, p+q)$.
In particular, if $\boldsymbol{\alpha}$ and $\boldsymbol{\beta}$ are 1-forms

$$\boldsymbol{\alpha} \wedge \boldsymbol{\beta} = \boldsymbol{\alpha} \otimes \boldsymbol{\beta} - \boldsymbol{\beta} \otimes \boldsymbol{\alpha}.$$

Properties: It follows from the definition that the exterior product is

associative $\quad (\boldsymbol{\alpha} \wedge \boldsymbol{\beta}) \wedge \boldsymbol{\gamma} = \boldsymbol{\alpha} \wedge (\boldsymbol{\beta} \wedge \boldsymbol{\gamma}),$

bilinear $\quad \boldsymbol{\alpha} \wedge (\boldsymbol{\beta} + \boldsymbol{\gamma}) = \boldsymbol{\alpha} \wedge \boldsymbol{\beta} + \boldsymbol{\alpha} \wedge \boldsymbol{\gamma}$

$$(\boldsymbol{\alpha} + \boldsymbol{\beta}) \wedge \boldsymbol{\gamma} = \boldsymbol{\alpha} \wedge \boldsymbol{\gamma} + \boldsymbol{\beta} \wedge \boldsymbol{\gamma}$$

$$f(\boldsymbol{\alpha} \wedge \boldsymbol{\beta}) = f\boldsymbol{\alpha} \wedge \boldsymbol{\beta} = \boldsymbol{\alpha} \wedge f\boldsymbol{\beta},$$

and in general not commutative

$$\boldsymbol{\alpha} \wedge \boldsymbol{\beta} = (-1)^{pq} \boldsymbol{\beta} \wedge \boldsymbol{\alpha} \qquad \text{if} \quad \boldsymbol{\alpha} \in \Lambda^p, \ \boldsymbol{\beta} \in \Lambda^q.$$

To show this last property notice that, in the process of interchanging $\boldsymbol{\beta}$ and $\boldsymbol{\alpha}$, comparable terms will be obtained by making p 1-forms jump q times, i.e. by pq permutations. It follows from this property that $(\wedge \boldsymbol{\omega})^k$ is identically zero if the degree of $\boldsymbol{\omega}$ is odd – but not otherwise.

Exercise: If $\boldsymbol{\theta}^1, \boldsymbol{\theta}^2, \ldots \boldsymbol{\theta}^p$ are 1-forms, show that

$$\boldsymbol{\theta}^1 \wedge \ldots \wedge \boldsymbol{\theta}^p = \epsilon^{1,2\ldots p}_{i_1 i_2 \ldots i_p} \boldsymbol{\theta}^{i_1} \otimes \cdots \otimes \boldsymbol{\theta}^{i_p}.$$

exterior algebra Grassman

The set of forms of all degrees on X together with the exterior product is called the **exterior (Grassman) algebra**; it is denoted $\Lambda(X)$ or simply Λ. Λ_x

[1]Some authors use the factor $1/(p+q)!$ rather than $1/p!q!$. Then if α and β are 1-forms

$$\boldsymbol{\alpha} \wedge \boldsymbol{\beta} = \tfrac{1}{2}(\boldsymbol{\alpha} \otimes \boldsymbol{\beta} - \boldsymbol{\beta} \otimes \boldsymbol{\alpha}).$$

is the space of forms at x. A 1-form is also called a **linear form**. A function f on X can be regarded either as an element of the ring on which the module $\Lambda(X)$ is defined, or as a 0-form. The product of a form $\boldsymbol{\alpha}$ and a function f is $\Lambda(X)$ linear form

$$f \wedge \boldsymbol{\alpha} = f\boldsymbol{\alpha}.$$

The algebra $\Lambda(X)$ is a **graded algebra**, that is, it is a collection $\{\Lambda^p(X)\}$ of modules indexed by the integers p together with an associative bilinear mapping $\Lambda(X) \times \Lambda(X) \to \Lambda(X)$. graded algebra

Local coordinates; basis for $\Lambda^p(X)$; strict components.

1) At a point $x \in X$. Let $(\boldsymbol{\theta}^i; i = 1, \ldots, n)$ be an arbitrary basis of $T_x^*(X^n)$. Let capital letters label ordered natural numbers: $I_j < I_{j+1}$.
A basis of $\Lambda_x^p(X^n)$ is the set of $\binom{n}{p}$ independent p-forms

$$\{\boldsymbol{\theta}^{I_1} \wedge \ldots \wedge \boldsymbol{\theta}^{I_p};\ I_j = 1, \ldots n\}.$$

Proof: The space of p-forms at x is a subspace of $\otimes^p T_x^*$. The set $\{\boldsymbol{\theta}^{i_1} \otimes \cdots \otimes \boldsymbol{\theta}^{i_p};\ i_j = 1, \ldots n\}$ is a basis for $\otimes^p T_x^*$. Hence a p-form can be written

$$\begin{aligned}
\boldsymbol{\alpha} &= \alpha_{i_1 \ldots i_p} \boldsymbol{\theta}^{i_1} \otimes \cdots \otimes \boldsymbol{\theta}^{i_p} \qquad \text{with } \alpha_{i_1 \ldots i_p} \text{ totally antisymmetric} \\
&= \frac{1}{p!} \epsilon^{k_1 \ldots k_p}_{i_1 \ldots i_p} \alpha_{k_1 \ldots k_p} \boldsymbol{\theta}^{i_1} \otimes \cdots \otimes \boldsymbol{\theta}^{i_p} \\
&= \frac{1}{p!} \alpha_{k_1 \ldots k_p} \boldsymbol{\theta}^{k_1} \wedge \ldots \wedge \boldsymbol{\theta}^{k_p} \\
&= \alpha_{K_1 \ldots K_p} \boldsymbol{\theta}^{K_1} \wedge \ldots \wedge \boldsymbol{\theta}^{K_p}.
\end{aligned}$$

The last two lines are both perfectly good expansions of $\boldsymbol{\alpha}$; however the set $\{\boldsymbol{\theta}^{k_1} \wedge \ldots \wedge \boldsymbol{\theta}^{k_p}\}$ is not a basis for $\Lambda_x^p(X)$ because the elements of the set are not independent, e.g. $\boldsymbol{\theta}^{k_1} \wedge \boldsymbol{\theta}^{k_2} \ldots \wedge \boldsymbol{\theta}^{k_p} = -\boldsymbol{\theta}^{k_2} \wedge \boldsymbol{\theta}^{k_1} \ldots \wedge \boldsymbol{\theta}^{k_p}$. In the space $\otimes^p T_x^*(X^n)$, the components of $\boldsymbol{\alpha}$ are $\{\alpha_{k_1 \ldots k_p}\}$; in the space $\Lambda_x^p(X^n)$ the components of $\boldsymbol{\alpha}$ are $\{\alpha_{K_1 \ldots K_p}\}$. The components with ordered indices are called the **strict components** of α. strict components

The dimension of the space $\Lambda_x^p(X^n)$ is equal to the binomial coefficient $\binom{n}{p} = n!/(n-p)!p!$. In particular an n-form $\boldsymbol{\alpha} \in \Lambda_x^n(X^n)$ has only one strict component $\alpha_{1 \ldots n}$. The sum of the dimension of the spaces of forms of all degrees is $\Sigma_{p=0}^n \binom{n}{p} = 2^n$.
A p-form with only one component $\boldsymbol{\alpha} = a\boldsymbol{\theta}^{i_1} \wedge \ldots \wedge \boldsymbol{\theta}^{i_p}$ is called a **monomial.** monomial

2) In the domain of a chart. Let $(\boldsymbol{\theta}^i)$ be an arbitrary basis for covariant vector fields in the domain of a chart. Then $(\boldsymbol{\theta}^i)$ is either the **natural** natural basis

local frame **basis** $(\mathbf{d}x^i)$ or a **local frame** $\{\boldsymbol{\theta}^i = a^i_j\, \mathbf{d}x^j\}$ where $[a^i_j(x)]$ is a regular (invertible) differentiable matrix. It follows that $\{\boldsymbol{\theta}^{I_1} \wedge \ldots \wedge \boldsymbol{\theta}^{I_p}\}$ is a basis for the p-forms in the chosen chart.

Exercise: Let $\boldsymbol{\omega} = \boldsymbol{\alpha} \wedge \boldsymbol{\beta}$; express the components of $\boldsymbol{\omega}$ in terms of the components of $\boldsymbol{\alpha}$ and $\boldsymbol{\beta}$. First let $\boldsymbol{\alpha}, \boldsymbol{\beta} \in \Lambda^1(X)$,

$$\boldsymbol{\alpha} = \alpha_i \boldsymbol{\theta}^i, \qquad \boldsymbol{\beta} = \beta_i \boldsymbol{\theta}^i \qquad i = 1, \ldots n.$$

Then

$$\begin{aligned} \boldsymbol{\alpha} \wedge \boldsymbol{\beta} &= \alpha_i \boldsymbol{\theta}^i \wedge \beta_j \boldsymbol{\theta}^j \\ &= \alpha_i \beta_j \boldsymbol{\theta}^i \wedge \boldsymbol{\theta}^j \\ &= (\alpha_I \beta_J - \alpha_J \beta_I) \boldsymbol{\theta}^I \wedge \boldsymbol{\theta}^J \qquad (I < J), \end{aligned}$$

which implies

$$(\boldsymbol{\alpha} \wedge \boldsymbol{\beta})_{IJ} = \alpha_I \beta_J - \alpha_J \beta_I. \qquad \blacksquare$$

Notice that

$$\alpha_i \beta_j \boldsymbol{\theta}^i \wedge \boldsymbol{\theta}^j \neq 2 \alpha_I \beta_J \boldsymbol{\theta}^I \wedge \boldsymbol{\theta}^J,$$

whereas

$$\boldsymbol{\omega} = \omega_{IJ} \boldsymbol{\theta}^I \wedge \boldsymbol{\theta}^J = \tfrac{1}{2} \omega_{ij} \boldsymbol{\theta}^i \wedge \boldsymbol{\theta}^j \quad \text{if } \omega_{ij} = -\omega_{ji}.$$

Notice also that the components of a form are always antisymmetric; $\alpha_i \beta_j$ is not a component because $\{\boldsymbol{\theta}^i \wedge \boldsymbol{\theta}^j\}$ is not a basis. These remarks are trivial but often forgotten on class tests.

Example: Let $\boldsymbol{\alpha}$ and $\boldsymbol{\beta}$ be two 1-forms in $\mathbb{R}^3$

$$\boldsymbol{\alpha} = \alpha_i\, \mathbf{d}x^i \qquad \boldsymbol{\beta} = \beta_i\, \mathbf{d}x^i.$$

Let (α_i) be the components of a vector $\boldsymbol{V}$ and (β_i) the components of a vector $\boldsymbol{U}$,

$$\begin{aligned} \boldsymbol{\alpha} \wedge \boldsymbol{\beta} = (\alpha_1 \beta_2 - \alpha_2 \beta_1)\, \mathbf{d}x^1 \wedge \mathbf{d}x^2 + (\alpha_1 \beta_3 - \alpha_3 \beta_1)\, \mathbf{d}x^1 \wedge \mathbf{d}x^3 \\ + (\alpha_2 \beta_3 - \alpha_3 \beta_2)\, \mathbf{d}x^2 \wedge \mathbf{d}x^3. \end{aligned}$$

The components of $\boldsymbol{\alpha} \wedge \boldsymbol{\beta}$ are related to the components of the vector product $\boldsymbol{V} \times \boldsymbol{U}$; the exact relation will be established later (p. 295). At present we notice that they would be equal if the basis of $\Lambda^2(\mathbb{R}^3)$ were $(\mathbf{d}x^2 \wedge \mathbf{d}x^3, \mathbf{d}x^3 \wedge \mathbf{d}x^1, \mathbf{d}x^1 \wedge \mathbf{d}x^2)$ instead of $(\mathbf{d}x^{I_1} \wedge \mathbf{d}x^{I_2})$.
Let now $\boldsymbol{\alpha} \in \Lambda^p(X^n)$, $\boldsymbol{\beta} \in \Lambda^q(X^n)$. Then

$$\begin{aligned} (\boldsymbol{\alpha} \wedge \boldsymbol{\beta})_{i_1 \ldots i_{p+q}} &= \frac{1}{p!} \frac{1}{q!} \epsilon^{j_1 \ldots j_p k_1 \ldots k_q}_{i_1 \ldots\ldots i_{p+q}} \alpha_{j_1 \ldots j_p} \beta_{k_1 \ldots k_q} \\ &= \epsilon^{J_1 \ldots J_p K_1 \ldots K_q}_{i_1 \ldots\ldots i_{p+q}} \alpha_{J_1 \ldots J_p} \beta_{K_1 \ldots K_q}, \end{aligned}$$

where the ordering of the upper case indices is a partial ordering in which the J subset and the K subset are both totally ordered but unrelated to each other.

Exercise: Let $\boldsymbol{\omega} = \boldsymbol{\alpha}^1 \wedge \ldots \wedge \boldsymbol{\alpha}^p$ with $\boldsymbol{\alpha}^j = \alpha^j_i \boldsymbol{\theta}^i \in \Lambda^1(X^n)$, that is α^j_i is the ith component of the 1-form $\boldsymbol{\alpha}^j$. Find the strict components of $\boldsymbol{\omega}$.

Answer:
$$\boldsymbol{\omega} = \alpha^1_{i_1}\alpha^2_{i_2} \ldots \alpha^p_{i_p} \boldsymbol{\theta}^{i_1} \wedge \ldots \wedge \boldsymbol{\theta}^{i_p}$$
$$= \epsilon^{j_1 \ldots j_p}_{I_1 \ldots I_p} \alpha^1_{j_1}\alpha^2_{j_2} \ldots \alpha^p_{j_p} \boldsymbol{\theta}^{I_1} \wedge \ldots \wedge \boldsymbol{\theta}^{I_p}. \quad ■$$

Change of basis. The transformation of the components of a p-form induced by the change of basis $(\boldsymbol{\theta}^i) \mapsto (\boldsymbol{\theta}^{i'} = a^{i'}_j \boldsymbol{\theta}^j)$ can be readily obtained from the above formula. Let $\boldsymbol{\omega} = \omega_{I'_1 \ldots I'_p} \boldsymbol{\theta}^{I'_1} \wedge \ldots \wedge \boldsymbol{\theta}^{I'_p}$ with $\omega_{i'_1 \ldots i'_p}$ totally antisymmetric. Then

$$\boldsymbol{\omega} = \frac{1}{p!} \omega_{i'_1 \ldots i'_p} \boldsymbol{\theta}^{i'_1} \wedge \ldots \wedge \boldsymbol{\theta}^{i'_p}$$

$$= \frac{1}{p!} \omega_{i'_1 \ldots i'_p} a^{i'_1}_{j_1} \ldots a^{i'_p}_{j_p} \boldsymbol{\theta}^{j_1} \wedge \ldots \wedge \boldsymbol{\theta}^{j_p}$$

$$= \frac{1}{p!} \omega_{i'_1 \ldots i'_p} \epsilon^{k_1 \ldots k_p}_{J_1 \ldots J_p} a^{i'_1}_{k_1} \ldots a^{i'_p}_{k_p} \boldsymbol{\theta}^{J_1} \wedge \ldots \wedge \boldsymbol{\theta}^{J_p}.$$

Therefore
$$\omega_{J_1 \ldots J_p} = \omega_{I'_1 \ldots I'_p} \epsilon^{k_1 \ldots k_p}_{J_1 \ldots J_p} a^{I'_1}_{k_1} \ldots a^{I'_p}_{k_p}.$$

The terms $\epsilon^{k_1 \ldots k_p}_{J_1 \ldots J_p} a^{I'_1}_{k_1} \ldots a^{I'_p}_{k_p}$ are the determinants of all the $p \times p$ matrices obtained from the $n \times n$ matrix $[a^{i'}_j]$ by eliminating $n - p$ rows and $n - p$ columns in all possible ways. It is often convenient to consider these determinants as the elements of a new matrix called the **p-compound matrix** derived from $[a^{i'}_j]$ (Problem 1, p. 270).

p-compound matrix

If the change of basis is a change of natural basis induced by the change of coordinates $x \mapsto \bar{x}(x)$, then $a^{i'}_i = \partial\bar{x}^{i'}/\partial x^i$ and

change of natural basis

$$\omega_{J_1 \ldots J_p}(x) = \omega_{I'_1 \ldots I'_p}(\bar{x}(x)) \frac{D(\bar{x}^{I'_1}, \ldots \bar{x}^{I'_p})}{D(x^{J_1}, \ldots x^{J_p})}.$$

$D(\bar{x}^{I'})/D(x^J)$ is the p-compound matrix derived from the jacobian matrix $[\partial\bar{x}^{i'}/\partial x^j]$. These relations give the rules for the change of variables of a p-integral in an n-dimensional space; a change of coordinates can be written as follows

$$\omega = \frac{1}{p!}\,\omega_{i_1\ldots i_p}(x)\,\mathbf{d}x^{i_1} \wedge \mathbf{d}x^{i_2} \ldots \wedge \mathbf{d}x^{i_p}$$

$$= \frac{1}{p!}\,\omega_{i'_1\ldots i'_p}(\bar{x}(x))\,\mathbf{d}\bar{x}^{i'_1}(x) \wedge \ldots \wedge \mathbf{d}\bar{x}^{i'_p}(x)$$

where $\mathbf{d}\bar{x}^{i'_j}(x) = (\partial \bar{x}^{i'_j}/\partial x^{k_j})\,\mathbf{d}x^{k_j}$.

2. EXTERIOR DIFFERENTIATION

Let $\boldsymbol{\alpha}$ be a differential p-form of class C^k ($\alpha_{I_1\ldots I_p}(x)$ are differentiable functions of x of class C^k).

exterior derivative coboundary d

The exterior differentiation operator d maps a p-form $\boldsymbol{\alpha}$ of class C^k into a $p+1$ form $\mathrm{d}\boldsymbol{\alpha}$ of class C^{k-1} called the **exterior derivative (coboundary)** of $\boldsymbol{\alpha}$. It has the following properties.

1) d is linear: $\mathrm{d}(\boldsymbol{\alpha}+\boldsymbol{\beta}) = \mathrm{d}\boldsymbol{\alpha} + \mathrm{d}\boldsymbol{\beta}$,
   if λ is a constant $\mathrm{d}(\lambda\boldsymbol{\alpha}) = \lambda\,\mathrm{d}\boldsymbol{\alpha}$.
2) $\mathrm{d}(\boldsymbol{\alpha} \wedge \boldsymbol{\beta}) = \mathrm{d}\boldsymbol{\alpha} \wedge \boldsymbol{\beta} + (-1)^p \boldsymbol{\alpha} \wedge \mathrm{d}\boldsymbol{\beta}$, $\quad p$ = degree of $\boldsymbol{\alpha}$.
3) $\mathrm{d}^2 = 0$.
4) If f is a 0-form, df is the ordinary differential $\mathbf{d}f$ of f.
5) The operation d is local: if $\boldsymbol{\alpha}$ and $\boldsymbol{\beta}$ coincide on an open set U, $\mathrm{d}\boldsymbol{\alpha} = \mathrm{d}\boldsymbol{\beta}$ on U; that is, the behavior of $\boldsymbol{\alpha}$ outside U does not affect $\mathrm{d}\boldsymbol{\alpha}$ on U, $(\mathrm{d}\boldsymbol{\alpha})|_U = \mathrm{d}(\boldsymbol{\alpha}|_U)$.

Theorem. Properties 1 *to* 4 *uniquely*[1] *define the operator* d.

Proof: Let d′ be an operator which satisfies the properties 1–4 and let $\boldsymbol{\alpha}$ be a differential p-form

$$\boldsymbol{\alpha} = \alpha_{I_1\ldots I_p} \wedge \mathbf{d}x^{I_1} \wedge \ldots \wedge \mathbf{d}x^{I_p}$$

where $\alpha_{I_1\ldots I_p}$ is treated here as a 0-form because in the course of the proof it will be operated on by d′. We shall show that d′$\boldsymbol{\alpha}$ is uniquely defined.

By property 2

$$\mathrm{d}'\boldsymbol{\alpha} = \mathrm{d}'\alpha_{I_1\ldots I_p} \wedge \mathbf{d}x^{I_1} \ldots \wedge \mathbf{d}x^{I_p} + \alpha_{I_1\ldots I_p}\,\mathrm{d}'(\mathbf{d}x^{I_1} \wedge \ldots \wedge \mathbf{d}x^{I_p}).$$

[1]If the exterior product is defined with $(p+q)!$ rather than $p!q!$ (p. 196), then the exterior derivative is modified accordingly. It becomes $\bar{\mathrm{d}} = \mathrm{d}/(p+1)$ when operating on a p-form. We chose the definition of the exterior product which leads to the simplest expression for the Stokes' formula.

By property 4

$$\mathrm{d}'\alpha_{I_1\ldots I_p} = \mathbf{d}\boldsymbol{\alpha}_{I_1\ldots I_p}, \qquad \mathrm{d}'x^i = \mathbf{d}\boldsymbol{x}^i.$$

By properties 2 and 3

$$\mathrm{d}'(\mathrm{d}'x^{I_1} \wedge \ldots \wedge \mathrm{d}'x^{I_p}) = 0.$$

Hence

$$\mathrm{d}'\boldsymbol{\alpha} = \mathbf{d}\boldsymbol{\alpha}_{I_1\ldots I_p} \wedge \mathbf{d}\boldsymbol{x}^{I_1} \wedge \ldots \wedge \mathbf{d}\boldsymbol{x}^{I_p}$$

and d′ is the uniquely defined operator d.
We shall check (1) that the operator d defined in local coordinates by this expression satisfies properties 1 to 5, (2) that its definition does not depend on the choice of coordinates, and (3) that $\mathbf{d}\boldsymbol{\alpha}$ given by this expression in all charts is a $(p+1)$-form defined over the whole manifold.

(1) From $\qquad \mathrm{d}\boldsymbol{\alpha} = \mathbf{d}\boldsymbol{\alpha}_{I_1\ldots I_p} \wedge \mathbf{d}\boldsymbol{x}^{I_1} \wedge \ldots \wedge \mathbf{d}\boldsymbol{x}^{I_p}$

we obtain the expansion of $\mathrm{d}\boldsymbol{\alpha}$

$$\begin{aligned}\mathrm{d}\boldsymbol{\alpha} &= \frac{\partial\alpha_{I_1\ldots I_p}}{\partial x^k}\mathbf{d}\boldsymbol{x}^k \wedge \mathbf{d}\boldsymbol{x}^{I_1} \wedge \ldots \wedge \mathbf{d}\boldsymbol{x}^{I_p} \\ &= \epsilon^{kI_1\ldots I_p}_{J_1\ldots J_{p+1}} \frac{\partial\alpha_{I_1\ldots I_p}}{\partial x^k}\mathbf{d}\boldsymbol{x}^{J_1} \wedge \ldots \wedge \mathbf{d}\boldsymbol{x}^{J_{p+1}}.\end{aligned}$$

From this definition of $\mathrm{d}\boldsymbol{\alpha}$, one can readily check properties 1, 2 and 5.

Property 3 follows from the symmetry in j and k of the second derivatives $\partial^2/\partial x^j \partial x^k$. Property 2 follows from the anticommutativity of two 1-forms. (*Hint*: in the regrouping of terms necessary to prove property 2, $\mathbf{d}\boldsymbol{\beta}_{I_1\ldots I_q}$ has to jump p terms $\mathbf{d}\boldsymbol{x}^{I_1} \wedge \ldots \wedge \mathbf{d}\boldsymbol{x}^{I_p}$.)

(2) Let us make the change of coordinates $(x^i) \mapsto (\bar{x}^{i'})$; let $(\mathrm{d}\boldsymbol{\alpha})_x$ be the expression for $\mathrm{d}\boldsymbol{\alpha}$ in the x coordinate system. From the rules obtained earlier for the transformation of a form under a change of coordinates we have

$$\boldsymbol{\alpha} = \frac{1}{p!}\alpha_{i_1\ldots i_p}\mathbf{d}\boldsymbol{x}^{i_1} \wedge \ldots \wedge \mathbf{d}\boldsymbol{x}^{i_p} = \frac{1}{p!}\alpha_{i'_1\ldots i'_p}(\bar{x}(x))\,\mathbf{d}\bar{\boldsymbol{x}}^{i'_1} \wedge \ldots \wedge \mathbf{d}\bar{\boldsymbol{x}}^{i'_p}.$$

Thus, after some calculation,

$$\begin{aligned}(\mathrm{d}\boldsymbol{\alpha})_x &= \frac{1}{p!}\frac{\partial\alpha_{i'_1\ldots i'_p}}{\partial\bar{x}^{k'}}\frac{\partial\bar{x}^{k'}}{\partial x^j}\mathbf{d}\boldsymbol{x}^j \wedge \mathbf{d}\bar{\boldsymbol{x}}^{i'_1} \wedge \ldots \wedge \mathbf{d}\bar{\boldsymbol{x}}^{i'_p} \\ &= \frac{1}{p!}\frac{\partial\alpha_{i'_1\ldots i'_p}}{\partial\bar{x}^{k'}}\mathbf{d}\bar{\boldsymbol{x}}^{k'} \wedge \mathbf{d}\bar{\boldsymbol{x}}^{i'_1} \wedge \ldots \wedge \mathbf{d}\bar{\boldsymbol{x}}^{i'_p} \\ &= (\mathrm{d}\boldsymbol{\alpha})_{\bar{x}}.\end{aligned}$$

(3) Let U_i and U_j be the domains of two maps, d_i and d_j the exterior derivative defined in U_i and U_j respectively. Then $\mathrm{d}_i = \mathrm{d}_j$ in $U_i \cap U_j$ by the

uniqueness theorem. This completes the proof of the existence and uniqueness of d. ■

Remark: The above proof shows that, whereas the $\partial\alpha_{I_1\dots I_p}/\partial x^k$ are not the components of a $p+1$ tensor, the antisymmetrized expressions $\epsilon^{kI_1\dots I_p}_{J_1\dots J_{p+1}}\,\partial\alpha_{I_1\dots I_p}/\partial x^k$ are a set of tensor components.
Remark: The exterior derivative of an n-form on X^n is zero.
Remark: d is an example of an antiderivation (see p. 206).

differential operator of degree s

Let $A = \{A^p\}$ be a graded algebra. A linear mapping $T: A \to A$ is a **differential operator of degree** s on A if

$$TT = 0 \quad \text{and } T: A^p \to A^{p+s}.$$

differential graded algebra DGA

Thus $\Lambda(X)$ together with d is a graded algebra (p. 197) with differential operator of degree 1 or a **differential graded algebra** (DGA).

In the following examples the forms are defined on $\mathbb{R}^3$.

Example 1: The exterior derivative of a 0-form f is the 1-form

$$\mathbf{d}f = \frac{\partial f}{\partial x^i}\,\mathbf{d}x^i.$$

The components of $\mathbf{d}f$ are the components of grad f.

Example 2: Exterior derivative of a 1-form $\boldsymbol{\alpha} = A\,\mathbf{d}x + B\,\mathbf{d}y + C\,\mathbf{d}z$. Let V be the vector with components (A, B, C).

$$\mathrm{d}\boldsymbol{\alpha} = \left(\frac{\partial C}{\partial y} - \frac{\partial B}{\partial z}\right)\mathbf{d}y \wedge \mathbf{d}z + \left(\frac{\partial A}{\partial z} - \frac{\partial C}{\partial x}\right)\mathbf{d}z \wedge \mathbf{d}x + \left(\frac{\partial B}{\partial x} - \frac{\partial A}{\partial y}\right)\mathbf{d}x \wedge \mathbf{d}y.$$

The components of $\mathrm{d}\boldsymbol{\alpha}$ with respect to the indicated basis, are the components of curl $\boldsymbol{V} \equiv \mathrm{rot}\,\boldsymbol{V} \equiv \nabla \times \boldsymbol{V}$.

Example 3: Exterior derivative of a 2-form

$$\boldsymbol{\omega} = P\,\mathbf{d}y \wedge \mathbf{d}z + Q\,\mathbf{d}z \wedge \mathbf{d}x + R\,\mathbf{d}x \wedge \mathbf{d}y,$$
$$\mathrm{d}\boldsymbol{\omega} = \left(\frac{\partial P}{\partial x} + \frac{\partial Q}{\partial y} + \frac{\partial R}{\partial z}\right)\mathbf{d}x \wedge \mathbf{d}y \wedge \mathbf{d}z.$$

Let $\boldsymbol{W}$ be the axial vector with components (P, Q, R); an axial vector is the set of the 3 unrelated components of an antisymmetric tensor of rank 2 defined on $\mathbb{R}^3$. In vector calculus notation $\boldsymbol{\omega} = \boldsymbol{W} \cdot \mathrm{d}\boldsymbol{\sigma}$ where $\mathrm{d}\boldsymbol{\sigma}$ is the "infinitesimal vector element of area".
The components of $\mathrm{d}\boldsymbol{\omega}$ are the components of div $\boldsymbol{W}$.

Example 4: From the property $d^2 = 0$ we conclude that

$$\mathrm{d}\,\mathbf{d}f \equiv \mathrm{curl}\ (\mathrm{grad}\ f) = 0,$$
$$\mathrm{d}\,\mathbf{d}\boldsymbol{\alpha} \equiv \mathrm{div}\ (\mathrm{curl}\ \boldsymbol{V}) = 0.$$

Exercise: Compute grad fg, curl $f\boldsymbol{V}$, div $f\boldsymbol{V}$, div $(\boldsymbol{V} \times \boldsymbol{U})$ where $\boldsymbol{V}$ and $\boldsymbol{U}$ are vectors in $\mathbb{R}^3$, using the relation

$$\mathrm{d}(\boldsymbol{\alpha} \wedge \boldsymbol{\beta}) = \mathbf{d}\boldsymbol{\alpha} \wedge \boldsymbol{\beta} + (-1)^p \boldsymbol{\alpha} \wedge \mathbf{d}\boldsymbol{\beta}, \quad p = \text{degree } \boldsymbol{\alpha}.$$

Answer:

$$\left.\begin{aligned} \mathrm{grad}\ fg &= g\ \mathrm{grad}\ f + f\ \mathrm{grad}\ g \\ \mathrm{curl}\ f\boldsymbol{V} &= \mathrm{grad}\ f \times \boldsymbol{V} + f\ \mathrm{curl}\ \boldsymbol{V} \\ \mathrm{div}\ f\boldsymbol{V} &= \mathrm{grad}\ f \cdot \boldsymbol{V} + f\ \mathrm{div}\ \boldsymbol{V} \end{aligned}\right\} \quad p = 0$$
$$\mathrm{div}\ (\boldsymbol{V} \times \boldsymbol{U}) = \boldsymbol{U} \cdot \mathrm{curl}\ \boldsymbol{V} - \boldsymbol{V} \cdot \mathrm{curl}\ \boldsymbol{U} \quad p = 1.$$

3. Reciprocal image of a form (pull back)

Reciprocal images have already been studied (pp. 116, 134, 138). However reciprocal images of forms play an important role in integration, both in integration of forms and in integration of exterior differential systems; we shall therefore study their properties more explicitly than before.

Let f be a differentiable mapping $X^n \to Y^q$, and let $\boldsymbol{v} \in T_x(X^n)$. Then

$$f'\boldsymbol{v} \in T_{f(x)}(Y^q).$$

Let f^* be the mapping induced by f on p-forms.

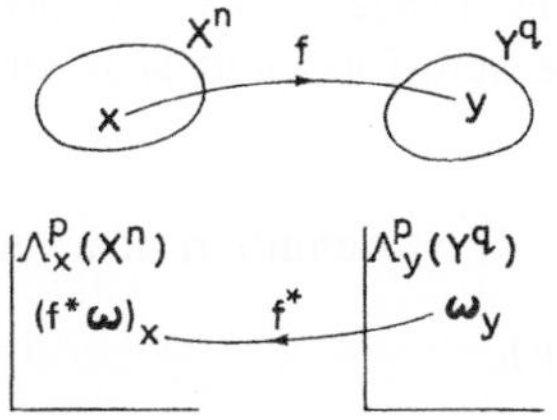

The **reciprocal image (pull back)** $f^*\boldsymbol{\omega}$ of the form $\boldsymbol{\omega}$ is defined by reciprocal image pull back

$$(f^*\boldsymbol{\omega})(\boldsymbol{v}_1, \ldots \boldsymbol{v}_p) = \boldsymbol{\omega}(f'\boldsymbol{v}_1, \ldots f'\boldsymbol{v}_p).$$

The reciprocal image $f^*\boldsymbol{\omega}$ is also called the form **induced** by f from $\boldsymbol{\omega}$. In local coordinates f is defined by $y^\alpha(x^i)$ with $i = 1, \ldots n$ and $\alpha = 1, \ldots q$.

Then

$$(f^*\omega)_{i_1\ldots i_p} = \frac{\partial y^{\alpha_1}}{\partial x^{i_1}} \cdots \frac{\partial y^{\alpha_p}}{\partial x^{i_p}} \omega_{\alpha_1\ldots\alpha_p}$$

$$= \frac{\mathrm{D}(y^{A_1}, \ldots y^{A_p})}{\mathrm{D}(x^{i_1}, \ldots x^{i_p})} \omega_{A_1\ldots A_p}$$

and since

$$\omega = \omega_{A_1\ldots A_p}(y)\, \mathbf{d}y^{A_1} \wedge \ldots \wedge \mathbf{d}y^{A_p} \qquad A_j = 1, \ldots q,$$

$$f^*\omega = \omega_{A_1\ldots A_p}(y(x))\, \mathbf{d}y^{A_1}(x) \wedge \ldots \wedge \mathbf{d}y^{A_p}(x)$$

$$= (f^*\omega)_{I_1\ldots I_p}(x)\, \mathbf{d}x^{I_1} \wedge \ldots \wedge \mathbf{d}x^{I_p} \qquad I_j = 1, \ldots n.$$

Thus we obtain $f^*\omega$ from ω simply by expressing y^α as a function of x^i, namely $y^\alpha(x^i)$, and hence $\mathbf{d}y^\alpha$ as the differential of $y^\alpha(x^i)$.

Theorem. The induced mapping f^* is a homomorphism on the differential graded algebra $\Lambda(Y^q)$.

Proof: We already know that

$f^*(\lambda\omega + \mu\theta) = \lambda f^*\omega + \mu f^*\theta$ homomorphism on the vector space $\Lambda^p(Y^q)$.

Simple algebraic manipulations show that

$f^*(\omega \wedge \theta) = f^*\omega \wedge f^*\theta$ homomorphism on the graded algebra $\Lambda(Y^q)$.

We shall now prove that

$f^*(\mathrm{d}\omega) = \mathrm{d}(f^*\omega)$ homomorphism on the DGA.

An arbitrary form is a finite sum of exterior products of functions and differentials of functions, hence we have to prove the theorem only for the cases where ω is a zero-form and ω is an exact one-form.

1) ω is a zero form g

$$f^*\,\mathbf{d}g = \mathbf{d}(g \circ f) \text{ by definition of a reciprocal image}$$

$$= \mathbf{d}(f^*g).$$

2) ω is an exact differential $\mathbf{d}g$

$$f^*(\mathrm{d}\,\mathbf{d}g) = 0.$$

On the other hand

$$\mathrm{d}(f^*\,\mathbf{d}g) = \mathrm{d}\,\mathbf{d}(f^*g) = 0. \qquad \blacksquare$$

Exercise: Prove $\mathrm{d} \circ f^* = f^* \circ \mathrm{d}$ using coordinate expressions – a long exercise in manipulation of coordinates.

Example: Anticipating section B.1, we state that if f is a diffeomorphism then

$$\int_{f(c)} \omega = \int_c f^* \omega.$$

Let c be the open rectangle in $\mathbb{R}^2$ defined by

$$0 < \varphi < 2\pi \qquad 0 < \theta < \pi.$$

Let g be the diffeomorphism $c \to (S^2 - \text{a set of measure } 0)$ defined by

$$g\colon (\theta, \varphi) \mapsto (\sin\theta\cos\varphi, \sin\theta\sin\varphi, \cos\theta).$$

Let i be the inclusion mapping from S^2 into $\mathbb{R}^3$

$$i\colon (\sin\theta\cos\varphi, \sin\theta\sin\varphi, \cos\theta) \mapsto (x^1, x^2, x^3).$$

Let $\boldsymbol{\omega}$ be a 2-form on $\mathbb{R}^3$

$$\boldsymbol{\omega} = P\, \mathbf{d}x^2 \wedge \mathbf{d}x^3 + Q\, \mathbf{d}x^3 \wedge \mathbf{d}x^1 + R\, \mathbf{d}x^1 \wedge \mathbf{d}x^2.$$

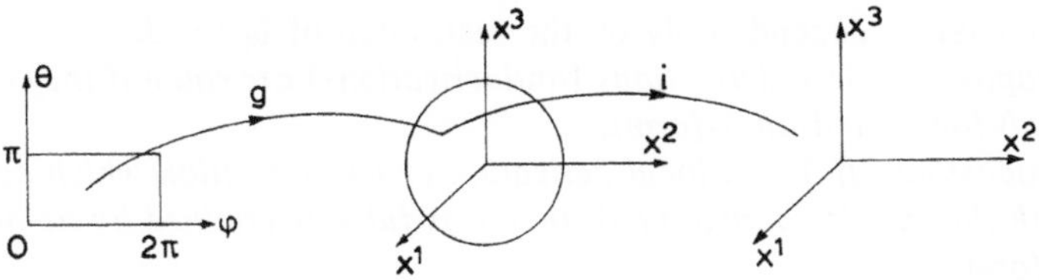

The reciprocal image of $\boldsymbol{\omega}$ on $\mathbb{R}^2$ is

$$(i \circ g)^* \boldsymbol{\omega} = (P \sin\theta\cos\varphi + Q\sin\theta\sin\varphi + R\cos\theta)\sin\theta\, \mathbf{d}\boldsymbol{\theta} \wedge \mathbf{d}\boldsymbol{\varphi}.$$

The integral $\int_c (i \circ g)^* \boldsymbol{\omega}$ can be computed readily whereas $\int_{i \circ g(c)} \boldsymbol{\omega}$ could be an insurmountable job. This example shows that once it is recognized that the integrand of a p-integral is a p-form, a change of variables in an n dimensional space becomes a trivial matter.

4. DERIVATION AND ANTIDERIVATION[1] ON $\Lambda(X)$

We have already mentioned that the Lie derivative is a derivation on the

[1]See [Lichnerowicz 1958].

space of tensor fields (p. 148) and that the exterior derivative is an antiderivation on $\Lambda(X)$ (p. 202). We wish now to give some general properties of derivation and antiderivation and to introduce another example of antiderivation, the interior product.

degree

Let T be a linear operator on $\Lambda(X)$, $T(\lambda\boldsymbol{\omega}+\mu\boldsymbol{\theta})=\lambda T\boldsymbol{\omega}+\mu T\boldsymbol{\theta}$, where λ, μ are constants; T is said to be of **degree** p if

$$T\colon \Lambda^r(X)\to\Lambda^{r+p}(X) \qquad \text{for every } r.$$

derivation

T is a **derivation** on $\Lambda(X)$ if its degree is even and if it obeys Leibniz rule

$$T(\boldsymbol{\omega}\wedge\boldsymbol{\theta})=T\boldsymbol{\omega}\wedge\boldsymbol{\theta}+\boldsymbol{\omega}\wedge T\boldsymbol{\theta}.$$

antiderivation

T is an **antiderivation** on $\Lambda(X)$ if its degree is odd and if it obeys the "antiLeibniz" rule

$$T(\boldsymbol{\omega}\wedge\boldsymbol{\theta})=T\boldsymbol{\omega}\wedge\boldsymbol{\theta}+(-1)^{\text{degree of }\omega}\boldsymbol{\omega}\wedge T\boldsymbol{\theta}.$$

Proposition. The commutator of two derivations is a derivation.
The anticommutator of two antiderivations is a derivation.
The commutator of a derivation with an antiderivation is an antiderivation.

local operator

An operator T on forms is said to be **local** if the restriction of $T\boldsymbol{\omega}$ to any open set U depends only on the restriction of $\boldsymbol{\omega}$ to U.
Proposition. Two derivations [antiderivations] are equal if they are equal on 0-forms and on 1-forms.
Proposition. If T is a local derivation or antiderivation which commutes with the exterior derivative d, *then it is fully determined by its action on 0-forms.*

Proof:

$$T(g\,\mathbf{d}f)=Tg\wedge\mathbf{d}f+gT(\mathbf{d}f)=Tg\wedge\mathbf{d}f+g\,\mathrm{d}(Tf). \qquad \blacksquare$$

Exercise: Show that the Lie derivative is a derivation on $\Lambda(X)$. It follows by antisymmetrization of (p. 148)

$$\mathcal{L}_v(\boldsymbol{s}\otimes\boldsymbol{u})=\mathcal{L}_v\boldsymbol{s}\otimes\boldsymbol{u}+\boldsymbol{s}\otimes\mathcal{L}_v\boldsymbol{u}$$

that

$$\mathcal{L}_v(\boldsymbol{\omega}\wedge\boldsymbol{\theta})=\mathcal{L}_v\boldsymbol{\omega}\wedge\boldsymbol{\theta}+\boldsymbol{\omega}\wedge\mathcal{L}_v\boldsymbol{\theta}. \qquad \blacksquare$$

The coordinate expression for the Lie derivative of a p-form can be written in a simple form, which however is not antisymmetric, namely

$\mathcal{L}_v\boldsymbol{\omega}$

$$\mathcal{L}_v\boldsymbol{\omega}=\frac{1}{p!}(v^k\partial_k\omega_{i_1\ldots i_p}+p\,\omega_{ki_2\ldots i_p}\partial_{i_1}v^k)\,\mathbf{d}x^{i_1}\wedge\ldots\wedge\mathbf{d}x^{i_p}.$$

The contracted multiplication or **interior product (inner product)** of a form $\boldsymbol{\omega}$ and a vector $\boldsymbol{v}$ is denoted $\mathrm{i}_v\boldsymbol{\omega}$ – also denoted $\boldsymbol{v} \lrcorner \boldsymbol{\omega}$ or $\mathrm{i}(\boldsymbol{v})\boldsymbol{\omega}$. The operator i_v is defined as follows.

interior product i_v

$v \lrcorner \omega$

1) i_v is an antiderivation.
2) $\mathrm{i}_v f = 0$.
3) $\mathrm{i}_v\, \mathbf{d}x^i = v^i$.

Then by the "antiLeibniz" rule

$$\mathrm{i}_v\boldsymbol{\omega} = \frac{1}{(p-1)!} v^j \omega_{ji_2\ldots i_p}\, \mathbf{d}x^{i_2} \wedge \ldots \wedge \mathbf{d}x^{i_p} = v^j \omega_{jI_1\ldots I_{p-1}} \mathbf{d}x^{I_1} \wedge \ldots \wedge \mathbf{d}x^{I_{p-1}}$$

$\mathrm{i}_v\omega$

where $\boldsymbol{\omega}$ is a p-form and its components have, if necessary, been antisymmetrized.

Properties:

1) $\mathrm{i}_v^2 = 0$.
2) $\mathrm{di}_v + \mathrm{i}_v\,\mathrm{d} = \mathscr{L}_v$ the Lie derivative with respect to $\boldsymbol{v}$.
3) $[\mathscr{L}_v, \mathrm{i}_w] \equiv \mathscr{L}_v \mathrm{i}_w - \mathrm{i}_w \mathscr{L}_v = \mathrm{i}_{[v,w]}$.
4) $\mathrm{d}\boldsymbol{\theta}(\boldsymbol{v}, \boldsymbol{w}) = \mathscr{L}_v(\boldsymbol{\theta}(\boldsymbol{w})) - \mathscr{L}_w(\boldsymbol{\theta}(\boldsymbol{v})) - \boldsymbol{\theta}([\boldsymbol{v}, \boldsymbol{w}])$, or more generally,

$$\mathrm{d}\boldsymbol{\omega}(\boldsymbol{v}_0, \boldsymbol{v}_1 \ldots \boldsymbol{v}_p) = \sum_{i=0}^{p} (-1)^i \boldsymbol{v}_i(\boldsymbol{\omega}(\boldsymbol{v}_0, \ldots \hat{\boldsymbol{v}}_i, \ldots \boldsymbol{v}_p))$$
$$+ \sum_{0 \le i < j \le p} (-1)^{i+j} \boldsymbol{\omega}([\boldsymbol{v}_i, \boldsymbol{v}_j], \boldsymbol{v}_0 \ldots \hat{\boldsymbol{v}}_i, \ldots \hat{\boldsymbol{v}}_j, \ldots \boldsymbol{v}_p).$$

5) $\mathscr{L}_{[v,w]} = [\mathscr{L}_v, \mathscr{L}_w]$.

Exercise: Prove these properties.

To establish the second property one can prove $[\mathscr{L}_v, \mathrm{d}] = 0$, for instance, by a calculation in local coordinates. Then because $\mathscr{L}_v f = \mathrm{i}_v\, \mathbf{d}f$ and because $\mathrm{i}_v f \overset{\mathrm{def}}{=} 0$ the operators $\mathscr{L}_v$ and $\mathrm{di}_v + \mathrm{i}_v\,\mathrm{d}$ are two derivations which commute with d and which coincide on 0-forms; hence they are identical.

To establish the third and the fourth property one can check them in local coordinates. It is sufficient to check the third identity on 0-forms and on 1-forms:

$$[\mathscr{L}_v, \mathrm{i}_w] f = 0, \qquad \mathrm{i}_{[v,w]} f = 0;$$

for $\boldsymbol{\alpha} = \alpha_i\, \mathbf{d}x^i$,

$$[\mathscr{L}_v, \mathrm{i}_w]\boldsymbol{\alpha} = \partial_l(w^i \alpha_i)v^l - w^i v^l \partial_l \alpha_i - w^l \alpha_i \partial_l v^i$$
$$= \mathrm{i}_{[v,w]}\boldsymbol{\alpha}.$$

The fourth property follows from (see footnote, p. 200)

$$\mathrm{d}\boldsymbol{\theta}(\boldsymbol{v}, \boldsymbol{w}) = (\partial_j\theta_i - \partial_i\theta_j)v^j w^i$$
$$\mathscr{L}_v(\boldsymbol{\theta}(\boldsymbol{w})) = v^j \partial_j(\theta_i w^i)$$
$$\boldsymbol{\theta}([\boldsymbol{v}, \boldsymbol{w}]) = \theta_i(v^j \partial_j w^i - w^j \partial_j v^i) = \mathrm{i}_{[v,w]}\boldsymbol{\theta}.$$

The fifth property follows from the fourth, which can be written

$$i_w i_v \, d\theta = \mathcal{L}_v i_w \theta - \mathcal{L}_w i_v \theta - i_{[v,w]}\theta$$

since $d\theta(v, w) = i_w \, d\theta(v, \cdot) = i_w i_v \, d\theta(\cdot, \cdot)$.

Set $\theta = df$, then

$$\mathcal{L}_{[v,w]} f = \mathcal{L}_v \mathcal{L}_w f - \mathcal{L}_w \mathcal{L}_v f.$$

The two derivations $\mathcal{L}_{[v,w]}$ and $[\mathcal{L}_v, \mathcal{L}_w]$ commute with d and coincide on 0-forms, hence they are identical. ∎

5. FORMS DEFINED ON A LIE GROUP

left, right invariant forms

We now consider a manifold G which is also a Lie group. A differential form ω on G is **left invariant** if[1] (pp. 150 and 155)

$$L_g^* \omega(L_g h) = \omega(h) \qquad \text{where } L_g h = gh, \quad g, h \in G.$$

As in the case of vector fields on G (p. 155), ω is left invariant if and only if $L_g^* \omega(g) = \omega(e)$ where e is the identity element of G.
The left invariant differential p-forms on G form a vector space of dimension $\binom{n}{p}$; for the proof see the argument given on pp. 155, 197.
A **right invariant** differential form is defined similarly.

If ω is a left [right] invariant differential form, then $d\omega$ is left [right] invariant:

$$L_g^* \, d\omega = dL_g^* \omega = d\omega.$$

Maurer–Cartan structure equations. The dual $\mathcal{G}^*$ of the Lie algebra $\mathcal{G}$ of G is the space of left invariant linear forms on G. Let (v_α) and (θ^β) be dual bases in $\mathcal{G}$ and $\mathcal{G}^*$ respectively. Since $d\theta^\alpha$ is also left invariant, it can be expressed in the basis $(\theta^B \wedge \theta^\Gamma)$ for the space of left invariant 2-forms (p. 197)

$$d\theta^\alpha = -c^\alpha_{B\Gamma} \theta^B \wedge \theta^\Gamma = -\tfrac{1}{2} c^\alpha_{\beta\gamma} \theta^\beta \wedge \theta^\gamma \quad \text{Maurer–Cartan structure equations.}$$

The strict components of $d\theta^\alpha$ are – up to a factor (-1) – the structure constants of the group G which have been defined, previously, by the relations

$$[v_\beta, v_\gamma] = c^\alpha_{\beta\gamma} v_\alpha.$$

[1]Here $\omega(g)$ means $\omega_g \in T_g^*(G)$.

Proof: The components of $\mathbf{d}\boldsymbol{\theta}^\alpha$ are

$$\begin{aligned}(\mathbf{d}\boldsymbol{\theta}^\alpha)_{\beta\gamma} &= \mathbf{d}\boldsymbol{\theta}^\alpha(\boldsymbol{v}_\beta, \boldsymbol{v}_\gamma)\\ &= \mathcal{L}_{\boldsymbol{v}_\beta}\boldsymbol{\theta}^\alpha(\boldsymbol{v}_\gamma) - \mathcal{L}_{\boldsymbol{v}_\gamma}\boldsymbol{\theta}^\alpha(\boldsymbol{v}_\beta) - \boldsymbol{\theta}^\alpha([\boldsymbol{v}_\beta, \boldsymbol{v}_\gamma])\\ &= -\boldsymbol{\theta}^\alpha([\boldsymbol{v}_\beta, \boldsymbol{v}_\gamma]) \qquad \text{using } \boldsymbol{\theta}^\alpha(\boldsymbol{v}_\gamma) = \delta^\alpha_\gamma\\ &= -\boldsymbol{\theta}^\alpha(c^\delta_{\beta\gamma}\boldsymbol{v}_\delta) = -c^\alpha_{\beta\gamma}.\end{aligned}$$

■

Notice that $c^\alpha_{\beta\gamma} = \mathbf{i}_{[\boldsymbol{v}_\beta, \boldsymbol{v}_\gamma]}\boldsymbol{\theta}^\alpha$.

***Exercise*: Derive the Jacobi identity (p. 156) from the relations $\mathbf{d}^2\boldsymbol{\theta}^\alpha = 0$.**

We now complete the proof of the following theorem stated in Chapter III (p. 158).

Theorem. A Lie group G has vanishing structure constants if and only if it is locally isomorphic to $\mathbb{R}^n$.

It remains to be proven that $c^\alpha_{\beta\gamma} = 0 \Rightarrow G \overset{\text{locally}}{\simeq} \mathbb{R}^n$.

Proof: Since $c^\alpha_{\beta\gamma} = 0$,

$$\mathbf{d}\boldsymbol{\theta}^\alpha = 0, \qquad \alpha = 1, \ldots, n.$$

It follows from the Poincaré lemma (p. 224) that in the domain U of any local chart there exist n independent differentiable functions φ^α such that

$$\boldsymbol{\theta}^\alpha = \mathbf{d}\varphi^\alpha.$$

These functions can be used as the coordinate functions of the chart. Since $\boldsymbol{\theta}^\alpha$ is left invariant

$$L^*_g\boldsymbol{\theta}^\alpha(L_g h) = \boldsymbol{\theta}^\alpha(h) = \mathbf{d}\varphi^\alpha(h) = \mathbf{d}h^\alpha.$$

On the other hand

$$L^*_g\boldsymbol{\theta}^\alpha(L_g h) = L^*_g\,\mathbf{d}\varphi^\alpha(L_g h) = L^*_g(\mathbf{d}(gh)^\alpha) = \mathbf{d}L^\alpha(g^\gamma, h^\beta) = \frac{\partial L^\alpha}{\partial h^\beta}\mathbf{d}h^\beta$$

where the $L^\alpha(g^\gamma, h^\beta)$ are the coordinates of gh; thus $\partial L^\alpha/\partial h^\beta$ must be the unit matrix

$$\partial L^\alpha/\partial h^\beta = \delta^\alpha_\beta.$$

Integration gives

$$L^\alpha(g^\gamma, h^\beta) = \Phi^\alpha(g^\gamma) + h^\alpha$$

where the Φ^α are arbitrary functions of g^γ. Now let U be a neighborhood of e and choose the Φ^α such that $\Phi^\alpha(e)=0$. Then

$$L^\alpha(g^\gamma,0)=\Phi^\alpha(g^\gamma)=g^\alpha$$

and

$$L^\alpha=g^\alpha+h^\alpha. \qquad \blacksquare$$

Exercise: Show that one can take $\Phi^\alpha(g^\gamma)=g^\alpha$ in a neighborhood U of any point by using the associativity of the group multiplication.

6. VECTOR VALUED DIFFERENTIAL FORMS

vector valued p form

An (exterior) p form φ at a point x of a differentiable manifold X **with values in a** given **vector space** V is a totally antisymmetric p linear map from T_xX into V,

$$\varphi_x:\quad (\times)^p T_xX\to V \quad \text{by} \quad (v_1,\dots v_p)\mapsto\varphi_x(v_1,\dots,v_p).$$

If V is finite dimensional – and let us say real – and if e_α, $\alpha=1,\dots r$ is a base of V we can write

$$\varphi_x(v_1,\dots,v_p)=\varphi_x^\alpha(v_1,\dots,v_p)e_\alpha$$

where $\varphi^\alpha(v_1,\dots,v_p)$ are real numbers. The mapping $(v_1,\dots v_p)\mapsto\varphi_x^\alpha(v_1,\dots v_p)$ is clearly p-linear and totally antisymmetric, therefore it defines an ordinary (scalar valued) p form φ_x^α at X and

$$\varphi_x=\varphi_x^\alpha\otimes e_\alpha.$$

vector valued p form on X

An **(exterior, differential)** p **form** φ **with values in a** given, **finite dimensional real vector space** V **on a manifold** X is an assignment $x\mapsto\varphi_x$, $x\in X$, φ_x a p-form at x with values in V. It can be written, if e_α is a basis of V,

$$\varphi=\varphi^\alpha\otimes e_\alpha$$

where the φ^α are the scalar valued p forms.
φ is of class C^k on X if the φ^α are of class C^k.

vector valued exterior differential

The **exterior differential** of φ is, by definition the $p+1$ form with values in V

$$d\varphi=d\varphi^\alpha\otimes e_\alpha.$$

It is clear that the definition does not depend on the choice of a basis in V, because if $e_\alpha=A_\alpha^{\alpha'}e_{\alpha'}$ and $\varphi=\varphi^\alpha\otimes e_\alpha$, then

$$\varphi=\varphi^{\alpha'}\otimes e_{\alpha'} \quad \text{with} \quad \varphi^{\alpha'}=A_\alpha^{\alpha'}\varphi^\alpha.$$

The $A_\alpha^{\alpha'}$ being constant numbers we have

$$d\varphi^{\alpha'} = A_\alpha^{\alpha'} \, d\varphi^\alpha$$

thus, if $d\varphi = d\varphi^\alpha \otimes e_\alpha$, we have also

$$d\varphi = d\varphi^{\alpha'} \otimes e_{\alpha'}.$$

Exercise: Show that the expression of $d\varphi$ can be deduced from the following hypothesis:

1) $d(\varphi + \psi) = d\varphi + d\psi$
2) $d^2\varphi = 0$
3) if a is an arbitrary linear form on V, a: $V \to \mathbb{R}$, then

$$d\langle a, \varphi\rangle = \langle a, d\varphi\rangle$$

where $\langle a, \varphi\rangle$ is the scalar valued p-form on X defined at $x \in X$ by

$$\langle a, \varphi\rangle_x : (v_1, \dots v_p) \mapsto \langle a, \varphi(v_1, \dots v_p)\rangle.$$

When the given vector space V is endowed with a Lie algebra structure by a bracket $[e_\alpha, e_\beta]$ between any pair of its elements, one defines the **bracket of a p-form and a q-form on X with values in V** as follows: bracket of vector valued forms

$$[\varphi, \psi] = (\varphi^\alpha \wedge \psi^\beta) \otimes [e_\alpha, e_\beta]$$

if

$$\varphi = \varphi^\alpha \otimes e_\alpha, \qquad \psi = \psi^\beta \otimes e_\beta.$$

It is easy to check that $[\varphi, \psi]$ does not depend on the basis chosen in V, and has the following properties:

1) If φ_1 and φ_2 are two p-forms

$$[\varphi_1 + \varphi_2, \psi] = [\varphi_1, \psi] + [\varphi_2, \psi].$$

2) If φ is a p-form and ψ a q-form

$$[\varphi, \psi] = (-1)^{pq+1}[\psi, \varphi].$$

3) If θ is an r-form, it results from the Jacobi identity in V that

$$(-1)^{pr}[\varphi, [\psi, \theta]] + (-1)^{pq}[\psi, [\theta, \varphi]] + (-1)^{qr}[\theta, [\varphi, \psi]] = 0.$$

4) The exterior differential of a bracket is

$$d[\varphi, \psi] = [d\varphi, \psi] + (-1)^p[\varphi, d\psi].$$

B. INTEGRATION

1. INTEGRATION

orientation, orientable, oriented

Two coordinate systems (x^i) and (y^j) on an open set of $\mathbb{R}^n$ are said to define the same **orientation** if the jacobian determinant $J = \mathrm{D}(x^i)/\mathrm{D}(y^j)$ is positive at all points of the set. A chart (U, φ) on a manifold X defines an orientation of U by means of the orientation provided by the coordinates $(\varphi^i(x) = x^i)$ on $\varphi(U) \in \mathbb{R}^n$. A differentiable manifold is said to be **orientable** if there exists an atlas such that on the overlap $U \cap V$ of any two charts (U, φ) and (V, ψ), $\mathrm{D}(\varphi^i)/\mathrm{D}(\psi^j) > 0$. A manifold defined in terms of such an atlas is said to be **oriented**.
An orientation at a point $x \in X$ can also be defined in terms of the orientation of the tangent vector space $T_x(X)$. If the manifold is orientable a frame transported along any path in the tangent bundle of the manifold comes back to its starting point with the same orientation.

Example 1: Orientation of a 2-sphere S^2 (see Problem III 6, p. 191).
Example 2: A Möbius band is not orientable.

Exercise: Show that a connected differentiable manifold is orientable if and only if its frame bundle consists of two disjoint sets.
Exercise: Let L^n be an n-dimensional vector space over the real numbers. The space $\Lambda^n(L^n)$ is one dimensional and is isomorphic to $\mathbb{R}$, but not canonically. To orient L^n is to decide which elements of $\Lambda^n(L^n)$ should be considered positive and which negative. Show that if L^n is oriented we may orient its dual in a natural way.[1]

odd forms

Originally the theory of integration was restricted to orientable manifolds. G. de Rham has extended the theory of integration to non-orientable manifolds by introducing forms of odd parity. Under a change of coordinates an **odd form** (e.g. the volume element, p. 294) transforms as follows:

$$\bar{\omega}_{i_1 \dots i_p} = \frac{J}{|J|} \omega_{j_1 \dots j_p} \frac{\partial x^{j_1}}{\partial \bar{x}^{i_1}} \cdots \frac{\partial x^{j_p}}{\partial \bar{x}^{i_p}}.$$

The difference between this transformation and the one for the previously introduced forms (even forms) is the sign of J. For integration on orientable manifolds it is sufficient to consider only the even forms; hence it is possible to ignore the parity of forms. Unless otherwise specified we shall limit our study to integration on orientable manifolds.

[1]See for instance [L. Schwartz 1955, p. 31].

Integration of a differential n-form on an n-dimensional paracompact oriented manifold X. To have a theory of integration on a smooth manifold X^n which parallels the Lebesgue integration on $\mathbb{R}^n$ (cf. p. 31) we shall extend the definition of a p-form (p-tensor field) on X^n to mappings $X^n \to \Lambda^p(T^*(X^n))$ defined almost everywhere, according to the following definitions.

A subset Y of X^n is of (Lebesgue) **measure zero** if it is a countable union of images in X^n, under inverse coordinate mappings, of sets of measure zero in $\mathbb{R}^n$. measure zero

A mapping $X^n \to Z$ is defined **almost everywhere** (a.e.) if it is defined except on a subset of measure zero. almost everywhere

We shall first define the **integral of an n-form with compact support in the domain U of a chart** with coordinates $(x^1, \dots x^n)$.
Let $\boldsymbol{\omega}$ be an n-form vanishing outside a compact set contained in U. By definition $\boldsymbol{\omega}$ is **integrable** on X if its component $\omega_{1\dots n}$ is integrable on $\mathbb{R}^n$ – that is Lebesgue measurable and integrable in the sense of Chapter I. D. The integral of $\boldsymbol{\omega}$ on U, which is also the integral on X, is then integrable

$$\int_X \boldsymbol{\omega} = \int_U \boldsymbol{\omega} = \int_{-\infty}^{+\infty} \cdots \int_{-\infty}^{+\infty} \omega_{12\dots n}(x^i)\, \mathrm{d}x^1\, \mathrm{d}x^2 \dots \mathrm{d}x^n.$$

Recall that the Lebesgue integral on $\mathbb{R}^n$ does not depend on the order of integrations, in accord with the notation in the last integral which does not include the wedge product. The definition does not depend on the choice of coordinates in U, provided they are compatible with the orientation. Indeed under a change of coordinates $(x^i) \mapsto (\tilde{x}^i)$, preserving the orientation $(\mathrm{D}(\tilde{x}^i)/\mathrm{D}(x^i) > 0)$ the definition of the integral of $\boldsymbol{\omega}$ on U becomes

$$\int_U \boldsymbol{\omega} = \int_{-\infty}^{+\infty} \cdots \int_{-\infty}^{+\infty} \tilde{\omega}_{12\dots n}(\tilde{x}^j)\, \mathrm{d}\tilde{x}^1 \dots \mathrm{d}\tilde{x}^n$$

with (cf. p. 199)

$$\tilde{\omega}_{1\dots n}(\tilde{x}^j) = \omega_{1\dots n}(x^i(\tilde{x}^j)) \frac{\mathrm{D}(\tilde{x}^l)}{\mathrm{D}(x^k)}$$

which corresponds to the classical formula for a change of variables in a multiple integral (p. 47).

We shall now define an **integral of an n-form with arbitrary support in X.**

partition of unity

The technique used to glue together the results obtained for each domain of a chart is called partition of unity. A **partition of unity** on X is a collection of functions $\theta_k \geq 0$ on X with the following properties.

1) The collection {supp θ_k} is locally finite (at each point $x \in X$ there are only a finite number of functions $\theta_k \neq 0$).
2) supp θ_k is compact.
3) $\Sigma_k \, \theta_k(x) = 1 \qquad \forall x \in X$.

The partition is said to be of class C^r if the θ_k are of class C^r. A partition of unity in $\mathbb{R}^n$ is easily constructed. Let $\{B_a^k\}$ be a covering of $\mathbb{R}^n$ by open balls of radius a centered at a countable number of points $x_{(k)}$. Then the collection of functions

$$\theta_k(x) = \frac{\theta_a(x - x_{(k)})}{\Sigma_j \, \theta_a(x - x_{(j)})},$$

with θ_a defined on p. 432, satisfies the above three conditions.

subordinate

A partition of unity $\{\theta_k\}$ is **subordinate** to a covering $\{U_i\}$ of X if one can find $U_i \supset$ supp θ_k for every k.

Theorem. If a space X is paracompact (p. 16), *it is always possible to find on it a partition of unity subordinate to any preassigned locally finite covering. If, moreover X is a C^r manifold this partition of unity can be required to be C^r.*

The easiest way to prove the theorem is actually to construct such a partition; we leave this construction as an exercise[1]. The importance of a partition of unity is not its practical usefulness in calculating integrals but its elegance and power for obtaining theoretical results. We shall always suppose in the following that the manifold X is paracompact. It can be proved that a connected manifold is paracompact if and only if it admits a countable atlas. Since a manifold is locally compact, the proof rests on the general theorem[2] which says that a connected locally compact space is paracompact if and only if it is a countable union of compact sets.

integral of ω on X

Let $\{\theta_k\}$ be a partition of unity subordinate to an atlas of X. The n-form ω is **integrable on** X if each n-form $\theta_k\omega$ is integrable and if the series $\Sigma_k \int_X \theta_k \omega$ converges. The integral of ω on X is then the sum of this series: $\int_X \omega = \Sigma_k \int_X \theta_k \omega$. One shows that this definition depends neither on the choice of atlas nor on the choice of partition of unity subordinate to it. If supp ω is compact – for instance if X is compact – the series has

[1][Choquet-Bruhat 1968, p. 69].
[2]Bourbaki, General topology, ch. I. §9.

only a finite number of terms. A consequence is that every continuous n-form is integrable on a compact n-manifold.

Rather obviously the definition of the integral of ω on a subset Y of X is

$$\int_Y \omega = \int_X \chi\omega$$

where χ is the characteristic function of Y in X (p. 1).

Properties: The integral is a linear functional on the space of integrable forms

$$\int_X \lambda\omega + \mu\theta = \lambda \int_X \omega + \mu \int_X \theta.$$

The additivity of the domains of integration follows from this linearity: Let X_1 and X_2 be two complementary subsets of X, let χ_1 and χ_2 be their respective characteristic functions in X; then

$$\int_X \omega = \int_X (\chi_1 + \chi_2)\omega = \int_X \chi_1\omega + \int_X \chi_2\omega = \int_{X_1} \omega + \int_{X_2} \omega.$$

These equalities suggest the notation

$$\int_X \omega = \langle X, \omega \rangle$$ $\langle X, \omega \rangle$

which is associated with bilinear functionals of elements in different spaces: we shall come back to this concept of duality in Chapter VI.

Calculation of integrals. The two following properties are very useful in the calculation of integrals:

1) The integral of an n-form ω on $A \subset X^n$ does not change if A is modified by a set of measure zero.

2) Let f be an orientation preserving *diffeomorphism* $X^n \to Y^n$ and ω an n-form on Y^n; then

$$\int_{X^n} f^*\omega = \int_{Y^n} \omega.$$

For the proof, use on Y^n an atlas which is the image by f of an atlas on X^n, then prove the equality of the integrals by the equality of their building blocks, namely the integrals in the domains of corresponding charts.

Example: Problem III 6, p. 191.

So far we have studied integrals of n-forms in n-dimensional manifolds. In the next paragraph we will extend the definition to integrals of p-forms on p-chains in n-dimensional manifolds.

2. STOKES' THEOREM

The forms considered in this paragraph are at least C^1. Stokes' formula is so beautiful that we first exhibit it, before we analyze it, prove it, exploit it:

$$\int_C \mathrm{d}\omega = \int_{\partial C} \omega$$

also written

$$\langle C, \mathrm{d}\omega\rangle = \langle \partial C, \omega\rangle$$

where ∂C is the boundary of C oriented coherently with C.
Note that the Ostrogradzky–Green and Riemann–Ampère–Stokes formulae are special cases of the Stokes formula. Indeed, when $\boldsymbol{V}$ is a vector in $\mathbb{R}^3$, it follows from Stokes' theorem that

$$\iiint_{\text{volume}} \nabla \cdot \boldsymbol{V}\, \mathrm{d}v = \iint_{\text{boundary}} \boldsymbol{V} \cdot \mathrm{d}\boldsymbol{\sigma} = \text{flux of } \boldsymbol{V} \text{ through a closed surface}$$

$$\iint_{\text{surface}} (\nabla \times \boldsymbol{V}) \cdot \mathrm{d}\boldsymbol{\sigma} = \int_{\text{boundary}} \boldsymbol{V} \cdot \mathrm{d}\boldsymbol{r} = \text{work of } \boldsymbol{V} \text{ along the boundary.}$$

Let us first give a precise definition of the domain of integration C or more generally of a chain C. The building blocks of p-chains can be either simplexes or rectangles in $\mathbb{R}^p$. In this book we shall use rectangles.

p-simplex A **p-simplex**[1] is a subset of $\mathbb{R}^n$ defined in terms of $p+1$ linearly independent points. This definition is a generalization, to higher or lower dimensions, of the definition of a closed triangle (2-simplex) in terms of three linearly independent points.

p-rectangle A **p-rectangle** P^p, or simply P, in $\mathbb{R}^p$ is a naturally oriented subset of $\mathbb{R}^p$ defined by

$$a^i \leq x^i \leq b^i, \qquad i = 1, \ldots, p.$$

[1]See for instance [Patterson, p. 91].

Simplexes and rectangles lead to the same definitions because a simplex can be decomposed into sets diffeomorphic to rectangles and vice-versa. Simplexes are more convenient in homology, rectangles more convenient in integration.
The definitions and proofs of theorems on integrals of p-forms on p-chains proceed along the following road:

```
             ↗ elementary p-chain → p-chain
rectangle                        ↓
             ↘ elementary domain of → domain of
                    integration        integration.
```

elementary p-chain

An **elementary p-chain** c on X^n is a pair (P, f) where P is a rectangle in $\mathbb{R}^p$ and $f: U \subset \mathbb{R}^p \to X^n$, with $P \subset U$, is a differentiable mapping. The domain of f is an open neighborhood U of P for convenience later so that the same mapping can be applied to the boundary of P.

supp c

elementary domain of integration

We will call **supp** c the image of P in X^n by f.
The elementary p-chain $c = (P, f)$ is called an **elementary p-domain of integration** if f is a diffeomorphism of U onto a differentiable p-submanifold (p. 239) of X. c or supp c is then also referred to as an elementary domain of integration.
All the results and theorems of the previous paragraph apply when considering the integral of a p-form over an elementary p-domain of integration. In particular if ω is a p-form on X and c an elementary domain of integration

$$\int_{\text{supp}\, c} \omega = \int_P f^*\omega.$$

p-chain C

A **p-chain** C on X^n is a formal linear combination of elementary p-chains with real coefficients

$$C = \sum_i \lambda_i c_i.$$

domain of integration

A formal locally finite linear combination, with coefficients ± 1, of elementary domains of integration is called a **domain of integration**.

Integrals of p-forms on p-chains. By definition, the integral of a p-form ω in X^n on an elementary p-chain c is

$$\int_c \omega = \int_P f^*\omega;$$

the integral of a p-form ω on a p-chain C is, if it exists,

$$\int_C \omega = \sum_i \lambda_i \int_{c_i} \omega.$$

When ω is continuous, it is always defined if C is finite, or, if ω has compact support and C is locally finite. The integral is bilinear in C and ω. We will denote in both cases $\int_C \omega = \langle C, \omega \rangle$.

equal chains

Two chains C and C' are said to be **equal** if $\int_C \omega = \int_{C'} \omega$ for all ω. The decomposition of a chain into elementary chains is not necessarily unique.

boundary

Boundaries. The **boundary** ∂P of a rectangle P in $\mathbb{R}^p$ is the $2p$ rectangles in R^{p-1} defined by the faces $x^i = a^i$ and $x^i = b^i$ of the rectangle; the $p-1$ coordinates of a point on these faces are $(x^1, \ldots \hat{x}^i, \ldots x^p)$ with $a^j \le x^j \le b^j$, $j = 1, \ldots \hat{i} \ldots p$.

coherently oriented

The boundary ∂P is said to be **coherently oriented** with P when the faces are given the following orientations

$$(x^1, \ldots \hat{x}^i, \ldots x^p) \quad \text{for} \quad \begin{cases} x^i = a^i \text{ and } i \text{ even} \\ x^i = b^i \text{ and } i \text{ odd} \end{cases}$$

$$\text{the opposite orientation} \quad \text{for} \quad \begin{cases} x^i = a^i \text{ and } i \text{ odd} \\ x^i = b^i \text{ and } i \text{ even.} \end{cases}$$

Exercise: Let ∂P be the coherently oriented boundary of a rectangle P in $\mathbb{R}^p$, where both P and $\mathbb{R}^p$ have the natural orientation. Let $(e_1, \ldots e_{p-1})$ be a frame, with orientation compatible with that of ∂P, at a point x in a face of P. Show that if $(e_1, \ldots e_{p-1}, e_p)$ is a frame at x, with orientation compatible with that of $\mathbb{R}^p$, then e_p is inside P for p even and outside P for p odd.

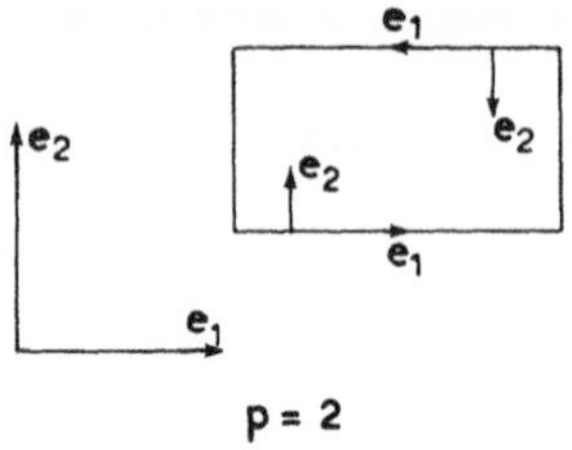

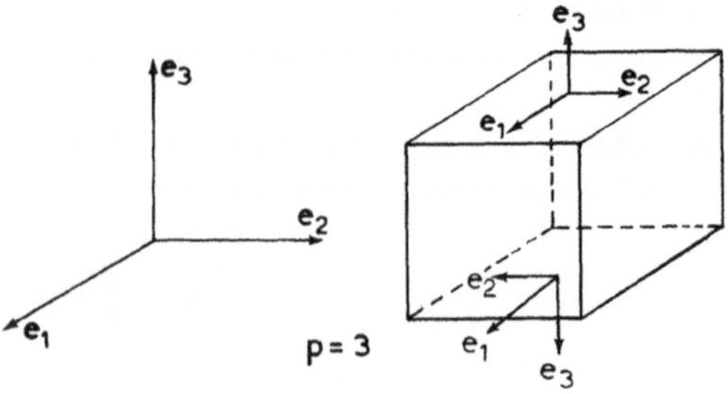

Use this result to deduce that for two adjacent elementary domains of integration c_1 and c_2, the intersection of supp ∂c_1 and supp ∂c_2 does not appear in supp $\partial(c_1 + c_2)$. Define "adjacent".

The boundary of an elementary p-chain (P, f) is related to the boundary of the rectangle P in an obvious fashion: it is a $(p-1)$-chain which is the sum of the elementary $(p-1)$-chains made up of the mapping f and the $2p$ coherently oriented rectangles in the boundary of P. The boundary of a chain $C = \Sigma_i \lambda_i c_i$ is by definition

$$\partial C = \sum_i \lambda_i \partial c_i.$$

The definition of a **manifold $\tilde{X}$ with boundary** is similar to the definition of a manifold given in Chapter III (p. 111); however there are now two different kinds of charts. Let $\mathbb{R}^{n+}$ denote the region of $\mathbb{R}^n$ for which $x^1 \geq 0$. Let $\{\tilde{U}_i, \varphi_i\}$ be an atlas on $\tilde{X}$. In some charts φ_i maps $\tilde{U}_i$ onto an open set in $\mathbb{R}^n$ homeomorphic to $\mathbb{R}^n$, in others onto $\mathbb{R}^{n+}$. The boundary $\partial\tilde{X}$ of $\tilde{X}$ is then defined as the set of all points of $\tilde{X}$ whose images under the coordinate mappings lie on the boundary $\mathbb{R}^{n-1}$ of $\mathbb{R}^{n+}$ defined by $x^1 = 0$. Clearly $\partial\tilde{X}$ is an $(n-1)$-dimensional manifold of the same class as $\tilde{X}$. manifold with boundary

An oriented manifold X^n is said to be **triangulable** if it can be decomposed into a union of adjacent n-dimensional elementary domains of integration with orientation compatible with that of X^n. All compact manifolds are triangulable, the triangulation being accomplished by a finite number of elementary domains of integration. If C is the domain of integration equal to the sum of the elementary domains of integration which triangulate X^n then triangulable

$$\int_{X^n} \omega = \int_C \omega$$

where ω is an n-form on X^n.

Mappings of chains. Let g: $X^n \to Y^m$ be a differentiable mapping and let $c = (P, f)$ be an elementary p-chain on X^n. By definition

$$g(c) = (P, g \circ f).$$

Therefore

$$\int_{g(c)} \omega = \int_P (g \circ f)^* \omega = \int_P f^* g^* \omega = \int_c g^* \omega$$

where ω is a p-form on Y^m. We now set, for $C = \Sigma\, \lambda_i c_i$,

$$g(C) = g(\Sigma_i\, \lambda_i c_i) = \Sigma_i\, \lambda_i g(c_i).$$

proper mapping

The mapping g is called **proper** if for any compact subset $V \subset Y^m$ the set $g^{-1}(V)$ is compact in X^n. Under such a mapping a locally finite chain remains locally finite and the reciprocal image of a form with compact support has compact support.

Theorem. Let g: $X^n \to Y^m$ be a proper differentiable mapping. If ω has compact support and C is locally finite, or if C is finite then

$$\int_{g(C)} \omega = \int_C g^*\omega.$$

The restrictions are made to assure that the integrals exist. Care must be taken in applying this theorem to integrals over domains of integration because the fact that c is an elementary domain of integration does not necessarily imply that $g(c)$ is.

Example: Let g: $S^1 \to S^1$ be a differentiable two–one mapping and let ω be a 1-form on S^1.

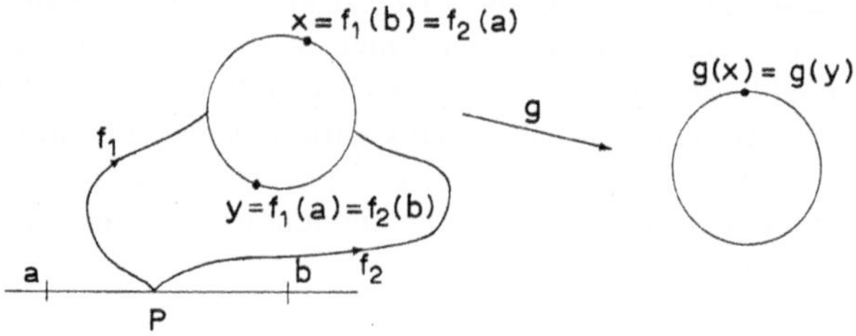

S^1 can be triangulated by two elementary domains of integration $c_1 = (P, f_1)$ and $c_2 = (P, f_2)$ as shown in the diagram

$$S^1 = \text{supp}\ c_1 \cup \text{supp}\ c_2.$$

By the theorem

$$\int_{S^1} g^*\omega = \int_{c_1+c_2} g^*\omega = \int_{g(c_1+c_2)} \omega.$$

However although supp $g(c_1 + c_2) = S^1$

$$\int_{g(c_1+c_2)} \omega \neq \int_{S^1} \omega$$

because $g(c_1 + c_2)$ is not a triangulation of S^1; it is not even a domain of integration. In fact

$$\int_{S^1} g^*\omega = 2\int_{S^1} \omega.$$

The general theorem can be stated in the following way.

Theorem. Let X^n and Y^n be connected and oriented and let $f\colon X^n \to Y^n$ be an orientation preserving proper differentiable mapping, then there exists an integer called the ***degree*** *of f such that* degree

$$\int_{X^n} f^*\omega = \deg(f) \int_{Y^n} \omega$$

for any n-form ω with compact support.

A partial proof of the theorem is given as an exercise in the following section (p. 228), elements for the proof will be given in section VIIB, p. 559.

Proof of Stokes' theorem. We are now ready to prove Stokes' theorem. We shall prove it for a rectangle; using the definitions the reader can easily extend this proof by himself to the case of a chain. We shall furthermore limit the proof to the particular case of a 2-rectangle; the generalization to an arbitrary rectangle involves only the labelling of an arbitrary number of coordinates[1].

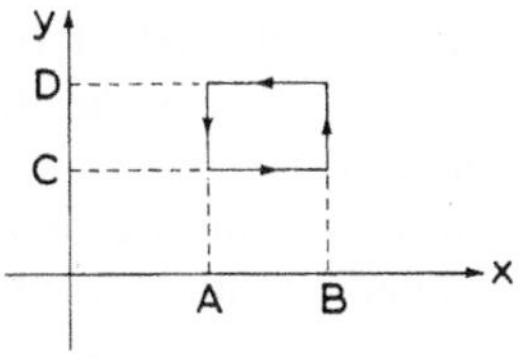

Let

$$\omega = a(x, y)\,\mathbf{d}x + b(x, y)\,\mathbf{d}y;$$

then

$$\mathbf{d}\omega = \left(\frac{\partial b}{\partial x} - \frac{\partial a}{\partial y}\right) \mathbf{d}x \wedge \mathbf{d}y,$$

[1]See for instance [Choquet-Bruhat 1968, p. 77].

and

$$\iint_P \mathrm{d}\boldsymbol{\omega} = \iint_P \frac{\partial b}{\partial x}\,\mathrm{d}x\,\mathrm{d}y - \iint_P \frac{\partial a}{\partial y}\,\mathrm{d}x\,\mathrm{d}y$$

$$= \int_C^D [b(B,y)-b(A,y)]\,\mathrm{d}y - \int_A^B [a(x,D)-a(x,C)]\,\mathrm{d}x$$

$$= \int_A^B a(x,C)\,\mathrm{d}x + \int_C^D b(B,y)\,\mathrm{d}y + \int_B^A a(x,D)\,\mathrm{d}x + \int_D^C b(A,y)\,\mathrm{d}y$$

$$= \int_{\partial P} \boldsymbol{\omega}. \qquad \blacksquare$$

Exercise: Let $\tilde{X}$ be a compact manifold with boundary; prove Stokes' theorem directly[1].

3. GLOBAL PROPERTIES

Using Stokes' formula we now show that $\mathrm{d}^2 = 0$ implies $\partial^2 = 0$.

Indeed: $\langle \partial^2 C, \boldsymbol{\omega}\rangle = \langle \partial C, \mathrm{d}\boldsymbol{\omega}\rangle = \langle C, \mathrm{d}^2\boldsymbol{\omega}\rangle = 0.$ ■

This result is an indication of the similarities and relationships between chains and forms.

Homology and cohomology. Both the set of forms $\Lambda(X)$ and the set of finite chains $C(X)$ on a manifold X have the structure of a graded vector space with differential operator:
$\Lambda(X)[C(X)]$ is a collection of vector spaces $\Lambda^p(X)[C_p(X)]$ over $\mathbb{R}$;
$\Lambda(X)$ $[C(X)]$ has a coboundary [boundary] operator such that

$$\text{coboundary} \quad \mathrm{d}\colon \Lambda(X)\to\Lambda(X), \quad \mathrm{d}\Lambda^p \subset \Lambda^{p+1}, \quad \mathrm{d}^2 = 0$$
$$\text{boundary} \quad \partial\colon C(X)\to C(X) \quad \partial C_p \subset C_{p-1}, \quad \partial^2 = 0.$$

closed, [co]cycle

A form $\boldsymbol{\omega}$ such that $\mathrm{d}\boldsymbol{\omega} = 0$ is called a **cocycle (closed form)**.
A finite chain C such that $\partial C = 0$ is called a **cycle**.

exact, [co]boundary

A form $\boldsymbol{\omega}$ such that $\boldsymbol{\omega} = \mathrm{d}\boldsymbol{\theta}$ is called a **coboundary (exact form)**.
A finite chain C such that $C = \partial B$ is called a **boundary**.

[1]See for instance [Choquet-Bruhat 1968, p. 80].

Note that

$$\boldsymbol{\omega} = \mathrm{d}\boldsymbol{\theta} \Rightarrow \mathrm{d}\boldsymbol{\omega} = 0 \quad \text{but usually} \quad \mathrm{d}\boldsymbol{\omega} = 0 \not\Rightarrow \boldsymbol{\omega} = \mathrm{d}\boldsymbol{\theta}$$
$$C = \partial B \Rightarrow \partial C = 0 \quad \text{but usually} \quad \partial C = 0 \not\Rightarrow C = \partial B.$$

Let $Z^p[Z_p]$ denote the vector space of cocycles [cycles] of degree [dimension] p; let $B^p[B_p]$ denote the vector space of coboundaries [boundaries] of degree [dimension] p. Since $\mathrm{d}^2 = 0$, $B^p \subset Z^p$; since $\partial^2 = 0$, $B_p \subset Z_p$.

The space $H^p = Z^p/B^p$ is the p-**cohomology vector space** $H^p(\Lambda(X))$ of X, often abbreviated $H^p(X)$. The elements of $H^p(X)$ are equivalence sets of cocycles. Two cocycles $\boldsymbol{\omega}_1$ and $\boldsymbol{\omega}_2$ belong to the same equivalence set, or are **homologous** ($\boldsymbol{\omega}_1 \sim \boldsymbol{\omega}_2$), if and only if they differ by a coboundary ($\boldsymbol{\omega}_1 - \boldsymbol{\omega}_2 = \mathrm{d}\boldsymbol{\theta}$). H^p is often called the de Rham cohomology group. Similarly $H_p = Z_p/B_p$ is the p-**homology vector space** $H_p(C(X))$ of X, often abbreviated $H_p(X)$. Two cycles C_1 and C_2 are **homologous** (belong to the same **homology class**) if and only if they differ by a boundary ($C_1 - C_2 = \partial B$).

p-cohomology
H^p

homologous
de Rham cohomology
p-homology
H_p

In general the coefficients of the elementary chains which make up a chain are allowed to be elements of any fixed group [See *Analysis, Manifolds and Physics, Part II*]; then H_p is a group but not necessarily a vector space. One therefore generally speaks of **[co]homology groups**.

[co]homology group

Exercise[1]: Prove that ∂ is a homomorphism on the space of chains $C(X)$. Prove that d is a homomorphism on the space of forms $\Lambda(X)$.
Let H be the quotient vector space $\operatorname{Ker} \partial/\operatorname{Im} \partial$, $[H^* \equiv \operatorname{Ker} \mathrm{d}/\operatorname{Im} \mathrm{d}]$; show that $H[H^*]$ is graded by the dimension [degree] of the chains [forms]

$$H(X) = \{H_p(X)\}, \qquad [H^*(X) = \{H^p(X)\}].$$

Exercise: Show that multiplication of forms give to $H^*(X)$ the structure of an algebra.

Conventions for 0-forms and 0-chains. $Z^0(X)$ is the space of differentiable functions on X with vanishing differential. Such functions are constant on each connected component of X. By convention $B^0(X) = \emptyset$: there are no forms of degree minus one. Thus a 0-form homologous to a closed form $\lambda \in \mathbb{R}$ is equal to λ. Hence, on a connected component of X, each element of $H^0(X)$ is in a one–one correspondence with $\mathbb{R}$ and the dimension of $H^0(X)$ is equal to the number of connected components of X.

0-forms
0-chains

[1] See, for instance, the summary with applications to physics in [Pham].

For the case of 0-chains see for instance [Patterson].

Betti numbers

The dimension of the p-cohomology [homology] group is called the **Betti number** $b^p[b_p]$ of X.

Euler–Poincaré characteristic

The Euler–Poincaré characteristic of a manifold X^n is the integer $\Sigma_{q=0}^n(-1)^q b_q$ where b_q is the qth Betti number of X^n.

Poincaré lemma. *On an open set U diffeomorphic to $\mathbb{R}^n$, all closed forms of degree $p \geq 1$ are exact; i.e. the Betti numbers $b^p = 0$ for $p = 1, \ldots n$.*

Proof: Let ω be a closed form, $d\omega = 0$; we shall determine θ such that $\omega = d\theta$. We shall first construct a linear operator T such that

1) $T: \Lambda^p(\mathbb{R}^n) \to \Lambda^{p-1}(\mathbb{R}^n)$
2) $T\,d + dT = \text{identity}$.

It follows that $dT\omega = \omega$, and the desired solution is $\theta = T\omega$.
The following operator, defined if ω is defined and differentiable on $\mathbb{R}^n$, satisfies the required condition for T:

$$T\omega(x) = \int_0^1 t^{p-1} i_x \omega(tx)\,dt, \qquad x \in \mathbb{R}^n$$

where $i_x\omega$ is the interior product of ω with the vector x

$$\begin{aligned}(T\,d + dT)\omega(x) &= \int_0^1 t^{p-1}(i_x\,d + d i_x)\omega(tx)\,dt \\ &= \int_0^1 t^{p-1}\mathcal{L}_x\omega(tx)\,dt \\ &= \int_0^1 \frac{d}{dt}(t^p\omega(tx))\,dt = \omega(x),\end{aligned}$$

where use has been made of

$$\begin{aligned}\frac{d}{dt}(t^p\omega(tx)) &= pt^{p-1}\omega(tx) + t^p\frac{\partial\omega}{\partial(tx)^i}\frac{d(tx^i)}{dt} \\ &= pt^{p-1}\omega(tx) + t^{p-1}\frac{\partial\omega}{\partial x^i}x^i.\end{aligned}$$

■

Exercise: Prove the Poincaré lemma for a monomial form. Obviously the previous proof applies to a monomial but a proof tailored to monomials is simple and interesting. Let $\boldsymbol{\omega}$ be a monomial

$$\boldsymbol{\omega} = \omega \, dx^1 \wedge \ldots \wedge dx^p,$$

$$d\boldsymbol{\omega} = 0 \Rightarrow \frac{\partial \omega}{\partial x^{p+1}} = \cdots = \frac{\partial \omega}{\partial x^n} = 0$$

ω depends only on $x^1, \ldots x^p$. $\boldsymbol{\theta} = \theta(x^1, \ldots x^p) \, dx^2 \wedge \ldots \wedge dx^p$ satisfies the equation $\boldsymbol{\omega} = d\boldsymbol{\theta}$ if $\partial\theta/\partial x^1 = \omega$.
On $\mathbf{R}^n$ there is an infinity of functions θ satisfying this differential equation. Let θ be one of them and $\boldsymbol{\theta}$ the corresponding form, all solutions can be written $\boldsymbol{\theta} + d\boldsymbol{\alpha}$. ■

Note that the Poincaré lemma does not apply if the form $\boldsymbol{\omega}$ fails to be differentiable at certain points of $\mathbf{R}^n$ because then the manifold on which $\boldsymbol{\omega}$ is differentiable is not homeomorphic to $\mathbf{R}^n$. For example, consider the closed form

$$\boldsymbol{\omega} = \frac{-x^2 \, \mathbf{dx}^1 + x^1 \, \mathbf{dx}^2}{(x^1)^2 + (x^2)^2} = d\left(\arctan \frac{x^2}{x^1}\right),$$

which is differentiable on $\mathbf{R}^2 - 0$. The form $\boldsymbol{\omega}$ is not exact because arctan (x^2/x^1) is not a differentiable function on $\mathbf{R}^2$.

Compact cohomology. Analogous definitions for groups H^p_c and Betti numbers b^p_c can be given, replacing $\Lambda(X)$ by the space $\hat{\mathscr{D}}(X)$ of C^∞ exterior differential forms with compact support and $C(X)$ by locally finite chains.
For the space $\hat{\mathscr{D}}(\mathbf{R}^n)$, a lemma analogous to the Poincaré lemma can be stated as follows: $\hat{\mathscr{D}}(\mathbf{R}^n)$

Poincaré lemma (compact support). In $\mathbf{R}^n$ a closed p-form with compact support is the coboundary of a $(p-1)$-form with compact support if $p \leq n-1$, and an n-form $\boldsymbol{\omega}$ with compact support is the coboundary of an $n-1$ form with compact support iff $\int_{\mathbf{R}^n} \boldsymbol{\omega} = 0$.

It follows that $b^p_c = 0$ for $0 \leq p \leq n-1$ and it can be shown that $b^n_c = 1$. The following table summarizes these results and also gives the Betti numbers for the space of forms $\Lambda(S^n)$ on the n-sphere and for the space of forms $\Lambda(\mathbf{T}^n)$ on the n-torus ($\mathbf{T}^n = \mathbf{R}^n/\mathbf{Z}^n$ = quotient space of $\mathbf{R}^n$ by the space of the integral lattice points in $\mathbf{R}^n$)

	b^0	b^1 ...	b^p ...	b^{n-1}	b^n	Remarks (pp. 226, 227)
$\Lambda(\mathbb{R}^n)$	1	0	0	0	0	
$\hat{\mathscr{D}}(\mathbb{R}^n)$	0	0	0	0	1	$b^p = b_c^{n-p}$.
$\Lambda(S^n), \hat{\mathscr{D}}(S^n)$	1	0	0	0	1	On compact manifolds
$\Lambda(\mathbb{T}^n), \hat{\mathscr{D}}(\mathbb{T}^n)$	1	n	$\binom{n}{p}$	n	1	$b^p = b_c^p$, hence $b^p = b^{n-p}$.

Exercise: Derive partially the results for S^n using S^2 as an example.[1] This table shows the striking difference between the space of forms with arbitrary support $\Lambda(\mathbb{R}^n)$ and the space of forms with compact support $\hat{\mathscr{D}}(\mathbb{R}^n)$. For a compact space X there is no difference between cohomology with compact support and cohomology with arbitrary support.

Duality. We already touched upon the duality of chains and forms when we stated that an integral is a bilinear functional

$$\int_C \omega = \langle C, \omega \rangle.$$

We shall in this paragraph make two independent, albeit similar, studies first with finite chains and forms with arbitrary supports, then with arbitrary locally finite chains and forms with compact support.

1) Finite chains, forms with arbitrary support.
The mapping $C(X) \times \Lambda(X) \to \mathbb{R}$ by $(C, \omega) \mapsto \langle C, \omega \rangle$ may be degenerate on $Z_p(X) \times Z^p(X)$, that is,

$$\langle C, \omega \rangle = 0 \quad \text{for all} \quad \omega \in Z^p$$

does not imply that $C = 0$ since the equation is true whenever $C = \partial B$. This suggests that a non-degenerate mapping will be obtained by passing to the quotient spaces H^p and H_p. This is de Rham's theorem.

de Rham theorem. The mapping

$$\mathscr{L}: H_p \times H^p \to \mathbb{R} \quad \text{by } ([C], [\omega]) \to \langle C, \omega \rangle \quad \text{for } C \in H_p,\ \omega \in H^p$$

is a bilinear non degenerate mapping which establishes the duality of the vector spaces H_p and H^p and the equality $b_p = b^p$ when $H_p[H^p]$ is finite dimensional.

[1][Choquet-Bruhat 1968, exercise p. 83].

It is easy to prove that $\mathscr{L}$ is a bilinear mapping from $H_p \times H^p$ into $\mathbb{R}$:

$C \in B_p \Rightarrow \langle C, \boldsymbol{\omega} \rangle = 0$ by Stokes' theorem because $C = \partial B$ and $\mathrm{d}\boldsymbol{\omega} = 0$

$\boldsymbol{\omega} \in B^p \Rightarrow \langle C, \boldsymbol{\omega} \rangle = 0$ by Stokes' theorem because $\boldsymbol{\omega} = \mathrm{d}\boldsymbol{\theta}$ and $\partial C = 0$.

To prove that $\mathscr{L}$ is non degenerate is to prove the converse of the above propositions[1]:

$$\langle C, \boldsymbol{\omega} \rangle = 0 \quad \text{for all } \boldsymbol{\omega} \in Z^p \Rightarrow C \sim 0$$

$$\langle C, \boldsymbol{\omega} \rangle = 0 \quad \text{for all } C \in Z_p \Rightarrow \boldsymbol{\omega} \sim 0.$$

2) Infinite chains, forms with compact support.
de Rham's theorem establishes similarly the duality of the compact cohomology H_c^p with the infinite chain homology IH_p for orientable manifolds. IH_p

The integral $\int_C \boldsymbol{\omega} = \langle C, \boldsymbol{\omega} \rangle$ of a cocycle $\boldsymbol{\omega}$ [closed form] over a cycle C is called a **period** of this form. It depends only on the cohomology class of $\boldsymbol{\omega}$ and the homology class of C. de Rham theorem can be reformulated as follows: period
There exists a closed form which has on X^n arbitrarily preassigned periods on linearly independent homology classes. This closed form is determined up to the addition of an exact form.

Corollary. A closed form is exact if and only if all its periods vanish.

Example: Let X^n be a compact connected manifold and $\boldsymbol{\omega}$ an n-form on X^n. We show that $\langle X^n, \boldsymbol{\omega} \rangle = 0 \Leftrightarrow \boldsymbol{\omega} = \mathrm{d}\boldsymbol{\theta}$. The dimension of $H_n(X^n)$ is one since $b_n = b^n = b^0 = 1$. X^n considered as a domain of integration is a cycle which is not a boundary; hence any element of $Z_n(X^n)$ can be written

$$\lambda X^n + \partial B$$

where λ is a real number. By de Rham's theorem

$$\begin{aligned} \boldsymbol{\omega} = \mathrm{d}\boldsymbol{\theta} &\Leftrightarrow \langle C, \boldsymbol{\omega} \rangle = 0 && \text{for all } C \in Z_p \\ &\Leftrightarrow \langle \lambda X^n + \partial B, \boldsymbol{\omega} \rangle = 0 && \text{for all } \lambda \in \mathbb{R} \\ &\Leftrightarrow \lambda \langle X^n, \boldsymbol{\omega} \rangle = 0 && \text{for all } \lambda \in \mathbb{R} \\ &\Leftrightarrow \langle X^n, \boldsymbol{\omega} \rangle = 0 \end{aligned}$$

where we have used the fact that all n-forms on X^n are cocycles. ■

[1]See proof for instance [de Rham].

Poincaré duality theorem. If X^n is orientable, the p-cohomology vector space H^p with arbitrary support is the dual of the compact $(n-p)$-cohomology vector space H_c^{n-p}; hence $b^p = b_c^{n-p}$.[1]

If X^n is compact, there is no difference between cohomology with compact supports and cohomology with arbitrary supports and

$$b^p = b^{n-p}.$$

Exercise: If X^n is a triangulable connected manifold then $IH_n(X^n)$ has at most dimension one.[2] In the case $X^n = \mathbb{R}^n$ this result can be read from the table p. 226 using $Ib_n = b_c^n$. Following the procedure of the above example, use this fact to show that for an n-form ω with compact support

$$\langle X^n, \omega \rangle = 0 \Leftrightarrow \omega = \mathrm{d}\theta.$$

Deduce from this property that an arbitrary n-form with compact support can be written

$$\lambda \varphi + \mathrm{d}\theta$$

where λ is a real number, φ is some fixed n-form with compact support such that

$$\langle X^n, \varphi \rangle \neq 0$$

and θ is an $(n-1)$-form with compact support. Show then that if $f: X^n \to Y^n$ is a proper differentiable mapping and if Y^n is also connected and triangulable, then there exists a number k such that

$$\int_{X^n} f^* \omega = k \int_{Y^n} \omega$$

for any ω with compact support. This constitutes a partial proof of a theorem quoted earlier (p. 221):

currents **Currents.** de Rham has unified the notions of chain and form in a much wider class of objects: the currents. We shall come back in Chapter VI to currents and, more generally, to tensor distributions on a manifold.

[1]See proof for instance [L. Schwartz 1966, Chap. IX].
[2]See for instance [Patterson].

C. EXTERIOR DIFFERENTIAL SYSTEMS

We turn now to the problem of finding integral manifolds of a system of exterior differential equations. Exterior differential systems made up of 1-forms, called **Pfaff systems**, are particularly simple. An integral manifold of a system S of forms of arbitrary degree may be described in terms of the integral manifolds of a Pfaff system associated with the closure of S. The Pfaff system is called the characteristic system of S. Pfaff system

Partial differential equations can be written as exterior differential systems. For example, a system of first order partial differential equations

$$F^\alpha\left(x^1,\ldots,x^n,z^1,\ldots,z^p,\frac{\partial z^1}{\partial x^1},\ldots\frac{\partial z^1}{\partial x^n},\ldots,\frac{\partial z^p}{\partial x^n}\right)=0$$

can be restated in terms of the following exterior differential system

$$\begin{cases} f^\alpha \equiv F^\alpha(x^i, z^I, p^I_i) = 0 \\ \boldsymbol{\theta}^I = \mathbf{d}z^I - p^I_i\,\mathbf{d}x^i = 0. \end{cases}$$

We shall treat in detail the case of a first order nonlinear partial differential equation (p. 250).

1. EXTERIOR EQUATIONS

We first consider quantities defined at a point $x \in X^n$; occasionally the subscript x is omitted for brevity in paragraphs 1, 2 and 3 when no confusion can arise. As a rule in this section the symbols $\boldsymbol{\theta}$, $\boldsymbol{\theta}^\alpha, \ldots$ designate 1-forms (Pfaff forms); $\boldsymbol{\omega}$, $\boldsymbol{\omega}^\alpha$, $\boldsymbol{\varphi}$, $\boldsymbol{\xi}$, $\boldsymbol{\zeta}, \ldots$ designate forms of arbitrary degree.

A **solution of the system of exterior equations** solution of a system of exterior equations

$$\{\boldsymbol{\omega}^\alpha_x = 0, \qquad \alpha = 1, 2, \ldots, N\}$$

is any subspace $P \subset T_x(X^n)$ such that $\boldsymbol{\omega}^\alpha_x(\boldsymbol{v}_1, \ldots \boldsymbol{v}_{p_\alpha}) = 0$ $\forall \alpha$ whenever $\boldsymbol{v}_1, \ldots \boldsymbol{v}_{p_\alpha} \in P$, i.e. any subspace P which **annuls** each $\boldsymbol{\omega}^\alpha_x$. annul

Note that if P is of dimension less than p, a p-form $\boldsymbol{\omega}_x$ vanishes on P trivially. It follows that, if all $\boldsymbol{\omega}^\alpha_x$ are of degree greater than or equal to p, then all subspaces of T_x of dimension less than p are solutions.

2. SINGLE EXTERIOR EQUATION $\boldsymbol{\omega}_x = 0$ WHERE $\boldsymbol{\omega}_x$ IS A p-FORM AT x

Let $\{\boldsymbol{\theta}^j;\ j = 1, \ldots, n\}$ be a basis in $T^*_x(X^n)$. The general expression for $\boldsymbol{\omega}_x$ includes all n elements of this basis. However it is often possible to find a linearly independent subset $\{\bar{\boldsymbol{\theta}}^\alpha;\ \alpha = 1, \ldots, r \text{ with } p \le r < n\}$ of T^*_x with less than n elements in terms of which $\boldsymbol{\omega}_x$ can be expressed:

$$\begin{aligned}\boldsymbol{\omega}_x &= \frac{1}{p!}\,\omega_{\alpha_1\ldots\alpha_p}\bar{\boldsymbol{\theta}}^{\alpha_1}\wedge\ldots\wedge\bar{\boldsymbol{\theta}}^{\alpha_p}\\ &= \omega_{A_1\ldots A_p}\bar{\boldsymbol{\theta}}^{A_1}\wedge\ldots\wedge\bar{\boldsymbol{\theta}}^{A_p}\end{aligned}$$

where $A_j = 1, \ldots, r$ and $A_1 < A_2 < \cdots < A_p$. Now let r be the minimum number of linearly independent 1-forms necessary to express $\boldsymbol{\omega}_x$; then $\boldsymbol{\omega}_x$ is said to be of **rank** r. Note that $p \le r \le n$. The space Q^*_x generated by the minimal set $\{\bar{\boldsymbol{\theta}}^\alpha;\ \alpha = 1, \ldots, r\}$ is called the **associated space** of $\boldsymbol{\omega}_x$; that is, Q^*_x is the smallest subspace of T^*_x such that[1] $\boldsymbol{\omega}_x \in \Lambda^p(Q^*_x)$.

rank

associated space Q^*_x

If $r = p$ the form $\boldsymbol{\omega}$ is monomial (consists of only one term $\omega_{1\ldots p}\bar{\boldsymbol{\theta}}^1\wedge\ldots\wedge\bar{\boldsymbol{\theta}}^p$) and conversely. A 1-form is always[2] of rank 1.

$\tilde{Q}_x$

Let $\tilde{Q}_x$ be the space of vectors $\boldsymbol{v}$ such that[3] $\mathrm{i}_v\boldsymbol{\theta} = 0$ for every $\boldsymbol{\theta} \in Q^*_x$. The space $\tilde{Q}_x$ is a subspace of T_x of dimension $n - r$ because the homogeneous equations $\bar{\theta}^\alpha_i v^i = 0$ for $\boldsymbol{v}$ have $n - r$ linearly independent solutions. Note that while T_x and T^*_x are dual spaces, $\tilde{Q}_x$ and Q^*_x are not dual.

The space $\tilde{Q}_x$ is a particularly interesting solution of the equation $\boldsymbol{\omega}_x = 0$. We shall call it the **maximal solution** of Q^*_x; every solution of the associated Pfaff system Q^*_x is a subspace of $\tilde{Q}_x$. The following theorem gives another characterization of the space $\tilde{Q}_x$.

maximal solution

Theorem. Let $\boldsymbol{\omega}_x$ be a p-form at $x \in X^n$ with associated space Q^*_x; let $\tilde{Q}_x$ be the space of vectors $\boldsymbol{v}$ such that $\mathrm{i}_v\boldsymbol{\theta} = 0$ for every $\boldsymbol{\theta} \in Q^*_x$. Then

$$\boldsymbol{v} \in \tilde{Q}_x \Leftrightarrow \mathrm{i}_v\boldsymbol{\omega}_x \equiv 0.$$

Proof:

1) $\boldsymbol{v} \in \tilde{Q} \Rightarrow \mathrm{i}_v\boldsymbol{\omega} = 0$ because if

$$\boldsymbol{\omega}_x = \omega_{A_1\ldots A_p}\bar{\boldsymbol{\theta}}^{A_1}\wedge\ldots\wedge\bar{\boldsymbol{\theta}}^{A_p}, \qquad A_j = 1, \ldots r$$

then (p. 207)

$$\mathrm{i}_v\boldsymbol{\omega}_x = \sum_j (\pm)\omega_{A_1\ldots A_p}\bar{\boldsymbol{\theta}}^{A_1}\wedge\ldots\mathrm{i}_v\bar{\boldsymbol{\theta}}^{A_j}\ldots\wedge\bar{\boldsymbol{\theta}}^{A_p} = 0.$$

2) $\mathrm{i}_{v'}\boldsymbol{\omega}_x \equiv 0 \Rightarrow \boldsymbol{v}' \in \tilde{Q}_x$.

Let $\tilde{Q}'_x$ be the space of vectors $\boldsymbol{v}'$ such that $\mathrm{i}_{v'}\boldsymbol{\omega} \equiv 0$; let Q'^*_x be the space

[1] Now that we have introduced $Q^*_x \subset T^*_x(X)$ we cannot use the abbreviated notation $\Lambda_x(X)$ any more but we have to use $\Lambda(T^*_xX)$ or $\Lambda(Q^*_x)$ as the case may be.

[2] If not identically zero.

[3] Both notations $\mathrm{i}_v\boldsymbol{\theta}$ and $\mathrm{i}(\boldsymbol{v})\boldsymbol{\theta}$ are used in this chapter.

of forms $\boldsymbol{\theta}'$ such that $\mathrm{i}_{v'}\boldsymbol{\theta}' = 0$, $\forall\, \boldsymbol{v}' \in \tilde{Q}'_x$. We shall prove that $Q^*_x \subset Q'^*_x$ which implies $\tilde{Q}'_x \subset \tilde{Q}_x$ and proves the proposition. Let $\{\boldsymbol{\theta}'^i;\ i = 1, \ldots n\}$ be a basis of T^*_x such that its first r' elements generate Q'^*_x. In this basis

$$\boldsymbol{\omega} = \omega'_{I_1 \ldots I_p} \boldsymbol{\theta}'^{I_1} \wedge \ldots \wedge \boldsymbol{\theta}'^{I_p}$$

and

$$\mathrm{i}_{v'}\boldsymbol{\omega} = \Big(\sum_{I_j \le r'} + \sum_{I_j \ge r'+1}\Big)(\pm)\, \omega'_{I_1 \ldots I_p} \boldsymbol{\theta}'^{I_1} \wedge \ldots \mathrm{i}_{v'}\boldsymbol{\theta}'^{I_j} \wedge \ldots \wedge \boldsymbol{\theta}'^{I_p}.$$

Since $\mathrm{i}_{v'}\boldsymbol{\theta}'^{i_j} = 0$ for $i_j \le r'$, the first sum is zero; moreover

$$\mathrm{i}_{v'}\boldsymbol{\theta}'^i = v'^i$$

where v'^i are the components of $\boldsymbol{v}'$ in the basis dual to $\{\boldsymbol{\theta}'^i\}$. Thus the first r' components of $\boldsymbol{v}'$ in this basis are zero, the remaining $n - r'$ ones are arbitrary. Since $\mathrm{i}_{v'}\boldsymbol{\omega} \equiv 0$

$$0 \equiv \mathrm{i}_{v'}\boldsymbol{\omega} \equiv \sum_{I_j \ge r'+1} (\pm)\, v'^{I_j} \omega_{I_1 \ldots I_j \ldots I_p} \boldsymbol{\theta}'^{I_1} \wedge \ldots \wedge \widehat{\boldsymbol{\theta}'^{I_j}} \wedge \ldots \wedge \boldsymbol{\theta}'^{I_p}$$

(the circumflex indicates omission) one concludes that

$$\omega_{I_1 \ldots I_j \ldots I_p} = 0 \text{ for } I_j \ge r' + 1.$$

Therefore $\boldsymbol{\omega}$ can be expressed in terms of the set $\{\boldsymbol{\theta}'^i;\ i = 1, \ldots, r'\}$, hence the associated space Q^*_x of $\boldsymbol{\omega}_x$ is contained in Q'^*_x. ■

The next theorem provides a construction for Q^*_x.

Theorem. Let $\{\boldsymbol{\theta}^i;\ i = 1, \ldots, n\}$ *be an arbitrary basis in* T^*_x

$$\boldsymbol{\omega}_x = \omega_{I_1 \ldots I_p} \boldsymbol{\theta}^{I_1} \wedge \ldots \wedge \boldsymbol{\theta}^{I_p}.$$

Set

$$\overset{\Delta}{\boldsymbol{\theta}}{}^\alpha = \omega_{i\alpha} \boldsymbol{\theta}^i = \omega_{iI_2 \ldots I_p} \boldsymbol{\theta}^i$$

where the labels $\alpha = 1, 2, \ldots, \binom{n}{p-1}$ *have been placed in one to one correspondence with the labels* $I_2 \ldots I_p$. *The set* $\{\overset{\Delta}{\boldsymbol{\theta}}{}^\alpha\}$ *spans* Q^*_x.

Proof:

$$\boldsymbol{v} \in \tilde{Q}_x \Big| \Rightarrow \mathrm{i}_v\boldsymbol{\omega} \equiv 0 \Rightarrow v^i \omega_{iI_2 \ldots I_p} = 0 \Rightarrow \mathrm{i}_v \overset{\Delta}{\boldsymbol{\theta}}{}^\alpha = 0.$$

Thus $\overset{\Delta}{\boldsymbol{\theta}}{}^\alpha \in Q^*_x$ because $\mathrm{i}_v \overset{\Delta}{\boldsymbol{\theta}}{}^\alpha = 0$, $\forall \boldsymbol{v} \in \tilde{Q}_x$. Since there are $n - r$ linearly independent vectors $\boldsymbol{v}$ satisfying the homogeneous equations $v^i \omega_{i\alpha} = 0$, the rank of the matrix $[\omega_{i\alpha}]$ is r. The rank of the set $\{\overset{\Delta}{\boldsymbol{\theta}}{}^\alpha\}$ is therefore r, hence it spans Q^*_x. ■

Example: The associated space of a 2-form $\boldsymbol{F} = F_{IJ}\boldsymbol{\theta}^I \wedge \boldsymbol{\theta}^J$ is spanned by the 1-forms $\boldsymbol{\theta}^\alpha \overset{\Delta}{=} F_{ij}\boldsymbol{\theta}^i$. The rank of $\boldsymbol{F}$ is the rank of the matrix $[F_{ij}]$. The rank of a square antisymmetric matrix is even; thus in an odd dimensional space X^n, the rank of $\boldsymbol{F}$ is strictly less than n: $2 \leq r < n$. In particular a 2-form in X^3 is at most of rank 2. For example if $P \neq 0$

$$\begin{aligned} \boldsymbol{F} &= P\ \mathbf{dx} \wedge \mathbf{dy} + Q\ \mathbf{dy} \wedge \mathbf{dz} + R\ \mathbf{dz} \wedge \mathbf{dx} \\ &= \left(\mathbf{dx} - \frac{Q}{P}\mathbf{dz}\right) \wedge (P\ \mathbf{dy} - R\ \mathbf{dz}) \\ &= \boldsymbol{\theta}^1 \wedge \boldsymbol{\theta}^2 \qquad \text{with obvious definitions for } \boldsymbol{\theta}^1 \text{ and } \boldsymbol{\theta}^2. \end{aligned}$$

3. SYSTEMS OF EXTERIOR EQUATIONS

Let $\{\boldsymbol{\omega}_x^\alpha = 0;\ \alpha = 1, \ldots, N\}$ be a system of N exterior equations in which the forms $\boldsymbol{\omega}_x^\alpha$ are not necessarily of the same degree but are all of degree *greater than zero*. Any attempt to find other systems with the same solutions as the system $\{\boldsymbol{\omega}_x^\alpha = 0\}$ leads naturally to the **ideal**

ideal I generated by $\{\boldsymbol{\omega}_x^\alpha\}$

$$I = \left\{\boldsymbol{\omega}_x = \sum_\alpha \boldsymbol{\xi}_x^\alpha \wedge \boldsymbol{\omega}_x^\alpha;\ \boldsymbol{\xi}_x^\alpha \in \Lambda(T_x^* X^n)\right\} \subset \Lambda(T_x^* X^n)$$

generated by the set $\{\boldsymbol{\omega}_x^\alpha\}$. This is because any subspace of $T_x X^n$ which annuls all elements of $\{\boldsymbol{\omega}_x^\alpha\}$ will also annul all exterior products $\boldsymbol{\xi}_x \wedge \boldsymbol{\omega}_x^\alpha$, where $\boldsymbol{\xi}_x \in \Lambda(T_x^*)$ is arbitrary, and all sums of such products (such that each term has the same degree). Since each $\boldsymbol{\omega}_x^\alpha$ is an element of I it is also true that any subspace of T_x which annuls each element of I will annul each $\boldsymbol{\omega}_x^\alpha$. Thus every solution of $\{\boldsymbol{\omega}_x^\alpha = 0\}$ annuls each element of I and conversely. Note that a zero form at a point is a number, the corresponding equation can have solutions only if the zero form is identically zero.

complete

A system $\{\boldsymbol{\omega}_x^\alpha\}$ [the ideal I generated by $\{\boldsymbol{\omega}^\alpha\}$] is called **complete** if any form which is annulled by every solution of $\{\boldsymbol{\omega}_x^\alpha = 0\}$ is an element of I.

Exercise: Show that every system of 1-forms is complete.

algebraically equivalent

Two systems $\{\boldsymbol{\omega}_x^\alpha = 0\}$ and $\{\boldsymbol{\omega}_x'^\beta = 0\}$ [sets $\{\boldsymbol{\omega}_x^\alpha\}$ and $\{\boldsymbol{\omega}_x'^\beta\}$], not necessarily with the same number of elements, are said to be **algebraically equivalent**, written $\{\boldsymbol{\omega}_x^\alpha = 0\} \sim \{\boldsymbol{\omega}_x'^\beta = 0\}$ $[\{\boldsymbol{\omega}_x^\alpha\} \sim \{\boldsymbol{\omega}_x'^\beta\}]$, if the sets $\{\boldsymbol{\omega}_x^\alpha\}$ and $\{\boldsymbol{\omega}_x'^\beta\}$ generate the same ideal. Thus two sets are algebraically equivalent if and only if each element of one can be expressed as a linear combination, with coefficients in Λ_x, of the elements of the other and vice versa. Clearly two algebraically equivalent systems have the same solutions. If a system is complete then it is algebraically equivalent to every other system with the same solutions.

Exercise: Show that the two following systems have the same solutions but do not generate the same ideal:

$$\begin{cases} \boldsymbol{\theta}^1 \wedge \boldsymbol{\theta}^3 = 0 \\ \boldsymbol{\theta}^1 \wedge \boldsymbol{\theta}^4 = 0 \\ \boldsymbol{\theta}^1 \wedge \boldsymbol{\theta}^2 - \boldsymbol{\theta}^3 \wedge \boldsymbol{\theta}^4 = 0 \end{cases} \quad \text{and} \quad \begin{cases} \boldsymbol{\theta}^1 \wedge \boldsymbol{\theta}^3 = 0 \\ \boldsymbol{\theta}^1 \wedge \boldsymbol{\theta}^4 = 0 \\ \boldsymbol{\theta}^1 \wedge \boldsymbol{\theta}^2 - \boldsymbol{\theta}^3 \wedge \boldsymbol{\theta}^4 = 0 \\ \boldsymbol{\theta}^1 \wedge \boldsymbol{\theta}^2 = 0 \end{cases}$$

where the $\boldsymbol{\theta}^i$ form a linearly independent set. *Hint*: See pp. 235–236.

Exercise: Show that the condition for two systems to be algebraically equivalent takes a simpler form when both systems are Pfaffian (systems of 1-forms). $\{\boldsymbol{\theta}^\alpha = 0\} \sim \{\boldsymbol{\theta}'^\beta = 0\}$ if the elements of each set can be expressed as linear combinations of the elements of the other, that is, if

$$\begin{cases} \boldsymbol{\theta}^\alpha = S^\alpha{}_\beta \boldsymbol{\theta}'^\beta \\ \boldsymbol{\theta}'^\beta = S'^\beta{}_\alpha \boldsymbol{\theta}^\alpha. \end{cases}$$

This is equivalent to saying that both sets generate the same vector space. If both sets are linearly independent the condition reduces further to $\boldsymbol{\theta}^\alpha = S^\alpha{}_\beta \boldsymbol{\theta}'^\beta$ where $S^\alpha{}_\beta$ is a regular (invertible) square matrix.

Let I be the ideal generated by the system $\{\boldsymbol{\omega}_x^\alpha\}$.
It is tempting to try to define the associated space of the system $\{\boldsymbol{\omega}_x^\alpha\}$ as the smallest subspace $Q_x^* \subset T_x^*$ such that $\boldsymbol{\omega}_x^\alpha \in \Lambda(Q_x^*), \forall \alpha$. However since any member of a system algebraically equivalent to $\{\boldsymbol{\omega}_x^\alpha\}$ is a linear combination, with coefficients in Λ_x, of the $\boldsymbol{\omega}_x^\alpha$, members of algebraically equivalent systems could fail to be elements of $\Lambda(Q_x^*)$ thus defined. The definition of Q_x^* must not depend on which system of the class of systems algebraically equivalent to $\{\boldsymbol{\omega}_x^\alpha\}$ is used to define it. The obvious definition is to say that Q_x^* is the smallest subspace of T_x^* from which one can construct a system algebraically equivalent to $\{\boldsymbol{\omega}_x^\alpha\}$. More precisely:

The **associated space** of the system $\{\boldsymbol{\omega}_x^\alpha\}$ [of the ideal I] is the smallest subspace $Q_x^* \subset T_x^*$ such that $\Lambda(Q_x^*)$ contains a subset $\{\boldsymbol{\omega}_x'^\beta\}$ which generates I (which is algebraically equivalent to $\{\boldsymbol{\omega}_x^\alpha\}$). associated space of I
The dimension of Q_x^* is called the **rank** of $\{\boldsymbol{\omega}_x^\alpha\}$ [of I]. rank of I

The systems which are subsets of $\Lambda(Q_x^*)$ are in a sense the simplest systems which generate I.

$\tilde{Q}_x$ As before the space $\tilde{Q}_x$ is defined to be the space of all vectors $\boldsymbol{v} \in T_x$ such that $\mathrm{i}_v\boldsymbol{\theta} = 0$ for all $\boldsymbol{\theta} \in Q_x^*$. The space $\tilde{Q}_x$ will be best characterized in terms of the ideal I because if $\{\boldsymbol{\omega}_x^\alpha\} \sim \{\boldsymbol{\omega}_x'^\beta\}$, then $\mathrm{i}_v\boldsymbol{\omega}_x^\alpha \equiv 0$, $\forall\,\alpha$ implies that

$$\mathrm{i}_v\boldsymbol{\omega}_x'^\beta = \sum_\alpha (\pm)\boldsymbol{\xi}^\alpha \wedge \mathrm{i}_v\boldsymbol{\omega}_x^\alpha + \sum_\alpha \mathrm{i}_v\boldsymbol{\xi}^\alpha \wedge \boldsymbol{\omega}_x^\alpha$$
$$= \sum_\alpha \mathrm{i}_v\boldsymbol{\xi}^\alpha \wedge \boldsymbol{\omega}_x^\alpha \in I,$$

but it does not necessarily follow that $\mathrm{i}_v\boldsymbol{\omega}_x'^\beta \equiv 0$. What is true is:

Theorem. Let I be the ideal generated by the system $\{\boldsymbol{\omega}_x^\alpha\}$; *let* Q_x^* *be the associated space of I. Then*

$$\boldsymbol{v} \in \tilde{Q}_x \Leftrightarrow \mathrm{i}_v\boldsymbol{\omega}_x \in I, \quad \forall \boldsymbol{\omega}_x \in I \quad [\text{equivalently } \mathrm{i}_v\boldsymbol{\omega}_x^\alpha \in I, \quad \forall \alpha].$$

Proof[1]: The proof is quite similar to that of the theorem on p. 230. Let $\tilde{Q}_x'$ be the space of vectors $\boldsymbol{v}'$ such that $\mathrm{i}_{v'}\boldsymbol{\omega} \in I$, $\forall \boldsymbol{\omega} \in I$. Let $\boldsymbol{\theta}' \in Q_x'^*$ be such that $\mathrm{i}_{v'}\boldsymbol{\theta}' = 0$, $\forall \boldsymbol{v}' \in \tilde{Q}_x'$. Let $\{\boldsymbol{e}_i'\}$ be a basis for T_x such that the last $n - r'$ vectors $\{\boldsymbol{e}_A'; A = r' + 1, \ldots, n\}$ span $\tilde{Q}_x'$; let $\{\boldsymbol{\theta}'^i\}$ be the dual basis for T_x^*. Let $\zeta^\alpha, \xi^\alpha \in \Lambda(T_x^*X^n)$.
1) It is easy to show that if $\boldsymbol{\omega} \in I$, then

$$\boldsymbol{v} \in \tilde{Q}_x \Rightarrow \mathrm{i}_v\boldsymbol{\omega} \in I.$$

Therefore $\tilde{Q}_x \subset \tilde{Q}_x'$, and $Q_x'^* \subset Q_x^*$.
2) To show that $\tilde{Q}_x' \subset \tilde{Q}_x$ we prove by induction that $\Lambda(Q_x'^*)$ contains a subset $\{\boldsymbol{\omega}'^\beta\}$ such that

$$\begin{cases} \boldsymbol{\omega}^\alpha = \sum_\beta \xi^\beta \wedge \boldsymbol{\omega}'^\beta, & \forall\,\alpha \\ \boldsymbol{\omega}'^\beta = \sum_\alpha \zeta^\alpha \wedge \boldsymbol{\omega}^\alpha, & \forall\,\beta, \end{cases}$$

that is, such that $\{\boldsymbol{\omega}^\alpha\}$ and $\{\boldsymbol{\omega}'^\beta\}$ generate the same ideal I. Then, since Q_x^* is the smallest such space, it follows that $Q_x^* \subset Q_x'^*$ and $\tilde{Q}_x' \subset \tilde{Q}_x$. Let $\boldsymbol{\omega} \in I$ be of degree 1. Now

$$\boldsymbol{v}' \in \tilde{Q}_x' \Rightarrow \mathrm{i}_{v'}\boldsymbol{\omega} \in I$$

and

$$\boldsymbol{\omega} \text{ of degree 1} \Rightarrow \mathrm{i}_{v'}\boldsymbol{\omega} \text{ of degree 0};$$

[1]Recall remark p. 229 concerning omission of subscript when no confusion can arise.

therefore $i_{v'}\boldsymbol{\omega} \equiv 0$, and $\boldsymbol{\omega} \in Q_x'^*$, because I contains no 0-forms other than 0.

Assume that there exist forms $\boldsymbol{\omega}'^{\beta} \in \cup_{q=1}^{p-1} \Lambda^q(Q_x'^*)$ such that for $\boldsymbol{\omega}$ of degree $p-1$

$$\boldsymbol{\omega} \in I \Leftrightarrow \boldsymbol{\omega} = \sum_\beta \boldsymbol{\xi}^\beta \wedge \boldsymbol{\omega}'^{\beta}.$$

Let $\boldsymbol{\omega}^\alpha$ be of degree p; then $i_{e_A'}\boldsymbol{\omega}^\alpha$ is of degree $p-1$ and can therefore be written

$$i_{e_A'}\boldsymbol{\omega}^\alpha = \sum_\beta \boldsymbol{\xi}^\beta \wedge \boldsymbol{\omega}'^{\beta}, \qquad A = r'+1, \ldots, n.$$

Since $\boldsymbol{\theta}'^A \wedge i_{e_A'}\boldsymbol{\omega}^\alpha$ is a p-form one can always construct the p-form $\boldsymbol{\omega}'^\alpha = \boldsymbol{\omega}^\alpha + \boldsymbol{\theta}'^A \wedge i_{e_A'}\boldsymbol{\omega}^\alpha$ and write

$$i_{e_A'}\boldsymbol{\omega}'^\alpha = i_{e_A'}\boldsymbol{\omega}^\alpha + i_{e_A'}\boldsymbol{\theta}'^A \wedge i_{e_A'}\boldsymbol{\omega}^\alpha - \boldsymbol{\theta}'^A \wedge i_{e_A'} i_{e_A'}\boldsymbol{\omega}^\alpha$$

Now

$$\begin{aligned} i_{e_A'}\boldsymbol{\omega}'^\alpha &= i_{e_A'}\boldsymbol{\omega}^\alpha + i_{e_A'}\boldsymbol{\theta}'^A \wedge i_{e_A'}\boldsymbol{\omega}^\alpha - \boldsymbol{\theta}'^A \wedge i_{e_A'} i_{e_A'}\boldsymbol{\omega}^\alpha \\ &\equiv 0. \end{aligned}$$

Therefore $i_{v'}\boldsymbol{\omega}'^\alpha \equiv 0$ for all $v' \in \tilde{Q}_x'$ and by the argument of the previous theorem $\boldsymbol{\omega}'^\alpha \in \Lambda^p(Q_x'^*)$. ■

Example: Consider the set S of forms in $\mathbb{R}^3$ at x:

$$\begin{cases} \boldsymbol{\omega}^1 = a\boldsymbol{\theta}^1 + b\boldsymbol{\theta}^2 + c\boldsymbol{\theta}^3 \\ \boldsymbol{\omega}^2 = P\boldsymbol{\theta}^2 \wedge \boldsymbol{\theta}^3 + Q\boldsymbol{\theta}^3 \wedge \boldsymbol{\theta}^1 + R\boldsymbol{\theta}^1 \wedge \boldsymbol{\theta}^2 \end{cases}$$

where the set $\{\boldsymbol{\theta}^i\}$ is a basis for $T_x^*(\mathbb{R}^3)$. Let I be the ideal generated by S;

$$\boldsymbol{\omega} \in I \Leftrightarrow \boldsymbol{\omega} = \boldsymbol{\xi}^1 \wedge \boldsymbol{\omega}^1 + \boldsymbol{\xi}^2 \wedge \boldsymbol{\omega}^2$$

with $\boldsymbol{\xi}^1, \boldsymbol{\xi}^2 \in \Lambda_x(\mathbb{R}^3)$. The elements $\boldsymbol{v} \in \tilde{Q}_x$ are determined by the conditions

$$i_v\boldsymbol{\omega}^1 \in I \text{ and } i_v\boldsymbol{\omega}^2 \in I.$$

1) Since $\boldsymbol{\omega}^1$ is a 1-form $i_v\boldsymbol{\omega}^1$ is a 0-form and since the only 0-form in I is 0

$$i_v\boldsymbol{\omega}^1 \in I \Leftrightarrow i_v\boldsymbol{\omega}^1 = 0.$$

2) Since $\boldsymbol{\omega}^2$ is a 2-form $i_v\boldsymbol{\omega}^2$ is a 1-form and therefore, looking at the structure of I,

$$i_v\boldsymbol{\omega}^2 \in I \Leftrightarrow i_v\boldsymbol{\omega}^2 = \lambda\boldsymbol{\omega}^1$$

where $\lambda \in \mathbb{R}$.

Let v^i be the components of $\boldsymbol{v}$ in the basis $\{\boldsymbol{e}_i\}$ of T_x dual to $\{\boldsymbol{\theta}^i\}$. The above two conditions on $\boldsymbol{v}$ are satisfied if and only if

$$b^i{}_j v^j = 0, \quad \text{where } [b^i{}_j] = \begin{bmatrix} a & b & c \\ aR & bR & -aP-bQ \\ bQ+cR & -bP & -cP \end{bmatrix}.$$

The rank of the matrix $[b^i{}_j]$ is equal to the rank of the system S. The 1-forms $\overset{\Delta}{\boldsymbol{\theta}}{}^i = b^i{}_j\boldsymbol{\theta}^j$ span the space Q^*_x. They are linearly independent if the rank of S is 3.

Suppose $c = aP + bQ = 0$; then $[b^i{}_j]$ is of rank 2, Q^*_x is generated by $\boldsymbol{\theta}^1$ and $\boldsymbol{\theta}^2$, and $\tilde{Q}$ is generated by e_3. Let $\boldsymbol{\omega}'^2 = \boldsymbol{\omega}^2 - \boldsymbol{\theta}^3 \wedge \mathrm{i}_{e_3}\boldsymbol{\omega}^2 = R\boldsymbol{\theta}^1 \wedge \boldsymbol{\theta}^2$. Then

$$\boldsymbol{\omega}'^2 = R\boldsymbol{\theta}^1 \wedge \boldsymbol{\theta}^2 \in \Lambda^2(Q^*_x)$$

$$\boldsymbol{\omega}^1 = a\boldsymbol{\theta}^1 + b\boldsymbol{\theta}^2 \in \Lambda^1(Q^*_x) = Q^*_x$$

and

$$\{\boldsymbol{\omega}^1, \boldsymbol{\omega}'^2\} \sim \{\boldsymbol{\omega}^1, \boldsymbol{\omega}^2\}.$$

4. EXTERIOR DIFFERENTIAL EQUATIONS

system of exterior differential equations

A **system of exterior differential equations** on X^n is a set of equations

$$\{\boldsymbol{\omega}^\alpha = 0, \qquad \alpha = 1, \ldots, N\}$$

Pfaff system
algebraically equivalent

where $\boldsymbol{\omega}^\alpha$ is a differential form on X^n. It is called a **Pfaff system** if all the $\boldsymbol{\omega}^\alpha$ are 1-forms. Two systems are **algebraically equivalent** if they generate the same ideal $I \subset \Lambda(X^n)$.

integral manifold

An **integral manifold** of the differential system $\{\boldsymbol{\omega}^\alpha = 0\}$ is a pair (Y, f) where Y is a differentiable manifold and f: $Y \to X^n$ is a differentiable mapping such that

$$f^*\boldsymbol{\omega}^\alpha \equiv 0, \qquad \forall\, \alpha.$$

This is equivalent to saying that for each $y \in Y$ the image $f' T_y(Y)$ in $T_{f(y)}(X^n)$ of the tangent space at y is a solution of the exterior equations $\{\boldsymbol{\omega}^\alpha_{f(y)} = 0\}$:

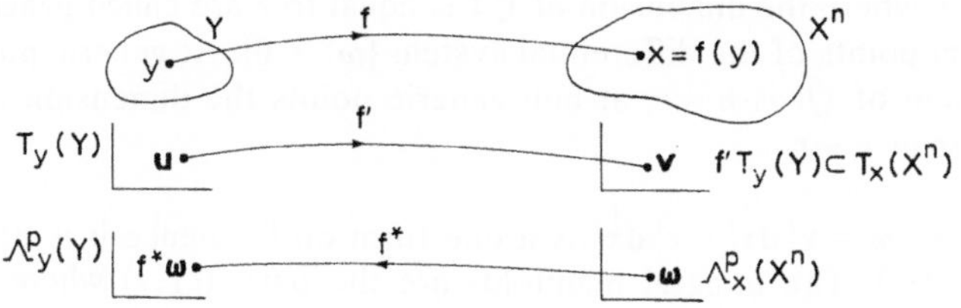

$$\begin{aligned} f^*\omega_y \equiv 0 &\Leftrightarrow f^*\omega_y(u_1, \dots, u_p) = 0 && \forall u_i \in T_y \\ &\Leftrightarrow \omega_x(f'u_1, \dots, f'u_p) = 0 && \forall u_i \in T_y \\ &\Leftrightarrow \omega_x(v_1, \dots, v_p) = 0 && \forall v_i \in f'T_y. \end{aligned}$$

Example: The surface

$$x = x(u, v), \quad y = y(u, v), \quad z = z(u, v), \quad u, v \in I \subset \mathbb{R}$$

in $\mathbb{R}^3$ is an integral manifold of the equation

$$\omega = P\, \mathbf{dx} \wedge \mathbf{dy} + Q\, \mathbf{dy} \wedge \mathbf{dz} + R\, \mathbf{dz} \wedge \mathbf{dx} = 0$$

if $f^*\omega \equiv 0$ where f: $(u, v) \mapsto (x, y, z)$ by the above equations.

$$f^*\omega = \left(f^*P\frac{D(x, y)}{D(u, v)} + f^*Q\frac{D(y, z)}{D(u, v)} + f^*R\frac{D(z, x)}{D(u, v)}\right) \mathbf{du} \wedge \mathbf{dv} \equiv 0.$$

All curves in $\mathbb{R}^3$ are integral manifolds of the equation. Note that the construction of an integral manifold must be done locally. In general a global construction is not possible.

Remark: When the mapping f is the inclusion mapping i, the integral manifold (Y, i) is abbreviated to Y.

Let U be the domain of a local chart in X^n. By the method of the preceeding paragraphs one can construct a set $\{\theta^\beta\}$ of differential 1-forms on U which generate the associated space Q^*_x of the exterior system $\{\omega^\alpha_x = 0\}$ at each point $x \in U$.
The set $\{\theta^\beta = 0\}$ is called the **associated Pfaff system** of the exterior differential system $\{\omega^\alpha = 0\}$. Note that the associated Pfaff system can only be constructed locally. associated Pfaff system

The set $\{\theta^\beta\}$ generates a submodule of the module $\Lambda^1(U)$. The rank r of the system $\{\omega^\alpha = 0\}$, and of the associated Pfaff system, is the **dimension** of this submodule. dimension
It can happen that, at certain points x, the dimension of Q^*_x is less than r.

generic points

Points x where the dimension of Q_x^* is equal to r are called **generic** (**non singular**) points of the differential system $\{\boldsymbol{\omega}^\alpha = 0\}$. At generic points the dimension of $\tilde{Q}_x$ is $n - r$; at non generic points the dimension of $\tilde{Q}_x$ is larger than $n - r$.

Example: $\boldsymbol{\omega} = x^2\,\mathbf{d}x^1 - x^1\,\mathbf{d}x^2$ is a one form on $\mathbb{R}^2$, hence it is of rank 1 (see p. 230). The integral manifolds are the pairs (c_k^1, i) where c_k^1 is a 1-cone defined by $x^2 = kx^1$ and i is the inclusion mapping. One checks readily that the tangent vectors $\boldsymbol{v}_x$ to c_k^1 at x satisfy the equation $\boldsymbol{\omega}_x(\boldsymbol{v}_x) = 0$, indeed: $\boldsymbol{v}_x = (a, ak)$ where a is arbitrary and $k = x^2/x^1$, $\boldsymbol{\omega}_x(\boldsymbol{v}_x) = ax^2 - kax^1 \equiv 0$. Everywhere, except at the origin, Q_x^* is of dimension $r = 1$; at the origin it is of dimension 0. Everywhere, except at the origin, $\tilde{Q}_x$ is of dimension $n - r = 1$; at the origin it is of dimesion 2. The origin is not a generic point; all the points in $\mathbb{R}^2 - \{0\}$ are generic points. Everywhere, except at the origin, $\tilde{Q}_x$ can be generated by a vector field $\boldsymbol{v}$: $x \mapsto \boldsymbol{v}_x = (a, (x^2/x^1)a)$; at the origin $\boldsymbol{v}_0$ is undefined. Notice also that there is exactly one integral manifold through each generic point.

Exercise: Let $\boldsymbol{\omega} = \mathrm{d}(x^2 + y^2) = 2x\,\mathbf{d}x + 2y\,\mathbf{d}y$; show that the integral manifolds of $\boldsymbol{\omega} = 0$ are the pairs (S_a^1, i) where S_a^1 is a 1-sphere of radius a and i is the inclusion mapping $i: S_a^1 \to \mathbb{R}^2$. *Hint*: The components of a tangent vector $\boldsymbol{v}$ to $i(S_a^1)$ at (x, y) are:

$$v^1 = -\alpha y, \qquad v^2 = \alpha x, \qquad \alpha \in \mathbb{R}.$$

closure

The differential system $\{\mathrm{d}\boldsymbol{\omega}^\alpha = 0, \boldsymbol{\omega}^\alpha = 0\}$ is called the **closure** of the system $\{\boldsymbol{\omega}^\alpha = 0\}$.

closed

A differential system is called **closed** if it is algebraically equivalent to its closure.

Let I be the ideal generated by the set $\{\boldsymbol{\omega}^\alpha\}$, that is,

$$\boldsymbol{\omega} \in I \Leftrightarrow \boldsymbol{\omega} = \sum_\alpha \boldsymbol{\xi}^\alpha \wedge \boldsymbol{\omega}^\alpha \qquad \text{with } \boldsymbol{\xi}^\alpha \in \Lambda(X^n).$$

$\mathrm{d}I$ is the ideal generated by the set $\{\boldsymbol{\omega}^\alpha, \mathrm{d}\boldsymbol{\omega}^\alpha\}$

$$\{\boldsymbol{\omega}^\alpha\} \text{ is closed} \Leftrightarrow \mathrm{d}\boldsymbol{\omega}^\alpha \in I, \forall\alpha \Leftrightarrow \mathrm{d}I \subset I.$$

It is of course always true that $I \subset \mathrm{d}I$. Closed systems have particularly nice properties. For this reason the following theorem is very convenient.

Theorem. A differential system $\{\boldsymbol{\omega}^\alpha = 0\}$ and its closure have the same integral manifolds.

Proof: Let (Y, f) be an integral manifold of $\{\omega^\alpha = 0\}$,

$$f^*\omega^\alpha \equiv 0 \text{ on } Y \qquad \forall \alpha.$$

Hence

$$0 \equiv \mathrm{d}f^*\omega^\alpha = f^*\,\mathrm{d}\omega^\alpha$$

and (Y, f) is an integral manifold of $\{\mathrm{d}\omega^\alpha = 0\}$. ■

Exercise: Show that if two systems are algebraically equivalent, then their closures are algebraically equivalent.

Exercise: Show that the closure of a differential system is closed; that is, show that if J is the ideal generated by $\{\mathrm{d}\omega^\alpha, \omega^\alpha\}$, then $\mathrm{d}J \subset J$.

An integral manifold is characterized by a pair (Y, f). The regularity of the integral manifold (Y, f) is discussed in terms of the regularity of $f(Y)$ as a subset of X^n. The subset $f(Y)$ is not in general a submanifold (see below). In the next section we shall impose successive restrictions on f until $f(Y)$ becomes a submanifold.

5. MAPPINGS OF MANIFOLDS

A subset Z of X^n is a **submanifold**[1] of X^n if every point $x \in Z$ is in the domain of a chart (U, φ) of X^n such that submanifold

$$\varphi\colon U \cap Z \to \mathbb{R}^q \times \{a\} \text{ by } \varphi(x) = (x^1, \ldots, x^q, a^1, \ldots, a^{n-q})$$

where a is a fixed element of $\mathbb{R}^{n-q}$. It is easy to check[2], that the charts $(\bar{U}, \bar{\varphi})$, where $\bar{U} = U \cap Z$ and $\bar{\varphi}\colon \bar{U} \to R^q$ by $\bar{\varphi}(x) = (x^1, \ldots, x^q)$, form an atlas on Z of the same class as the atlas $\{(U, \varphi)\}$ of X^n.
If Z already has a manifold structure, it is called a submanifold of X if it can be given a submanifold structure which is equivalent to the already existing structure.

[1]The reader is cautioned that there are subtle variations of these definitions among various authors. For example what is here called a submanifold is sometimes called a proper or regular submanifold thus allowing a submanifold to be somewhat less well behaved.
[2][Choquet-Bruhat 1968, p. 12].

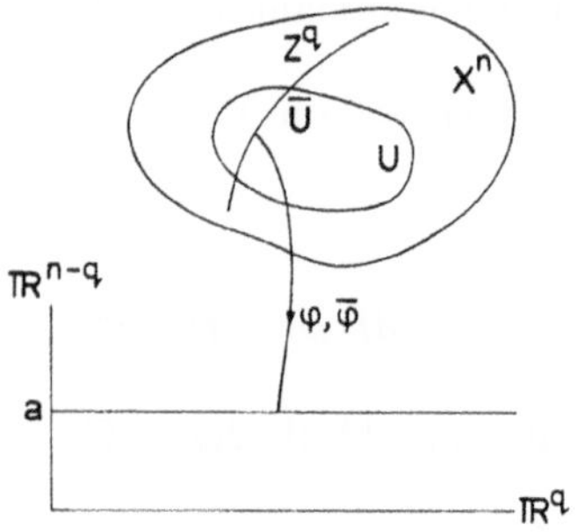

Submanifolds defined by a system of equations.

Theorem. Let a subset Z of X^n be defined by a system of p equations; $Z = \{x \in X^n; f^\alpha(x) = 0, \alpha = 1, \ldots, p\}$ *such that the* $f^\alpha(x)$ *are differentiable functions of x and such that the mapping* $X^n \to \mathbb{R}^p$ *defined by* $x \mapsto (f^1(x), \ldots, f^p(x))$ *is of rank p for every* $x \in Z$. *Then Z is a submanifold of dimension* $n - p$.

Proof: Let $x_0 \in Z$ and (U, φ) be a chart whose domain contains x_0. The mapping $x \mapsto (f^1(x), \ldots, f^p(x))$ being of rank p, there is one $p \times p$ matrix $[\partial f^\alpha / \partial x^i]$ such that its determinant at x_0 is different from zero, say the matrix corresponding to $i = 1, \ldots, p$. Then there exists a neighborhood $U' \subset U$ of x_0 where the following change of coordinates is admissible:

$$x'^i = f^i(x), \quad x'^{p+j} = x^j \qquad i = 1, \ldots, p; \; j = 1, \ldots, n-p.$$

The local chart (U', φ') is such that

$$\varphi'(U' \cap Z) = (0, \ldots 0, x'^{p+1}, \ldots x'^n).$$

A similar chart exists for every point of Z; hence Z is a submanifold of X^n. ∎

We can apply this theorem to show that an $(n-1)$-sphere S^{n-1} is a submanifold of $\mathbb{R}^n$,

$$S^{n-1} = \left\{ x \in \mathbb{R}^n;\; \sum_i (x^i)^2 - 1 = 0 \right\};$$

the mapping $x \mapsto \Sigma_i (x^i)^2 - 1$ is of rank 1 at each point $x \in S^n$.
It is clear that the cone

$$C^{n-1} = \left\{ x \in \mathbb{R}^n;\; (x^1)^2 - \sum_{i \neq 1} (x^i)^2 = 0 \right\}$$

is not a submanifold of $\mathbb{R}^n$. The mapping $x \mapsto (x^1)^2 - \Sigma_{i\neq 1}(x^i)^2$ is of rank 0 at the point $x = 0$ but of rank 1 on $\mathbb{R}^n - \{0\}$. Hence the cone without the point $x = 0$ is a submanifold of $\mathbb{R}^n$.

The most general integral manifold of a differential system is a pair (Y^q, f) where Y^q is a differentiable manifold and $f: Y^q \to X^n$ is a differentiable mapping. If the differentiable mapping $f: Y^q \to X^n$ is of rank q for every point $y \in Y^q$, then it is called an **immersion**. immersion

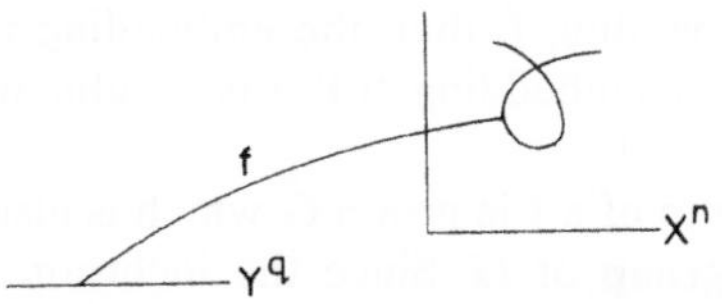

Example of a non-injective immersion.

An immersion is not necessarily injective; hence $f(Y^q)$ is not necessarily a manifold. Notice, however, that the induced mapping f' on the tangent bundle is injective. In the infinite dimensional case (p. 549) immersions are characterized through this property. Another characterization of immersions is given by the following theorem.

Theorem.[1] *A differentiable mapping $f: Y^q \to X^n$ is an immersion if and only if the set of germs g of differentiable functions on Y^q coincides with the reciprocal images under f of the germs $\bar{g}$ of differentiable functions on X^n i.e. $g = \bar{g} \circ f$.*

An injective immersion is an **embedding**. The set $f(Y^q)$ with the differentiable structure induced by the embedding f is a manifold. The differentiable structure on $f(Y^q)$ induced by f is the set of charts $\{(F(V_i), \psi_i \circ F^{-1})\}$ where $\{(V_i, \psi_i)\}$ is an atlas on Y^q and $F: Y^q \to f(Y^q)$ by $F(y) = f(y)$; the mapping F differs from f in that it is surjective. For the proof that this set of charts does define a differentiable structure see [Choquet-Bruhat 1968, p. 15]. embedding

However the manifold structure induced by f on $f(Y^q)$ may not be equivalent to a submanifold structure on $f(Y^q)$. Consider, for instance, the open interval $I \subset \mathbb{R}$ and the curve $f: I \to \mathbb{R}^2$ which comes arbitrarily close to itself without ever touching itself. If $\bar{U}$ is an open subset (domain of a chart in the differentiable structure induced by f) of $f(I)$ which

[1] For the proof see [Choquet-Bruhat 1968, p. 14].

contains the point of approach, there is no open subset of $U \subset \mathbb{R}^2$ such that $\bar{U} = U \cap F(I)$.

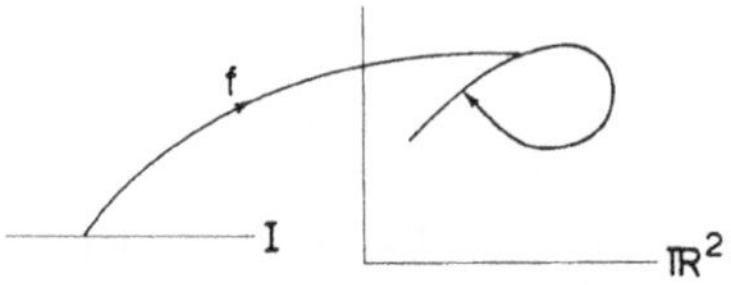

regular embedding

If $f(Y^q)$ has a submanifold structure equivalent to the manifold structure induced by the embedding f, then the embedding is said to be **regular**. Thus if f is a regular embedding $f(Y^q)$ is a submanifold of X^n.

Lie subgroup

Example: A subgroup of a Lie group G which is also a submanifold of G is called a **Lie subgroup** of G. Since the inclusion mapping is differentiable, every Lie subgroup is itself a Lie group. Consider the orthogonal group $O(n)$ which is the subgroup of the Lie group GL (n) made up of all $n \times n$ matrices A such that $A^{\mathrm{T}}A = \mathbb{1}$ where A^{T} is the transpose of A. Since the relation $A^{\mathrm{T}}A = \mathbb{1}$ is equivalent to a set of $n(n+1)/2$ independent numerical equations, the subgroup $O(n)$ is a submanifold of GL(n).

submersion

For completeness we include the definition of a submersion. If $f: Y^q \to X^n$ is a differentiable mapping of rank n for every $y \in Y^q$, then f is called a **submersion.** Clearly if f is a submersion then $q \geq n$.

6. PFAFF SYSTEMS

We now consider the particular case of Pfaff systems of exterior differential equations. In the next section we show how an arbitrary system can be discussed in terms of the associated Pfaff system of its closure (the characteristic system).

The associated space Q_x^* of a Pfaff system $\{\boldsymbol{\theta}^\alpha = 0\}$ at a point x is simply the vector space generated by the set $\{\boldsymbol{\theta}_x^\alpha\}$. In the domain of a chart U any Pfaff system of rank r is algebraically equivalent to a set of r linearly independent differential 1-forms. We can therefore without loss of generality assume that any Pfaff system of rank r on U has r linearly independent elements.

The space $\tilde{Q}_x$ contains all solutions of the system $\{\boldsymbol{\theta}_x^\alpha = 0\}$. An integral manifold of $\{\boldsymbol{\theta}^\alpha = 0\}$ is therefore a pair (Y, f) such that $f'T_y(Y) \subset \tilde{Q}_{f(y)}$ for all $y \in Y$.

An integral manifold which is a submanifold of X^n thus has dimension at most $n-r$ in the neighborhood of a generic point.

Example: Let $\boldsymbol{\theta} = \theta_i\, \mathbf{d}x^i$ be a 1-form in $\mathbb{R}^3$. The integral manifolds of the equation $\boldsymbol{\theta} = 0$ are

1) the curve $x^i = x^i(t)$ such that $\theta_i(x^j(t))\, \mathrm{d}x^i/\mathrm{d}t = 0$
2) the surface, if it exists, $x^i = x^i(u, v)$, $u, v \in I \subset \mathbb{R}$ such that

$$f^*\boldsymbol{\theta} = \theta_i(x^j(u, v)) \frac{\partial x^i}{\partial u}\, \mathbf{d}u + \theta_i(x^j(u, v)) \frac{\partial x^i}{\partial v}\, \mathbf{d}v \equiv 0,$$

that is, such that $\theta_i(x^j(u, v)) \dfrac{\partial x^i}{\partial u} \equiv 0$ and $\theta_i(x^j(u, v)) \dfrac{\partial x^i}{\partial v} \equiv 0.$

In general there is no function $x^i(u, v)$ which satisfies these conditions. An integral submanifold of maximum dimension of a given Pfaff system exists (theorem, p. 249) if the system is completely integrable.

The Pfaff system $\{\boldsymbol{\theta}^\alpha = 0\}$ of rank r is **completely integrable in** U if there exist r independent differentiable functions y^α on U such that

completely integrable in U

$$\{\boldsymbol{\theta}^\alpha = 0\} \sim \{\mathbf{d}y^\alpha = 0\}.$$

Since the mapping $x \mapsto (y^\beta(x))$ is of rank r in a neighborhood of a generic point x_0, the equations

$$y^\beta(x) = y^\beta(x_0)$$

define a submanifold Y^{n-r} of X^n in a neighborhood of x_0, and (Y^{n-r}, i) is an integral manifold of the system $\{\boldsymbol{\theta}^\alpha = 0\}$. The functions $y^\alpha(x)$ are called the **first integrals** of the system.

first integrals

A system is said to be **completely integrable on the manifold** X^n if there exists a covering of X^n by open sets U_i in each of which the system is completely integrable. This definition does not imply the existence of a global solution.

completely integrable on X

Under what conditions is a Pfaff system completely integrable, i.e. when is it algebraically equivalent to a system of exact differentials? Any system of exact differentials is closed. So is any system algebraically equivalent to it. The next theorem states that this condition is also sufficient.

Frobenius theorem. A Pfaff system $\{\boldsymbol{\theta}^\alpha = 0\}$ is completely integrable in a neighborhood of a generic point if and only if it is closed in this neighborhood.

Proof of sufficiency: Let I be the ideal generated by the set $\{\boldsymbol{\theta}^\alpha\}$ of r linearly independent 1-forms. We must show that in the neighborhood of a generic point

$$\mathbf{d}\boldsymbol{\theta}^\alpha \in I \quad \forall \alpha$$

implies that there exist functions z^α such that $\{\boldsymbol{\theta}^\alpha = 0\} \sim \{\mathbf{d}z^\alpha = 0\}$. The proof is by recurrence on the dimension of the space. In $\mathbb{R}^r$ the result is trivial: $\boldsymbol{\theta}^\alpha = \theta_i^\alpha\, \mathbf{d}x^i$ where $[\theta_i^\alpha]$ is a regular (invertible) square matrix; thus $\{\boldsymbol{\theta}^\alpha = 0\} \sim \{\mathbf{d}x^i = 0\}$. We assume the result is true in $\mathbb{R}^{n-1}$ and show that it is true in $\mathbb{R}^n$.

Let $\boldsymbol{\theta}^\alpha = \theta_i^\alpha\, \mathbf{d}x^i$ with $\alpha = 1, \ldots, r$ and $i = 1, \ldots, n$; and define r linearly independent 1-forms $\{\bar{\boldsymbol{\theta}}^\alpha\}$ in $\mathbb{R}^{n-1}$ by

$$\boldsymbol{\theta}^\alpha \equiv \bar{\boldsymbol{\theta}}^\alpha + \theta_n^\alpha\, \mathbf{d}x^n$$

with the variable x^n treated as a parameter. Now

$$\mathbf{d}\boldsymbol{\theta}^\alpha = \sum_{i,j=1}^{n-1} \frac{\partial \theta_i^\alpha}{\partial x^j}\, \mathbf{d}x^j \wedge \mathbf{d}x^i + \sum_{i=1}^{n-1} \left(\frac{\partial \theta_i^\alpha}{\partial x^n}\, \mathbf{d}x^n \wedge \mathbf{d}x^i + \frac{\partial \theta_n^\alpha}{\partial x^i}\, \mathbf{d}x^i \wedge \mathbf{d}x^n \right)$$

$$\equiv \bar{\mathbf{d}}\bar{\boldsymbol{\theta}}^\alpha + \boldsymbol{\xi} \wedge \mathbf{d}x^n$$

where $\bar{\mathbf{d}}$ designates differentiation in $\mathbb{R}^{n-1}$. Let $\bar{I}$ be the ideal generated by $\{\bar{\boldsymbol{\theta}}^\alpha\}$. As $\mathbf{d}\boldsymbol{\theta}^\alpha \in I$, this equation says that $\bar{\mathbf{d}}\bar{\boldsymbol{\theta}}^\alpha \in \bar{I}$; that is, the system $\{\bar{\boldsymbol{\theta}}^\alpha\}$ is closed in $\mathbb{R}^{n-1}$. Thus there exist by assumption functions y^α of the variables $x^1, \ldots, x^{n-1}$ and the parameter x^n such that $\{\bar{\boldsymbol{\theta}}^\alpha = 0\} \sim (\bar{\mathbf{d}}y^\alpha = 0\}$, i.e. such that

$$\bar{\boldsymbol{\theta}}^\alpha = S_\beta^\alpha \bar{\mathbf{d}}y^\beta$$

where $[S_\beta^\alpha]$ is a regular square matrix. Now construct a system $\{\boldsymbol{\theta}'^\alpha\}$ in $\mathbb{R}^n$, algebraically equivalent to the system $\{\boldsymbol{\theta}^\alpha\}$, by

$$\begin{aligned}
\boldsymbol{\theta}'^\beta &= (S^{-1})_\alpha^\beta (\bar{\boldsymbol{\theta}}^\alpha + \theta_n^\alpha\, \mathbf{d}x^n) \\
&= \bar{\mathbf{d}}y^\beta + (S^{-1})_\alpha^\beta \theta_n^\alpha\, \mathbf{d}x^n \\
&= \mathbf{d}y^\beta + \left(-\frac{\partial y^\beta}{\partial x^n} + (S^{-1})_\alpha^\beta \theta_n^\alpha \right) \mathbf{d}x^n \equiv \mathbf{d}y^\beta + b^\beta(x^1, \ldots x^n)\, \mathbf{d}x^n
\end{aligned}$$

where b^β has been introduced to simplify the notation. The next step is to show that the b^β depend only on the $r+1$ independent variables (y^α, x^n). This follows from the fact that the system $\{\boldsymbol{\theta}'^\beta\}$ is closed:

$$\mathbf{d}\boldsymbol{\theta}'^\beta = \mathbf{d}b^\beta \wedge \mathbf{d}x^n = \boldsymbol{\varphi}_\alpha^\beta \wedge \boldsymbol{\theta}'^\alpha = \boldsymbol{\varphi}_\alpha^\beta \wedge (\mathbf{d}y^\alpha + b^\alpha\, \mathbf{d}x^n)$$

for some 1-forms $\boldsymbol{\varphi}_\alpha^\beta$. The last equality can be true only if the $\boldsymbol{\varphi}_\alpha^\beta$ are proportional to $\mathbf{d}x^n$ (otherwise $\varphi_\alpha^\beta \wedge \mathbf{d}y^\alpha$ cannot be of the form $\mathbf{d}b^\beta \wedge \mathbf{d}x^n$). Set $\varphi_\alpha^\beta = c_\alpha^\beta\, \mathbf{d}x^n$. Then

$$\mathbf{d}b^\beta \wedge \mathbf{d}x^n = c_\alpha^\beta\, \mathbf{d}x^n \wedge \mathbf{d}y^\alpha.$$

Thus b^β can be expressed in terms of the y^α and x^n. The system

$$\{\boldsymbol{\theta}'^\alpha = 0\} = \{\mathbf{d}y^\alpha + b^\alpha(y^\beta, x^n)\,\mathbf{d}x^n = 0\}$$

is therefore a system of ordinary differential equations. If $z^\alpha(y^\beta, x^n)$ are r first integrals of the system then

$$\{\boldsymbol{\theta}^\alpha = 0\} \sim \{\boldsymbol{\theta}'^\alpha = 0\} \sim \{\mathbf{d}z^\alpha = 0\}. \qquad \blacksquare$$

The proof of the theorem provides a means to construct the first integrals of a completely integrable system.

Example: Consider the form in $\mathbb{R}^3$.

$$\boldsymbol{\theta} = yz\,\mathbf{d}x + xz\,\mathbf{d}y + \mathbf{d}z.$$

The system $\boldsymbol{\theta} = 0$ is closed because $\mathbf{d}\boldsymbol{\theta} = (-y\,\mathbf{d}x - x\,\mathbf{d}y) \wedge \boldsymbol{\theta}$. Let $\bar{\boldsymbol{\theta}} = zy\,\mathbf{d}x + zx\,\mathbf{d}y$. A first integral u of $\bar{\boldsymbol{\theta}} = 0$ in $\mathbb{R}^2$ is easily found

$$\bar{\boldsymbol{\theta}} = xyz\,\bar{\mathbf{d}}(\ln(xy)) \equiv S\,\bar{\mathbf{d}}u.$$

Since $\bar{\mathbf{d}}u = \mathbf{d}u$

$$\begin{aligned}\boldsymbol{\theta}' &= \mathbf{d}u + (xyz)^{-1}\,\mathbf{d}z = \mathbf{d}u + \mathrm{e}^{-u}z^{-1}\,\mathbf{d}z\\ &= \mathrm{e}^{-u}\,\mathrm{d}(\mathrm{e}^u + \ln z) = \mathrm{e}^{-u}\,\mathrm{d}(\ln z\mathrm{e}^{xy})\\ &= \mathrm{e}^{-u}\mathrm{e}^{-xy}z^{-1}\,\mathrm{d}(z\mathrm{e}^{xy}).\end{aligned}$$

Thus a first integral of $\boldsymbol{\theta}' = 0$ is $z\mathrm{e}^{xy}$.

A particularly useful criterion of complete integrability can be derived from the following theorem. integrability criterion

Frobenius theorem. A Pfaff system $\{\boldsymbol{\theta}^\alpha = 0;\ \alpha = 1, \ldots, r\}$ is completely integrable if and only if

$$\mathbf{d}\boldsymbol{\theta}^\alpha \wedge \boldsymbol{\theta}^1 \wedge \ldots \wedge \boldsymbol{\theta}^r \equiv 0 \qquad \textit{for } \alpha = 1, \ldots, r.$$

Proof: We shall prove the more general result that a form $\boldsymbol{\omega}$ is an element of the ideal generated by r independent Pfaff forms $\{\boldsymbol{\theta}^\alpha\}$ if and only if

$$\boldsymbol{\omega} \wedge \boldsymbol{\theta}^1 \wedge \ldots \wedge \boldsymbol{\theta}^r = 0.$$

Obviously $\boldsymbol{\omega} \in I \;\Rightarrow\; \boldsymbol{\omega} \wedge \boldsymbol{\theta}^1 \wedge \ldots \wedge \boldsymbol{\theta}^r = 0$ because this product involves at least two identical one-forms.

To prove the converse, let us choose a basis $\{\boldsymbol{\theta}^i, i = 1, \ldots, n\}$ for the 1-forms in a neighborhood U such that $\boldsymbol{\theta}^i = \boldsymbol{\theta}^\alpha$ for $i = 1, \ldots, r$; in this basis

$$\boldsymbol{\omega} = \omega_{I_1 \ldots I_p} \boldsymbol{\theta}^{I_1} \wedge \ldots \wedge \boldsymbol{\theta}^{I_p}$$

$$\boldsymbol{\omega} \wedge \boldsymbol{\theta}^1 \wedge \ldots \wedge \boldsymbol{\theta}^r \equiv 0 \Rightarrow \omega_{I_1 \ldots I_p} = 0 \qquad \text{if } \{I_1, \ldots, I_p\} \cap \{1, \ldots r\} = \emptyset.$$

Hence $\boldsymbol{\omega} \in I$. Thus the integrability criterion stated above is equivalent to the statement $\mathrm{d}\boldsymbol{\theta}^\alpha \in I$. ■

Application: A system $\{\boldsymbol{\theta}^\alpha = 0\}$ of rank n or $n-1$ in an n-dimensional space is always completely integrable.

Proof: a) System of rank n (see p. 233). Let us choose for a basis of the space of 1-forms in a neighborhood of x the n independent Pfaff forms $\{\boldsymbol{\theta}^\alpha\}$; $\mathrm{d}\boldsymbol{\theta}^\alpha$ being a sum of products of two elements of the basis,

$$\mathrm{d}\boldsymbol{\theta}^\alpha \wedge \boldsymbol{\theta}^1 \wedge \ldots \wedge \boldsymbol{\theta}^n \equiv 0.$$

An integral manifold is of dimension zero, it reduces to a point.

b) System of rank $n-1$. Let us choose for a basis the $n-1$ independent Pfaff forms $\{\boldsymbol{\theta}^\alpha\}$ completed by an nth linearly independent forms $\boldsymbol{\theta}$; $\mathrm{d}\boldsymbol{\theta}^\alpha$ is a sum of products each containing at least one of the Pfaff forms $\boldsymbol{\theta}^\alpha$, hence

$$\mathrm{d}\boldsymbol{\theta}^\alpha \wedge \boldsymbol{\theta}^1 \wedge \ldots \wedge \boldsymbol{\theta}^{n-1} \equiv 0.$$

The maximum dimension of an integral manifold is one. This result can also be obtained directly:

$$\{\boldsymbol{\theta}^\alpha = \theta_i^\alpha \,\mathbf{d}x^i = 0; \quad \alpha = 1, \ldots n-1,\ i = 1, \ldots n\} \sim \left\{\frac{\mathrm{d}x^\alpha}{\mathrm{d}x^n} = A^\alpha(x^i)\right\}.$$

If necessary, the coordinates can be relabelled so that the $(n-1) \times (n-1)$ matrix $[\theta_\beta^\alpha]$ be invertible. It is known that if θ_i^α is lipschitzian (a fortiori differentiable) A^α is lipschitzian and the system of ordinary differential equations

$$\left\{\frac{\mathrm{d}x^\alpha}{\mathrm{d}x^n} = A^\alpha(x^i)\right\}$$

is integrable; an integral manifold is a curve given by the $n-1$ equations $x^\alpha = x^\alpha(x^n)$. ■

Example[1]: In thermodynamics, a confusing symbol "đ" is often used to label a 1-form which is not an exact differential; for example

$$\text{đ}\boldsymbol{Q} = \mathbf{d}\boldsymbol{U} + p\,\mathbf{d}\boldsymbol{V}$$

[1]For more on this subject see [Souriau].

where U is the internal energy, p the pressure and V the volume of the system.
Set

$$\boldsymbol{\theta} = \mathbf{d}U + p\,\mathbf{d}V$$

and choose p, V for independent variables,

$$\boldsymbol{\theta} = \frac{\partial U}{\partial p}\,\mathbf{d}p + \left(\frac{\partial U}{\partial V} + p\right)\mathbf{d}V.$$

The equation $\boldsymbol{\theta} = 0$ is completely integrable (one equation in a 2-dimensional space):

$$\mathbf{d}\boldsymbol{\theta} \wedge \boldsymbol{\theta} = (\mathbf{d}p \wedge \mathbf{d}V) \wedge \left(\frac{\partial U}{\partial p}\,\mathbf{d}p + \left(\frac{\partial U}{\partial V} + p\right)\mathbf{d}V\right) \equiv 0.$$

Thus, $\{\boldsymbol{\theta} = 0\} \sim \{\mathbf{d}y = 0\}$ with $\boldsymbol{\theta} = A\,\mathbf{d}y$. Indeed $\boldsymbol{\theta} = T\,\mathrm{d}S$ where T is the temperature and S the entropy; S is the first integral of the exterior differential equation $\mathbf{d}U + p\,\mathbf{d}V = 0$.

Exercise: In $\mathbb{R}^3$, when is the Pfaff equation $\boldsymbol{\theta} = \theta_i\,\mathbf{d}x^i = 0$ integrable? (see p. 243). It is completely integrable in a neighborhood U of $x_0 \in \mathbb{R}^3$ if and only if

$$\mathbf{d}\boldsymbol{\theta} \wedge \boldsymbol{\theta} \equiv 0$$

i.e. if and only if

$$\theta_1\left(\frac{\partial \theta_3}{\partial x^2} - \frac{\partial \theta_2}{\partial x^3}\right) + \theta_2\left(\frac{\partial \theta_1}{\partial x^3} - \frac{\partial \theta_3}{\partial x^1}\right) + \theta_3\left(\frac{\partial \theta_2}{\partial x^1} - \frac{\partial \theta_1}{\partial x^2}\right) \equiv 0,$$

also written

$$\boldsymbol{\theta} \cdot \operatorname{curl} \boldsymbol{\theta} \equiv 0.$$

Off hand one might have said $\boldsymbol{\theta} = 0$ is integrable if (θ_i) is the gradient of an arbitrary function. This condition is sufficient but not necessary. If $\boldsymbol{\theta} = 0$ is completely integrable in a neighborhood U, then through each generic point $x_0 \in U$ there is exactly one 2-dimensional integral manifold

$$\varphi(x^1, x^2, x^3) = \varphi(x_0^1, x_0^2, x_0^3)$$

defined in a neighborhood of x_0.

Exercise: Integrable constraints[1]. Show that the constraints of a vertical disk rolling on a horizontal plane are not integrable.
The velocity of the center of the disk is $v = a\dot{\phi}$, its direction is per-

[1]See for instance [Goldstein, 2nd ed., pp. 14–15].

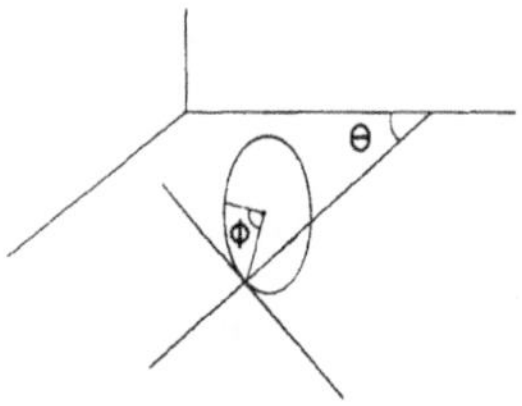

pendicular to the axis of the disk

$$\dot{x} = v \sin\theta = a\dot{\phi}\sin\theta$$
$$\dot{y} = -v\cos\theta = -a\dot{\phi}\cos\theta.$$

The constraints are given by the Pfaff system of 2 equations in a 4 dimensional space

$$\begin{cases} \boldsymbol{\omega}^1 \equiv \mathbf{dx} - a\sin\theta\,\mathbf{d\phi} = 0 \\ \boldsymbol{\omega}^2 \equiv \mathbf{dy} + a\cos\theta\,\mathbf{d\phi} = 0. \end{cases}$$

It is not completely integrable because the two expressions

$$\mathrm{d}\boldsymbol{\omega}^1 \wedge \boldsymbol{\omega}^1 \wedge \boldsymbol{\omega}^2 = -a\cos\theta\,\mathbf{d\theta}\wedge\mathbf{d\phi}\wedge\mathbf{dx}\wedge\mathbf{dy}$$
$$\mathrm{d}\boldsymbol{\omega}^2 \wedge \boldsymbol{\omega}^1 \wedge \boldsymbol{\omega}^2 = -a\sin\theta\,\mathbf{d\theta}\wedge\mathbf{d\phi}\wedge\mathbf{dx}\wedge\mathbf{dy}$$

are not identically zero. There are no integral manifolds of dimension $n-r=2$.

Dual expression of the Frobenius theorem. A Pfaff system of rank r can be specified by giving a set of $n-r$ linearly independent vector fields $\{\boldsymbol{v}_A\}$ which generate $\tilde{Q}_x$ at each generic point x. Such a set is called a **Pfaff system of vector fields**.

Pfaff system of vector fields

Theorem. A Pfaff system is completely integrable in a neighborhood U of a generic point if and only if the Lie bracket $[\boldsymbol{v}_A, \boldsymbol{v}_B]_x \in \tilde{Q}_x$ for all $x \in U$ and all elements of $\{\boldsymbol{v}_A\}$; i.e. $[\boldsymbol{v}_A, \boldsymbol{v}_B] \wedge \boldsymbol{v}_1 \wedge \ldots \boldsymbol{v}_{n-r} \equiv 0$, $\forall A, B = 1, \ldots n-r$.

Proof: Let I be the ideal generated by the Pfaff system. The system is completely integrable if and only if $\mathrm{d}\boldsymbol{\theta} \in I$, $\forall \boldsymbol{\theta} \in Q_x^*$. Since a Pfaff system is complete (p. 232), for every $\boldsymbol{\theta} \in Q_x^*$

$$\mathrm{d}\boldsymbol{\theta} \in I \quad \Leftrightarrow \quad \mathrm{d}\boldsymbol{\theta}(\boldsymbol{v}, \boldsymbol{w}) = 0, \quad \forall \boldsymbol{v}, \boldsymbol{w} \in \tilde{Q}_x.$$

On the other hand (p. 207)

$$\mathrm{d}\boldsymbol{\theta}(\boldsymbol{v}, \boldsymbol{w}) = \mathcal{L}_v(\boldsymbol{\theta}(\boldsymbol{w})) - \mathcal{L}_w(\boldsymbol{\theta}(\boldsymbol{v})) - \boldsymbol{\theta}([\boldsymbol{v}, \boldsymbol{w}])$$
$$= -\boldsymbol{\theta}([\boldsymbol{v}, \boldsymbol{w}]) \text{ when } \boldsymbol{v}, \boldsymbol{w} \in \tilde{Q}_x, \boldsymbol{\theta} \in Q_x^*.$$

Hence for $\boldsymbol{v}$, $\boldsymbol{w} \in \tilde{Q}_x$ and $\boldsymbol{\theta} \in Q_x^*$

$$\mathrm{d}\boldsymbol{\theta}(\boldsymbol{v}, \boldsymbol{w}) = 0 \quad \Leftrightarrow \quad \boldsymbol{\theta}([\boldsymbol{v}, \boldsymbol{w}]) = 0,$$
$$\Leftrightarrow \quad [\boldsymbol{v}, \boldsymbol{w}] \in \tilde{Q}_x. \qquad \blacksquare$$

Exercise: Show that if $\boldsymbol{v}_A$, $\boldsymbol{v}_B \in \tilde{Q}_x$, $[\boldsymbol{v}_A, \boldsymbol{v}_B]$ is not necessarily in $\tilde{Q}_x$. Let, for example, $n = 3$ and $\dim \tilde{Q}_x = 2$. Then

$$[\boldsymbol{v}_A, \boldsymbol{v}_B] \in \tilde{Q}_x \quad \Leftrightarrow \quad [\boldsymbol{v}_A, \boldsymbol{v}_B] = \lambda \boldsymbol{v}_A + \mu \boldsymbol{v}_B$$
$$\Leftrightarrow \quad v_A^i v_{B,i}^j - v_B^i v_{A,i}^j = \lambda v_A^j + \mu v_B^j.$$

This system of 3 equations in the 2 unknowns λ, μ is compatible if the determinant of the 3×3 matrix $[v_A^i v_{B,i}^j - v_B^i v_{A,i}^j, v_A^j, v_B^j]$ is zero. This condition is precisely the condition $\mathrm{d}\boldsymbol{\theta}(\boldsymbol{v}_A, \boldsymbol{v}_B) = 0$; indeed the compatibility of the three equations $\boldsymbol{\theta}(\boldsymbol{v}_A) = 0$, $\boldsymbol{\theta}(\boldsymbol{v}_B) = 0$, $\mathrm{d}\boldsymbol{\theta}(\boldsymbol{v}_A, \boldsymbol{v}_B) = 0$, i.e. the compatibility of

$$0 = \theta_j v_A^j$$
$$0 = \theta_j v_B^j$$
$$0 = \tfrac{1}{2}(\theta_{j,i} - \theta_{i,j}) v_A^i v_B^j = -\tfrac{1}{2}\theta_j (v_A^i v_{B,i}^j - v_B^i v_{A,i}^j)$$

is equivalent to the compatibility of the 3 equations which determine λ and μ.

Sufficient condition for complete integrability. It was shown at the beginning of this section that a completely integrable Pfaff system has an integral submanifold of maximal dimension in a neighborhood of any generic point. The following theorem is a converse of this result.

Theorem. A Pfaff system of rank r on X^n is completely integrable in a neighborhood of a generic point x_0 if there exists a diffeomorphism φ of a neighborhood U of x_0 onto a product $T^r \times Y^{n-r}$ such that $(\varphi^{-1}(\{t\} \times Y), i)$ is, for every $t \in T$, an integral submanifold of dimension $n - r$.

Proof: Let (t^α, y^A) be a system of local coordinates on U such that $(t^\alpha; \alpha = 1, \ldots, r)$ and $(y^A; A = 1, \ldots, n - r)$ are respectively local coordinates on T^r and on Y^{n-r}. The Pfaff system $\{\boldsymbol{\theta}^\alpha\}$ can then be written

$$\{\boldsymbol{\theta}^\alpha \equiv \theta_\beta^\alpha \, \mathbf{d}t^\beta + \theta_A^\alpha \, \mathbf{d}y^A = 0\}.$$

However since the form $i^*\boldsymbol{\theta}^\alpha \equiv \theta_A^\alpha \, \mathbf{d}y^A$ induced on the integral manifold $\{t\} \times Y^{n-r}$ is identically zero, it follows that $\theta_A^\alpha = 0$ for all α and all A. Thus $\boldsymbol{\theta}^\alpha \equiv \theta_\beta^\alpha \, \mathbf{d}t^\beta$ and

$$\{\boldsymbol{\theta}^\alpha \equiv \theta_\beta^\alpha \, \mathbf{d}t^\beta = 0\} \sim \{\mathbf{d}t^\beta = 0\}$$

because the matrix $[\theta_\beta^\alpha]$ is a regular square matrix. $\blacksquare$

7. CHARACTERISTIC SYSTEM

characteristic system

A powerful tool in the investigation of an exterior differential system S is its characteristic system C. We first define the characteristic system of a system $S = \{\boldsymbol{\omega}^\alpha = 0\}$ from which 0-forms are excluded. Let $\bar{S}$ be the closure of S

$$\bar{S} = \{\boldsymbol{\omega}^\alpha = 0, \mathrm{d}\boldsymbol{\omega}^\alpha = 0\}.$$

The **characteristic system** of S is the associated Pfaff system of $\bar{S}$.
If $S = \{\boldsymbol{\omega}^\alpha = 0, f^\beta = 0\}$, where f^β are 0-forms, the characteristic system C of S is by definition

$$C = \begin{cases} \text{the associated Pfaff system of } \{\boldsymbol{\omega}^\alpha = 0,\ \mathrm{d}\boldsymbol{\omega}^\alpha = 0,\ \mathbf{d}f^\beta = 0\} \\ \text{the equations } f^\beta = 0. \end{cases}$$

class

characteristic manifold

characteristic vector field

The rank r of the associated Pfaff system of $\{\boldsymbol{\omega}^\alpha, \mathrm{d}\boldsymbol{\omega}^\alpha, \mathbf{d}f^\beta\}$ is called the **class** of S. An integral manifold of C of dimension $n - r$ is called a **characteristic manifold**. The elements of the space $\tilde{Q}$ of vector fields associated with $\bar{S}$ are called **characteristic vector fields**.

The reason for defining C in terms of a Pfaff system associated with $\bar{S}$ rather than S is that it might have the following properties.

In a neighborhood of a generic point C is completely integrable.

Characteristic manifolds can be used to construct integral manifolds of S; in particular, they can be used to solve the Cauchy problem, i.e., to construct an integral manifold which satisfies some initially given data (boundary conditions, Cauchy data).

A differential form $\boldsymbol{\omega}$ is a Poincaré integral invariant for its characteristic system; that is,

$$\int_c \boldsymbol{\omega} = \int_{c'} \boldsymbol{\omega}$$

where c' is the transform of c under the one parameter group of transformations generated by a characteristic vector field.

Example: 1st order partial differential equations. We begin with an example, namely the construction of the characteristic system of the most general first order partial differential equation[1] in $\mathbf{R}^n$:

$$F\left(x^1, \ldots, x^n, z, \frac{\partial z}{\partial x^1}, \ldots, \frac{\partial z}{\partial x^n}\right) = 0.$$

[1]Cf. [Courant and Hilbert, Chapt. 2].

The solutions are the differentiable functions $z(x^i)$ which satisfy $F = 0$. The problem can be restated in terms of the following exterior differential system S defined in the space $\mathbb{R}^{2n+1}$ of $(2n+1)$-tuples

$$(x^a;\, a = 1, \ldots 2n+1) \equiv (x^i, z, p_i;\, i = 1, \ldots, n,\, z = x^{n+1}, p_i = x^{n+1+i}):$$

$$S \equiv \begin{cases} f \equiv F(x^i, z, p_i) = 0 \\ \boldsymbol{\theta} \equiv \mathbf{d}z - p_i\, \mathbf{d}x^i = 0. \end{cases}$$

The solutions $z(x^i)$ are then determined by the n-dimensional integral manifolds $(U^n \subset \mathbb{R}^n, g)$ of S, where $g\colon U^n \to \mathbb{R}^{2n+1}$ by $(x^i) \mapsto (x^i, z(x^i), p_j(x^i))$. We shall follow the procedure given at the beginning of this paragraph to determine the characteristic system C of S.
1) Closure $\bar{S}$ of S,

$$\bar{S} = \begin{cases} f = 0, & \mathbf{d}f = \left(\dfrac{\partial F}{\partial x^i} + p_i \dfrac{\partial F}{\partial z}\right) \mathbf{d}x^i + \dfrac{\partial F}{\partial p_i}\, \mathbf{d}p_i = 0 \\ \boldsymbol{\theta} = 0, & \mathbf{d}\boldsymbol{\theta} = -\,\mathbf{d}p_i \wedge \mathbf{d}x^i = 0. \end{cases}$$

2) Characteristic system C of S. Set $\{\boldsymbol{\omega}^\alpha = 0\} \equiv \{\mathbf{d}f = 0, \boldsymbol{\theta} = 0, \mathbf{d}\boldsymbol{\theta} = 0\}$

$$C = \begin{cases} \text{Pfaff system } P = \{\bar{\boldsymbol{\theta}}^\beta = 0\} \text{ associated with } \{\boldsymbol{\omega}^\alpha = 0\} \\ f = 0. \end{cases}$$

The forms $\bar{\boldsymbol{\theta}}^\beta$ are determined by the method of the theorem and example on pp. 234–236. Let $(v^a) = (v^i, v^{n+1}, u_i)$ be the components of the vector $\boldsymbol{v}$ with respect to the basis dual to the basis $\mathbf{d}x^a = (\mathbf{d}x^i, \mathbf{d}z, \mathbf{d}p_i)$ in $\mathbb{R}^{2n+1}$. Let I be the ideal generated by $\{\boldsymbol{\omega}^a\}$. The condition that $\boldsymbol{v}$ be a characteristic vector field,

$$\mathrm{i}_v \boldsymbol{\omega}^\alpha \in I \qquad \forall \alpha,$$

can be written in the form $v^a b_{a\beta} = 0$ for some $b_{a\beta}$. The forms $\bar{\boldsymbol{\theta}}^\beta$ are then given by

$$\bar{\boldsymbol{\theta}}^\beta = \mathbf{d}x^a b_{a\beta}.$$

Since $\boldsymbol{\theta}$ and $\mathbf{d}f$ are 1-forms we can choose $\bar{\boldsymbol{\theta}}^1 = \boldsymbol{\theta}$, $\bar{\boldsymbol{\theta}}^2 = \mathbf{d}f$. Then the only condition left to be considered is $\mathrm{i}_v\, \mathbf{d}\boldsymbol{\theta} \in I$, or

$$\mathrm{i}_v\, \mathbf{d}\boldsymbol{\theta} \equiv -\,u_i\, \mathbf{d}x^i + v^i\, \mathbf{d}p_i = \lambda(x^a)\boldsymbol{\theta} + \mu(x^a)\, \mathbf{d}f$$

where λ and μ are arbitrary functions. Since $\mathrm{i}_v\, \mathbf{d}\boldsymbol{\theta}$ does not have a $\mathbf{d}z$ term, $\lambda(x^a) = 0$. The condition $\mathrm{i}_v\, \mathbf{d}\boldsymbol{\theta} = \mu(x^a)\, \mathbf{d}f$ can be written

$$\frac{v^1}{\partial F/\partial p_1} = \cdots = \frac{v^n}{\partial F/\partial p_n} = \frac{-u_1}{\partial F/\partial x^1 + p_1(\partial F/\partial z)} = \cdots = \frac{-u_n}{\partial F/\partial x^n + p_n(\partial F/\partial z)}$$

which is a set of $2n-1$ equations of the form

$$v^\alpha b_{\alpha\beta} = 0.$$

The equations $\{\bar{\boldsymbol{\theta}}^\beta \equiv \mathbf{d}x^\alpha b_{\alpha\beta} = 0;\ \beta = 3, \ldots, 2n+1\}$ can thus be written

$$\frac{\mathrm{d}x^1}{\partial F/\partial p_1} = \cdots = \frac{\mathrm{d}x^n}{\partial F/\partial p_n} = \frac{-\mathrm{d}p_1}{\partial F/\partial x^1 + p_1(\partial F/\partial z)} = \cdots = \frac{-\mathrm{d}p_n}{\partial F/\partial x^n + p_n(\partial F/\partial z)}.$$

These equations together with $\boldsymbol{\theta} = 0$, $\mathbf{d}f = 0$ and $f = 0$ form the characteristic system C of S. It can be written

$$C = \begin{cases} \dfrac{\mathrm{d}x^1}{\partial F/\partial p_1} = \cdots = \dfrac{\mathrm{d}x^n}{\partial F/\partial p_n} = \dfrac{-\mathrm{d}p_1}{\partial F/\partial x^1 + p_1(\partial F/\partial z)} = \cdots \\ \qquad = \dfrac{-\mathrm{d}p_n}{\partial F/\partial x^n + p_n(\partial F/\partial z)} = \dfrac{\mathrm{d}z}{\Sigma_i\, p_i(\partial F/\partial p_i)} \\ f = 0 \end{cases}$$

$\boldsymbol{\theta} = 0$ is replaced by the last equation on the second line, since

$$\frac{p_1\,\mathrm{d}x^1}{p_1(\partial F/\partial p_1)} = \cdots = \frac{p_n\,\mathrm{d}x^n}{p_n(\partial F/\partial p_n)} = \frac{\Sigma\, p_i\,\mathrm{d}x^i}{\Sigma\, p_i(\partial F/\partial p_i)};$$

the equation $\mathbf{d}f = 0$ is automatically satisfied if the first $2n-1$ equations of the characteristic system are satisfied.

The rank of P (the class of S) is $2n$. The characteristic manifolds of S are curves in $\mathbb{R}^{2n+1}$ which are integral manifolds of P and which satisfy $f = 0$. Equivalently they can be considered as curves in $\mathbb{R}^{n+1}$, the space of $(n+1)$-tuples (x^i, z), *together with*, at each point, the n-dimensional tangent vector space which satisfies $\boldsymbol{\theta} = 0$; these curves are called **characteristic strips** or **bicharacteristics**.

characteristic strips bicharacteristics

Geometric interpretation. At a fixed point $A = (x^i, z)$ in $\mathbb{R}^{n+1}$ each set of values for the p_i, satisfying the equation $F(x^i, z, p_i) = 0$, determines an n-dimensional tangent plane through A. The family of planes so determined envelops a cone with vertex A called the **Monge cone**. A solution of the partial differential equation $F(x^i, z, \partial z/\partial x^i) = 0$ is a hypersurface in $\mathbb{R}^{n+1}$ tangent to the Monge cone at each point of the hypersurface. A bicharacteristic curve through A is tangent, at A, to a generator of the Monge cone.

Monge cone

Complete integrability of the characteristic system.

Theorem. In a neighborhood U of a generic point:

1) *The characteristic system C of an exterior differential system S is completely integrable.*

2) *If* $(y^\lambda; \lambda = 1, \ldots, r)$ *is a complete independent set of first integrals (p. 243) of C, then the closure of S is algebraically equivalent to a system whose members can be expressed in terms of the* y^λ *and their differentials* $\mathbf{d}y^\lambda$.

Proof: 1) Let $\bar{I}$ be the ideal generated by $\bar{S}$. The relations

$$\mathrm{i}_{[v,w]} = \mathcal{L}_v \mathrm{i}_w - \mathcal{L}_w \mathrm{i}_v$$

and

$$\mathcal{L}_v = \mathrm{d}\mathrm{i}_v + \mathrm{i}_v \mathrm{d}$$

may be used to write

$$\mathrm{i}_{[v,w]}\boldsymbol{\omega} = \mathrm{d}\mathrm{i}_v\mathrm{i}_w\boldsymbol{\omega} + \mathrm{i}_v\,\mathrm{d}\mathrm{i}_w\boldsymbol{\omega} - \mathrm{d}\mathrm{i}_w\mathrm{i}_v\boldsymbol{\omega} - \mathrm{i}_w\,\mathrm{d}\mathrm{i}_v\boldsymbol{\omega}.$$

Since $\bar{S}$ is closed, if v and w are characteristic vector fields then

$$\mathrm{i}_{[v,w]}\boldsymbol{\omega} \in \bar{I} \text{ for all } \boldsymbol{\omega} \in \bar{I}.$$

Therefore $[v, w]$ is also a characteristic vector field. By the theorem on p. 248, C is completely integrable.

2) Let $(y^A; A = r+1, \ldots, n)$ be a set of independent functions such that $(y^i) = (y^\lambda, y^A)$ is a set of coordinate functions on U. Let (e_i) be the basis for $T(U)$ dual to $(\mathbf{d}y^i)$. The set (e_A) generates $\tilde{Q}$; the set $(\mathbf{d}y^\lambda)$ generates Q^*. By the definition of Q^* there is a system algebraically equivalent to $\bar{S}$ whose elements $\boldsymbol{\omega}^\alpha$ can be written

$$\boldsymbol{\omega}^\alpha = \frac{1}{p!}\,\omega^\alpha_{\lambda_1\ldots\lambda_p}(y^i)\,\mathbf{d}y^{\lambda_1} \wedge \ldots \wedge \mathbf{d}y^{\lambda_p}, \quad \lambda_k = 1, \ldots, r.$$

We must show that the algebraically equivalent system can be chosen so that the $\omega^\alpha_{\lambda_1\ldots\lambda_p}$ do not depend on the y^A. First note that since $\bar{S}$ is closed and e_A is a characteristic vector field

$$\mathrm{i}_{e_A}\,\mathrm{d}\boldsymbol{\omega}^\alpha \in \bar{I};$$

that is,

$$\frac{\partial \omega^\alpha_{\lambda_1\ldots\lambda_p}}{\partial y^A}\,\mathbf{d}y^{\lambda_1} \wedge \ldots \wedge \mathbf{d}y^{\lambda_p} = \boldsymbol{\xi}^\alpha{}_\beta \wedge \boldsymbol{\omega}^\beta$$

for some $\boldsymbol{\xi}^\alpha_\beta \in \Lambda(U)$. The proof is by induction on the degree of $\boldsymbol{\omega}^\alpha$. We illustrate using a system made up only of 1-forms $\{\boldsymbol{\theta}^\alpha; \alpha = 1, \ldots, r_1\}$ and

2-forms $\{\boldsymbol{\phi}^\beta;\ \beta = 1, \ldots, r_2\}$. For a 1-form the above equations become

$$\frac{\partial \theta^\alpha_\lambda}{\partial y^A} = f^\alpha_\gamma \theta^\gamma_\lambda \quad \begin{cases} \lambda = 1, \ldots, r \\ \gamma = 1, \ldots, r_1. \end{cases}$$

The general solution of the equations is

$$\theta^\alpha_\lambda = \bar{\theta}^{(k)}_\lambda \bar{z}^\alpha_{(k)}$$

where the functions $\bar{z}^\alpha_{(k)}$, $k = 1, \ldots, r_1$ are r_1 independent solutions of the equations

$$\frac{\partial z^\alpha}{\partial y^A} = f^\alpha_\gamma(y^i) z^\gamma,$$

and the $\bar{\theta}^{(k)}_\lambda$ are arbitrary functions which are independent of y^A. Thus the set $\{\bar{\theta}^{(k)}_\lambda \mathbf{d}y^\lambda\}$ is algebraically equivalent to the set $\{\boldsymbol{\theta}^\alpha\}$ and does not involve y^A. Repetition of the above procedure for each value of A will result in a set $\{\bar{\boldsymbol{\theta}}^{(k)}\}$ algebraically equivalent to $\{\boldsymbol{\theta}^\alpha\}$, which does not involve any y^A.

For a 2-form the equations are

$$\frac{\partial \phi^\beta_{\lambda\mu}}{\partial y^A} = g^\beta_\delta \phi^\delta_{\lambda\mu} + \xi^\beta_{\delta\lambda} \theta^\delta_\mu.$$

The general solution is

$$\phi^\beta_{\lambda\mu} = \bar{\phi}^{(l)}_{\lambda\mu} \bar{w}^\beta_{(l)}$$

where the functions $\bar{w}^\beta_{(l)}$, $l = 1, \ldots, r_2$ are r_2 independent solutions of the equations

$$\frac{\partial w^\beta}{\partial y^A} = g^\beta_\delta w^\delta,$$

and the $\bar{\phi}^{(l)}_{\lambda\mu}$ satisfy the equations

$$\frac{\partial \bar{\phi}^{(l)}_{\lambda\mu}}{\partial y^A} \bar{w}^\beta_{(l)} = \xi^\beta_{\delta\lambda} \theta^\delta_\mu.$$

Since we can assume that the θ^δ_μ do not depend on the y^A, the equations can be integrated to give

$$\bar{\phi}^{(l)}_{\lambda\mu} = \bar{\bar{\phi}}^{(l)}_{\lambda\mu} + \zeta^{(l)}_{\delta\lambda} \theta^\delta_\mu$$

where the $\bar{\bar{\phi}}^{(l)}_{\lambda\mu}$ do not depend on y^A. Then

$$\boldsymbol{\phi}^\beta = \bar{w}^\beta_{(l)} \bar{\bar{\boldsymbol{\phi}}}^{(l)} + \bar{w}^\beta_{(l)} \zeta^{(l)}_{\delta\lambda} \mathbf{d}y^\lambda \wedge \boldsymbol{\theta}^\delta.$$

Therefore

$$\{\boldsymbol{\theta}^\alpha, \boldsymbol{\phi}^\beta\} \sim \{\bar{\boldsymbol{\theta}}^{(k)}, \bar{\bar{\boldsymbol{\phi}}}^{(l)}\}. \qquad \blacksquare$$

Construction of integral manifolds. Let S be a system of rank r in X^n. Suppose (W^m, f) is an integral manifold of S. If all points in $f(W^m)$ are generic points, (W^m, f) can be used together with the characteristic manifolds of S to generate an integral manifold of S of dimension $m+n-r$.

Example: Let S be the equation $\boldsymbol{\omega}=0$ where $\boldsymbol{\omega}$ is a 2-form in $\mathbb{R}^3$ defined by

$$\omega_{12}=\omega(x^1,x^2),\quad \omega_{13}=\omega_{23}=0$$

$$\boldsymbol{\omega}=\omega(x^1,x^2)\,\mathbf{d}x^1\wedge\mathbf{d}x^2.$$

The characteristic system C of S is $\{\mathbf{d}x^1=0,\ \mathbf{d}x^2=0\}$. The characteristic manifolds are $x^1=x_0^1$, $x^2=x_0^2$ where x_0^1, x_0^2 are constants; that is, the characteristic manifolds are vertical lines. Any curve

$$f: t\mapsto(x^1(t),x^2(t),x^3(t)),\quad t\in I\subset\mathbb{R}$$

is an integral manifold of S:

$$f^*\boldsymbol{\omega}=\omega\frac{\mathrm{d}x^1}{\mathrm{d}t}\mathbf{d}t\wedge\frac{\mathrm{d}x^2}{\mathrm{d}t}\mathbf{d}t\equiv0.$$

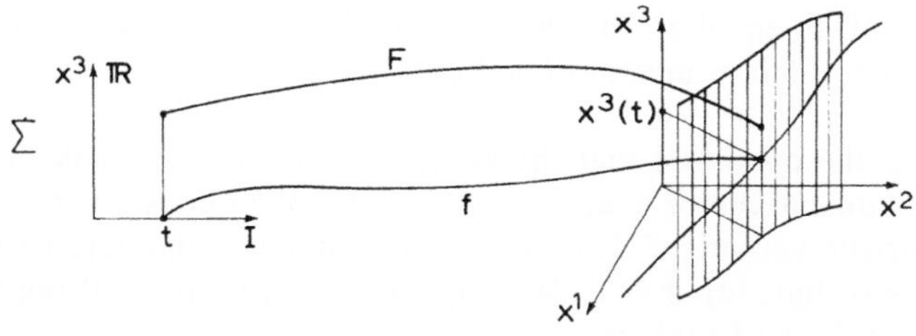

The pair (Σ, F), where $\Sigma=I\times\mathbb{R}$ and where $F(\Sigma)$ is the cylinder obtained by drawing a vertical line at each point of $f(I)$, is an integral manifold of S.

$$F:\Sigma\to\mathbb{R}^3\quad\text{by}\quad(t,x^3)\mapsto(x^1(t),x^2(t),x^3).$$

Since $f(I)\subset F(\Sigma)$, (Σ, F) is said to contain (I, f).

Theorem. Given an integral manifold (W^m, f) of a system S of rank r for which all points $x\in f(W^m)$ are generic, the pair (Σ, F), defined as follows, is also an integral manifold.

$\Sigma=W^m\times\mathbb{R}^{n-r}$ with coordinates (u^α,u^A), $\alpha=1,\ldots,m$, $A=r+1,\ldots,n$

$$F:\Sigma\to X^n\ by\begin{cases}y^\lambda=f^\lambda(u^\alpha), & \lambda=1,\ldots,r,\\ y^A=u^A, & A=r+1,\ldots,n;\end{cases}$$

i.e. choose a coordinate system on X^n such that the first integrals define the ***first r axes.***

The coordinates $(y^i; i=1,\ldots,n)$ *of a point* $x \in X^n$ *are such that the characteristic system* C *of* S *is algebraically equivalent to the system* $\{\mathbf{d}y^\lambda = 0\}$; *the functions* $y^\lambda = f^\lambda(u^\alpha)$ *are the coordinate representation of* f *in the* (y^i) *system of coordinates.*
Moreover, if $f(W^m)$ *is a submanifold of* X^n *and if the matrix* $[\partial y^\lambda/\partial u^\alpha]$ *is of rank* m, $F(\Sigma)$ *is also a submanifold of* X^n *in a neighborhood of* $f(W^m)$.

Proof: By the previous theorem (p. 253) we can assume that any form $\boldsymbol{\omega}^\alpha$ in the closure of S can be written

$$\boldsymbol{\omega}^\alpha = \frac{1}{p!}\omega^\alpha_{\lambda_1\ldots\lambda_p}(y^\lambda)\,\mathbf{d}y^{\lambda_1}\wedge\ldots\wedge\mathbf{d}y^{\lambda_p} \qquad \lambda, \lambda_k = 1,\ldots,r.$$

Thus (Σ, F) is an integral manifold of $\bar{S}$ and of S because

$$F^*\boldsymbol{\omega}^\alpha = f^*\boldsymbol{\omega}^\alpha \equiv 0.$$

Since $f(W^m)$ is a submanifold of X^n, the matrix $[\partial y^i/\partial u^\alpha]$ is of rank m. The condition that the matrix $[\partial y^\lambda/\partial u^\alpha]$ be of rank m is required to make the rank of F equal to $m+n-r$. Thus F is an immersion. It is also regular and 1–1 in a neighborhood of $f(W^m)$. ■

Note that the condition that the matrix $[\partial y^\lambda/\partial u^\alpha]$ be of rank m implies that the tangent vector space at a point of $f(W^m)$ does not contain any characteristic vectors of S. This is because if $\boldsymbol{v}$ is a characteristic vector then $v^\lambda = 0$; but, for $\boldsymbol{w} \in T_x(W^m)$, $(f'\boldsymbol{w})^\lambda = (\partial y^\lambda/\partial u^\alpha)w^\alpha = 0$ implies $\boldsymbol{w} = 0$ if $[\partial y^\lambda/\partial u^\alpha]$ is of rank m.

This theorem is used in the next paragraph.

Cauchy problem for a first order partial differential equation.
The problem is to find a solution for a first order P.D.E. $z = h(x^i)$ of the equation $F(x^1,\ldots,x^n, z, \partial z/\partial x^1,\ldots,\partial z/\partial x^n) = 0$ which contains a given $(n-1)$-dimensional surface $(W^{n-1}, \bar{g})$ where W^{n-1} is an open set of $\mathbb{R}^{n-1}$,

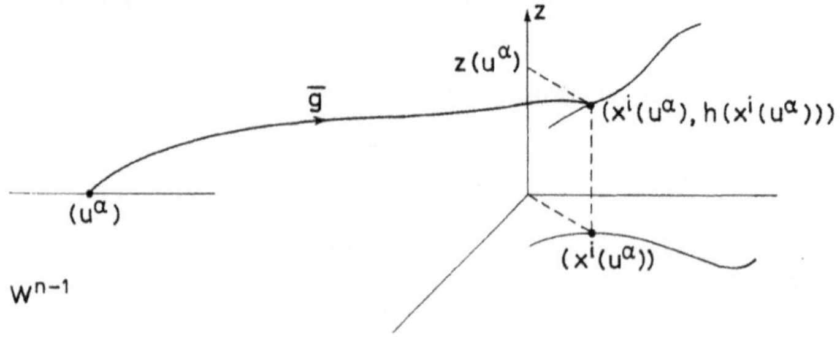

$(u^\alpha; \alpha = 1, \ldots, n-1) \in W^{n-1}$, and $\bar{g}$: $W^{n-1} \to \mathbb{R}^{n+1}$ by $(u^\alpha) \mapsto (x^i(u^\alpha), z(u^\alpha) \equiv h(x^i(u^\alpha)))$. (Problems IV 4 and IV 5.)

The equation $F = 0$ can be replaced by the system

$$S = \begin{cases} f \equiv F(x^i, z, p_i) = 0 \\ \boldsymbol{\theta} \equiv \mathbf{d}z - p_i\, \mathbf{d}x^i = 0 \end{cases}$$

in $\mathbb{R}^{2n+1}$. Suppose we can construct a mapping g which extends the range of $\bar{g}$ to $\mathbb{R}^{2n+1}$,

$$g: W^{n-1} \to R^{2n+1} \quad \text{by} \quad (u^\alpha) \mapsto (x^i(u^\alpha), z(u^\alpha), p_i(u^\alpha)),$$

and such that (W^{n-1}, g) is an integral manifold of S. The integral manifold (W^{n-1}, g) contains $(W^{n-1}, \bar{g})$; it is called an **initial integral manifold**. According to the previous theorem, an integral manifold of S of dimension n which contains (W^{n-1}, g) can then be generated by characteristic manifolds through each point of $g(W^{n-1})$ if all points of $g(W^{n-1})$ are generic points of S.

initial integral manifold

The construction of the desired solution to S according to this blueprint proceeds as follows. The pair (W^{n-1}, g) is to be an integral manifold of S. Hence $g^*f \equiv 0$ and $g^*\boldsymbol{\theta} \equiv 0$ on W^{n-1}, i.e.

$$\begin{cases} F(x^i(u^\alpha), z(u^\alpha), p_i(u^\alpha)) \equiv 0 \\ \left(\dfrac{\partial z}{\partial u^\alpha} - p_i(u^\alpha)\dfrac{\partial x^i}{\partial u^\alpha}\right) \mathbf{d}u^\alpha \equiv 0. \end{cases}$$

The implicit function theorem says that these n equations have a differentiable solution $\{p_i(u^\alpha)\}$ in the neighborhood of a point (u_0^α) if the determinant

$$\Delta = \det \begin{bmatrix} \dfrac{\partial F}{\partial p_1} & \cdots & \dfrac{\partial F}{\partial p_n} \\ \dfrac{\partial x^1}{\partial u^1} & \cdots & \dfrac{\partial x^n}{\partial u^1} \\ \cdot & & \cdot \\ \dfrac{\partial x^1}{\partial u^{n-1}} & \cdots & \dfrac{\partial x^n}{\partial u^{n-1}} \end{bmatrix} \neq 0.$$

Note that $(\partial x^i/\partial u^1)$ are the components of a tangent vector to the curve $x^i(u^1)$, $u^2 \ldots u^{n-1}$ being constant; if a row $(\partial x^i/\partial u^\alpha$; α fixed, $i = 1, \ldots n)$ is zero at a point, $\bar{g}(W^{n-1})$ has a vertical tangent space at that point.

A necessary condition for the above matrix to be of rank n is that the matrix $[\partial x^i/\partial u^\alpha]$ be of rank $n-1$. This implies that $\bar{g}$ is an immersion and moreover that the tangent space at any point of $\bar{g}(W^{n-1})$ is not "vertical", that is, it does not contain any vector whose only non-vanishing component is the z-component.

Let $(y^\lambda(x^i, z, p_i), y^{2n+1}(x^i, z, p_i))$ be a set of coordinate functions on $\mathbb{R}^{2n+1}$ where the y^λ are a set of $2n$ independent first integrals of the characteristic system of S. Let $U^{n-1} \subset W^{n-1}$ be such that all points of $g(U^{n-1})$ are generic. Then the pair (Σ^n, G), where $\Sigma^n = U^{n-1} \times \mathbb{R}$ and

$$G: (u^\alpha, t) \mapsto (y^\lambda = g^\lambda(u^\alpha), y^{2n+1} = t),$$

is an integral manifold of S. The mapping G can also be expressed in terms of the coordinates (x^i, z, p_i)

$$G: (u^\alpha, t) \mapsto (x^i(u^\alpha, t), z(u^\alpha, t), p_i(u^\alpha, t)).$$

For fixed u^α, G defines a characteristic curve in $\mathbb{R}^{2n+1}$ passing through the corresponding point of $g(W^{n-1})$. Thus when $t = 0$

$$x^i(u^\alpha, 0) = x^i(u^\alpha)$$
$$z(u^\alpha, 0) = z(u^\alpha)$$
$$p_i(u^\alpha, 0) = p_i(u^\alpha).$$

If the equations $x^i = x^i(u^\alpha, t)$ can be solved for the differentiable functions $t = t(x^i)$ and $u^\alpha = u^\alpha(x^i)$ in a neighborhood of the point $(u^\alpha, t = 0)$, then substitution in $z = z(u^\alpha, t)$ gives the differentiable function $z = h(x^i)$ which is the desired solution of the original partial differential equation.

According to the inverse function theorem a necessary and sufficient condition for the existence of $t(x^i)$ and $u^\alpha(x^i)$ is that the Jacobian determinant $\mathrm{D}(x^i)/\mathrm{D}(u^\alpha, t)|_{t=0}$ not vanish. Because of the characteristic equations

$$\frac{\mathbf{d}x^1}{\partial F/\partial p_1} = \cdots = \frac{\mathbf{d}x^n}{\partial F/\partial p_n}$$

and the fact that in a neighborhood of a point $(u^\alpha, t = 0)$

$$\mathbf{d}x^i = \frac{\partial x^i}{\partial t}\mathbf{d}t + \frac{\partial x^i}{\partial u^\alpha}\mathbf{d}u^\alpha,$$

the determinant $\mathrm{D}(x^i)/\mathrm{D}(u^\alpha, t)|_{t=0}$ is proportional to Δ. The above construction proves the following theorem.

Theorem. If $\Delta \neq 0$ at every point of the integral manifold (W^{n-1}, g) of the exterior system S, then there exists an integral manifold (Σ, G) of S which passes through (W^{n-1}, g) and which gives rise to a solution of the Cauchy problem for the partial differential equation $F(x^i, z, \partial z/\partial x^i) = 0$.

An integral manifold (W^{n-1}, g) of the system S is called **Cauchy characteristic** if $\Delta = 0$ at each point of W^{n-1}. Cauchy characteristic manifold

Example: Consider the equation $F(x^1, x^2, z, \partial z/\partial x^1, \partial z/\partial x^2) = 0$. One looks for a surface $z = h(x^1, x^2)$ in $\mathbb{R}^3$ which passes through the curve

$$u \mapsto (x^1(u),\ x^2(u),\ z(u)), \qquad u \in I \subset \mathbb{R}.$$

The condition that

$$\det\begin{bmatrix} \dfrac{\partial F}{\partial p_1} & \dfrac{\partial F}{\partial p_2} \\ \dfrac{\partial x^1}{\partial u} & \dfrac{\partial x^2}{\partial u} \end{bmatrix} \neq 0$$

at some point says that, when projected onto $\mathbb{R}^2$, the directions of the tangent vector of the curve and the bicharacteristic vector do not coincide at that point. This is because the first two components v^1, v^2 of a bicharacteristic vector satisfy the equation

$$\frac{v^1}{\partial F/\partial p_1} = \frac{v^2}{\partial F/\partial p_2}.$$

Example: Problem VI 11, p. 535.

Example: Quasi-linear equation in $\mathbb{R}^3$. For the system

$$a^1(x^1, x^2, z)p_1 + a^2(x^1, x^2, z)p_2 = b(x^1, x^2, z)$$
$$\mathbf{d}z = p_1\,\mathbf{d}x^1 + p_2\,\mathbf{d}x^2$$

the Monge cone (p. 252) reduces to a straight line. A solution is a surface whose tangent plane at any point contains this line. The bicharacteristic system is

$$\frac{\mathrm{d}x^1}{a^1(x^i, z)} = \frac{\mathrm{d}x^2}{a^2(x^i, z)} = \frac{\mathrm{d}z}{b(x^i, z)}.$$

Since the equations do not involve the p_i, they form a differential system in $\mathbb{R}^3$. In order to solve the Cauchy problem, it is sufficient to integrate this system. (See the next example.) The Cauchy problem defined by the differentiable curve

$$C: \qquad u \mapsto (x^1(u), x^2(u), z(u)), \qquad u \in I \subset \mathbb{R}$$

has a solution if the condition

$$\frac{\partial x^1}{\partial u}a^2 - \frac{\partial x^2}{\partial u}a^1 = 0$$

is not satisfied for any $u \in I$. If, on the other hand, the condition is satisfied for all $u \in I$, then the linear equations

$$a^1 p_1 + a^2 p_2 = b$$

$$p_1 \frac{\partial x^1}{\partial u} + p_2 \frac{\partial x^2}{\partial u} = \frac{\partial z}{\partial u}$$

which determine p_1 and p_2 have infinitely many solutions, if in addition

$$b \frac{\partial x^1}{\partial u} - a^1 \frac{\partial z}{\partial u} = 0$$

for all $u \in I$. These two conditions are equivalent to saying that C is a characteristic curve.

Example: The characteristic system of the quasi-linear equation

$$zp_1 + p_2 = 1, \qquad \mathbf{d}z = p_1\, \mathbf{d}x^1 + p_2\, \mathbf{d}x^2$$

in $\mathbb{R}^3$ is

$$\frac{\mathrm{d}x^1}{z} = \frac{\mathrm{d}x^2}{1} = \frac{\mathrm{d}z}{1} = \frac{-\mathrm{d}p_1}{p_1^{\,2}} = \frac{-\mathrm{d}p_2}{p_1 p_2}.$$

In terms of any parameter t such that $\mathbf{d}z = \mathbf{d}t$, the bicharacteristic curve through the point (x_0^1, x_0^2, z_0) is

$$\begin{aligned} x^1 &= \tfrac{1}{2}t^2 + z_0 t + x_0^1 \\ x^2 &= t + x_0^2 \\ z &= t + z_0. \end{aligned}$$

The surface generated by the bicharacteristic curves through the differentiable curve

$$u \mapsto (x^1(u), x^2(u), z(u)) \qquad u \in I \subset \mathbb{R}$$

is the pair $(I \times \mathbb{R}, G)$ where G: $(u, t) \mapsto (x^1, x^2, z)$ is such that

$$\begin{aligned} x^1(u, t) &= \tfrac{1}{2}t^2 + z(u)t + x^1(u) \\ x^2(u, t) &= t + x^2(u) \\ z(u, t) &= t + z(u). \end{aligned}$$

Example: The characteristic system of the equation

$$a^i(x^j)p_i = 0, \quad \mathbf{d}z = p_i\, \mathbf{d}x^i \qquad i, j = 1, \ldots, n$$

is

$$\frac{\mathrm{d}x^i}{a^i(x^j)} = \frac{\mathrm{d}z}{0} = \frac{-\mathrm{d}p_i}{\Sigma_j p_j \partial a^j / \partial x^i} \qquad i, j = 1, \ldots, n.$$

The Cauchy problem can be solved by integrating the first n equalities. These equalities form a differential system in $\mathbb{R}^{n+1}$ since they do not involve the p_i. One first integral is $z = \text{const}$. Let (y^λ, y^n, z) be a set of coordinate functions on $\mathbb{R}^{n+1}$ where the $y^\lambda(x^i)$ are $n-1$ independent first integrals of the first $n-1$ equalities above. The solution of the Cauchy problem is determined by the pair $(W^{n-1} \times \mathbb{R}, G)$ which contains $(W^{n-1}, \bar{g})$ and where the mapping

$$\bar{G}: (u^\alpha, t) \mapsto (y^\lambda, y^n, z)$$

is defined by

$$y^\lambda = \bar{g}^\lambda(u^\alpha)$$
$$z = z(u^\alpha)$$
$$y^n = t.$$

When the first set of equations is solved for the u^α, substitution in the second equation gives the solution of the Cauchy problem as a function of the y^λ. The general solution of the equation $a^i(x^j)p_i = 0$ is thus an arbitrary function of the $n-1$ independent first integrals $y^\lambda(x^i)$.

8. INVARIANTS

Invariance with respect to a Pfaff system. Let $P = \{\boldsymbol{\theta}^\alpha = 0\}$ be a Pfaff system; let $\tilde{P}$ be the associated space of vector fields

$$\tilde{P} = \{\boldsymbol{v};\ \boldsymbol{\theta}^\alpha(\boldsymbol{v}) = 0 \quad \forall \alpha\}.$$

The space $\tilde{P}$ is a module over the ring $C^\infty(X)$ of C^∞ differentiable functions on X.

A form $\boldsymbol{\omega}$ is **invariant with respect to the Pfaff system** P if it is invariant under the local one parameter group of diffeomorphisms generated by each element of $\tilde{P}$. In other words $\boldsymbol{\omega}$ is invariant with respect to P if and only if $\mathcal{L}_v \boldsymbol{\omega} = 0$ for all $\boldsymbol{v} \in \tilde{P}$.

invariant with respect to P

Theorem. The form $\boldsymbol{\omega}$ is invariant with respect to the Pfaff system P if and only if

$$\mathrm{i}_v \boldsymbol{\omega} = 0 \quad \text{and} \quad \mathrm{i}_v\, \mathbf{d}\boldsymbol{\omega} = 0$$

for all $\boldsymbol{v} \in \tilde{P}$.

Proof: Since $\tilde{P}$ is a module over the ring $C^\infty(X)$, if $\boldsymbol{v} \in \tilde{P}$ then $f\boldsymbol{v} \in \tilde{P}$ for any $f \in C^\infty(X)$. Let $\boldsymbol{v} \in \tilde{P}$. The result follows from the relations (p. 207)

$$\mathcal{L}_v \boldsymbol{\omega} = \mathbf{d}\mathrm{i}_v \boldsymbol{\omega} + \mathrm{i}_v\, \mathbf{d}\boldsymbol{\omega}$$

$$\mathcal{L}_{fv}\boldsymbol{\omega} = \mathbf{d}f \wedge \mathrm{i}_v\boldsymbol{\omega} + f\,\mathrm{d}\mathrm{i}_v\boldsymbol{\omega} + f\mathrm{i}_v\,\mathrm{d}\boldsymbol{\omega}$$

$$= \mathbf{d}f \wedge \mathrm{i}_v\boldsymbol{\omega} + f\mathcal{L}_v\boldsymbol{\omega}. \qquad \blacksquare$$

If $\boldsymbol{\omega}$ is a closed form then it is invariant with respect to its associated Pfaff system Q^*, that is, $\mathcal{L}_v\boldsymbol{\omega} = 0$ for all $v \in \tilde{Q}$. Note that $\tilde{Q}$ is non-trivial if the rank of $\boldsymbol{\omega}$ is less than n (p. 230).

Exercise: Prove the following partial converse of the above statement. If $\boldsymbol{\omega}$ is of degree $n-1$ and rank $n-1$ then it is closed.

Proof: Let $y^1, \ldots, y^{n-1}$ be $n-1$ independent first integrals of the characteristic system of $\boldsymbol{\omega}$. Then $\boldsymbol{\omega}$ can be written (p. 253)

$$\boldsymbol{\omega} = \frac{1}{(n-1)!}\,\omega_{\lambda_1\ldots\lambda_{n-1}}(y^1, \ldots, y^{n-1})\,\mathbf{d}y^{\lambda_1} \wedge \ldots \wedge \mathbf{d}y^{\lambda_{n-1}} \qquad \lambda = 1, \ldots, n-1.$$

Hence $\mathrm{d}\boldsymbol{\omega} = 0$. $\blacksquare$

Exercise: If $\boldsymbol{\omega}$ is invariant with respect to P, $\mathrm{d}\boldsymbol{\omega}$ and $(\wedge\boldsymbol{\omega})^p$ are invariant with respect to P. If $\boldsymbol{\omega}$ and $\boldsymbol{\theta}$ are invariant with respect to P, $\boldsymbol{\omega} \wedge \boldsymbol{\theta}$ is invariant with respect to P (Problem IV 4, p. 272).

Suppose $\boldsymbol{\omega}$ is invariant with respect to the Pfaff system P. It is interesting to analyze the fact that $v \in \tilde{P}$ implies $fv \in \tilde{P}$ for all $f \in C^\infty(X)$. This reflects the fact that previously (p. 144) we were considering parametrized curves and now we are considering geometric curves.

$$\frac{\mathrm{d}\sigma(t)}{\mathrm{d}t} = v(\sigma(t)) \text{ defines a parametrized curve } \sigma: I \subset \mathbb{R} \to X^n.$$

$$\frac{\mathrm{d}x^1}{v^1(x^i)} = \frac{\mathrm{d}x^2}{v^2(x^i)} = \cdots = \frac{\mathrm{d}x^n}{v^n(x^i)}$$

defines a geometric curve; i.e. a submanifold of X^n whose tangent space at each point x contains v_x; one of the coordinates, for instance x^n if $v^n(x^i) \neq 0$, can be used as a parameter to solve this system of $n-1$ first order differential equations. Geometric curves are equivalence classes of parametrized curves; two parametrized curves σ and $\bar{\sigma}$ belong to the same class if $\sigma(t) \equiv \bar{\sigma}(tf(x) + g(x))$ where $f(x)$ is a non vanishing function on X^n,

$$\frac{\mathrm{d}\bar{\sigma}}{\mathrm{d}t} = f(\bar{\sigma}(t))v(\bar{\sigma}(t)).$$

Example (Problem V 1, p. 335). Let $\boldsymbol{F} = \frac{1}{2}F_{ij}\, \mathbf{d}x^i \wedge \mathbf{d}x^j$, with $i, j = 0, 1, 2, 3$, be a closed 2-form in $\mathbb{R}^4$ with

$$[F_{ij}] = \begin{bmatrix} 0 & E_1 & E_2 & E_3 \\ -E_1 & 0 & -H_3 & H_2 \\ -E_2 & H_3 & 0 & -H_1 \\ -E_3 & -H_2 & H_1 & 0 \end{bmatrix}.$$

In order to have non-trivial solutions of the equation

$$\mathrm{i}_{\boldsymbol{v}}\boldsymbol{F} = 0$$

the rank of $\boldsymbol{F}$ must be less than 4; in other words we require that $\det[F_{ij}] = (\boldsymbol{E} \cdot \boldsymbol{H})^2 = 0$. If $\boldsymbol{F}$ is not identically zero its rank must therefore be 2 (Example, p. 232). There are therefore two linearly independent solutions of the equations

$$v^i F_{ij} = 0.$$

For example,

$$v_1^0 = 0 \qquad v_1^\alpha = H_\alpha \qquad \alpha = 1, 2, 3$$

and

$$v_2^0 = \sum_\alpha (H_\alpha)^2 \qquad v_2^\alpha = (\boldsymbol{E} \times \boldsymbol{H})_\alpha.$$

The most general vector field $\boldsymbol{v}$ such that $\mathrm{i}_{\boldsymbol{v}}\boldsymbol{F} = 0$ is therefore

$$\boldsymbol{v} = f(x)\boldsymbol{v}_1 + g(x)\boldsymbol{v}_2$$

where $f, g \in C^0(\mathbb{R}^4)$. Thus $\boldsymbol{F}$ is invariant with respect to the Pfaff system P which is such that $\tilde{P}$ is generated by $\boldsymbol{v}_1$ and $\boldsymbol{v}_2$.

Remark: Let $S = \{\boldsymbol{\omega}^\alpha = 0\}$ be a differential system in X. The real vector space

$$\{\boldsymbol{v} \in T(X);\ \mathscr{L}_{\boldsymbol{v}}\boldsymbol{\omega}^\alpha = 0 \qquad \forall \alpha\}$$

is a Lie algebra. The vector fields $\boldsymbol{v}$ are called **isovector fields**[1] of S. isovector

Integral invariants. A p-form $\boldsymbol{\omega}$ is an **absolute integral invariant** of a Pfaff system P if its integral on a p-chain is invariant under the transformations of the chain generated by each element of $\tilde{P}$. Let $\{\sigma_t\}$ be the local one-parameter group of transformations of X^n generated by an element $\boldsymbol{v}$ of $\tilde{P}$. Let c be an elementary p-chain (B, f), where B is a rectangle and $f: B \subset \mathbb{R}^p \to X^n$. The chain c^t defined by $(B, f_t \equiv \sigma_t \circ f)$ is called the **transform** of c under σ_t.

absolute integral invariant

transform

Transformations of elementary p-chains are extended to transformations of chains by linearity.

[1]See [Estabrook].

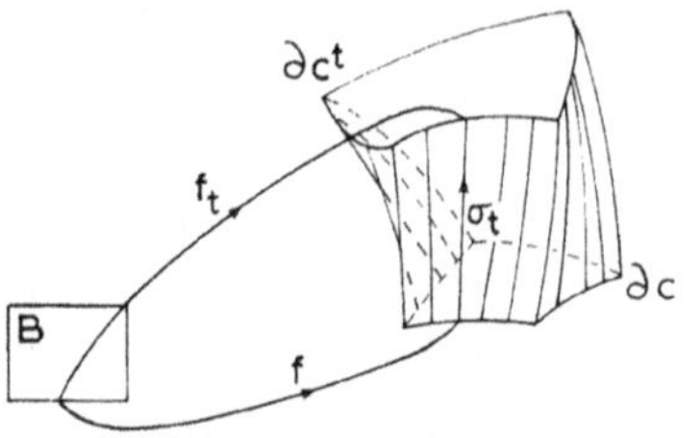

In fact, if $v \in \tilde{P}$, the integral of ω over a chain is unchanged if the chain is deformed in any smooth way by moving it along the integral curves of v; this is because $v \in \tilde{P}$ implies $fv \in \tilde{P}$ for any $f \in C^\infty(X)$.

Theorem. The three following statements are equivalent:

1) *ω is invariant with respect to P* i.e. $i_v\omega = 0$, $i_v\, d\omega = 0 \quad \forall v \in \tilde{P}$.
2) *$\mathcal{L}_v\omega = 0$ for all $v \in \tilde{P}$* i.e. *$\sigma_t^*\omega \equiv \omega$ for each σ_t generated by an element of $\tilde{P}$.*
3) *ω is an absolute invariant of P* i.e. *$\int_c \omega = \int_{c^t} \omega$ for each transformation generated by an element of $\tilde{P}$.*

Proof: We have already shown the equivalence of the first two statements (p. 261); we shall prove that 2 implies 3. By definition

$$\int_c \omega = \int_B f^*\omega$$

and by 2

$$\int_{c^t} \omega = \int_B f^*\sigma^*{}_t\omega = \int_B f^*\omega.$$

Conversely 3 implies 2

$$\int_{c^t} \omega = \int_c \omega \Rightarrow \int_B f^*\sigma^*{}_t\omega = \int_B f^*\omega \text{ for every block B}$$

$$\Rightarrow f^*(\sigma^*{}_t\omega - \omega) = 0 \text{ for every } f$$

$$\Rightarrow \sigma^*{}_t\omega = \omega \text{ for every } \sigma_t. \qquad \blacksquare$$

relative integral invariant

A p-form is a **relative integral invariant** of a Pfaff system P if its integral on a p-boundary is invariant under the transformations of the p-boundary generated by each element of $\tilde{P}$,

$$\int_{\partial c} \omega = \int_{\partial c^t} \omega.$$

The following statements are equivalent, 1) and 2) by Stokes' theorem.

1) $\boldsymbol{\omega}$ is a relative invariant of P i.e. $\int_{\partial c} \boldsymbol{\omega} = \int_{\partial c^t} \boldsymbol{\omega}$
2) $\mathrm{d}\boldsymbol{\omega}$ is an absolute invariant of P i.e. $\int_c \mathrm{d}\boldsymbol{\omega} = \int_{c^t} \mathrm{d}\boldsymbol{\omega}$
3) $\mathrm{d}\boldsymbol{\omega}$ is invariant with respect to P i.e. $\mathrm{i}_{\boldsymbol{v}}\, \mathrm{d}\boldsymbol{\omega} = 0 \quad \forall \boldsymbol{v} \in \tilde{P}$.

9. INTEGRAL INVARIANTS OF CLASSICAL DYNAMICS

Let the manifold X^n, with coordinates (q^i), be the **configuration space** of a system with n degrees of freedom. The manifold $X^n \times \mathbb{R}$, with coordinates (q^i, t) is called the **configuration space time** of the system. The natural coordinates (pp. 119, 136) on $T(X^n)$ and $T^*(X^n)$ will be denoted $(q^i, \dot{q}^i)$ and (q^i, p_i) respectively. The cotangent bundle $T^*(X^n)$ is called the **phase space** of the system. The manifold $T^*(X^n) \times \mathbb{R}$ is called the **state space** of the system. The lagrangian is a function L: $T(X^n) \times \mathbb{R} \to \mathbb{R}$ by $L(q^i, \dot{q}^i, t) \in \mathbb{R}$; the vector with components $(p_i = \partial L / \partial \dot{q}^i)$ is a covariant vector on X^n (Problem III 1). The hamiltonian is a function H: $T^*(X^n) \times \mathbb{R} \to \mathbb{R}$ by $H(q^i, p_i, t) \in \mathbb{R}$.

configuration space
configuration space time
phase space
state space

Theorem. The 1-*form* $\boldsymbol{\omega} = p_i\, \mathbf{d}q^i - H\, \mathbf{d}t$ *on* $T^*(X^n) \times \mathbb{R}$ *is a relative integral invariant of the hamiltonian Pfaff system* (p. 274)

$$\frac{\mathrm{d}q^i}{\partial H / \partial p_i} = \mathrm{d}t = \frac{-\mathrm{d}p_i}{\partial H / \partial q^i};$$

i.e., $\mathrm{d}\boldsymbol{\omega}$ *is an absolute integral invariant.*

Proof: The space of vector fields associated with the Hamiltonian system is 1-dimensional and is generated by the **hamiltonian vector field**

hamiltonian vector field

$$\boldsymbol{v} = \frac{\partial H}{\partial p_i} \frac{\partial}{\partial q^i} + \frac{\partial}{\partial t} - \frac{\partial H}{\partial q^i} \frac{\partial}{\partial p_i}.$$

The form $\mathrm{d}\boldsymbol{\omega}$ can be written

$$\mathrm{d}\boldsymbol{\omega} = -\frac{\partial H}{\partial q^i} \mathbf{d}q^i \wedge \mathbf{d}t - \mathbf{d}q^i \wedge \mathbf{d}p_i + \frac{\partial H}{\partial p_i} \mathbf{d}t \wedge \mathbf{d}p_i.$$

One proves that $\mathrm{i}_{\boldsymbol{v}}\, \mathrm{d}\boldsymbol{\omega} = 0$ by a straight forward calculation. ∎

Exercise: Carry out the calculation for a system with one degree of freedom.

The components of $\boldsymbol{v}$ are then $v^1 = \partial H / \partial p$, $v^2 = 1$, $v^3 = -\partial H / \partial q$; the components of $\mathrm{d}\boldsymbol{\omega}$ are then $(12) = -\partial H / \partial q$, $(13) = -1$, $(23) = \partial H / \partial p$, where we have set $(\mathbf{d}\boldsymbol{\omega})_{ij} = (ij)$. Then

$$\mathrm{i}_v\,\mathbf{d}\boldsymbol{\omega} = (v^2(21) + v^3(31))\,\mathbf{d}\boldsymbol{q} + (v^1(12) + v^3(32))\,\mathbf{d}\boldsymbol{t} + (v^1(13) + v^2(23))\,\mathbf{d}\boldsymbol{p} = 0.$$

$\mathbf{d}\boldsymbol{\omega}$ is often called "the" integral invariant of dynamics. It serves to build other integral invariants such as $(\wedge\,\mathbf{d}\boldsymbol{\omega})^n$ (see Exercise p. 262). For example the form $\boldsymbol{\alpha} = \mathbf{d}\boldsymbol{p}_1 \wedge \ldots \wedge \mathbf{d}\boldsymbol{p}_n \wedge \mathbf{d}\boldsymbol{q}^1 \wedge \ldots \wedge \mathbf{d}\boldsymbol{q}^n$ on $T^*(X^n)$ can be obtained from

$$(\wedge\,\mathbf{d}\boldsymbol{\omega})^n = (-1)^{n-1} n!\,\boldsymbol{\alpha} + \boldsymbol{\theta} \wedge \mathbf{d}\boldsymbol{t} \qquad \text{on } T^*(X^n) \times \mathbb{R}$$

by the projection $(x, t) \mapsto x$ with $x \in T^*(X^n)$.

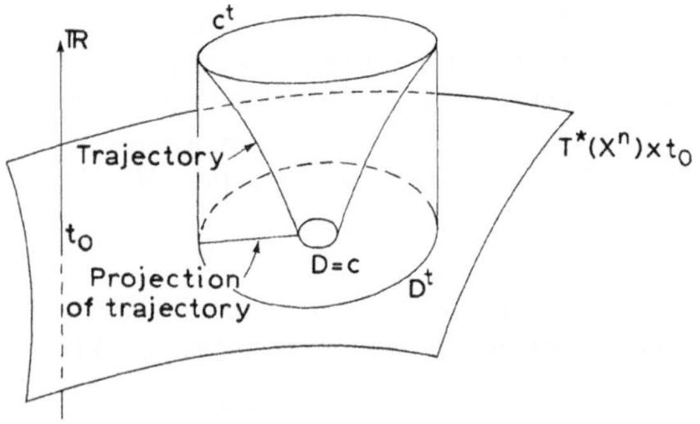

Liouville's theorem can be obtained from the absolute invariance of $(\wedge\mathbf{d}\boldsymbol{\omega})^n$ with respect to Hamilton's equations. Indeed

$$\int_c (\wedge\mathbf{d}\boldsymbol{\omega})^n = \int_{c^t} (\wedge\mathbf{d}\boldsymbol{\omega})^n$$

where c^t is a chain on $T^*(X^n) \times \mathbb{R}$ generated from c by the trajectories of Hamilton's equations.

Let $c = D$ be a chain on $T^*(X^n) \times t_0$, let D^t be the projection of c^t on $T^*(X^n) \times t_0$; then we have the **Liouville theorem**

Liouville theorem

$$\int_D \boldsymbol{\alpha} = \int_{D^t} \boldsymbol{\alpha}.$$

A coordinate transformation

$$Q^i = Q^i(q^j, p_j, t)$$
$$P_i = P_i(q^j, p_j, t)$$
$$t = t$$

on $U \times \mathbb{R} \subset T^*(X^n) \times \mathbb{R}$ is called **a canonical transformation** if there exists a function $\bar{H}$ on $U \times \mathbb{R}$ such that

canonical transformations

$$\mathbf{d}\boldsymbol{\Omega} = \mathbf{d}\boldsymbol{\omega}$$

where $\boldsymbol{\omega}$ and $\boldsymbol{\Omega}$ are the 1-forms given respectively in the coordinates (q^i, p_i, t) and (Q^i, P_i, t) by

$$\boldsymbol{\omega} = p_i\, \mathbf{d}q^i - H\, \mathbf{d}t$$
$$\boldsymbol{\Omega} = P_i\, \mathbf{d}Q^i - \bar{H}\, \mathbf{d}t.$$

It follows that if the transformation is canonical and if U is homeomorphic to $\mathbb{R}^n$, then $\boldsymbol{\omega} = \boldsymbol{\Omega} + \mathbf{d}\varphi$ where $\varphi\colon U \times \mathbb{R} \to \mathbb{R}$ is an arbitrary differentiable function. The function φ is called the **generating function** of the transformation. The canonical transformations form a local pseudo group.

generating function

Theorem. A canonical transformation conserves the form of the Hamilton equations.

Proof. The hamiltonian vector field $\boldsymbol{v}$ satisfies $\mathrm{i}_{\boldsymbol{v}}\, \mathrm{d}\boldsymbol{\omega} = 0$. If $(q^i, p_i, t) \mapsto (Q^i, P_i, t)$ is canonical then there exists a function H such that $\mathrm{d}\boldsymbol{\Omega} = \mathrm{d}\boldsymbol{\omega}$ for $\boldsymbol{\Omega} = P_i\, \mathbf{d}Q^i - \bar{H}\, \mathbf{d}t$. Let $\boldsymbol{V} = s\boldsymbol{v}$ have coordinates (V^i, V^t, V_i) with respect to the basis $(\partial/\partial \boldsymbol{Q}^i, \partial/\partial t, \partial/\partial \boldsymbol{P}_i)$ for $T(T^*X^n)$. By appropriate choice of s we may take $V^t = 1$. Then

$$0 = \mathrm{i}(\boldsymbol{V})\, \mathrm{d}\boldsymbol{\Omega} = \left(\frac{\partial \bar{H}}{\partial Q^i} + V_i\right) \mathrm{d}Q^i - \left(V^i \frac{\partial \bar{H}}{\partial Q^i} + V_i \frac{\partial \bar{H}}{\partial P_i}\right) \mathrm{d}t$$
$$+ \left(-V^i + \frac{\partial \bar{H}}{\partial P_i}\right) \mathrm{d}P^i.$$

Hence $V^i = \partial\bar{H}/\partial P_i$, $V_i = -\partial\bar{H}/\partial Q^i$; and the corresponding Pfaff system is

$$\frac{\mathbf{d}Q^i}{\partial\bar{H}/\partial P_i} = \mathbf{d}t = \frac{-\mathbf{d}P_i}{\partial\bar{H}/\partial Q^i}.$$

This is a particular case of the theorem proved on p. 269. ■

10. SYMPLECTIC STRUCTURES AND HAMILTONIAN SYSTEMS[1]

We give here some of the results about dynamical systems with time independent hamiltonian in the modern language of symplectic geometry.

[1]See for instance [Abraham, Godbillon, Souriau 1970]. For a geometric formalism and recent general results in the case of time independent hamiltonians see [Lichnerowicz 1974 and 1975].

symplectic form

Let X be a smooth manifold of even dimension $2n$. A **symplectic form** F on X is an exterior differential closed 2-form of rank $2n$.

canonical 1-form

Example: The **canonical 1-form** $\boldsymbol{\theta}$ of a dynamical system with an n dimensional configuration manifold X^n is a 1-form on T^*X^n (phase space) defined in natural coordinates (p. 137) by:

$$\boldsymbol{\theta} = p_i \, \mathrm{d}q^i.$$

canonical 2-form

The **canonical 2-form**, $\boldsymbol{F} = \mathrm{d}\boldsymbol{\theta}$, is a symplectic form on T^*X^n, which reads in natural coordinates

$$\boldsymbol{F} = \mathrm{d}p_i \wedge \mathrm{d}q^i.$$

The following theorem states that a symplectic form can generally be expressed in this way.

Darboux theorem. If F is a symplectic form on the 2n-dimensional manifold X, there exists about each point coordinates $(x^1, \ldots x^n, y_1, \ldots y_n)$, *called canonical, in which the symplectic form has the expression*

$$F = \sum_{i=1}^{n} \mathrm{d}y_i \wedge \mathrm{d}x^i.$$

Proof: See Problem IV 6, p. 281.

symplectomorphism

A diffeomorphism f of X is said to be **symplectic** (a **symplectomorphism**) if it leaves invariant the symplectic form $\boldsymbol{F}$:

$$f^*\boldsymbol{F} = \boldsymbol{F}.$$

The hamiltonian H of a dynamical system with configuration manifold X^n is a function on T^*X^n. The hamilton equations, which determine the trajectories of the system, are a differential system on T^*X^n which reads in natural coordinates[1]

$$\frac{\mathrm{d}q^i}{\mathrm{d}t} = \frac{\partial H}{\partial p_i}, \qquad \frac{\mathrm{d}p_i}{\mathrm{d}t} = -\frac{\partial H}{\partial q^i}.$$

The vector field $\boldsymbol{v}$ on T^*X^n given in natural coordinates[2] by ($v^i = \partial H/\partial p_i$, $v_i = -\partial H/\partial q^i$) satisfies clearly the equation

$$\mathrm{i}(\boldsymbol{v})\boldsymbol{F} = -\mathrm{d}H$$

and also, since $\boldsymbol{F}$ is closed, the relation

$$\mathcal{L}_{\boldsymbol{v}}\boldsymbol{F} = 0.$$

[1]We still denote by H the expression of H in a local chart.
[2]Note the position of i in v^i and v_i to distinguish the two sets of components.

Therefore the flow[1] of (the group of transformations generated by) the vector field $\boldsymbol{v}$ leaves invariant the canonical 2-form.

A vector field $\boldsymbol{v}$ on a symplectic manifold (X, F) is said to be **hamiltonian** if $\mathrm{i}(\boldsymbol{v})\boldsymbol{F}$ is a closed form: hamiltonian vector field

$$\mathrm{di}(\boldsymbol{v})\boldsymbol{F} = 0 \text{ or equivalently } \mathcal{L}_{\boldsymbol{v}}\boldsymbol{F} = 0.$$

As a consequence of the definition we have the following theorem.

Theorem[2]. *The flow $f = \{f_t\}$ of a hamiltonian vector field on $(X, \boldsymbol{F})$ is a group of symplectomorphisms of X.*

If $\mathrm{i}(\boldsymbol{v})F$ is not only closed but also an exact differential,

$$\mathrm{i}(\boldsymbol{v})F = -\mathrm{d}H.$$

The vector field $\boldsymbol{v}$ is said to be **globally hamiltonian** and the function H on X (determined up to an additive constant) is called the **hamiltonian**. Sometimes a vector field such that $\mathrm{i}(\boldsymbol{v})F$ is a closed form is called locally hamiltonian, and the name "hamiltonian" is applied only to these vector fields that we have called "globally hamiltonian". globally hamiltonian hamiltonian

Theorem. The set of hamiltonian vector fields is a Lie subalgebra of the Lie algebra of the set of vector fields.

Proof: If $\boldsymbol{v}_1$ and $\boldsymbol{v}_2$ are hamiltonian, $[\boldsymbol{v}_1, \boldsymbol{v}_2]$ is hamiltonian, since:

$$\mathcal{L}_{[\boldsymbol{v}_1, \boldsymbol{v}_2]}\boldsymbol{F} = [\mathcal{L}_{\boldsymbol{v}_1}, \mathcal{L}_{\boldsymbol{v}_2}]\boldsymbol{F}.$$

The other properties are trivial to verify. ■

Given a function H on X, with differential $\mathrm{d}H$, there exists one, and only one, hamiltonian vector field $\boldsymbol{v}$ on X with hamiltonian H, i.e. such that $\mathrm{i}(\boldsymbol{v})\boldsymbol{F} = -\mathrm{d}H$. Indeed since the rank of $\boldsymbol{F}$ equals the dimension of X, we have, more generally

Theorem. The mapping $\boldsymbol{v} \mapsto \mathrm{i}(\boldsymbol{v})\boldsymbol{F}$ is an isomorphism of the module of vector fields onto the module of 1-forms.

Given a 1-form $\boldsymbol{\theta}$, we define a vector field $\boldsymbol{v}_{\boldsymbol{\theta}}$ by the equation

$$\mathrm{i}(\boldsymbol{v}_{\boldsymbol{\theta}})\boldsymbol{F} = \boldsymbol{\theta}.$$

[1]Or local flow, if $\boldsymbol{v}$ has no global flow (cf. p. 145).
[2]To be adapted for local flows.

Poisson bracket

We then define the **Poisson bracket** of two 1-forms $\boldsymbol{\theta}$ and $\boldsymbol{\psi}$ on a symplectic manifold by:

$$(\boldsymbol{\theta}, \boldsymbol{\psi}) = \mathrm{i}([\boldsymbol{v}_{\theta}, \boldsymbol{v}_{\psi}])\boldsymbol{F}$$

and the Poisson bracket of two differentiable functions f and g on a symplectic manifold by

$$(f, g) = -\boldsymbol{F}(\boldsymbol{v}_{\mathbf{d}f}, \boldsymbol{v}_{\mathbf{d}g}).$$

The components in canonical coordinates (q^i, p_i) of $\boldsymbol{v}_{\mathbf{d}f}$, defined by $\mathrm{i}(\boldsymbol{v}_{\mathbf{d}f})\boldsymbol{F} = \mathbf{d}f$, are $(-\partial f/\partial p_i, \partial f/\partial q^i)$. Hence

$$(f, g) = \frac{\partial f}{\partial q^i}\frac{\partial g}{\partial p_i} - \frac{\partial g}{\partial q^i}\frac{\partial f}{\partial p_i}.$$

Poisson brackets and Lie brackets obey similar identities. Poisson brackets are widely used in the study of integrability conditions and integral manifolds.

PROBLEMS AND EXERCISES

PROBLEM 1. PROPERTIES OF p-COMPOUND MATRICES

Let $A = (a^i_j)$ be an $n \times n$ matrix. The determinants of all the $p \times p$ matrices obtained from A are $\epsilon^{k_1 \dots k_p}_{j_1 \dots j_p} a^{l_1}_{k_1} \dots a^{l_p}_{k_p}$. It is often convenient to consider these determinants as the elements of a new matrix called the p-compound matrix derived from A.

In particular if $p = n$, the n-compound matrix has only one element equal to the determinant of A.

Let A be an $m \times n$ matrix, let $P_A = (p^H_K)$ be the p-compound matrix of A

$$p^H_K = \epsilon^{k_1 \dots k_p}_{K_1 \dots K_p} a^{H_1}_{k_1} a^{H_2}_{k_2} \dots a^{H_p}_{k_p}.$$

Let B be an $n \times q$ matrix; note that AB is an $m \times q$ matrix, P_A is an $\binom{m}{p} \times \binom{n}{p}$ matrix, P_B is an $\binom{n}{p} \times \binom{q}{p}$ matrix, and P_{AB} is an $\binom{m}{p} \times \binom{q}{p}$ matrix.

Prove the following properties:

$P_{AB} = P_A P_B$.

The p-compound of a unit matrix is a unit matrix.

$P_{A^{-1}} = P_A^{-1}$.

$P_{A^{\mathrm{T}}} = P_A^{\mathrm{T}}$, *where* T *stands for transposed.*

The p-compound matrix of a symmetric matrix is a symmetric matrix.

PROBLEM 2. POINCARÉ LEMMA. FIRST SET OF MAXWELL EQUATIONS. WORMHOLES (see Problems V 1 and Vbis 2)

Show that an electromagnetic field F defined on an arbitrary manifold M can not necessarily be derived from a potential A. Let M be a spacetime manifold with wormholes; show that the wormholes will appear to be electric charges.

Answer: The first set of Maxwell equations (see example p. 263)

$$F_{ij,k} + F_{jk,i} + F_{ki,j} = 0 \equiv \begin{cases} \operatorname{div} \boldsymbol{H} = 0 & \text{for } i, j, k, = 1, 2, 3 \\ \operatorname{curl} \boldsymbol{E} + \dot{\boldsymbol{H}} = 0 & \text{for either } i, j, \text{ or } k = 0 \end{cases}$$

stands for $\mathrm{d}\boldsymbol{F} = 0$, where $\boldsymbol{F}$ is a 2-form on spacetime. The equations $\boldsymbol{H} = \operatorname{curl} \vec{A}$, $\boldsymbol{E} = -\dot{\vec{A}} - \nabla A_0$ stand for $\boldsymbol{F} = \mathrm{d}\boldsymbol{A}$, with $\boldsymbol{A} = (A_0, \vec{A})$. The Maxwell equations $\mathrm{d}\boldsymbol{F} = 0$ imply $\boldsymbol{F} = \mathrm{d}\boldsymbol{A}$ on spaces homeomorphic to $\mathbb{R}^n$.

The electromagnetic field F_{ij} on an arbitrary manifold – even on an open set of $\mathbb{R}^3$ not homeomorphic to $\mathbb{R}^3$ – is not necessarily derived from a potential A_i. See Problem Vbis 2 for the properties of an electromagnetic field which cannot be derived globally from a potential. In his theory of geometrodynamics, Wheeler has proposed the existence of holes, called "wormholes" in the structure of space and explained electric charges as a trapping of electric lines of force by the wormholes. This phenomenon is related to the fact that $\mathrm{d}\,{}^*\boldsymbol{F} = 0$ (second set of Maxwell equations p. 335) does not imply ${}^*\boldsymbol{F} = \mathrm{d}\boldsymbol{U}$ in a space with wormholes.

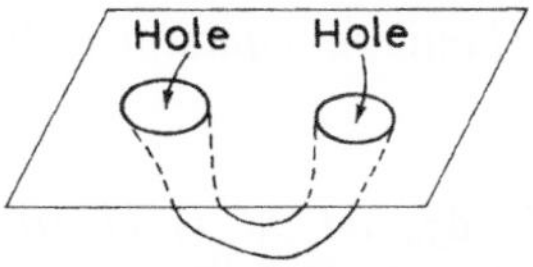

Reference: J.A. Wheeler, in: *Battelle Rencontres*, eds. C.M. DeWitt and J.A. Wheeler (Benjamin, 1968) p. 266.

PROBLEM 3. INTEGRAL MANIFOLDS

Show that two systems which have the same integral manifolds are not necessarily algebraically equivalent.

Answer: $\boldsymbol{\omega} = P\,\mathbf{d}x + Q\,\mathbf{d}y + R\,\mathbf{d}z$ is not algebraically equivalent to $(\boldsymbol{\omega}, \mathbf{d}\boldsymbol{\omega})$ unless $\mathbf{d}\boldsymbol{\omega}$ is of the form $\boldsymbol{\omega} \wedge \boldsymbol{\theta}$. Nevertheless $\boldsymbol{\omega} = 0$ and $(\boldsymbol{\omega} = 0, \mathbf{d}\boldsymbol{\omega} = 0)$ have the same integral manifolds. Indeed:
Let $z = f(x, y)$ be an integral manifold of $\boldsymbol{\omega} = 0$, then

$$(P(x, y, f) + R(x, y, f) f'_x(x, y))\,\mathbf{d}x + (Q(x, y, f) + R(x, y, f) f'_y(x, y))\,\mathbf{d}y = 0.$$

Since $\mathbf{d}f'_x \wedge \mathbf{d}x + \mathbf{d}f'_y \wedge \mathbf{d}y = 0$, it follows that

$$(\mathbf{d}P + \mathbf{d}Rf'_x + R\,\mathbf{d}f'_x) \wedge \mathbf{d}x + (\mathbf{d}Q + \mathbf{d}Rf'_y + R\,\mathbf{d}f'_y) \wedge \mathbf{d}y = 0.$$

On the other hand

$$\mathbf{d}\omega = \mathbf{d}P \wedge \mathbf{d}x + \mathbf{d}Q \wedge \mathbf{d}y + \mathbf{d}R \wedge \mathbf{d}z$$
$$\mathbf{d}\omega = 0 \quad \text{for} \quad z = f(x, y).$$

PROBLEM 4. FIRST ORDER PARTIAL DIFFERENTIAL EQUATION; HAMILTON–JACOBI EQUATION; HAMILTON EQUATIONS; HAMILTON VECTOR FIELD; INTEGRAL MANIFOLD; LAGRANGIAN MANIFOLD (see Problems III 1, II 2, II 3)

a) *Let* M *be the configuration space of a system with n degrees of freedom, let* $x \in \mathsf{M}$. *Let* $\mathsf{W} = \mathsf{M} \times \mathbb{R}$, *let* $(x, x^0) = \tilde{x} \in \mathsf{W}$. *Let* $\mathsf{X}^\nu = (T^*\mathsf{W}) \times \mathbb{R}$; *note* $\nu = 2n + 3$. *Introduce the coordinates* (x^i) *on* M, $i = 1, \dots n$, *the fiber coordinates* (x^α, y_α) *on* $T^*\mathsf{W}$, $\alpha = 1, \dots n, 0$, *and* (x^α, y_α, z) *on* X^ν.
Let L *be the lagrangian of the system and let* H *be its hamiltonian. Let* $\bar{S}$ *be a function on* W *satisfying the Hamilton–Jacobi equation; in local coordinates*

$$\frac{\partial \bar{S}}{\partial x^0} + H\left(x^\alpha, \frac{\partial \bar{S}}{\partial x^i}\right) = 0.$$

Prove that the characteristic system C *of the Hamilton–Jacobi equation contains the Hamilton equations. Which other equation is contained in* C?

b) *Show that* $\mathbf{d}\theta \equiv \mathbf{d}x^\alpha \wedge \mathbf{d}y_\alpha$ *on* X^ν *is invariant under the group of transformations generated by the Hamiltonian vector field on* X^ν *defined by*

$$\boldsymbol{V} = \frac{\partial}{\partial \boldsymbol{x}^0} + \frac{\partial H}{\partial y_i}\frac{\partial}{\partial \boldsymbol{x}^i} - \frac{\partial H}{\partial x^\alpha}\frac{\partial}{\partial \boldsymbol{y}_\alpha} + \left(y_i \frac{\partial H}{\partial y_i} - H\right)\frac{\partial}{\partial \boldsymbol{z}}.$$

c) *Let* (L, f) *be the integral manifold of the Hamilton–Jacobi equation for the initial Cauchy data* $(\mathbb{C}, g)$: $g(\mathbb{C}) \subset f(\mathsf{L}) \subset \mathsf{X}^\nu$, *where* $f(\mathsf{L})$ *is an* $n+1$ *manifold, solution of the exterior differential system S, that contains the n manifold* $g(\mathbb{C})$.
Let q be an arbitrary path on M, *let* $p(x^0) \overset{\text{def}}{=} \partial L / \partial \dot{q}(x^0)$; *the solution of the Hamilton equations induce curves on* $T^*\mathsf{M}$, $T^*\mathsf{W}$, X^ν. *We shall use the notations*:

$$(\bar{q}(x_0), \bar{p}(x_0)) \in T^*\mathsf{M}, \quad (\bar{Q}(\lambda), \bar{P}(\lambda)) \in T^*\mathsf{W}, \quad \bar{\sigma}(\lambda) \in \mathsf{X}^\nu.$$

Use the solutions of the Hamilton equations to construct $f(\mathbb{L})$. Set $\boldsymbol{\Omega} \overset{\text{def}}{=} y_\alpha\, \mathbf{d}x^\alpha$*, the Liouville form on* $f(\mathbb{L})$*; show that* $\boldsymbol{\Omega}$ *is a closed form. When is it an exact form?*

Answer: a) The exterior differential system S corresponding to the Hamilton–Jacobi equation is

$$S = \begin{cases} F \equiv y_0 + H(x^\alpha, y_i) = 0 \\ \boldsymbol{\theta} \equiv \mathbf{d}z - y_\alpha\, \mathbf{d}x^\alpha = 0 \end{cases}$$

where $z = \bar{S}(x^\alpha)$ and $y_\alpha = \partial\bar{S}/\partial x^\alpha$.
The characteristic system C of the exterior system S on $\mathbb{X}^\nu$ consists of

$$C = \begin{cases} F = 0 \\ \text{the associated Pfaff system } \{\bar{\boldsymbol{\theta}}^\beta = 0\} \text{ of } \{\boldsymbol{\theta} = 0,\ \mathbf{d}F = 0,\ \mathbf{d}\boldsymbol{\theta} = 0\}. \end{cases}$$

Let Q_X^* be the subspace of $T_X^*\mathbb{X}^\nu$ spanned by $\{\bar{\boldsymbol{\theta}}_X^\beta\}$; let $\tilde{Q}_X$ be the solution of the set of exterior equations $\{\bar{\boldsymbol{\theta}}_X^\beta = 0\}$, i.e.

$V \in \tilde{Q}_X \subset T_X\mathbb{X}^\nu \iff \mathrm{i}_V\bar{\boldsymbol{\theta}}_X^\beta = 0$ for every element of the set $\{\bar{\boldsymbol{\theta}}_X^\beta\}$.

Let I be the ideal generated by $\{\boldsymbol{\theta}, \mathbf{d}F, \mathbf{d}\boldsymbol{\theta}\}$

$$V \in \tilde{Q}_X \iff \mathrm{i}_V\, \mathrm{d}\theta \in I$$

$$\mathrm{i}_V\, \mathbf{d}\boldsymbol{\theta} \in I \iff \mathrm{i}_V\, \mathbf{d}\boldsymbol{\theta} = \mu\, \mathbf{d}F + \lambda\boldsymbol{\theta} \quad \text{for some numbers } \lambda, \mu.$$

Using the natural basis $(\mathbf{d}x^\alpha, \mathbf{d}y_\alpha, \mathbf{d}z)$ in $T_X^*\mathbb{X}^\nu$ and $(\partial/\partial \boldsymbol{x}^\alpha, \partial/\partial \boldsymbol{y}_\alpha, \partial/\partial \boldsymbol{z})$ in $T_X\mathbb{X}^\nu$ we write[1]

$$\boldsymbol{V} = V^\alpha \frac{\partial}{\partial \boldsymbol{x}^\alpha} + V_\alpha \frac{\partial}{\partial \boldsymbol{y}_\alpha} + V^z \frac{\partial}{\partial \boldsymbol{z}};$$

in these coordinates the equation $\mathrm{i}_V\, \mathbf{d}\boldsymbol{\theta} = \mu\, \mathbf{d}F + \lambda\boldsymbol{\theta}$ reads

$$V^\alpha\, \mathrm{d}y_\alpha - V_\alpha\, \mathrm{d}x^\alpha = \mu\left(\mathrm{d}y_0 + \frac{\partial H}{\partial x^\alpha}\mathrm{d}x^\alpha + \frac{\partial H}{\partial y_i}\mathrm{d}y_i\right) + \lambda(\mathrm{d}z - y_\alpha\, \mathrm{d}x^\alpha).$$

It is satisfied for $\lambda = 0$ (i.e. this equation does not determine V^z) and

$$\mu = V^0 = \frac{V^i}{\partial H/\partial y_i} = \frac{-V_\alpha}{\partial H/\partial x^\alpha}.$$

These equations are satisfied by all $\boldsymbol{V} \in \tilde{Q}_X$; hence the following $2n+1$ Pfaff forms $\boldsymbol{\theta}^i$ and $\boldsymbol{\theta}_\alpha$ which satisfy $\boldsymbol{\theta}^i(\boldsymbol{V}) = \boldsymbol{\theta}_\alpha(\boldsymbol{V}) = 0$ are in Q_X^*

$$\boldsymbol{\theta}^i \equiv \mathbf{d}x^i - \frac{\partial H}{\partial y_i}\mathbf{d}x^0 \quad \text{and} \quad \boldsymbol{\theta}_\alpha \equiv \mathbf{d}y_\alpha + \frac{\partial H}{\partial x^\alpha}\mathbf{d}x^0.$$

The system $\{\boldsymbol{\theta}, \mathbf{d}F, \boldsymbol{\theta}^i, \boldsymbol{\theta}_\alpha\}$ is not linearly independent, indeed:

$$\mathbf{d}F = \frac{\partial H}{\partial x^i}\boldsymbol{\theta}^i + \frac{\partial H}{\partial y_\alpha}\boldsymbol{\theta}_\alpha.$$

[1]Note the position of α in V^α and V_α to distinguish the two sets of components.

The system $\{\bar{\boldsymbol{\theta}}^{\beta}\} = \{\boldsymbol{\theta}, \boldsymbol{\theta}^{i}, \boldsymbol{\theta}_{\alpha}\}$ is a system of $2n+2$ linearly independent Pfaff forms on X^{ν} algebraically equivalent to $\{\boldsymbol{\theta}, \mathbf{d}F, \mathbf{d}\boldsymbol{\theta}\}$; its rank is $2n+2$. The class of the system S is $2n+2$.

The characteristic system C of S consists of the following equations

$$C = \begin{cases} F \equiv y_0 + H(x^{\alpha}, y_i) = 0 \\ \boldsymbol{\theta} \equiv \mathbf{d}z - y_{\alpha}\, \mathbf{d}x^{\alpha} = 0 \\ \left.\begin{array}{l} \boldsymbol{\theta}^{i} \equiv \mathbf{d}x^{i} - \dfrac{\partial H}{\partial y_i}\mathbf{d}x^{0} = 0 \\ \boldsymbol{\theta}_{\alpha} \equiv \mathbf{d}y_{\alpha} + \dfrac{\partial H}{\partial x^{\alpha}}\mathbf{d}x^{0} = 0 \end{array}\right\} \text{Hamiltonian equations.} \end{cases}$$

Remark 1: The equation $F=0$ is an algebraic equation in y_{α} which may admit N sets of solutions $\{(y_{\alpha}^{A});\ A = 1, \dots N\}$. By virtue of the equation $\mathbf{d}\bar{S}^{A}(x) - \omega_{\alpha}^{A}(x)\,\mathbf{d}x^{\alpha} = 0$, the vector $(k, -k\omega_{\alpha}^{A}(x))$, where k is an arbitrary number, is perpendicular to the surface $S^{A}(x)$. It determines the tangent space to $S^{A}(x)$ at x. The set of N vectors $\{(k, -k\omega_{\alpha}^{A}(x))\}$ lie on a cone C, the N corresponding tangent spaces envelop the dual cone Γ (Monge cone). In the following we shall consider any one set of solutions $(\omega_{\alpha}^{A}(x))$ to $F=0$ but we shall suppress the index A.

Remark 2: When $\partial H/\partial x^{0} = 0$, then $\mathrm{d}y_0 = 0$ and $H(x^{i}, y_i)$ is a constant of the motion (first integral).

Remark 3: The Hamilton equations do not depend on $\bar{S}$; they can be solved independently of the equation $\boldsymbol{\theta} = 0$. Using the solutions of the Hamilton equations, one can then solve the equation $\boldsymbol{\theta} = 0$ by a simple quadrature

$$\mathbf{d}\bar{S}(x^{\alpha}) = \left(\sum y_i \frac{\partial H}{\partial y_i} - H\right)\mathbf{d}x^{0} = L(\bar{q}(x^{0}), \dot{\bar{q}}(x^{0}), x^{0})\,\mathbf{d}x^{0}.$$

The function $\bar{S}$ is the integral of the Lagrangian at the classical path. When there is more than one classical path between two points $x_1^{\alpha}, x_2^{\alpha}$, $\bar{S}$ is a multivalued function of x_1^{α} and x_2^{α}.

Remark 4: The characteristic system can be written

$$C = \begin{cases} F = 0 \\ \mathbf{d}x^{0} = \dfrac{\mathbf{d}x^{i}}{\partial H/\partial y_i} = \dfrac{-\mathbf{d}y_{\alpha}}{\partial H/\partial x^{\alpha}} = \dfrac{\mathbf{d}z}{\sum y_i(\partial H/\partial y_i) - H}. \end{cases}$$

Its solutions are the trajectories of the Hamilton vector field

$$\boldsymbol{V} = \frac{\partial}{\partial \boldsymbol{x}^{0}} + \frac{\partial H}{\partial y_i}\frac{\partial}{\partial \boldsymbol{x}^{i}} - \frac{\partial H}{\partial x^{\alpha}}\frac{\partial}{\partial \boldsymbol{y}_{\alpha}} + \left(y_i\frac{\partial H}{\partial y_i} - H\right)\frac{\partial}{\partial \boldsymbol{z}}.$$

trajectory A **trajectory** is an equivalence class of parametrized curves; indeed the

vector fields given at each point X in $\mathbb{X}^\nu$ by $f(X)\boldsymbol{V}(X)$, where f is an arbitrary non vanishing differentiable function, define the same trajectory as $\boldsymbol{V}$. In other words we can solve the system of $2n+2$ ordinary differential equations in C and give the solutions as $2n+1$ functions of x^0. But x^0 is not the only parameter which can be used. We can, in general, consider the system of ν ordinary differential equations obtained by adding to C the equation $\mathrm{d}x^0 = f(X)\,\mathrm{d}\lambda$, i.e. $x^0 = f(X)\lambda + g(X)$, and express the solution as a curve $\bar{\sigma}: \mathbb{R} \to X^\nu$ parametrized by λ.
It is easy to check that the parametrized curves in the same equivalence class are obtained one from another by a change of parameter $\lambda \to \lambda' = f(X)\lambda + g(X)$. Indeed let $\bar{\sigma}_1$ be the curve parametrized by $\lambda' = x^0$, then $\mathrm{d}\bar{\sigma}_1(\lambda')/\mathrm{d}\lambda' = V(\bar{\sigma}_1(\lambda'))$. Set $\bar{\sigma}_1(\lambda') = \bar{\sigma}_1(f(X)\lambda + g(X)) \overset{\text{def}}{=} \bar{\sigma}(\lambda)$, then

$$\frac{\mathrm{d}\bar{\sigma}}{\mathrm{d}\lambda} = \frac{\mathrm{d}\bar{\sigma}_1}{\mathrm{d}\lambda'}\frac{\mathrm{d}\lambda'}{\mathrm{d}\lambda} = f(\bar{\sigma}(\lambda))\,V(\bar{\sigma}(\lambda)).$$

b) Under the group of transformations generated by $\boldsymbol{V}$, $X \mapsto X'$ where $X' = \bar{\sigma}(\lambda, X)$ is a point on the trajectory $\bar{\sigma}(\lambda)$ of the Hamilton vector field $\boldsymbol{V}$ going through X. $\mathbf{d}\boldsymbol{\theta}$ is invariant under the mapping $\bar{\sigma}(\lambda, .)$ if $\mathbf{d}\boldsymbol{\theta}_X = \bar{\sigma}^*(\lambda, .)\,\mathbf{d}\boldsymbol{\theta}_{X'}$. By the theorem given on p. 261, a form is invariant with respect to its characteristic system, thus $\boldsymbol{\theta}$, $\mathbf{d}\boldsymbol{F}$, $\mathbf{d}\boldsymbol{\theta}$ are invariant with respect to the group of transformations generated by $\boldsymbol{V}$. It is also easy to check directly that $\mathrm{i}_V\,\mathbf{d}\boldsymbol{\theta} = 0$:

$$\mathrm{i}_V\,\mathbf{d}\boldsymbol{\theta} = V^0\left(\frac{\partial H}{\partial x^0} + \frac{\partial H}{\partial y_i}\frac{\partial H}{\partial x^i} - \frac{\partial H}{\partial x^i}\frac{\partial H}{\partial y_i} - \frac{\partial H}{\partial x^0}\right)\mathbf{d}\boldsymbol{x}^0 \equiv 0,$$

$\mathrm{i}_V\boldsymbol{\theta}$ and $\mathrm{i}_V\,\mathbf{d}\boldsymbol{F}$ are identically zero.

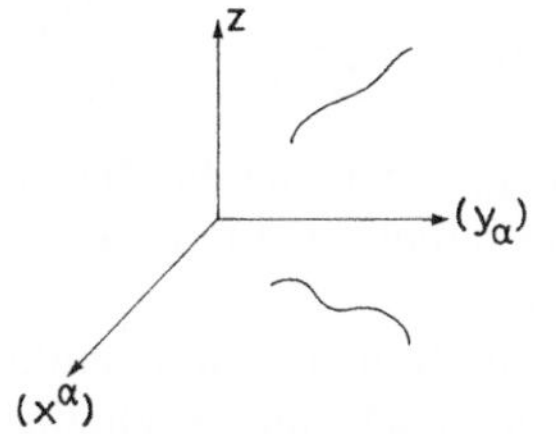

c) The Cauchy data $(\mathbb{C}, g)$ define an n dimensional space in $\mathbb{X}^\nu$

$$g: \mathbb{C} \to \mathbb{X}^\nu \text{ by } (u^i) \mapsto g(u) = (g^\alpha(u), g_\alpha(u), g^z(u))$$

where the components $g_\alpha(u)$ have been obtained by solving the system of algebraic equations (assumed soluble)

$$\begin{cases} g_0(u) + H(g^\alpha(u), g_i(u)) = 0 \\ \partial g^z(u)/\partial u^i - g_\alpha(u)\partial g^\alpha(u)/\partial u^i = 0. \end{cases}$$

The unknown function $\bar{S}$ must contain the initial manifold, i.e.

$$\bar{S}(x^\alpha) = g^z(u) \quad \text{for } x^\alpha = g^\alpha(u).$$

Let $\bar{\sigma}(\lambda, u)$ be the solution of the Hamilton equations satisfying the initial conditions

$$\bar{\sigma}(0, u) = g(u).$$

The integral manifold $(\mathbb{L}, f)$ for the Cauchy data $(\mathbb{C}, g)$ is defined by the $(n+1)$ manifold $\mathbb{L} = \mathbb{R}^n \times \mathbb{R}$ and the mapping

$$f: \mathbb{L} \to \mathbb{X}^\nu \quad \text{by} \quad (u^i, \lambda) = \bar{\sigma}(\lambda, u).$$

The integral manifold is generated from the initial manifold by the trajectories of the Hamilton equations.

On $f(\mathbb{L})$, $y_\alpha \, \mathbf{d}x^\alpha = \mathbf{d}z$, hence $\mathbf{d}\boldsymbol{\Omega} = \mathbf{d}^2 z = 0$, and $\boldsymbol{\Omega}$ is closed.

Let $\bar{f}(\mathbb{L})$ be the projection of $f(\mathbb{L})$ on $T^*\mathbb{W}$; $\bar{f}(\mathbb{L})$ is an $(n+1)$ manifold contained in the $2(n+1)$ manifold $T^*\mathbb{W}$ such that the symplectic form $\mathbf{d}\boldsymbol{\Omega} = \Sigma \, \mathbf{d}y_\alpha \wedge \mathbf{d}x^\alpha$ is zero on $\bar{f}(\mathbb{L})$. Such a manifold is called a **lagrangian manifold**.

lagrangian manifold

$\boldsymbol{\Omega}$ is an exact form $\mathbf{d}\bar{S}$ when $y_\alpha = \partial \bar{S}(x^\alpha)/\partial x^\alpha$. On $f(\mathbb{L})$, y_α and x^α are solutions of the Hamilton equations

$$\mathrm{d}x^0 = \frac{\mathrm{d}x^i}{\partial H/\partial y_i} = -\frac{\mathrm{d}y_\alpha}{\partial H/\partial x^\alpha}.$$

Let $x^\alpha = \bar{\sigma}^\alpha(\lambda, u)$ and $y_\alpha = \bar{\sigma}_\alpha(\lambda, u)$ be the solution of the Hamilton equations satisfying the initial conditions $\bar{\sigma}(0, u) = g(u)$. $\boldsymbol{\Omega}$ is an exact form on $f(\mathbb{L})$ if and only if

1) there exist global solutions of the Hamilton equations,
2) the $n+1$ variables (λ, u) can be eliminated between the $2n+2$ equations $x^\alpha = \bar{\sigma}^\alpha(\lambda, u)$, $y_\alpha = \bar{\sigma}_\alpha(\lambda, u)$ giving y_α as a function of x^α.

This last condition is satisfied only if the jacobian matrix

$$\frac{\partial(\bar{\sigma}^\alpha, \bar{\sigma}_\alpha)}{\partial(\lambda, u)} \text{ is of rank } n+1.$$

PROBLEM 5. FIRST ORDER PARTIAL DIFFERENTIAL EQUATION; CATASTROPHES: SOAP FILM PROBLEM (see Problems II 3 and IV 4)

In Problem II 3 *we found the area of a soap film held by two wire circles of radius r and R, at a distance l apart, by finding the critical value of S*

$$S: X \to \mathbf{R} \text{ by } S(q) = 2\pi \int_0^l q(x)(1+\dot{q}(x)^2)^{1/2}\,dx$$

$$= \int_0^l L(q(x), \dot{q}(x))\,dx$$

where $q \in X$ *is a* C^1 *function on the interval* $(0, l)$ *such that* $q(0) = r$, $q(l) = R$. *We shall now compute the area* $\bar{S}$ *as a function of R and l by the method of the previous problem and study the global properties of the integral manifold.*

Answer: a) *Hamilton–Jacobi equation*: We first compute the Hamilton–Jacobi equation for the function $\bar{S}(l, R)$; we have

$$\frac{\partial L}{\partial \dot{q}(x)} = \frac{2\pi\dot{q}(x)q(x)}{(1+\dot{q}^2(x))^{1/2}} = p(x)$$

$$\dot{q}(x) = p(x)/(4\pi^2 q^2(x) - p^2(x))^{1/2}$$

$$H(q(x), p(x)) = -(4\pi^2 q^2(x) - p^2(x))^{1/2}$$

and the Hamilton–Jacobi equation

$$\frac{\partial \bar{S}}{\partial l} + H\left(R, \frac{\partial \bar{S}}{\partial R}\right) = \frac{\partial \bar{S}}{\partial l} - \left(4\pi^2 R^2 - \left(\frac{\partial \bar{S}}{\partial R}\right)^2\right)^{1/2} = 0.$$

Set $\bar{S}(l, R) = z$, $\partial \bar{S}/\partial l = y_l$, $\partial \bar{S}/\partial R = y_R$. We note that $\partial H/\partial l = 0$ hence $H(R, y_R)$ is a first integral; set $H(R, y_R) = -\alpha$.

b) *Characteristic system C*: It is readily written down; using the first integral one has

$$C = \begin{cases} y_l - \alpha = 0 \\ \alpha\dot{q}(l) = p(l), \quad \alpha\dot{p}(l) = 4\pi^2 q(l), \quad \alpha\dot{z}(l) = 4\pi^2 q^2(l) \end{cases}$$

where $R = q(l)$, $y_R = p(l)$.

Eliminating $\dot{p}(l)$, we obtain

$$\alpha^2\ddot{q}(l) - 4\pi^2 q(l) = 0$$

$$q(l) = \frac{\alpha}{2\pi}\cosh\frac{2\pi}{\alpha}(l - D).$$

The solutions of the Hamilton equations are indeed the solutions of the Euler–Lagrange equation. Set $\alpha = 2\pi C$ for quick comparisons with the results of Problem II 3.
We can easily complete the integration of the characteristic system:

$$R = q(l) = C\cosh\frac{l - D}{C} \overset{\text{def}}{=} h(l, D, C)$$

$$y_R = p(l) = 2\pi C\sinh\frac{l - D}{C} \overset{\text{def}}{=} k(l, D, C)$$

$$z = z_0 + 2\pi C\int_0^l \cosh^2\frac{t - D}{C}\,\mathrm{d}t$$

$$= z_0 + \pi C^2\left(\sinh\frac{l - D}{C}\cosh\frac{l - D}{C} + \frac{l - D}{C} + \sinh\frac{D}{C}\cosh\frac{D}{C} + \frac{D}{C}\right)$$

$$\overset{\text{def}}{=} f(l, D, C, z_0).$$

c) *Cauchy problem*: We shall use the same notation as in the previous problem. We want to find the integral manifold $f(\mathbb{L})$ such that $g(\mathbb{C}) \subset f(\mathbb{L}) \subset \mathbb{X}^\nu$. Here $\mathbb{C}$ is one dimensional; the Cauchy data is a curve $g: \mathbb{C} \to \mathbb{X}^\nu$ by

$$u \mapsto \{l = g^l(u),\quad R = g^R(u),\quad y_l = 2\pi C,\quad y_R = g_R(u),\quad z = g^z(u)\}.$$

We can choose arbitrarily $g^l(u)$, $g^R(u)$; $g^z(u)$, we obtain C and $g_R(u)$ by solving the algebraic equations:

$$\partial g^z/\partial u - 2\pi C\,\partial g^l/\partial u - g_R(u)\,\partial g^R/\partial u = 0$$

and

$$4\pi^2 C^2 = 4\pi^2(g^R(u))^2 - (g_R(u))^2,\quad C \geq 0.$$

The constants of integration C, D, z_0 of the characteristic system are determined by the requirement that $f(\mathbb{L})$ contains $g(\mathbb{C})$, namely

$$g^R(u) = h(g^l(u), D, C)$$

$$g_R(u) = k(g^l(u), D, C)$$

$$g^z(u) = f(g^l(u), D, C, z_0).$$

This system of 3 equations for the 3 unknowns D, C, z_0 is soluble only if $g^R(u) \geq R_0(g^l(u))$ where R_0 is the function determined in Problem II 3; another calculation of R_0 is given in paragraph d below.

In practice, we may be interested in the surface area $\bar{S}(R, l)$ of a soap film held by a fixed wire circle of radius r and a wire circle of variable radius R at a variable distance l from the r-circle. The integral manifold Σ is then defined by

$$z = f(l, D, C, z_0)$$
$$R = h(l, D, C)$$
$$l = l$$

together with $r = h(0, D, C)$ and $z_0 = \pi|r^2 - R^2|$. Σ contains the following Cauchy data

$$u \mapsto \{l = g^l(u), \quad R = g^R(u), \quad y_l = 2\pi C, \quad y_R = g_R(u),$$
$$z = f(g^l(u), D, C, z_0), \quad r = h(0, D, C) \quad \text{and} \quad z_0 = \pi|r^2 - R^2|\},$$

i.e. we can still choose arbitrarily $g^l(u)$, and $g^R(u)$ but $g^z(u)$ is determined by the requirement that $g(u) \in \Sigma$.

d) *Global properties of Σ*: In Problem II 3 we showed that for a given set (l, R, r) the Euler–Lagrange equation $S'(q_0) = 0$ has two solutions q_0^1 and q_0^2 if R is larger than a certain function R_0 of l, one solution q_0^0 if $R = R_0(l)$, and no solution if $R < R_0(l)$. On the other hand,

$$S(q_0) = f(l, D, C, \pi|r^2 - R^2|) = z.$$

If we were to write $z = \bar{S}(l, R)$, $\bar{S}$ would not be a function in the sense we give to that word since it is double valued for $R > R_0(l)$. We note that on the curve $\sigma(l) = (l, R_0(l), \bar{S}(l, R_0(l))$ where $\bar{S}$ is single valued, the tangent space to the manifold Σ is vertical (parallel to the z axis).

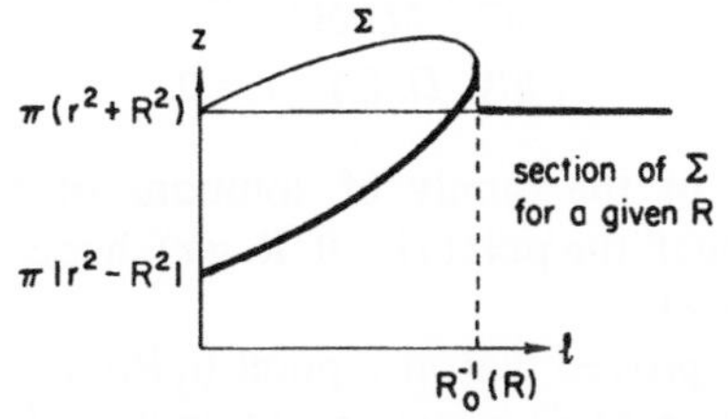

The lower sheet of Σ is the "physical" sheet, it gives the value of the area of the soap bubble generated by the catenary 1 of the first figure of Problem II 3. For a given R, the area increases with l until the breaking value $l = R_0^{-1}(R)$. The area $\pi(r^2 + R^2)$ of the two discs after the break is less than the area of the film before the break. The discontinuity of the area as l varies continuously is a catastrophe. The upper sheet of Σ has

no physical interpretation per se but it has an important physical consequence: the soap films breaks when the upper sheet coalesces with the lower sheet.
We shall compute $R_0(l)$ by computing $\sigma(l)$ as follows. Let χ be the projection map $(l, R, z) \mapsto (l, R)$; its derivative $\chi'(l, R, z)$ maps $T_{(l,R,z)}\Sigma$ into $\mathbb{R}^2$.

$$T_{(l,R,z)}\Sigma \text{ vertical} \quad \Leftrightarrow \quad \chi'(l, R, z) \text{ not one–one } (\chi \text{ singular})$$

($\chi'(l, R, z)$ maps several vectors in $T_{(l,R,z)}\Sigma$ into the same vector in $\mathbb{R}^2$). Hence the curve $\sigma(l)$ is given by the singular points of the projection map χ i.e. the points where χ' is not one–one.
Consider the mapping $G \circ F$ which coordinatizes Σ

$$(C, l) \overset{F}{\mapsto} (D, C, l) \overset{G}{\mapsto} (z, R, l) \overset{\chi}{\mapsto} (R, l).$$

χ is singular if and only if $(\chi \circ G \circ F)'(C, l)$ is of rank less than 2; i.e. if and only if its determinant vanishes:

$$\det \frac{\partial(R, l)}{\partial(C, l)} = \frac{\partial h}{\partial C} + \frac{\partial h}{\partial D}\frac{\partial D}{\partial C}$$
$$= \sinh \frac{l-D}{C}\left(\operatorname{cotgh}\frac{l-D}{C} - \frac{l-D}{C} + \operatorname{cotgh}\frac{D}{C} - \frac{D}{C}\right) = 0$$

where $D = C \operatorname{arg\,cosh}(r/C)$.

The curve $R = R_0(l)$ is given by the set of equations

$$\begin{cases} R - h(l, D, C) = 0 \\ \dfrac{\partial h}{\partial C} + \dfrac{\partial h}{\partial D}\dfrac{\partial D}{\partial C} = 0 \\ h(0, D, C) - r = 0. \end{cases}$$

It is the envelope of the family of solutions of the Euler–Lagrange equation going through the point $(l = 0, R = r)$, hence the points $(l, R_0(l))$ are conjugate to $(0, r)$.
In Problem II 3 we proved that, if a point $(l, R_0(l))$ is conjugate to $(0, r)$ along a curve q_0^0, then $S(q_0^0)$ is not an extremum. Here we have proved that the singular values $\chi(\sigma(l))$ of χ are the conjugate points of $(0, r)$. Hence the singular values of χ determine solutions of the Euler–Lagrange equation which do not extremise S, i.e. they determine degenerate critical values of S.

e) *Catastrophes*: A system whose behavior is determined by the critical values of a function S is called a gradient system; for instance, a system whose equilibrium position minimizes the potential energy, or the soap bubble whose area is the critical value of S. The Thom theory of catastrophes analyzes the discontinuous behavior of a gradient system which is caused by continuous changes in the factors which control the system.

Here continuous changes in R and l bring a discontinuous change in the film area when the point (l, R) moves from the domain $R > R_0(l)$ into the domain $R < R_0(l)$. The $\mathbb{R}^2$ space of pairs (l, R) is called the control space. The space $\mathbb{Q}$ of C^1 paths q mapping the interval $(0, l)$ into $\mathbb{R}$ such that $q(0) = r, q(l) = R$ is called the behaviour space or the state space.

Catastrophes are jumps from a critical value of S to another one characterized by the singularities of the projection map $\chi: \Sigma \to$ control space. In general let f be a smooth function on the product space, control space $\times$ behaviour space

$$f: \mathbb{R}^k \times \mathbb{Q} \to \mathbb{R}$$

such that the behaviour of the system is a solution of

$$f'_{q_0}(x, q_0) = 0 \text{ where } x \in \mathbb{R}^k.$$

It can be shown that the equation $f'_{q_0}(x, q_0) = 0$ determines a k-dimensional manifold, say Σ. Catastrophes are the singularities of the projection map $\chi: \Sigma \to \mathbb{R}^k$. The nature of the catastrophes, if any, of a gradient system is determined by the dimension of its control space. This explains the variety of applications of catastrophe theory. The complexity of the system, i.e. the dimension of its state space $\mathbb{Q}$, infinite for instance in the case of lagrangian systems, does not affect the nature of the catastrophes.

Reference: R. Thom, Morphogenesis and Structural Stability (Benjamin, 1972).

PROBLEM 6. DARBOUX THEOREM

a) *Let $\boldsymbol{F}$ be a skew symmetric 2-form on an m dimensional vector space E. Show that $\boldsymbol{F}$ can have rank m only if m is even. Show that if $m = 2n$ and $\boldsymbol{F}$ has rank $2n$ it can be written, in a convenient basis on E as*

$$(F_{ij}) = \begin{pmatrix} 0 & I \\ -I & 0 \end{pmatrix}$$

where I is in the $n \times n$ unit matrix.

b) *Let $\boldsymbol{F}$ be a symplectic form on a smooth manifold X. Prove that for each $x_0 \in X$ there is a local coordinate system about x_0 in which $\boldsymbol{F}$ is constant.*

Answer: a) Elementary algebra.

b) Consider a coordinate system about x_0; denote by $\bar{\boldsymbol{\omega}}$ the expression of a form $\boldsymbol{\omega}$ in these coordinates. Denote by $\boldsymbol{F}_1$ the form which, in these coordinates, is constant and such that $\bar{\boldsymbol{F}}_1 = \bar{\boldsymbol{F}}(x_0)$. We shall construct a change of coordinates $f_1: \tilde{x} \mapsto x$ such that $f_1^* \bar{\boldsymbol{F}}_1 = \bar{\boldsymbol{F}}$, i.e. the expression $\tilde{\boldsymbol{F}}$ of $\boldsymbol{F}$ in the new coordinates is equal to $\bar{\boldsymbol{F}}_1$.

Let $\bar{\boldsymbol{F}}_t = \bar{\boldsymbol{F}} + t(\bar{\boldsymbol{F}}_1 - \bar{\boldsymbol{F}})$. $\boldsymbol{F}_t$ is non degenerate on some neighborhood of the origin for all $t \in [0, 1]$. By Poincaré lemma, we can set $\bar{\boldsymbol{F}}_1 - \bar{\boldsymbol{F}} = \mathbf{d}\bar{\boldsymbol{\omega}}$ for some 1-form $\boldsymbol{\omega}$ on this neighborhood. Assume $\boldsymbol{\omega}(x_0) = 0$.

Let $\boldsymbol{v}_t$ be the vector field defined by $\mathrm{i}(v_t)\boldsymbol{F}_t = -\boldsymbol{\omega}$. It is indeed possible to define a vector field by this equation because the mapping $\boldsymbol{v}_t \mapsto \mathbf{i}(\boldsymbol{v}_t)\boldsymbol{F}_t$ is an isomorphism (linear mapping by a non degenerate form). Let f_t be the local flow of $\boldsymbol{v}_t$ at the point x_0.
We shall prove first an interesting property of the flow f_t of a t-independent vector field $\boldsymbol{v}$. Namely

$$(\mathrm{d}/\mathrm{d}t) f_t^* \boldsymbol{\alpha} = f^*(\mathcal{L}_v \boldsymbol{\alpha})$$

where $\boldsymbol{\alpha}$ is an arbitrary form on X. Indeed

$$(f_t^* \boldsymbol{\alpha})(\boldsymbol{w}_1, \boldsymbol{w}_2, \ldots \boldsymbol{w}_k)(x) = \boldsymbol{\alpha}(\boldsymbol{u}_1, \boldsymbol{u}_2, \ldots \boldsymbol{u}_k)(f_t(x))$$

with $\boldsymbol{u}_j = f_t' \boldsymbol{w}_j$. Set $f_t(x) = f(t, x)$, then $u^i = (\partial f^i(t, x)/\partial x^j) w^j$. From the flow equation $\mathrm{d}f^i/\mathrm{d}t = v^i(f(t, x))$, we obtain

$$\frac{\mathrm{d}}{\mathrm{d}t}(\alpha_{i_1 \ldots i_k}(f(t, x)) u_1^{i_1} \ldots u_k^{i_k}) = f^*(\mathcal{L}_v \boldsymbol{\alpha})(\boldsymbol{w}_1, \ldots \boldsymbol{w}_k). \quad \blacksquare$$

Here the vector field is t-dependent, thus

$$\frac{\mathrm{d}}{\mathrm{d}t}(f_t^* \boldsymbol{F}_t) = f_t^*(\mathcal{L}_{v_t} \boldsymbol{F}_t) + f_t^* \frac{\mathrm{d}\boldsymbol{F}_t}{\mathrm{d}t}$$

$$= f_t^* \,\mathrm{d}\mathrm{i}(\boldsymbol{v}_t)\boldsymbol{F}_t + f_t^*(\boldsymbol{F}_1 - \boldsymbol{F})$$

$$= f_t^*(-\mathrm{d}\boldsymbol{\omega}) + f_t^*(\mathrm{d}\boldsymbol{\omega}) = 0.$$

Therefore $f_1^* \bar{\boldsymbol{F}}_1 = f_0^* \bar{\boldsymbol{F}} = \bar{\boldsymbol{F}}$. The mapping f_1 is the change of coordinates transforming $\bar{\boldsymbol{F}}$ into the constant form $\bar{\boldsymbol{F}}_1$.
When X is finite dimensional, the Darboux theorem says that locally about

each point there are coordinates $(x^1, \dots x^n, y_1, \dots y_n)$ such that $\boldsymbol{F} = \sum_{i=1}^n \mathbf{d}y_i \wedge \mathbf{d}x^i$.

References: See for instance P.R. Chernoff and J.E. Marsden, *Properties of Infinite Dimensional Hamiltonian Systems*, Lecture Notes in Mathematics No. 425 (Springer Verlag, 1974). N.M.J. Woodhouse, *Lectures on Geometric Quantization*, Lecture Notes in Physics No. 53 (Springer Verlag, 1976).

PROBLEM 7. TIME DEPENDENT HAMILTONIANS

Let T^*M, with symplectic structure $\boldsymbol{F} = \mathbf{d}q^i \wedge \mathbf{d}p_i$ be the phase space of a dynamical system whose hamiltonian H is time dependent. A trajectory of the dynamical system is now considered as a trajectory of the vector field $\boldsymbol{w}$ on $T^*M \times \mathbb{R}$ with components in canonical coordinates $(\partial H/\partial p_i, -\partial H/\partial q^i, 1)$.
Let f be a symplectic diffeomorphism on T^*M. Show that the diffeomorphism

$$(f, \mathrm{Id}): T^*M \times \mathbb{R} \to T^*M \times \mathbb{R}$$

leaves the trajectories of the dynamical system invariant.

Answer: It is straightforward, using the product structure to prove that (f, Id) leaves invariant the vector field $\boldsymbol{w}$.

V. RIEMANNIAN MANIFOLDS. KÄHLERIAN MANIFOLDS

A. THE RIEMANNIAN STRUCTURE

1. PRELIMINARIES[1]

A **riemannian manifold** is a smooth manifold X together with a continuous 2-covariant tensor field $\boldsymbol{g}$, called the **metric tensor**, such that

riemannian manifold
metric tensor

1) $\boldsymbol{g}$ is symmetric
2) for each $x \in X$, the bilinear form $\boldsymbol{g}_x$ is non-degenerate; since X is finite dimensional in this chapter this means $\boldsymbol{g}_x(\boldsymbol{v}, \boldsymbol{w}) = 0$ for all $\boldsymbol{v} \in T_x$ if and only if $\boldsymbol{w} = 0$.

Such a manifold is said to possess a riemannian structure. A riemannian manifold [riemannian structure] is called **proper** if

proper

$$\boldsymbol{g}_x(\boldsymbol{v}, \boldsymbol{v}) > 0 \qquad \forall \boldsymbol{v} \in T_x, \quad \boldsymbol{v} \neq 0, \quad x \in X.$$

Otherwise the manifold is called **pseudo-riemannian** or is said to possess an **indefinite metric**. In terms of a moving frame $(\boldsymbol{\theta}^i)$ (p. 134) the tensor $\boldsymbol{g}$, sometimes denoted $\mathrm{d}s^2$, is written

pseudo-riemannian
indefinite metric

$$\boldsymbol{g} = \mathrm{d}s^2 = g_{ij}\boldsymbol{\theta}^i\boldsymbol{\theta}^j.$$

The notation $\boldsymbol{\theta}^i\boldsymbol{\theta}^j$ is used rather than $\boldsymbol{\theta}^i \otimes \boldsymbol{\theta}^j$ because g is symmetric, i.e.

$$\boldsymbol{\theta}^i\boldsymbol{\theta}^j = \tfrac{1}{2}(\boldsymbol{\theta}^i \otimes \boldsymbol{\theta}^j + \boldsymbol{\theta}^j \otimes \boldsymbol{\theta}^i).$$

The tensor $\boldsymbol{g}$ endows each vector space T_x with an **inner product (scalar product)** $(\boldsymbol{v}|\boldsymbol{w})$ defined by

inner scalar product

$$(\boldsymbol{v}|\boldsymbol{w}) = \boldsymbol{g}_x(\boldsymbol{v}, \boldsymbol{w}) \qquad \forall\, \boldsymbol{v}, \boldsymbol{w} \in T_x.$$

It follows that

$$g_{ij} = (\boldsymbol{e}_i|\boldsymbol{e}_j)$$

if $(\boldsymbol{e}_i)$ is the basis dual to $(\boldsymbol{\theta}^i)$.

The norm $\|\boldsymbol{v}\|$ of a vector $\boldsymbol{v} \in T_x$ is defined by

$$\|\boldsymbol{v}\|^2 = \boldsymbol{g}_x(\boldsymbol{v}, \boldsymbol{v}) = g_{ij}v^iv^j.$$

[1]In this chapter $T_x(X)$ is abbreviated to T_x. For alternate sign and index positioning conventions, see for example [Misner, Thorne and Wheeler 1973].

This is a norm in the usual sense in the case of a proper riemannian manifold. Otherwise $\|v\|$ may be a real or imaginary number. If $\|v\| = 0$ the vector v is called a **null (isotropic)** vector. At each point $x \in X$ the null vectors form a cone in T_x called the **null cone (light cone)**; the null cone consists of all vectors v such that

isotropic (null) vector
light (null) cone

$$g_{ij}v^i v^j = 0.$$

A null vector is orthogonal to itself.

canonical isomorphism

An inner product on any vector space defines a **canonical isomorphism** between the space and its dual. For a fixed u the mapping

$$(u|\cdot): T_x \to \mathbb{R} \qquad \text{by } v \mapsto (u|v)$$

is an element of T_x^*. The canonical isomorphism is the mapping

$$T_x \to T_x^* \qquad \text{by } u \mapsto (u|\cdot) \equiv u^*.$$

In coordinate language

$$(u|v) = g_{ij}u^i v^j.$$

Therefore

$$u^* = g_{ij}u^i \theta^j.$$

In general the same symbol u is used to denote both the contravariant vector u and the covariant vector u^* which is the image of u under the canonical isomorphism. The u^i are called the **contravariant components** of u; the u_i are called the **covariant components** of u; they are related by

contravariant, covariant components

$$u_i = g_{ij}u^j \qquad u^i = g^{ij}u_j$$

where the g^{ij} are the elements of the inverse of the matrix $[g_{ij}]$

$$g^{ij}g_{jk} = \delta^i_k.$$

inner product on T^*

An **inner product** is defined **on T^*** by means of the canonical isomorphism

$$(u^*|v^*) = (u|v)$$
$$= g^{ik}u_i v_k.$$

Similarly there are canonical isomorphisms between the spaces $\otimes^p T_x$, $\otimes^p T_x^*$, $\otimes^q T_x \otimes^{p-q} T_x^*$, etc. The contravariant and covariant components of the tensor t are related by

$$t_{i_1 \ldots i_p} = g_{i_1 j_1} g_{i_2 j_2} \ldots g_{i_p j_p} t^{j_1 \ldots j_p}$$

$$t^{i_1 \ldots i_p} = g^{i_1 j_1} \ldots g^{i_p j_p} t_{j_1 \ldots j_p}.$$

raising and lowering indices

The g^{ij} are the contravariant components of the tensor g. One says that the indices are **raised** and **lowered** by means of the tensor g. The **mixed**

components of a tensor are obtained in the same way, for example, — mixed components

$$t^i{}_j = g^{ik}t_{kj}.$$

With a suitable choice of basis $(e_{i'})$ for T_x the quadratic form $g_x(v, v) = g_{ij}v^iv^j$ can be written as a sum of k positive and $(n-k)$ negative squares

$$g_{ij}v^iv^j = g_{i'j'}v^{i'}v^{j'} = \sum_{i'=1}^{k}(v^{i'})^2 - \sum_{i'=k+1}^{n}(v^{i'})^2.$$

The number k is called the **index** of the quadratic form, it does not depend on the basis (e_i); the number $k-(n-k)$ is called the **signature**[1] of the quadratic form. In terms of the basis $(\theta^{i'})$ dual to $(e_{i'})$ — index, signature

$$g_x = \sum_{i'=1}^{k}\theta^{i'}\theta^{i'} - \sum_{i'=k+1}^{n}\theta^{i'}\theta^{i'}.$$

Since g is continuous the index of $g_x(v, v)$ will be the same at each point x of a connected manifold. One therefore speaks of the signature or the index of the riemannian manifold.

The index of a proper riemannian manifold X^n is n. On such a manifold a basis (frame) (e_i) is called **orthonormal** if — orthonormal

$$(e_i|e_j) = \delta_{ij}.$$

There is a bijection between the set of orthonormal frames in T_x and the group $O(n)$ of orthogonal $n\times n$ matrices. In other words given any orthonormal frame (e_i) any other frame $(e_{i'})$ can be obtained from it by means of an orthogonal matrix

$$e_{i'} = a^j_{i'}e_j.$$

The space $O(X)$ of pairs (x, τ_x), where τ_x is an orthonormal frame of T_x, can be given the structure of a principal fibre bundle with typical fibre and structural group $O(n)$. It is a reduction of the principal fibre bundle of frames (p. 131). — $O(X)$

Exercise: Spell out the topology and the atlas on $O(X^n)$.

We now restrict our attention to the case of fundamental interest in physics, the 4-dimensional pseudo-riemannian manifold of index 1. Such a manifold is called a **hyperbolic manifold**. The metric is called **lorentzian**. It is conventional to label the coordinates with greek indices which take the values 0, 1, 2, 3; latin indices take the values 1, 2, 3. A basis (e_α) on a — hyperbolic manifold, lorentzian metric

[1]These two terms are sometimes defined in a slightly different way.

orthonormal

hyperbolic manifold is called orthonormal if

$$(e_\alpha | e_\beta) = 0 \qquad \alpha \neq \beta$$
$$(e_0 | e_0) = 1 \qquad (e_i | e_i) = -1.$$

In terms of an orthonormal basis (θ^α)

$$g = \theta^0 \theta^0 - \sum_{i=1}^{3} \theta^i \theta^i.$$

With respect to an orthonormal frame the equation for the null cone C_x in T_x is

$$(v^0)^2 - \sum_{i=1}^{3} (v^i)^2 = 0.$$

A vector $w \in T_x$ such that

$$g_{\alpha\beta} w^\alpha w^\beta < 0,$$

spacelike vector

i.e. outside the null cone, is called **spacelike**. A vector $w \in T_x$ such that

$$g_{\alpha\beta} w^\alpha w^\beta > 0,$$

timelike vector

i.e. inside the null cone, is called **timelike**.

future directed

past directed

time-orientable

The null cone C_x is made up of two half-cones. If one of these half-cones is singled out and called the future half-cone C_x^+ then T_x is said to be time-oriented. A timelike vector inside C_x^+ is said to be **future directed**; a timelike vector inside the past null half-cone C_x^- is said to be **past directed**. If T_x can be time-oriented in a continuous fashion at each point $x \in X$, then X is said to be **time-orientable**. A hyperbolic manifold may be orientable but not time-orientable.

Example: Let V^3 be the region of E^3 such that $-a \le x^2 \le a$ with the points (x^0, x^1, a) and $(-x^0, -x^1, -a)$ identified. Let $X^4 = V^3 \times \mathbb{R}$. Let

$$g = \mathbf{d}x^0\, \mathbf{d}x^0 - \mathbf{d}x^1\, \mathbf{d}x^1 - \mathbf{d}x^2\, \mathbf{d}x^2 - \mathbf{d}x^3\, \mathbf{d}x^3.$$

The manifold X^4 is orientable but not time-orientable.

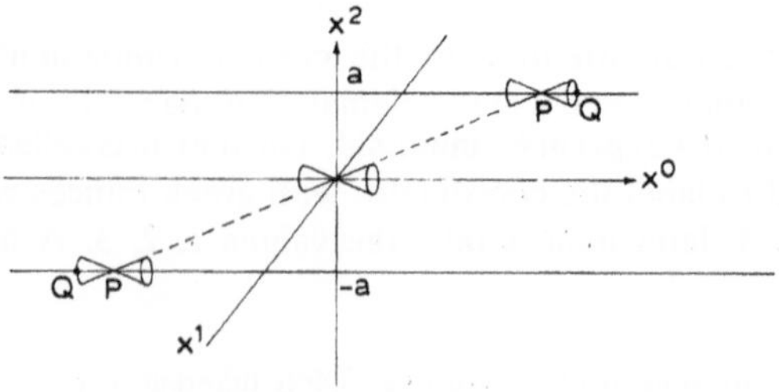

It is usual to choose a coordinate system $(x^\alpha) = (x^0, x^i)$ on a hyperbolic manifold in such a way that $\partial/\partial x^0$ is a timelike vector and the $\partial/\partial x^i$ are spacelike vectors.

Exercise: Show that the three following statements are equivalent
a) $\partial/\partial x^0$ is timelike and $\partial/\partial x^i$ is spacelike.
b) $g_{00} > 0$ and $g^{ij}v_iv_j$ is negative definite.
c) $g^{00} > 0$ and $g_{ij}v^iv^j$ is negative definite.

Theorem. Let $\boldsymbol{u}$ *be a timelike vector field such that*

$$g_{\alpha\beta}u^\alpha u^\beta = 1.$$

There exist, in a coordinate neighborhood U, three spacelike vector fields which together with $\boldsymbol{u}$ *form an orthonormal moving frame in U.*

Proof: Setting

$$\boldsymbol{\theta}^0 = \boldsymbol{u} = u_\alpha \, \mathbf{d}x^\alpha \text{ and } \gamma_{\alpha\beta} = g_{\alpha\beta} - u_\alpha u_\beta$$

one can write

$$ds^2 = g_{\alpha\beta} \, \mathbf{d}x^\alpha \, \mathbf{d}x^\beta = \boldsymbol{\theta}^0\boldsymbol{\theta}^0 + \gamma_{\alpha\beta} \, \mathbf{d}x^\alpha \, \mathbf{d}x^\beta.$$

Since the manifold is hyperbolic the quadratic form $\gamma_{\alpha\beta}v^\alpha v^\beta$ can be decomposed into a sum of three negative squares.
Therefore $\gamma_{\alpha\beta} \, \mathbf{d}x^\alpha \, \mathbf{d}x^\beta$ can be written

$$\gamma_{\alpha\beta} \, \mathbf{d}x^\alpha \, \mathbf{d}x^\beta = -\sum_{i=1}^{3} \boldsymbol{\theta}^i\boldsymbol{\theta}^i.$$

The set $(\boldsymbol{\theta}^0, \boldsymbol{\theta}^i)$ is thus an orthonormal moving frame (p. 134). ■

Exercise: Show that if $\boldsymbol{u} = \frac{1}{U}\frac{\partial}{\partial x^0}$, where $U^2 = g_{00}$, then

$$\boldsymbol{\theta}^0 = U \, \mathbf{d}x^0 + \frac{g_{i0}}{U} \, \mathbf{d}x^i$$

$$\gamma_{\alpha\beta} \, \mathbf{d}x^\alpha \, \mathbf{d}x^\beta = \left(-\frac{g_{i0}g_{j0}}{g_{00}} + g_{ij}\right) \mathbf{d}x^i \, \mathbf{d}x^j.$$

Answer: This follows from the fact that the components of $\boldsymbol{u}$ are then

$$u^0 = 1/U \qquad u^i = 0$$
$$u_0 = g_{00}u^0 = U \qquad u_i = g_{i0}u^0 = g_{i0}/U.$$ ■

Let $(\boldsymbol{e}_\alpha)$ be an orthonormal frame. We denote by $\eta^{\alpha\beta}$ the components of

g_x with respect to this frame. Thus

$$[\eta_{\alpha\beta}] = [\eta^{\alpha\beta}] = \begin{bmatrix} 1 & 0 & 0 & 0 \\ 0 & -1 & 0 & 0 \\ 0 & 0 & -1 & 0 \\ 0 & 0 & 0 & -1 \end{bmatrix}.$$

complete Lorentz group

The **complete Lorentz group**, denoted $L(4)$ or $O(1,3)$, is the group of all linear transformations L: $(e_\beta) \mapsto (a^\alpha{}_\beta e_\alpha)$ which leave the form $\eta_{\alpha\beta}v^\alpha v^\beta$ invariant, that is[1]

$$a^\alpha{}_\mu \eta_{\alpha\beta} a^\beta{}_\nu = \eta_{\mu\nu},$$

where the matrices $[\eta_{\alpha\beta}]$ and $[\eta^{\alpha\beta}]$ are used to raise and lower indices. It follows that

$$\det [a^\alpha{}_\beta] = \pm 1.$$

orthochronous Lorentz group

The Lorentz transformations for which $a^0{}_0 > 0$ form a group called the **orthochronous Lorentz group**. An orthochronous Lorentz transformation preserves the sense of time-like vectors.

proper Lorentz group

The Lorentz transformations for which $a^0{}_0 > 0$ and $\det [a^\alpha{}_\beta] = +1$ form a group called the **proper Lorentz group**, denoted L_0 or SO(1, 3).
The transformations of spatial reflection $s: (e_0, e_i) \mapsto (e_0, -e_i)$ and time reflection $t: (e_0, e_i) \mapsto (-e_0, e_i)$ have determinant -1. The complete Lorentz group is made up of the transformations L_0, sL_0, tL_0, and stL_0. The Lorentz group is a Lie subgroup of the Lie group GL(4), see pp. 173 and 174. The group manifold of $L(4)$ is not connected.

On a hyperbolic manifold there is a bijection between the set of orthonormal frames in T_x and the Lorentz group $L(4)$. In other words, given any orthonormal frame (e_α) any other frame $(e_{\beta'})$ can be obtained from it by means of a Lorentz transformation

$$e_{\beta'} = a^\alpha{}_{\beta'} e_\alpha.$$

$L(X)$

The space $L(X)$ of pairs (x, τ_x) where τ_x is an orthonormal frame of T_x, is a principal fibre bundle over X^4 with typical fibre and structural group $L(4)$. It is a reduction of the principal fibre bundle of frames (p. 131).

2. GEOMETRY OF SUBMANIFOLDS

Induced metric. Let $f: X^n \to Y^m$ be a differentiable mapping and let Y^m be a riemannian manifold with metric tensor g. The reciprocal image f^*g of g is a 2-covariant symmetric tensor on X^n defined by (p. 138)

$$(f^*g)_x\,(v, w) = g_y(f'v, f'w).$$

[1]Note that if one defines $a^{\alpha\rho} = \eta^{\rho\mu} a^\alpha{}_\mu$ and $a_{\alpha\nu} = \eta_{\alpha\beta} a^\beta{}_\nu$, then one can write the Lorentz transformation condition as $a^{\alpha\rho} a_{\alpha\nu} = \delta^\rho{}_\nu$.

The induced metric f^*g is sometimes called the **first fundamental form** of X^n. first fundamental form

Theorem. If Y^m is a proper riemannian manifold with metric g and if the mapping $f: X^n \to Y^m$ is an immersion (of rank n) then the tensor f^*g defines a proper riemannian structure on X^n.

Proof: (1) f^*g is clearly symmetric.
(2) f^*g is non-degenerate because the following statements are equivalent:

a) $(f^*g)_x(v, w) = 0$ for every v,
b) $g_y(f'v, f'w) = 0$ for every v,
c) $f'w = 0$,
d) $w = 0$, since f is of rank n.

(3) $(f^*g)_x(v, v)$ is positive definite because

$$g_y(f'v, f'v) > 0 \text{ for every } v$$

implies

$$(f^*g)_x(v, v) > 0 \text{ for every } v.$$

■

The theorem does not hold if Y^m is not proper riemannian[1], or if f is not an immersion.

Example: Let $f: \mathbb{R} \to \mathbb{R}^2$ by $x \mapsto (f^1(x), f^2(x))$. Let $g_y(w, w) = (w^1)^2 + (w^2)^2$. Then, for $w = f'v$, $w^1 = (\mathrm{d}f^1/\mathrm{d}x)v$ and $w^2 = (\mathrm{d}f^2/\mathrm{d}x)v$

$$\begin{aligned}(f^*g)_x(v, v) &= g_y(w, w) \\ &= v^2((\mathrm{d}f^1/\mathrm{d}x)^2 + (\mathrm{d}f^2/\mathrm{d}x)^2).\end{aligned}$$

The tensor f^*g is a metric on $\mathbb{R}$ only if there is no point at which both $\mathrm{d}f^1/\mathrm{d}x$ and $\mathrm{d}f^2/\mathrm{d}x$ are zero, that is only if f is of rank 1.

Example: Let the mapping $f: X^3 \to Y^4$ be given in terms of local coordinates by

$$(x^1, x^2, x^3) \mapsto (y^0, y^1, y^2, y^3) = (x^1, x^1, x^2, x^3).$$

Let

$$g_y(w, w) = (w^0)^2 - (w^1)^2 - (w^2)^2 - (w^3)^2.$$

Then

$$(f^*g)_x(v, v) = -(v^2)^2 - (v^3)^2$$

[1]For theorems valid if Y^m is not proper riemannian see for instance (Hawking, Ellis, pp. 44–45].

and

$$(f^*g)_x(u, v) = -u^2v^2 - u^3v^3.$$

The form $(f^*g)_x(u, v)$ is thus degenerate and therefore does not define a riemannian structure on X^3.

null submanifold

Let $W \subset X^4$ be a submanifold of the hyperbolic manifold X^4. Let i: $W \to X^4$ be the inclusion mapping. W is called a **null submanifold** of X^4 if the form i^*g is degenerate.

normal

Exercise: Let f: $X^{n-1} \to Y^n$ be an imbedding. The non-zero form $n \in T^*_{f(x)}(Y)$ such that $\langle n, f'v \rangle = 0$, for any vector $v \in T_x(X)$, defines a **normal** to $f(X)$ which is unique, up to a normalizing factor. Let g be a lorentzian metric on Y. Show[1] that

$$f^*g \text{ is lorentzian if } g^{\alpha\beta}n_\alpha n_\beta < 0$$

$$f^*g \text{ is degenerate if } g^{\alpha\beta}n_\alpha n_\beta = 0$$

$$f^*g \text{ is negative definite if } g^{\alpha\beta}n_\alpha n_\beta > 0.$$

Second fundamental form. See p. 312.

3. EXISTENCE OF A RIEMANNIAN STRUCTURE

Proper structure.

Theorem. A smooth (C^1) manifold X can be given a proper riemannian structure if and only if it is paracompact.

Proof: a) A riemannian manifold is a metric space (p. 326) and therefore (p. 16) paracompact. b) Let $\{\theta_i\}$ be a partition of unity (p. 214) subordinate to an atlas $\{(U_i, \varphi_i)\}$ on X. Let g_i be the 2-covariant tensor on U_i whose components with respect to the natural basis are

$$(g_i)_{hk} = \delta_{hk}.$$

Then the 2-covariant tensor $\Sigma_i\, \theta_i g_i$ defines a proper riemannian structure on X. This is because the sum of two positive definite quadratic forms is again a positive definite quadratic form. ■

This construction cannot be used to build a hyperbolic riemannian structure because the sum of two quadratic forms of index 1 need not have index 1.

[1]See for instance [Hawking, Ellis, p. 45] where other conventions are used.

Hyperbolic structure. A **line element (direction)** at $x \in X$ is a 1-dimensional vector subspace of T_x. line element

Theorem. On a paracompact C^1 manifold X the existence of a continuous line element field is equivalent to the existence of a hyperbolic riemannian structure on X.

Proof: (1) Let X be a manifold with a continuous line element field. Since X is paracompact it can be given a proper riemannian structure $\boldsymbol{g}$. Let $\boldsymbol{u} \in T_x$ be a unit vector in the line element at x; the vector $\boldsymbol{u}$ is uniquely determined up to a sign. Consider the quadratic form

$$\bar{g}_x(\boldsymbol{v}, \boldsymbol{v}) = 2(g_x(\boldsymbol{v}, \boldsymbol{u}))^2 - g_x(\boldsymbol{v}, \boldsymbol{v}) \qquad \boldsymbol{v} \in T_x.$$

In local coordinates

$$\bar{g}_x(\boldsymbol{v}, \boldsymbol{v}) = (2u_\alpha u_\beta - g_{\alpha\beta})v^\alpha v^\beta.$$

The form $\bar{g}_x(\boldsymbol{v}, \boldsymbol{v})$ is hyperbolic. This can be verified by choosing an orthonormal basis $(\boldsymbol{e}_\alpha)$ for T_x such that $\boldsymbol{e}_0 = \boldsymbol{u}$. Then

$$u^0 = u_0 = 1, \quad u^i = u_i = 0, \quad g_{\alpha\beta} = \delta_{\alpha\beta}$$

and

$$\bar{g}_x(\boldsymbol{v}, \boldsymbol{v}) = (v^0)^2 - (v^1)^2 - (v^2)^2 - (v^3)^2.$$

The corresponding tensor $\bar{\boldsymbol{g}}$ therefore defines a hyperbolic riemannian structure on X.
(2) Let X be a manifold with a hyperbolic riemannian structure $\bar{\boldsymbol{g}}$. Let $\boldsymbol{g}$ define a proper riemannian structure on X. At each point $x \in X$ one of the eigenvectors $\boldsymbol{u}$ of the eigenvalue problem $\bar{g}_{\alpha\beta}u^\beta = \lambda g_{\alpha\beta}u^\beta$ is timelike. This timelike eigenvector determines a continuous line element field on X. ∎

Theorem. (1) *Any non-compact paracompact C^1 manifold can be given a hyperbolic riemannian structure.*
(2) *A connected compact C^1 manifold can be given a hyperbolic riemannian structure if its Euler–Poincaré characteristic (p. 224) is zero.*
(3) *If a compact connected orientable manifold can be given a hyperbolic riemannian structure then its Euler–Poincaré characteristic is zero.*

Proof: The proof[1] is in terms of the existence of a continuous line element field.

[1]See for instance [Choquet-Bruhat 1968, p. 102], [Markus], [Samelson].

4. VOLUME ELEMENT. THE STAR OPERATOR

Throughout this section X^n will be an oriented (p. 212) riemannian manifold. Let g denote the determinant of the matrix $[g_{ij}]$, that is,

$$g = \epsilon^{i_1 \dots i_n}_{1 \dots n} g_{i_1 1} \dots g_{i_n n}.$$

With respect to a different basis $(\boldsymbol{e}_{i'})$, with

$$\boldsymbol{e}_{k'} = A^i{}_{k'} \boldsymbol{e}_i,$$

the components of $\boldsymbol{g}$ are

$$g_{k'l'} = A^i{}_{k'} A^j{}_{l'} g_{ij}.$$

Therefore

$$g' = \Delta^2 g$$

where Δ is the determinant of the matrix $[A^i{}_{j'}]$. It follows that under an orientation preserving transformation $\sqrt{|g|}$ transforms like the strict component of an n-form

$$\sqrt{|g'|} = |\Delta| \sqrt{|g|}$$

where $|\ |$ denotes the absolute value. Let

$$\begin{aligned} \boldsymbol{\tau} &= \frac{1}{n!} \tau_{i_1 \dots i_n} \mathbf{d}x^{i_1} \wedge \dots \wedge \mathbf{d}x^{i_n} \\ &= \sqrt{|g|}\, \mathbf{d}x^1 \wedge \dots \wedge \mathbf{d}x^n. \end{aligned}$$

Then

$$\tau_{i_1 \dots i_n} = \epsilon^{1 \dots n}_{i_1 \dots i_n} \tau_{1 \dots n} = \epsilon^{1 \dots n}_{i_1 \dots i_n} \sqrt{|g|}.$$

volume element — The n-form $\boldsymbol{\tau}$ is called the **volume element** of X^n. In terms of an orthonormal basis $(\boldsymbol{\theta}^i)$

$$\boldsymbol{\tau} = \boldsymbol{\theta}^1 \wedge \dots \wedge \boldsymbol{\theta}^n.$$

Exercise: Let $t_{1 \dots n}$ be the strict component of an n-form on X^n. Show that the tensor with components $t^{i_1 \dots i_n}$ is antisymmetric and has strict component

$$t^{1 \dots n} = \frac{1}{g} t_{1 \dots n}.$$

Infer that

$$\tau^{i_1 \dots i_n} = \frac{\operatorname{sign} g}{\sqrt{|g|}} \epsilon^{i_1 \dots i_n}_{1 \dots n}.$$

Answer:

$$\begin{aligned} t^{1 \dots n} &= g^{1 i_1} \dots g^{n i_n} t_{i_1 \dots i_n} = g^{1 i_1} \dots g^{n i_n} \epsilon^{1 \dots n}_{i_1 \dots i_n} t_{1 \dots n} \\ &= \frac{1}{g} t_{1 \dots n}. \end{aligned}$$

■

The inner product on T_x^* may be used to define an **inner product on** Λ_x^p as follows

inner product on Λ_x^p

$$(\mathbf{d}x^{i_1} \wedge \cdots \wedge \mathbf{d}x^{i_p} | \mathbf{d}x^{j_1} \wedge \cdots \wedge \mathbf{d}x^{j_p}) = \det [(\mathbf{d}x^{i_k} | \mathbf{d}x^{j_l})]$$
$$= \epsilon^{i_1 \ldots i_p}_{k_1 \ldots k_p} g^{j_1 k_1} \ldots g^{j_p k_p}.$$

By linearity, for

$$\alpha = \frac{1}{p!} \alpha_{i_1 \ldots i_p} \mathbf{d}x^{i_1} \wedge \ldots \wedge \mathbf{d}x^{i_p},$$

$$\beta = \frac{1}{p!} \beta_{i_1 \ldots i_p} \mathbf{d}x^{i_1} \wedge \ldots \wedge \mathbf{d}x^{i_p},$$

$$(\alpha | \beta) = \frac{1}{p!} \alpha_{i_1 \ldots i_p} \beta^{i_1 \ldots i_p} = \alpha_{I_1 \ldots I_p} \beta^{I_1 \ldots I_p}.$$

The **star operator** defines an isomorphism $\Lambda^p(X^n) \to \Lambda^{n-p}(X^n)$ by $\beta \mapsto * \beta$. The **dual** $* \beta$ of β denotes the unique $(n-p)$-form such that

star operator dual

$$\tau(\alpha | \beta) = \alpha \wedge * \beta \quad \text{for every } \alpha \in \Lambda^p(X^n).$$

By choosing, for example, $\alpha = \mathbf{d}x^1 \wedge \ldots \wedge \mathbf{d}x^p$ the components of $* \beta$ can be found:

$$\tau(\mathbf{d}x^1 \wedge \cdots \wedge \mathbf{d}x^p | \beta) = \tau \beta^{1 \ldots p}$$
$$= \frac{1}{(n-p)!} \tau_{1 \ldots p i_{p+1} \ldots i_n} \mathbf{d}x^1 \wedge \cdots \wedge \mathbf{d}x^p \wedge \mathbf{d}x^{i_{p+1}} \wedge \cdots \wedge \mathbf{d}x^{i_n} \beta^{1 \ldots p}$$
$$= \mathbf{d}x^1 \wedge \cdots \wedge \mathbf{d}x^p \wedge \frac{1}{(n-p)!} (* \beta)_{i_{p+1} \ldots i_n} \mathbf{d}x^{i_{p+1}} \wedge \cdots \wedge \mathbf{d}x^{i_n}.$$

Therefore

$$\tau_{1 \ldots p i_{p+1} \ldots i_n} \beta^{1 \ldots p} = (* \beta)_{i_{p+1} \ldots i_n}.$$

Summing over all possible permutations of the indices $1, \ldots, p$ one has

$$(* \beta)_{i_{p+1} \ldots i_n} = \frac{1}{p!} \tau_{i_1 \ldots i_n} \beta^{i_1 \ldots i_p}.$$

Although this formula has been proved only for the special case where $i_1 \ldots i_p$ and $i_{p+1} \ldots i_n$ are permutations of $1 \ldots p$ and $(p+1) \ldots n$, respectively, it is clear that the formula is valid for any permutation $i_1 \ldots i_n$ of $1 \ldots n$.

Exercise: Let α and β be two 1-forms in $\mathbb{R}^3$

$$\alpha = \alpha_i \, \mathbf{d}x^i \qquad \beta = \beta_i \, \mathbf{d}x^i.$$

Let (α_i) be the components of a vector $\boldsymbol{V}$ and (β_i) be the components of a vector $\boldsymbol{U}$. Show that the components of $*(\boldsymbol{\alpha} \wedge \boldsymbol{\beta})$ are the components of the vector $\boldsymbol{V} \times \boldsymbol{U}$ (cf. p. 198).

Exercise: Show that the star isomorphism $\Lambda^p \to \Lambda^{n-p}$ preserves the inner product up to sign, that is that $(\boldsymbol{\alpha}|\boldsymbol{\beta}) = \text{sign } g(*\boldsymbol{\alpha}|*\boldsymbol{\beta})$.

Answer:

$$(*\boldsymbol{\alpha}|*\boldsymbol{\beta}) = \frac{1}{(n-p)!}\frac{1}{p!}\frac{1}{p!}\tau_{i_1\ldots i_n}\alpha^{i_1\ldots i_p}\tau^{j_1\ldots j_p i_{p+1}\ldots i_n}\beta_{j_1\ldots j_p}$$

$$= \frac{\text{sign } g}{(n-p)!}\frac{1}{p!}\frac{1}{p!}\epsilon^{1\ldots n}_{i_1\ldots i_n}\alpha^{i_1\ldots i_p}\epsilon^{j_1\ldots j_p i_{p+1}\ldots i_n}_{1\ldots n}\beta_{j_1\ldots j_p}$$

$$= \frac{\text{sign } g}{p!}\alpha^{i_1\ldots i_p}\beta_{i_1\ldots i_p} = \text{sign } g(\boldsymbol{\alpha}|\boldsymbol{\beta}). \qquad \blacksquare$$

$*^{-1}$ The inverse $*^{-1}$ of the operator $*$ is the operator $(-1)^{p(n-p)}$ sign $g\,*$. This is clear from the identities

$$\begin{aligned}\boldsymbol{\tau}(\boldsymbol{\beta}|\boldsymbol{\alpha}) = \boldsymbol{\beta} \wedge *\boldsymbol{\alpha} &= (-1)^{p(n-p)} *\boldsymbol{\alpha} \wedge \boldsymbol{\beta}\\ &= \text{sign } g\,\boldsymbol{\tau}(*\boldsymbol{\beta}|*\boldsymbol{\alpha}) = \text{sign } g\,\boldsymbol{\tau}(*\boldsymbol{\alpha}|*\boldsymbol{\beta})\\ &= \text{sign } g *\boldsymbol{\alpha} \wedge **\boldsymbol{\beta}.\end{aligned}$$

δ The star operator can be used to define a linear differential operator δ of degree -1 (p. 202) on the graded algebra $\Lambda(X)$

$$\delta\boldsymbol{\omega} = (-1)^p *^{-1} \mathrm{d} * \boldsymbol{\omega} \qquad \boldsymbol{\omega} \in \Lambda^p(X^n).$$

It is clear that δ is of degree -1; we will show that $\delta\delta = 0$

$$\delta\delta\boldsymbol{\omega} = (-1)^{p-1}(-1)^p *^{-1} \mathrm{d} * *^{-1} \mathrm{d} * \boldsymbol{\omega} = - *^{-1} \mathrm{dd} * \boldsymbol{\omega} = 0.$$

Exercise: Let $\boldsymbol{\alpha} = \alpha_i\,\mathbf{d}x^i$ be a 1-form in $\mathbb{R}^n$. Let (α_i) be the components of a vector $\boldsymbol{V}$. Show that $\delta\boldsymbol{\alpha} = -\text{div } \boldsymbol{V}$.

global inner product If the support of either $\boldsymbol{\alpha}$ or $\boldsymbol{\beta}$, $\boldsymbol{\alpha}, \boldsymbol{\beta} \in \Lambda^p(X^n)$, is compact the **global inner product** $[\boldsymbol{\alpha}|\boldsymbol{\beta}]$ of $\boldsymbol{\alpha}$ and $\boldsymbol{\beta}$ is by definition

$$[\boldsymbol{\alpha}|\boldsymbol{\beta}] = \int_{X^n} (\boldsymbol{\alpha}|\boldsymbol{\beta})\boldsymbol{\tau}.$$

By the definition of the star operator, we obtain

$$[\boldsymbol{\alpha}|\boldsymbol{\beta}] = \int_{X^n} \boldsymbol{\alpha} \wedge * \boldsymbol{\beta}.$$

Let T be a linear operator of degree s on the graded algebra $\Lambda_0(X)$ of differential forms with compact support. The **metric transpose** T' of T is defined by the relation metric transpose

$$[T\boldsymbol{\alpha}|\boldsymbol{\beta}] = [\boldsymbol{\alpha}|T'\boldsymbol{\beta}] \qquad \boldsymbol{\alpha} \in \Lambda_0^p, \quad \boldsymbol{\beta} \in \Lambda_0^{p+s}.$$

Theorem. d *and* δ *are the metric transposes of each other.*

Proof: Since

$$\begin{aligned}[][\mathrm{d}\boldsymbol{\alpha}|\boldsymbol{\beta}] &= \int_{X^n} \mathrm{d}\boldsymbol{\alpha} \wedge * \boldsymbol{\beta} \qquad \boldsymbol{\alpha} \in \Lambda_0^{p-1}, \boldsymbol{\beta} \in \Lambda_0^p \\ &= \int_{X^n} \mathrm{d}(\boldsymbol{\alpha} \wedge * \boldsymbol{\beta}) + (-1)^p \int_{X^n} \boldsymbol{\alpha} \wedge \mathrm{d} * \boldsymbol{\beta} \\ &= \int_{X^n} \mathrm{d}(\boldsymbol{\alpha} \wedge * \boldsymbol{\beta}) + (-1)^p \int_{X^n} \boldsymbol{\alpha} \wedge * *^{-1} \mathrm{d} * \boldsymbol{\beta}\end{aligned}$$

it follows from Stokes' formula that δ is the metric transpose of d,

$$[\mathrm{d}\boldsymbol{\alpha}|\boldsymbol{\beta}] = \int_{X^n} \boldsymbol{\alpha} \wedge * \delta\boldsymbol{\beta} = [\boldsymbol{\alpha}|\delta\boldsymbol{\beta}].$$

d is the metric transpose of δ since

$$[\delta\boldsymbol{\alpha}|\boldsymbol{\beta}] = [\boldsymbol{\beta}|\delta\boldsymbol{\alpha}] = [\mathrm{d}\boldsymbol{\beta}|\boldsymbol{\alpha}] = [\boldsymbol{\alpha}|\mathrm{d}\boldsymbol{\beta}]. \qquad \blacksquare$$

Note that for a compact manifold with boundary the relation $[\mathrm{d}\boldsymbol{\alpha}|\boldsymbol{\beta}] = [\boldsymbol{\alpha}|\delta\boldsymbol{\beta}]$ does not hold.

Exercise: Show that $[*\boldsymbol{\alpha}|\boldsymbol{\beta}] = \operatorname{sign} g[\boldsymbol{\alpha}|*^{-1}\boldsymbol{\beta}]$.

Answer: Let $\boldsymbol{\alpha} \in \Lambda^p$, $\boldsymbol{\beta} \in \Lambda^{n-p}$, then

$$\begin{aligned}[][*\boldsymbol{\alpha}|\boldsymbol{\beta}] &= \int (*\boldsymbol{\alpha}|\boldsymbol{\beta})\boldsymbol{\tau} = \int (\boldsymbol{\beta}|*\boldsymbol{\alpha})\boldsymbol{\tau} = \int \boldsymbol{\beta} \wedge * * \boldsymbol{\alpha} \\ &= (-1)^{p(n-p)} \int * * \boldsymbol{\alpha} \wedge \boldsymbol{\beta} \\ &= \operatorname{sign} g \int \boldsymbol{\alpha} \wedge * *^{-1} \boldsymbol{\beta} = \operatorname{sign} g[\boldsymbol{\alpha}|*^{-1}\boldsymbol{\beta}].\end{aligned}$$

The differential operator δ can be used to give an intrinsic expressio the second set of Maxwell's equations (Problem 1, p. 335).

5. ISOMETRIES

isometry

Let X^n and Y^n be two smooth manifolds with riemannian structures g and γ, respectively. The mapping $f: X^n \to Y^n$ is called an **isometry** if f is a diffeomorphism and

$$f^*\gamma = g.$$

Two manifolds are said to be isometric if there exists an isometry of one onto the other.

local isometry

The mapping $f: X^n \to Y^n$ is called a **local isometry** if for each $x \in X^n$ there exist neighborhoods U of x and V of $f(x)$ such that f is an isometry of U onto V.

The isometries of X^n onto itself form a group $\mathcal{I}(X^n)$.

Killing vector field on (X^n, g)

Let v be a vector field on X^n. The vector field v generates a one parameter pseudogroup of local isometries (is a Killing vector field relative to this pseudogroup (p. 154)) if and only if $\mathcal{L}_v g = 0$ (p. 152). Since isometries are the most important transformations on a riemannian manifold (X^n, g), a **Killing vector field** on (X^n, g) is assumed to be relative to isometries unless otherwise specified.

Theorem: Suppose there exists an isometry $f: X^n \to X^n$; then for each chart (U, φ) there exists a chart (U', φ') such that $\varphi(U) = \varphi'(U')$ and

$$[(\varphi^{-1})^* g](x^i) = [(\varphi'^{-1})^* g](x^i) \qquad \forall (x^i) \in \varphi(U) \subset \mathbb{R}^n. \tag{1}$$

The converse is true locally.

Proof: (1) Suppose $f: X^n \to X^n$ is an isometry. Given a chart (U, φ) on X^n, let $(U', \varphi') = (f(U), \varphi \circ f^{-1})$. In other words the coordinates on U' are chosen so as to make $f: x \mapsto x'$ such that $\varphi(x) = \varphi'(x') = (x^i)$. Then

$$(\varphi'^{-1})^* g = ((\varphi \circ f^{-1})^{-1})^* g = (\varphi^{-1})^* f^* g = (\varphi^{-1})^* g$$

since $f^* g = g$.

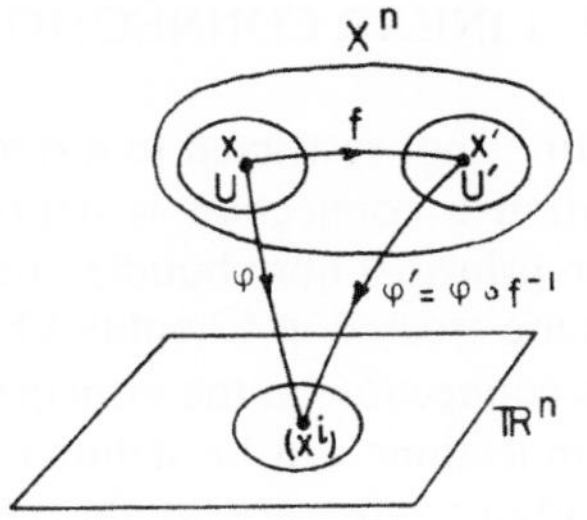

In terms of the special coordinates the condition (1) becomes

$$\bar{g}_{ij}(x^k) = \bar{g}'_{ij}(x^k) \qquad \forall (x^k) \in \varphi(U) = \varphi'(U')$$

where $\bar{g} \equiv (\varphi^{-1})^* g$ and $\bar{g}' \equiv (\varphi'^{-1})^* g$ are the induced metrics on $\varphi(U) = \varphi'(U') \subset \mathbb{R}^n$.

(2) Suppose conversely that given a chart (U, φ) there exists a chart (U', φ') such that both $\varphi(U) = \varphi'(U')$ and (1) hold. Then $f \equiv \varphi'^{-1} \circ \varphi : U \to U'$ is a local isometry since

$$f^* g = (\varphi'^{-1} \circ \varphi)^* g = \varphi^* (\varphi'^{-1})^* g = \varphi^* (\varphi^{-1})^* g = g.$$

f is not necessarily extendible to a global isometry $X^n \to X^n$. ■

Any riemannian manifold isometric to the riemannian manifold $\mathbb{R}^n$ with chart $(\mathbb{R}^n, \mathrm{Id})$ and metric flat space

$$\mathrm{d}s^2 = \sum_{i=1}^{n} \epsilon_i \, \mathbf{d}x^i \, \mathbf{d}x^i, \quad \epsilon_i = \pm 1,$$

is called a **flat space**. If $\epsilon_i = +1$, $\forall i$ the space is called **Euclidean** and denoted E^n. Euclidean space

The **Minkowski space** is the flat space with metric Minkowski space

$$\mathrm{d}s^2 = \mathbf{d}x^0 \, \mathbf{d}x^0 - \sum_{i=1}^{n-1} \mathbf{d}x^i \, \mathbf{d}x^i.$$

A **locally flat** space is a space locally isometric to a flat space. locally flat

B. LINEAR CONNECTIONS

The idea of a connection is not restricted to a riemannian manifold. In its most abstract formulation a connection is defined on a principal fibre bundle. Connections on principal fibre bundles and their important recent applications to physics are treated in Chapter Vbis.

linear connection

A **linear connection** is a connection on the principal fibre bundle of frames. Alternatively linear connections can be defined directly in terms of the covariant derivative without reference to the bundle of frames. In this chapter we shall follow this simpler more direct approach to linear connections.

1. LINEAR CONNECTIONS

In euclidean space two vectors of different origin are compared by parallel translating one or both vectors to the same origin, and the derivative of a vector v defined along a curve given in cartesian coordinates by $(x^i(t))$ is the vector of components $(\partial v^i/\partial x^j)\,dx^j/dt$. There is no canonical way to generalize the two related concepts, parallel transport and derivative, to vectors on an arbitrary differentiable manifold. Indeed to speak of the equality of the components of two vectors based at two different points $a \in X$ and $b \in X$, one must be able to assign a uniquely defined frame at b given a frame at a, and to speak of derivative one must be able to speak of parallel transport.

We note also that under an arbitrary change of variables, $x^{j'} = x^{j'}(x^i)$, the set $(\partial v^i/\partial x^j)$ does not transform as the set of components of a tensor.

The definition of a connection makes it possible to define the parallel transport of a vector along a curve and to define a derivative of a vector which is itself a tensor.

In this chapter "differentiable" is understood as "infinitely differentiable" for simplicity. It is easy to figure out corresponding statements for finite orders of differentiability.

linear connection

A **linear connection** on a smooth manifold X is a mapping $v \mapsto \nabla v$ from the germs (p. 118) of smooth vector fields on X into the germs of differentiable tensor fields of type $(1, 1)$ on X such that

$$\nabla(v + w) = \nabla v + \nabla w$$

and

$$\nabla(fv) = \mathbf{d}f \otimes v + f\nabla v$$

where f is a germ of a differentiable function on X.

The tensor $\nabla \boldsymbol{v}$ is called the **covariant derivative** of $\boldsymbol{v}$ **(absolute derivative)**. The **connection coefficients** γ^j_{ki} are defined by the relation

covariant derivative
absolute derivative
connection coefficients

$$\nabla \boldsymbol{e}_i = \gamma^j_{ki} \boldsymbol{\theta}^k \otimes \boldsymbol{e}_j$$

where $(\boldsymbol{e}_i)$ and $(\boldsymbol{\theta}^i)$ are dual frames.
Then

$$\begin{aligned}\nabla \boldsymbol{v} &= \nabla(v^i \boldsymbol{e}_i) = \mathbf{d}v^i \otimes \boldsymbol{e}_i + v^i \nabla \boldsymbol{e}_i \\ &= (\mathbf{d}v^i + v^j \gamma^i_{kj} \boldsymbol{\theta}^k) \otimes \boldsymbol{e}_i \\ &= (\boldsymbol{e}_k(v^i) + \gamma^i_{kj} v^j)\boldsymbol{\theta}^k \otimes \boldsymbol{e}_i.\end{aligned}$$

The components of the tensor $\nabla \boldsymbol{v}$ are denoted $\nabla_k v^i$ or $v^i{}_{;k}$

$$\nabla \boldsymbol{v} = \nabla_k v^i \boldsymbol{\theta}^k \otimes \boldsymbol{e}_i = v^i{}_{;k} \boldsymbol{\theta}^k \otimes \boldsymbol{e}_i.$$

In terms of the **connection forms** $\boldsymbol{\omega}^j_i = \gamma^j_{ki} \boldsymbol{\theta}^k$

connection forms

$$\nabla \boldsymbol{v} = (\mathbf{d}v^i + \boldsymbol{\omega}^i_j v^j) \otimes \boldsymbol{e}_i.$$

We use the fact that $\nabla \boldsymbol{v}$ is a tensor to find the transformation law of the connection coefficients. Under the change of basis

$$\boldsymbol{e}_i = a^{j'}_i \boldsymbol{e}_{j'}, \quad \boldsymbol{\theta}^k = a^k_{l'} \boldsymbol{\theta}^{l'}, \quad a^{j'}_i a^k_{j'} = \delta^k_i, \quad v^i = a^i_{j'} v^{j'},$$

we have

$$\mathbf{d}v^i = a^i_{l'} \mathbf{d}v^{l'} + v^{l'} \mathbf{d}a^i_{l'} = a^i_{l'} \mathbf{d}v^{l'} + v^{l'} \boldsymbol{e}_{h'}(a^i_{l'})\boldsymbol{\theta}^{h'}.$$

Therefore

$$\begin{aligned}\nabla \boldsymbol{v} &= (\mathbf{d}v^i + v^j \gamma^i_{kj} \boldsymbol{\theta}^k) \otimes a^{j'}_i \boldsymbol{e}_{j'} \\ &= (\mathbf{d}v^{j'} + a^{j'}_i v^{l'} \mathbf{d}a^i_{l'} + a^m_{l'} v^{l'} \gamma^i_{km} a^{j'}_i a^k_{h'} \boldsymbol{\theta}^{h'}) \otimes \boldsymbol{e}_{j'} \\ &= (\mathbf{d}v^{j'} + v^{l'} \gamma^{j'}_{h'l'} \boldsymbol{\theta}^{h'}) \otimes \boldsymbol{e}_{j'}\end{aligned}$$

and

$$\gamma^{j'}_{h'l'} = a^{j'}_i \boldsymbol{e}_{h'}(a^i_{l'}) + a^m_{l'} \gamma^i_{km} a^{j'}_i a^k_{h'}.$$

Note that the connection coefficients are not the components of a tensor.
In terms of the natural basis

$$\nabla \boldsymbol{v} = \left(\frac{\partial v^i}{\partial x^k} + \Gamma^i_{kl} v^l\right) \mathbf{d}\boldsymbol{x}^k \otimes \frac{\partial}{\partial \boldsymbol{x}^i}$$

where the Γ^i_{kl} are the **connection coefficients** in the natural basis, often called **Christoffel symbols**.

connection coefficient
Christoffel symbol

Under a change of natural basis induced by a change of coordinates $(x^i) \mapsto (x^{i'})$:

$$\Gamma^{j'}_{h'l'} = \frac{\partial x^{j'}}{\partial x^i} \frac{\partial}{\partial x^{h'}} \left(\frac{\partial x^i}{\partial x^{l'}}\right) + \frac{\partial x^m}{\partial x^{l'}} \frac{\partial x^k}{\partial x^{h'}} \frac{\partial x^{j'}}{\partial x^i} \Gamma^i_{km}.$$

On a C^k $(k \geq 2)$ manifold a connection is said to be of class C^r if, in all charts of an atlas, the Γ_{ij}^k are of class C^r. If $r \leq k-2$ the definition is coherent and does not depend on the atlas. If $\boldsymbol{v}$ is C^{k-1} and the connection C^{k-2}, then $\nabla \boldsymbol{v}$ is C^{k-2}.

covariant derivative in the direction of $\boldsymbol{u}$

Parallel translation (transport). The **covariant derivative** $\nabla_u \boldsymbol{v}$ **of** $\boldsymbol{v}$ **in the direction of** $\boldsymbol{u}$ is by definition

$$\nabla_u \boldsymbol{v} = (\nabla \boldsymbol{v})(\boldsymbol{u}) \quad \text{or} \quad \nabla_u \boldsymbol{v}(\cdot) = \nabla \boldsymbol{v}(\boldsymbol{u}, \cdot).$$

That is to say

$$\begin{aligned} \nabla_u \boldsymbol{v} &= u^k (e_k(v^i) + \gamma_{kj}^i v^j) e_i \\ &= (\boldsymbol{u}(v^i) + \gamma_{kj}^i u^k v^j) e_i. \end{aligned}$$

In particular

$$\nabla_{e_i} e_k = \gamma_{ik}^j e_j.$$

Notice that $\nabla_u \boldsymbol{v}$ is linear in $\boldsymbol{u}$ over the ring of functions on X

$$\nabla_{fu_1+gu_2} v = f\nabla_{u_1} v + g\nabla_{u_2} v \qquad f, g : X \to \mathbb{R}.$$

parallel along a curve

A vector $\boldsymbol{v}$ is said to be **parallel along a curve** $C: t \mapsto C(t)$ if

$$\nabla_u \boldsymbol{v} = 0, \qquad \boldsymbol{u} = C' \frac{\mathrm{d}}{\mathrm{d}t}, \qquad u^i = \frac{\mathrm{d}C^i}{\mathrm{d}t}, \qquad \text{for all points of } C.$$

The vector $\nabla_u \boldsymbol{v}$ is sometimes denoted $\mathrm{D}\boldsymbol{v}/\mathrm{d}t$.
The vector $\boldsymbol{u} = \mathrm{d}C/\mathrm{d}t$ is defined only at points along the curve C. It can however be extended to a vector field on a neighborhood of any point of C.[1] When written in terms of local coordinates it is clear that the vector $\nabla_u \boldsymbol{v}$ is independent of the extension.

affine geodesic

An **affine geodesic** on X is a curve $C: t \mapsto C(t)$ such that

$$\nabla_u \boldsymbol{u} = \lambda(t) \boldsymbol{u}, \qquad \boldsymbol{u} = \mathrm{d}C/\mathrm{d}t$$

geodesic

for some function λ on $\mathbb{R}$. If $\nabla_u \boldsymbol{u} = 0$ the curve is a **geodesic**.
In a local chart (U, φ) using the natural basis, with $C^i(t) = \varphi^i \circ C(t)$ and $u^i(t) = u^i \circ C(t)$, one has

$$u^i = \boldsymbol{u}(\varphi^i) = \left(C' \frac{\mathrm{d}}{\mathrm{d}t}\right)(\varphi^i) = \frac{\mathrm{d}}{\mathrm{d}t}(\varphi^i \circ C(t)) = \frac{\mathrm{d}C^i}{\mathrm{d}t}$$

and

$$\boldsymbol{u}(u^i) = \left(C' \frac{\mathrm{d}}{\mathrm{d}t}\right)(u^i) = \frac{\mathrm{d}}{\mathrm{d}t}(u^i \circ C(t)) = \frac{\mathrm{d}u^i(t)}{\mathrm{d}t}.$$

[1] See for instance [Helgason, p. 28].

Therefore

$$(\nabla_u u)^i = u(u^i) + u^k \Gamma^i_{kj} u^j$$
$$= \frac{d^2 C^i}{dt^2} + \Gamma^i_{kj} \frac{dC^k}{dt} \frac{dC^j}{dt}.$$

Exercise: Consider the change of parameter $s = \tau(t)$, $C(t) = \bar{C}(s) = \bar{C} \circ \tau(t)$. Set $\boldsymbol{u} = dC/dt$ and $\bar{\boldsymbol{u}} = d\bar{C}/ds$. If C is an affine geodesic, i.e. if $\nabla_u \boldsymbol{u} = \lambda(t)\boldsymbol{u}$, show that

$$\nabla_{\bar{u}} \bar{\boldsymbol{u}} = \left[\lambda(t)\frac{d\tau}{dt} - \frac{d^2\tau}{dt^2}\right]\left(\frac{d\tau}{dt}\right)^{-2} \bar{\boldsymbol{u}}.$$

Answer: Since $\dfrac{dC^i}{dt} = \dfrac{d\bar{C}^i}{ds}\dfrac{d\tau}{dt}$ we have

$$\frac{d}{dt}\frac{d\bar{C}^i}{ds}\frac{d\tau}{dt} + \Gamma^i_{kj}\frac{d\bar{C}^k}{ds}\frac{d\bar{C}^j}{ds}\left(\frac{d\tau}{dt}\right)^2 = \lambda(t)\frac{d\tau}{dt}\frac{d\bar{C}^i}{ds}$$

or

$$\left(\frac{d\tau}{dt}\right)^2\frac{d^2\bar{C}^i}{ds^2} + \Gamma^i_{jk}\frac{d\bar{C}^k}{ds}\frac{d\bar{C}^j}{ds}\left(\frac{d\tau}{dt}\right)^2 = \left[\lambda(t)\frac{d\tau}{dt} - \frac{d^2\tau}{dt^2}\right]\frac{d\bar{C}^i}{ds}.$$

If $\nabla_{\bar{u}}\bar{\boldsymbol{u}} = 0$ then s is called an **affine parameter**. Note that if s is affine then so is $as + b$, with $a, b \in \mathbb{R}$, since $(d^2/ds^2)(as + b) = 0$. affine parameter

Theorem. Let $x \in X$ and $\boldsymbol{v} \in T_x(X)$, $\boldsymbol{v} \neq 0$. There exists a unique, up to a change in parameter, maximal affine geodesic $C: t \mapsto C(t)$ such that

$$C(0) = x \quad \text{and} \quad \left.\frac{dC}{dt}\right|_{t=0} = \boldsymbol{v}.$$

Proof: The proof depends on theorems on the existence and uniqueness of the solution of a system of ordinary differential equations[1]. ■

The covariant derivative is extended to germs of tensors of arbitrary type by requiring that the directional covariant derivative satisfy

1) $\nabla_v f = \boldsymbol{v}(f) \qquad f \in C^1(X)$
2) $\nabla_v(\boldsymbol{t} + \boldsymbol{s}) = \nabla_v \boldsymbol{t} + \nabla_v \boldsymbol{s}$
3) $\nabla_v(\boldsymbol{t} \otimes \boldsymbol{s}) = \nabla_v \boldsymbol{t} \otimes \boldsymbol{s} + \boldsymbol{t} \otimes \nabla_v \boldsymbol{s}$
4) ∇_v commutes with the operation of contracted multiplication.

[1]See for instance [Helgason, p. 31].

Then if t is a tensor of type (q, p), the covariant derivative ∇t is the tensor of type $(q+1, p)$ defined by

$$(\nabla t)(v, v_1, \ldots, v_q, \omega_1, \ldots, \omega_p) = (\nabla_v t)(v_1, \ldots, v_q, \omega_1, \ldots, \omega_p).$$

We use 4) to find the covariant derivative of the 1-form α. The covariant derivative in the direction v of the contraction of $\alpha \otimes u$ is equal to the contraction of $\nabla_v \alpha \otimes u + \alpha \otimes \nabla_v u$

$$\nabla_v[\alpha(u)] = (\nabla_v \alpha)(u) + \alpha(\nabla_v u)$$

or

$$(\nabla_v \alpha)(u) = \nabla_v[\alpha(u)] - \alpha(\nabla_v u).$$

Take $u = e_i$ to get

$$\begin{aligned}(\nabla_v \alpha)_i &= v(\alpha_i) - \alpha(\gamma^j_{ki} v^k e_j)\\ &= v(\alpha_i) - \gamma^j_{ki} v^k \alpha_j.\end{aligned}$$

Thus

$$\nabla_v \alpha = v^k(e_k(\alpha_i) - \gamma^j_{ki}\alpha_j)\theta^i$$

and

$$\begin{aligned}\nabla \alpha &= (e_k(\alpha_i) - \gamma^j_{ki}\alpha_j)\theta^k \otimes \theta^i\\ &= (\mathbf{d}\alpha_i - \alpha_j \omega^j_i) \otimes \theta^i.\end{aligned}$$

In particular

$$\nabla \theta^i = -\omega^i_j \otimes \theta^j = -\gamma^i_{kj}\theta^k \otimes \theta^j$$

and

$$\nabla_v \theta^i = -v^k \gamma^i_{kj}\theta^j.$$

This result is used together with 2) and 3) to find the covariant derivative of an arbitrary tensor. Consider for example the type (2, 1) tensor

$$t = t^i{}_{kl} e_i \otimes \theta^k \otimes \theta^l.$$

The covariant derivative in the direction v is

$$\begin{aligned}\nabla_v t &= v(t^i{}_{kl}) e_i \otimes \theta^k \otimes \theta^l\\ &\quad + t^i{}_{kl} \nabla_v e_i \otimes \theta^k \otimes \theta^l\\ &\quad + t^i{}_{kl} e_i \otimes \nabla_v \theta^k \otimes \theta^l\\ &\quad + t^i{}_{kl} e_i \otimes \theta^k \otimes \nabla_v \theta^l\\ &= v^j(e_j(t^i{}_{kl}) + \gamma^i_{jm} t^m{}_{kl} - \gamma^m_{jk} t^i{}_{ml}\\ &\quad - \gamma^m_{jl} t^i{}_{km}) e_i \otimes \theta^k \otimes \theta^l.\end{aligned}$$

Exercise: Show that the components

$$\nabla_j t^i{}_{kl} = e_j(t^i{}_{kl}) + \gamma^i_{jm} t^m{}_{kl} - \gamma^m_{jk} t^i{}_{ml} - \gamma^m_{jl} t^i{}_{km}$$

of ∇t transform like the components of a tensor of type (3, 1) under a change of basis.

Exercise: Show that the formula for the components of the covariant derivative of a product is identical with the usual formula for the derivative of a product. This exercise is to be contrasted with the next one.

Answer: Notice that $\nabla_j t^i{}_{kl} = (\nabla t)_j{}^i{}_{kl} = (\nabla_{e_j} t)^i{}_{kl}$. Consider the product

$$s \otimes t = (s^i{}_k e_i \otimes \theta^k) \otimes t_l \theta^l = s^i{}_k t_l e_i \otimes \theta^k \otimes \theta^l.$$

By 3)

$$\nabla_{e_j}(s \otimes t) = (\nabla_{e_j} s) \otimes t + s \otimes (\nabla_{e_j} t)$$

so that

$$(\nabla_{e_j}(s \otimes t))^i{}_{kl} = (\nabla_{e_j} s)^i{}_k t_l + s^i{}_k (\nabla_{e_j} t)_l.$$

Therefore

$$(\nabla(s \otimes t))_j{}^i{}_{kl} = (\nabla s)_j{}^i{}_k t_l + s^i{}_k (\nabla t)_{jl}$$

or

$$\nabla_j (s^i{}_k t_l) = t_l \nabla_j s^i{}_k + s^i{}_k \nabla_j t_l.$$

Exercise: Show that $\nabla(s \otimes t) \neq \nabla s \otimes t + s \otimes \nabla t$.

Answer: $\nabla s \otimes t + s \otimes \nabla t$ is meaningless because one can add tensor products only if the corresponding factors in each term have the same rank. The components relationship of the previous exercise does not imply that $\nabla(s \otimes t)$ splits into $\nabla s \otimes t$ plus $s \otimes \nabla t$ because the j index does not occur in the same place in these two terms. The covariant derivative does not obey Leibniz rule.

Torsion and curvature. The **torsion operation** τ and **curvature operation** ρ are defined by

torsion operation
curvature operation

$$\tau(u, v) = \nabla_u v - \nabla_v u - [u, v]$$

$$\rho(u, v) = \nabla_u \nabla_v - \nabla_v \nabla_u - \nabla_{[u, v]}$$

where $u, v \in \mathscr{X}(X)$, the Lie algebra of C^∞ vector fields on X (p. 133). Note that

$$\tau(u, v) = -\tau(v, u)$$

$$\rho(u, v) = -\rho(v, u)$$

$$\tau(fu, gv) = fg\tau(u, v)$$

$$\rho(fu, gv)hw = fgh\rho(u, v)w$$

for all $f, g, h \in C(X)$ and $u, v, w \in \mathscr{X}(X)$.

torsion tensor
curvature tensor

The **torsion tensor** T, and **curvature tensor** R are defined by

$$T(\alpha, u, v) = \alpha(\tau(u, v))$$
$$R(w, \alpha, u, v) = \alpha(\rho(u, v)w)$$

where $\alpha \in \Lambda^1(X)$.
In a moving frame the components of T and R are

$$T^i{}_{kl} = T(\theta^i, e_k, e_l) = \gamma^i_{kl} - \gamma^i_{lk} - c^i_{kl}$$

$$R_i{}^j{}_{kl} = R(e_i, \theta^j, e_k, e_l)$$
$$= e_k(\gamma^j_{li}) - e_l(\gamma^j_{ki}) + \gamma^j_{km}\gamma^m_{li} - \gamma^j_{lm}\gamma^m_{ki} - c^m_{kl}\gamma^j_{mi}$$

where the c^i_{kl} are defined through the bracket of the vector fields e_k, e_l by

$$[e_k, e_l] = c^i_{kl}e_i.$$

structure coefficients

The c^i_{kl} are differentiable functions on X which are called the **structure coefficients** of the moving frame (e_i). Using the natural frame one has $[\partial/\partial x^k, \partial/\partial x^l] = 0$.

$$T^i_{kl} = \Gamma^i_{kl} - \Gamma^i_{lk}$$

and
$$R_i{}^j{}_{kl} = \partial_k\Gamma^j_{li} - \partial_l\Gamma^j_{ki} + \Gamma^j_{km}\Gamma^m_{li} - \Gamma^j_{lm}\Gamma^m_{ki}.$$

Note that T and R measure the "symmetry" of the tensor $\nabla\nabla t$; that is, in particular,

$$\nabla_k\nabla_l v^j - \nabla_l\nabla_k v^j = R_i{}^j{}_{kl}v^i - T^i{}_{kl}\nabla_i v^j$$
$$\nabla_k\nabla_l f - \nabla_l\nabla_k f = -T^i_{kl}\nabla_i f.$$

Exercise: Prove these relations.

torsion forms

Due to the antisymmetry of $T^i{}_{kl}$ and $R_i{}^j{}_{kl}$ in the last two indices it is possible to define the **torsion forms**

$$\Theta^i = \tfrac{1}{2}T^i{}_{kl}\theta^k \wedge \theta^l$$

curvature forms

and the **curvature forms**

$$\Omega^j_i = \tfrac{1}{2}R_i{}^j{}_{kl}\theta^k \wedge \theta^l.$$

Theorem. Cartan structural equations. ω^j_i *being the connection forms* $\gamma^j_{ki}\theta^k$,

$$\Theta^i = \tfrac{1}{2}T^i{}_{kl}\theta^k \wedge \theta^l = d\theta^i + \omega^i_l \wedge \theta^l$$
$$\Omega^j_i = \tfrac{1}{2}R_i{}^j{}_{kl}\theta^k \wedge \theta^l = d\omega^j_i + \omega^j_m \wedge \omega^m_i.$$

Proof: Using the relation (p. 207)

$$\mathbf{d}\boldsymbol{\theta}(u, v) = \mathcal{L}_u(\boldsymbol{\theta}(v)) - \mathcal{L}_v(\boldsymbol{\theta}(u)) - \boldsymbol{\theta}([u, v])$$

we have

$$\mathbf{d}\boldsymbol{\theta}^i(e_k, e_l) = -\boldsymbol{\theta}^i([e_k, e_l]) = -c^i_{kl}.$$

Therefore

$$\mathbf{d}\boldsymbol{\theta}^i = -\tfrac{1}{2}c^i_{kl}\boldsymbol{\theta}^k \wedge \boldsymbol{\theta}^l.$$

Also

$$\boldsymbol{\omega}^i_l \wedge \boldsymbol{\theta}^l = \gamma^i_{kl}\boldsymbol{\theta}^k \wedge \boldsymbol{\theta}^l = \tfrac{1}{2}(\gamma^i_{kl} - \gamma^i_{lk})\boldsymbol{\theta}^k \wedge \boldsymbol{\theta}^l$$

which proves the first equation. ∎

Exercise: Prove the second equation.

Differentiation of the Cartan structural equations gives

$$\mathbf{d}\boldsymbol{\Theta}^k = \mathbf{d}\boldsymbol{\omega}^k_i \wedge \boldsymbol{\theta}^i - \boldsymbol{\omega}^k_i \wedge \mathbf{d}\boldsymbol{\theta}^i = \boldsymbol{\Omega}^k_i \wedge \boldsymbol{\theta}^i - \boldsymbol{\omega}^k_i \wedge \boldsymbol{\Theta}^i$$

$$\mathbf{d}\boldsymbol{\Omega}^k_i = \mathbf{d}\boldsymbol{\omega}^k_l \wedge \boldsymbol{\omega}^l_i - \boldsymbol{\omega}^k_l \wedge \mathbf{d}\boldsymbol{\omega}^l_i = \boldsymbol{\Omega}^k_l \wedge \boldsymbol{\omega}^l_i - \boldsymbol{\omega}^k_l \wedge \boldsymbol{\Omega}^l_i.$$

In terms of completely antisymmetric components and with respect to a natural frame the first equation becomes

$$\frac{1}{3!}\Big(\sum_{(jli)} \partial_j T^k_{li}\Big)\,\mathbf{d}x^j \wedge \mathbf{d}x^l \wedge \mathbf{d}x^i = \frac{1}{3!}\Big(\sum_{(jli)} (R_i{}^k{}_{jl} - \Gamma^k_{jm}T^m_{li})\Big)\,\mathbf{d}x^j \wedge \mathbf{d}x^l \wedge \mathbf{d}x^i$$

where $\Sigma_{(jli)}$ denotes the sum over cyclic permutations of the three indices. Now

$$\sum_{(jli)} \nabla_j T^k_{li} = \sum_{(jli)} (\partial_j T^k_{li} + \Gamma^k_{jm}T^m_{li} - \Gamma^m_{jl}T^k_{mi} - \Gamma^m_{ji}T^k_{lm})$$

$$= \sum_{(jli)} (\partial_j T^k_{li} + \Gamma^k_{jm}T^m_{li} + T^m_{ji}T^k_{ml})$$

thus we obtain the identities

$$\sum_{(jli)} R_i{}^k{}_{jl} = \sum_{(jli)} (\nabla_j T^k_{li} - T^m_{ji}T^k_{ml})$$

which are true with respect to any frame.

A similar manipulation of the second equation leads to the **Bianchi identities** Bianchi identities

$$\sum_{(jkl)} \nabla_j R_i{}^m{}_{kl} = \sum_{(jkl)} T^h_{kj}R_i{}^m{}_{hl}.$$

2. RIEMANNIAN CONNECTION

Theorem. On a riemannian manifold there exists a unique linear connection such that
1) The torsion tensor $\boldsymbol{T} = 0$,
2) The covariant derivative of $\boldsymbol{g}$ vanishes, $\nabla \boldsymbol{g} = 0$.

Proof: We demonstrate the existence and uniqueness of the connection by deriving an explicit expression for the connection coefficients. In local coordinates the first condition becomes (see p. 306)

$$\mathbf{d}\boldsymbol{\theta}^i + \boldsymbol{\omega}_j^{\ i} \wedge \boldsymbol{\theta}^j = 0 \quad \text{or} \quad \gamma^i_{\ lj} - \gamma^i_{\ jl} = c^i_{\ lj}$$

and the second becomes

$$\mathbf{d}g_{ij} = \boldsymbol{\omega}_i^{\ l} g_{lj} + \boldsymbol{\omega}_j^{\ l} g_{il} \quad \text{or} \quad e_k(g_{ij}) = \gamma^l_{\ ki} g_{lj} + \gamma^l_{\ kj} g_{il}.$$

The two equations can be combined to give

$$\tfrac{1}{2}[e_k(g_{ij}) + e_i(g_{jk}) - e_j(g_{ki})] = \tfrac{1}{2}(c^l_{\ ki} g_{lj} + c^l_{\ kj} g_{li} + c^l_{\ ij} g_{lk}) + \gamma^l_{\ ik} g_{lj}.$$

Therefore

$$\gamma^m_{\ ik} = \tfrac{1}{2} g^{jm}[e_k(g_{ij}) + e_i(g_{jk}) - e_j(g_{ki})] - \tfrac{1}{2}(c^m_{\ ki} + g^{jm} g_{li} c^l_{\ kj} + g^{jm} g_{lk} c^l_{\ ij}).$$

Defining $\gamma_{mik} = g_{mj}\gamma^j_{\ ik}$ and $c_{mik} = g_{mj} c^j_{\ ik}$ we have

$$\gamma_{mik} = -\tfrac{1}{2}(c_{mki} + c_{ikm} + c_{kim}) + \tfrac{1}{2}(e_k(g_{im}) + e_i(g_{mk}) - e_m(g_{ki})). \qquad \blacksquare$$

The connection defined by these connection coefficients is called the **riemannian connection.**

riemannian connection

In the case of a natural frame

$$\Gamma^m_{ik} = \tfrac{1}{2} g^{jm}(\partial_k g_{ij} + \partial_i g_{kj} - \partial_j g_{ik}) \quad \text{and} \quad \Gamma^m_{ik} = \Gamma^m_{ki}.$$

In the case of an orthonormal frame

$$\gamma_{mik} = \tfrac{1}{2}(c_{mik} + c_{imk} + c_{kmi})$$

since in this case $g_{ij} = \pm\delta_{ij}$, the signs depending on the index of the metric. Thus

$$\gamma_{mik} = -\gamma_{kim}$$

for an orthonormal frame.

Exercise: Use this fact to show that $\nabla \boldsymbol{\tau} = 0$ where $\boldsymbol{\tau}$ is the volume element on X (p. 294).

Answer: In terms of an orthonormal frame $(\boldsymbol{\theta}^i)$

$$\boldsymbol{\tau} = \boldsymbol{\theta}^1 \wedge \cdots \wedge \boldsymbol{\theta}^n = \epsilon^{1\ldots n}_{i_1\ldots i_n} \boldsymbol{\theta}^{i_1} \otimes \cdots \otimes \boldsymbol{\theta}^{i_n}.$$

Then

$$\nabla_k \tau_{i_1 \ldots i_n} = -\gamma^s{}_{ki_1} \epsilon^{1 \ldots n}_{si_2 \ldots i_n} - \gamma^s{}_{ki_2} \epsilon^{1 \ldots n}_{i_1 s \ldots i_n} - \cdots$$
$$= -\gamma^{i_1}{}_{ki_1} \epsilon^{1 \ldots n}_{i_1 \ldots i_n} - \gamma^{i_2}{}_{ki_2} \epsilon^{1 \ldots n}_{i_1 \ldots i_n} - \cdots$$

where no summation occurs over the repeated indices $i_1, \ldots, i_n$. But

$$\gamma^{i_1}{}_{ki_1} = \pm \gamma_{i_1 k i_1} = 0$$

because of the antisymmetry in the first and last indices. ∎

Remark: The parallel translation defined by the riemannian connection preserves the scalar product.

Since $\nabla_k g_{ij} = 0$, it follows from the relation $g_{ij} g^{jk} = \delta_i^k$ that $\nabla_k g^{ij} = 0$. This insures that

$$\nabla_k v_i = 0 \quad \Leftrightarrow \quad \nabla_k v^i = 0$$

and similarly for tensors of higher order.

Curvature tensor, Ricci tensor, curvature scalar. The components of the curvature tensor of a riemannian connection, called the **Riemann tensor**, satisfy the following identities

Riemann tensor

1) $R_i{}^j{}_{kl} = -R_i{}^j{}_{lk}$
2) $\sum_{(ikl)} R_i{}^j{}_{kl} = 0$
3) $\sum_{(mkl)} \nabla_m R_i{}^j{}_{kl} = 0$ Bianchi identities
4) $R_{ijkl} = -R_{jikl}$
5) $R_{ijkl} = R_{klij}$.

The first identity comes from the definition of the curvature tensor. The second and third are the identities of the previous section with $\boldsymbol{T} = 0$. If we define $\boldsymbol{\omega}_{ij} = g_{jk} \boldsymbol{\omega}_i^k$ and $\boldsymbol{\Omega}_{ij} = g_{jk} \boldsymbol{\Omega}_i^k$ then the fourth identity follows from the fact that in an orthonormal frame

$$\mathbf{d} g_{ij} = 0 = \boldsymbol{\omega}_{ij} + \boldsymbol{\omega}_{ji}$$

$$\boldsymbol{\Omega}_{ij} = \mathbf{d}\boldsymbol{\omega}_{ij} + \boldsymbol{\omega}_{mj} \wedge \boldsymbol{\omega}_i{}^m = -\boldsymbol{\Omega}_{ji}.$$

The identities 1), 2) and 4) then imply 5).

The **Ricci tensor** is a contraction of the curvature tensor. Its components are by definition

Ricci tensor

$$R_{ik} = R_i{}^j{}_{kj}.$$

Clearly $R_{ik} = R_{ki}$.

With respect to a natural frame

$$R_{ik} = \partial_k \Gamma^j_{ji} - \partial_j \Gamma^j_{ki} + \Gamma^j_{km} \Gamma^m_{ji} - \Gamma^j_{jm} \Gamma^m_{ki}.$$

Note that

$$\Gamma^j_{jk} = \tfrac{1}{2} g^{jm} (\partial_k g_{mj}) = \frac{1}{2g} \frac{\partial g}{\partial x^k} = \frac{\partial}{\partial x^k} \ln |g|^{1/2}$$

since

$$g^{ij} = \frac{1}{g} \frac{\partial g}{\partial g_{ij}}.$$

Riemann curvature scalar

The **Riemann curvature scalar** is by definition

$$R = g^{ij} R_{ij}.$$

Contraction of the Bianchi identities gives

$$\nabla_m R_{ik} - \nabla_k R_{im} + \nabla_j R_i{}^j{}_{mk} = 0.$$

Multiplication by g^{im} gives

$$\nabla_i R^i_k - \nabla_k R + \nabla_j R^j_k = 0.$$

contracted Bianchi identity

This last identity, called the **contracted Bianchi identity**, reads

$$\nabla_j (R^j_k - \tfrac{1}{2} \delta^j_k R) = 0,$$

Einstein tensor

and $R^j_k - \frac{1}{2}\delta^j_k R$ is called the **Einstein tensor.**

On a Riemannian manifold the Cartan structure equations are

$$\mathrm{d}\boldsymbol{\theta}^l = \boldsymbol{\theta}^j \wedge \boldsymbol{\omega}_j{}^l$$

$$\tfrac{1}{2} R_i{}^j{}_{kl} \boldsymbol{\theta}^k \wedge \boldsymbol{\theta}^l = \boldsymbol{\Omega}_i{}^j = \mathrm{d}\boldsymbol{\omega}_i{}^j - \boldsymbol{\omega}_i{}^l \wedge \boldsymbol{\omega}_l{}^j.$$

In the case of a manifold with a fair amount of symmetry these equations can be used efficiently to compute the curvature tensor (cf. Problem 2, p. 341).

Locally flat manifolds.

Theorem. A riemannian manifold X^n is locally flat (p. 299) if and only if the curvature tensor $\boldsymbol{R} \equiv 0$.

Proof: It is obvious that $\boldsymbol{R}$ vanishes if the manifold is locally flat. This can be seen by choosing, in a neighborhood of each point, a local chart such that

$$\mathrm{d}s^2 = \sum_{i=1}^{n} \epsilon_i \, \mathbf{d}x^i \, \mathbf{d}x^i, \qquad \epsilon_i = \pm 1.$$

Let $(\boldsymbol{\theta}^i)$ be an orthonormal moving frame on a chart U in X^n with coordinates (x^a). Now suppose $\boldsymbol{R} \equiv 0$ and consider the Pfaff system of rank n^2

$$\boldsymbol{\Lambda}_i^j = \mathbf{d}z_i^j - \boldsymbol{\omega}_i^m z_m^j$$

on the space $U \times \mathbb{R}^{n^2}$ where $(z_i^j) \in \mathbb{R}^{n^2}$. Since $\boldsymbol{\Omega}_i^j = 0$

$$\begin{aligned} \mathrm{d}\boldsymbol{\Lambda}_i^j &= -(\boldsymbol{\Lambda}_m^j + \boldsymbol{\omega}_m^k z_k^j) \wedge \boldsymbol{\omega}_i^m - \boldsymbol{\omega}_i^l \wedge \boldsymbol{\omega}_l^m z_m^j - \boldsymbol{\Omega}_i^m z_m^j \\ &= \boldsymbol{\omega}_i^m \wedge \boldsymbol{\Lambda}_m^j. \end{aligned}$$

By the Frobenius integration theorem (p. 243) the system $\boldsymbol{\Lambda}_i^j = 0$ is therefore completely integrable. Let the functions $f_i^j(x^a, z_k^l)$ be a set of n^2 independent first integrals of the system. The equations

$$f_i^j(x^a, z_k^l) = f_i^j(x_0^a, z_{0k}^l)$$

determine an integral manifold of the system through the point (x_0^a, z_{0k}^l). We will choose the point $(z_{0k}^l) \in \mathbb{R}^{n^2}$ to be such that the matrix $[z_{0k}^l]$ is an element of the group $O(n)[L(n)]$. It is clear from the form of the $\boldsymbol{\Lambda}_i^j$ that a vector $\boldsymbol{v}$ tangent to an integral manifold must be such that $\pi'_U \boldsymbol{v} = 0$, where $\pi_U : U \times \mathbb{R}^{n^2} \to U$ is the projection mapping. Therefore these equations can be solved for the z_i^j to give

$$z_i^j = a_i^j(x^a).$$

The matrix $[a_i^j(x_0^a)]$ is an element of the group $O(n)[L(n)]$. Since the $\boldsymbol{\Lambda}_i^j$ vanish identically on an integral manifold

$$\mathbf{d}a_i^j = \boldsymbol{\omega}_i^m a_m^j.$$

The n 1-forms $\boldsymbol{\alpha}^j = a_i^j \boldsymbol{\theta}^i$ are therefore such that

$$\nabla \boldsymbol{\alpha}^j = 0;$$

that is to say, the vectors $\boldsymbol{\alpha}^j$ are parallel along any curve in U. The forms $\boldsymbol{\alpha}^j$ are closed

$$\mathrm{d}\boldsymbol{\alpha}^j = \mathbf{d}a_i^j \wedge \boldsymbol{\theta}^i + a_i^j \boldsymbol{\theta}^k \wedge \boldsymbol{\omega}_k^i = 0.$$

Thus by the Poincaré lemma (p. 224) there exist n linearly independent functions u^j on U such that $\boldsymbol{\alpha}^j = \mathbf{d}u^j$. These n functions may be used as a coordinate system on U.

Now the matrix $[a_i^j(x^a)]$ is in fact an element of $O(n)[L(n)]$ at each point $(x^a) \in U$. This is because, since $\boldsymbol{\omega}_{ij} = -\boldsymbol{\omega}_{ji}$ and $\mathbf{d}g_{ij} = 0$,

$$\begin{aligned} \mathrm{d}(a_i^j a^i{}_m) &= \boldsymbol{\omega}_i^l a_l^j a^i{}_m + a_i^j \boldsymbol{\omega}^{il} a_{lm} \\ &= \boldsymbol{\omega}_i^l a_l^j a^i{}_m - \boldsymbol{\omega}_l^i a_i^j a^l{}_m = 0 \end{aligned}$$

where $a'^{i}_{m} = g^{il} a_l^k g_{km}$. Thus since $a_i^j(x_0^a) a'^{i}_{m}(x_0^a) = \delta^j_m$ it follows that $a_i^j a'^{i}_{m} = \delta^j_m$ at every point of U. Therefore since $\mathbf{d}u^j = a_i^j \boldsymbol{\theta}^i$ the basis $(\mathbf{d}u^j)$ is orthonormal. ∎

3. SECOND FUNDAMENTAL FORM

Let S be a p-dimensional submanifold of an n dimensional riemannian manifold X with metric g. The induced metric (p. 290) $\bar{g}$ on S is the pull back of g by the inclusion mapping $i: S \to X$. The tangent space T_xS is a p-dimensional vector subspace of T_xX, for $x \in S \subset X$. We suppose that T_xS is not null [isotropic] for g (p. 292), and we denote by $(T_xS)^\perp$ its g-orthogonal complement.

If u is a tangent vector to S at x, the covariant derivative $(\nabla_u v)_x$, in the riemannian connection ∇ of the metric g of a differentiable vector field v on the submanifold S is a well defined vector of the tangent space T_xX. We denote by $(\nabla_u v)_x^\parallel$ its component along T_xS and $(\nabla_u v)_x^\perp$ its normal component, i.e. its component along $(T_xS)^\perp$.

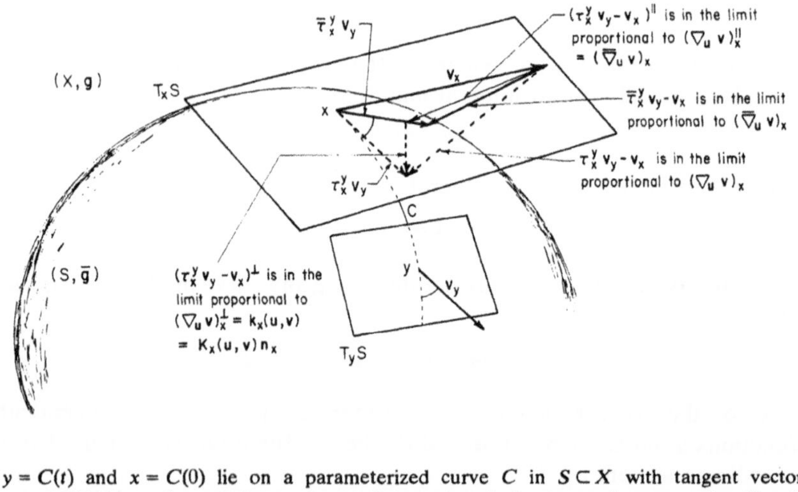

$y = C(t)$ and $x = C(0)$ lie on a parameterized curve C in $S \subset X$ with tangent vector $u_x = \mathrm{d}C(t)/\mathrm{d}t|_{t=0}$. $\tau^y_x v_y$ and $\bar{\tau}^y_x v_y$ are the parallel translates in the riemannian connections ∇ on X and $\bar{\nabla}$ on S, respectively, of the vector v_y to T_xX and T_xS (see p. 358). The limit referred to in the figure is at $t = 0$; the differences referred to should, of course, be divided by t before taking the limit. We have chosen C to be a geodesic so that the parallelism of v_y and $\bar{\tau}^y_x v_y$ could be indicated by the equality of the angles they make respectively with the tangents (not drawn) to C at y and x. Figure contributed by G. Grunberg.

Theorem: 1. $(\nabla_u v)_x^{\|}$ *is the covariant derivative of* v *in the riemannian connection of the metric* $\bar{g}$ *on* S *induced by* g *on* X:

$$(\nabla_u v)_x^{\|} = (\bar{\nabla}_u v)_x.$$

2. $(\nabla_u v)_x^{\perp} = (\nabla_v u)_x^{\perp}$, *and* $(\nabla_u v)_x^{\perp}$ *depends only on the vectors* u_x *and* v_x.

Definition: The (symmetric) mapping

$$k_x\colon T_xS \times T_xS \to (T_xS)^{\perp} \quad \text{by} \quad (u_x, v_x) \mapsto k_x(u_x, v_x) \equiv (\nabla_u v)_x^{\perp}$$

is called the **second fundamental form** of the submanifold S of X. second fundamental form

Proof: 1. To show that $(\nabla_u v)^{\|} = \bar{\nabla}_u v$ we show that $(\nabla_u v)^{\|}$ satisfies all the properties of a covariant derivative in the riemannian connection of $\bar{g}$ (cf. pp. 303 and 308).
a) $(\nabla_u v)^{\|}$ depends linearly on v and $(\nabla_u(fv))^{\|} = (\nabla_u f)v^{\|} + f(\nabla_u v)^{\|}$. These properties are immediate consequence of the properties of ∇_u and the linearity of the projection operator on T_xS. The operator $v \mapsto (\nabla_u v)^{\|}$ defines therefore a covariant derivative in a connection $\nabla^{\|}$ on S.
b) The above connection has no torsion. Since ∇ has no torsion, $\nabla_{u'}v' - \nabla_{v'}u' = [u', v']$ where u' and v' are vector fields on X which, on S, are identical with u and v. Then on S, $u'^{\|} = u$, $v'^{\|} = v$ and $[u', v']^{\|} = [u, v]$, so

$$\nabla_u^{\|} v - \nabla_v^{\|} u \equiv (\nabla_u v)^{\|} - (\nabla_v u)^{\|} = (\nabla_{u'} v' - \nabla_{v'} u')^{\|} = [u', v']^{\|} = [u, v].$$

c) The connection $\nabla^{\|}$ on S respects the metric $\bar{g}$. Indeed we shall have $\nabla^{\|}\bar{g} = 0$ if

$$\nabla_u^{\|}(\bar{g}(v, w)) = \bar{g}(\nabla^{\|}_u v, w) + \bar{g}(v, \nabla^{\|}_u w)$$

for all vector fields u, v, w on S: the left hand side is the derivative in the direction u of the function $\bar{g}(v, w)$, derivative also denoted (p. 303) $u(\bar{g}(v, w))$, and we have the identity (cf. p. 303 the Leibniz rule and derivation of a contracted product)

$$u(\bar{g}(v, w)) \equiv (\nabla_u \bar{g})(v, w) + \bar{g}(\nabla^{\|}_u v, w) + \bar{g}(v, \nabla^{\|}_u w)$$

which reduces to the above formula for every v, w on S if and only if $\nabla_u \bar{g} = 0$. By the definition of $\bar{g}$

$$\bar{g}(v, w) = g(v, w), \qquad \forall v, w \in T_xS \subset T_xX$$

thus $u(\bar{g}(v, w)) = u(g(v, w))$, $\forall u, v, w$ vector fields on S, while

$$\bar{g}(\nabla^{\|}_u v, w) \equiv g(\nabla_u v - (\nabla_u v)^{\perp}, w) = g(\nabla_u v, w)$$

since $g((\nabla_u v)^{\perp}, w) = 0$, because $(\nabla_u v)^{\perp}$ and w are g-orthogonal. Now, since

$\nabla \boldsymbol{g} = 0$ we have

$$\boldsymbol{u}(\boldsymbol{g}(\boldsymbol{v}, \boldsymbol{w})) = \boldsymbol{g}(\nabla_{\boldsymbol{u}}\boldsymbol{v}, \boldsymbol{w}) + \boldsymbol{g}(\boldsymbol{v}, \nabla_{\boldsymbol{u}}\boldsymbol{w}).$$

Comparing the various equalities gives

$$(\nabla^{\parallel}{}_{\boldsymbol{u}}\bar{\boldsymbol{g}})(\boldsymbol{v}, \boldsymbol{w}) = 0 \qquad \text{for all vector fields } \boldsymbol{u}, \boldsymbol{v}, \boldsymbol{w} \text{ on } S$$

thus $\nabla^{\parallel}\boldsymbol{g} = 0$. By uniqueness, $\nabla^{\parallel}$ must be the riemannian connection $\bar{\nabla}$.
2. The symmetry property $(\nabla_{\boldsymbol{u}}\boldsymbol{v})^{\perp} = (\nabla_{\boldsymbol{v}}\boldsymbol{u})^{\perp}$ is a consequence of b) of the proof of 1, which gives

$$\nabla_{\boldsymbol{u}}\boldsymbol{v} - \nabla_{\boldsymbol{v}}\boldsymbol{u} = \bar{\nabla}_{\boldsymbol{u}}\boldsymbol{v} - \bar{\nabla}_{\boldsymbol{v}}\boldsymbol{u}.$$

We show that $(\nabla_{\boldsymbol{u}}\boldsymbol{v})_x^{\perp}$ depends only on the vectors $\boldsymbol{u}_x$ and $\boldsymbol{v}_x$. Let f be a differentiable function on S; we have

$$\nabla_{f\boldsymbol{u}}\boldsymbol{v} = f\nabla_{\boldsymbol{u}}\boldsymbol{v} = f(\bar{\nabla}_{\boldsymbol{u}}\boldsymbol{v} + (\nabla_{\boldsymbol{u}}\boldsymbol{v})^{\perp}) = \bar{\nabla}_{f\boldsymbol{u}}\boldsymbol{v} + f(\nabla_{\boldsymbol{u}}\boldsymbol{v})^{\perp}$$

from which we deduce, using also the symmetry of $\boldsymbol{k}$,

$$\boldsymbol{k}(f\boldsymbol{u}, h\boldsymbol{v}) = fh\boldsymbol{k}(\boldsymbol{u}, \boldsymbol{v})$$

if h is another differentiable function on S. The implication from this formula that $\boldsymbol{k}_x(\boldsymbol{u}, \boldsymbol{v})$ depends only on $\boldsymbol{u}_x$ and $\boldsymbol{v}_x$ can be easily seen by using local coordinates (x^i), then

$$\boldsymbol{u} = u^i(\partial/\partial x^i), \qquad \boldsymbol{v} = v^i(\partial/\partial x^i)$$

where u^i and v^i are functions in a neighborhood of x in S, thus

$$\boldsymbol{k}_x(\boldsymbol{u}, \boldsymbol{v}) = u^i(x)v^j(x)\boldsymbol{k}_x(\partial/\partial x^i, \partial/\partial x^j)$$

depends only on the values at x of u^i and v^i. ■

$\boldsymbol{k}_x$ is a quadratic form on T_xS, which values in $(T_xS)^{\perp}$. In the case where X has dimension $p+1$, i.e. where S is an hypersurface of X, $(T_xS)^{\perp}$ is one dimensional, it is generated by the unit normal $\boldsymbol{n}$ to S. One then sets

$$\boldsymbol{k}_x(\boldsymbol{u}, \boldsymbol{v}) = \boldsymbol{K}_x(\boldsymbol{u}, \boldsymbol{v})\boldsymbol{n}$$

extrinsic curvature

mean extrinsic curvature

and $\boldsymbol{K}_x$ is an ordinary symmetric covariant 2-tensor. This tensor is called the **extrinsic curvature** of S as a submanifold of X. The trace of $\boldsymbol{K}$ in the metric $\bar{\boldsymbol{g}}$ is called the **mean extrinsic curvature** of S. An hypersurface with **tr $\boldsymbol{K} = \bar{g}^{ij}K_{ij} = 0$ in a proper riemannian metric is a minimal hypersurface; in a lorentzian metric it is a maximal hypersurface. They realize local minima [resp. maxima] of the n-dimensional volume.**

Exercise 1: Compute the extrinsic curvature $\boldsymbol{K}$ of a hypersurface $S \subset X$ in local coordinates. Show that $K = -\frac{1}{2}(\mathcal{L}_n g)^{\|}$ where g is the metric on X and n is the unit normal to the embedded hypersurface defined as follows. Let f: $S \to X$ by $s \mapsto x$ be the embedding of S in X. The **unit normal** $n \in T^*_{f(s)}X$ is defined, up to a sign, by

unit normal

$$\begin{cases} \langle n, f'v\rangle = 0 & \text{for all } v \in T_sS \\ g^{\alpha\beta} n_\alpha n_\beta = \pm 1. \end{cases}$$

$\mathcal{L}_n g$ is to be understood as the Lie derivative of g with respect to any vector field equal to n at $x = f(s)$. Its projection $(\mathcal{L}_n g)^{\|}$ on $T_{f(s)}f(S)$ is

$$(\mathcal{L}_n g)^{\|}_{\alpha\beta} = h^{\gamma}_{\alpha} h^{\delta}_{\beta} (\mathcal{L}_n g)_{\gamma\delta}$$

where $h_{\alpha\beta} = g_{\alpha\beta} \mp n_\alpha n_\beta$, $h^{\beta}_{\alpha} = h_{\alpha\gamma} g^{\gamma\beta}$.

Answer: Let (s^i), $i = 1, \ldots, p$ be local coordinates in S, and let (x^α), $\alpha = 0, 1, \ldots, n$ be local coordinates in X. In terms of these coordinates the embedding is $(x^\alpha) = (x^0, x^i) = (f^\alpha(s)) = (0, s^i)$, so $x^0 = 0$ is the local equation of $f(S)$. Let (v^i) be the components of a vector $v \in T_sS$, then the components of $f'v$ are $(0, v^i)$ and the components of n are $n_i = 0$, $n_0^2 = (g^{00})^{-1}$. By definition

$$K(e_i, e_j) n^\alpha = (\nabla_{e_i} e_j)^{\perp\alpha} = n^\alpha n_\beta (\nabla_{e_i} e_j)^\beta \quad \text{for } e_i, e_j \in T_{f(s)}f(S)$$

since, for $\omega \in T_{f(s)}X$, we can write $\omega^\alpha = \delta^\alpha_\beta \omega^\beta = (h^\alpha_\beta + n^\alpha n_\beta)\omega^\beta \equiv (\omega^{\|})^\alpha + (\omega^{\perp})^\alpha$. Hence for (e_i) the natural basis in $T_{f(s)}f(S)$

$$K_{ij} = K(e_i, e_j) = n_\beta (\nabla_{e_i} e_j)^\beta = n_\beta \Gamma^\beta_{ij} = n_0 \Gamma^0_{ij}.$$

On the other hand

$$\begin{aligned} -\tfrac{1}{2}(\mathcal{L}_n g)^{\|}_{ij} &= -\tfrac{1}{2} h^\alpha_i h^\beta_j (\mathcal{L}_n g)_{\alpha\beta} = -\tfrac{1}{2}(\mathcal{L}_n g)_{ij} && \text{in the chosen coordinates} \\ &= -\tfrac{1}{2}(n_{i;j} + n_{j;i}) = \Gamma^0_{ij} n_0 && \text{in the chosen coordinates.} \quad \blacksquare \end{aligned}$$

Exercise 2: The **Gauss–Codazzi relations**. Let X be a riemannian manifold with metric g, $S \subset X$ a hypersurface with induced metric $\bar{g}$ and extrinsic curvature K. Set $G_{\mu\nu} = R_{\mu\nu} - \frac{1}{2} g_{\mu\nu} R$ (Einstein tensor). Prove that:

Gauss–Codazzi relations

$$2G^0_0 = -\bar{R} - K^i_j K^j_i + (K^i_i)^2 \quad \text{Gauss relation}$$

$$G^0_i = \bar{\nabla}_j K^j_i - \bar{\nabla}_i \ \mathrm{tr}\, K \quad \text{Codazzi relation.}$$

Answer: Using Gauss normal coordinates. Let (x^α), $\alpha = 0, 1, \ldots, p$ be local coordinates in X with $x^0 = 0$ the local equation of S. The **Gauss normal coordinate metric** is

Gauss normal coordinates

$$ds^2 = (dx^0)^2 + g_{ij}(x^\mu)\, dx^i\, dx^j.$$

With this metric $g_{00}=1$, $g_{0i}=0$, $g_{ij}|_S=\bar{g}_{ij}$, $g^{00}=1$, $g^{0i}=0$, $g^{ij}g_{jk}=\delta^i_k$;

$$\Gamma^0_{00}=\Gamma^0_{0i}=\Gamma^i_{00}=0, \qquad \Gamma^l_{0k}=-g^{lm}\Gamma^0_{mk}, \qquad \Gamma^i_{jk}|_S=\bar{\Gamma}^i_{jk},$$

$$G^0_0=\tfrac{1}{2}(R_{00}-g^{ij}R_{ij}), \qquad G^0_i=g^{00}R_{0i};$$

$$R_{00}=\partial_0\Gamma^k_{k0}+\Gamma^k_{0m}\Gamma^m_{k0}, \qquad R_{0j}=\partial_j\Gamma^k_{k0}-\partial_k\Gamma^k_{j0}+\Gamma^k_{jm}\Gamma^m_{k0}-\Gamma^k_{km}\Gamma^m_{j0},$$

$$R_{ij}=\bar{R}_{ij}-\partial_0\Gamma^0_{ij}+\Gamma^0_{jm}\Gamma^m_{0i}+\Gamma^k_{j0}\Gamma^0_{ki}-\Gamma^k_{k0}\Gamma^0_{ji}.$$

It follows from exercise 1 that in Gauss normal coordinates $\Gamma^0_{ij}=K_{ij}$. Hence $\partial_0\Gamma^l_{0k}=-\partial_0 g^{lm}\Gamma^0_{mk}=-2K^l_jK^j_k-g^{lm}\partial_0K_{mk}$, and similar calculations for $\partial_j\Gamma^k_{k0}$ and $\partial_k\Gamma^k_{j0}$. The results follow.
See another method in [Hawking, Ellis, pp. 44–47].

4. DIFFERENTIAL OPERATORS

linear differential operator

A mapping D: $C^\infty(X^n)\to C^\infty(X^n)$ is called a **linear differential operator of order** m if on each local chart (U,φ) there exist functions $a_j\in C^\infty(U)$, $j=(j_1,\ldots,j_n)$, such that

$$(\mathrm{D}f)(x)=\sum_{|j|=0}^{m} a_j\mathrm{D}^j(f\circ\varphi^{-1})(\varphi(x)), \quad f\in C^\infty(X^n), \quad x\in U,$$

where $\mathrm{D}^j=(\partial/\partial x^1)^{j_1}\ldots(\partial/\partial x^n)^{j_n}$.
A linear differential operator is local: for an open subset $V\subset X^n$, $(\mathrm{D}f_1)|_V=(\mathrm{D}f_2)|_V$ if $f_1|_V=f_2|_V$. A linear differential operator therefore defines an operator on germs of functions. Moreover, for any $f\in C^\infty(X^n)$

$$\operatorname{supp}\mathrm{D}f\subset\operatorname{supp}f.$$

invariant with respect to Φ

A linear differential operator D is said to be **invariant with respect to a diffeomorphism** Φ of X if

$$\boldsymbol{\Phi}^*(\mathrm{D}f)=\mathrm{D}(\boldsymbol{\Phi}^*f)$$

for all $f\in C^\infty(X)$.

The concept of a differential operator can be extended to differential forms, or more generally to sections of a vector bundle over X (cf. pp. 482–486). We give expressions in terms of covariant derivatives for the following commonly used differential operators.

Exterior derivative d. Let

$$\omega=\frac{1}{p!}\omega_{i_1\ldots i_p}\mathbf{d}x^{i_1}\wedge\cdots\wedge\mathbf{d}x^{i_p}.$$

Then

$$\mathrm{d}\boldsymbol{\omega} = \frac{1}{p!}\,\partial_j\omega_{i_1\ldots i_p}\,\mathbf{d}\boldsymbol{x}^j \wedge \mathbf{d}\boldsymbol{x}^{i_1} \wedge \cdots \wedge \mathbf{d}\boldsymbol{x}^{i_p}.$$

Since the torsion tensor vanishes on a riemannian manifold

$$\begin{aligned}\mathrm{d}\boldsymbol{\omega} &= \frac{1}{p!}\,\nabla_j\omega_{i_1\ldots i_p}\,\mathbf{d}\boldsymbol{x}^j \wedge \mathbf{d}\boldsymbol{x}^{i_1} \wedge \cdots \wedge \mathbf{d}\boldsymbol{x}^{i_p}\\ &= \frac{1}{p!(p+1)!}\,\epsilon^{ji_1\ldots i_p}_{k_1\ldots k_{p+1}}\nabla_j\omega_{i_1\ldots i_p}\,\mathbf{d}\boldsymbol{x}^{k_1}\wedge\cdots\wedge\mathbf{d}\boldsymbol{x}^{k_{p+1}}.\end{aligned}$$

The operator δ (see Problem 1, p. 335). Using the expression above we have

$$\begin{aligned}\mathrm{d}({}^*\boldsymbol{\omega}) &= \frac{1}{p!(n-p)!}\,\tau_{i_1\ldots i_n}\nabla_j\omega^{i_1\ldots i_p}\,\mathbf{d}\boldsymbol{x}^j\wedge\mathbf{d}\boldsymbol{x}^{i_{p+1}}\wedge\cdots\wedge\mathbf{d}\boldsymbol{x}^{i_n}\\ &= \frac{1}{(n-p+1)!p!(n-p)!}\,\epsilon^{ji_{p+1}\ldots i_n}_{k_p\ldots k_n}\tau^{i_1\ldots i_p}{}_{i_{p+1}\ldots i_n}\nabla_j\omega_{i_1\ldots i_p}\,\mathrm{d}x^{k_p}\wedge\cdots\wedge\mathrm{d}x^{k_n}.\end{aligned}$$

Then the δ operator defined p. 296 is

$$\delta\boldsymbol{\omega} = \frac{(-1)^p(-1)^{(n-p+1)(p-1)}\,\mathrm{sign}\,g}{(p-1)!(n-p+1)!p!(n-p)!}\,\tau_{k_p\ldots k_n k_1\ldots k_{p-1}}\epsilon^{k_p\ldots k_n}_{ji_{p+1}\ldots i_n}$$

$$\times\,\tau^{i_1\ldots i_n}\nabla^j\omega_{i_1\ldots i_p}\,\mathbf{d}\boldsymbol{x}^{k_1}\wedge\cdots\wedge\mathbf{d}\boldsymbol{x}^{k_{p-1}}$$

and by the properties of the Kronecker symbol

$$\begin{aligned}\delta\boldsymbol{\omega} &= \frac{(-1)^p(-1)^{(n-p+1)(p-1)}}{(p-1)!p!(n-p)!}\,\epsilon^{1\ldots n}_{ji_{p+1}\ldots i_n k_1\ldots k_{p-1}}\epsilon^{i_1\ldots i_n}_{1\ldots n}\\ &\qquad\times\nabla^j\omega_{i_1\ldots i_p}\,\mathbf{d}\boldsymbol{x}^{k_1}\wedge\cdots\wedge\mathbf{d}\boldsymbol{x}^{k_{p-1}}\\ &= \frac{(-1)^p(-1)^{(n-p+1)(p-1)}}{(p-1)!p!(n-p)!}(-1)^{(p-1)(n-p)}\epsilon^{i_1\ldots i_n}_{jk_1\ldots k_{p-1}i_{p+1}\ldots i_n}\\ &\qquad\times\nabla^j\omega_{i_1\ldots i_p}\,\mathbf{d}\boldsymbol{x}^{k_1}\wedge\cdots\wedge\mathbf{d}\boldsymbol{x}^{k_{p-1}}\\ &= \frac{-1}{(p-1)!p!}\,\epsilon^{i_1\ldots i_p}_{jk_1\ldots k_{p-1}}\nabla^j\omega_{i_1\ldots i_p}\,\mathbf{d}\boldsymbol{x}^{k_1}\wedge\cdots\wedge\mathbf{d}\boldsymbol{x}^{k_{p-1}}.\end{aligned}$$

Thus finally

$$\delta\boldsymbol{\omega} = \frac{-1}{(p-1)!}\,\nabla^j\omega_{jk_1\ldots k_{p-1}}\,\mathbf{d}\boldsymbol{x}^{k_1}\wedge\cdots\wedge\mathbf{d}\boldsymbol{x}^{k_{p-1}}.$$

The divergence (cf. p. 296). If $\boldsymbol{v}$ is a vector with components v^i then

$$\mathrm{div}\,\boldsymbol{v} = \nabla_i v^i.$$

Since $\Gamma^i_{ki} = \partial_k \ln |g|^{1/2}$ we can write

$$\operatorname{div} v = \frac{1}{|g|^{1/2}} \partial_i(|g|^{1/2} v^i).$$

Exercise: Show that, $\boldsymbol{\tau}$ being the volume element (p. 294),

$$\mathrm{d} * \boldsymbol{v} = \boldsymbol{\tau} \operatorname{div} \boldsymbol{v}.$$

Infer from Stoke's formula (p. 216) that

$$\int_c (\operatorname{div} \boldsymbol{v})\boldsymbol{\tau} = \int_{\partial c} * \boldsymbol{v}.$$

Answer:

$$\begin{aligned} * \boldsymbol{v} &= \frac{1}{(n-1)!} \tau_{i_1 i_2 \ldots i_n} v^{i_1} \, \mathbf{dx}^{i_2} \wedge \cdots \wedge \mathbf{dx}^{i_n} \\ &= \frac{1}{(n-1)!} |g|^{1/2} \epsilon^{1 \ldots n}_{i_1 \ldots i_n} v^{i_1} \, \mathbf{dx}^{i_2} \wedge \cdots \wedge \mathbf{dx}^{i_n} \\ &= |g|^{1/2} \sum_{i=1}^{n} (-1)^{i-1} v^i \, \mathbf{dx}^1 \wedge \cdots \wedge \widehat{\mathbf{dx}^i} \wedge \cdots \wedge \mathbf{dx}^n \end{aligned}$$

$$\mathrm{d} * \boldsymbol{v} = \sum_{i=1}^{n} (-1)^{i-1} \partial_k (|g|^{1/2} v^i) \, \mathbf{dx}^k \wedge \mathbf{dx}^1 \wedge \cdots \wedge \widehat{\mathbf{dx}^i} \wedge \cdots \wedge \mathbf{dx}^n.$$

Since all terms of the summation for which $k \neq i$ vanish

$$\begin{aligned} \mathrm{d} * \boldsymbol{v} &= \partial_k (|g|^{1/2} v^k) \, \mathbf{dx}^1 \wedge \cdots \wedge \mathbf{dx}^n \\ &= \frac{1}{|g|^{1/2}} \partial_k (|g|^{1/2} v^k) \boldsymbol{\tau} = \boldsymbol{\tau} \operatorname{div} \boldsymbol{v}. \end{aligned}$$ ■

Note that $* \boldsymbol{v} = \mathrm{i}_v \boldsymbol{\tau}$.

laplacian Δ

The laplacian. The linear differential operator $\Delta: \Lambda^p(X) \to \Lambda^p(X)$ defined by

$$\Delta = \mathrm{d}\delta + \delta \mathrm{d}$$

is called the laplacian.

harmonic

If the form $\boldsymbol{\omega}$ is such that $\Delta \boldsymbol{\omega} = 0$, it is called **harmonic**.

The operator Δ is self-adjoint; that is,

$$[\Delta \boldsymbol{\alpha} | \boldsymbol{\beta}] = [\delta \boldsymbol{\alpha} | \delta \boldsymbol{\beta}] + [\mathrm{d}\boldsymbol{\alpha} | \mathrm{d}\boldsymbol{\beta}] = [\boldsymbol{\alpha} | \Delta \boldsymbol{\beta}]$$

where $\boldsymbol{\alpha}$ and $\boldsymbol{\beta}$ are p-forms with compact support.

On a proper riemannian manifold Δ is positive; that is,

$$[\Delta\omega|\omega] \geq 0 \quad \forall\, \omega \text{ with compact support}$$

since

$$[\Delta\omega|\omega] = [\delta\omega|\delta\omega] + [\mathrm{d}\omega|\mathrm{d}\omega].$$

Exercise: Show that

$$*\Delta = \Delta *$$
$$\delta\Delta = \Delta\delta$$
$$\mathrm{d}\Delta = \Delta\mathrm{d}.$$

Exercise:[1] Let $\boldsymbol{\omega} = \frac{1}{p!}\omega_{i_1\ldots i_p}\,\mathbf{d}\boldsymbol{x}^{i_1}\wedge\cdots\wedge\mathbf{d}\boldsymbol{x}^{i_p}$, show that in local coordinates

$$\begin{aligned}(\Delta\boldsymbol{\omega})_{i_1\ldots i_p} &= -g^{ij}\nabla_i\nabla_j\omega_{i_1\ldots i_p}\\ &+\sum_q(-1)^q R^h_{i_q}\,\omega_{hi_1\ldots\hat{i}_q\ldots i_p}\\ &+2\sum_{\substack{r,q\\ r<q}}(-1)^{r+q}R^{h\ \ j}_{\ i_q\ i_r}\,\omega_{hi_1\ldots\hat{i}_r\ldots\hat{i}_q\ldots i_p}.\end{aligned}$$

In the case of a 0-form f

$$\begin{aligned}\Delta f = \delta\,\mathrm{d}f &= -g^{ij}\nabla_j\nabla_i f\\ &= -g^{ij}\partial_i\partial_j f + g^{ij}\Gamma^l_{ij}\partial_l f\\ &= -\frac{1}{|g|^{1/2}}\,\partial_i(g^{ij}|g|^{1/2}\partial_j f).\end{aligned}$$

Note that the sign is the opposite of the sign of the laplacian as it is usually defined. In this book the symbol Δ on a 0-form is used to mean $\delta\mathrm{d}$ or $g_{ij}(\partial/\partial x^i)(\partial/\partial x^j)$ according to convenience.

Theorem: *Let $\Phi: X \to X$ be a diffeomorphism. Then Φ is an isometry (p. 298) if and only if*

$$\Phi^*(\Delta f) = \Delta(\Phi^* f) \qquad \forall f \in C^\infty(X).$$

Proof: (1) Assume that Φ is an isometry. Let (U, φ) be a local chart about $x \in X$, and choose $(U', \varphi') = (\Phi(U), \varphi\circ\Phi^{-1})$ as a local chart about $x' = \Phi(x)$. Then (see proof of theorem, p. 297)

$$\bar{g}_{\hat{i}\hat{j}} = \bar{g}'_{\hat{i}\hat{j}}, \quad \text{i.e. } g_{ij}\circ\varphi^{-1} = g_{i'j'}\circ\varphi'^{-1} \quad \text{on } \varphi(U) = \varphi'(U').$$

Here $\boldsymbol{g}$ is the metric on X, and $\bar{\boldsymbol{g}} = (\varphi^{-1})^*\boldsymbol{g}$, $\bar{\boldsymbol{g}}' = (\varphi'^{-1})^*\boldsymbol{g}$ are the (equal) induced metrics on $\varphi(U) = \varphi'(U') \subset \mathbb{R}^n$, while unprimed components, primed components are with respect to the coordinate systems (U, φ), (U', φ') respectively, and circumflexed components are with respect to the

[1]See, for instance [Lichnerowicz 1958, pp. 1–5].

natural coordinate system on $\mathbb{R}^n$. Hence

$$g_{ij}(x) = (g_{i'j'} \circ \varphi'^{-1} \circ \varphi)(x) = (g_{i'j'} \circ \Phi)(x) = g_{i'j'}(x').$$

Since (see proof of theorem, p. 308) the g_{ij}'s [$g_{i'j'}$'s] determine $\Gamma^l_{ij}(x)$ [$\Gamma^{l'}_{i'j'}(x')$], it follows that $\Gamma^l_{ij}(x) = \Gamma^{l'}_{i'j'}(x')$. Also

$$\frac{\partial f}{\partial x^{j'}}(x') = \frac{\partial (f \circ \Phi)}{\partial x^j}(x) \quad \text{and} \quad \frac{\partial^2 f}{\partial x^{i'}\, \partial x^{j'}}(x') = \frac{\partial^2 (f \circ \Phi)}{\partial x^i\, \partial x^j}(x). \tag{2}$$

From

$$\Delta(\Phi^* f)(x) = \Delta(f \circ \Phi)(x) = -g^{ij}(x) \frac{\partial^2 f \circ \Phi}{\partial x^i\, \partial x^j}(x) + g^{ij}(x)\, \Gamma^l_{ij}(x) \frac{\partial f \circ \Phi}{\partial x^l}(x)$$

and

$$\begin{aligned} \Phi^*(\Delta f)(x) &= (\Delta f) \circ \Phi(x) = \Delta f(x') \\ &= -g^{i'j'}(x') \frac{\partial^2 f}{\partial x^{i'}\, \partial x^{j'}}(x') + g^{i'j'}(x')\, \Gamma^{l'}_{i'j'}(x') \frac{\partial f}{\partial x^{l'}}(x') \end{aligned}$$

it then follows that $\Phi^*(\Delta f) = \Delta(\Phi^* f)$.

(2) Conversely, assume that $\Phi^*(\Delta f) = \Delta(\Phi^* f)$. Use the same coordinate systems as before. Since equation (2) is independent of the metric,

$$\begin{aligned} 0 &= \Delta(\Phi^* f)(x) - \Phi^*(\Delta f)(x) \\ &= -[g^{ij}(x) - g^{i'j'}(x')] \frac{\partial^2 f \circ \Phi}{\partial x^i\, \partial x^j}(x) + [g^{ij}(x)\, \Gamma^l_{ij}(x) - g^{i'j'}(x')\, \Gamma^{l'}_{i'j'}(x')] \frac{\partial f \circ \Phi}{\partial x^l}(x). \end{aligned}$$

The arbitrariness of f then implies that the coefficients in brackets vanish; in particular $g^{ij}(x) = g^{i'j'}(x')$, or equivalently $\boldsymbol{g} = \Phi^* \boldsymbol{g}$. ■

C. GEODESICS

1. ARC LENGTH

On a (smooth) proper riemannian manifold X the presence of the metric $\boldsymbol{g}$ allows the definition of the **length $J(C)$ of a differentiable curve** (path) $C: [a, b] \to X$ by $t \mapsto C(t)$

arc length

$$J(C) = \int_a^b \left(\frac{\mathrm{d}C(t)}{\mathrm{d}t} \middle| \frac{\mathrm{d}C(t)}{\mathrm{d}t} \right)^{1/2} \mathrm{d}t.$$

We remark that all paths (curves) deduced from C by a change of parameter – that is $\gamma = C \circ \psi$ with $\psi: \tau \mapsto t = \psi(\tau)$ – have the same

length. The length is a property of the geometrical curve, defined by the equivalence class of parametrized paths.
If the parameter t is such that $(\mathrm{d}C/\mathrm{d}t|\mathrm{d}C/\mathrm{d}t)$ is constant along C the curve is said to be parametrized proportionally to arc length.
The distance between two points x and y of X is by definition

$$\mathrm{d}(x, y) = \inf_C J(C)$$

where C runs over all piecewise differentiable curves from x to y. It will be proved that this distance turns the manifold X into a metric space (cf. p. 326). The resulting topology coincides with the topology of X as a manifold.
On a proper riemannian manifold a curve of minimum length between two nearby points is a geodesic. In order to generalize to the case of a pseudo-riemannian manifold it is convenient to consider the **energy** of the curve C energy

$$E(C) = \int_a^b \left(\frac{\mathrm{d}C(t)}{\mathrm{d}t}\middle|\frac{\mathrm{d}C(t)}{\mathrm{d}t}\right) \mathrm{d}t.$$

This integral is not invariant with respect to the transformation $C \mapsto C \circ \psi$ (change of parameter).

2. VARIATIONS

J and E are functions on the space $\Omega(x, y)$ of differentiable paths from x to y, that is of differentiable mappings $C: [a, b] \to X$, $C(a) = x$, $C(b) = y$. The space $\Omega(x, y)$ is not a vector space; we cannot apply directly the results of Chapter II to study the extrema of J or E. In fact $\Omega(x, y)$ can be endowed with a structure of an infinite dimensional riemannian manifold (cf. Chapter VII). However, to avoid a rather cumbersome formalism we shall study directly the stationary (critical) points of E, by a variation method described below. A critical point of E will be called a **geodesic**. One can show, in the proper riemannian case, that a critical point of E is also a critical point of J, parametrized proportionally to arc length, and that any critical point of J is equivalent to some path which is also a critical point of E. geodesic

Let $C: [a, b] \to X$ by $t \mapsto C(t)$ be a C^1 path. A **variation (with fixed end points)** of C is a one parameter family of paths $C_u: [a, b] \to X$ by $t \mapsto \psi(u, t)$ such that variation

a) $\psi(u, t)$ is continuously differentiable for $-\epsilon \leq u \leq \epsilon$, $a \leq t \leq b$,
b) $C_0 = C$,
c) $\psi(u, a) = C(a) = x$ and $\psi(u, b) = C(b) = y$, $-\epsilon \leq u \leq \epsilon$.

stationary A function I on the space of paths $\Omega(x, y)$ is said to be **stationary** at C if for each variation C_u of C

$$\frac{\mathrm{d}}{\mathrm{d}u} I(C_u)\bigg|_{u=0} = 0.$$

We consider in particular the function

$$I(C) = \int_a^b L \circ \tilde{C}(t)\, \mathrm{d}t,$$

where $L: T(X) \times \mathbb{R} \to \mathbb{R}$ is a smooth function called the lagrangian. The curve $\tilde{C}: t \mapsto (C(t), \mathrm{d}C(t)/\mathrm{d}t, t) \equiv \tilde{C}(t) \in T(X) \times \mathbb{R}$ is called the lift of the curve $C: t \mapsto C(t)$ on X. In this case

$$\frac{\mathrm{d}}{\mathrm{d}u} I(C_u) = \frac{\mathrm{d}}{\mathrm{d}u} \int_a^b L \circ \tilde{\psi}(u, t)\, \mathrm{d}t.$$

By the definition of the differential of a function on a manifold we have

$$\frac{\mathrm{d}}{\mathrm{d}u}(L \circ \tilde{\psi}(u, t)) = \mathrm{d}L \circ \frac{\partial \tilde{\psi}}{\partial u}(u, t)$$

where $\mathrm{d}L$, the differential of L on TX, is a linear mapping

$$\mathrm{d}L: T(TX) \to \mathbb{R}$$

and $\partial\tilde{\psi}/\partial u$ is a tangent vector to TX, $\partial\tilde{\psi}/\partial u \in T(TX)$, since

$$\tilde{\psi}: (u, t) \mapsto \left(\psi(u, t), \frac{\partial}{\partial t}\psi(u, t)\right).$$

Thus C will be stationary if and only if

$$\int_a^b \left\{\mathrm{d}L \circ \frac{\partial \tilde{\psi}}{\partial u}(u, t)\right\}_{u=0} \mathrm{d}t = 0.$$

Assume that for $t_i \leq t \leq t_{i+1}$ the range of C is in the domain U_i of a chart of X, with local coordinates x^i. Denote by $\bar{L}$ the expression for L in this domain. For such values of t

$$\left\{\mathrm{d}L \circ \frac{\partial \tilde{\psi}}{\partial u}\right\}_{u=0} = \left\{\frac{\partial \bar{L}}{\partial x^i}\frac{\partial \psi^i}{\partial u} + \frac{\partial \bar{L}}{\partial v^i}\frac{\partial^2 \psi^i}{\partial u \partial t}\right\}_{u=0},$$

where $\partial\bar{L}/\partial x^i$ actually means $(\partial\bar{L}/\partial x^i)\circ\bar{\psi}$, etc. Therefore

$$\int_{t_i}^{t_{i+1}} \left\{\mathrm{d}L \circ \frac{\partial\bar{\psi}}{\partial u}\right\}_{u=0} \mathrm{d}t = \int_{t_i}^{t_{i+1}} \left\langle \mathscr{E}_L, \frac{\partial\psi}{\partial u}\right\rangle_{u=0} \mathrm{d}t + \left[\left\langle \mathbf{d}_v\boldsymbol{L}, \frac{\partial\psi}{\partial u}\right\rangle_{u=0}\right]_{t_i}^{t_{i+1}}$$

where $\mathscr{E}_L$ and $\mathbf{d}_v\boldsymbol{L}$ are covariant vectors on X (cf. Problem III 1, p. 170) whose components with respect to the coordinates (x^i) at the point $C^i(t)$ are

$$\mathscr{E}_i \equiv \frac{\partial\bar{L}}{\partial x^i}(C^k(t), C'^k(t)) - \frac{\mathrm{d}}{\mathrm{d}t}\left(\frac{\partial\bar{L}}{\partial v^i}(C^k(t), C'^k(t))\right),$$

$$(\mathrm{d}_vL)_i \equiv \frac{\partial\bar{L}}{\partial v^i}(C^k(t), C'^k(t)).$$

Thus, finally

$$\left\{\frac{\mathrm{d}}{\mathrm{d}u} I(C_u)\right\}_{u=0} = \int_a^b \left\langle \mathscr{E}_L, \frac{\partial\psi}{\partial u}\right\rangle_{u=0} \mathrm{d}t$$

and the necessary and sufficient condition for C to be a stationary (critical) point of I is that the Euler equations

$$\mathscr{E}_L = 0, \quad \text{i.e. } \mathscr{E}_i = \frac{\partial\bar{L}}{\partial x^i} - \frac{\mathrm{d}}{\mathrm{d}t}\frac{\partial\bar{L}}{\partial v^i} = 0, \quad i = 1, \ldots n,$$

be satisfied on C.

An integral curve of the Euler equations is called a **geodesic of the lagrangian** L. **geodesic of the lagrangian**

Exercise: Show that under a regularity condition on L a piecewise C^1 curve C is stationary with respect to piecewise C^1 variations if and only if C is C^1 and satisfies Euler equations.

Answer: There is a subdivision $a = t_0 < t_1 \cdots < t_k = b$ of $[a, b]$ such that C is C^1 in each interval $[t_i, t_{i+1}]$.

A variation $C_u: t \mapsto \psi(u, t)$ of C is defined by a mapping ψ which is C^1 on each strip $[-\epsilon, \epsilon] \times [t_i, t_{i+1}]$. The preceding computation shows that

$$\left\{\frac{\mathrm{d}}{\mathrm{d}u} I(C_u)\right\}_{u=0} = \int_a^b \left\langle \mathscr{E}_L, \frac{\partial\psi}{\partial u}\right\rangle \mathrm{d}t + \sum_{i=1}^{k-1} \left\langle [\mathbf{d}_v\boldsymbol{L}]_i, \left[\frac{\partial\psi}{\partial u}\right]_i\right\rangle = 0$$

where $[\quad]_i$ denotes the magnitude of the discontinuity at $t = t_i$.
The conditions $\mathscr{E}_L = 0$ and $[\mathbf{d}_v L]_i = 0$ are then shown to be necessary by choosing appropriate variations ψ. They are obviously sufficient. The condition $[\mathbf{d}_v L]_i \equiv [L'_v]_i = 0$ implies $[C']_i = 0$ if the second derivative L''_{vv} is a non degenerate quadratic form. ■

Let $C: t \mapsto C(t)$ be a geodesic of L. If $\partial L/\partial t = 0$, then at each point of C

$$\frac{\mathrm{d}L}{\mathrm{d}t} = \frac{\partial \bar{L}}{\partial x^i}\frac{\mathrm{d}C^i}{\mathrm{d}t} + \frac{\partial \bar{L}}{\partial v^i}\frac{\mathrm{d}^2C^i}{\mathrm{d}t^2} = \left(\frac{\mathrm{d}}{\mathrm{d}t}\frac{\partial \bar{L}}{\partial v^i}\right)\frac{\mathrm{d}C^i}{\mathrm{d}t} + \frac{\partial \bar{L}}{\partial v^i}\frac{\mathrm{d}^2C^i}{\mathrm{d}t^2}$$
$$= \frac{\mathrm{d}}{\mathrm{d}t}\left(\frac{\partial \bar{L}}{\partial v^i}\frac{\mathrm{d}C^i}{\mathrm{d}t}\right).$$

Thus

$$v^i(\partial \bar{L}/\partial v^i) - \bar{L} = \text{constant}$$

is a first integral of the Euler equations provided that L does not depend explicitly on t. This first integral is called the **energy integral**.

energy integral

We now consider the special case of the energy function E. Here $L(x, v_x, t) = g_x(v_x, v_x)$ or $\bar{L}(x^i, v^i, t) = g_{ij}(x^k)v^iv^j$. The Euler equations become

$$\frac{\partial g_{ij}}{\partial x^k}\frac{\mathrm{d}C^i}{\mathrm{d}t}\frac{\mathrm{d}C^j}{\mathrm{d}t} - 2\frac{\mathrm{d}}{\mathrm{d}t}\left(g_{kj}\frac{\mathrm{d}C^j}{\mathrm{d}t}\right) = 0$$

or

$$g_{kj}\frac{\mathrm{d}^2C^j}{\mathrm{d}t^2} + \frac{\partial g_{kj}}{\partial x^m}\frac{\mathrm{d}C^m}{\mathrm{d}t}\frac{\mathrm{d}C^j}{\mathrm{d}t} - \frac{1}{2}\frac{\partial g_{ij}}{\partial x^k}\frac{\mathrm{d}C^i}{\mathrm{d}t}\frac{\mathrm{d}C^j}{\mathrm{d}t} = 0.$$

Multiplication by g^{lk} gives

$$\frac{\mathrm{d}^2C^l}{\mathrm{d}t^2} + \Gamma^l_{mj}\frac{\mathrm{d}C^m}{\mathrm{d}t}\frac{\mathrm{d}C^j}{\mathrm{d}t} = 0.$$

We see that C is an affine geodesic (p. 302) and that t is an affine parameter. Note that $\partial^2 L/\partial v^i \partial v^j = g_{ij}$, hence the geodesics are C^1 curves. They are C^∞ curves if the metric is C^∞ due to the differential equation they satisfy.
The energy integral is

$$g_{ij}\frac{\mathrm{d}C^i}{\mathrm{d}t}\frac{\mathrm{d}C^j}{\mathrm{d}t} = \text{constant};$$

the tangent vector has constant length along a geodesic. In particular on a hyperbolic manifold if a geodesic has a timelike [spacelike] [null] tangent

vector at one point then it has a timelike [spacelike] [null] tangent vector at all points.

Exercise: Show using the energy integral, that a geodesic of E with non vanishing tangent vector is also a geodesic of J.

3. EXPONENTIAL MAPPING, NORMAL COORDINATES

We consider the first order differential system for the **geodesic flow** on $T(X)$ (margin: geodesic flow)

$$\frac{\mathrm{d}C^i}{\mathrm{d}t} = v^i, \quad \frac{\mathrm{d}v^i}{\mathrm{d}t} = -\Gamma^i_{kj}v^k v^j.$$

According to the classical existence and uniqueness theorems, for any $x_0 \in X$ there exists an $\epsilon > 0$ and an open neighborhood V of 0 in T_{x_0} such that there is, for all $w \in V$, a unique geodesic

$$C_w : [-\epsilon, \epsilon] \to X \text{ by } t \mapsto C_w(t)$$

with $C^i_w(0) = x^i_0$ and $\mathrm{d}C^i_w/\mathrm{d}t|_{t=0} = w^i$. Moreover the C^i_w depend differentiably on the w^i.

The fact that the equations for a geodesic are invariant with respect to an affine transformation allows a refinement of this result. If $C(t)$ is a geodesic then so is $C(\lambda t)$. And

$$\left.\frac{\mathrm{d}C^i(\lambda t)}{\mathrm{d}t}\right|_{t=0} = \lambda \left.\frac{\mathrm{d}C^i(t)}{\mathrm{d}t}\right|_{t=0}.$$

This means that the domain of any curve through x_0 can be extended to the interval $[-1, 1]$ provided that the neighborhood of 0 in T_{x_0} is suitably chosen. For example take $\{(w^i) \in T_{x_0};\ w/\epsilon \in V\}$. We assume in what follows that W_x is such an open neighborhood of 0 in T_x. Let C_w be the geodesic through x with tangent vector w at x.

The mapping $\exp: W_x \to X$ by $w \mapsto C_w(1) \equiv \exp_x w$ is the **exponential mapping**. Notice that $\exp tw = C_{tw}(1) = C_w(t)$ and $\mathrm{d}(\exp tw)/\mathrm{d}t|_{t=0} = w$. (margin: exponential mapping)

Theorem. The mapping $w \mapsto \exp_x w$ *is a diffeomorphism of an open neighborhood of* 0 *in* T_x *onto an open neighborhood of* x.

Proof: Since exp is differentiable on T_x we have only to show that $\exp' : T_0(T_x) \to T_x(X)$ is an isomorphism at x. Of course $T_x(X)$ and

$T_0(T_x)$ have the same dimension. The curve $t \mapsto tv$ on T_x is mapped by exp onto the geodesic $t \mapsto \exp tv = C_v(t)$. Therefore $\exp'(v) = v$. ∎

(geodetic) normal coordinates

Let (U, φ) be a coordinate system on X^n with $\varphi^{-1}(0) = x_0$. The coordinate system (U, φ) is **normal** with respect to x_0 if the inverse images under φ of straight lines through the origin in $\mathbb{R}^n$ are geodesics on X^n. Since the mapping $T_x(X^n) \to \mathbb{R}^n$ by $v \mapsto (v^i)$ is a diffeomorphism it follows that the exponential mapping can be used to define a normal coordinate system on a neighborhood of any point of X^n.

If (x^i) are normal coordinates on $U \subset X^n$ with respect to x_0, then every geodesic through x_0 is of the form

$$C^i(t) = a^i t \qquad (a^i) \in \mathbb{R}^n.$$

Moreover $\Gamma^i_{jk}(x_0) = 0$. This follows from the fact that

$$\Gamma^i_{jk}(a^i t) a^j a^k = 0 \qquad \forall (a^i) \in \mathbb{R}^n.$$

Exercise: Let $x = \exp_{x_0} tv$. Prove[1] that, in normal coordinates,

$$g_{ij}(x) = g_{ij}(x_0) + \tfrac{1}{3} t^2 R_{ikjl}(x_0) v^k v^l + o(t^2).$$

Normal coordinates are used in proving the following theorem.

Theorem[2]. *Every point of a riemannian manifold has a neighborhood U such that for any two points in U there is a unique geodesic which joins the points and lies in U.*

geodesically complete

The manifold X is **geodesically complete at x_0** if $\exp_{x_0} v$ is defined for all $v \in T_{x_0}$. The manifold X is **geodesically complete** if $\exp_x v$ is defined for all $v \in T_x$ and all $x \in X$. This is equivalent to saying that every geodesic on X can be extended to an infinite geodesic.
The euclidean space E^n is geodesically complete. So is the sphere $S^n = \{(x^i) \in E^{n+1};\ \Sigma_i (x^i)^2 = 1\}$ with the metric induced from E^{n+1}.

[1]See for instance O. Veblen, *Invariants of Quadratic Differential Forms* (Cambridge University Press, 1927) pp. 94–100, and L. Parker, in: *Recent Developments in Gravitation*, eds. M. Levy and S. Deser (Plenum Press N.Y., 1979) p. 228.
[2]See for example [Kobayashi and Nomizu, Ch. III §8]; [E. Cartan, Ch. X §1].

4. GEODESICS ON A PROPER RIEMANNIAN MANIFOLD

Throughout this section X is a proper riemannian manifold. Consider the function $d: X \times X \to \mathbb{R}$ by

$$d(x, y) = \inf_C J(C)$$

where C runs over all piecewise differentiable curves from x to y. There is not necessarily a curve between x and y of length $d(x, y)$. For example on the manifold $E^n - \{0\}$ the two points (x^i) and $(-x^i)$ cannot be connected by a curve of length $2\sqrt{\Sigma_i (x^i)^2}$.

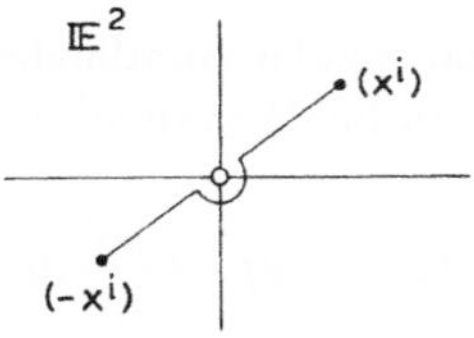

Theorem. The function d is a metric (p. 23) *on X.*

Proof: It is clear that $d(x, y) \geq 0$ and $d(x, y) = d(y, x)$ for all $x, y \in X$. Let $x, y, z \in X$. We know that there exist curves $C_1: [a, b] \to X$ and $C_2: [b, c] \to X$ connecting x to y and y to z such that

$$J(C_1) - d(x, y) < \epsilon$$

$$J(C_2) - d(y, z) < \epsilon.$$

If $C: [a, c] \to X$ where $C|_{[a,b]} = C_1$ and $C|_{[b,c]} = C_2$ then

$$J(C) \leq J(C_1) + J(C_2).$$

But

$$J(C_1) + J(C_2) < d(x, y) + d(y, z) + 2\epsilon.$$

Therefore $d(x, y) + d(y, z) \geq d(x, z)$.
The following theorem establishes that $d(x, y) = 0$ if and only if $x = y$. ∎

Theorem. Each point $x \in X$ has a neighborhood U such that for every $y \in U$ there is a unique geodesic which joins x and y and lies in U. The length of this geodesic is $d(x, y)$.

Proof: Let (z^i) be a set of coordinates on a neighborhood of $x \in X$ such that the natural basis for T_x is orthonormal. If the exponential mapping is used to construct a normal coordinate system around x the natural basis of T_x with respect to the new system will still be orthonormal. Let (y^i) be such a set of normal coordinates on a neighborhood U of x. We introduce "polar coordinates" (t, φ^α) on U

$$y^i = tc^i(\varphi^\alpha) \qquad \sum_i [c^i(\varphi^\alpha)]^2 = 1$$

where the φ^α are the usual angular coordinates on S^{n-1}. The set (t, φ^α) is of course not a "legal" coordinate system[1] on U. Since

$$\mathbf{d}y^i = c^i\,\mathbf{d}t - t\frac{\partial c^i}{\partial \varphi^\alpha}\mathbf{d}\varphi^\alpha$$

we have

$$\begin{aligned} \mathrm{d}s^2 &= g_{ij}\,\mathbf{d}y^i\,\mathbf{d}y^j \\ &= \gamma_{nn}\,\mathbf{d}t^2 + 2\gamma_{\alpha n}\,\mathbf{d}t\,\mathbf{d}\varphi^\alpha + \gamma_{\alpha\beta}\,\mathbf{d}\varphi^\alpha\,\mathbf{d}\varphi^\beta \end{aligned}$$

where

$$\begin{aligned} \gamma_{nn} &= g_{ij}c^ic^j \\ \gamma_{\alpha n} &= tc^ig_{ij}(\partial c^j/\partial \varphi^\alpha) \\ \gamma_{\alpha\beta} &= t^2 g_{ij}\frac{\partial c^i}{\partial \varphi^\alpha}\frac{\partial c^j}{\partial \varphi^\beta}. \end{aligned}$$

The curves $t \mapsto c^it$, $c^i = \text{constant}$, are geodesics on U. The energy integral gives

$$g_{ij}c^ic^j = \text{constant}.$$

But $g_{ij}(x)c^ic^j = \delta_{ij}c^ic^j = 1$. Therefore

$$\gamma_{nn} = g_{ij}c^ic^j = 1.$$

In terms of the coordinates (t, φ^α) the curves $t \mapsto (t, \varphi^\alpha)$, $\varphi^\alpha = \text{constant}$,

[1]See for instance p. 190 a correct coordinate system on S^2.

are geodesics. The Euler equations give

$$\Gamma^n_{nn} = \Gamma^\alpha_{nn} = 0$$

or

$$\gamma^{n\beta}\frac{\partial \gamma_{\beta n}}{\partial t} = 0 \quad \text{and} \quad \gamma^{\alpha\beta}\frac{\partial \gamma_{\beta n}}{\partial t} = 0.$$

Now provided that $t \neq 0$ the rank of the matrix $\begin{bmatrix} \gamma_{nn} & \gamma_{n\alpha} \\ \gamma_{\beta n} & \gamma_{\beta\alpha} \end{bmatrix}$ is n. Therefore $\partial\gamma_{\beta n}/\partial t = 0$ for $t \neq 0$ and also for $t = 0$ since $\gamma_{\beta n}$ is continuously differentiable. Thus $\gamma_{\beta n} = 0$ at $t = 0$. We have now shown that

$$\mathrm{d}s^2 = \mathbf{d}t^2 + \gamma_{\alpha\beta}\,\mathbf{d}\varphi^\alpha\,\mathbf{d}\varphi^\beta$$

on U.

Let $y_1 = (t_1, \varphi_1^\alpha) \in U$. The only geodesic in U connecting x and y_1 is $t \mapsto (t, \varphi_1^\alpha)$. The length of this geodesic is

$$\int_0^{t_1} \sqrt{\mathrm{d}s^2} = \int_0^{t_1} \mathrm{d}t = t_1.$$

The length of any other curve $c: t \mapsto (t, \varphi^\alpha(t))$ from x to y_1 is

$$J(c) = \int_0^{t_1} \left(1 + \gamma_{\alpha\beta}\frac{\mathrm{d}\varphi^\alpha}{\mathrm{d}t}\frac{\mathrm{d}\varphi^\beta}{\mathrm{d}t}\right)^{1/2} \mathrm{d}t.$$

Since the $\gamma_{\alpha\beta}$ determine a positive definite quadratic form, $J(c) > t_1$. Thus $d(x, y_1)$ is equal to the length of the geodesic fórm x to y_1. ■

It is clear that $x \neq y_1$ implies $d(x, y_1) \neq 0$ for $y_1 \in U$. Suppose $x_1 \notin U$. Let $B \subset U$ be the open ball $B = \{y; d(x, y) < r\}$. If C is a piecewise differentiable curve from x to x_1 let C' be the portion of this curve from x to its first intersection with the boundary of B. Then

$$d(x, x_1) = \inf_C J(C) \geq \inf_{C'} J(C') = r > 0.$$

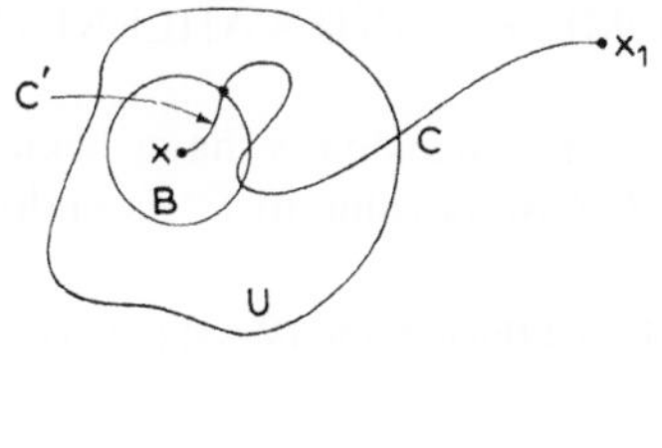

The topology given by d is equivalent to the topology of X as a manifold. This follows from the fact that, given r such that $B = \{(y^i);\ \Sigma_i\,(y^i)^2 < r\} \subset U$, there exist positive numbers k and K such that

$$\{y;\, d(x, y) < k\} \subset B \subset \{y;\, d(x, y) < K\} \subset U$$

where we have used the notation of the preceding theorem.

The following theorems describe some global properties of proper riemannian manifolds.

Hopf–Rinow theorem. On a connected proper riemannian manifold X the following conditions are equivalent.

a) *X is geodesically complete.*

b) *X is geodesically complete at a point $x_0 \in X$.*

c) *X is a complete metric space with metric d.*

d) *All closed subsets of X which are bounded with respect to d are compact.*

Corollary. Any compact proper riemannian manifold is geodesically complete.

Theorem[1]. *On a complete connected proper riemannian manifold any two points x and y can be connected by a geodesic of length $d(x, y)$.*

Notice that the geodesic is not necessarily unique: two diametrically opposite points on a sphere can be joined by infinitely many geodesics of the same length. More on this subject in Problem 3.

5. GEODESICS ON A HYPERBOLIC MANIFOLD

We refer the reader to the following references: [Choquet-Bruhat 1968], [Lichnerowicz 1968] and [Hawking and Ellis, Ch. 4].

D. ALMOST COMPLEX AND KÄHLERIAN MANIFOLDS

Complex manifolds Z of dimension n have been defined p. 112. The tangent vector space T_zZ isomorphic to $\mathbb{C}^n$ is endowed with a canonical

[1]See for example [Kobayashi and Nomizu, Ch. IV, §4] or [Spivak, Vol. I].

automorphism J by multiplication by the number $i \in \mathbb{C}$:

$$J\colon T_zZ \to T_zZ \quad \text{by} \quad \boldsymbol{v} \mapsto i\boldsymbol{v}.$$

This automorphism is such that

$$J^2 = -\mathbb{1}.$$

The complex manifold Z is also a real (analytic) manifold X of dimension $2n$. If $z^i = x^i + iy^i$ are local complex coordinates for Z, then the $2n$ real numbers x^i, y^i are admissible real coordinates for X and the automorphism J gives an automorphism of the tangent space to X, with $J^2 = -\mathbb{1}$. In the canonical coordinates (x^i, y^i), J becomes the canonical automorphism J_0 of $\mathbf{R}^{2n}$ given by

$$J_0(\partial/\partial x^i) = \partial/\partial y^i, \qquad J_0(\partial/\partial y^i) = -\partial/\partial x^i.$$

One is thus led to the following definition:

Definition: An **almost complex manifold** is a real differentiable manifold X together with a field of linear automorphisms J of the tangent spaces, called an **almost complex structure**, such that at each point $x \in X$, $J^2 = -\mathbb{1}$. almost complex

Proposition. An almost complex manifold is of even dimension and orientable.

Proof: The tangent space at an arbitrary point $x \in X$ can be given a complex vector space structure by the following definition of multiplication of a vector by a complex number

$$(a + ib)\boldsymbol{v} = a\boldsymbol{v} + bJ\boldsymbol{v}.$$

It follows from $(\operatorname{Det} J)^2 = \operatorname{Det} J^2 = \operatorname{Det}(-\mathbb{1}) = (-1)^n$ that n is even. There exists for T_xX a basis of the form $(\boldsymbol{v}_1, \ldots, \boldsymbol{v}_{n/2}, J\boldsymbol{v}_1, \ldots, J\boldsymbol{v}_{n/2})$ which can be shown to be of the same orientation as the one given by the coordinates of an admissible atlas. ■

To a complex manifold is naturally associated a $2n$-real manifold with an almost complex structure. The following theorem gives a necessary and sufficient condition for the converse to be true, namely for the almost complex structure J to coincide with the canonical operator J on a complex n-dimensional manifold associated with a real $2n$-dimensional manifold. The almost complex structure J is then said to be a **complex structure**. complex structure

Theorem. An almost complex structure J corresponds to a complex structure if and only if it satisfies the following condition:

$$\boldsymbol{N}(\boldsymbol{u}, \boldsymbol{v}) \equiv 2\{[J\boldsymbol{u}, J\boldsymbol{v}] - [\boldsymbol{u}, \boldsymbol{v}] - J[\boldsymbol{u}, J\boldsymbol{v}] - J[J\boldsymbol{u}, \boldsymbol{v}]\} = 0$$

that is, in coordinates

$$N^i_{jk} \equiv 2(J^h_j \partial_h J^i_k - J^h_k \partial_h J^i_j - J^i_h \partial_j J^h_k + J^i_h \partial_k J^h_j) = 0.$$

torsion

The 3-tensor $\boldsymbol{N}$ is called the **torsion of the almost complex structure.**

Proof: a) it is easy to see that the condition is necessary: in the local coordinates (x^i, y^i) of X corresponding to the complex coordinates $(z^i = x^i + \mathrm{i}y^i)$ of Z the tensor J is identical to J_0, therefore has vanishing partial derivatives.
b) The sufficiency rests on the Frobenius theorem generalized to tensor fields[1]. An almost complex structure is integrable if its torsion vanishes.

The existence of an almost complex structure on a $2n$ dimensional manifold X rests on the possibility of reducing its frame bundle, which has $\mathrm{GL}(2n, \mathbb{R})$ as structure group, to a bundle with group $\mathrm{GL}(n, \mathbb{C})$, considered as a subgroup of $\mathrm{GL}(2n, \mathbb{R})$ as follows:

real representation

Lemma with definition. $\mathrm{GL}(n, \mathbb{C})$ *can be identified with a subgroup of* $\mathrm{GL}(2n, \mathbb{R})$ *by the following homomorphism*, called the **real representation of** $\mathrm{GL}(n, \mathbb{C})$:

$$A + \mathrm{i}B \mapsto \begin{pmatrix} A & B \\ -B & A \end{pmatrix}, \qquad A + \mathrm{i}B \in \mathrm{GL}(n, \mathbb{C})$$

A and B are real $n \times n$ matrices.
The subgroup of $\mathrm{GL}(2n, \mathbb{R})$ *which is identifiable with* $\mathrm{GL}(n, \mathbb{C})$ *is the subgroup which commutes with the canonical complex structure* J_0 *of* $\mathbb{R}^{2n}$.

Proof: It is a simple calculation to check that

$$A + \mathrm{i}B \mapsto \begin{pmatrix} A & B \\ -B & A \end{pmatrix}$$

is a group homomorphism from $\mathrm{GL}(n, \mathbb{C})$ into $\mathrm{GL}(2n, \mathbb{R})$ and that $SJ_0 = J_0 S$ if

$$S = \begin{pmatrix} A & B \\ -B & A \end{pmatrix} \quad \text{and} \quad J_0 = \begin{pmatrix} 0 & \mathbb{1} \\ -\mathbb{1} & 0 \end{pmatrix},$$

where $\mathbb{1}$ denotes the $n \times n$ identity matrix.

Complex linear frame bundle.

Let X be an almost complex manifold of dimension $2n$, with almost complex

[1] See for instance [Lichnerowicz 1976].

structure J. We can consider a linear frame at x (p. 128) as a non singular linear mapping $\rho_x \colon \mathbb{R}^{2n} \to T_x X$. We say that this linear frame is a **complex linear frame** if it is compatible with the almost complex structure J, that is if

complex linear frame

$$\rho \circ J_0 = J \circ \rho$$

where J_0 is the canonical automorphism of $\mathbb{R}^{2n}$ defined on p. 330. It is easy to see that ρ is then a non singular linear mapping of $\mathbb{C}^n$ (identified canonically with $\mathbb{R}^{2n}$) onto $T_x X$. The set of complex linear frames over X forms a principal fibre bundle $\mathbb{C}(X)$, with base X and group $\mathrm{GL}(n, \mathbb{C})$.

Proposition. Given a 2n-dimensional manifold X there exists a natural bijective correspondence between the almost complex structures and the reductions of the bundle F(X) of linear frames over X to $\mathrm{GL}(n, \mathbb{C})$.

a) Consider a reduction (p. 131) $P(X)$ of the bundle $F(X)$ of linear frames over X to $\mathrm{GL}(n, \mathbb{C})$ (identified with its real representation). We can define an almost complex structure on X by choosing at each point $x \in X$ a linear frame $\rho_x \in P(X)$, i.e. a linear isomorphism $\rho_x \colon \mathbb{R}^{2n} \to T_x X$ and defining the mapping J as

$$J = \rho J_0 \rho^{-1}$$

where J_0 is the canonical complex structure on $\mathbb{R}^{2n}$. The mapping J is independent of the choice of ρ because another frame $\rho' \in P(X)$ at x differs from ρ by right multiplication by an element of $\mathrm{GL}(n, \mathbb{C})$.

b) Consider an almost complex structure J on X. It defines the complex linear frame bundle $\mathbb{C}(X)$ which is a reduction of $F(X)$ to $\mathrm{GL}(n, \mathbb{C})$.

Example: S^2 admits a complex structure, S^6 admits an almost complex structure[1] (it is not known if it admits a complex one). The other spheres S^n do not admit almost complex structures.

Connections in almost complex manifolds.

Definition: A **connection in the principal bundle $\mathbb{C}(X)$ of complex linear frames on** X is an almost complex connection on an almost complex manifold. We shall see (p. 380) that, given a riemannian manifold X with metric g, a linear connection Γ on X comes from a connection in the bundle $O(X)$ of orthogonal frames if and only if g has vanishing covariant derivative with respect to Γ. The proof of the following theorem is analogous.

[1]Cf. for instance [Lichnerowicz 1976], [Kobayashi, Nomizu II, p. 143].

Theorem. A linear connection Γ on an almost complex manifold X is an almost complex connection if and only if the almost complex structure J, mixed 2-tensor, has vanishing covariant derivative with respect to Γ.

It can be proved[1] that every almost complex manifold admits an almost complex connection such that its torsion T is given by

$$N = 8T$$

where N is the torsion previously defined for the almost complex structure J of X.

hermitian metric

A **hermitian metric** on an almost complex manifold X is a riemannian metric invariant by the almost complex structure J, that is

$$\boldsymbol{g}(Ju, Jv) = \boldsymbol{g}(u, v)$$

for any vector fields $\boldsymbol{u}$ and $\boldsymbol{v}$. The manifold X, together with J and $\boldsymbol{g}$ is called an **almost hermitian manifold**.

almost hermitian manifold

Theorem. Every almost complex manifold admits a hermitian metric provided it is paracompact.

Proof: If X is paracompact it admits a riemannian metric $\boldsymbol{g}$ (cf. p. 292). We construct a hermitian metric $\boldsymbol{h}$ by setting

$$\boldsymbol{h}(u, v) = \boldsymbol{g}(u, v) + \boldsymbol{g}(Ju, Jv). \qquad \blacksquare$$

Let $(X, J, \boldsymbol{g})$ be an almost hermitian manifold. We define a 2-form $\boldsymbol{\phi}$ on X, called the **fundamental 2-form**, by setting for any vector fields $\boldsymbol{u}$ and $\boldsymbol{v}$

fundamental 2-form

$$\boldsymbol{\phi}(u, v) = \boldsymbol{g}(u, Jv).$$

Since $\boldsymbol{g}$ is invariant by J, so is ϕ

$$\boldsymbol{\phi}(Ju, Jv) = \boldsymbol{\phi}(u, v).$$

If the fundamental 2-form $\boldsymbol{\phi}$ is closed the hermitian metric $\boldsymbol{g}$ on the almost complex manifold (X, J) is called a **Kähler metric**.

Kähler metric

An almost complex manifold (X, J) with a Kähler metric $\boldsymbol{g}$ is called an **almost Kähler manifold**. If moreover the almost complex structure J is integrable, that is if (X, J) corresponds to a complex manifold Z, $\boldsymbol{g}$ and $\boldsymbol{\phi}$ being respectively as before hermitian and closed, the manifold is said to be a **Kähler manifold**.

almost Kähler manifold

Kähler manifold

[1]Proof for instance in [Kobayashi, Nomizu II, pp. 138–140].

We note that the fundamental 2-form $\boldsymbol{\phi}$ of a hermitian metric $\boldsymbol{g}$ on an almost complex manifold (X, J) is non degenerate – i.e. has maximum rank $2n$, since $\boldsymbol{g}$ is non degenerate and J is non singular. Therefore,

Theorem. ***An almost Kähler manifold has a symplectic structure*** *(p. 267), defined by the 2-form* $\boldsymbol{\phi}$.
Conversely, let X be a 2n-dimensional manifold, with a 2-form $\boldsymbol{\phi}$, *of rank 2n at each point. It is always possible to construct on X an almost complex structure and a hermitian metric such that* $\boldsymbol{\phi}$ *is the corresponding fundamental form*[1].
To perform calculation on Kähler manifolds it is convenient to use either the coordinates (x^i, y^i) adapted to the complex structure in which the automorphism J has the canonical form J_0, or the complex coordinates $z^i = x^i + \mathrm{i}y^i$ and their complex conjugates $\bar{z}^i = x^i - \mathrm{i}y^i$. In these coordinates the automorphism J is given by

$$\mathrm{i}\begin{pmatrix} \mathbb{1} & 0 \\ 0 & -\mathbb{1} \end{pmatrix}.$$

The Kähler metric can then be written

$$\mathrm{d}s^2 = g_{i\bar{j}}\, \mathrm{d}z^i\, \mathrm{d}\bar{z}^j, \qquad i, j = 1, \ldots n$$

where $g_{i\bar{j}}$ is an $n \times n$ hermitian metric ($g_{i\bar{j}} = a_{ij} + \mathrm{i}b_{ij} = a_{ji} - \mathrm{i}b_{ji}$). The fundamental 2-form is then

$$\boldsymbol{\phi} = -\mathrm{i}g_{i\bar{j}}\, \mathrm{d}z^i \wedge \mathrm{d}\bar{z}^j.$$

Example: The space $\mathbb{C}^n$, with the hermitian metric

$$\mathrm{d}s^2 = \sum \mathrm{d}z^i\, \mathrm{d}\bar{z}^i$$

is a Kähler manifold, with the closed 2-form

$$\phi = -\mathrm{i} \sum \mathrm{d}z^i \wedge \mathrm{d}\bar{z}^i.$$

References to some physical systems defined on complex and almost complex manifolds: R. Penrose, Proceedings of the International Congress of Mathematicians (Helsinki 1978); E.J. Flaherty, Hermitian and Kählerian Geometry in Relativity, Springer Verlag lectures in Physics 46 (Springer Verlag, 1976); G.W. Gibbons and C.N. Pope, "CP^2 as a gravitational instanton", Commun. Math. Phys. 61 (1978) 239–248; R.O. Wells, Differential Analysis on Complex Manifolds (Springer Verlag, 1980).

[1]See proof in [Lichnerowicz, 1955, 1976].

PROBLEMS AND EXERCISES

PROBLEM 1. MAXWELL EQUATIONS; GRAVITATIONAL RADIATION

How should a gravitational radiation field be identified? Electromagnetic radiation is identified in terms of energy and in terms of sources. These concepts are ambiguous in gravitation theory. It is therefore necessary to look for other properties of electromagnetic radiation which can be generalized to gravitational radiation. See the properties of the wave fronts in Problem VI 11, *p.* 535. *See also Problem* 2, *p.* 341, *example on p.* 263.

1) *Maxwell equations.*

It has been shown in Problem IV 2, *p.* 271 *that the first set of Maxwell equations can be written* $\mathrm{d}\boldsymbol{F} = 0$. *Show that the second set, namely*

$$\frac{1}{\sqrt{|g|}}(\sqrt{|g|}F^{\alpha\beta})_{,\beta} = J^{\alpha} \equiv \begin{cases} \operatorname{div} \boldsymbol{E} = \rho & \text{for } \alpha = 0 \\ \operatorname{curl} \boldsymbol{H} - \dot{\boldsymbol{E}} = j & \text{for } \alpha \neq 0 \end{cases}$$

can be written $\delta \boldsymbol{F} + \boldsymbol{J} = 0$, *where* $\boldsymbol{J} = -j_i\, \mathbf{d}\boldsymbol{x}^i + \rho\, \mathbf{d}\boldsymbol{x}^0$.

Answer:

$$* \boldsymbol{F} = (* \boldsymbol{F})_{B_1B_2}\, \mathbf{d}\boldsymbol{x}^{B_1} \wedge \mathbf{d}\boldsymbol{x}^{B_2}$$

$$(* \boldsymbol{F})_{B_1B_2} = \epsilon^{1\,2\,3\,4}_{A_1A_2B_1B_2}\sqrt{|g|}\,g^{A_1\mu_1}g^{A_2\mu_2}F_{\mu_1\mu_2} = \epsilon^{1\,2\,3\,4}_{A_1A_2B_1B_2}\sqrt{|g|}\,F^{A_1A_2}$$

$$\mathrm{d} * \boldsymbol{F} = \epsilon^{1\,2\,3\,4}_{A_1A_2B_1B_2}(\sqrt{|g|}F^{A_1A_2})_{,\lambda}\, \mathbf{d}\boldsymbol{x}^{\lambda} \wedge \mathbf{d}\boldsymbol{x}^{B_1} \wedge \mathbf{d}\boldsymbol{x}^{B_2}$$

$$= \epsilon^{1\,2\,3\,4}_{A_1A_2B_1B_2}\epsilon^{\lambda\, B_1 B_2}_{\Sigma_1\Sigma_2\Sigma_3}(\sqrt{|g|}F^{A_1A_2})_{,\lambda}\, \mathbf{d}\boldsymbol{x}^{\Sigma_1} \wedge \mathbf{d}\boldsymbol{x}^{\Sigma_2} \wedge \mathbf{d}\boldsymbol{x}^{\Sigma_3}$$

$$\delta \boldsymbol{F} = (-1)^2 *^{-1} \mathrm{d} * \boldsymbol{F} = - * \mathrm{d} * \boldsymbol{F}$$

$$= \frac{-1}{3!}\,\epsilon^{1\,2\,3\,4}_{\mu_1\mu_2\mu_3\sigma}\epsilon^{1\,2\,3\,4}_{A_1A_2B_1B_2}\epsilon^{\lambda\, B_1B_2}_{\sigma_1\sigma_2\sigma_3}\sqrt{|g|}\,g^{\mu_1\sigma_1}g^{\mu_2\sigma_2}g^{\mu_3\sigma_3}(\sqrt{|g|}F^{A_1A_2})_{,\lambda}\, \mathbf{d}\boldsymbol{x}^{\sigma}$$

$$= (\delta \boldsymbol{F})_{\sigma}\mathbf{d}\boldsymbol{x}^{\sigma}$$

$$(\delta \boldsymbol{F})^{\mu} = g^{\mu\sigma}(\delta \boldsymbol{F})_{\sigma}$$

$$= -(\sqrt{|g|}F^{A_1A_2})_{,\lambda}\,\epsilon^{1\,2\,3\,4}_{A_1A_2B_1B_2}\epsilon^{\lambda\, B_1B_2}_{\sigma_1\sigma_2\sigma_3}\sqrt{|g|}\,\frac{1}{3!}\,\epsilon^{1\,2\,3\,4}_{\mu_1\mu_2\mu_3\sigma}g^{\mu_1\sigma_1}g^{\mu_2\sigma_2}g^{\mu_3\sigma_3}g^{\mu\sigma}$$

$$= -(\sqrt{|g|}F^{A_1A_2})_{,\lambda}\,\epsilon^{1\,2\,3\,4}_{A_1A_2B_1B_2}\epsilon^{\lambda\, B_1B_2}_{\sigma_1\sigma_2\sigma_3}\sqrt{|g|}\,\frac{1}{3!}\,\epsilon^{\sigma_1\sigma_2\sigma_3\mu}_{1\,2\,3\,4}\,\frac{1}{g}$$

$$= (\sqrt{|g|}F^{A_1A_2})_{,\lambda}\,\epsilon^{1\,2\,3\,4}_{A_1A_2B_1B_2}\epsilon^{\lambda B_1B_2\mu}_{1\,2\,3\,4}\,\frac{1}{\sqrt{|g|}}, \quad g < 0$$

$$= (\sqrt{|g|}F^{A_1A_2})_{,\lambda}\frac{1}{\sqrt{|g|}}\epsilon^{\lambda\mu}_{A_1A_2}$$

$$= \frac{1}{2!}(\sqrt{|g|}F^{\alpha_1\alpha_2})_{,\lambda}\frac{1}{\sqrt{|g|}}(\delta^{\lambda}_{\alpha_1}\delta^{\mu}_{\alpha_2} - \delta^{\mu}_{\alpha_1}\delta^{\lambda}_{\alpha_2})$$

$$= -\frac{1}{\sqrt{|g|}}(\sqrt{|g|}F^{\mu\lambda})_{,\lambda}.$$

Also

$$J^{\mu} = g^{\mu\lambda}J_{\lambda}.$$

In summary, the Maxwell equations which in arbitrary local coordinates read

$$\nabla_{\gamma}F_{\alpha\beta} + \nabla_{\alpha}F_{\beta\gamma} + \nabla_{\beta}F_{\gamma\alpha} = 0$$
$$-\nabla_{\alpha}F^{\alpha\beta} + J^{\beta} = 0$$

can be written intrinsically $\mathrm{d}\boldsymbol{F} = 0$, $\delta\boldsymbol{F} + \boldsymbol{J} = 0$.
The conservation law $\delta\boldsymbol{J} = -\nabla_{\alpha}J^{\alpha} = 0$ follows from $\delta^2 = 0$.
The equation $\mathrm{d}\boldsymbol{F} = 0$ implies that there exists *locally* a vector potential $\boldsymbol{A}$ such that $\boldsymbol{F} = \mathrm{d}\boldsymbol{A}$. Hence

$$\delta\,\mathrm{d}\boldsymbol{A} + \boldsymbol{J} = 0, \text{ or in coordinates } -\nabla^{\alpha}(\partial_{\alpha}A_{\beta} - \partial_{\beta}A_{\alpha}) + J_{\beta} = 0.$$

The vector potentials $\boldsymbol{A}$ are defined by $\boldsymbol{F}$ modulo an additive exact differential. A change of potential $\boldsymbol{A} \mapsto \bar{\boldsymbol{A}} = \boldsymbol{A} + \mathbf{d}\boldsymbol{B}$ is called a **gauge transformation**. The Lorentz auxiliary condition $\delta\boldsymbol{A} = 0$ restricts further the choice of gauge to functions B such that $\delta\,\mathbf{d}\boldsymbol{B} = \Delta B = 0$. Although the Lorentz condition does not determine the gauge uniquely, one calls a potential $\boldsymbol{A}$ such that $\delta\boldsymbol{A} = 0$ "a potential in the **Lorentz gauge**". In the Lorentz gauge

gauge transformation

Lorentz gauge

$$0 = \delta\,\mathrm{d}\boldsymbol{A} + \mathrm{d}\delta\boldsymbol{A} + \boldsymbol{J} = \Delta\boldsymbol{A} + \boldsymbol{J}.$$

2) *Physical interpretation.*
In Minkowski space $\mathrm{d}s^2 = (\mathrm{d}x^0)^2 - \Sigma\,(\mathrm{d}x^i)^2$, *the volume element is* $\boldsymbol{\tau} = \mathbf{d}x^0 \wedge \mathbf{d}x^1 \wedge \mathbf{d}x^2 \wedge \mathbf{d}x^3$, $\overset{*}{F}_{\alpha\beta} = \frac{1}{2}\epsilon^{0123}_{\gamma\delta\alpha\beta}F^{\gamma\delta}$ *and*

$$F_{\alpha\beta} = \begin{bmatrix} 0 & E_1 & E_2 & E_3 \\ -E_1 & 0 & -H_3 & H_2 \\ -E_2 & H_3 & 0 & -H_1 \\ -E_3 & -H_2 & H_1 & 0 \end{bmatrix}, \qquad F^{\alpha\beta} = \begin{bmatrix} 0 & -E_1 & -E_2 & -E_3 \\ E_1 & 0 & -H_3 & H_2 \\ E_2 & H_3 & 0 & -H_1 \\ E_3 & -H_2 & H_1 & 0 \end{bmatrix},$$

$$\overset{*}{F}_{\alpha\beta} = \begin{bmatrix} 0 & -H_1 & -H_2 & -H_3 \\ H_1 & 0 & -E_3 & E_2 \\ H_2 & E_3 & 0 & -E_1 \\ H_3 & -E_2 & E_1 & 0 \end{bmatrix}.$$

In an arbitrary riemannian manifold, we can define the electric and magnetic components of a 2-form F, *relative to a time like unit vector* u *by*

$$E_\alpha(u) = F_{\alpha\beta}u^\beta \qquad \text{and} \qquad H_\alpha(u) = -\overset{*}{F}_{\alpha\beta}u^\beta.$$

The vectors (E^α) *and* (H^α) *are space like*

$$g_{\alpha\beta}E^\alpha u^\beta = 0, \qquad g_{\alpha\beta}H^\alpha u^\beta = 0.$$

The Maxwell tensor $\tau^{\alpha\beta} = \frac{1}{4}g^{\alpha\beta}F^{\gamma\delta}F_{\gamma\delta} - F^{\alpha\gamma}F^\beta{}_\gamma$ *is a useful tool to discuss the physical properties of the Maxwell field. Use its value in Minkowski space to identify the various components.*

Answer:

$\tau^{00} = \frac{1}{2}(E^2 + H^2)$ is the energy density W.
$\tau^{0i} = (E \times H)_i$ is the Poynting vector P_i.
$\tau^{ij} = -E_iE_j - H_iH_j + \frac{1}{2}\delta_{ij}(E^2 + H^2)$ is the Maxwell stress tensor.

The Lorentz force $\dot{f}_\alpha = F_{\alpha\beta}J^\beta$ is equal by virtue of the Maxwell equations to

$$f^\alpha = \nabla_\beta \tau^{\beta\alpha}.$$

3) *Transformations which leave* F *invariant. Other invariants.*
The closed form F *is invariant under the group of transformations generated by the vector field* v *if* (*notation p.* 230 *footnote*)

$$\mathrm{i}(v)F = 0.$$

In order that this equation have non trivial solutions the determinant of F *must vanish. When* F *is an antisymmetric 2-form on* $\mathbb{R}^2$, *the most general solution is* (*p.* 263)

$$v = f(x)v_1 + g(x)v_2$$

where f *and* g *are arbitrary continuous functions on* $\mathbb{R}^4$; *the components of the vector fields* v_1 *and* v_2 *in an inertial* (*Lorentz*) *frame are*

$$v_1^0 = 0, \quad v_1^i = H_i \qquad \textit{and} \qquad v_2^0 = \sum H_i^2, \quad v_2^i = (E \times H)_i.$$

The general case is obtained by expressing E_i, H_i *in terms of* F_{ij}.
Similarly the most general vector field u *such that* $\mathrm{i}(u)\overset{*}{F} = 0$ *is* $u = f(x)u_1 + g(x)u_2$ *where the components of* u_1 *and* u_2 *in an inertial frame are*

$$u_1^0 = 0, \quad u_1^i = E_i \qquad \textit{and} \qquad u_2^0 = \sum E_i^2, \quad u_2^i = (E \times H)_i.$$

If F *is invariant under the group of transformations generated by* v, *so is* $F \wedge \overset{*}{F} = \frac{1}{2}\tau F_{\alpha\beta}\overset{*}{F}{}^{\alpha\beta} = \overset{*}{F} \wedge \overset{*}{F}$. *In an inertial frame* $F_{\alpha\beta}\overset{*}{F}{}^{\alpha\beta} = 4(E \cdot H)$. *Note that* $\det F = \det \overset{*}{F} = (\frac{1}{4}F_{\alpha\beta}\overset{*}{F}{}^{\alpha\beta})^2$.

If F and $\overset{*}{F}$ are invariant under the group of transformations generated by v, so is $F \wedge \overset{*}{F} = \frac{1}{2}\tau F_{\alpha\beta}\overset{*}{F}{}^{\alpha\beta} = \frac{1}{2}\tau \overset{*}{F}_{\alpha\beta}\overset{*}{F}{}^{\alpha\beta}$. In an inertial frame $F_{\alpha\beta}F^{\alpha\beta} = 2(E^2 - H^2)$.
An electromagnetic field F is called a **singular field** *if there exists a vector field l such that both F and $\overset{*}{F}$ are invariant under the local group of transformations generated by l. l is called a* **fundamental vector** *of F. In other words F is called singular if the intersection of the spaces Q and $\overset{*}{Q}$ associated respectively with F and $\overset{*}{F}$ is non empty, i.e. if*

singular field

fundamental vector

$$\mathrm{i}(l)F = 0 \qquad \textit{together with} \qquad \mathrm{i}(l)\overset{*}{F} = 0$$

or equivalently $l^\alpha F_{\alpha\beta} = 0$ together with $l_\alpha F_{\beta\gamma} + l_\beta F_{\gamma\alpha} + l_\gamma F_{\alpha\beta} = 0$.
Prove that an eigenvector l of the non zero form F is also an eigenvector of $\overset{*}{F}$ if and only if it is a null vector.

Answer: $l^\alpha F_{\alpha\beta} = al_\beta$ and $l^\alpha \overset{*}{F}_{\alpha\beta} = bl_\beta$ imply $al_\beta l^\beta = 0$ and $bl_\beta l^\beta = 0$.

If l is not isotropic then $a = b = 0$, and $l^\alpha F_{\alpha\beta} = l^\alpha \overset{*}{F}_{\alpha\beta} = 0$ implies $F = 0$. Conversely, an isotropic eigenvector of F is an isotropic eigenvector of $\tau_{\alpha\beta}$ hence an isotropic eigenvector of $\overset{*}{F}$ whether F is regular or singular.

4) *Pure radiation field.*
A **pure radiation field** *is a non zero solution of the Maxwell equations in the absence of matter, $J = 0$, satisfying any one of the following three equivalent properties.*

pure radiation field

a) *F is a singular field.*

b) $F_{\alpha\beta}F^{\alpha\beta} = 0$ *and* $F_{\alpha\beta}\overset{*}{F}{}^{\alpha\beta} = 0$.

c) *The Poynting vector relative to an arbitrary time like direction does not vanish. (Check that unless $E^2 = H^2$ and $E \cdot H = 0$, there exists an inertial frame where the Poynting vector $(E \times H)$ vanishes.)*

5) *Einstein equation.*
We shall construct various quantities analogous to the quantities used in the case of the electromagnetic field to define a pure radiation field. The components of the Riemann tensor can be separated into two sets analogous respectively to the electric and magnetic components of the electromagnetic tensor

$$E_{ij} = R_{i0j0} \quad \textit{and} \quad H_{ij} = \tfrac{1}{2}\epsilon_{ikl}R_{klj0}.$$

We can define intrinsically these components by considering the Riemann tensor as a double 2-form and defining its left adjoint (cf. p. 295)

$$(*R)_{\alpha\beta\gamma\delta} = \tfrac{1}{2}\tau_{\alpha\beta\rho\sigma}R^{\rho\sigma}{}_{\gamma\delta}.$$

We shall need later the right adjoint $(R*)_{\alpha\beta\gamma\delta} = \frac{1}{2}\tau_{\gamma\delta\rho\sigma}R_{\alpha\beta}{}^{\rho\sigma}$ *and the bi-adjoint* $(*R*)_{\alpha\beta\gamma\delta} = \frac{1}{4}\tau_{\alpha\beta\eta\nu}\tau_{\gamma\delta\rho\sigma}R^{\eta\nu\rho\sigma}$.
What are the properties of the electric and magnetic components relative to a time like unit vector $\boldsymbol{u}$ *defined as follows*

$$E_{\alpha\beta}(\boldsymbol{u}) = R_{\alpha\gamma\beta\delta}u^{\gamma}u^{\delta}$$
$$H_{\alpha\beta}(\boldsymbol{u}) = -(*R)_{\alpha\gamma\beta\delta}u^{\gamma}u^{\delta}?$$

Answer: a) $E_{\alpha\beta}$ is obviously symmetric, $H_{\alpha\beta}$ is symmetric if and only if $R_{\alpha\beta} = \lambda g_{\alpha\beta}$, i.e. if and only if the space is an Einstein space (Problem 4, p. 355). In that case $*R_{\alpha\beta\gamma\delta} = *R_{\gamma\delta\alpha\beta}$.
b) Set $X_{\alpha\beta} = (*R*)_{\alpha\gamma\beta\delta}u^{\gamma}u^{\delta}$ and $Y_{\alpha\beta} = (R*)_{\alpha\gamma\beta\delta}u^{\gamma}u^{\delta}$ and let $\boldsymbol{Z}$ be any one of the tensors, $\boldsymbol{E}$, $\boldsymbol{H}$, $\boldsymbol{X}$ or $\boldsymbol{Y}$.

$$Z_{\alpha\beta}u^{\beta} = 0,\ Z \text{ is a spacelike tensor.}$$

$Z^{\alpha\beta}Z_{\alpha\beta} > 0$ unless $Z_{\alpha\beta} = 0$. Indeed in an orthonormal frame using $\boldsymbol{u}$ as a time axis

$$Z^{\alpha\beta}Z_{\alpha\beta} = Z^{00}Z_{00} + Z^{0\alpha}Z_{0\alpha} + Z^{\alpha 0}Z_{\alpha 0} + Z^{\alpha\beta}Z_{\alpha\beta} = \sum (Z_{\alpha\beta})^2.$$

c) Set $A = \frac{1}{8}R^{\alpha\beta\gamma\delta}R_{\alpha\beta\gamma\delta}$ and $B = \frac{1}{8}R^{\alpha\beta\gamma\delta}(*R)_{\alpha\beta\gamma\delta}$. A and B are given in terms of the electric and magnetic components by

$$A = E^{\alpha\beta}E_{\alpha\beta} - H^{\alpha\beta}H_{\alpha\beta}, \quad B = -2E^{\alpha\beta}H_{\alpha\beta}.$$

What are the properties of the Bel–Robinson tensor $\boldsymbol{T}$ *defined by*

$$T^{\alpha\beta\gamma\delta} = R^{\alpha\eta\gamma\nu}R^{\beta}{}_{\eta}{}^{\delta}{}_{\nu} + (*R*)^{\alpha\eta\gamma\nu}(*R*)^{\beta}{}_{\eta}{}^{\delta}{}_{\nu} - Ag^{\alpha\beta}g^{\gamma\delta} + (\gamma \sim \delta)$$

where $(\gamma \sim \delta)$ *are the terms obtained from the previous ones by interchanging* γ *and* δ?

Answer:

a) $W \overset{\text{def}}{=} T_{\alpha\beta\gamma\delta}u^{\alpha}u^{\beta}u^{\gamma}u^{\delta} \geq 0$ and vanishes only when $R_{\alpha\beta\gamma\delta} = 0$.
b) $\nabla_{\alpha}T^{\alpha\beta\gamma\delta} = 0$ when $R_{\alpha\beta} = \lambda g_{\alpha\beta}$.

Proof: a) The tensor $\boldsymbol{T}$ can also be written

$$T^{\alpha\beta\gamma\delta} = R^{\alpha\eta\gamma\nu}R^{\beta}{}_{\eta}{}^{\delta}{}_{\nu} + (*R*)^{\alpha\eta\gamma\nu}(*R*)^{\beta}{}_{\eta}{}^{\delta}{}_{\nu}$$
$$+ (*R)^{\alpha\eta\delta\nu}(*R)^{\beta}{}_{\eta}{}^{\gamma}{}_{\nu} + (R*)^{\alpha\eta\delta\nu}(R*)^{\beta}{}_{\eta}{}^{\gamma}{}_{\nu}.$$

Hence $W = E^{\alpha\beta}E_{\alpha\beta} + H^{\alpha\beta}H_{\alpha\beta} + X^{\alpha\beta}X_{\alpha\beta} + Y^{\alpha\beta}Y_{\alpha\beta}$ and the result follows.

b) The proof is fairly involved and we shall only give the main steps.

$$(*R*)_{\alpha\beta\gamma\delta} = -R_{\alpha\beta\gamma\delta} + g_{\alpha\gamma}B_{\beta\delta} - g_{\beta\gamma}B_{\alpha\delta} + g_{\beta\delta}B_{\alpha\gamma} - g_{\alpha\delta}B_{\beta\gamma}$$

where $B_{\alpha\beta} = R_{\alpha\beta} - \frac{1}{4}Rg_{\alpha\beta}$. Note that $B_{\alpha\beta} = 0$ when $R_{\alpha\beta} = \lambda g_{\alpha\beta}$, i.e. when the space is an Einstein space

$$R^{\alpha\beta\gamma\delta}R_{\alpha\beta\epsilon\delta} + (*R*)^{\alpha\beta\gamma\delta}(*R*)_{\alpha\beta\epsilon\delta} - 4Ag^{\gamma}_{\epsilon} = 0$$

$$\nabla_\alpha T_{\beta\gamma\delta\epsilon} = 0$$

$$\nabla_\alpha(*R*)^{\beta\alpha\gamma\delta} = 0.$$ ■

The natural generalization of the 3 properties which can be used to characterize a pure electromagnetic radiation field are the following.

a) The Riemann tensor is a singular double 2-form

$Sl_\gamma R_{\alpha\beta\lambda\eta} = 0$, where S is summation after circular permutation of $\alpha\beta\gamma$

$l^\alpha R_{\alpha\beta\lambda\eta} = 0.$

b) The Poynting vector relative to a time like unit vector does not vanish

$$P^\alpha(u) = (g^{\rho\alpha} - u^\rho u^\alpha)T_{\rho\beta\lambda\eta}u^\beta u^\lambda u^\eta.$$

c) The four fundamental scalars vanish

$$R^{\alpha\beta}{}_{\lambda\eta}R^{\lambda\eta}{}_{\alpha\beta} = 0, \qquad R^{\alpha\beta}{}_{\lambda\eta}(*R)^{\lambda\eta}{}_{\alpha\beta} = 0$$

$$R^{\alpha\beta}{}_{\lambda\eta}R^{\lambda\eta}{}_{\rho\sigma}R^{\rho\sigma}{}_{\alpha\beta} = 0, \qquad R^{\alpha\beta}{}_{\lambda\eta}R^{\lambda\eta}{}_{\rho\sigma}(*R)^{\rho\sigma}{}_{\alpha\beta} = 0.$$

Bel has shown that these four generalizations are not equivalent. The first one however is more than a formal generalization and is justified by the study of the discontinuities of the Riemann tensor across a gravitational wave front. The first condition implies the others but not vice versa.

References: A. Lichnerowicz, Annali di Matematica Pura ed Applicata ser 4 vol. 50 (1960) pp. 1–95. A. Trautman, F.A.E. Pirani and H. Bondi, *Lectures on General Relativity*, Vol. I (Prentice-Hall Inc., 1965).

PROBLEM 2. THE SCHWARZSCHILD SOLUTION

Seek on a hyperbolic manifold X^4 a metric of the form

$$ds^2 = e^{\nu(r,t)}\,dt^2 - e^{\lambda(r,t)}\,dr^2 - r^2\,d\theta^2 - r^2\sin^2\theta\,d\phi^2$$

Einstein equation

which satisfies the **Einstein empty space field equations**

$$R_{\mu\nu} = 0.$$

Answer: The calculations can be done conveniently with respect to the orthonormal frame

$$\begin{aligned}\boldsymbol{\theta}^0 &= e^{\nu/2}\,\mathbf{d}t\\ \boldsymbol{\theta}^1 &= e^{\lambda/2}\,\mathbf{d}r\\ \boldsymbol{\theta}^2 &= r\,\mathbf{d}\theta\\ \boldsymbol{\theta}^3 &= r\sin\theta\,\mathbf{d}\phi.\end{aligned}$$

Then letting $\lambda' = \partial\lambda/\partial r$, $\nu' = \partial\nu/\partial r$, $\dot{\lambda} = \partial\lambda/\partial t$, $\dot{\nu} = \partial\nu/\partial t$, we have

$$\begin{aligned}\mathbf{d}\boldsymbol{\theta}^0 &= \tfrac{1}{2}e^{-\lambda/2}\nu'\boldsymbol{\theta}^1\wedge\boldsymbol{\theta}^0\\ \mathbf{d}\boldsymbol{\theta}^1 &= \tfrac{1}{2}e^{-\lambda/2}\dot{\lambda}\boldsymbol{\theta}^0\wedge\boldsymbol{\theta}^1\\ \mathbf{d}\boldsymbol{\theta}^2 &= \frac{1}{r}e^{-\lambda/2}\boldsymbol{\theta}^1\wedge\boldsymbol{\theta}^2\\ \mathbf{d}\boldsymbol{\theta}^3 &= \frac{1}{r}e^{-\lambda/2}\boldsymbol{\theta}^1\wedge\boldsymbol{\theta}^3 + \frac{1}{r}\cot\theta\,\boldsymbol{\theta}^2\wedge\boldsymbol{\theta}^3.\end{aligned}$$

We can obtain the connection forms by solving, by inspection, the Cartan structural equation $\mathbf{d}\boldsymbol{\theta}^\alpha + \boldsymbol{\omega}_\beta{}^\alpha\wedge\boldsymbol{\theta}^\beta = 0$

$$\begin{aligned}\boldsymbol{\omega}_1{}^0 &= \boldsymbol{\omega}_{10} = -\boldsymbol{\omega}_{01} = \boldsymbol{\omega}_0{}^1 = \tfrac{1}{2}e^{-\lambda/2}\nu'\boldsymbol{\theta}^0 + \tfrac{1}{2}e^{-\nu/2}\dot{\lambda}\boldsymbol{\theta}^1\\ \boldsymbol{\omega}_1{}^2 &= -\boldsymbol{\omega}_2{}^1 = \frac{1}{r}e^{-\lambda/2}\boldsymbol{\theta}^2\\ \boldsymbol{\omega}_1{}^3 &= -\boldsymbol{\omega}_3{}^1 = \frac{1}{r}e^{-\lambda/2}\boldsymbol{\theta}^3\\ \boldsymbol{\omega}_2{}^3 &= -\boldsymbol{\omega}_3{}^2 = \frac{1}{r}\cot\theta\,\boldsymbol{\theta}^3\\ \boldsymbol{\omega}_0{}^2 &= \boldsymbol{\omega}_2{}^0 = \boldsymbol{\omega}_0{}^3 = \boldsymbol{\omega}_3{}^0 = 0.\end{aligned}$$

Then

$$\begin{aligned}\mathbf{d}\boldsymbol{\omega}_1{}^0 &= \tfrac{1}{2}[e^{-\lambda}(\tfrac{1}{2}\nu'(\nu'-\lambda')+\nu'') - e^{-\nu}(\tfrac{1}{2}\dot{\lambda}(\dot{\lambda}-\dot{\nu})+\ddot{\lambda})]\boldsymbol{\theta}^1\wedge\boldsymbol{\theta}^0\\ \mathbf{d}\boldsymbol{\omega}_1{}^2 &= -\frac{1}{2r}e^{-(\lambda+\nu)/2}\dot{\lambda}\boldsymbol{\theta}^0\wedge\boldsymbol{\theta}^2 - \frac{1}{2r}e^{-\lambda}\lambda'\boldsymbol{\theta}^1\wedge\boldsymbol{\theta}^2\\ \mathbf{d}\boldsymbol{\omega}_1{}^3 &= \frac{1}{r^2}e^{-\lambda/2}\cot\theta\,\boldsymbol{\theta}^2\wedge\boldsymbol{\theta}^3 - \frac{1}{2r}e^{-(\lambda+\nu)/2}\dot{\lambda}\boldsymbol{\theta}^0\wedge\boldsymbol{\theta}^3 - \frac{1}{2r}e^{-\lambda}\lambda'\boldsymbol{\theta}^1\wedge\boldsymbol{\theta}^3\\ \mathbf{d}\boldsymbol{\omega}_2{}^3 &= -\frac{1}{r^2}\boldsymbol{\theta}^2\wedge\boldsymbol{\theta}^3.\end{aligned}$$

We use the second Cartan structural equation to compute the Riemann tensor. The non-vanishing curvature forms are thus

$$\begin{aligned}
\Omega_0{}^1 &= \Omega_1{}^0 = \mathrm{d}\omega_0{}^1 \\
&= \tfrac{1}{4}[\mathrm{e}^{-\lambda}(\tfrac{1}{2}\nu'(\nu'-\lambda')+\nu'') - \mathrm{e}^{-\nu}(\tfrac{1}{2}\dot\lambda(\dot\lambda-\dot\nu)+\ddot\lambda)](\theta^1\wedge\theta^0-\theta^0\wedge\theta^1) \\
\Omega_2{}^0 &= -\omega_0{}^1\wedge\omega_1{}^2 \\
&= -\frac{1}{4r}\mathrm{e}^{-\lambda}\nu'(\theta^0\wedge\theta^2-\theta^2\wedge\theta^0) - \frac{1}{4r}\mathrm{e}^{-(\lambda+\nu)/2}\dot\lambda(\theta^1\wedge\theta^2-\theta^2\wedge\theta^1) \\
\Omega_0^3 &= \Omega_3^0 = -\omega_0{}^1\wedge\omega_1{}^3 \\
&= -\frac{1}{4r}\mathrm{e}^{-\lambda}\nu'(\theta^0\wedge\theta^3-\theta^3\wedge\theta^0) - \frac{1}{4r}\mathrm{e}^{-(\lambda+\nu)/2}\dot\lambda(\theta^1\wedge\theta^3-\theta^3\wedge\theta^1) \\
\Omega_1^2 &= -\Omega_2^1 = \mathrm{d}\omega_1{}^2 - \omega_1{}^3\wedge\omega_3{}^2 \\
&= \frac{1}{4r}\mathrm{e}^{-(\lambda+\nu)/2}\dot\lambda(\theta^2\wedge\theta^0-\theta^0\wedge\theta^2) - \frac{1}{4r}\mathrm{e}^{-\lambda}\lambda'(\theta^1\wedge\theta^2-\theta^2\wedge\theta^1) \\
\Omega_1^3 &= -\Omega_3^1 = \mathrm{d}\omega_1{}^3 - \omega_1{}^2\wedge\omega_2{}^3 \\
&= \frac{1}{4r}\mathrm{e}^{-\lambda}\lambda'(\theta^3\wedge\theta^1-\theta^1\wedge\theta^3) + \frac{1}{4r}\mathrm{e}^{-(\lambda+\nu)/2}\dot\lambda(\theta^3\wedge\theta^0-\theta^0\wedge\theta^3) \\
\Omega_2^3 &= -\Omega_3^2 = \mathrm{d}\omega_2{}^3 - \omega_2{}^1\wedge\omega_1{}^3 \\
&= \frac{1}{2r^2}(\mathrm{e}^{-\lambda}-1)(\theta^2\wedge\theta^3-\theta^3\wedge\theta^2).
\end{aligned}$$

The components of the curvature tensor can now be read off from these equations and used to calculate the components of the Ricci tensor

$$\begin{aligned}
R_{00} &= R_0{}^1{}_{01} + R_0{}^2{}_{02} + R_0{}^3{}_{03} \\
&= -\tfrac{1}{2}\mathrm{e}^{-\lambda}(\tfrac{1}{2}\nu'(\nu'-\lambda')+\nu'') + \tfrac{1}{2}\mathrm{e}^{-\nu}(\tfrac{1}{2}\dot\lambda(\dot\lambda-\dot\nu)+\ddot\lambda) - \frac{1}{r}\mathrm{e}^{-\lambda}\nu' \\
R_{01} &= R_0{}^2{}_{12} + R_0{}^3{}_{13} = -\frac{1}{r}\mathrm{e}^{-(\lambda+\nu)/2}\dot\lambda \\
R_{11} &= R_1{}^0{}_{10} + R_1{}^2{}_{12} + R_1{}^3{}_{13} \\
&= \tfrac{1}{2}\mathrm{e}^{-\lambda}(\tfrac{1}{2}\nu'(\nu'-\lambda')+\nu'') - \tfrac{1}{2}\mathrm{e}^{-\nu}(\tfrac{1}{2}\dot\lambda(\dot\lambda-\dot\nu)+\ddot\lambda) - \frac{1}{r}\mathrm{e}^{-\lambda}\lambda' \\
R_{22} &= R_2{}^0{}_{20} + R_2{}^1{}_{21} + R_2{}^3{}_{23} \\
&= \frac{1}{2r}\mathrm{e}^{-\lambda}\nu' + \frac{1}{r^2}(\mathrm{e}^{-\lambda}-1) - \frac{1}{2r}\mathrm{e}^{-\lambda}\lambda' \\
R_{33} &= R_3{}^0{}_{30} + R_3{}^1{}_{31} + R_3{}^2{}_{32} \\
&= \frac{1}{2r}\mathrm{e}^{-\lambda}\nu' + \frac{1}{r^2}(\mathrm{e}^{-\lambda}-1) - \frac{1}{2r}\mathrm{e}^{-\lambda}\lambda' \\
R_{02} &= R_{03} = R_{12} = R_{13} = R_{23} = 0.
\end{aligned}$$

It follows from Einstein equations and the expression for R_{01} that $\dot{\lambda} = 0$. The remaining equations for ν and λ are

$$\tfrac{1}{2}\nu'(\nu' - \lambda') + \nu'' + \frac{2}{r}\nu' = 0$$

$$\tfrac{1}{2}\nu'(\nu' - \lambda') + \nu'' - \frac{2}{r}\lambda' = 0$$

$$e^{-\lambda}(\nu' - \lambda') + \frac{2}{r}(e^{-\lambda} - 1) = 0.$$

The difference of the first two equations gives $\nu' + \lambda' = 0$. The last equation can then be solved for λ to give

$$e^{-\lambda} = 1 - K/r$$

where K is a constant. It follows that

$$e^{\nu} = F(t)(1 - K/r)$$

where F is an arbitrary function of t. The equation obtained by adding the first two equations is the same as the equation obtained by differentiating the last equation after multiplying by r. It is possible to eliminate the dependence of ν on t by making a change of coordinates $(t, r, \theta, \varphi) \mapsto (\hat{t}(t), r, \theta, \varphi)$ such that

$$\frac{d\hat{t}}{dt} = [F(t)]^{1/2}.$$

We therefore take $F(t) = 1$ and write

$$ds^2 = \left(1 - \frac{K}{r}\right) \mathbf{d}t^2 - \frac{\mathbf{d}r^2}{(1 - K/r)} - r^2\, \mathbf{d}\theta^2 - r^2 \sin^2\theta\, \mathbf{d}\varphi^2.$$

For physical reasons it is assumed that $K > 0$.

PROBLEM 3. GEODETIC MOTION; EQUATION OF GEODETIC DEVIATION; EXPONENTIATION; CONJUGATE POINTS

Generalize Problem II 2, *p.* 100 *to the case of* C^1 *functions* $q: \bar{\Omega} \to M$ *where* M *is a riemannian manifold with metric* g.

Answer: Let $\mathscr{C}^1(\bar{\Omega})$ be the space of C^1 mappings q on the interval $\bar{\Omega} = [a, b] \subset \mathbb{R}$ into M as defined p. 29; let $\mathscr{C}^1(\bar{\Omega}, a) \subset \mathscr{C}^1(\bar{\Omega})$ be the space of mappings q such that $q(a) = A$; let $\mathscr{C}^1(\bar{\Omega}, a, b) \subset \mathscr{C}^1(\bar{\Omega}, a)$ be the space of mappings q such that $q(a) = A$ and $q(b) = B$.

a) *Fermi geodetic system of coordinates.*

We shall use the **Fermi geodetic system of coordinates** defined as follows.

Fermi geodetic coordinates

Let $\{P_\alpha\}$ be a set of orthonormal vector fields parallel propagated along q. The set $\{P_\alpha\}$ is a basis for the module of continuous vector fields along q. $P_\alpha(q(x))$ will be abbreviated to $P_\alpha(x)$ and $g(q(x))$ to $g(x)$ when possible without ambiguity.

$$(P_\alpha(x), P_\beta(x)) = \delta_{\alpha\beta}.$$

Let ∇_x denote the covariant derivative along the path $x \mapsto q(x)$ at the point $q(x)$:

$$\nabla_x P_\alpha(x) = 0.$$

Let J be a two point contravariant tensor along q.

$J(q(x), q(y)) \in T_{q(x)}M \otimes T_{q(y)}M$ will be abbreviated to $J(x, y)$; its coordinate expression is

$$J(x, y) = P_\alpha(x) J^{\alpha\beta}(x, y) P_\beta(y).$$

The basis $\{P^\alpha\}$ dual to $\{P_\alpha\}$ is defined by

$$\langle P^\alpha(x), P_\beta(x)\rangle = \delta^\alpha_\beta.$$

In this system of coordinates

$$\nabla_x J(x, y) = P_\alpha(x) \frac{\mathrm{d}}{\mathrm{d}x} J^{\alpha\beta}(x, y) P_\beta(y)$$

and the results of Problem II 2, p. 100 are readily generalized.

b) *Geodesics. Geodetic deviation.*

Geodesics are the critical points q_0 of the action, then often called the **energy function** $S(q) = \frac{1}{2}\int_\Omega (\dot q(x)|\dot q(x))\,\mathrm{d}x$; $S'(q_0) = 0$.

Let $\bar\alpha(u)$ be a **one parameter variation** of q_0 keeping the end points fixed, $u \in U$.

energy function
one parameter variation

$$\text{Set } \bar\alpha(u)(x) = \alpha(u, x), \ \frac{\partial\alpha}{\partial u}(0, x) = h(x) \text{ and } \frac{\partial^2\alpha}{\partial u^2}(0, x) = k(x),$$

then

$$\frac{\mathrm{d}(S \circ \bar\alpha)}{\mathrm{d}u}(0) = S'(q_0)h$$

$$\frac{\mathrm{d}^2(S \circ \bar\alpha)}{\mathrm{d}u^2}(0) = S''(q_0)hh + S'(q_0)k.$$

Use the following equations.

α)
$$\frac{\partial}{\partial u}\left(\frac{\partial}{\partial x}\alpha(u, x)\middle|\frac{\partial}{\partial x}\alpha(u, x)\right) = 2\left(\nabla_u \frac{\partial}{\partial x}\alpha(u, x)\middle|\frac{\partial}{\partial x}\alpha(u, x)\right)$$

and

$$\frac{\partial}{\partial x}\left(\frac{\partial\alpha}{\partial u}\middle|\frac{\partial\alpha}{\partial x}\right)=\left(\nabla_x\frac{\partial\alpha}{\partial u}\middle|\frac{\partial\alpha}{\partial x}\right)+\left(\frac{\partial\alpha}{\partial u}\middle|\nabla_x\frac{\partial\alpha}{\partial x}\right).$$

β)
$$\nabla_u\frac{\partial}{\partial x}\alpha(u,x)=\nabla_x\frac{\partial}{\partial u}\alpha(u,x).$$

γ) Let v be a vector field on the parametrized surface

$$\alpha: U\times\bar{\Omega}\to M$$

$$\nabla_u\nabla_x v=\nabla_x\nabla_u v+R\left(\frac{\partial\alpha}{\partial x},\frac{\partial\alpha}{\partial u}\right)v$$

where R is the Riemann tensor.
The first differential (first variation)

$$S'(q_0)h=-\int_\Omega\left(h(x)|\nabla_x\frac{\mathrm{d}q_0}{\mathrm{d}x}\right)\mathrm{d}x$$

gives the geodesic equation

$$S'(q_0)=-\nabla_x\,\mathrm{d}q_0/\mathrm{d}x=0.$$

The second differential (second variation)

$$S''(q_0)hh=-\int_\Omega\left(h(x)|\nabla_x^2h+R\left(\frac{\mathrm{d}q_0}{\mathrm{d}x},h\right)\frac{\mathrm{d}q_0}{\mathrm{d}x}\right)\mathrm{d}x$$

geodetic deviation

gives the **equation of geodetic deviation**

$$J_{q_0}h_0=\nabla_x^2h_0+R\left(\frac{\mathrm{d}q_0}{\mathrm{d}x},h_0\right)\frac{\mathrm{d}q_0}{\mathrm{d}x}=0.$$

Remark: It is easy to generalize these equations to paths which are only piecewise differentiable (see Milnor).

c) *Solutions of the geodetic deviation equation.*
It is convenient to introduce the Jacobi fields $J(\cdot,a)$ and $K(\cdot,a)$ defined as follows:

$$J(a,a)=0\quad\text{and}\quad\nabla_xJ(x=a,a)=g^{-1}(a)$$
$$K(a,a)=g^{-1}(a)\quad\text{and}\quad\nabla_xK(x=a,a)=0.$$

Using the notation of Problem II 2, p. 100 we can write the solution h_0 of the geodetic deviation equation satisfying Dirichlet data or Cauchy data as before,

$$h_0(x)=-J(x,a)M(a,b)h_0(b)-J(x,b)M(b,a)h_0(a)\quad\text{(Dirichlet data)}$$
$$h_0(x)=J(x,a)g(a)\nabla h_0(a)+K(x,a)g(a)h_0(a)\quad\text{(Cauchy data).}$$

Note that the above expressions are contracted products. For instance, using the basis $\{P_\alpha\}$ and $\{P^\alpha\}$ introduced in paragraph a, we have

$$M(a, b) = P^\alpha(a)M_{\alpha\beta}(a, b)P^\beta(b)$$

$$J(x, a)M(a, b)h_0(b) = P_\alpha(x)J^{\alpha\beta}(x, a)M_{\beta\gamma}(a, b)h_0^\gamma(b) = P_\alpha(x)h_0^\alpha(x).$$

d) *Conjugate points as the critical points of the exponential map.* In Problem II 1, p. 98, we have given three equivalent characterizations of conjugate points.

α) Two points $A = q_0(a)$ and $B = q_0(b)$ are conjugate along q_0, if and only if there exists a nonzero Jacobi field vanishing at A and B.

β) Let $\bar{\beta}(u)$ be a variation through geodesics of q_0 keeping the point A fixed; A and B are conjugate along q_0 if and only if $\beta(u, b) - \beta(0, b)$ is at least of order u^2.

γ) A and B are conjugate along q_0 if and only if

$$\det J(b, a) = 0.$$

Recall that $J(b, a)$ is the inverse of the VanVleck matrix.

We shall use the fact that a geodesic can be obtained by exponentiation to obtain a fourth characterization of conjugate points. Let $\bar{\beta}(u)$ be a variation through geodesics of q_0 keeping the point A fixed. $\bar{\beta}(u)$ can be obtained by exponentiating a vector $v(u)$ in the tangent plane to M at A.

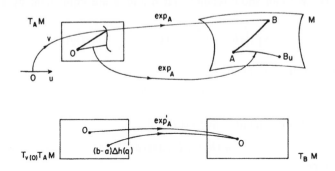

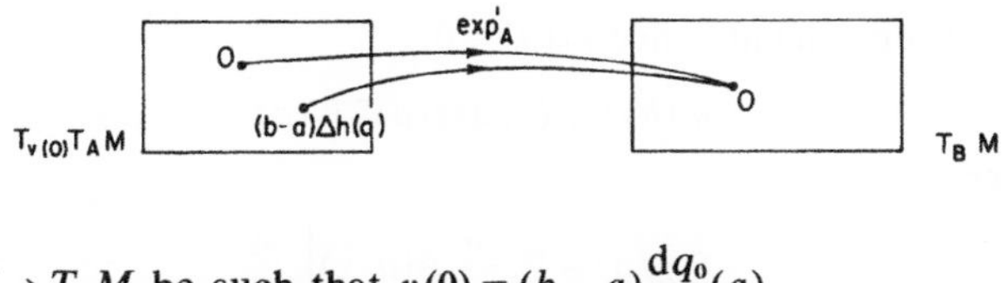

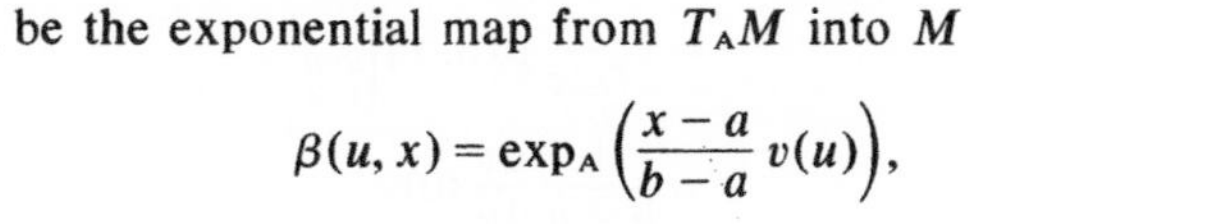

Let $v\colon U \to T_A M$ be such that $v(0) = (b - a)\dfrac{dq_0}{dx}(a)$.

Let $\exp_A$ be the exponential map from $T_A M$ into M

$$\beta(u, x) = \exp_A\left(\frac{x - a}{b - a}\, v(u)\right),$$

conversely

$$v(u) = (b - a)\frac{\partial \beta}{\partial x}(u, a).$$

The derivative of the exponential mapping at $v(0)$ maps $T_{v(0)}(T_A M)$ into $T_B M$.

$$\frac{\partial\beta}{\partial u}(0,x) = \exp'_A\left(\frac{x-a}{b-a}\,v(0)\right)\frac{x-a}{b-a}\frac{\mathrm{d}v}{\mathrm{d}u}(0).$$

We have shown in Problem II 1, p. 98, that

$$\frac{\partial\beta}{\partial u}(0,x) = h_0(x);$$

hence

$$h_0(b) = \frac{\partial\beta}{\partial u}(0,b) = \exp'_A(v(0))\frac{\mathrm{d}v}{\mathrm{d}u}(0).$$

Since $\bar{\beta}(u)$ is a variation keeping A fixed, $h_0(a) = 0$.
Hence if there exists a vector V such that $\exp'_A(v(0))V = 0$, we can construct a nonzero Jacobi field $(\partial\beta/\partial u)(0,x)$ vanishing at a and b, it suffices to choose v so that $(\mathrm{d}v/\mathrm{d}u)(0) = V$. Conversely if there is a nonzero Jacobi field vanishing on the boundary $\exp'_A(v(0))$ is not one–one. In conclusion we obtain a fourth characterization of conjugate points.

δ) Two points $A = q_0(a)$ and $B = q_0(b)$ are conjugate along q_0 if and only if $(b-a)(\mathrm{d}q_0/\mathrm{d}x)(a)$ is a critical point of $\exp_A$.

Remark: We can show readily that δ and γ are equivalent by proving that

$$\exp'_A\left((b-a)\frac{\mathrm{d}q_0}{\mathrm{d}x}(a)\right) = \frac{1}{b-a}J(b,a)g(a).$$

Indeed on the one hand

$$h_0(b) = \frac{\partial\beta}{\partial u}(0,b)$$

and on the other hand, since $h_0(a) = 0$

$$h_0(b) = J(b,a)g(a)\nabla h_0(a).$$

Moreover

$$\begin{aligned}\nabla h_0(a) &= \nabla_x \frac{\partial}{\partial u}\beta(u,x)\Big|_{\substack{u=0\\x=a}} \\ &= \nabla_u \frac{\partial}{\partial x}\beta(u,x)\Big|_{\substack{u=0\\x=a}} \\ &= \frac{1}{b-a}\frac{\mathrm{d}v}{\mathrm{d}u}(0).\end{aligned}$$

Remark: We can show readily that α and β are equivalent by making a

Taylor expansion of $\beta(u, x)$. In a chart at x we write

$$\beta(u, x) = \beta(0, x) + u\frac{\partial\beta}{\partial u}(0, x) + \tfrac{1}{2}u^2\frac{\partial^2\beta}{\partial u^2}(0, x) + \cdots.$$

Since $h_0(x) = (\partial\beta/\partial u)(0, x)$ we can write

$$\begin{aligned}\beta(u, x) &= \exp_{q_0(x)} u h_0(x)\\ &= q_0(x) + u \exp'_{q_0(x)}(0) h_0(x) + \text{terms of order } u^2.\end{aligned}$$

$\beta(u, b) - \beta(0, b)$ is of order u^2 if and only if $h_0(b) = 0$.

e) *Natural coordinate systems.*
We give below the equation of geodetic deviation, both in the Fermi geodetic system of coordinates and in the natural system of coordinates. Letters from the first part of the greek alphabet label coordinates in the Fermi system, letters from the second part label coordinates in the natural system

$$\begin{aligned}h_0(x) &= h_0^{\alpha}(x)P_{\alpha}(x) = h_0{}^{\mu}(x)\partial_{\mu}\\ (\partial_{\mu}|\partial_{\nu}) &= g_{\mu\nu}\\ (P_{\alpha}|P_{\beta}) &= \delta_{\alpha\beta}\end{aligned}$$

$g_{\mu\nu}$ and $g^{\mu\nu}$ lower and raise μ or ν; $\delta_{\alpha\beta}$ and $\delta^{\alpha\beta}$ lower and raise α or β

$$h_0{}^{\mu}(x) = h_0{}^{\alpha}(x)P_{\alpha}{}^{\mu}(x), \qquad h_0{}^{\alpha}(x) = h_0{}^{\mu}(x)P_{\mu}{}^{\alpha}(x).$$

The equation of geodetic deviation reads

$$\nabla_x{}^2 h_0{}^{\mu} + R_{\rho\sigma\nu}{}^{\mu} v^{\rho} h_0{}^{\sigma} v^{\nu} = 0 \quad \text{where } v = \mathrm{d}q_0/\mathrm{d}x$$

or

$$\frac{\mathrm{d}^2 h_0{}^{\alpha}}{\mathrm{d}x^2} + R_{\beta\gamma\delta}{}^{\alpha} v^{\beta} h_0{}^{\gamma} v^{\delta} = 0.$$

One deduces from this equation that a riemannian manifold with non-positive curvature has no conjugate point. For more on this subject see [Kobayashi, Nomizu, vol. II, Ch. VIII], [Milnor 1969, p. 101].

Reference: J. Milnor, *Morse Theory*, Annals of Mathematic Studies 51 (Princeton University Press, 1963, 1969).
We use here Milnor's choice of the Riemannian tensor to facilitate further reading in Milnor's book:

$$R_{\alpha\beta\gamma}{}^{\delta} = \partial_{\beta}\Gamma^{\delta}_{\alpha\gamma} - \partial_{\alpha}\Gamma^{\delta}_{\beta\gamma} + \Gamma^{\mu}_{\alpha\gamma}\Gamma^{\delta}_{\beta\mu} - \Gamma^{\mu}_{\beta\gamma}\Gamma^{\delta}_{\alpha\mu}.$$

PROBLEM 4. CAUSAL STRUCTURES: CONFORMAL SPACES: WEYL TENSOR

Introduction: A local causal structure on a manifold X is an assignment $x \mapsto c(x)$ for every $x \in X$ of a closed convex cone[1] $c(x) \subset T_x(M)$ which is locally defined by a finite number of inequalities on continuous functions on the tangent bundle. The causal future $\mathscr{E}^+(x_0)$ of a point is determined, through a closure operation, by the set of oriented differentiable paths issuing from x_0, whose tangents at each point x are in $c(x)$. The structure is causal if $y \in \mathscr{E}^+(x)$ defines a partial ordering. For instance, let M be the Minkowski space with metric $\eta_{\alpha\beta}$ (p. 290), and let $c(x)$ be the forward time cone

$$c(x) = \{y \in M\,;\, \eta_{\alpha\beta}(y^\alpha - x^\alpha)(y^\beta - x^\beta) \geq 0,\ y^0 - x^0 \geq 0,$$

then $\mathscr{E}^+(x) = c(x)$.

causal riemannian

A time oriented hyperbolic riemannian manifold with a causal structure induced by its metric[2] is called a **riemannian causal manifold**. A mapping $f\colon X \to Y$ is said to preserve the causal structure if, Y being in the causal future of x, $f(y)$ is in the causal future of $f(x)$;

$$x \lessdot y \quad \text{implies} \quad f(x) \lessdot f(y).$$

1) Conformal automorphisms; Weyl conformal curvature tensor

conformal automorphisms

A smooth mapping σ of a riemannian space (X, g) into itself is said to be **conformal** *if it satisfies the equation*

$$\sigma^*(g(\sigma(x))) = \exp(2\Phi(x))g(x)$$

where Φ is a smooth function on X.
The defining relations for the future light cone are invariant under the conformal mappings σ. It follows that a conformal mapping preserves the local causal structure induced by the metric,

a) *Set* $\bar{g}(x) \overset{\text{def}}{=} \sigma^*(g(\sigma(x)))$, *compute the connection $\bar{\Gamma}$ and the Riemann tensor $\bar{\boldsymbol{R}}$ of $\bar{g}$ in terms of the connection Γ and the Riemann tensor $\boldsymbol{R}$ of g.*
b) *Find a 4-tensor invariant by conformal transformations and give some of its properties.*

[1]Cone means here "half cone" in the sense of p. 288.
[2]$c(x) = c^+(x)$ in the notation p. 288.

Answer:

a) $\bar{\Gamma}^l_{ij} = \Gamma^l_{ij} + \delta^l_i \Phi_{,j} + \delta^l_j \Phi_{,i} - g_{ij} g^{lm} \Phi_{,m}$ with $\Phi_{,i} = \partial\Phi/\partial x^i$.
Set $\Phi_{ij} = \Phi_{,ij} - \Phi_{,i}\Phi_{,j}$, $\Delta_1\Phi = g^{ij}\Phi_{,i}\Phi_{,j}$ and $\Delta_2\Phi = g^{ij}\Phi_{,ij}$,

we have, for an n-dimensional space

$$\bar{R}^h{}_{ijk} = R^h{}_{ijk} + \delta^h_k \Phi_{ij} - \delta^h_j \Phi_{ik} + g^{hl}(g_{ij}\Phi_{lk} - g_{ik}\Phi_{lj}) + (\delta^h_k g_{ij} - \delta^h_j g_{ik})\Delta_1\Phi$$
$$\bar{R}_{ij} = \bar{g}^{hk}\bar{R}_{hijk} = R_{ij} + (n-2)\Phi_{ij} + g_{ij}(\Delta_2\Phi + (n-2)\Delta_1\Phi)$$
$$\bar{R} = e^{-2\Phi}(R + 2(n-1)\Delta_2\Phi + (n-1)(n-2)\Delta_1\Phi).$$

b) We can eliminate Φ_{ij} from the pair of equations $\bar{R}_{ij} = \ldots$ and $\bar{g}_{ij}\bar{R} = \ldots$; substituting Φ_{ij} into the equation $\bar{R}^h{}_{ijk} = \ldots$ we obtain an equation which can be written $\bar{C}^h{}_{ijk} = C^h{}_{ijk}$ where

$$C^{ij}{}_{km} = R^{ij}{}_{km} - \frac{1}{n-2}(R^i_k\delta^j_m - R^j_k\delta^i_m + R^j_m\delta^i_k - R^i_m\delta^j_k) + \frac{1}{(n-1)(n-2)} R\epsilon^{ij}_{km}.$$

The tensor $\boldsymbol{C}$ was first considered by Weyl who called it the conformal curvature tensor. If $n \geq 4$ a space with vanishing Weyl tensor is conformally flat. If $n = 3$ the tensor $\boldsymbol{C}$ vanishes identically. Another tensor (see for instance Eisenhart) can be used to determine if a space is conformally flat. If $n = 2$ or $n = 1$, all spaces are conformally flat.
Properties of the Weyl tensor.

1) Invariant under conformal mappings $\sigma^*\boldsymbol{C} = \boldsymbol{C}$.
2) The Weyl tensor is a traceless tensor $C^{is}{}_{ks} = 0$.
3) Equal to the Riemann tensor in Ricci flat spaces $R_{ij} = 0$.

The Weyl tensor is used to classify gravitational fields.

c) *Show that the equation*

$$\Delta_2 u + \frac{n-2}{4(n-1)} Ru = 0$$

is invariant by the transformation

$$\boldsymbol{g} \mapsto \exp 2\Phi \boldsymbol{g} = \bar{\boldsymbol{g}}, \qquad u \mapsto \exp\frac{2\Phi}{p} u = \bar{u}$$

if $p = -4/(n-2)$.

Answer:

$$\bar{\Delta}_2\bar{u} = \bar{g}^{ij}\partial^2_{ij}\bar{u} - \bar{g}^{ij}\bar{\Gamma}^l_{ij}\partial_l\bar{u}$$
$$= \exp\left(-\frac{n+2}{2}\Phi\right)\left\{\Delta_2 u + \left(\frac{4}{p^2}\Delta_1\Phi + \frac{2}{p}\Delta_2\Phi\right)u + \right.$$

$$+\frac{4}{p}\partial_i\Phi\partial_j u - g^{ij}\left(2\frac{\partial_l\Phi}{p}u + \partial_l u\right)(\delta_i^l\partial_j\Phi + \delta_j^l\partial_i\Phi - g_{ij}g^{lm}\partial_m\Phi)\Big\}$$

which reduces to, if $p = -4/(n-2)$

$$\bar{\Delta}_2\bar{u} = \exp\left(-\frac{n+2}{2}\Phi\right)\left\{\Delta_2 u - \frac{n-2}{2}\left(\Delta_2\Phi + \frac{n-2}{2}\Delta_1\Phi\right)u\right\}$$

which gives by a result p. 351

$$\bar{\Delta}_2\bar{u} + \frac{n-2}{4(n-1)}\bar{R}\bar{u} \equiv \exp\left(-\frac{n+2}{2}\Phi\right)\left\{\Delta_2 u + \frac{n-2}{4(n-1)}Ru\right\}$$

if

$$\bar{u} = \exp\frac{2\Phi}{p}u = \exp\left(-\frac{n-2}{2}\Phi\right)u.$$

2) Causal groups; the conformal group

The conformal diffeomorphisms of a riemannian space X into itself form a Lie group, called a conformal group or a causal group. The group of conformal diffeomorphisms on the compactified Minkowski space $(\mathbb{M}, \eta) \cup \{\infty\}$ is called the conformal group and labelled $C(3, 1)$. Construct the Lie algebra of the conformal group.

conformal Killing vector field

Answer: Let $\{\sigma_t\}$ be a one parameter subgroup of the conformal group; the vector field $\boldsymbol{v}$ generated by $\{\sigma_t\}$ is called a **conformal Killing vector field,**

$$\mathcal{L}_v\eta = 2\Phi\eta. \tag{1}$$

The subgroup of $C(3, 1)$ such that $\Phi = 0$ is the group of isometries; the subgroup such that $\Phi =$ constant is the group of isometries and dilations. The conformal Killing vector fields of $C(3, 1)$ satisfy the system of partial differential equations

$$(\mathcal{L}_v\eta)_{\alpha\beta} = \eta_{\gamma\beta}\partial_\alpha v^\gamma + \eta_{\gamma\alpha}\partial_\beta v^\gamma = 2\phi\eta_{\alpha\beta}$$

where $\partial_\alpha = \partial/\partial x^\alpha$ and where ϕ is obtained by contraction $2\phi = \frac{1}{2}\partial_\alpha v^\alpha$.

The system $\partial_\alpha v_\beta + \partial_\beta v_\alpha = \frac{1}{2}\partial_\rho v^\rho \eta_{\alpha\beta}$ can be solved exactly: its integrability conditions show that the Taylor expansion of $v^\alpha(x)$ terminates at third order. Indeed, set $\psi = \frac{1}{2}\partial_\rho v^\rho$, the integrability condition

$$\partial_\lambda\partial_\alpha v_\beta - \partial_\alpha\partial_\lambda v_\beta = 0 \quad \text{gives}$$

$$\partial_\lambda\partial_\beta v_\alpha - \partial_\alpha\partial_\beta v_\lambda = \partial_\lambda\psi\eta_{\alpha\beta} - \partial_\alpha\psi\eta_{\lambda\beta}. \tag{2}$$

The integrability condition $\partial_\rho\partial_\beta(\partial_\lambda v_\alpha - \partial_\alpha v_\lambda) - \partial_\beta\partial_\rho(\partial_\lambda v_\alpha - \partial_\alpha v_\lambda) = 0$ gives after contraction with $\eta^{\rho\alpha}$

$$2\partial_\beta\partial_\lambda\psi + \eta^{\rho\alpha}\partial_\alpha\partial_\rho\psi\eta_{\beta\lambda} = 0$$

which in turn gives by contraction $\eta^{\rho\alpha}\partial_\alpha\partial_\rho\psi = 0$, hence $\partial_\beta\partial_\lambda\psi = 0$. It follows that the Taylor expansion of ψ terminates at second order

$$\psi(x) = 2b + 4b_\alpha x^\alpha$$

where (b, b_α) are five independent constants. Substituting ψ into (2) we obtain, after integration with respect to x^β

$$\partial_\lambda v_\alpha - \partial_\alpha v_\lambda = 4(b_\lambda x_\alpha - b_\alpha x_\lambda) + 2a_{\alpha\lambda} \tag{3}$$

where $a_{\alpha\lambda} = -a_{\lambda\alpha}$ are 6 independent constants. Finally adding (3) and (1) and integrating with respect x^λ we obtain the general solution

$$v^\alpha(x) = a^\alpha + a^\alpha{}_\beta x^\beta + bx^\alpha + 2b^\beta\eta_{\beta\gamma}x^\gamma x^\alpha - b^\alpha\eta_{\beta\gamma}x^\beta x^\gamma \tag{4}$$

where a^α are 4 independent constants which together with $a_{\alpha\beta}$, b and b_α make 15 independent constants. The conformal group $C(3, 1)$ is a 15 dimensional group.

Lie algebra of the conformal group. The set of conformal Killing vector fields on X is a representation of the Lie algebra of the conformal group on X. The natural basis for the vector fields in X is $\{\partial_\alpha\}$. The fifteen linearly independent conformal Killing vector fields on M can be labelled

$$\boldsymbol{p}_\alpha = \partial_\alpha, \qquad \boldsymbol{m}_{\alpha\beta} = x_\alpha\partial_\beta - x_\beta\partial_\alpha$$

$$\boldsymbol{d} = x^\alpha\partial_\alpha, \qquad \boldsymbol{k}_\alpha = 2x_\alpha x^\beta\partial_\beta - x^2\partial_\alpha.$$

The Lie brackets are $[\boldsymbol{p}_\alpha, \boldsymbol{p}_\beta] = 0$, $[\boldsymbol{m}_{\alpha\beta}, \boldsymbol{p}_\gamma] = \eta_{\beta\gamma}\boldsymbol{p}_\alpha - \eta_{\alpha\gamma}\boldsymbol{p}_\beta$,

$[\boldsymbol{m}_{\alpha\beta}, \boldsymbol{m}_{\gamma\sigma}] = \eta_{\beta\gamma}\boldsymbol{m}_{\alpha\delta} + \eta_{\alpha\delta}\boldsymbol{m}_{\beta\gamma} - \eta_{\alpha\gamma}\boldsymbol{m}_{\beta\delta} - \eta_{\beta\delta}\boldsymbol{m}_{\alpha\gamma}$,

$[\boldsymbol{d}, \boldsymbol{m}_{\alpha\beta}] = 0$, $[\boldsymbol{d}, \boldsymbol{p}_\alpha] = -\boldsymbol{p}_\alpha$, $[\boldsymbol{d}, \boldsymbol{k}_\alpha] = \boldsymbol{k}_\alpha$, $[\boldsymbol{m}_{\alpha\beta}, \boldsymbol{k}_\gamma] = \eta_{\beta\gamma}\boldsymbol{k}_\alpha - \eta_{\alpha\gamma}\boldsymbol{k}_\beta$,

$[\boldsymbol{p}_\alpha, \boldsymbol{k}_\beta] = -\boldsymbol{m}_{\alpha\beta} + \eta_{\alpha\beta}\boldsymbol{d}$, $[\boldsymbol{k}_\alpha, \boldsymbol{k}_\beta] = 0$.

Flows of the conformal Killing vector fields on the Minkowski space. The flows are the solutions of the equation

$$\mathrm{d}x^\alpha(t)/\mathrm{d}t = v^\alpha(x(t)) \tag{5}$$

where v^α is given by equation (4). To obtain the transformations generated by each of the fifteen linearly independent Killing vector fields we shall set all the constants of integrations in equation (4) equal to zero except one:

$a^\alpha \neq 0$,	flow of p_α:	$x^\alpha(t) = a^\alpha t + x^\alpha(t_0)$	translations
$b \neq 0$,	flow of d:	$x^\alpha(t) = x^\alpha(t_0)\exp(bt)$	dilations
$a^\alpha_\beta \neq 0$,	flow of $m_{\alpha\beta}$:	$x^\alpha(t) = (\exp(a))^\alpha_\beta x^\beta(t_0)$	Lorentz rotation

where $a = (a^\alpha_\beta)$.

To solve equation (5) when $b^\alpha \neq 0$ we shall rewrite the infinitesimal transformation $x^\alpha(t_0) \mapsto x^\alpha(t_0 + \epsilon)$ as follows

$$x^\alpha(t_0 + \epsilon) = QPQx^\alpha(t_0)$$

where $Qx^\alpha(t_0) = x^\alpha(t_0)/x^2(t_0)$ and $Px^\alpha(t_0) = x^\alpha(t_0) + \epsilon b^\alpha$; we have then

$$\begin{aligned} x^\alpha(t) &= Q\exp(P)Qx^\alpha(t_0) \\ &= (x^\alpha(t_0) + tb^\alpha x^2(t_0))(1 + 2x^\alpha b_\alpha t + b^2x^2(t_0)t^2)^{-1}. \end{aligned}$$

This transformation is called a proper conformal transformation. Translation, dilations, Lorentz rotations and proper conformal transformations are transformations of the conformal group in the connected component of the identity. One can check that $Q: x^\alpha \mapsto x^\alpha/|x|^2$ is a causal transformation. It is called the conformal inversion. The proper conformal transformations and the conformal inversion map infinity, which is not a point of Minkowski space into a finite point and vice versa. The conformal group is defined on the so called conformally compactified Minkowski space.

3) **Show that the Maxwell equations aré invariant under conformal transformations**

Answer: We want to show that the Maxwell equations $dF = 0$, $\delta F = 0$ are the same whether they are computed with respect to g or to $\bar{g}$ when $\bar{g}(x) = \exp(2\Phi(x))g(x)$.
Since $dF = 0$ does not depend on the metric, we have only to compute $\bar{\nabla}_\alpha \bar{g}^{\alpha\mu}\bar{g}^{\beta\nu}F_{\mu\nu}$. Using the expression of $\bar{\Gamma}$ computed in 2) we obtain $\bar{\nabla}_\alpha \bar{g}^{\alpha\mu}\bar{g}^{\beta\nu}F_{\mu\nu} = \exp(-4\Phi(x))\nabla_\alpha F^{\alpha\beta}$ where ∇_α is the covariant derivative with respect to the metric g. ■

Note that $\bar{g}^{\beta\alpha}J_\alpha = \exp(-2\Phi(x))J^\beta$, hence $\delta F + J$ is not invariant under conformal transformations.

4). **Metric rescalings; conformal spaces**

Two riemannian spaces (X, g) and $(\bar{X}, \bar{g})$ are said to be conformal if there exists a mapping $f: X \to \bar{X}$ such that $f^*\bar{g} = \exp(2\Phi)g$ where Φ is a *smooth function on X. The mapping f is sometimes called a metric rescaling: it rescales the metric* $\mathbf{g}$ *into* $\exp(2\Phi)\mathbf{g}$. *It is not a conformal mapping in the same sense as 2).*

a) *Show that the n-sphere with the metric induced by the euclidean metric in* $\mathbb{R}^{n+1}$ *is conformally flat.* (See Problem Vbis 2.)

Answer: Let $\boldsymbol{\eta}$ be the euclidean metric on $\mathbb{R}^n$; let (r, Ω), where Ω stands for $(n-1)$ angles, be the polar coordinates on $\mathbb{R}^n$; let the pull back of $\boldsymbol{\eta}$ induced by the change of coordinates $f: (r, \Omega) \mapsto (x_1, \ldots, x_n)$ be $\mathbf{g} = f^*\boldsymbol{\eta} = \mathrm{d}r^2 + r^2\,\mathrm{d}\Omega^2$. For instance in $\mathbb{R}^3$, $\mathrm{d}\Omega^2 = \mathrm{d}\theta^2 + \sin^2\theta\,\mathrm{d}\varphi^2$. The metric $\bar{\mathbf{g}}$ on the n-sphere $x_1{}^2 + \cdots + x_{n+1}^2 = a^2$ induced by the euclidean metric on $\mathbb{R}^{n+1}$ is

$$\bar{\mathbf{g}} = a^2(a^2 - r^2)^{-1}\,\mathrm{d}r^2 + r^2\,\mathrm{d}\Omega^2.$$

Let $k: (\rho, \Omega) \mapsto (r, \Omega)$ by $r = 4a^2\rho/(4a^2 + \rho^2)$, then

$$k^*\bar{\mathbf{g}} = (1 + \rho^2/4a^2)^{-2}(\mathrm{d}\rho^2 + \rho^2\,\mathrm{d}\Omega^2).$$ ■

b) 1) *Show that the Friedman–Robertson–Walker (F.R.W.) isotropic, open universe*

$$\bar{\mathbf{g}} = R^2(\eta)(\mathrm{d}\eta^2 - \mathrm{d}\rho^2 - \sinh^2\rho\,\mathrm{d}\Omega^2), \quad \mathrm{d}\Omega^2 = \sin^2\theta\,\mathrm{d}\varphi^2 + \mathrm{d}\theta^2$$

is conformally flat.

2) *Same question for the Einstein space (closed universe with constant radius R)*

$$\mathbb{R} \times S^3 \text{ with } \bar{\mathbf{g}} = R^2(\mathrm{d}\eta^2 - \mathrm{d}\rho^2 - \sin^2\rho\,\mathrm{d}\Omega^2)$$

and for the Minkowski space

$$\mathbf{g} = \mathrm{d}t^2 - \mathrm{d}r^2 - r^2\,\mathrm{d}\Omega^2.$$

Answer:

1) Let $l: (\eta, \rho, \Omega) \mapsto (t, r, \Omega)$ by $r = A\exp(\eta)\sinh\rho$ and $t = A\exp(\eta)\cosh\rho$. Then

$$l^*\mathbf{g} = A^2\exp(2\eta)(R(\eta))^{-2}\bar{\mathbf{g}}.$$

2) Let $l: (\eta, \rho, \Omega) \mapsto (t, r, \Omega)$ by $t + r = 2R\tan\frac{1}{2}(\eta + \rho)$ and $t - r = 2R\tan\frac{1}{2}(\eta - \rho)$. Then

$$l^*\mathbf{g} = 4(\cos\eta + \cos\rho)^{-2}\bar{\mathbf{g}}$$

$$l^{-1*}\bar{\mathbf{g}} = \left(1 + \frac{1}{4R^2}(t + r)^2\right)^{-1}\left(1 + \frac{1}{4R^2}(t - r)^2\right)^{-1}\mathbf{g}.$$ ■

5) **Einstein spaces; the Einstein space**

A space such that $R_{ij} = Rg_{ij}/n$ with R constant is called an Einstein space. Prove the following.

a) *Any two-dimensional riemannian space is an Einstein space.*

b) *A space of constant curvature K is an Einstein space with $R = K(1-n)n$.*

c) *If an Einstein space is conformally flat, it is a space of constant curvature.*

Example: The space-time V^4 with metric

$$g = (1 + \tfrac{1}{3}ar^2 + c/r)\,dt^2 - (1 + \tfrac{1}{3}ar^2 + c/r)^{-1}\,dr^2 - r^2\,d\Omega^2$$

is an Einstein space. It is actually the most general spherically symmetric solution, up to a constant scale factor, of the **Einstein equation with a cosmological constant a in empty space**

Einstein equation

$$R_{\mu\nu} - \tfrac{1}{2}g_{\mu\nu}R - ag_{\mu\nu} = 0.$$

Einstein Universe

The **Einstein Universe** $\mathbb{R} \times S^3$ is not an Einstein space.

References: E. E. Fairchild Jr, *Applications of the Conformal Group in Physics*, Ph.D. Thesis University of Texas, Austin 1975; L. P. Eisenhart, *Riemannian Geometry* (Princeton University Press, 1926); L. Infeld and A. E. Schild, Phys. Rev. 68 (1945) 250–272 and 70 (1946) 410–425; G. Rosen, Am. J. Phys. 40 (1972) 1023–1027; D. R. Brill, A simple Derivation of the General Redshift Formula, in: *Methods of Local and Global Differential Geometry in General Relativity*, eds. D. Farmsworth, J. Fink, J. Porter and A. Thompson, Lecture Notes in Physics # 14 (Springer Verlag, 1972); I. E. Segal, *Mathematical Cosmology and Extragalactic Astronomy* (Academic Press N.Y., 1976); F. A. E. Pirani, Gravitational Radiation, in: *Gravitation, an introduction to current research*, ed. L. Witten (John Wiley and Sons, 1962); R. Penrose, Structure of Space Time, in: Battelle Rencontres 1967 Lectures in Mathematics and Physics, eds. C. M. DeWitt and J. A. Wheeler (W. A. Benjamin Inc., 1968).

Vbis. CONNECTIONS ON A PRINCIPAL FIBRE BUNDLE

The past years have seen the gradual emergence of the fundamental role played by gauge fields in physics and the possibility of describing gauge fields in terms of connections on principal fibre bundles. Therefore, in the revised edition, we have thought it desirable to remove the section on connections on principal fibre bundles from Chapter V on riemannian manifolds and develop it in a full chapter which, in a physics book, would probably have been titled *The Geometry of Gauge Fields* (see Problem Vbis 1). This leaves in Chapter V the theory of linear connections (connections on the frame bundle) presented directly from the concept of covariant derivatives.

The phrase "gauge transformation" was introduced in 1918 by Hermann Weyl[1] in a geometrically tantalizing, but unphysical,[2] attempt at constructing a unified theory for the electromagnetic field and the gravitational field, the two long range force fields. Nowadays, with a somewhat different meaning attached to the word "gauge", gauge theories remain at the core of the unifying schemes proposed for the four known fundamental interactions: gravitation, electromagnetism, weak and strong interactions in particle physics. The Salam–Weinberg model is a unified theory of electromagnetism with the weak interactions. It is based on two key concepts: non-abelian gauge field and broken symmetry. For an analysis of the role of gauge fields and symmetry in the development of physics and for the applications of gauge theory to the fundamental interactions of physics, we refer the reader to the excellent articles of Yang[3] and O'Raifeartaigh[4], respectively.

A. CONNECTIONS ON A PRINCIPAL FIBRE BUNDLE

1. DEFINITIONS

In a principal fibre bundle (P, X, π, G) each fibre is diffeomorphic to the typical fibre G. However, the diffeomorphism is not canonical; it depends

[1]For Weyl a gauge transformation was a local change of scale of the unit length.

[2]Einstein pointed out that in Weyl's scheme the atomic spectra would depend upon the world line of the atom:
"Weyl became convinced that this theory was not true as a theory of gravitation; but still it was so beautiful that he did not wish to abandon it and so he kept it alive for the sake of its beauty." S. Chandrasekhar, Physics Today, July 1979.

[3][Yang 1979].

[4][O'Raifeartaigh].

on the covering $\{U_i\}$ of X and on the choice of mappings ϕ_i: $\pi^{-1}(U_i)\to U_i\times G$. A connection on P leads to a correspondence between any two fibres along a curve C in X. One says that a point p of the fibre over a point x of the curve is **parallel translated** along the curve by means of this correspondence. We also say that the curve $\hat{C}$ described by the **parallel transport** of p is a **horizontal lift** of the curve C.

parallel translated

parallel transport

horizontal lift

To obtain the horizontal lifts of the curves C of X, it will be sufficient to define its "infinitesimal" counterpart, that is to associate with each tangent vector v_x at any $x\in X$, and each point p of the fibre $G_x\equiv\pi^{-1}(x)$, a tangent vector v_p to P at p, called a **horizontal vector**, which projects by π' onto v_x. The horizontal lift of a curve C in X, through a point $p\in P$, will then be obtained as the integral curve through p of these horizontal vectors which projects by π onto C.

$G_x\equiv\pi^{-1}(x)$

horizontal vector

We want the parallel transport to be compatible with the differentiable and the principal fibre bundle structure of P; the mapping between $T_x(X)$ and $T_p(P)$ must preserve their vector structures and be differentiable with respect to x; moreover, we have seen that the structure group G has a global right action on P denoted $\tilde{R}_g$, which preserves, and acts transitively on each fibre (p. 129). We wish therefore that two different horizontal lifts $\hat{C}_1$ and $\hat{C}_2$ of a curve C in X, for the same connection but through different points p_1 and p_2 in $\pi^{-1}(x)$, be related by the transformation $\tilde{R}_g$ which brings p_1 onto p_2. These requirements lead to the following definition of a connection.

connection on a principal bundle

Definition (a): A **connection on the principal bundle** (P, X, π, G) is a mapping σ_p: $T_x(X)\to T_p(P)$, $x=\pi(p)$ for every $p\in P$ such that

(1) σ_p is linear,

(2) $\pi'\sigma_p$ is the identity mapping on $T_x(X)$,

(3) σ_p depends differentiably on p,

(4) $\sigma_{\tilde{R}_g p}=\tilde{R}'_g\sigma_p$, $g\in G$.

Let C: $I\subset\mathbb{R}\to X$ by $t\mapsto C(t)$ be a curve in X passing through the point $x_0=C(0)$ and let p_0 be in the fibre at x_0. The parallel translation of p_0 along C is given by the curve $\hat{C}$: $I\subset\mathbb{R}\mapsto P$ by $t\mapsto\hat{C}(t)$ defined by

$$\frac{\mathrm{d}\hat{C}(t)}{\mathrm{d}t}=\sigma_p\frac{\mathrm{d}C(t)}{\mathrm{d}t},\qquad \hat{C}(0)=p_0,\qquad \hat{C}(t)=p.$$

Remark: The classical existence theorem for differential equations (p. 94) gives the existence of the horizontal lift $\hat{C}$ of C only in a neighborhood of p_0. However, it is possible here, by using a covering of C by a finite number of arcs and using the properties of σ_p to obtain a uniform estimate, and thus to show the existence of the horizontal lift $\hat{C}$ of the entire curve C. A change of parametrization of C changes the parametrization of $\hat{C}$, but the geometrical curve $\hat{C}$ is independent of the parametrization of C.

The properties of the parallel transport for which we were looking follow from the definition (a), namely

(1) $\hat{C}$ projects by π onto C:

$$\pi\hat{C}(t) = C(t),$$

(2) the right action of G on P commutes with parallel transport: two horizontal lifts $\hat{C}_2$ and $\hat{C}_1$ of the same curve C in X satisfy $\hat{C}_2(t) = \tilde{R}_g\hat{C}_1(t)$ for all t, with the same g, if they satisfy it for $t = 0$, since both sides, due to property (4), satisfy the same differential equation.

Since σ_p is linear the space of horizontal vectors at p

$$H_p \equiv \sigma_p(T_x(X)), \qquad x = \pi(p)$$

is a vector subspace of $T_p(P)$. Due to the property (a2) we have also

$$\pi' H_p = T_x(X), \qquad x = \pi(p)$$

thus H_p is isomorphic to $T_x(X)$, by the linear mapping π'.
We can express the definition (a) in terms of these horizontal vector spaces H_p, as follows:

Definition (b): A **connection on the principal bundle** (P, X, π, G) is a field of vector spaces H_p, $H_p \subset T_p(P)$, such that connection

(1) π': $H_p \to T_x(X)$, $x = \pi(p)$, is an isomorphism,

(2) H_p depends differentiably on p,

(3) $H_{\tilde{R}_g p} = \tilde{R}'_g H_p$.

The elements of H_p are the horizontal vectors at p. The elements of the tangent space $T_p(G_x) \equiv V_p$ to the fibre G_x at p are called **vertical vectors**. Since $\pi' V_p = 0$, then $T_p(P) = H_p \oplus V_p$, that is, any $v \in T_p(P)$ can be written, uniquely vertical vector

$$v = v_{\mathrm{H}} + v_{\mathrm{V}} \qquad v_{\mathrm{H}} \in H_p, \qquad v_{\mathrm{V}} \in V_p.$$

Note that v_{V} (also denoted ver v) depends like v_{H} (also denoted hor v) on the choice of H_p. ver v hor v

Canonical isomorphisms between the Lie algebra $\mathscr{G}$ of G and the vertical spaces V_p. First note that, since G acts effectively on P by $\tilde{R}_g$, there is a natural vector isomorphism between the Lie algebra $\mathscr{G}$ of G and the space of

Killing vector fields $\{v^K\}$ on P relative to G (p. 154), defined by $\hat{v}_{(\alpha)} \leftrightarrow v^K_{(\alpha)}$ where $\hat{v}_{(\alpha)} \in \mathscr{G}$ and $v^K_{(\alpha)} \in \{v^K\}$ are both generated by $v_{(\alpha)}(e) = \mathrm{d}g(s)/\mathrm{d}s|_{s=0} \in T_e(G)$:

$$\hat{v}_{(\alpha)}(g) = L'_g(e)\, v_{(\alpha)}(e), \qquad v^K_{(\alpha)}(p) = \mathrm{d}(\tilde{R}_{g(s)}p)/\mathrm{d}s|_{s=0}. \tag{1}$$

Next, because p and $\tilde{R}_g p$ lie in the same fibre, any Killing vector $v^K(p)$ is a vertical vector. In addition a Killing vector field does not vanish at any point unless it corresponds to the zero element of the Lie algebra. Since the correspondence is linear, the dimension of the space $\{v^K(p)\}$ is equal to the dimension of the space $\{v^K\}$, which is equal to the dimension of the group G and of V_p. In conclusion, let $\hat{v} \in \mathscr{G}$ correspond to $\mathrm{d}g(s)/\mathrm{d}s|_{s=0} \in T_e(G)$; the equation

canonical isomorphism between $\mathscr{G}$ and V_p

$$v(p) = \mathrm{d}(\tilde{R}_{g(s)}p)/\mathrm{d}s|_{s=0}$$

defines the **canonical isomorphism between $\mathscr{G}$ and V_p**,

$v \leftrightarrow \hat{v}$

$$v(p) \leftrightarrow \hat{v}, \qquad v(p) \in V_p. \tag{2}$$

fundamental vector field

Given an element $\gamma \in \mathscr{G}$, this isomorphism defines a vector field v_γ on P, called a **fundamental vector field**, by $\widehat{v_\gamma(p)} = \gamma$. We also write $\check{\gamma} = v_\gamma(p)$.

When we are given the field of horizontal subspaces H_p we have, for each $p \in P$, a well-defined family of linear mappings

$$T_p(P) \to \mathscr{G} \quad \text{by} \quad v \mapsto \widehat{\mathrm{ver}\, v}, \qquad p \in P. \tag{3}$$

In agreement with previous definitions (p. 210) we call the family of mappings (3) a 1-form ω on P with values in the vector space $\mathscr{G}$, the Lie algebra of G:

$$\omega(v) = \widehat{\mathrm{ver}\, v}, \qquad \text{and thus } \omega(\mathrm{hor}\, v) = 0, \qquad \forall v \in T_p(P).$$

Note that if (e_α) is a basis for $\mathscr{G}$ and if (θ^i) is a basis for $T^*_p(P)$, then ω can be written[1]

$$\omega = \omega^\alpha \otimes e_\alpha = \omega^\alpha_i \theta^i \otimes e_\alpha$$

where the ω^α are 1-forms on P.

[1] We have $\omega(v) = (\omega(v))^\alpha e_\alpha$ and we write $(\omega(v))^\alpha = \omega^\alpha(v)$.

It results from the property 2 of the definition (b) that the ω^α are differentiable 1-forms on P, and thus that ω is a differentiable 1-form on P with values in $\mathscr{G}$. The equivariance (property 3 of definition (b)) of the horizontal subspaces H_p insures that R'_g preserves the decomposition of any tangent vector v to P into a horizontal and a vertical part:

$$\tilde{R}'_g v = (\tilde{R}'_g v)_{\mathrm{H}} + (\tilde{R}'_g v)_{\mathrm{V}} = \tilde{R}'_g v_{\mathrm{H}} + \tilde{R}'_g v_{\mathrm{V}} \quad \text{if} \quad v = v_{\mathrm{H}} + v_{\mathrm{V}}.$$

If we compute the pull back of ω by $\tilde{R}_g$ we find, by the definitions

$$(\tilde{R}^*_g \omega)(v) = \omega(\tilde{R}'_g v) = \omega((\tilde{R}'_g v)_{\mathrm{V}}) = \omega(R'_g v_{\mathrm{V}}).$$

The restriction of ω to a fibre $G_x = \pi^{-1}(x)$ defines a 1-form on G_x (which we shall also call ω) by

$$\omega(v_{\mathrm{V}}) = \widehat{v_{\mathrm{V}}}, \qquad v_{\mathrm{V}} \in T_p(G_x) \equiv V_p.$$

This form can be identified with the Maurer–Cartan canonical 1-form (p. 168) on G, through the identification $\mathscr{I}$: $G \to G_x$ obtained by choosing a point p_0 in G_x, and setting $\mathscr{I}(h) = p$ iff $p = \tilde{R}_h p_0$.
We deduce then from the transformation law of this canonical 1-form (p. 169) that

$$(\tilde{R}^*_g \omega)(v) = \mathrm{Ad}(g^{-1})\omega(v)$$

where Ad is the adjoint representation defined on p. 167.

We arrive thus at the third definition.

Definition (c): A **connection on the principal fibre bundle** (P, X, π, G) is a 1-form ω on P with values in the vector space $\mathscr{G}$, such that connection
(1) $\omega_p(u) = \hat{u}$ where $u \in V_p$ and $\hat{u} \in \mathscr{G}$ are related by the canonical isomorphism (2),
(2) ω_p depends differentiably on p,
(3) $\omega_{\tilde{R}_g p}(\tilde{R}'_g v) = \mathrm{Ad}(g^{-1})\omega_p(v)$.
If a connection is given by the definition (c), we define the horizontal subspaces as the kernels of the mappings ω_p: $T_p(P) \to \mathscr{G}$, namely

$$H_p = \{v \in T_p(P);\ \omega_p(v) = 0\};$$

one verifies easily that these spaces satisfy the properties of the definition (b), and thus the equivalence of the two definitions.

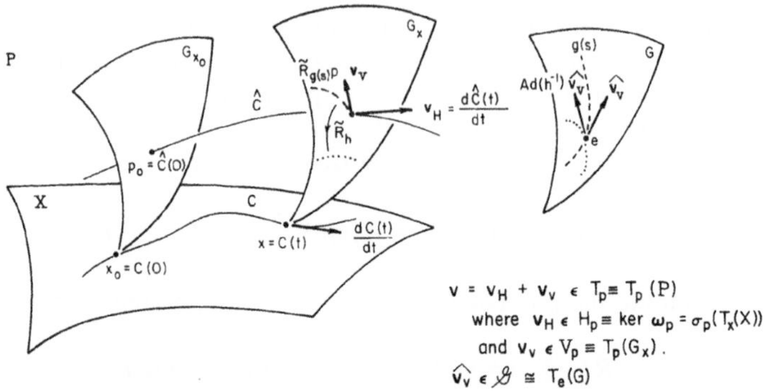

$\boldsymbol{\omega}$ is a form on $T_p(P)$ with values in $\mathcal{G}$ such that $\boldsymbol{\omega}(\boldsymbol{v}_V) = \widehat{\boldsymbol{v}_V}$; given $\boldsymbol{v}$ and given $\boldsymbol{\omega}$, the vector $\boldsymbol{\omega}(\boldsymbol{v}) \in \mathcal{G}$ determines uniquely $\boldsymbol{v}_V$; the horizontal vector $\boldsymbol{v}_H$ and hence the curve $\hat{C}(t)$ is uniquely determined by $\boldsymbol{v}_H = \boldsymbol{v} - \boldsymbol{v}_V$.

$\tilde{R}'_h \boldsymbol{v}_V$ generates the curve $\tilde{R}_h \circ \tilde{R}_{g(t)} \circ \tilde{R}_h^{-1} p = \tilde{R}_{h^{-1}g(t)h} p$ (p. 147), hence $\boldsymbol{\omega}_{\tilde{R}_h p}(\tilde{R}'_h \boldsymbol{v}_V) = \mathrm{Ad}\,(h^{-1})\widehat{\boldsymbol{v}_V}$. Conditions 2 and 3 of the definition ensure that $\boldsymbol{\omega}$ is compatible with the structure of P: given $\boldsymbol{\omega}_p$, condition 3 determines $\boldsymbol{\omega}$ at another point on the same fibre; condition 2 ensures that $\hat{C}(t)$ is differentiable.

2. LOCAL CONNECTION 1-FORMS ON THE BASE MANIFOLD

For a given connection $\boldsymbol{\omega}$ we shall now associate with each differentiable local section of $\pi^{-1}(U) \subset P$, $U \subset X$, a 1-form on U with values in $\mathcal{G}$. Let

$$f\colon U \subset X \to f(U) \subset P, \qquad \pi f = \mathrm{Id}$$

be a local section of P; we define a 1-form $f^*\boldsymbol{\omega}$ on U with values in $\mathcal{G}$ by the pull back of $\boldsymbol{\omega}$ by f: that is, if $\boldsymbol{u} \in T_x(X)$, $x \in U$

$$(f^*\boldsymbol{\omega})_x(\boldsymbol{u}) = \boldsymbol{\omega}_{f(x)}(f'\boldsymbol{u}).$$

Conversely:

Theorem. Given a differentiable 1-*form* $\bar{\boldsymbol{\omega}}$ *on* U *with values in* $\mathcal{G}$, *and a differentiable section* f *of* $\pi^{-1}(U)$, *there exists one and only one connection* $\boldsymbol{\omega}$ *on* $\pi^{-1}(U)$ *such that* $f^*\boldsymbol{\omega} = \bar{\boldsymbol{\omega}}$.

Proof: (1) Let $\boldsymbol{v} \in T_{p_0}(P)$, $p_0 = f(x)$; we write

$$\boldsymbol{v} = \boldsymbol{v}_1 + \boldsymbol{v}_2$$

where $\boldsymbol{v}_1 = (f' \circ \pi')\boldsymbol{v}$, and therefore $\boldsymbol{v}_2$ is vertical: we have $\pi'\boldsymbol{v}_2 = 0$ since $\pi' \circ f' = \mathrm{Id}$.

We define a 1-form $\boldsymbol{\omega}$ with values in $\mathcal{G}$ at each point $p_0 \in f(U)$ of the section by

$$\boldsymbol{\omega}_{p_0}(\boldsymbol{v}) = \bar{\boldsymbol{\omega}}_x(\pi' \boldsymbol{v}) + \hat{\boldsymbol{v}}_2$$

where $\hat{\boldsymbol{v}}_2$ is obtained from $\boldsymbol{v}_2$ by the canonical isomorphism $V_p \to \mathcal{G}$.
(2) We extend $\boldsymbol{\omega}$ to all the points of $\pi^{-1}(U)$ by the action of G, that is we set

$$\boldsymbol{\omega}_p(\boldsymbol{v}) = \mathrm{Ad}(g^{-1})\boldsymbol{\omega}_{p_0}(\tilde{R}_g^{-1'}\boldsymbol{v}) \quad \text{if} \quad p = \tilde{R}_g p_0.$$

The 1-form thus defined enjoys the properties 1–2–3 of the definition (c), and it is the only one such that $f^*\boldsymbol{\omega} = \bar{\boldsymbol{\omega}}$. ∎

This construction can be extended to the case where $\bar{\boldsymbol{\omega}}$ is a differentiable 1-form on the whole base X and leads to the following theorem.

Theorem: *There exist on each principal bundle with paracompact base X infinitely many connections.*

Proof: Let $\bar{\boldsymbol{\omega}}$ be a differentiable 1-form on X with values in $\mathcal{G}$, and $\{\theta_i\}$ a partition of unity subordinate to the covering $\{U_i\}$ of X associated with the fibre bundle structure of P. Denote by $\boldsymbol{\omega}_i$ the connection form on $\pi^{-1}(U_i)$, obtained from $\bar{\boldsymbol{\omega}}$ by a choice of a section f_i over U_i. The 1-form on P:

$$\boldsymbol{\omega} = \sum_i (\theta_i \circ \pi)\boldsymbol{\omega}_i$$

is a connection on P, since it is easily seen that it satisfies the conditions of the definition (c), if it is so for the $\boldsymbol{\omega}_i$'s. ∎

Note that, if P is not a trivial bundle, the 1-form $\bar{\boldsymbol{\omega}}$ is not the pull back of $\boldsymbol{\omega}$ by a differentiable section f of P since such a section $X \to P$ does not exist in that case (see Problem Vbis 2).

We shall prove the converse of the above theorem, namely, given a trivialization $\{U_i, \phi_i\}$ of the bundle P and a connection $\boldsymbol{\omega}$ on P, there corresponds a unique family $\{\bar{\boldsymbol{\omega}}_i\}$ of connection 1-forms on the base manifold. First, we define the **section s_i of $\pi^{-1}(U_i)$ canonically associated with the trivialization ϕ_i.** section associated to a trivialization

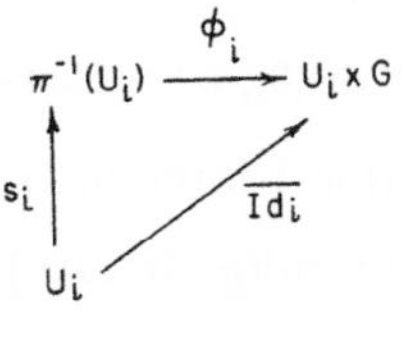

Let $\overline{\mathrm{Id}}: U_i \to U_i \times G$ by $x \mapsto (x, e)$. A trivialization ϕ_i defines a section s_i, and vice versa, through the equation

$$s_i = \phi_i^{-1} \circ \overline{\mathrm{Id}}.$$

connection form in a trivialization

Let $\bar{\boldsymbol{\omega}}_i = s_i^* \boldsymbol{\omega}$, the form $\bar{\boldsymbol{\omega}}_i$ on U_i is called the **connection form in the local trivialization ϕ_i.**

potential
local gauge

Potentials. In the Yang–Mills theories of physics, the 1-forms $\bar{\boldsymbol{\omega}}_i$ are usually called **potentials (gauge potentials)** and the trivializations ϕ_i are called **local gauges**. The $\bar{\boldsymbol{\omega}}_i$ are related to the traditional potential $\boldsymbol{A}$ by multiplicative constants.

Theorem. At a point $x \in U_i \cap U_j$ the connection forms $\bar{\boldsymbol{\omega}}_i$ and $\bar{\boldsymbol{\omega}}_j$ in the local gauges ϕ_i and ϕ_j corresponding to the same connection $\boldsymbol{\omega}$ on P are linked by the relation

$$\bar{\boldsymbol{\omega}}_{i,x} = \mathrm{Ad}(g_{ji}^{-1}(x))\bar{\boldsymbol{\omega}}_{j,x} + (g_{ji}^* \boldsymbol{\theta}_{\mathrm{MC}})_x \tag{4}$$

where g_{ij} is the transition mapping (cf. pp. 125, 126):

$$g_{ij}: U_i \cap U_j \to G \quad by \quad x \mapsto g_{ij}(x) = \overset{\Delta}{\phi}_{i,x} \circ \overset{\Delta}{\phi}{}_{j,x}^{-1} \in G$$

and $g_{ij}^* \boldsymbol{\theta}_{\mathrm{MC}}$ denotes the pull-back on $U_i \cap U_j$ of the Maurer–Cartan 1-form on G (cf. p. 168) *by this transition mapping.*

Proof: The pull back $\phi_i^{-1*}\boldsymbol{\omega}: TU_i \times TG \to \mathscr{G}$ is a $\mathscr{G}$-valued 1-form on $U_i \times G$. Let $(\boldsymbol{v}, \boldsymbol{w}) \in T_xU_i \times T_gG$;

$$(\phi_i^{-1*}\boldsymbol{\omega})(\boldsymbol{v}, \boldsymbol{w}) = (\phi_i^{-1*}\boldsymbol{\omega})(\boldsymbol{v}, \boldsymbol{0}_g) + (\phi_i^{-1*}\boldsymbol{\omega})(\boldsymbol{0}_x, \boldsymbol{w})$$

where $\boldsymbol{0}_g$ and $\boldsymbol{0}_x$ are the zero vectors at g and x respectively.
Using $\boldsymbol{0}_g = R'_g\boldsymbol{0}_e$ and the definition of a connection form (p. 361) we obtain

$$\begin{aligned}(\phi_i^{-1*}\boldsymbol{\omega})(\boldsymbol{v}, \boldsymbol{0}_g) &= \mathrm{Ad}(g^{-1})(\phi_i^{-1*}\boldsymbol{\omega})(\boldsymbol{v}, \boldsymbol{0}_e)\\ &= \mathrm{Ad}(g^{-1})(\phi_i^{-1} \circ \overline{\mathrm{Id}})^*\boldsymbol{\omega}(\boldsymbol{v})\\ &= \mathrm{Ad}(g^{-1})s_i^*\boldsymbol{\omega}(\boldsymbol{v}).\end{aligned}$$

Since $(\boldsymbol{0}_x, \boldsymbol{w})$ is a vertical vector,

$$(\phi_i^{-1*}\boldsymbol{\omega})(\boldsymbol{0}_x, \boldsymbol{w}) = \boldsymbol{\theta}_{\mathrm{MC}}(\boldsymbol{w}) \tag{5}$$

where $\boldsymbol{\theta}_{\mathrm{MC}}$ is the Maurer–Cartan 1-form on G. Finally

$$(\phi_i^{-1*}\boldsymbol{\omega})(\boldsymbol{v}, \boldsymbol{w}) = \mathrm{Ad}(g^{-1})s_i^*\boldsymbol{\omega}(\boldsymbol{v}) + \boldsymbol{\theta}_{\mathrm{MC}}(\boldsymbol{w}).$$

To obtain $\bar{\boldsymbol{\omega}}_i = s_i^* \boldsymbol{\omega}$ in terms of $\bar{\boldsymbol{\omega}}_j = s_j^* \boldsymbol{\omega}$, we compute[1]

$$((\phi_j^{-1} \circ \phi_j \circ \phi_i^{-1})^* \boldsymbol{\omega})(v, w) = (\phi_j^{-1*} \boldsymbol{\omega})((\phi_j \circ \phi_i^{-1})'(x, g)(v, w)) \qquad (6)$$
$$= (\phi_j^{-1*} \boldsymbol{\omega})((\mathrm{id}_X, L_{g_{ji}(x)})'(x, g)(v, w)).$$

Since

$$(\mathrm{id}_X, L_{g_{ji}(x)}): (U_i \cap U_j) \times G \to (U_i \cap U_j) \times G \quad \text{by} \quad (x, g) \mapsto (x, g_{ij}(x) g),$$

we have

$$(\mathrm{id}_X, L_{g_{ji}(x)})'(x, g): T_x(U_i \cap U_j) \times T_g G \to T_x(U_i \cap U_j) \times T_{g_{ji}(x)g} G$$

by

$$\begin{pmatrix} v \\ w \end{pmatrix} \mapsto \begin{pmatrix} \partial x / \partial x & \partial x / \partial g \\ \partial L_{g_{ji}(x)} g / \partial x & \partial L_{g_{ji}(x)} g / \partial g \end{pmatrix} \begin{pmatrix} v \\ w \end{pmatrix}$$

Hence

$$(\phi_j^{-1*} \boldsymbol{\omega})(\mathrm{id}_X, L_{g_{ji}(x)})'(x, g)(v, w)$$
$$= (\phi_j^{-1*} \boldsymbol{\omega})(v, R_g'(g_{ji}(x)) g_{ji}'(x)(v) + L_{g_{ji}(x)}'(g)(w))$$
$$= (\phi_j^{-1*} \boldsymbol{\omega})(v, 0_{g_{ji}(x)g}) + (\phi_j^{-1*} \boldsymbol{\omega})(0_x, R_g'(g_{ji}(x)) g_{ji}'(x) v + L_{g_{ji}(x)}'(g) w$$
$$= \mathrm{Ad}\, g^{-1}(\mathrm{Ad}(g_{ji}(x))^{-1} \bar{\boldsymbol{\omega}}_j(v) + \boldsymbol{\theta}_{\mathrm{MC}}(g_{ji}'(x) v)) + \boldsymbol{\theta}_{\mathrm{MC}}(w) \quad \text{by (5)}$$
$$= (\phi_i^{-1*} \boldsymbol{\omega})(v, w) \quad \text{by (6)}.$$

Since
$$(\phi_i^{-1*} \boldsymbol{\omega})(v, w) = \mathrm{Ad}(g^{-1}) \bar{\boldsymbol{\omega}}_i(v) + \boldsymbol{\theta}_{\mathrm{MC}}(w),$$

it follows that

$$\bar{\boldsymbol{\omega}}_i(v) = \mathrm{Ad}(g_{ji}(x)^{-1}) \bar{\boldsymbol{\omega}}_j(v) + \boldsymbol{\theta}_{\mathrm{MC}}(g_{ji}'(x) v). \qquad (7) \blacksquare$$

Remark: If G is a subgroup of $\mathrm{GL}(n)$ acting on TX by linear transformations and if we identify $\mathscr{G}$ with the tangent space $T_e G$ at the identity, the above transformation may be written with the natural identification of the linear mappings L_g and R_g with their derivatives (see p. 163):

$$\bar{\boldsymbol{\omega}}_i(v) = g_{ji}^{-1}(x)\, \bar{\boldsymbol{\omega}}_j(v)\, g_{ji}(x) + g_{ji}^{-1}(x)\, g_{ji}'(x)\, v. \qquad (8)$$

Formula (8) is also used, in the general case, as an abbreviated notation for formula (7).

[1]The following calculation is fairly involved and the reader may prefer the notation $T_x f(v)$ for $f'(x)(v)$; he should then write $\boldsymbol{\omega}(T_{(x,g)}(\mathrm{id}_X, L_{g_{ij}^{-1}})(v, w))$ etc., in the lines which follow. We prefer not to change the notation used throughout this book. A prime denotes the derivative with respect to the argument. Thus $g_{ij}'(x)\colon T_x X \to T_{g_{ij}(x)} G$ and $\mathrm{d}L_{g_{ij}(x)}(g)/\mathrm{d}x = \mathrm{d}\boldsymbol{R}_g(g_{ij}(x))/\mathrm{d}x = \boldsymbol{R}_g'(g_{ij}(x)) g_{ij}'(x)$. For a heuristic proof see p. 407.
In checking the equations leading to (6), recall that $w \in T_g G$ and $g_{ji}'(x) v \in T_{g_{ji}(x)}$.

Example: **Electromagnetic potential**. Consider a principal bundle with structure group $U(1)$, i.e., the group of complex numbers z of unit norm written $z = \exp ia$. A transition function from $U_i \cap U_j$ into $U(1)$ is then given by $g_{ij}(x) = \exp i\phi(x)$, $g_{ji}(x) = \exp(-i\phi(x))$ where $x \mapsto \phi(x)$ is a real valued function on $U_i \cap U_j$. A connection in a local gauge is a 1-form $\bar{\boldsymbol{\omega}}_i$ on $U_i \subset X$ with values in the Lie algebra $\mathscr{U}(1)$ of $U(1)$. If we choose $\bar{\boldsymbol{\omega}}_i(x) = -ie\boldsymbol{A}_i(x)$, the real function $\boldsymbol{A}_i(x)$ is the electromagnetic potential. It follows from (5) or (6) that the transformation law of the potentials under a change of gauge is given by

$$\boldsymbol{A}_i(\boldsymbol{v}) = \boldsymbol{A}_j(\boldsymbol{v}) + \frac{1}{e}\,\phi'(x)\boldsymbol{v},$$

i.e., in coordinates

$$A_{i\mu}v^{\mu} = A_{j\mu}v^{\mu} + \frac{1}{e}\,\partial_{\mu}\phi(x)v^{\mu}.$$

Note that this transformation law is valid on $U_i \cap U_j$, that is when the potentials are defined on open sets diffeomorphic to $\mathbb{R}^n$.

Example: **Christoffel symbols.** Consider the frame bundle with structure group GL(n). A connection in a local gauge is a 1-form with values in the Lie algebra $\mathscr{GL}(n)$. If the connection is riemannian (p. 308) its components are the Christoffel symbols $\Gamma^{\mu}_{\nu\rho} \equiv \{^{\mu}_{\nu\rho}\}$, and $\Gamma^{\mu}_{\nu\rho}v^{\nu}$ are the components of a matrix of $\mathscr{GL}(n)$. Equation (6) gives the usual transformation of the Christoffel symbols under a change of coordinates.

Exercise: Prove directly that the mapping σ_p: $T_x(X) \to T_p(P)$ in the definition (a), together with a given local trivialization of P, defines a one-form with values in T_eG equal, up to a sign, to the local connection $\bar{\boldsymbol{\omega}}$ in the given trivialization, provided one identifies T_eG and $\mathscr{G}$.

Answer:
(1) Let ϕ: $\pi^{-1}(U) \to U \times G$ be a local trivialization such that $\phi(p_0) = (x, e)$.
Let ϕ'_{p_0}: $T_{p_0}P \to T_xX \times T_eG$ then $\phi'_{p_0} \circ \sigma_{p_0}$: $T_xX \to T_xX \times T_eG$ by $\boldsymbol{u} \mapsto (\boldsymbol{u}, \boldsymbol{w})$. Hence $\phi'_{p_0} \circ \sigma_{p_0} = (\mathrm{id}, \boldsymbol{\alpha})$ where id is the identity on T_xX and where $\boldsymbol{\alpha}$ is a 1-form on U with value in T_eG.
(2) We shall show that $\boldsymbol{\alpha} = -s * \boldsymbol{\omega}$ where $s = \phi^{-1} \circ \overline{\mathrm{Id}}$: $U \to U \times G \to \pi^{-1}(U)$ by $x \mapsto (x, e) \mapsto p_0$.
Indeed

$$(\boldsymbol{u}, \boldsymbol{0}_e) = \phi'_{p_0}(s'(\boldsymbol{u})) = \phi'_{p_0}(\mathrm{hor}\ s'(\boldsymbol{u}) + \mathrm{ver}\ s'(\boldsymbol{u}))$$

where $\boldsymbol{0}_e$ is the zero vector in T_eG.

Set $[\boldsymbol{\omega}(s'(\boldsymbol{u}))]^{\vee}$ the vertical vector at $s(x)$ canonically associated to the element of the Lie algebra $\boldsymbol{\omega}(s'(\boldsymbol{u}))$. The $\vee$ operation is the inverse of the hat operation which maps a vertical vector v at $p \in G_x$ into $\hat{v} \in \mathcal{G}$. With this notation

$$\phi'_{p_0}(\text{hor } s'(\boldsymbol{u}) + \text{ver } s'(\boldsymbol{u})) = \phi'_{p_0}(\sigma_{p_0}(\boldsymbol{u}) + [\boldsymbol{\omega}(s'(\boldsymbol{u}))]^{\vee}) = (\boldsymbol{u}, \boldsymbol{0}_e).$$

This last equation is the basic relation between σ_{p_0} and $\boldsymbol{\omega}$. More explicitly it gives

$$(\boldsymbol{u}, \boldsymbol{0}_e) = \phi'(\sigma_{p_0}(\boldsymbol{u})) + \phi'([(s^*\boldsymbol{\omega})(\boldsymbol{u})]^{\vee}) \equiv (\boldsymbol{u}, \boldsymbol{\alpha}(\boldsymbol{u})) + (\boldsymbol{0}_x, s^*\boldsymbol{\omega}(\boldsymbol{u})).$$

Hence $\boldsymbol{\alpha}(\boldsymbol{u}) = -s^*\boldsymbol{\omega}(\boldsymbol{u})$. ∎

3. COVARIANT DERIVATIVE

A connection on a principal bundle (P, X, π, G) defines the covariant derivative of sections of some vector bundles, the so called associated bundles to the principal bundle. For example, a linear connection, i.e. a connection on the frame bundle of the differentiable manifold X, defines the covariant derivative of vector and tensor fields.

Definition: A **vector bundle** (E, X, π_1, F, G) with base X, typical fibre F and structural group G is said to be **associated to the principal bundle** (P, X, π, G) by the representation ρ of G on the vector space F (p. 163) if its transition functions are the images under ρ of the corresponding transition functions of the principal bundle P; i.e., let the local trivializations of P relative to the covering $\{U_i\}$ of X be

associated vector bundle

$$\Phi_i: \pi^{-1}(U_i) \to U_i \times G,$$

$$\overset{\Delta}{\Phi}_{i,x}: \pi^{-1}(x) \to G, \qquad x \in U_i.$$

Then the transition functions of P are

$$g_{ji}: U_j \cap U_i \to G \quad \text{by} \quad x \mapsto g_{ji}(x) = \overset{\Delta}{\Phi}_{j,x} \circ \overset{\Delta}{\Phi}{}^{-1}_{i,x} \in G;$$

the transition functions of E are relative to the same covering of X, and given by

$$\phi_i: \pi_1^{-1}(U_i) \to U_i \times F,$$

$$\overset{\Delta}{\phi}_{i,x}: \pi_1^{-1}(x) \to F, \qquad x \in U_i$$

with

$$\overset{\Delta}{\phi}_{j,x} \circ \overset{\Delta}{\phi}{}^{-1}_{i,x} = \rho(g_{ji}(x)) \in \mathscr{L}(F, F),$$

i.e., if $\boldsymbol{V}_i$ and $\boldsymbol{V}_j$ are elements of F which are images of the same $\boldsymbol{v}_x \in F_x$ by the trivializations ϕ_i and ϕ_j, respectively, then

$$\boldsymbol{V}_i = \rho(g_{ij}(x))\boldsymbol{V}_j.$$

This definition can easily be extended to define associated fibre bundles, not necessarily vector bundles, to a principal bundle; ρ is then a realization (p. 162) of G on a differentiable manifold F. Conversely, given a fibre bundle $B = (B, X, \pi, G)$ of typical fibre Y (p. 125), one can construct a principal bundle $P(B) \equiv (P(B), X, P(\pi), G)$ associated with B as follows.

admissible

A map u: $Y \to B_x$ from the typical fibre to the fibre at x is said to be **admissible** iff there is some $g \in G$ such that $\overset{\Delta}{\phi}_{i,x} \circ u = g$. This definition is independent of the choice of U_i, i.e., if $x \in U_i \cap U_j$ then

$$\overset{\Delta}{\phi}_{j,x} \circ u = (\overset{\Delta}{\phi}_{j,x} \circ \overset{\Delta}{\phi}{}^{-1}_{i,x}) \circ (\overset{\Delta}{\phi}_{i,x} \circ u) = g'g \in G,$$

for some $g' \in G$.

To construct $P(B)$ set each fibre $P(B)_x = \{u$: $Y \to B_x$ such that u is admissible$\}$ and set $P(B) = \cup P(B)_x$. Define $P(\pi)$: $P(B) \to X$ by $P(\pi)(P(B)_x) = x$ and define $P(\phi_i)$: $(P(\pi))^{-1}(U_i) \to U_i \times G$ by $u \mapsto (P(\pi)(u), \overset{\Delta}{\phi}_{i,x} \circ u)$. Then $P(\phi_i)$ is a bijection and one may give $P(B)$ the unique differentiable structure with respect to which each $P(\phi_i)$ is a diffeomorphism. The transition maps $P(\overset{\Delta}{\phi_i})_{,x} \circ P(\overset{\Delta}{\phi_j})^{-1}_{,x}$ are the same as those of B, so that B and $P(B)$ are associated as advertised. The right action of G on $P(B)$ can be defined by $(u, g) \mapsto u \circ g \equiv ug$. The map $u \circ g$ is another admissible map as required. (See Example p. 404.)

Example: If $Y = \mathbb{R}^n$ and $B = T(X^n)$ for some n dimensional manifold X^n then $P(T(X^n))$ is equivalent (p. 126) to the frame bundle $F(X^n)$ (p. 128).

Example: If π: $E \to X$ is a vector bundle with typical fibre F and group $\rho(G) \subset \mathrm{GL}(F)$, where G is an abstract group and ρ is a faithful representation of G on F, then the above construction can be slightly modified to yield a principal bundle $P(E)$ of abstract group G to which E is associated by the representation ρ. The modification consists of defining $P(\phi_i)$ by $u \mapsto (P(\pi)(u), \rho^{-1}(\overset{\Delta}{\phi}_{i,x} \circ u))$ instead of as above. The right action of G on $P(E)$ may then be defined by $ug \equiv u \circ \rho(g)$.

The definition of an associated bundle E to the principal bundle P leads naturally to the concept of parallel transport of an element of E along a curve C from x_1 to x_2 assumed to be sufficiently close for both of them to be in an open set $U_i \subset X$.

Definition: The **parallel transport** of $v_1 \in E$, $\pi_1(v_i) = x_1$ along a curve C joining x_1 and x_2 is an element $v_2 \in E$, $\pi_1(v_2) = x_2$, such that, suppressing the x label on $\overset{\Delta}{\Phi}_{i,x}$ and $\overset{\Delta}{\phi}_{i,x}$, parallel transport

$$\overset{\Delta}{\phi}_i(v_2) = \rho(\overset{\Delta}{\Phi}_i(p_2)\overset{\Delta}{\Phi}_i(p_1)^{-1})\overset{\Delta}{\phi}_i(v_1)$$

where p_2 is the parallel transport of $p_1 \in P$, $\pi(p_1) = x_1$, along the curve C from x_1 to x_2 (p. 358).

Lemma. The element v_2 parallel transported from v_1 along C does not depend on the choice of the trivialization nor the choice of p_1.

Proof:
(1) Set

$$\overset{\Delta}{\Phi}_i(p_1) = g_{1,i} \in G, \qquad \overset{\Delta}{\Phi}_i(p_2) = g_{2,i} \in G$$

$$\overset{\Delta}{\phi}_i(v_1) = V_{1,i}, \qquad \overset{\Delta}{\phi}_i(v_2) = V_{2,i}.$$

Since $g_{2,i} = g_{ij}(x_2)g_{2,j}$ and $g_{1,i}^{-1} = g_{1,j}^{-1}g_{ij}^{-1}(x_1)$ we have, according to the definition of parallel transport,

$$V_{2,i} = \rho(g_{ij}(x_2)g_{2,j}g_{1,j}^{-1}g_{ij}^{-1}(x_1))\,V_{1,i}.$$

Since ρ is a representation of G

$$\rho(g_{ij}^{-1}(x_2))V_{2,i} = \rho(g_{2,j}g_{1,j}^{-1})\rho(g_{ij}^{-1}(x_1))\,V_{1,i}$$

hence, according to the definition of associated bundles

$$V_{2,j} = \rho(g_{2,j}g_{1,j}^{-1})V_{1,j}$$

which proves that the definition of parallel transport does not depend on the choice of trivialization.
(2) It does not depend either on the choice of p_1, provided $\pi(p_1) = x_1$, since parallel transport in P commutes with the action $\tilde{R}_g$ of G on the fibres: the parallel transport p_2' of p_1' is

$$p_2' = \tilde{R}_h p_2 \quad \text{if} \quad p_1' = \tilde{R}_h p_1,$$

with p_2 parallel transported from p_1 and, in a given trivialization, according to the definition of $\tilde{R}_h$

$$g_{2,i}' = g_{2,i}h \quad \text{if} \quad g_{1,i}' = g_{1,i}h,$$

hence

$$\rho(g'_{2,i}g'^{-1}_{1,i}) = \rho(g_{2,i}g^{-1}_{1,i}). \qquad \blacksquare$$

Covariant derivative (see Problem Vbis 1, p. 404 for another definition of covariant derivative). Let $\boldsymbol{v}$ be (the germ of (p. 118)) a differentiable section of a vector bundle (E, X, π_1, F, G) associated to the principal bundle (P, X, π, G). Let $\boldsymbol{u}$ be a vector at $x_0 \in X$ tangent to a curve C: $t \mapsto C(t)$ on X with $C(0) = x_0$. The **covariant derivative** of $\boldsymbol{v}$ **in the** $\boldsymbol{u}$ **direction** at x_0 is by definition the vector

directional covariant derivative

$$(\nabla_{\boldsymbol{u}}\boldsymbol{v})(x_0) \equiv (\nabla_{\boldsymbol{u}}\boldsymbol{v})_{x_0} = \lim_{t=0} (\tau_0^t \boldsymbol{v}_t - \boldsymbol{v}_0)/t$$

where $\tau_0^t \boldsymbol{v}_t$ is the vector $\boldsymbol{v}_t \equiv \boldsymbol{v}_{x(t)}$ of the given vector field $\boldsymbol{v}$ parallel transported along C from $x(t)$ to x_0, the only requirement on C being $\dot{x}(0) = \boldsymbol{u}$. We shall simultaneously show the existence of this limit and give its expression in a given trivialization (U, Φ) of P, $x_0 \in U \subset X$, Φ: $\pi^{-1}(U) \to U \times G$. According to the definition of parallel transport in E, we have in the chosen trivialization

$$(\tau_0^t \boldsymbol{v}_t)_U = \rho(g(0)g^{-1}(t))(\boldsymbol{v}_t)_U$$

where $(\boldsymbol{v}_t)_U \equiv \boldsymbol{V}_t$ is the element of F representing $\boldsymbol{v}_t$

$$\boldsymbol{V}_t \overset{\Delta}{=} \phi_{x(t)}\boldsymbol{v}_t$$

and $g(0)$ and $g(t)$ are the elements of G representing, respectively a point $p_0 \in G_{x_0}$ and its parallel transport $p_t \in G_{x(t)}$ along C from x_0 to $x(t)$. We choose p_0 such that $g(0) = e$ for simplicity.
We can now compute

$$(\nabla_{\boldsymbol{u}}\boldsymbol{v})_U(x_0) = \lim_{t=0} (\rho(g^{-1}(t))\boldsymbol{V}_t - \boldsymbol{V}_0)/t$$

$$= \mathrm{d}(\rho(g^{-1}(t))\boldsymbol{V}_t)/\mathrm{d}t|_{t=0}.$$

The Leibniz formula and the derivation of the compound mapping $\mathbb{R} \to G \to \mathscr{L}(F, F)$ by $t \mapsto g^{-1}(t) \mapsto \rho(g^{-1}(t))$ gives

$$(\nabla_{\boldsymbol{u}}\boldsymbol{v})_U(x_0) = \mathrm{d}\boldsymbol{V}_t/\mathrm{d}t|_{t=0} - \rho'(e)\,\mathrm{d}g(t)/\mathrm{d}t|_{t=0}\boldsymbol{V}_0.$$

Note that with the notation of the exercise p. 366, $\Phi'_{p_0} \circ \sigma_{p_0} = (\mathrm{id}, \boldsymbol{\alpha})$ and $\mathrm{d}g(t)/\mathrm{d}t|_{t=0} = \boldsymbol{\alpha}(\mathrm{d}x(t)/\mathrm{d}t|_{t=0}) = \boldsymbol{\alpha}(\boldsymbol{u})$, hence

$$(\nabla_{\boldsymbol{u}}\boldsymbol{v})_U(x_0) = \mathrm{d}\boldsymbol{V}_t/\mathrm{d}t|_{t=0} - \rho'(e)\boldsymbol{\alpha}(\boldsymbol{u})\boldsymbol{V}_0.$$

As expected, $\nabla_{\boldsymbol{u}}\boldsymbol{v}$ depends on the curve C only through $\mathrm{d}x/\mathrm{d}t|_{t=0}$.

One can also define the **absolute differential** ∇ **of** v by

absolute differential

$$(\nabla_{u}\boldsymbol{v})_U(x_0) = (\nabla\boldsymbol{v})_U(\boldsymbol{u})(x_0) = \frac{\partial \boldsymbol{V}}{\partial x^{\mu}}u^{\mu} - \rho'(e)\alpha_{\mu}u^{\mu}V_0 \equiv (\nabla_{\mu}\boldsymbol{v})_U u^{\mu}(x_0).$$

Remark on the signs in the covariant derivatives on the base X:
If the local connection on the base is defined by

$$\phi' \circ \sigma_p(u) = (\mathrm{id}, \boldsymbol{\alpha})$$

then,

$$(\nabla_{u}\boldsymbol{v})_U = \mathrm{d}\boldsymbol{V}/\mathrm{d}t - \rho'(e)\boldsymbol{\alpha}(\boldsymbol{u})\boldsymbol{V}.$$

If the local connection on the base is defined by

$$\bar{\boldsymbol{\omega}} = s^*\boldsymbol{\omega}$$

then (p. 367),

$$(\nabla_{u}\boldsymbol{v})_U = \mathrm{d}\boldsymbol{V}/\mathrm{d}t + \rho'(e)\bar{\boldsymbol{\omega}}(\boldsymbol{u})\boldsymbol{V}.$$

Example 1: **Complex scalar field on a differentiable manifold** X.
Let ψ be a section of a vector bundle on X of typical fibre $\mathbb{C}$, structural group $U(1)$ associated to a principal fibre bundle with group $U(1)$ in the representation

$$\rho(g)z = gz, \qquad g \in U(1) \subset \mathbb{C}, \qquad z \subset \mathbb{C}.$$

Let $\bar{\boldsymbol{\omega}}$ be a local connection 1-form on $U \subset X$,

$$(\nabla_{u}\psi)_U = (\partial\psi_U/\partial x^{\mu})u^{\mu} + \bar{\omega}_{\mu}u^{\mu}\psi_U$$

where ψ_U and $(\nabla_u\psi)_U$ are the complex numbers corresponding to ψ and $\nabla_u\psi$ in the given gauge. The conventional choice of the electromagnetic potential is such that

$$\nabla_{\mu} = \partial_{\mu} - \mathrm{i}eA_{\mu}.$$

Example 2: **Covariant derivatives of tensor fields in the linear connection.**
Let $\boldsymbol{\omega}$ be a linear connection on the frame bundle $F(X^n)$ of a differentiable manifold X of dimension n. The tangent bundle $T(X^n)$ is associated to $F(X^n)$ by the canonical representation of GL $(n, \mathbb{R})$ on $\mathbb{R}^n$, i.e., the group of isomorphisms of $\mathbb{R}^n$. Then $\rho = \mathrm{id}$ and $\rho'_e = \mathrm{id}$. A local trivialization of $T(X^n)$ is given by a system of local coordinates (x^{α}), the representative of a vector $\boldsymbol{v} \in T(X^n)$ is the vector $(v^{\alpha}) \in \mathbb{R}^n$. The 1-form $\bar{\boldsymbol{\omega}}$ is

$$\bar{\boldsymbol{\omega}}(\boldsymbol{u}) = \Gamma^{\alpha}_{\mu\beta}u^{\mu}\boldsymbol{E}^{\beta}_{\alpha}$$

where E^β_α is a basis of $\mathscr{GL}(n, \mathbb{R})$ identified with $T_e\,\mathrm{GL}(n, \mathbb{R})$.
Choose for E^β_α the $n \times n$ matrix whose only nonvanishing element is in the α-row and the β-column:

$$(E^\beta_\alpha v)^\lambda = \delta^\lambda_\alpha v^\beta.$$

It follows that, in local coordinates

$$(\nabla_u v)^\lambda = \frac{\partial v^\lambda}{\partial x^\mu} u^\mu + \Gamma^\lambda_{\mu\beta} u^\mu v^\beta,$$

a formula already derived pp. 301–302.

If we now consider the tensor bundle on X of a given type, one obtains the formulae established p. 304. For example, consider the tensor bundle of twice contravariant tensors $t \in T_xX \otimes T_xX$, $F = \mathbb{R}^{n^2}$, $G = \mathrm{GL}(n, \mathbb{R})$. Let g be the $n \times n$ matrix S of elements S^μ_ν, then $\rho(S)$ is the $n^2 \times n^2$ matrix $S \otimes S$ of elements $S^\mu_\nu S^\rho_\sigma$, i.e., $\rho(S)$: $t \mapsto \bar{t}$ by $\bar{t}^{\mu\rho} = S^\mu_\nu S^\rho_\sigma t^{\nu\sigma}$. We have $\rho'(S)|_{S=1}$: $T_e(\mathrm{GL}(n, \mathbb{R})) \to T_1L(\mathbb{R}^{n^2}; \mathbb{R}^{n^2}) \simeq L(\mathbb{R}^{n^2}; \mathbb{R}^{n^2})$ by $T \mapsto \mathrm{Id} \otimes T + T \otimes \mathrm{Id}$, or in components

$$(\rho'(\mathbb{1})T)^{\mu\rho}_{\nu\sigma} = \frac{\partial}{\partial S^\alpha_\beta} (S^\mu_\nu S^\rho_\sigma)\Big|_{S=1} T^\alpha_\beta = \delta^\mu_\nu T^\rho_\sigma + T^\mu_\nu \delta^\rho_\sigma$$

$$\rho'(\mathbb{1})T\colon\ t \mapsto \bar{t} \quad \text{by} \quad \bar{t}^{\mu\rho} = T^\rho_\sigma t^{\mu\sigma} + T^\mu_\nu t^{\nu\rho}.$$

Similarly the covariant derivative of a covariant vector is obtained from the representation

$$\rho(S)\colon\ t \to \bar{t} \quad \text{by} \quad \bar{t}_\rho = (S^{-1})^\sigma_\rho t_\sigma.$$

4. CURVATURE

Definitions: Let (P, X, π, G) be a principal fibre bundle with connection H_p defined by the 1-form ω on P with values in $\mathscr{G}$. Let $h\colon T_p(P) \to H_p$ by $v \mapsto v_h$. The **exterior covariant derivative** $\mathrm{D}\phi$ of an r-form $\phi = \phi^\alpha \otimes e_\alpha$ on P with values in some vector space with basis (e_α) is defined by the relation

exterior covariant derivative

$$\mathrm{D}\phi(v_1, \ldots, v_{r+1}) = \mathrm{d}\phi(hv_1, \ldots, hv_{r+1})$$

where $\mathrm{d}\phi = (\mathrm{d}\phi^\alpha) \otimes e_\alpha$ (see p. 210).

curvature form

The 2-form $\Omega = \mathrm{D}\omega$ with values in $\mathscr{G}$ is called the **curvature form** of the connection ω (curvature form of the connection H_p).

A connection $\boldsymbol{\omega}$ is said to be **flat** if $\boldsymbol{\Omega} \equiv 0$. flat

It will be useful to introduce the following definitions. A differentiable r-form $\boldsymbol{\alpha}$ on a principal fibre bundle P is said to be a **horizontal form** if $\boldsymbol{\alpha}(\boldsymbol{v}_1, \ldots \boldsymbol{v}_r) = 0$ whenever at least one of the vectors $\boldsymbol{v}_1, \ldots \boldsymbol{v}_r$ is vertical. horizontal form

A differentiable r-form $\boldsymbol{\alpha}$ on P with values in a vector space E is said to be of **type** (ρ, E) if

$$\tilde{R}_g^* \boldsymbol{\alpha} = \rho(g^{-1})\boldsymbol{\alpha}, \qquad \forall g \in G,$$

where ρ is a representation of G on E.
One also says that $\boldsymbol{\alpha}$ is **equivariant** under the right action R_g by the representation ρ. equivariant
If in addition $\boldsymbol{\alpha}$ is horizontal it is said to be **tensorial of type** (ρ, E). tensorial

Lemma. The curvature form $\boldsymbol{\Omega}$ *is a tensorial form of type* $(\mathrm{Ad}, \mathscr{G})$:

$$(\tilde{R}_g^* \boldsymbol{\Omega})(\boldsymbol{u}, \boldsymbol{v}) = \mathrm{Ad}(g^{-1})\boldsymbol{\Omega}(\boldsymbol{u}, \boldsymbol{v}).$$

Proof: $\boldsymbol{\Omega}$ is a tensorial form by definition. By the properties 1 and 3 of the H_p spaces we have

$$\tilde{R}_g' h = h\tilde{R}_g', \qquad g \in G, \qquad h \equiv \text{hor}.$$

Thus

$$\begin{aligned}
(\tilde{R}_g^* \boldsymbol{\Omega})(\boldsymbol{u}, \boldsymbol{v}) &= \boldsymbol{\Omega}(\tilde{R}_g' \boldsymbol{u}, R_g' \boldsymbol{v}) \\
&= \mathrm{d}\boldsymbol{\omega}(h\tilde{R}_g' \boldsymbol{u}, h\tilde{R}_g' \boldsymbol{v}) = \mathrm{d}\boldsymbol{\omega}(\tilde{R}_g' h\boldsymbol{u}, \tilde{R}_g' h\boldsymbol{v}) \\
&= (\tilde{R}_g^* \, \mathrm{d}\boldsymbol{\omega})(h\boldsymbol{u}, h\boldsymbol{v}) \\
&= (\mathrm{d}\tilde{R}_g^* \boldsymbol{\omega})(h\boldsymbol{u}, h\boldsymbol{v}) \\
&= (\mathrm{d}(\mathrm{Ad}(g^{-1})\boldsymbol{\omega}))(h\boldsymbol{u}, h\boldsymbol{v}) = \mathrm{Ad}(g^{-1}) \, \mathrm{d}\boldsymbol{\omega}(h\boldsymbol{u}, h\boldsymbol{v})
\end{aligned}$$

since $\mathrm{Ad}\, g^{-1}$ is a given linear map in the space of values of $\boldsymbol{\omega}$. ■

Cartan structural equation

Cartan structural equation. *If* $\boldsymbol{\omega}$ *is a connection on* P *and* $\mathrm{D}\boldsymbol{\omega} = \boldsymbol{\Omega}$ *then*[1]

$$\boldsymbol{\Omega}(\boldsymbol{u}, \boldsymbol{v}) = \mathrm{d}\boldsymbol{\omega}(\boldsymbol{u}, \boldsymbol{v}) + [\boldsymbol{\omega}(\boldsymbol{u}), \boldsymbol{\omega}(\boldsymbol{v})].$$

Proof: $\boldsymbol{\Omega}(\boldsymbol{u}_p, \boldsymbol{v}_p) = \mathrm{d}\boldsymbol{\omega}(\boldsymbol{u}_p - \text{ver}\, \boldsymbol{u}_p, \boldsymbol{v}_p - \text{ver}\, \boldsymbol{v}_p)$. Extend ver $\boldsymbol{u}_p$ and ver $\boldsymbol{v}_p$ to Killing vector fields ver $\boldsymbol{u}$ and ver $\boldsymbol{v}$ in a neighborhood of p.

[1]See p. 306. If one uses the other definition of the exterior derivative (p. 200), then $\bar{\boldsymbol{\Omega}}(\boldsymbol{u}, \boldsymbol{v}) = \bar{\mathrm{d}}\boldsymbol{\omega}(\boldsymbol{u}, \boldsymbol{v}) + \frac{1}{2}[\boldsymbol{\omega}(\boldsymbol{u}), \boldsymbol{\omega}(\boldsymbol{v})]$ and $\boldsymbol{\Omega} = 2\bar{\boldsymbol{\Omega}}$.

a) By property 4 (p. 207) we have

$$\mathrm{d}\boldsymbol{\omega}(\text{ver } \boldsymbol{u}, \text{ver } \boldsymbol{v}) = \mathscr{L}_{\text{ver } \boldsymbol{u}}(\boldsymbol{\omega}(\text{ver } \boldsymbol{v})) - \mathscr{L}_{\text{ver } \boldsymbol{v}}(\boldsymbol{\omega}(\text{ver } \boldsymbol{u})) - \boldsymbol{\omega}[\text{ver } \boldsymbol{u}, \text{ver } \boldsymbol{v}].$$

Since $\boldsymbol{\omega}(\text{ver } \boldsymbol{v}) = \widehat{\text{ver } \boldsymbol{v}}$ = constant, we have

$$\begin{aligned} \mathrm{d}\boldsymbol{\omega}(\text{ver } \boldsymbol{u}, \text{ver } \boldsymbol{v}) &= -\boldsymbol{\omega}([\text{ver } \boldsymbol{u}, \text{ver } \boldsymbol{v}]) = -\widehat{[\text{ver } \boldsymbol{u}, \text{ver } \boldsymbol{v}]} \\ &= -[\widehat{\text{ver } \boldsymbol{u}}, \widehat{\text{ver } \boldsymbol{v}}] = -[\boldsymbol{\omega}(\text{ver } \boldsymbol{u}), \boldsymbol{\omega}(\text{ver } \boldsymbol{v})]. \end{aligned}$$

b) Using again property 4 (p. 207) and $\boldsymbol{\omega}(\text{hor } \boldsymbol{v}) = 0$ we have $\mathrm{d}\boldsymbol{\omega}(\text{ver } \boldsymbol{u}, \text{hor } \boldsymbol{v}) = -\boldsymbol{\omega}[\text{ver } \boldsymbol{u}, \text{hor } \boldsymbol{v}]$. Note that $[\text{ver } \boldsymbol{u}, \text{hor } \boldsymbol{v}]$ is horizontal since

$$[\text{ver } \boldsymbol{u}, \text{hor } \boldsymbol{v}] = \mathscr{L}_{\text{ver } \boldsymbol{u}} \text{hor } \boldsymbol{v} = \lim_{t=0} \frac{1}{t} (\tilde{R}'_{g(t)^{-1}} \text{hor } \boldsymbol{v} - \text{hor } \boldsymbol{v})$$

and $\tilde{R}'_{g(t)^{-1}} \text{hor } \boldsymbol{v}$ is horizontal. Hence $\mathrm{d}\boldsymbol{\omega}(\text{ver } \boldsymbol{u}, \text{hor } \boldsymbol{v}) = 0$.
c) $\mathrm{d}\boldsymbol{\omega}(\boldsymbol{u}, \text{ver } \boldsymbol{v}) = \mathrm{d}\boldsymbol{\omega}(\text{ver } \boldsymbol{u}, \text{ver } \boldsymbol{v}) + \mathrm{d}\boldsymbol{\omega}(\text{hor } \boldsymbol{u}, \text{ver } \boldsymbol{v}) = -[\boldsymbol{\omega}(\text{ver } \boldsymbol{u}), \boldsymbol{\omega}(\text{ver } \boldsymbol{v})]$. Combining (a) and (c) gives the structural equation. ■

Exercise: Show that

$$[\boldsymbol{\omega}(\boldsymbol{u}), \boldsymbol{\omega}(\boldsymbol{v})] = \tfrac{1}{2}[\boldsymbol{\omega}, \boldsymbol{\omega}](\boldsymbol{u}, \boldsymbol{v}) = \tfrac{1}{2} c^{\gamma}_{\alpha\beta} \boldsymbol{e}_{\gamma} (\boldsymbol{\omega}^{\alpha} \wedge \boldsymbol{\omega}^{\beta})(\boldsymbol{u}, \boldsymbol{v})$$

where $(\boldsymbol{e}_{\alpha})$ is a basis of $\mathscr{G}$ and $\boldsymbol{\omega} = \boldsymbol{\omega}^{\alpha} \otimes \boldsymbol{e}_{\alpha}$.

Answer: With the notation defined in the footnote p. 360,

$$[\boldsymbol{\omega}, \boldsymbol{\omega}] = (\boldsymbol{\omega}^{\alpha} \wedge \boldsymbol{\omega}^{\beta}) \otimes [\boldsymbol{e}_{\alpha}, \boldsymbol{e}_{\beta}] \qquad \text{(cf. p. 211)}$$

$$\begin{aligned} [\boldsymbol{\omega}, \boldsymbol{\omega}](\boldsymbol{u}, \boldsymbol{v}) &= (\boldsymbol{\omega}^{\alpha}(\boldsymbol{u})\boldsymbol{\omega}^{\beta}(\boldsymbol{v}) - \boldsymbol{\omega}^{\beta}(\boldsymbol{u})\boldsymbol{\omega}^{\alpha}(\boldsymbol{v}))[\boldsymbol{e}_{\alpha}, \boldsymbol{e}_{\beta}] \qquad \text{(cf. p. 196)} \\ &= 2[\boldsymbol{\omega}(\boldsymbol{u}), \boldsymbol{\omega}(\boldsymbol{v})]. \end{aligned}$$

The second equality follows from

$$[\boldsymbol{\omega}, \boldsymbol{\omega}](\boldsymbol{u}, \boldsymbol{v}) = (\boldsymbol{\omega}^{\alpha} \wedge \boldsymbol{\omega}^{\beta})(\boldsymbol{u}, \boldsymbol{v}) c^{\gamma}_{\alpha\beta} \boldsymbol{e}_{\gamma}.$$

Exercise: The collection of horizontal vector fields (p. 359) on a principal bundle is a Pfaff system of vector fields (p. 248). Show that this Pfaff system is completely integrable if and only if, the connection is flat.

Answer: Let $\boldsymbol{u}$, $\boldsymbol{v}$ be any two horizontal vector fields, then $\boldsymbol{\omega}(\boldsymbol{u}) = \boldsymbol{\omega}(\boldsymbol{v}) = 0$ and $\boldsymbol{\Omega}(\boldsymbol{u}, \boldsymbol{v}) = \mathrm{d}\boldsymbol{\omega}(\boldsymbol{u}, \boldsymbol{v}) = -\boldsymbol{\omega}[\boldsymbol{u}, \boldsymbol{v}]$ (p. 248). Hence $\boldsymbol{\Omega}(\boldsymbol{u}, \boldsymbol{v}) = 0 \Leftrightarrow [\boldsymbol{u}, \boldsymbol{v}]$ is a horizontal vector field. The Frobenius condition (p. 248) for a Pfaff system to be completely integrable is satisfied.

Local curvature on the base manifold. In a local trivialization (U_i, ϕ_i) the 2-form $\boldsymbol{\Omega}$ on $\pi^{-1}(U_i)$ is represented by the 2-form $\bar{\boldsymbol{\Omega}}_i$ on U_i, defined through the corresponding cross section s_i by

$$\bar{\boldsymbol{\Omega}}_i = s_i^* \boldsymbol{\Omega}.$$

It results from the Cartan structure equation and the commutation of the pull back with d that

$$\bar{\boldsymbol{\Omega}}_i = \mathrm{d}\bar{\boldsymbol{\omega}}_i + \tfrac{1}{2}[\bar{\boldsymbol{\omega}}_i, \bar{\boldsymbol{\omega}}_i].$$

Example: In a bundle with the commutative structure group $U(1)$, $\bar{\boldsymbol{\Omega}}_i = \mathrm{d}\bar{\boldsymbol{\omega}}_i$.

In Yang–Mills theories a 2-form $\bar{\boldsymbol{\Omega}}_i$ is called a **field strength in the gauge** ϕ_i, and is usually labelled $\boldsymbol{F}_i$ up to a multiplicative constant. field strength

A calculation similar to the one carried out on p. 364 for the transformation of a local connection under a change of trivialization, and to the proof of the previous lemma, gives the transformation law of $\bar{\boldsymbol{\Omega}}_i$ at a point $x \in U_i \cap U_j$ under a change of trivialization. Let (U_i, ϕ_i) and (U_j, ϕ_j) be two local trivializations. Then

$$\bar{\boldsymbol{\Omega}}_i(\boldsymbol{u}, \boldsymbol{v}) = \mathrm{Ad}(g_{ji}(x)^{-1})\bar{\boldsymbol{\Omega}}_j(\boldsymbol{u}, \boldsymbol{v}).$$

If G is a subgroup of $\mathrm{GL}(n, \mathbb{R})$ this formula may be written without abuse of notation

$$\bar{\boldsymbol{\Omega}}_i(\boldsymbol{u}, \boldsymbol{v}) = g_{ji}^{-1}(x)\bar{\boldsymbol{\Omega}}_j(\boldsymbol{u}, \boldsymbol{v})g_{ji}(x).$$

Coordinate expressions for the potentials and the field strengths. Let $(\boldsymbol{e}_\alpha)$ denote a basis of $\mathcal{G}$ and $c^\alpha_{\beta\gamma}$ the structure constants (p. 156) given by $[\boldsymbol{e}_\alpha, \boldsymbol{e}_\beta] = c^\gamma_{\alpha\beta}\boldsymbol{e}_\gamma$. Let $(\boldsymbol{e}_\mu)$ denote a basis of T_xX for $x \in U$. Then the components $\bar{\omega}^\alpha_\mu$ and $\bar{\Omega}^\alpha_{\mu\nu}$ are defined by $\bar{\boldsymbol{\omega}}(\boldsymbol{e}_\mu) = \bar{\omega}^\alpha_\mu \boldsymbol{e}_\alpha$, $\bar{\boldsymbol{\Omega}}(\boldsymbol{e}_\mu, \boldsymbol{e}_\nu) = \bar{\Omega}^\alpha_{\mu\nu}\boldsymbol{e}_\alpha$, while the structural equation gives

$$\bar{\Omega}^\alpha_{\mu\nu} = \partial_\mu \bar{\omega}^\alpha_\nu - \partial_\nu \bar{\omega}^\alpha_\mu + c^\alpha_{\beta\gamma}\bar{\omega}^\beta_\mu \bar{\omega}^\gamma_\nu.$$

Bianchi identities. In terms of the basis $(\boldsymbol{e}_\alpha)$ of $\mathcal{G}$ the Cartan structural equation (p. 373) becomes

$$\boldsymbol{\Omega}^\alpha = \mathrm{d}\boldsymbol{\omega}^\alpha + \tfrac{1}{2}c^\alpha_{\beta\gamma}\boldsymbol{\omega}^\beta \wedge \boldsymbol{\omega}^\gamma.$$

Differentiation gives

$$\mathrm{d}\boldsymbol{\Omega}^\alpha = \tfrac{1}{2}c^\alpha_{\beta\gamma}\,\mathrm{d}\boldsymbol{\omega}^\beta \wedge \boldsymbol{\omega}^\gamma - \tfrac{1}{2}c^\alpha_{\beta\gamma}\boldsymbol{\omega}^\beta \wedge \mathrm{d}\boldsymbol{\omega}^\gamma.$$

Thus, since $\boldsymbol{\omega}$ vanishes on horizontal vectors

$$\mathrm{D}\boldsymbol{\Omega}^\alpha(\boldsymbol{v}, \boldsymbol{u}, \boldsymbol{w}) = \mathrm{d}\boldsymbol{\Omega}^\alpha(h\boldsymbol{v}, h\boldsymbol{u}, h\boldsymbol{w}) = 0 \qquad \forall \boldsymbol{u}, \boldsymbol{v}, \boldsymbol{w}.$$

Therefore we have the **Bianchi identities** Bianchi identities

$$\mathrm{D}\boldsymbol{\Omega} = 0.$$

See on p. 307 expressions for the Bianchi identities in terms of components for the case of a linear connection.

5. LINEAR CONNECTIONS

linear connection

A **linear connection** on a smooth manifold X is a connection on the principal fibre bundle $F(X)$ of frames on X.

Let Φ_p: $T_x(X^n) \to \mathbb{R}^n$, $p = (x, \rho_x)$, be the mapping which maps[1] a vector in $T_x(X^n)$ into its components with respect to the frame ρ_x. In other words if $\rho_x = (e_i)$ then $\Phi_p u = (\theta^i(u))$ where (θ^i) is the basis dual to (e_i). The 1-form θ on $F(X)$ with values in $\mathbb{R}^n$ defined by

$$\theta_p(v) = \Phi_p \pi' v, \qquad v \in T_p F(X)$$

soldering form

is called the **soldering (canonical) form** of X.

torsion form

The 2-form $\Theta = D\theta$ is called the **torsion form** of the linear connection on X.

Property: The soldering form is tensorial of type $(\rho, \mathbb{R}^n)$ with $\rho(g) = g$, $g \in GL(n)$ (p. 373). To prove $(\tilde{R}^*_g \theta)(v) = g^{-1}\theta(v)$, we calculate

$$(\tilde{R}^*_g \theta)(v) = \theta(\tilde{R}'_g v) = \Phi_{\tilde{R}_g p} \pi' v = g^{-1} \Phi_p \pi' v = g^{-1} \theta_p(v).$$

The tensorial property follows from the definition.

The torsion form Θ on $F(X)$ is tensorial of type $(\rho(g) = g, \mathbb{R}^n)$.

Cartan structural equation

Cartan structural equation (see also pp. 306, 373).

$$\Theta(u, v) \equiv D\theta = d\theta(u, v) + \omega(u)\theta(v) - \omega(v)\theta(u)$$

where ω is the connection on $F(X)$ defining the covariant exterior derivatives, and $\omega(u)\theta(v)$ denotes the action of $\omega(u) \in \mathcal{GL}(n)$ on $\theta(v) \in \mathbb{R}^n$.

Proof:

$$\Theta(u, v) = d\theta(\text{hor } u, \text{hor } v)$$

$$= d\theta(u, v) - d\theta(\text{ver } u, v) - d\theta(u, \text{ver } v) + d\theta(\text{ver } u, \text{ver } v)$$

$$d\theta(\text{ver } u, v) = \mathcal{L}_{\text{ver } u}(\theta(v)) - \mathcal{L}_v(\theta(\text{ver } u)) - \theta([\text{ver } u, v]) \quad \text{(p. 207)}$$

$$\theta([\text{ver } u, v]) = i_{[\text{ver } u, v]}\theta = [\mathcal{L}_{\text{ver } u}, i_v]\theta \quad \text{(p. 207)}$$

[1]If frame ρ_x is considered as a mapping from $\mathbb{R}^n$ into $T_x(X^n)$, then $\Phi_p = \rho_x^{-1}$.

hence

$$\mathrm{d}\boldsymbol{\theta}(\mathrm{ver}\,\boldsymbol{u}, \boldsymbol{v}) = \mathrm{i}_{\boldsymbol{v}}\mathcal{L}_{\mathrm{ver}\,\boldsymbol{u}}\boldsymbol{\theta}$$

$$\mathcal{L}_{\mathrm{ver}\,\boldsymbol{u}}\boldsymbol{\theta} = \frac{\mathrm{d}}{\mathrm{d}t}\tilde{R}^*_{g(t)}\boldsymbol{\theta}\Big|_{t=0} \quad \text{with} \quad \mathrm{d}g(t)/\mathrm{d}t\Big|_{t=0} = \widehat{\mathrm{ver}\,\boldsymbol{u}} = \boldsymbol{\omega}(\boldsymbol{u})$$

$$\mathrm{i}_{\boldsymbol{v}}\mathcal{L}_{\mathrm{ver}\,\boldsymbol{u}}\boldsymbol{\theta} = \frac{\mathrm{d}}{\mathrm{d}t}g^{-1}(t)\Big|_{t=0}\boldsymbol{\theta}(\boldsymbol{v}) = -\boldsymbol{\omega}(\boldsymbol{u})\boldsymbol{\theta}(\boldsymbol{v}).$$

The structural equation follows. ■

The torsion structural equation can be written

$$\boldsymbol{\Theta} = \mathrm{d}\boldsymbol{\theta} + [\boldsymbol{\omega}, \boldsymbol{\theta}],$$

where [,] denotes the Lie bracket in the Lie algebra of the affine group (p. 157): If $(\boldsymbol{v}^{\beta}_{\alpha}, \boldsymbol{v}_{\gamma})$ is a basis of this Lie algebra, then $\boldsymbol{\omega}(\boldsymbol{u}) = \omega^{\alpha}_{\beta}(\boldsymbol{u})\boldsymbol{v}^{\beta}_{\alpha}$, $\boldsymbol{\theta}(\boldsymbol{v}) = \theta^{\alpha}(\boldsymbol{v})\boldsymbol{v}_{\alpha}$, $[\boldsymbol{v}^{\alpha}_{\beta}, \boldsymbol{v}_{\gamma}] = \delta^{\alpha}_{\gamma}\boldsymbol{v}_{\beta}$ and it is easy to show that both forms of the torsion are equivalent.[1]

Exercise: Prove that $\mathrm{D}\boldsymbol{\Theta} = \boldsymbol{\Omega} \wedge \boldsymbol{\theta}$, where $\boldsymbol{\Omega} = \mathrm{D}\boldsymbol{\omega}$.

Answer: Follows by taking the derivative of the structural equation.

Coordinate expression of the soldering form. Let (x^{α}) be coordinates on $U \subset X$, let $\partial_{\alpha} = \partial/\partial x^{\alpha}$, $(\rho_x)_{\alpha} = a^{\beta}_{\alpha}\partial_{\beta}$ and $\boldsymbol{\theta} = \theta^{\alpha}\boldsymbol{e}_{\alpha}$ where $(\boldsymbol{e}_{\alpha})$ is the natural basis on $\mathbb{R}^n$. If $\pi'\boldsymbol{v} = \boldsymbol{u}$, then

$$\boldsymbol{\theta}_p(\boldsymbol{v}) = \Phi_p(u) = \Phi_p(\mathbf{d}x^{\beta}(\boldsymbol{u})\partial_{\beta}) = \Phi_p(\mathbf{d}x^{\beta}(\boldsymbol{u})(a^{-1})^{\gamma}_{\beta}a^{\delta}_{\gamma}\partial_{\delta}) = \mathbf{d}x^{\beta}(\boldsymbol{u})(a^{-1})^{\gamma}_{\beta}\boldsymbol{e}_{\gamma}.$$

So

$$[\boldsymbol{\theta}_p(\boldsymbol{v})]^{\gamma} = (a^{-1})^{\gamma}_{\beta}\,\mathbf{d}x^{\beta}(\pi'\boldsymbol{v}).$$

Exercise: Given a connection on $F(X)$ and a vector $\boldsymbol{u} \in \mathbb{R}^n$, show that the soldering form $\boldsymbol{\theta}$ can be used to construct a horizontal vector field on the frame bundle.

Answer: Given a fixed vector $\boldsymbol{u}$ in $\mathbb{R}^n$, let $\boldsymbol{v}_{\boldsymbol{u}}(p)$ be the horizontal vector at p such that

$$\boldsymbol{\theta}(\boldsymbol{v}_{\boldsymbol{u}}(p)) = \boldsymbol{u}, \qquad p \in F(x), \qquad \boldsymbol{u} \in \mathbb{R}^n.$$

[1] For a further understanding of this interpretation of $\boldsymbol{\Theta}$ see [Kobayashi and Nomizu, pp. 125–130].

The horizontal vector field v_u defined by this equation is unique. It is called the **standard horizontal (basic) vector field associated with** u.

standard horizontal (basic) vector field

Curvature and torsion on the base manifold of the frame bundle $F(X)$. We shall first give the form on the base manifold X canonically related to an arbitrary tensorial form $\boldsymbol{\alpha}$ on $F(X)$ of type (ρ, E) (p. 373).

Theorem. There is a bijective correspondence between the space of differential forms on X with values in the space of tensor fields of a given type on the one hand and the space of tensorial forms of type $(\rho, \mathbb{R}^{n^q})$ *on the frame bundle* $F(X)$ *on the other hand,* ρ *and* q *being determined by the type of the tensor field.*

Proof: Let $\boldsymbol{\alpha}$ be a differential r-form on $F(X)$ with values in $\mathbb{R}^{n^q} = E$, which is tensorial (p. 373) of type (ρ, E).

Let now Φ_p, $p = (x, \rho_x)$, map a tensor at x into its components in the frame defined by ρ_x on T_xX, i.e., we generalize to the tensorial case the symbol Φ_p introduced for the vectorial case. Let $\boldsymbol{u}_i \in T_xX$ and $\boldsymbol{v}_i \in T_pF(X)$ be such that $\boldsymbol{u}_i = \pi'(p)\boldsymbol{v}_i$. We shall show that the form $\bar{\boldsymbol{\alpha}}$ on X defined by

$$\bar{\boldsymbol{\alpha}}_x(\boldsymbol{u}_1, \ldots \boldsymbol{u}_r) = \Phi_p^{-1}(\boldsymbol{\alpha}(\boldsymbol{v}_1, \ldots \boldsymbol{v}_r)), \qquad \boldsymbol{u}_i = \pi'(p)\boldsymbol{v}_i$$

does not depend on the choice of ρ_x nor the choice of $(\boldsymbol{v}_i)$ provided that their projections be $(\boldsymbol{u}_i)$.

(1) If $\boldsymbol{v}_i$ and $\boldsymbol{v}'_i$ project onto the same $\boldsymbol{u}_i$, then $\boldsymbol{v}_i - \boldsymbol{v}'_i$ is vertical, hence the value of $\bar{\boldsymbol{\alpha}}$ is unchanged when one replaces $\boldsymbol{v}_i$ by $\boldsymbol{v}'_i$.

(2) Let $p'_x = (x, \rho'_x)$, then there exists $g \in G$ such that $\rho' = \tilde{R}_g p$. Let $\boldsymbol{v}'_i = \tilde{R}'_g \boldsymbol{v}_i$, then $\pi' \boldsymbol{v}'_i = \pi \boldsymbol{v}_i = \boldsymbol{u}_i$. On the other hand $\boldsymbol{\alpha}_{p'}(\boldsymbol{v}'_1, \ldots, \boldsymbol{v}'_r) = (\tilde{R}^*_g \boldsymbol{\alpha})_p(\boldsymbol{v}_1, \ldots, \boldsymbol{v}_r) = \rho(g^{-1})\boldsymbol{\alpha}_p(\boldsymbol{v}_1, \ldots, \boldsymbol{v}_p)$ and $\Phi_{p'}^{-1} = \Phi_p^{-1}\rho(g)$; hence the value of $\bar{\boldsymbol{\alpha}}_x$ does not depend on the choice of ρ_x and the form $\bar{\boldsymbol{\alpha}}$ is uniquely defined. ∎

We shall now use this theorem to construct the forms on X canonically associated with the soldering from $\boldsymbol{\theta}$, the torsion form $\boldsymbol{\Theta} = \mathrm{D}\boldsymbol{\theta}$ and the curvature form $\boldsymbol{\Omega} = \mathrm{D}\boldsymbol{\omega}$. The forms canonically associated with $\boldsymbol{\Theta}$ and $\boldsymbol{\Omega}$ are the torsion and the curvature forms introduced in Chapter V (p. 306). Note that a connection form is not tensorial, hence does not define a tensor on X.

(1) The soldering form $\boldsymbol{\theta}$ on $F(X)$ is tensorial, of type $(\rho(g) = g, \mathbb{R}^n)$. It defines a differential 1-form $\bar{\boldsymbol{\theta}}$ on X with values in the space of contravariant vector fields, i.e., it defines a (1-covariant 1-contravariant) tensor whose components are $\bar{\theta}^\alpha_\beta = \delta^\alpha_\beta$.

(2) The torsion form $\boldsymbol{\Theta}$ on $F(X)$ is tensorial, of type $(\rho(g)=g, \mathbb{R}^n)$. It defines a differential 2-form $\bar{\boldsymbol{\Theta}}$ on X with values in the space of contravariant vector fields, i.e., it defines a (2-covariant 1-contravariant) tensor. $\bar{\boldsymbol{\Theta}}$ is equal to the torsion form $\boldsymbol{\Theta}$ on X defined p. 306. Indeed it follows from the structural equation (p. 376) that

$$\bar{\boldsymbol{\Theta}}(\pi'\boldsymbol{u}, \pi'\boldsymbol{v}) = \mathrm{d}\bar{\boldsymbol{\theta}}(\pi'\boldsymbol{u}, \pi'\boldsymbol{v}) + \bar{\boldsymbol{\omega}}(\pi'\boldsymbol{u})\bar{\boldsymbol{\theta}}(\pi'\boldsymbol{v}) - \bar{\boldsymbol{\omega}}(\pi'\boldsymbol{v})\bar{\boldsymbol{\theta}}(\pi'\boldsymbol{u})$$

which reads in components

$$\bar{\boldsymbol{\Theta}}^\alpha(\pi'\boldsymbol{u}, \pi'\boldsymbol{v}) = \mathrm{d}\bar{\boldsymbol{\theta}}^\alpha(\pi'\boldsymbol{u}, \pi'\boldsymbol{v}) + \bar{\boldsymbol{\omega}}^\alpha_\beta(\pi'\boldsymbol{u})\bar{\boldsymbol{\theta}}^\beta(\pi'\boldsymbol{v}) - \bar{\boldsymbol{\omega}}^\alpha_\beta(\pi'\boldsymbol{v})\bar{\boldsymbol{\theta}}^\beta(\pi'\boldsymbol{u}).$$

One can write this expression in abbreviated notation

$$\boldsymbol{\Theta} = \mathrm{d}\boldsymbol{\theta} + \boldsymbol{\omega} \wedge \boldsymbol{\theta}.$$

This is indeed the torsion form defined p. 306 since $(\bar{\boldsymbol{\theta}}^\beta)$ is a basis on T^*_xX dual to the basis defined by ρ_x on T_xX.

(3) The curvature form $\boldsymbol{\Omega}$ is tensorial of type $(\mathrm{Ad}, \mathbb{R}^{n^2})$. It defines a differential 2-form $\boldsymbol{\Omega}$ on X with values in the space of (1-covariant 1-contravariant) tensor field, i.e., it defines a (3-covariant 1-contravariant) tensor. $\bar{\boldsymbol{\Omega}}$ is equal to the curvature form $\boldsymbol{\Omega}$ on X defined p. 306. Indeed it follows from the structural equation (p. 373) that

$$\bar{\boldsymbol{\Omega}}(\pi'\boldsymbol{v}, \pi'\boldsymbol{u}) = \mathrm{d}\bar{\boldsymbol{\omega}}(\pi'\boldsymbol{v}, \pi'\boldsymbol{u}) + [\bar{\boldsymbol{\omega}}(\pi'\boldsymbol{v}), \bar{\boldsymbol{\omega}}(\pi'\boldsymbol{u})].$$

On the other hand, if $(\boldsymbol{E}^\beta_\alpha)$ is a basis of $\mathscr{GL}(n, \mathbb{R})$,

$$[\boldsymbol{E}^\alpha_\beta, \boldsymbol{E}^\gamma_\delta] = \delta^\alpha_\delta \boldsymbol{E}^\gamma_\beta - \delta^\gamma_\beta \boldsymbol{E}^\alpha_\delta \qquad \text{(see p. 157)}$$

we can write $\bar{\boldsymbol{\omega}}(\pi'\boldsymbol{v}) = \bar{\boldsymbol{\omega}}^\alpha_\beta(\pi'\boldsymbol{v})\boldsymbol{E}^\beta_\alpha$ where $\bar{\boldsymbol{\omega}}^\alpha_\beta(\pi'\boldsymbol{v})$ is a numerical matrix element. Then

$$\begin{aligned}[\bar{\boldsymbol{\omega}}(\pi'\boldsymbol{v}), \bar{\boldsymbol{\omega}}(\pi'\boldsymbol{u})] &= \bar{\boldsymbol{\omega}}^\alpha_\beta(\pi'\boldsymbol{v})\bar{\boldsymbol{\omega}}^\gamma_\delta(\pi'\boldsymbol{u})[\boldsymbol{E}^\beta_\alpha, \boldsymbol{E}^\delta_\gamma] \\ &= (\bar{\boldsymbol{\omega}}^\alpha_\gamma \wedge \bar{\boldsymbol{\omega}}^\gamma_\beta)(\pi'\boldsymbol{v}, \pi'\boldsymbol{u})\boldsymbol{E}^\beta_\alpha.\end{aligned}$$

Therefore $\bar{\boldsymbol{\Omega}} = \bar{\boldsymbol{\Omega}}^\alpha_\beta \boldsymbol{E}^\beta_\alpha$ is given by

$$\bar{\boldsymbol{\Omega}}^\alpha_\beta = \mathrm{d}\bar{\boldsymbol{\omega}}^\alpha_\beta + \bar{\boldsymbol{\omega}}^\alpha_\gamma \wedge \bar{\boldsymbol{\omega}}^\gamma_\beta.$$

This is indeed the curvature form defined (p. 306) since it has been shown in Example 2, p. 371, that, in natural coordinates, $(\bar{\boldsymbol{\omega}}(\pi'\boldsymbol{v}))^\alpha_\beta = \Gamma_{\beta\rho}^{\ \alpha} v^\rho$.

bundle homomorphism

Connections induced by bundle homomorphisms.
F: $(P_1, X_1, \pi_1, G_1) \to (P_2, X_2, \pi_2, G_2)$ is said to be a **bundle homomorphism** if it is a bundle morphism (p. 127)

$$\begin{cases} F \text{ is fibre preserving} \\ F \text{ induces a diffeomorphism } f\colon X_1 \to X_2 \end{cases}$$

and if, in addition there exists a homomorphism ϕ: $G_1 \to G_2$ such that

$$F(\tilde{R}_{g_1} p_1) = \tilde{R}_{\phi(g_1)} F(p_1), \quad p_1 \in P_1 \quad \text{and} \quad g_1 \in G_1 \tag{1}$$

i.e., with the simplified notation, $\tilde{R}_g p = pg$,

$$F(p_1 g_1) = F(p_1)\phi(g_1).$$

Theorem. If there exists a homomorphism from the principal bundle P_1 to the principal bundle P_2, a connection on P_1 determines a unique connection on P_2.

Proof: We shall construct the horizontal subspaces $H_{p_2} \subset T_{p_2}(P_2)$ in terms of the horizontal spaces $H_{p_1} \subset T_{p_1}(P_1)$. Given $p_2 \in P_2$, choose $p_1 \in P_1$ and $g_2 \in G_2$ such that

$$p_2 = F(p_1) g_2.$$

Set $H_{p_2} \equiv \tilde{R}'_{g_2}(F'(p_1) H_{p_1})$. We shall show that H_{p_2} thus defined does not depend on the choice of p_1, g_2. Indeed let p'_1 and g'_2 be such that $p_2 = \tilde{R}_{g'_2} F(p'_1)$. Hence there exists g_1 such that $p'_1 = \tilde{R}_{g_1} p_1$. Moreover $\tilde{R}_{g_2} F(p_1) = R_{g'_2} F(R_{g_1} p_1)$ implies $g_2 = \phi(g_1) g'_2$. Thus if $H'_{p_2} = \tilde{R}'_{g'_2} H_{F(p'_1)}$, then

$$H'_{p_2} = \tilde{R}'_{g'_2} \tilde{R}'_{\phi(g_1)} H_{F(p_1)} = \tilde{R}'_{g_2} H_{F(p_1)} = H_{p_2}.$$

It is easy to show that the horizontal spaces H_{p_2} satisfy all the properties of a connection as defined on p. 359. ∎

Corollary. If (P, X, π, G) is a reduction (p. 131) of the frame bundle $F(X)$, then a connection on P determines a connection on $F(X)$ (a linear connection).

This is a straightforward application of the theorem with F being defined as the inclusion on P and ϕ as the inclusion of G into GL(n).

Given a riemannian manifold X, a linear connection on $F(X)$ which is determined by a connection on the bundle of orthonormal frames $O(X)$ with structural group $O(n)$ (p. 287) is called a **metric connection**. metric connection

Theorem. The covariant derivative of the metric tensor on a riemannian manifold, with respect to a metric connection vanishes.

This theorem follows from the fact that, under parallel transport defined by a metric connection, an orthonormal frame remains orthonormal.

The uniqueness of the riemannian connection (p. 308) can then be restated:

Theorem. On a riemannian manifold there exists a unique metric connection such that the torsion vanishes.

If X is a four-dimensional pseudo riemannian space, a connection on the bundle of Lorentz frames $L(X)$ with structural group $L(4)$ (p. 290) determines a unique connection on $F(X)$. Again, the covariant derivative of the metric tensor on X with respect to this connection vanishes.

B. HOLONOMY

When a tangent vector at some point x of a manifold X endowed with a linear connection is parallel transported along a closed curve, we say "a loop" from x to x, the resulting vector is in general different from the original one, in contradistinction to the standard parallelism in euclidean space. The difference is related to the nonvanishing of the curvature of the connection, as we have already seen on an infinitesimal level by the very definition of the curvature tensor through the non commutativity of covariant derivatives in two different directions.

We shall in this section study the parallel transport along loops in a general principal fibre bundle. Its relation to curvature will be given by the Ambrose–Singer theorem. We have first to give some properties of reduction of principal fibre bundle, which are also of interest in themselves.

In this section manifolds are supposed to be differentiable of some class C^k, $1 \leq k \leq +\infty$, all groups are Lie groups or Lie subgroups. Curves will be piecewise C^k, $1 \leq k \leq +\infty$.

1. REDUCTION

We have said (cf. p. 132) that a (differentiable) principal fibre bundle (P,

X, π, G) is reducible to a principal fibre bundle (P_1, X, π_1, G_1) if G_1 is a (Lie) subgroup of (the Lie group) G, and P_1 a submanifold of P such that the injection f: $P_1 \to P$ is a bundle morphism, i.e.

$$\pi f(p) = \pi_1 p, \qquad \forall p \in P_1$$

which commutes with the action of G_1, i.e.

$$f(\tilde{R}_g p) = \tilde{R}_g(f(p)), \qquad \forall p \in P_1, \qquad g \in G_1.$$

We will now prove the theorem.

Theorem. A principal fibre bundle (P, X, π, G) *is reducible to* (P_1, X, π_1, G_1), *with* G_1 *a subgroup of* G *if and only if it admits a family of local trivializations whose transition functions take their values in* G_1.

Proof: 1) Suppose (P, X, π, G) is reducible to (P_1, X, π_1, G_1). Consider an admissible family of local trivializations of P_1:

$$\bigcup_{i \in I} U_i = X, \qquad \varphi_i: \pi_1^{-1} U_i \to U_i \times G_1$$

the corresponding transition functions are (cf. pp. 125, 126)

$$g_{ij}: U_i \cap U_j \to G_1 \text{ by } g_{ij} = \overset{\Delta}{\varphi}_{i,x} \circ \overset{\Delta}{\varphi}{}^{-1}_{j,x}.$$

We identify P_1 with its image by f in P and we extend the local trivialization of P_1 to P as follows:
If $p \in \pi^{-1}(x)$, $x \in U_i$, and if we chose some $p_1 \in \pi_1^{-1}(x)$, there exists $g \in G$ such that $p = \tilde{R}_g p_1$. We set $\phi_i(p) = \tilde{R}_g \varphi_i(p_1)$, it is easily seen that $\phi_i(p)$ does not depend on the choice of p_1, and that $p \mapsto \phi_i(p)$ is a diffeomorphism of $\pi^{-1}(U_i)$ onto $U_i \times G$. The transition functions $\overset{\Delta}{\phi}_{i,x} \circ \overset{\Delta}{\phi}{}^{-1}_{j,x} = \overset{\Delta}{\varphi}_{i,x} \circ \overset{\Delta}{\varphi}{}^{-1}_{j,x}$ take their values in G_1.
2) Conversely, assume P admits a family of local trivialization whose transition functions g_{ij} have their values in G_1, and are therefore differentiable maps $U_i \cap U_j \to G_1$. We construct P_1 as the quotient of $\bigcup_{i \in I} U_i \times G_1$ by the equivalence relation $(x, g_i) \in U_i \times G_1 \sim (x, g_j) \in U_j \times G_1$ if $g_i = g_j g_{ij}(x)$.
It is not difficult[1] to endow P_1 with a principal fibre bundle structure with group G_1, base X, projection π_1; with $\pi_1^{-1}(U_i)$ diffeomorphic to $U_i \times G_1$ and transition functions g_{ij}.

[1]Cf. [Chevalley, p. 95, proposition 1] and [Kobayashi, Nomizu I, p. 11].

Each $U_i \times G_1$ is naturally embedded in the corresponding $U_i \times G$; we have for each index i an embedding:

$$f_i\colon \pi_1^{-1}(U_i) \to U_i \times G_1 \to U_i \times G \to \pi^{-1}(U_i).$$

On $\pi_1^{-1}(U_i \cap U_j)$ we have $f_i = f_j$, since the transition function is the same for P and P_1, thus a well defined embedding $f\colon P_1 \to P$, and P_1 is a reduction of P. ∎

We have defined on p. 367 vector bundles associated with a principal fibre bundle. More generally we can associate with the principal fibre bundle (P, X, π, G) fibre bundles whose typical fibre F is a manifold on which G acts on the left: $(g, u) \in G \times F \to \sigma_g u \in F$, with $g \mapsto \sigma_g$ a realization of G (cf. p. 162).

Definition: The fibre bundle (E, X, π_E, F, G) is said to be **associated** with the principal fibre bundle (P, X, π, G), through the realization of G as a group of diffeomorphisms of the typical fibre F if, given an admissible family (U_i, ϕ_i) of local trivializations of P: associated bundle

$$\phi_i\colon \pi^{-1}(U_i) \to U_i \times G, \qquad \bigcup_{i \in I} U_i = X$$

$$g_{ij}\colon U_i \cap U_j \to G \quad \text{by} \quad g_{ij}(x) = \overset{\Delta}{\phi}_{i,x} \circ \overset{\Delta}{\phi}{}_{j,x}^{-1}$$

there exists an associated family (U_i, φ_i) of local trivializations of E, namely:

$$\varphi_i\colon \pi_E^{-1}(U_i) \to U_i \times F$$

such that:

$$\sigma_{g_{ij}}\colon U_i \cap U_j \to \{\sigma_g;\, g \in G\} \quad \text{by} \quad \sigma_{g_{ij}}(x) = \overset{\Delta}{\varphi}_{i,x} \circ \overset{\Delta}{\varphi}{}_{j,x}^{-1}.$$

Exercise: Show that if F is a manifold on which G acts on the left through the realization $\{\sigma_g,\ g \in G\}$ one can associate with the principal fibre bundle (P, X, π, G) a fibre bundle (E, X, π_E, F, G) by taking the quotient of $P \times F$ under the following action τ_g, $g \in G$:

$$\tau_g\colon P \times F \to P \times F \quad \text{by} \quad (p, u) \mapsto (\tilde{R}_g p, \sigma_g^{-1} u).$$

Answer: It is easy to check that τ_g is a group action: $\tau_{g_1 g_2} = \tau_{g_1} \circ \tau_{g_2}$. Let us denote by $E = P \times_G F$ the quotient space of $P \times F$ under the indicated action, i.e. under the equivalence relation: $\boldsymbol{E = P \times_G F}$

$$(p, u) \simeq (p_1, u_1) \quad \text{if} \quad \exists g \in G \quad \text{such that} \quad (p, u) = (\tilde{R}_g p_1, \sigma_g^{-1} u_1).$$

We denote an element of E by $v = \mathcal{R}(p, u)$.
To the projection π: $P \to X$ by $p \mapsto x = \pi(p)$ we associate the mapping π_E: $\mathcal{R}(p, u) \mapsto x = \pi(p)$; it is a projection $E \to X$.
We define an equivalence relation $\bar{\mathcal{R}}$ on $G \times F$ by

$$(a, u) \simeq (a_1, u_1) \quad \text{if} \quad \exists g \quad \text{with} \quad (a, u) = (a_1 g, \sigma_g^{-1} u_1).$$

The quotient of $G \times F$ by $\bar{\mathcal{R}}$ can be identified with F, for instance by setting

$$\bar{\mathcal{R}}(a, u) = (e, \sigma_a u).$$

Let now

$$\varphi: \pi^{-1}(U) \to U \times G \quad \text{by} \quad p \mapsto (x, \overset{\Delta}{\varphi}(p))$$

be a local trivialization of P. We define a bijection ϕ

$$\phi: \pi_E^{-1}(U) \to U \times F \quad \text{by} \quad \mathcal{R}(p, u) \mapsto (x, \bar{\mathcal{R}}(\overset{\Delta}{\varphi}(p), u)).$$

We endow $\pi_E^{-1}(U)$ with the differentiable structure, such that ϕ is a diffeomorphism.
The family of local trivializations (U_i, φ_i) induces a differentiable fibre bundle structure on E. The transition functions on $U_i \cap U_j$ are such that

$$\sigma_{g_{ij}}(x) = \overset{\Delta}{\phi}_{i,x} \circ \overset{\Delta}{\phi}{}_{j,x}^{-1} = \sigma_{g_{ij}(x)} \quad \text{where} \quad g_{ij}(x) = \overset{\Delta}{\varphi}_{i,x} \circ \overset{\Delta}{\varphi}{}_{j,x}^{-1}. \qquad ■$$

Theorem. If G_1 is a subgroup of G the quotient P/G_1 of P under the right action of G_1, where (P, X, π, G) is a principal fibre bundle, has a fibre bundle structure with base X, typical fibre $F = G/G_1$ and structure group G.

Proof: The space $E = P/G_1$ is the space of equivalence classes $[p]$ of $p \in P$ with

$$p \sim p' \quad \text{if} \quad p' = \tilde{R}_{g_1} p,\ g_1 \in G_1,\ p \in P,\ p' \in P.$$

The projection π: $P \to X$ induces a projection π_E: $E \to X$, and the diffeomorphism

$$\varphi: \pi_E^{-1}(U) \to U \times G \quad \text{by} \quad p \mapsto (x, \overset{\Delta}{\varphi}(p)),\ x = \pi(p)$$

induces a bijection:

$$\phi^{-1}(U): \pi_E^{-1}(U) \to U \times F \quad \text{by} \quad [p] \mapsto (x, [\overset{\Delta}{\varphi}(p)])$$

where $[\overset{\Delta}{\varphi}(p)]$ is the element of G/G_1 corresponding to $\overset{\Delta}{\varphi}(p)$; i.e. the right coset $\overset{\Delta}{\varphi}(p)G_1$. The bijection ϕ is a diffeomorphism if $\pi_E^{-1}(U)$ is endowed with the appropriate differentiable structure. The family (ϕ_i, U_i) associated with an admissible family (φ_i, U_i) for the fibered structure of P

defines a fibered structure on E, whose transition functions on $U_i \cap U_j$ have their values in G: indeed

$$\phi_i \circ \phi_j^{-1} = \varphi_i \circ \varphi_j^{-1}$$

since if

$$g_{ij}(x)\colon G \to G \quad \text{by} \quad g \mapsto g_{ij}(x)g$$

then also

$$g_{ij}(x)\colon G/G_1 \to G/G_1 \quad \text{by} \quad gG_1 \mapsto g_{ij}(x)gG_1. \qquad \blacksquare$$

Theorem. The structure group G of a principal fibre bundle (P, X, π, G) is reducible to a subgroup $G_1 \subset G$ if and only if the fibre bundle P/G_1 admits a cross section.

Proof: 1) Suppose that P/G_1 admits a cross section $\sigma\colon X \to P/G_1$. We denote by μ the projection $P \to P/G_1$ by $p \mapsto [p] = \{\tilde{R}_{g_1}p, g_1 \in G_1\}$. We denote by P_1 the subspace of P, $P_1 = (\mu^{-1} \circ \sigma)(X)$. The injection $P_1 \to P$ is a bundle morphism which commutes with the action of G_1.
2) Suppose G is reducible to G_1, and let P_1 denote a reduced bundle with injection $f\colon P_1 \to P$. Let μ be the projection $P \to P/G_1$. The mapping $\mu \circ f$ is constant on the fibres of P_1, since f commutes with the right action of G_1, thus $\sigma = \mu \circ f \circ \pi_1^{-1}$ is a well defined mapping $X \to P/G_1$. ■

We have already mentioned (p. 131) that every fibre bundle with base manifold $\mathbf{R}^n$ is equivalent to a trivial fibre bundle. We now give a theorem in the case where the fibre is $\mathbf{R}^m$:

Theorem. Every fibre bundle (E, X, π, F, G) such that the base manifold X is paracompact and the fibre F is diffeomorphic to $\mathbb{R}^m$ admits infinitely many cross sections.

Proof: It is easy to show that when E is a vector bundle, i.e. when F_x is a vector space, it admits infinitely many cross sections. Indeed, let $\{\theta_i\}$ be a partition of unity on X subordinate to a locally finite covering by open sets V_i such that $V_i \subset U_j$, some open set of an atlas of X over which E is trivializable. Let σ_i be an arbitrary cross section over V_i. Then the element of F_x given by the finite sum

$$\sigma(x) = \sum_i \theta_i(x)\sigma_i(x)$$

is a cross section over X. The theorem is stronger because it does not require a canonical identification of a point of F_x with the origin of $\mathbb{R}^m$, nor the group G to be linear. For the proof[1] one uses the property that a

[1]Cf. [R. Godement, *Theorie des Faisceaux* p. 151, Hermann Paris, 1958] or [Kobayashi and Nomizu loc. cit. Vol. I, p. 58].

differentiable function defined on a closed set of $\mathbb{R}^m$ can be extended to the whole of $\mathbb{R}^m$, together with Zorn's lemma, to show that every cross section defined over a closed set $\bar{Y} \subset X$ can be extended to a cross section over X.

More generally one can prove that a principal fibre bundle over a paracompact manifold with structure group a connected Lie group G can be reduced to a principal fibre bundle with structure group any maximal compact subgroup G_1 of G because, by Iwasawa[1] theorem, G is diffeomorphic with the direct product of G_1 and a space $\mathbb{R}^m$.

An application of the previous theorem is that all tensor bundles over a manifold X have cross sections. A cross section of the tangent bundle is a vector field; it can be zero at some points. These points are called **singular points** of the vector field.

singular points

As an application of the theorem we shall now prove:

Theorem. The principal bundle $F(X)$ of linear frames of a paracompact manifold X can be reduced to a principal bundle with structure group $O(n)$. To each reduction of $F(X)$ to $O(n)$ corresponds a riemannian metric on X.

Proof: The quotient $GL(n, \mathbb{R})/O(n)$ is diffeomorphic to[2] the space $\mathbb{R}^{n(n+1)/2}$. The fibre bundle $F/O(n)$, where $(F, X, \pi, GL(n, \mathbb{R}))$ is the bundle of linear frames, admits therefore a cross section, and F is reducible to a bundle with structure group $O(n)$. To each cross section of $F/O(n)$ – i.e. to each reduction $(Q, X, \pi, O(n))$ of F – corresponds a riemannian metric g on X by setting

$$g_x(u, v) = (\rho_x^{-1} u | \rho_x^{-1} v)$$

where $(\,|\,)$ denotes the euclidean inner product in $\mathbb{R}^n$ and ρ_x^{-1} the linear isomorphism $T_xX \to \mathbb{R}^n$ corresponding to an element of Q, $\rho_x \in \pi^{-1}_{(x)}$, identified with a linear frame via the bundle imbedding $f\colon Q \to F$. The invariance of $(\,|\,)$ by $O(n)$ implies that $g_x(u, v)$ is independent of the choice of ρ_x^{-1}. Conversely a riemannian metric on X defines a bundle of orthogonal frames which is obviously a reduction of F. This fact has already been stated in part A (pp. 380–381). ■

2. HOLONOMY GROUPS

loop

Let c be a closed curve starting and ending at the point x, called a **loop** at x, on the base manifold X of a principal fibre bundle (P, X, ω, G) with connection ω. The parallel displacement along c of the elements p of the fibre $G_x = \pi^{-1}(x)$ defines a mapping $G_x \to G_x$, which we still denote by c. We have:

[1]Cf. Iwasawa, Ann. of Maths. 50 (1949) 507–558.
[2]Cf. for instance [Chevalley, p. 16].

Theorem. The mapping c: $G_x \to G_x$ *defined by parallel transport along the loop* c *is a diffeomorphism which commutes with* $\tilde{R}_g$.

Proof: It is an immediate consequence of the properties (p. 358) of a connection. The inverse c^{-1} of the mapping c is obtained by parallel transport along the same geometric curve, with the opposite orientation.

Theorem and definition. The set C_x *of mappings* c *corresponding to the loops at* x *forms a group, which is called the* **holonomy group** *of the connection* ω *with reference point* x.

holonomy group

Proof: We endow C_x with a group structure by defining the product c_1c_2 as the mapping $G_x \to G_x$ obtained by parallel transporting first along c_2, then along c_1. ∎

The subgroup of the holonomy group consisting of the parallel displacement along loops which are homotopic to zero in X is called the **restricted holonomy group**. In a simply connected manifold it coincides with the holonomy group.

restricted holonomy group

The holonomy group at x can be identified with a subgroup of the structure group G. Indeed let $p \in P$, $\pi(p) = x$. By parallel transport along a loop c at x, we get a point $c(p) \in \pi^{-1}(x)$, thus there exists $a \in G$ with $c(p) = \tilde{R}_a p$. If a_1, $a_2 \in G$ correspond to the loops c_1 and c_2 then a_1a_2 correspond to c_1c_2 since $\tilde{R}_{a_1}\tilde{R}_{a_2} = \tilde{R}_{a_1a_2}$.
Hence we have defined, once $p \in \pi^{-1}(x)$ is chosen, an isomorphism between the holonomy group at x [the restricted holonomy group if we consider only loops homotopic to zero] and a subgroup $\mathcal{H}(p)$ [$\mathcal{H}_0(p)$] of G, called the [restricted] holonomy group with reference point $p \in P$. If we change the point $p \in \pi^{-1}(x)$ to $p' = \tilde{R}_g p$ we change $\mathcal{H}(p)$ for the isomorphic subgroup $\mathcal{H}(p') = g\mathcal{H}(p)g^{-1}$ [$\mathcal{H}_0(p') = g\mathcal{H}_0(p)g^{-1}$].

Theorem[1]. *If* X *is connected* $\mathcal{H}(p)$ *is a Lie subgroup of* G *whose connected component of the identity is the restricted holonomy group* $\mathcal{H}_0(p)$, *while* $\mathcal{H}(p)/\mathcal{H}_0(p)$ *is countable.*

Let $p_0 \in P$ be chosen. We denote by $P(p_0)$ the set of points of P which can be obtained by parallel transport from p_0, i.e. joined to p_0 by an horizontal curve of the connection. $P(p_0)$ is called the **holonomy bundle** at p_0. If we replace p_0 by a point $p \in P(p_0)$ we obtain, by parallel transport, the same holonomy bundle. Thus we see that P is the disjoint union of holonomy bundles of points which cannot be joined by a horizontal curve in P.

holonomy bundle

[1]Cf. [Lichnerowicz 1976]; [Kobayashi, Nomizu I, p. 74].

Exercise: Show that the holonomy group $\mathcal{H}(p)$ is the same subgroup of G at each point of an holonomy bundle $P(p_0)$.

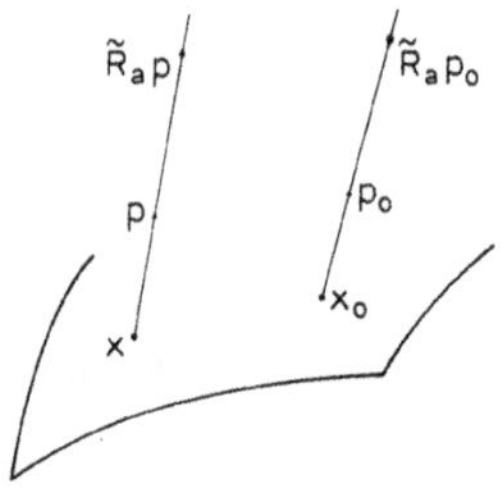

Answer: If $a \in \mathcal{H}(p_0)$ there is a loop c at x_0 such that, by parallel transport of p_0 along c, $c(p_0) = \tilde{R}_a p_0$. Now if p is obtained by parallel transport from p_0 along some curve γ from x_0 to x, the same is true of $\tilde{R}_a p$ from $\tilde{R}_a p_0$: thus $\tilde{R}_a p$ is obtained by parallel transport from p along the loop $\gamma \circ c \circ \gamma^{-1}$, thus $a \in \mathcal{H}(p)$. Conversely, if $a \in \mathcal{H}(p)$, then $a \in \mathcal{H}(p_0)$ therefore $\mathcal{H}(p) = \mathcal{H}(p_0)$. The same reasoning with loops homotopic to zero gives $\mathcal{H}_0(p) = \mathcal{H}_0(p_0)$.

The following theorem gives a link between the topological structure of a principal bundle and its holonomy bundle relative to some connection defined on it.

Theorem. Let (P, X, π, G) be a principal fibre bundle with a connection $\boldsymbol{\omega}$, where X is connected, then each holonomy bundle $P(p_0)$ is a reduced bundle of P, with structure group $\mathcal{H}(p_0)$.

Proof: 1) The projection $\pi: P \to X$ maps also $P(p_0)$ onto X if X is connected, since $p_0 \in \pi^{-1}(x_0)$ can be parallel transported above any point $x \in X$, to some $p \in \pi^{-1}(x)$.
2) If $p \in P(p_0)$ and $a \in \mathcal{H}(p_0)$ then $\tilde{R}_a p \in P(p_0)$; conversely if p, $p' \in P(p_0)$, $\pi(p) = \pi(p')$ there exists $a \in \mathcal{H}(p_0)$ such that $p' = \tilde{R}_a p$. Therefore if $\varphi: \pi^{-1}(U) \to U \times G$ is a local trivialization of P, the restriction $\varphi|_{P(p_0)}$ defines a bijection $\phi: P(p_0) \cap \pi^{-1}(U) \to U \times \mathcal{H}(p_0)$. The properties 1, 2 show that $(P(p_0), X, \pi, \mathcal{H}(p_0))$ is a principal fibre bundle which is a reduction of (P, X, π, G). ■

reducible connection

Definition: Let $f = (P_1, X, \pi_1, G_1) \to (P, X, \pi, G)$ be a reduction of a principal fibre bundle. A connection $\boldsymbol{\omega}$ on P is **reducible** by f if there exists a connection $\boldsymbol{\omega}_1$ on P_1 such that $\boldsymbol{\omega}$ is its image by f.
We have already proved (p. 380) that if such a connection $\boldsymbol{\omega}_1$ exists, then the horizontal subspaces of $\boldsymbol{\omega}_1$ are mapped into horizontal subspaces of $\boldsymbol{\omega}$ by f. Moreover we have

Theorem. A connection $\boldsymbol{\omega}$ *on P is reducible by f if the* 1*-form* $f^*\boldsymbol{\omega}$ *on* P_1 *has its values in* $\mathscr{G}_1$*, the Lie algebra of* G_1 *identified with a Lie subalgebra of G.*

Proof: It is easy to check that $\boldsymbol{\omega}_1 = f^*\boldsymbol{\omega}$ is a connection on P_1. ■

Theorem. A connection $\boldsymbol{\omega}$ *on a principal fibre bundle is reducible to a connection in any of its holonomy bundles.*

Proof: The bundle (P, X, π, G) is reducible to the holonomy bundle $(P(p), X, \pi, \mathscr{H}(p))$. We can define a connection in $P(p)$ by setting its horizontal subspaces equal to the horizontal spaces of the original connection of P, at the points of $P(p)$, since these spaces are tangent to $P(p)$ by its very definition – it is easy to check that they satisfy the required properties to define a connection $\boldsymbol{\omega}_1$, reduction of $\boldsymbol{\omega}$.

Theorem (Ambrose–Singer). *Let* (P, X, π, G) *be a principal fibre bundle, with a connection* $\boldsymbol{\omega}$*, X connected. The Lie algebra of an holonomy group* $\mathscr{H}(p_0)$ *is equal to the subspace of the Lie algebra* $\mathscr{G}$ *of G which is spanned by all elements of the form* $\boldsymbol{\Omega}_p(\boldsymbol{u}, \boldsymbol{v})$ *where* $\boldsymbol{\Omega}_p$ *is the curvature* 2*-form of* $\boldsymbol{\omega}$ *at any arbitrary point* $p \in P(p_0)$*, the holonomy bundle, and* $\boldsymbol{u}$*,* $\boldsymbol{v}$ *are arbitrary horizontal vectors at p.*

Proof: By virtue of the construction on p. 388, we may assume $P = P(p_0)$, $G = \mathscr{H}(p_0)$. We denote by V the Lie algebra, subspace of $\mathscr{G}$, spanned by all elements of the form $\boldsymbol{\Omega}_p(\boldsymbol{u}, \boldsymbol{v})$, $p \in P(p_0)$, $\boldsymbol{u}$, $\boldsymbol{v}$ horizontal vectors at p. At each point $p \in P$ we consider the subspace S_p of T_pP spanned by the horizontal subspace H_p and by the subspace of the vertical subspace corresponding to V (i.e. tangent Killing vectors corresponding to transformations of the fibre generated by elements in V). The subspace $S_p \subset T_pP$ has dimension $n + r$, with $r = \dim V$, $n = \dim X$. It depends differentiably on p, and can be proven[1] to satisfy the Frobenius condition (p. 248) for complete integrability, using the facts that V is a Lie subalgebra of $\mathscr{G}$, that the bracket of a Killing vertical vector field with a horizontal field is horizontal (p. 374), and that the bracket of two horizontal vector fields has a vertical component which belongs to V by the structure equation $\boldsymbol{\omega}([\boldsymbol{u}, \boldsymbol{v}]) = -\boldsymbol{\Omega}(\boldsymbol{u}, \boldsymbol{v})$ if $\boldsymbol{u}$, $\boldsymbol{v}$ are horizontal vector fields.
The family of hyperplanes S_p admits by the Frobenius theorem an integral manifold S of dimension $n + r$, passing through p_0. It certainly contains all horizontal curves starting from p_0, and thus coincide with P, since $\dim P = n + \dim \mathscr{G}$ we have $\dim \mathscr{G} = r$, i.e. $V = \mathscr{G}$. ■

[1]For more details see for instance [Lichnerowicz 1976], [Kobayashi, Nomizu I, p. 89].

C. CHARACTERISTIC CLASSES AND INVARIANT CURVATURE INTEGRALS

1. CHARACTERISTIC CLASSES

It can be shown that some quantities constructed with the curvature $\boldsymbol{\Omega}$ of a connection $\boldsymbol{\omega}$ on a principal differentiable fibre bundle P are in fact topological invariants of this bundle, that is
a) they do not depend on the choice of the connection $\boldsymbol{\omega}$
b) they are conserved by a bundle diffeomorphism $P \to P'$.
Examples of these quantities are the Euler, Chern and Pontrijagin characteristic classes. A characteristic class is the cohomology class (p. 223) of a certain closed form on the base manifold X of P.

We first construct closed forms, $f(\Omega)$ on X, whose cohomology class in the de Rham algebra $H^*(X)$ is independent of the choice of a connection.

Definition: A symmetric multilinear mapping from the Lie algebra $\mathcal{G}$ of G into $\mathbf{R}$

$$f\colon \mathcal{G} \times \mathcal{G} \cdots \times \mathcal{G} \to \mathbf{R}$$

Ad G invariant

is said to be **invariant** by G **(Ad G invariant)** if

$$f(\text{Ad}\, g V_1, \ldots, \text{Ad}\, g V_k) = f(V_1, \ldots, V_k), \quad \forall g \in G, \forall V_1, \ldots, V_k \in \mathcal{G}.$$

Theorem 1. *Let f be an invariant k-linear symmetric mapping $\mathcal{G} \times \cdots \times \mathcal{G} \to \mathbf{R}$, and $P(X, \pi, G)$ be a principal fibre bundle with base X, group G, and projection π. Let $\boldsymbol{\omega}$ be a connection on P and $\boldsymbol{\Omega}$ its curvature. The exterior differential form $f(\boldsymbol{\Omega})$ of degree $2k$ defined by*

$$f(\boldsymbol{\Omega})(v_1, \ldots, v_{2k}) = \frac{1}{(2k)!} \sum_\sigma \text{sign}\,\sigma\, f(\boldsymbol{\Omega}(v_{\sigma(1)}, v_{\sigma(2)}), \ldots, \boldsymbol{\Omega}(v_{\sigma(2k-1)}, v_{\sigma(2k)}))$$

for $v_1, \ldots, v_{2k} \in T_pP$, sum over all permutations σ, is then such that
a) *$f(\boldsymbol{\Omega})$ projects to a unique closed $2k$-form, say $\bar{f}(\boldsymbol{\Omega})$, on X, i.e. there exists a unique $\bar{f}(\boldsymbol{\Omega})$, such that*

$$f(\boldsymbol{\Omega}) = \pi^* \bar{f}(\boldsymbol{\Omega}).$$

b) *The element of the de Rham cohomology $H^*(X)$ associated with $\bar{f}(\boldsymbol{\Omega})$ is independent of the choice of $\boldsymbol{\omega}$.*

Proof: a) Let $\boldsymbol{\alpha}$ be a q-form on P; its projection $\bar{\boldsymbol{\alpha}}$ on X, if it exists, must be such that

$$\bar{\boldsymbol{\alpha}}_x(u_1, \ldots u_q) = \boldsymbol{\alpha}_p(v_1, \ldots v_q) \qquad \forall u_i = \pi' v_i \in T_xX,\ x = \pi(p);$$

the form $\bar{\boldsymbol{\alpha}}$ will exist and be unique if the right hand side depends neither on p, nor on the choice $v_i, \ldots v_q$. These conditions are satisfied if $\boldsymbol{\alpha}$ is invariant under the right action of G in P:

$$R_g^* \boldsymbol{\alpha} = \boldsymbol{\alpha}$$

and if α vanishes whenever one of the vectors v_i is vertical. These conditions are satisfied by $f(\boldsymbol{\Omega})$ since

α) $R_g^* \boldsymbol{\Omega} = \mathrm{Ad}(g^{-1})\boldsymbol{\Omega}$ and f is invariant by G,

β) $\boldsymbol{\Omega}$ vanishes on vertical vectors.

b) The proof that $\bar{f}(\boldsymbol{\Omega})$ is closed rests on the fact that $\mathrm{d}f(\boldsymbol{\Omega}) = \mathrm{D}f(\boldsymbol{\Omega}) = 0$:

$$\begin{aligned} \mathrm{d}f(\boldsymbol{\Omega})(v_1, \ldots, v_n) &= \pi^* \,\mathrm{d}\bar{f}(\boldsymbol{\Omega})(v_1, \ldots, v_n) \\ &= \mathrm{d}\bar{f}(\boldsymbol{\Omega})(\pi' \operatorname{hor} v_1, \ldots, \pi' \operatorname{hor} v_n) \\ &= \mathrm{d}f(\boldsymbol{\Omega})(\operatorname{hor} v_1, \ldots, \operatorname{hor} v_n) = \mathrm{D}f(\boldsymbol{\Omega})(v_1, \ldots, v_n). \end{aligned}$$

c) Consider two connection forms $\boldsymbol{\omega}_0$ and $\boldsymbol{\omega}_1$ on P. We want to show that their curvatures are such that $f(\boldsymbol{\Omega}_1) - f(\boldsymbol{\Omega}_0)$ is an exact differential. We shall use a homotopy type argument, setting:

$$\boldsymbol{\phi} = \boldsymbol{\omega}_1 - \boldsymbol{\omega}_0, \qquad \boldsymbol{\omega}_t = \boldsymbol{\omega}_0 + t\boldsymbol{\phi}.$$

$\boldsymbol{\phi}$ is a tensorial 1-form of type Ad, and $\boldsymbol{\omega}_t$ satisfies the properties (p. 361) required for being a connection on P. Let D_t and $\boldsymbol{\Omega}_t$ denote the corresponding covariant exterior differentiation and curvature form. We have

$$\boldsymbol{\Omega}_t = \mathrm{d}\omega_t + \tfrac{1}{2}[\omega_t, \omega_t].$$

Therefore

$$\mathrm{d}\boldsymbol{\Omega}_t/\mathrm{d}t = \mathrm{d}\boldsymbol{\phi} + \tfrac{1}{2}[\boldsymbol{\phi}, \boldsymbol{\omega}_t] + \tfrac{1}{2}[\omega_t, \boldsymbol{\phi}] = \mathrm{D}_t\boldsymbol{\phi}$$

by a calculation similar to the proof of the structural equation (p. 373). To show that $f(\boldsymbol{\Omega}_1) - f(\boldsymbol{\Omega}_0)$ is exact, write

$$\begin{aligned} f(\boldsymbol{\Omega}_1, \ldots, \boldsymbol{\Omega}_1) - f(\boldsymbol{\Omega}_0, \ldots, \boldsymbol{\Omega}_0) &= \int_0^1 (\mathrm{d}f(\boldsymbol{\Omega}_t, \ldots, \boldsymbol{\Omega}_t)/\mathrm{d}t)\,\mathrm{d}t, \quad \text{identity} \\ &= k \int_0^1 f(\mathrm{d}\boldsymbol{\Omega}_t/\mathrm{d}t, \ldots, \boldsymbol{\Omega}_t)\,\mathrm{d}t, \quad f \text{ symmetric} \\ &= k \int_0^1 f(\mathrm{D}_t\boldsymbol{\phi}, \boldsymbol{\Omega}_t, \ldots, \boldsymbol{\Omega}_t)\,\mathrm{d}t, \quad \text{see above} \end{aligned}$$

$$= k\int_0^1 D_t f(\boldsymbol{\phi}, \boldsymbol{\Omega}_t, \ldots, \boldsymbol{\Omega}_t)\,dt, \quad \text{using Bianchi } D_t\boldsymbol{\Omega}_t = 0$$

$$= k\int_0^1 df(\boldsymbol{\phi}, \boldsymbol{\Omega}_t, \ldots, \boldsymbol{\Omega}_t)\,dt, \quad \text{tensorial form}$$

$$= d\int_0^1 kf(\boldsymbol{\phi}, \boldsymbol{\Omega}_t, \ldots, \boldsymbol{\Omega}_t)\,dt$$

$$\equiv d\boldsymbol{\Phi}.$$

This, together with the fact that $\boldsymbol{\Phi}$ being invariant under the right action of G on P and vanishing on a vertical vector projects to a form $\bar{\boldsymbol{\Phi}}$ on X, gives

$$\bar{f}(\boldsymbol{\Omega}_1) - \bar{f}(\boldsymbol{\Omega}_0) = d\bar{\boldsymbol{\Phi}}. \qquad \blacksquare$$

characteristic class

The element in $H^*(X)$ represented by $\bar{f}(\boldsymbol{\Omega})$ is called a **characteristic class.** It depends on the chosen principal bundle and on f.

$I(G)$

We now make the set of formal sums of symmetric multilinear mappings: $\mathscr{G}\times\cdots\times\mathscr{G}\to\mathbb{R}$ into a commutative algebra $I(G)$ by setting

$$fg(\boldsymbol{V}_1, \ldots, \boldsymbol{V}_{k+l}) = ((k+l)!)^{-1}\sum_\sigma f(\boldsymbol{V}_{\sigma(1)}, \ldots, \boldsymbol{V}_{\sigma(k)}) \times g(\boldsymbol{V}_{\sigma(k+1)}, \ldots, \boldsymbol{V}_{\sigma(k+l)}).$$

The following theorem says that the characteristic classes of a principal fibre bundle are a subalgebra of the cohomology algebra of the base space.

Theorem 2. *The mapping $I(G)\to H^*(X)$ by $f\mapsto\bar{f}(\Omega)$ is an algebra homomorphism.*

Weil homomorphism

It is called the **Weil homomorphism**.
The proof is a direct consequence of the definitions of the algebra $H^*(X)$ (p. 223) recalling that the exterior product of forms of even degree is commutative.

polynomial on $\mathscr{G}$

Invariant polynomials. A symmetric multilinear mapping f from a vector space, say $\mathscr{G}$, into $\mathbf{R}$, is also called a **polynomial** on $\mathscr{G}$. If we choose a basis in $\mathscr{G}$, and set

$$f(\boldsymbol{V}) \equiv f(\boldsymbol{V}, \ldots, \boldsymbol{V})$$

$f(\boldsymbol{V})$ is a polynomial of degree k in the components of $\boldsymbol{V}$, if f is k-multilinear.

The algebra $I(G)$ of Ad G-invariant symmetric multilinear mapping is then identified with the algebra of Ad G-invariant polynomials.

In the case where G is the complex linear group $\mathrm{GL}(m, \mathbb{C})$, one obtains m invariant polynomials $f_1, \ldots, f_m$ on $\mathrm{GL}(m, \mathbb{C})$ by writing the characteristic polynomial of a matrix $V \in \mathscr{GL}(m, \mathbb{C})$, namely

$$\det\left(\lambda \mathbb{1}_m - \frac{1}{2\mathrm{i}\pi} V\right) = \sum_{k=0}^{m} f_k(V)\lambda^{m-k}.$$

The normalization of the eigenvalues is chosen such that the Chern numbers (p. 396) are integer.

Chern classes: We consider a complex vector bundle E over a (real) differentiable manifold X, with typical fibre $\mathbb{C}^m$ and structure group $\mathrm{GL}(m, \mathbb{C})$, and its associated principal fibre bundle $P(X, \pi, \mathrm{GL}(m, \mathbb{C}))$. Let $\boldsymbol{\Omega}$ be the curvature form of some connection $\boldsymbol{\omega}$ on P.

Definition: The **k-th Chern class** $C_k(E)$ of a complex vector bundle over X is the cohomology class of the closed $2k$-form γ_k on X which is such that Chern class

$$\pi^* \gamma_k = f_k(\boldsymbol{\Omega}).$$

The form γ_k exists and is closed by Theorem 1, p. 390.

If we express the curvature form by a matrix valued 2-form $(\bar{\boldsymbol{\Omega}}^i_j)$ on X then the $2k$-form γ_k representing the Chern class $C_k(E)$ is given by:

$$\gamma_k = \frac{(-1)^k}{(2\pi\mathrm{i})^k (k!)} \epsilon^{j_1 \cdots j_k}_{i_1 \cdots i_k} \bar{\boldsymbol{\Omega}}^{i_1}_{j_1} \wedge \cdots \wedge \bar{\boldsymbol{\Omega}}^{i_k}_{j_k}$$

(see Problems Vbis 2 and 3).

It can be shown that the Chern classes generate the whole algebra of characteristic classes of the considered complex vector bundle E.
Of special importance, due to its appearance in the Atiyah–Singer index theorem is the **Chern character**, denoted $\mathrm{ch}(E)$, of the complex vector bundle E; it is the cohomology class defined by the closed form $\boldsymbol{c}$ such that Chern character

$$\pi^* \boldsymbol{c} = \mathrm{trace}(\exp(-\boldsymbol{\Omega}/2\pi\mathrm{i})) .$$

This equation is to be understood as follows: Substitute $\boldsymbol{\Omega}$ for $V \in \mathscr{GL}(m, \mathbb{C})$ in

$$\mathrm{trace}(\exp(-V/2\pi\mathrm{i})) \equiv \mathrm{trace} \sum_{k=0}^{\infty} \frac{1}{k!} (-2\mathrm{i}\pi)^{-k} V^k \equiv \sum_{k=0}^{\infty} f_k(V)$$

and use the definition of $f(\boldsymbol{\Omega})$ given on p. 390.
The series giving $\boldsymbol{c}$ has in fact only a finite number of terms since Ω^k projects on a $2k$-form on X, which vanishes if $2k > \dim X$.

Euler class

Euler class. The Euler class of a differentiable oriented manifold X of even dimensions $2n$ is the characteristic class of the tangent bundle represented by the following closed $2n$ form on X:

$$\gamma = \frac{(-1)^n}{(4\pi)^n n!} \epsilon^{1\ldots 2n}_{i_1\ldots i_{2n}} \bar{\boldsymbol{\Omega}}^{i_1}_{i_2} \wedge \cdots \wedge \bar{\boldsymbol{\Omega}}^{i_{2n-1}}_{i_{2n}}$$

where $\bar{\boldsymbol{\Omega}}^i_j$ is the curvature 2-form of a riemannian connection on X.
To verify that γ defines a characteristic class of the tangent bundle we consider the corresponding polynomial in the Lie algebra $\mathcal{O}(2n)$

$$f(\boldsymbol{V}) = \frac{(-1)^n}{(4\pi)^n n!} \epsilon^{1\ldots 2n}_{i_1\ldots i_{2n}} V^{i_1}_{i_2} \cdots V^{i_{2n-1}}_{i_{2n}}$$

$$\boldsymbol{V} = (V^i_j) \in \mathcal{O}(2n), \quad \text{i.e.} \quad V^i_j = -V^j_i \qquad \text{(cf. p. 174);}$$

by classical calculus we see that

$$f^2(\boldsymbol{V}) = (2\pi)^{-2n} \det \boldsymbol{V}$$

and thus f is invariant by $\mathrm{Ad}(O(2n))$.

The definition of the Euler class can be extended to fibre bundles over X whose typical fibre[1] is $\mathbb{R}^{2m}$.
Pontrijagin classes. They are to real vector bundles what the Chern classes are to complex vector bundles.
One defines $\mathrm{Ad}(\mathrm{GL}(m,\mathbb{R}))$ invariant polynomial functions $g_0, g_1, \ldots, g_m$ on $\mathcal{GL}(m,\mathbb{R})$ by

$$\det\left(\lambda \mathbb{1}_m - \frac{1}{2\pi}\boldsymbol{V}\right) \equiv \sum_{k=0}^{m} \lambda^{m-k} g_k(\boldsymbol{V}).$$

Pontrijagin class

Definition: The k-th **Pontrijagin class** $p_k(E)$ of a differentiable vector bundle E, with base X, typical fiber $\mathbb{R}^m$ and structure group $\mathrm{GL}(m,\mathbb{R})$ is the element of $H^{4k}(X)$ with representant $\boldsymbol{\beta}_k$ the $4k$-closed form on X such that

$$\pi^* \boldsymbol{\beta}_k = g_{2k}(\boldsymbol{\Omega})$$

where $\boldsymbol{\Omega}$ is the curvature 2-form of a connection on the associated principal bundle.

Exercise: Show that the closed $2k$-form $\boldsymbol{\alpha}_k$ on X such that $\pi^*\boldsymbol{\alpha}_k = g_k(\boldsymbol{\Omega})$, with k an odd number is always cohomologous to zero.
Answer: If $\boldsymbol{V} \in \mathcal{O}(m)$, the Lie algebra of $O(m)$ we have (cf. p. 173)

[1] Cf. for instance [Kobayashi, Nomizu, pp. 314–320].

$V = -{}^tV$ thus

$$\det\left(\lambda \mathbb{1}_m - \frac{1}{2\pi} V\right) = \det{}^t\left(\lambda \mathbb{1}_m - \frac{1}{2\pi} V\right) = \det\left(\lambda \mathbb{1}_m + \frac{1}{2\pi} V\right)$$

thus

$$\sum_{k=0}^{m} \lambda^{m-k} g_k(V) = \sum_{k=0}^{m} (-1)^k \lambda^{m-k} g_k(V) \quad \text{for} \quad V \in \mathcal{O}(m)$$

which implies $g_k(V) = 0$ for odd k.
On the other hand we know that a bundle P with structure group $\mathrm{GL}(m, \mathbb{R})$ can always be reduced to $O(m)$, there is always a connection on P whose curvature 2-forms takes its values in the Lie subalgebra $\mathcal{O}(m)$ of $\mathcal{GL}(m, \mathbb{R})$.

Remark: The k-th Pontrijagin class is the $2k$-th Chern class of the complexified $E^c(X, \mathbb{C}^m, \mathrm{GL}(m, \mathbb{C}))$ of the vector bundle $E(X, \mathbb{R}^m, \mathrm{GL}(m, \mathbb{R}))$. If we write $\boldsymbol{\Omega}$ as a, real, matrix valued 2-form $(\boldsymbol{\Omega}^i_j)$, then

$$g_{2k}(\boldsymbol{\Omega}) = \frac{1}{(2\pi)^{2k}} \frac{1}{(2k!)^2} \epsilon^{j_1 \cdots j_{2k}}_{i_1 \cdots i_{2k}} \boldsymbol{\Omega}^{i_1}_{j_1} \wedge \cdots \wedge \boldsymbol{\Omega}^{i_{2k}}_{j_{2k}}.$$

It can be shown that the algebra of the characteristic classes of the vector bundle E is generated by its Pontrijagin classes.

2. GAUSS–BONNET THEOREM AND CHERN NUMBERS

A characteristic class is a topological invariant of the bundle to which it is related, the same is true of the integral of its representant over a cycle (p. 222) of the base manifold X. This integral is certainly defined if the cycle is compact, and depends only on the homology class of the cycle.
The best known invariant of this type is the **Euler number** of an even dimensional manifold X which is the integral over X of the Euler class of its tangent bundle. Euler number

Theorem (Gauss–Bonnet–Chern–Avez). *Let X be a $2n$ dimensional oriented compact, riemannian or pseudo riemannian manifold; let γ be its Euler class and χ its Euler number*

$$\chi(X) = \int_X \gamma,$$

the Euler number of X is equal to its Euler Poincaré characteristic (p. 224).

Proof (cf. [Chern], [Guillemin and Pollack, p. 196]). ■

Note 1: The coefficient in front of the Euler class has been chosen such that the Euler number is an integer.
Note 2: If X is 2 dimensional then $\gamma = (1/2\pi)\Omega_1^2 = (1/2\pi)K\tau$ where K is the **gaussian curvature**, and τ the volume element. For the 2-sphere $\chi(S^2) = 2$, for the 2-torus $\chi(T^2) = 0$. One proves also $\chi(S^{2n}) = 2$, $\chi(T^{2n}) = 0$. It can be shown that

Poincaré–Hopf theorem. The Euler number of a compact manifold X is also equal to the sum of the indices of the zeros of any vector field on X which has only isolated zeros[1].

Chern number

Important topological invariants for the complex vector bundles over X are the Chern numbers. It can be shown that the integral of a representant of a Chern class C_k over a $2k$-cycle in X, with the given choice of the coefficients in front of γ_k is an integer. The name **Chern number** however is usually given only to an integral over the whole n-dimensional manifold X of a representant of a characteristic class of degree n; such a characteristic class is an element of the algebra generated by the Chern classes. In the physical case $n = 4$ there are two Chern numbers

$$\int_X \gamma_2(\Omega) \quad \text{and} \quad \int_X \gamma_1(\Omega) \wedge \gamma_1(\Omega).$$

See in Problems Vbis 2 and 3 the computations of Chern numbers.

Pontrijagin number

Also interesting is the Pontrijagin number of a real vector bundle over a compact 4-dimensional manifold X. This **Pontrijagin number** is equal to the integral over X of a representant of the first, and unique, Pontrijagin class of the bundle – i.e. to the Chern number $\int_X \gamma_2$ of the complexified bundle.

3. THE ATIYAH–SINGER INDEX THEOREM

An interesting application of characteristic classes is their use in the theorem of Atiyah and Singer, which gives the equality between the analytic index and the topological index of an elliptic complex over a compact manifold X.
A linear differential operator D: $C^\infty(E) \to C^\infty(F)$ is a linear mapping from C^∞ sections of a vector bundle E over a base manifold X into C^∞ sections

[1]The index of a vector field v on X, at an isolated zero y, is the degree of the map (cf. Ch VII, p. 557) $(x^i) \mapsto (v^i(x^j))/\{\Sigma(v^i(x^j))^2\}^{1/2}$ of a small sphere of center (y^i) into the unit sphere – it does not depend on the choice of coordinates. Cf. for instance [Milnor 1965].

of another vector bundle F over X which reads, in local coordinates

$$(\mathrm{D}u)_I \equiv \sum_{|k|=0}^{m} a_{kI}^{A}\, \mathrm{D}^k u_A,$$

$$k = (k_1, \ldots, k_n), \mathrm{D}^k = \left(\frac{\partial}{\partial x^1}\right)^{k_1} \cdots \left(\frac{\partial}{\partial x^n}\right)^{k_n}, |k| = k_1 + \cdots + k_n$$

where the a_{kI}^{A} are C^∞ functions of the coordinates $(x^1, \ldots, x^n)$.

The **principal symbol of** D at a point $x \in X$ is the linear map $\sigma_x(\xi)$ (D): $E_x \to F_x$ from the fibre E_x of E at x into the fibre F_x of F at x defined by the matrix principal symbol

$$(\sigma_x(\xi)(\mathrm{D}))_I^A = \sum_{|k|=m} a_{kI}^{A} \xi^k, \qquad \xi \in T_x^* X.$$

The operator D is **elliptic** on X if for each $x \in X$ and $\xi \neq 0$ the linear map $\sigma_x(\xi)$ (D) is an isomorphism from E_x onto F_x. elliptic operator

Let now $\{E_p\}$, $p = 1, \ldots N$, be a finite sequence of vector bundles over X and let:

$$\mathrm{D}_p\colon C^\infty(E_p) \to C^\infty(E_{p+1})$$

be a sequence of linear differential operators. This sequence is a **complex** if $\mathrm{D}_{p+1}\mathrm{D}_p = 0$, that is if complex

$$\text{Image } \mathrm{D}_p \subset \text{kernel } \mathrm{D}_{p+1}.$$

The sequence $\{\mathrm{D}_p\}$ is an **elliptic complex** if in addition, for each x and $\xi \neq 0$ the sequence of linear maps elliptic complex

$$\{\sigma_x(\xi)(\mathrm{D}_p)\colon E_{p,x} \to E_{p+1,x}\}$$

is an **exact sequence** that is if exact sequence of operators

$$\ker \sigma_x(\xi)(\mathrm{D}_{p+1}) = \text{Image } \sigma_x(\xi)(\mathrm{D}_p)$$

for $p = 1, \ldots, N$.

We identify, via a scalar product in the fibres, possibly deduced from a riemannian metric $\boldsymbol{g}$ on X, the bundles E and F with the dual bundles E^* and F^*. The adjoint D^* of D is the mapping $C^\infty(F) \to C^\infty(E)$ such that, for all sections $u \in C^\infty(E)$, $v \in C^\infty(F)$, u and v with compact support we have, $(\,|\,)$ being the inner product in the fibres and $\boldsymbol{\tau}$ the volume element of X,

$$\int_X (\mathrm{D}u|v)\boldsymbol{\tau} = \int_X (u|\mathrm{D}^*v)\boldsymbol{\tau}.$$

D^* is given in local coordinates by

$$(\mathrm{D}^*v)^A = \sum (-1)^{|k|}\mathrm{D}^k(a_{kI}^{A}|\det \boldsymbol{g}|^{1/2}v^I).$$

Exercise: Show that if $\{D_p, E_p\}$ is an elliptic complex the complex $\{D_p^*, E_{p+1}\}$, with a reverse order for the sequence, is also elliptic.
Answer: a) $D_{p+1} D_p = 0$ implies $D_p^* D_{p+1}^* = (D_{p+1} D_p)^* = 0$.
b) If σ_p^* denotes the principal symbol of D_p^* we have, where $\perp$ means orthogonality with respect to the scalar product in $E_{p+1,x}$:

$$(b|\sigma_p^* a) = (\sigma_p b|a) \qquad \forall a \in E_{p+1,x},\ b \in E_{p,x}$$

thus

$$\operatorname{Im} \sigma_{p-1}^* \perp \ker \sigma_{p-1} \quad \text{and} \quad \ker \sigma_p^* \perp \operatorname{Im} \sigma_p.$$

Theorem. The complex $\{D_p, E_p\}$ *is an elliptic complex if and only if the operators* Δ_p: $E_p \to E_p$ *given by*

$$\Delta_p \equiv D_p^* D_p + D_{p-1} D_{p-1}^*$$

laplacians of an elliptic complex

are elliptic operators, called **laplacians of the complex**.

Proof: We denote $\sigma_x(D_p) = \sigma_p$. We have

$$A_p \equiv \sigma_x(\Delta_p) = \sigma_p^* \sigma_p + \sigma_{p-1} \sigma_{p-1}^*$$

where $\sigma_p^* = \sigma_x(D_p^*)$ is the adjoint linear map $E_{p+1,x} \to E_{p,x}$ of the linear map σ_p.
a) We suppose that $\ker \sigma_p = \operatorname{Im} \sigma_{p-1}$, and we show that the linear map $\sigma_x(\Delta_p)$: $E_{p,x} \to E_{p,x}$ is injective, and therefore bijective. The equality

$$(\sigma_p^* \sigma_p + \sigma_{p-1} \sigma_{p-1}^*) a = 0, \qquad a \in E_{p,x}$$

implies, if we denote by $(\,|\,)$ the scalar product in the fibre $E_{p,x}$:

$$((\sigma_p^* \sigma_p + \sigma_{p-1} \sigma_{p-1}^*) a | a)$$

therefore

$$(\sigma_p a|\sigma_p a) + (\sigma_{p-1}^* a|\sigma_{p-1}^* a) = 0$$

and thus

$$\sigma_p a = 0, \qquad \sigma_{p-1}^* a = 0$$

from the exactness of the sequence of symbols we deduce that there exists $b \in E_{p-1,x}$ such that

$$a = \sigma_{p-1} b$$

and, since $\sigma_{p-1}^* a = 0$

$$\sigma_{p-1}^* \sigma_{p-1} b = 0$$

which in turn implies

$$(\sigma_{p-1}b, \sigma_{p-1}b) = 0, \qquad \sigma_{p-1}b = 0$$

i.e. $a = 0$.

b) We suppose A_p is injective, and $\sigma_p\sigma_{p-1} = 0$ and we prove $\ker \sigma_p = \operatorname{Im} \sigma_{p-1}$. We can always write

$$E_{p,x} = \operatorname{Im} \sigma_{p-1} \oplus \operatorname{Im} \sigma_p^* \oplus H_p.$$

The first two spaces are orthogonal by the property $\sigma_p\sigma_{p-1} = 0$, and H_p denotes their orthogonal complement on $E_{p,x}$. If $a \in H_p$ then $\sigma_{p-1}^* a = 0$ and $\sigma_p a = 0$, hence $A_p a = 0$, and $a = 0$ if A_p is injective. Since $\ker \sigma_p$ and $\operatorname{Im} \sigma_{p-1}$ are both the orthogonal complement of $\operatorname{Im} \sigma_p^*$ in $E_{p,x}$, they are identical. ∎

Decomposition theorem. If $\{D_p, E_p\}$ *is an elliptic complex over a compact manifold X each space* $C^\infty(E_p)$ (p. 396) *can be written as a direct, topological,* L^2 *orthogonal sum*

$$C^\infty(E_p) = \text{Range } D_{p-1} \oplus \text{Range } D_p^* \oplus \ker \Delta_p$$

i.e. each $f_p \in C^\infty(E_p)$ admits a unique decomposition, depending continuously on f_p:

$$f_p = D_{p-1}f_{p-1} + D_p^* f_{p+1} + h_p, \quad \text{with} \quad \Delta_p h_p = 0.$$

Proof: It follows formally the same lines that the proof given for the finite dimensional space $E_{p,x}$ given above. Its validity in the new context relies on the fact that the kernel of an elliptic operator on a compact manifold is finite dimensional. ∎

Example: Let E_p be the vector bundle of exterior forms of degree p on X; $C^\infty(E_p) = \Lambda^p(X)$ (p. 196) let $D_p = d$ be the operator of exterior differentiation. The sequence $\{E_p, D_p\}$ is a complex since $d^2 = 0$. It is called the **de Rham complex**. — de Rham complex

The adjoint D_p^*: $\Lambda^{p+1}(X) \to \Lambda^p(X)$ of d is the operator δ defined p. 296. The laplacian Δ_p of the complex is the laplacian on p-forms $d\delta + \delta d$ defined p. 318. Its symbol is the matrix $-g^{ij}\xi_i\xi_j I$, where I is the identity mapping $\Lambda_x^p(X) \to \Lambda_x^p(X)$, therefore Δ_p is an elliptic operator and the complex is elliptic. A p-form γ such that $\Delta_p\gamma = 0$ is called a **harmonic** ***p*-form**. — harmonic p-form

A p-form $\boldsymbol{\omega}$ on a compact manifold X can be written as a sum:

$$\boldsymbol{\omega} = d\boldsymbol{\alpha} + \delta\boldsymbol{\beta} + \boldsymbol{\omega}_h \quad \text{with} \quad \Delta\boldsymbol{\omega}_h = 0.$$

This decomposition goes back to Hodge and is known under the name **Hodge decomposition**. — Hodge decomposition

In the Hodge decomposition of $\boldsymbol{\omega}$ the harmonic form $\boldsymbol{\omega}_{\mathrm{h}}$, and the closed and coclosed forms $\mathrm{d}\boldsymbol{\alpha}$ and $\delta\boldsymbol{\beta}$ are determined in a unique way. If $\boldsymbol{\omega}$ is closed [resp. coclosed] then $\delta\boldsymbol{\beta} = 0$ [resp. $\mathrm{d}\boldsymbol{\alpha} = 0$], because $\boldsymbol{\omega}$ is then L^2-orthogonal to all coclosed [resp. closed] forms. Thus the harmonic component $\boldsymbol{\omega}_{\mathrm{h}}$ of $\boldsymbol{\omega}$ determines its cohomology class, and we have:

Theorem (Hodge). *The space of harmonic p forms on a compact manifold X is isomorphic to the cohomology space $H^p(X)$. In particular*

$$b_p = \dim \ker \Delta_p$$

where b_p is the p-th Betti number of X.

Cohomology of elliptic complexes. The cohomology space $H^p(X)$ is defined (cf. p. 223) as the quotient space of the space of closed p-forms under the equivalence relation $\boldsymbol{\omega}_1 \simeq \boldsymbol{\omega}_2$ if $\boldsymbol{\omega}_1 - \boldsymbol{\omega}_2 = \mathrm{d}\boldsymbol{\varphi}$. Since, in a general complex

$$\ker \mathrm{D}_p \supset \operatorname{Im} \mathrm{D}_{p-1}$$

one may also define a cohomology space $H^p(E, \mathrm{D})$ for the complex as

$$H^p(E, \mathrm{D}) = \ker \mathrm{D}_p / \operatorname{Im} \mathrm{D}_{p-1}.$$

The decomposition theorem proves that

$$\dim H^p(E, \mathrm{D}) = \dim \ker \Delta_p$$

because, as in the case of the de Rham complex, if $\mathrm{D}_p f_p = 0$ then $f_p = \mathrm{D}_{p-1} f_{p-1} + h_p$.

The Hodge theorem gives a relation between an analytic property of the differential operator Δ_p on X, the dimension of its kernel, and a topological property of X. A further relation is, since $\Sigma_p(-1)^p b_p = \chi(X)$, the Euler number of X,

$$\chi(X) = \sum_p (-1)^p \dim \ker \Delta_p.$$

It is a relation of this type that the Atiyah–Singer index theorem generalizes to an arbitrary elliptic complex.

analytical index of an elliptic complex

Definitions[1]: The **analytical index of an elliptic complex** $\{\mathrm{D}_p, E_p\}$ is

$$\operatorname{index} \{\mathrm{D}_p, E_p\} = \sum_p (-1)^p \dim \ker \Delta_p.$$

symbol bundle

To define the topological index one must define the **symbol bundle** $\Sigma(\mathrm{D})$

[1]Cf. Atiyah and Singer, Bull. A.M.S. 69 (1963) 422–433, [Palais 1965], Shanahan, Springer Lecture Notes 638, 1978.

relative to the elliptic complex, which is a complex vector bundle over the appropriately compactified cotangent bundle $\psi(X)$ of X.

The **topological index of the elliptic complex** is then

topological index of an elliptic complex

$$\text{top index}\{D_p, E_p\} = \int_{\psi(X)} \text{ch}(\Sigma(D)) \wedge \rho^* \text{tod}\, X$$

where $\text{ch}(\Sigma(D))$ is the Chern character of the symbol bundle, ρ the projection $\psi(X) \to X$, and $\text{tod}(X)$ the **Todd class** of the complexified tangent bundle of X, that is the element of the algebra of characteristic classes of this bundle corresponding to the polynomial $\Pi_{j=1}^{k} x_j/\{1-e^{-x_j}\}$. When the Euler class $e(X)$ of X is not zero, and has no zero divisor the formula for the topological index can be written:

Todd class

$$\text{top Ind}\{D_p, E_p\} = (-1)^{n(n+1)/2} \int_X \text{Ch}\,(\oplus(-1)^p E_p) \frac{\text{tod}(X)}{e(X)}.$$

Theorem (Atiyah–Singer). *If* $\{D_p, E_p\}$ *is an elliptic complex over a compact manifold* X *then*

$$\text{index}\{D_p, E_p\} = \text{top. index}\{D_p, E_p\}.$$

Remarks:

1. If the complex reduces to one elliptic operator $D: C^\infty(E_1) \to C^\infty(E_2)$ $\Delta_1 = D^*D$ and $\Delta_2 = DD^*$ are both elliptic, $\ker \Delta_1 = \ker D$ and $\ker \Delta_2 = \ker D^*$. The index of (E, D) coincides with the classical definition

$$\text{index}\,(E, D) = \dim \ker D - \dim \ker D^*$$

of the index of an elliptic operator. It is zero if the operator is self adjoint.

2. The index of an elliptic complex over an odd dimensional manifold is always zero.

Note that the **cokernel of** D is **coker** $D \equiv C^\infty(E_2)\backslash\text{range } D \simeq \ker D^*$.

A linear operator D is a **Fredholm operator** if ker D and coker D are finite dimensional. An elliptic operator on a compact manifold is a Fredholm operator.

cokernel of D

Fredholm operator

PROBLEMS AND EXERCISES

PROBLEM 1. THE GEOMETRY OF GAUGE FIELDS[1]

This problem develops results obtained in Section A and introduces

[1]Contributed by T. Jacobson and G. Sammelmann.

expressions seen in physics. The following notation is used. Let $P = (P, X, \pi, G)$ be a principal bundle over space-time X. Quantities defined on P: connection $\boldsymbol{\omega}$, curvature $\boldsymbol{\Omega}$. Indices i, j, k refer to trivializations ϕ_i, ϕ_j, ϕ_k, henceforth called local gauges; s_i is the section defined by the local gauge ϕ_i, $s_i = \phi_i^{-1} \circ \overline{\mathrm{Id}}$ (p. 364); the right action $\tilde{R}_g$ on P is written $\tilde{R}_g u = ug$.

Quantities defined on the base X: gauge potentials $\bar{\boldsymbol{\omega}}_i$ in the local gauge ϕ_i, $\bar{\boldsymbol{\omega}}_i = s_i^* \boldsymbol{\omega}$, gauge field $\bar{\boldsymbol{\Omega}}_i = s_i^* \boldsymbol{\Omega}$. To recover the standard expressions used in physics, one defines the electromagnetic field $F_{\mu\nu}$ by $-\frac{1}{2}ieF_{\mu\nu}\, \mathbf{d}x^\mu \wedge \mathbf{d}x^\nu = \bar{\boldsymbol{\Omega}}_i$, and the Yang–Mills field $F^a_{\mu\nu}$ by $\frac{1}{2}F^a_{\mu\nu}\boldsymbol{E}_a\, \mathbf{d}x^\mu \wedge \mathbf{d}x^\nu = \bar{\boldsymbol{\Omega}}_i$, with $\boldsymbol{E}_a \in \mathscr{G}$. Greek indices refer to coordinates on X, latin indices in the first part of the alphabet refer to coordinates on G. The transition functions are

$$g_{ij}(x) = \overset{\Delta}{\phi}_{i,x} \circ \overset{\Delta}{\phi}{}^{-1}_{j,x} \in G.$$

a) **Gauge principle and gauge field equations.** A brief historical note will serve to introduce the gauge principle in the simpler context of the early fifties. Protons and neutrons were known as different states of the same particles, the nucleons, and their interactions were explained in terms of π mesons (pions) exchange. The total number of nucleons was observed to be conserved in the course of their interactions. Noether's theorem[1] suggested that this conserved quantity corresponds to a symmetry under unitary "gauge" transformations of the nucleon field. That is, if one takes the nucleon field to be a doublet ψ of fields and describes the nucleon-pion interactions by the action of the antihermitian generators $i\boldsymbol{\sigma} = (i\sigma_1, i\sigma_2, i\sigma_3)$ of SU(2) on this doublet, then the lagrangian would be invariant under the gauge transformation[2] $\psi(x) \mapsto \exp(i\boldsymbol{\alpha}(x) \cdot \boldsymbol{\sigma})\psi(x)$. A lagrangian is constructed out of fields and their derivatives so that derivatives of $\boldsymbol{\alpha}(x)$ will in general appear when one performs this transformation. This is analogous to what happens when one subjects an ordinary derivative of a tensor field to a non-linear coordinate transformation. In that case one can define a covariant derivative using a linear connection in order to render the derivative a coordinate invariant concept. The same procedure must be followed in the case at hand if one is to construct a lagrangian invariant under local gauge transformations, only here the connection is not on the principal bundle of linear frames of the spacetime manifold but rather on a principal bundle over spacetime whose group is the **gauge group**, in this example SU(2). The pull back of the connection 1-form by a local section is a **gauge potential** and will in

gauge group
gauge potential

[1]To an invariance in the lagrangian one can associate a locally conserved four-vector current. One defines the charge as the integral over 3-space of the 4th component of the current. It follows from the current conservation that the time derivative of the charge vanishes. See for instance [Itzykson and Zuber, pp. 23–29].

[2]If $\boldsymbol{\alpha}(x)$ is a constant, this gauge transformation is called **global**, otherwise it is called **local**.

global
local
gauge transformation

the quantum field theory give rise to "gauge bosons". The pull back of the corresponding curvature is a **gauge field**. Sections of associated vector bundles are called **matter fields**. When the coupling of the gauge potential with the matter fields is exclusively through the **gauge covariant derivative** it is called a **minimal coupling**.

gauge field
matter field
gauge covariant derivative
minimal coupling

Inspired by electromagnetic theory in which the gauge potential is the electromagnetic vector potential, one assumes, in a gauge theory, that the gauge potential has a dynamics of its own, that is, that the lagrangian includes a kinetic energy term for the gauge potential. One requires this kinetic term to be independent of the matter fields, invariant under Lorentz and gauge transformations, and to contain terms quadratic in the derivatives of the potential but no higher. It is then determined uniquely up to a constant factor. The choice of this factor determines the coupling constant g and it would be natural to write the kinetic energy term $-\mathrm{Tr}F_{\mu\nu}F^{\mu\nu}/4g^2$ where $F_{\mu\nu}$, the pullback of the curvature form, is treated as taking values in the adjoint representation of the Lie algebra of the gauge group, rather than in the abstract Lie algebra. We shall, however, use the standard physics convention, factor g out of the potential, writing it as $g\boldsymbol{A}$, and write the Yang–Mills lagrangian

$$\mathscr{L}(\phi, \boldsymbol{A}) = \mathscr{L}_0(\mathrm{D}_\mu\phi, \phi) - \mathrm{Tr}F_{\mu\nu}F^{\mu\nu}/4$$

where ϕ represents the matter fields, D_μ the covariant derivative[1], and $F^{\mu\nu} = \eta^{\mu\rho}\eta^{\nu\sigma}F_{\rho\sigma}$ where $\boldsymbol{\eta}$ is the Minkowski metric.

The SU(2) gauge theory of strong interactions, suggested by Yang and Mills as an analogy to the $U(1)$ gauge theory of electromagnetism discovered by Weyl, turns out to be physically untenable. The gauge principle, however, together with the concept of hidden symmetry, has been used successfully in building a unified theory of electromagnetism and the weak interactions in which the gauge group is $\mathrm{SU}(2)\times U(1)$, and is guiding present attempts to understand and unify all the fundamental interactions.

Derive the Yang–Mills field vacuum equations. Show that dual and antiself-dual fields satisfy Yang–Mills vacuum equations.

Answer: In the absence of matter fields, the Yang–Mills lagrangian is

$$\mathscr{L} = -\tfrac{1}{4}\mathrm{Tr}\,F_{\mu\nu}F^{\mu\nu} = -\tfrac{1}{4}F^a_{\mu\nu}F^{\mu\nu}_a \quad \text{with} \quad F^a_{\mu\nu} = \partial_\mu A^a_\nu - \partial_\nu A^a_\mu + ec^a_{bc}A^b_\mu A^c_\nu.$$

[1]Labelled ∇_μ in the text. Physicists often use D_μ for gauge covariant derivatives, keeping ∇_μ for covariant derivatives in a linear connection.

The Yang–Mills field vacuum equations are (cf. Euler's equation, p. 78)

$$0 = \partial_\beta \frac{\partial \mathcal{L}}{\partial A^a_{\alpha,\beta}} - \frac{\partial \mathcal{L}}{\partial A^a_\alpha} = -\partial_\beta F^{\beta\alpha}_a + e c^b_{ca} A^c_\beta F^{\beta\alpha}_b = -\mathrm{D}_\beta F^{\beta\alpha}_a .$$

This can be shown to be the coordinate expression of the pullback of $\mathrm{D}^*\boldsymbol{\Omega} = 0$. (See p. 336, the calculation of $^*\boldsymbol{F}$ for the electromagnetic case.) If a field is self-dual or antiself-dual $^*\boldsymbol{\Omega} = \pm\boldsymbol{\Omega}$, the field equation $\mathrm{D}^*\boldsymbol{\Omega} = 0$ is satisfied by virtue of the Bianchi identity $\mathrm{D}\boldsymbol{\Omega} = 0$.

b) **Matter fields**. For simplicity we consider only scalar fields, referred to as Higgs fields, so that transformation behavior under spacetime coordinate changes is trivial. We now describe two equivalent ways in which one may view a matter field.

1. Let ρ: $G \to \mathrm{GL}(F)$ be a representation of the gauge group on the vector space F. F is usually chosen to be $\mathbb{C}^n$ for some n and ρ is chosen to be a unitary representation of G on $\mathbb{C}^n$. A **Higgs field** of type (ρ, F) may be defined as a section ψ of a vector bundle with typical fibre F associated with P by the representation ρ. See p. 367 for associated vector bundles, covariant derivatives of their sections and examples.

Higgs field

2. An equivalent definition of a **Higgs field**[1] of type (ρ, F) is a map $\tilde{\psi}$: $P \to F$ which is equivariant under G, i.e., $\tilde{\psi}(ug) = \rho(g^{-1})\tilde{\psi}(u)$ for all $u \in P$, $g \in G$.

Higgs field

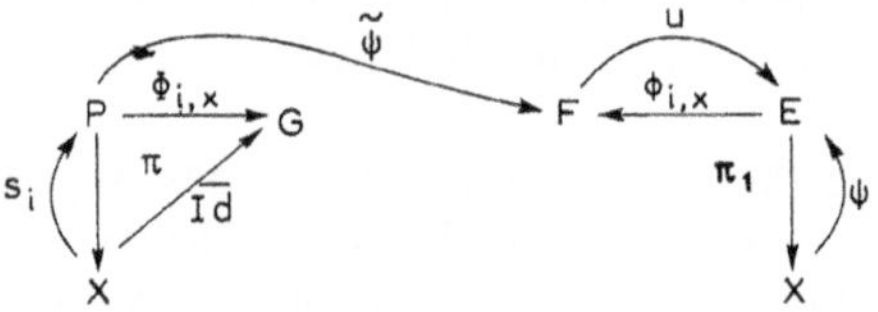

u is both an admissible map: $F \to \pi_1^{-1}(x)$ and a point in $\pi^{-1}(x)$. The representations ψ and $\tilde{\psi}$ of a Higgs field are related by $\tilde{\psi}(u) = u^{-1} \circ \psi(x)$, $x = \pi(u)$.

Show that the covariant derivative $\mathrm{D}\tilde{\psi}$: $TP \to F$ *defined by*

$$\mathrm{D}\tilde{\psi}(\boldsymbol{v}) \equiv \tilde{\psi}'(\mathrm{hor}\ \boldsymbol{v})$$

is essentially equivalent to the covariant derivative of a section of an associated vector bundle as defined via parallel transport (p. 369).

Answer[2]: A vector field $\boldsymbol{v}$ on P can be decomposed into its vertical and horizontal components by the connection one form $\boldsymbol{\omega}$

$$\boldsymbol{v} = (\boldsymbol{\omega}(\boldsymbol{v}))^* + \mathrm{hor}\ \boldsymbol{v}$$

[1][Trautman].
[2]Contributed by T. Jacobson.

where $(\boldsymbol{\omega}(\boldsymbol{v}))^*$ is the fundamental vector field defined by $\boldsymbol{\omega}(\boldsymbol{v}) \in \mathscr{G}$. According to the notation p. 360, if $\boldsymbol{\omega}(\boldsymbol{v}) = \hat{v}$, then $\hat{v}^* = \text{ver}\, \boldsymbol{v}$. Now, by definition

$$\mathrm{D}\tilde{\psi}(\boldsymbol{v}) \equiv \tilde{\psi}'(\boldsymbol{v} - (\boldsymbol{\omega}(\boldsymbol{v}))^*)$$

$$\mathrm{D}\tilde{\psi}_u(\boldsymbol{v}) = \tilde{\psi}'(\boldsymbol{v}) - \tilde{\psi}'\left(\frac{\mathrm{d}}{\mathrm{d}t}\tilde{R}_{g(t)}u\bigg|_{t=0}\right)$$

for $g(t) = \exp t\hat{v}$ where we have identified $\mathscr{G}$ and T_eG,

$$\begin{aligned}\tilde{\psi}'\left(\frac{\mathrm{d}}{\mathrm{d}t}\tilde{R}_{g(t)}u\bigg|_{t=0}\right) &= \frac{\mathrm{d}}{\mathrm{d}t}(\tilde{\psi}(\tilde{R}_{g(t)}u))\bigg|_{t=0}\\ &= \frac{\mathrm{d}}{\mathrm{d}t}(\rho(\exp(-t\hat{v}))\tilde{\psi}(u))\bigg|_{t=0} \qquad \text{since } \tilde{\psi} \text{ is equivariant}\\ &= \rho_e'(-\hat{v})\tilde{\psi}(u).\end{aligned}$$

Hence

$$\mathrm{D}\tilde{\psi}(\boldsymbol{v}) = \tilde{\psi}'(\boldsymbol{v}) + \rho_e'(\boldsymbol{\omega}(\boldsymbol{v}))\tilde{\psi}(u)$$

and the local expression on the manifold is

$$(s_i^*\mathrm{D}\tilde{\psi})(\boldsymbol{v}_x) = \tilde{\psi}'(s_i'\boldsymbol{v}_x) + \rho_e'(\boldsymbol{\omega}(s_i'\boldsymbol{v}_x))\tilde{\psi}(s_i(x)).$$

If we write $\psi = \tilde{\psi} \circ s_i$, and $s_i^*\mathrm{D}\tilde{\psi} \equiv \mathrm{D}\psi$, then

$$\mathrm{D}\psi(\boldsymbol{v}_x) = \psi'(\boldsymbol{v}_x) + \rho_e'(\bar{\omega}_i(\boldsymbol{v}_x))\psi(x)$$

which is to be compared with the expression obtained on p. 371. ■

c) **General gauge transformations**. A gauge transformation corresponds to a change of local sections or to a global automorphism of the principal bundle. The first viewpoint, sometimes called the passive viewpoint, has been developed in Section A. We shall develop the active viewpoint in this section and compare both viewpoints.

Let $\mathscr{F}$ be a **vertical automorphism** of P, i.e.,

vertical automorphism

$$\mathscr{F}\colon \pi^{-1}(x) \to \pi^{-1}(x), \qquad \mathscr{F}(ug) = \mathscr{F}(u)g, \qquad u \in P, \qquad ug = \tilde{R}_g u.$$

Transformations induced by vertical automorphisms are sometimes called **general gauge transformations**.

general gauge transformations

Let Φ_i and Φ_j be two different trivializations of $\pi^{-1}(U)$, $U \subset X$, chosen so that

$$\overset{\Delta}{\Phi}_j(u) = \overset{\Delta}{\Phi}_i(\mathscr{F}(u)).$$

active viewpoint
passive viewpoint

We can then consider $\overset{\Delta}{\Phi}_i(u)$ and $\overset{\Delta}{\Phi}_i(\mathscr{F}(u))$ as originating from two different points of P **(active viewpoint)**, or $\overset{\Delta}{\Phi}_i(u)$ and $\overset{\Delta}{\Phi}_j(u)$ as two different trivializations of the same point **(passive viewpoint)**.

1. **General gauge transformations of connections and potentials.**
Let $\mathscr{F}$ be a vertical automorphism of P and $\boldsymbol{\omega}$ a connection. Show that

$$\mathscr{F}^*\boldsymbol{\omega}(v) = \mathrm{Ad}(\check{\phi}(u)^{-1})\boldsymbol{\omega}(v) + \boldsymbol{\theta}_{\mathrm{MC}}(\check{\phi}'(u)v), \qquad v \in T_uP,$$

where $\check{\phi}$: $P \to G$ such that $\mathscr{F}(u) = u\check{\phi}(u)$. Set $\mathbf{A}_i = s_i^*\boldsymbol{\omega}$ and $\mathbf{A}_i' = s_i^*(\mathscr{F}^*\boldsymbol{\omega})$. Give the relationship between $\mathbf{A}_i$ and A_i'.

Answer: Since the right action of G acts freely on $\pi^{-1}(x)$, there exists a unique map $\check{\phi}$: $P \to G$ such that $\mathscr{F}(u) = u\check{\phi}(u)$. The vertical automorphism $\mathscr{F}$ is stated in terms of a group action on $\pi^{-1}(x)$. A proof similar to the proof of equation (4) (p. 364) gives the stated expression for $\mathscr{F}^*\boldsymbol{\omega}$. It follows that

$$A_i'(v) = \mathrm{Ad}((\check{\phi} \circ s_i(x))^{-1})A_i(v) + \boldsymbol{\theta}_{\mathrm{MC}}((\check{\phi} \circ s_i)'(x)v), \qquad v \in T_xX. \quad \blacksquare$$

2. **General gauge transformations of the local expressions of matter fields.**
In a physics equation, a matter field is usually given by its local expression in a given trivialization. A section s_i on P can be used to determine both a local trivialization on P and a local trivialization on E:

$$\overset{\Delta}{\boldsymbol{\Phi}}_{i,x} \circ s_i(x) = \overline{\mathrm{Id}}$$

$$\overset{\Delta}{\phi}_{i,x} \circ s_i(x) = \mathrm{Id}_F.$$

Then

$$\overset{\Delta}{\phi}_{i,x} \circ \psi(x) = \tilde{\psi}(s_i(x)).$$

Show that

$$\mathscr{F}^*\tilde{\psi}(s_i(x)) = \rho((\check{\phi}(s_i(x)))^{-1})\tilde{\psi}(s_i(x)).$$

Answer:

$$\mathscr{F}^*\tilde{\psi}(u) = \tilde{\psi}(\mathscr{F}(u)) = \tilde{\psi}(u\check{\phi}(u)) = \rho(\check{\phi}(u)^{-1})\tilde{\psi}(u). \quad \blacksquare$$

Compare general gauge transformations from the active and passive viewpoints.
Active viewpoint: two points, u, $\mathscr{F}u \in P$, one trivialization Φ_i

the section	$s_i \to s_i$
the trivialization of E	$\overset{\Delta}{\phi}_{i,x} \to \overset{\Delta}{\phi}_{i,x}$

the connection on P	$\boldsymbol{\omega} \to \mathscr{F}^* \boldsymbol{\omega}$
the potential	$s_i^* \boldsymbol{\omega} \to s_i^*(\mathscr{F}^* \boldsymbol{\omega})$
the matter field	$\tilde{\psi} \to \mathscr{F}^* \tilde{\psi}$.

Passive viewpoint: one point $u \in P$, two trivializations Φ_i, Φ_j such that $\overset{\Delta}{\Phi}_j(u) = \overset{\Delta}{\Phi}_i(\mathscr{F}(u))$

the section	$s_i(x) \to \mathscr{F}(s_i(x)) = s_i(x)\tilde{\phi}(s_i(x))$
the trivialization of E	$\overset{\Delta}{\phi}_{i,x} \to (\overset{\Delta}{\phi}(s_i(x)))^{-1}\overset{\Delta}{\phi}_{i,x}$
the connection on P	$\boldsymbol{\omega} \to \boldsymbol{\omega}$
the potential	$s_i^* \boldsymbol{\omega} \to (\mathscr{F} \circ s_i)^* \boldsymbol{\omega}$
the matter field	$\tilde{\psi} \to \tilde{\psi}$.

When can a potential be transformed to zero by a gauge transformation?

Answer: Let $\boldsymbol{A}_i$ and $\boldsymbol{A}_j$ be the potentials in the local gauges Φ_i and Φ_j corresponding to the same connection, then (Eq. (6), p. 365)

$$\boldsymbol{A}_i(\boldsymbol{v}) = g_{ji}^{-1}(x)\boldsymbol{A}_j(\boldsymbol{v})g_{ji}(x) + g_{ji}^{-1}(x)g'_{ji}(x)\boldsymbol{v}.$$

$\boldsymbol{A}_i(\boldsymbol{v}) \equiv 0$, iff $\boldsymbol{A}_j(\boldsymbol{v}) = -g'_{ji}(x)\boldsymbol{v}g_{ji}^{-1}(x)$. To know if a given potential $\boldsymbol{A}_j$ is of this form, we write down the integrability conditions for the equation $A_{j\mu}(x) = -\partial_\mu g_{ji}(x)g_{ji}^{-1}(x)$. Omitting the latin subscripts temporarily,

$$\begin{aligned} 0 = (\partial_\mu\partial_\nu - \partial_\nu\partial_\mu)g(x) &= \partial_\mu(A_\nu(x)g(x)) - \partial_\nu(A_\mu(x)g(x)) \\ &= (\partial_\mu A_\nu - \partial_\nu A_\mu + [A_\mu, A_\nu])g(x). \end{aligned}$$

In conclusion, a potential is locally gauge equivalent to zero if and only if the field $\boldsymbol{F}$ is zero, i.e., when the connection $\boldsymbol{A}$ is flat.

Remark: Note that when $\boldsymbol{F} \neq 0$, two potentials giving the same $\boldsymbol{F}$ are not in general locally gauge equivalent, unless the group is abelian. Therefore, in a non-abelian gauge theory there is more information in the gauge potentials than in the gauge fields.

Heuristic comment: Let $\phi(x)$ be some "multiplet field" which transforms tensorially under the action of a group G. Let us define parallel transport by a rule

$$\phi(x) \to \phi(x + \delta x) = \phi(x) + \boldsymbol{A}\delta x\phi(x)$$

where δx is an infinitesimal displacement and $\boldsymbol{A}\delta x$ is an operator on ϕ. If we consider x-dependent transformations of the field

$$\bar{\phi}(x) = g(x)\phi(x) \qquad \text{(strictly speaking } \bar{\phi}(x) = \rho(g(x))\phi(x))$$

the definition of parallel transport is "gauge independent" if

$$g(x)\phi(x) + \bar{A}\delta x g(x)\phi = g(x + \delta x)(\phi + A\delta x \phi),$$

i.e.,

$$\bar{A} = gAg^{-1} + g'g^{-1}.$$

References: H. Weyl, Z. Phys. 56 (1929) 330; C.N. Yang and R.L. Mills, "Conservation of Isotopic Spin and Isotopic Gauge Invariance", Phys. Rev. 96 (1954) 191–195; M. Ikeda and Y. Miyachi, "On an extended framework for the description of elementary particles", Prog. Theor. Phys. 16 (1956) 537–547; B.S. DeWitt, *Dynamical Theory of Groups and Fields* (Gordon and Breach, New York, 1965); R. Stora, "Continuum Gauge Theories", in *New Developments in Quantum Field Theory and Statistical Mechanics*, eds. M. Lévy and P. Mitter (Plenum Press, New York, 1977); A. Trautman, "The geometry of gauge fields", Czech. J. Phys. B29 (1979) 107–116; M. Daniel and C.M. Viallet, "The geometrical setting of gauge theories of the Yang–Mills type", Rev. Mod. Phys. 52 (1980) 175–197; L.D. Faddeev and A.A. Slavnov, *Gauge fields* (Benjamin, Reading MA, 1980); C. Itzykson and J.B. Zuber, *Quantum Field Theory* (McGraw-Hill, New York, 1980).

PROBLEM 2. CHARGE QUANTIZATION. MONOPOLES[1]

a) *Show that, in the context of a $U(1)$ gauge theory of electromagnetism, the discreteness of the unitary representations of $U(1)$ implies the quantization of electric charge.*
b) *Can Maxwell equations accommodate the existence of monopoles (isolated magnetic charges) which have not yet been observed?*
c) *Can the field $\boldsymbol{H}$ of a magnetic monopole be described by a vector potential $\boldsymbol{A}$?*
d) *Identify the monopole charges with the Chern numbers characterizing the principal $U(1)$ bundles over $\mathbb{R}^2 \times S^2$.*
e) *Show that the magnetic field defined by the following electromagnetic potentials is the field created by a monopole of charge n at the origin*:

$$\boldsymbol{A}_+ = \tfrac{1}{2}n(\cos\theta - 1)\,\mathbf{d}\boldsymbol{\phi} \quad \text{on} \quad U_+ = \mathbb{R}^4 - \{r = 0 \text{ or } z < 0\}$$

$$\boldsymbol{A}_- = \tfrac{1}{2}n(\cos\theta + 1)\,\mathbf{d}\boldsymbol{\phi} \quad \text{on} \quad U_- = \mathbb{R}^4 - \{r = 0 \text{ or } z > 0\}$$

where r, θ, ϕ are the polar coordinates on $\mathbb{R}^3$, $z = r\cos\theta$.

Answer:
a) Let ρ be a one-dimensional unitary representation of $U(1)$ on $\mathbb{C}$

$$\rho\colon U(1) \to U(1) \quad \text{by} \quad \exp it \mapsto \exp iat.$$

The condition $a(t + 2\pi) = at + 2\pi n$ where n is an integer gives $a = n$.
If we assume that the coupling of the electromagnetic field A_μ with the matter field ψ of type $(\rho, \mathbb{C})$ is the minimal coupling provided by the covariant derivative (p. 371).

[1]Contributed by Ted Jacobson.

$$D_\mu\psi = \partial_\mu\psi - \rho'_e eA_\mu\psi = \partial_\mu\psi - \mathrm{i}\, neA_\mu\psi,$$

then the charge of the matter field ψ is ne where e is fixed since the same electromagnetic potential couples to all charged fields. In general the minimal coupling of a gauge potential $s^*\omega$ with a matter field ψ is through the covariant derivative

$$(s^*D\psi)(v) = (\psi \circ s)'(v) + \rho'_e(s^*\omega(v))(\psi \circ s)(x), \quad v \in T_xX$$

so that the possible minimal couplings of different matter fields correspond to the possible choices of the Lie algebra homomorphism $\rho'(e)$: $\mathscr{G} \to \mathscr{U}(n)$, the group acting on ψ being restricted to $U(n)$.
b) A magnetic monopole of charge g at the origin of space would create a magnetic field $\boldsymbol{H}$ satisfying the equation $\operatorname{div} \boldsymbol{H} = 4\pi g\delta_0$. Maxwell's equations (p. 271) are satisfied on a space from which the origin has been deleted.
c) If the magnetic field is defined globally by a potential, i.e., if $\boldsymbol{H} = \operatorname{curl} \boldsymbol{A}$ (or $\boldsymbol{F} = \mathrm{d}\boldsymbol{A}$), then the magnetic flux through any surface S enclosing the monopole vanishes: if ∂S is empty $\int_S \boldsymbol{F} = \int_{\partial S} \boldsymbol{A} = 0$, i.e.

$$\int_S \boldsymbol{H} \cdot \mathrm{d}\boldsymbol{\sigma} = \int_S \operatorname{curl} \boldsymbol{A} \cdot \mathrm{d}\boldsymbol{\sigma} = \oint_C \boldsymbol{A} \cdot \mathrm{d}s - \oint_C \boldsymbol{A} \cdot \mathrm{d}s = 0$$

where C is any closed curve on S. Thus if the field of a magnetic monopole is to be described by a potential, at least one of the assumptions justifying this calculation must fail to hold:
In Dirac's description of a monopole[1], the string singularity in the potential invalidates the application of Stokes' Theorem in the above calculation.

Viewing the vector potential as given by a connection 1-form ω defined globally on a principal $U(1)$ bundle over spacetime provides an alternate mathematical description[2]: there is no globally defined $\boldsymbol{A}$ such that $\boldsymbol{F} = \mathrm{d}\boldsymbol{A}$. The bundle must be non-trivializable, since the existence of a global section (gauge) s would provide a way to define the potential $\boldsymbol{A} = s^*\omega$ globally on spacetime. All fibre bundles over a contractible paracompact base space are trivializable (p. 131); we have rendered spacetime non-contractible, however, by deleting the world line of the pole. Since, all principal fibre bundles (over a paracompact base) with a contractible structural group are trivializable (p. 385), a magnetic monopole can be described only if we choose $U(1)$ as the electromagnetic gauge group as opposed to $(\mathbb{R}, +)$.

[1][Dirac 1931, 1948].
[2][Lubkin] and [Yang 1977].

To construct a $U(1)$ bundle appropriate for the description of a particular magnetic pole we define potentials on patches of an open cover of $\mathbb{R}^4$-{pole world line} in such a way that on the overlaps the potentials differ by gauge transformations, and so that the magnetic field defined at any given point is the field that would arise from a pole located at the origin. These gauge transformations can be used to define the transition functions of a principal $U(1)$ bundle over $\mathbb{R}^4$-{pole world line}; the potentials are then interpreted as pull-backs by local sections of a globally defined connection 1-form on the bundle.

Since $\mathbb{R}^4$-{line} is homeomorphic to $\mathbb{R}^2 \times S^2$, and $\mathbb{R}^2$ is contractible, classification of such bundles reduces to classification of $U(1)$ bundles over S^2. Covering S^2 by two patches which overlap in an arbitrarily thin band around the equator one sees that the choice of transition functions essentially amounts to the choice of a continuous map $S^1 \to U(1)$. The bundles are thus classified by the homotopically different ways one can choose this map, i.e., by the **winding numbers**. One can show that other choices for the open covering of S^2 lead to no new possibilities.
d) The k-th Chern class $C_k(E)$ is the cohomology class of γ_k given on p. 393:

$$\gamma_0 = 1, \qquad \gamma_1 = -(2\pi\mathrm{i})^{-1}\,\mathrm{tr}\,\bar{\boldsymbol{\Omega}}, \quad \gamma_2 = \tfrac{1}{2}(2\pi\mathrm{i})^{-2}((\mathrm{tr}\,\bar{\boldsymbol{\Omega}})^2 - \mathrm{tr}(\bar{\boldsymbol{\Omega}} \wedge \bar{\boldsymbol{\Omega}})), \ldots$$

Here

$$\bar{\boldsymbol{\Omega}} = \mathrm{i}\boldsymbol{F}, \qquad \gamma_1 = -\boldsymbol{F}/2\pi, \qquad \gamma_n = 0 \quad \text{for} \quad n \geq 2.$$

The Chern number

$$C_1 = \int_{S^2} \gamma_1 = -\frac{1}{2\pi}\int_{U_+} \boldsymbol{F}_+ - \frac{1}{2\pi}\int_{U_-} \boldsymbol{F}_- = -\frac{1}{2\pi}\int_{S^1} (\boldsymbol{A}_+ - \boldsymbol{A}_-) = \frac{1}{2\pi}\int_{S^1} n\,\mathbf{d}\boldsymbol{\phi}$$
$$= n.$$

Here U_+ and U_- have a "very thin" overlap, namely their boundary S^1.
e) On $U_+ \cap U_-$, the gauge potentials $\mathrm{i}\boldsymbol{A}_+$ and $\mathrm{i}\boldsymbol{A}_-$ are related by a gauge transformation, $\mathrm{i}\boldsymbol{A}_+ = \mathrm{i}\boldsymbol{A}_- - \mathrm{i}n\,\mathbf{d}\boldsymbol{\phi}$. The electromagnetic potentials $\boldsymbol{A}_+$ and $\boldsymbol{A}_-$ define the following electromagnetic field which satisfies the required conditions,

$$\boldsymbol{F} = -\tfrac{1}{2}n \sin\theta\, \mathbf{d}\boldsymbol{\theta} \wedge \mathbf{d}\boldsymbol{\phi} \qquad \text{on} \qquad U_+ \cup U_- = \mathbb{R}^4 - \{\text{pole world line}\}$$
$$= \tfrac{1}{2}nr^{-3}(x\,\mathbf{dz} \wedge \mathbf{dy} + y\,\mathbf{dx} \wedge \mathbf{dz} + z\,\mathbf{dy} \wedge \mathbf{dx})$$

in cartesian coordinates. Note (p. 263)

$$\boldsymbol{H} = n\boldsymbol{r}/r^3.$$

References: P.A.M. Dirac, "Quantized singularities in the electromagnetic field", Proc. Roy. Soc. A133 (1931) 60–72; "The theory of magnetic poles", Phys. Rev. 74 (1948) 817–830; E. Lubkin, "Geometric definition of gauge invariance", Ann. Phys. 23 (1963) 233–283.

PROBLEM 3. INSTANTON SOLUTION OF EUCLIDEAN SU(2) YANG–MILLS THEORY (CONNECTION ON A NONTRIVIAL SU(2) BUNDLE OVER S^4)

(a) *Construct the stereographic coordinates on the 4 sphere. Notation:* $U_+ \equiv S^4 - \{\text{North pole}\}$, $U_- \equiv S^4 - \{\text{South pole}\}$, $(\rho_\pm, \Omega_\pm) \equiv$ *stereographic coordinates on* U_+, U_- *respectively, where* $\Omega = (\theta, \phi, \psi)$ *are the Euler angles* (pp. 181–190). $a \equiv$ *radius of* S^4 *when embedded in* $\mathbb{R}^5$.
(b) *Construct an* SU(2) *bundle over* S^4. *Let* $M(s)$, $s \in U_+ \cap U_-$ *be the transition function corresponding to a rotation of Euler angles* $\Omega = (\theta, \phi, \psi)$ *by the correspondence defined p.* 188. *Compute the Maurer–Cartan form* $\boldsymbol{\theta}_{\mathrm{MC}}(M(s))$ *as a function of* Ω. *Compute* $\bar{\boldsymbol{\theta}}_{\mathrm{MC}}(M(s)) = \mathrm{Ad}\,(M(s))\boldsymbol{\theta}_{\mathrm{MC}}(M(s))$. *Let the basis on* $\mathscr{SU}(2)$ *be* $(\mathrm{i}\sigma_\alpha/2)$ *where* (σ_α) *are the Pauli matrices defined on p.* 183.
(c) *Show that the following potentials are related by a gauge transformation:*

$$\begin{cases} \boldsymbol{A}_+(s)(\boldsymbol{v}) = -4a^2(\rho^2 + 4a^2)^{-1} M'(s)(\boldsymbol{v}) M^{-1}(s) & \text{on } U_+ \\ \boldsymbol{A}_-(s)(\boldsymbol{v}) = \rho^2(\rho^2 + 4a^2)^{-1} M^{-1}(s) M'(s)(\boldsymbol{v}) & \text{on } U_-. \end{cases}$$

Compute the Yang–Mills field $\boldsymbol{F}_+$ *and* $\boldsymbol{F}_-$ *given by* $\boldsymbol{A}_+$ *and* $\boldsymbol{A}_-$ *respectively. Do* $\boldsymbol{F}_+$ *and* $\boldsymbol{F}_-$ *satisfy the Yang–Mills field equations? Consider the same questions for*

$$\begin{cases} \bar{\boldsymbol{A}}_+(s)(\boldsymbol{v}) = -\rho^2(\rho^2 + 4a^2)^{-1} M(s) M^{-1\prime}(s)(\boldsymbol{v}) \\ \bar{\boldsymbol{A}}_-(s)(\boldsymbol{v}) = 4a^2(\rho^2 + 4a^2)^{-1} M^{-1\prime}(s)(\boldsymbol{v}) M(s). \end{cases}$$

(d) *Compute the Chern numbers characterizing the* SU(2) *bundles over* S^4.

Answer:
(a) **Geometry of the 4-sphere** (see also p. 355). Let $(u^1, \ldots, u^5) \in \mathbb{R}^5$. The metric $\bar{g}$ on S^4 induced by the euclidean metric $\Sigma_{i=1}^5 (\mathbf{d}u^i)^2$ on $\mathbb{R}^5$ is

$$\bar{g} = a^2(a^2 - r^2)^{-1}\, \mathbf{d}r^2 + r^2\, \mathbf{d}\Omega^2 \tag{1}$$

where a is the radius of S^4, $\Sigma_{i=1}^5 (u^i)^2 = a^2$, and where (r, Ω) are some polar coordinates on $\mathbb{R}^4$. We call **polar coordinates** on $\mathbb{R}^4$ any system of coordinates such that, if $r^2 = \Sigma_{i=1}^4 (u^i)^2$ and if Ω stands for a triple of angles, the pull back of the euclidean metric η on $\mathbb{R}^4$ under the change of coordinates F: $(r, \Omega) \mapsto (u^1, \ldots, u^4)$ is $f^*\eta = \mathbf{d}r^2 + r^2\, \mathbf{d}\Omega^2$. polar coordinates

Proof of equation (1): on S^4, $(u^5)^2 = a^2 - r^2$, hence $\bar{g} = f^*\eta + (\mathbf{d}\sqrt{a^2 - r^2})^2 = a^2(a^2 - r^2)^{-1}\, \mathbf{d}r^2 + r^2\, \mathbf{d}\Omega^2$, $r < a$. Now let k: $(\rho, \Omega) \mapsto (r, \Omega)$ by $r = 4a^2\rho(4a^2 + \rho^2)^{-1}$, then

$$g = k^*\bar{g} = (1 + \rho^2/4a^2)^{-2}\, (\mathbf{d}\rho^2 + \rho^2\, \mathbf{d}\Omega^2).$$

stereographic coordinates

(ρ, Ω) are **stereographic coordinates**: See figure, let $s \in S^4$. Let (ρ, Ω) be the coordinates of the intersection of Ns with $\mathbb{R}^4$ and (r, Ω) be the coordinates of the vertical projection of s. Then $(\rho - r)/\rho = (a \pm (a^2 - r^2)^{1/2})/2a$, hence $r = 4a^2\rho(4a^2 + \rho^2)^{-1}$. We can take for Ω the Euler angles (pp. 181–190) (θ, ϕ, ψ) which parametrize the 3-sphere.

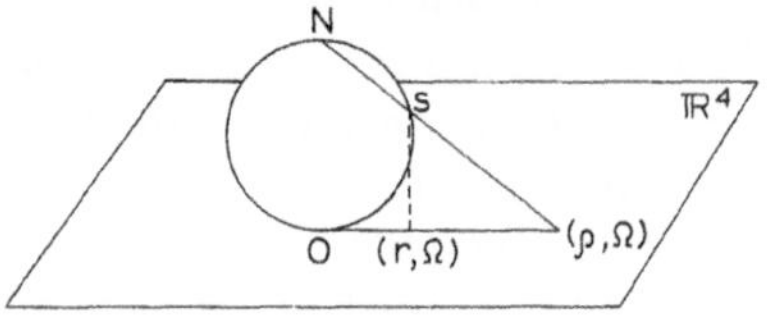

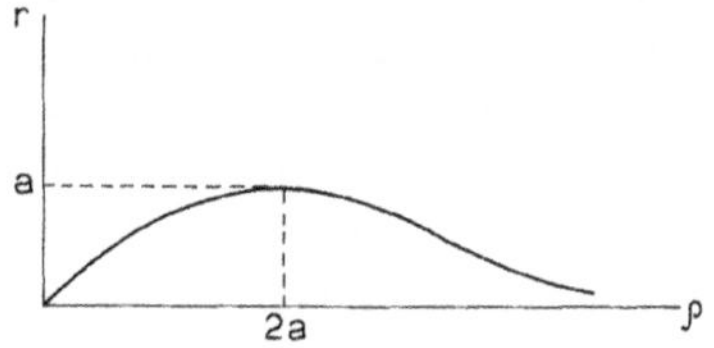

$$f\colon (r, \Omega) \mapsto (u^1, \ldots, u^4) \text{ by } \begin{cases} u^1 = r \cos \tfrac{1}{2}\theta \cos \tfrac{1}{2}(\psi + \phi) \\ u^2 = r \cos \tfrac{1}{2}\theta \sin \tfrac{1}{2}(\psi + \phi) \\ u^3 = r \sin \tfrac{1}{2}\theta \sin \tfrac{1}{2}(\psi - \phi) \\ u^4 = r \sin \tfrac{1}{2}\theta \cos \tfrac{1}{2}(\psi - \phi) \end{cases}$$

and

$$\mathbf{d}\boldsymbol{\Omega}^2 = \tfrac{1}{4}(\mathbf{d}\boldsymbol{\theta}^2 + \mathbf{d}\boldsymbol{\phi}^2 + \mathbf{d}\boldsymbol{\psi}^2 + 2\cos\theta\, \mathbf{d}\boldsymbol{\phi}\, \mathbf{d}\boldsymbol{\psi}).$$

(b) **SU(2) bundle over S^4**. Let ϕ_+ and ϕ_- define the trivializations on $\pi^{-1}(U_+)$ and $\pi^{-1}(U_-)$ respectively and let the transition function $\overset{\Delta}{\phi}_+(s) \circ \overset{\Delta}{\phi}{}_-^{-1}(s) = M(s) \in \mathrm{SU}(2)$, $s \in U_+ \cap U_-$. The matrix M is of the form

$$M = \begin{pmatrix} a + ib & c + id \\ -c + id & a - ib \end{pmatrix}.$$

If $a = u^1/r$, $b = u^2/r$, $c = u^3/r$, $d = u^4/r$, M corresponds to a rotation of Euler angles (θ, ϕ, ψ) by the correspondence defined p. 188.
Note that M is a mapping on $U_+ \cap U_-$ which is parametrized by $(\rho, \theta, \phi, \psi)$ but M is independent of ρ. It is convenient to consider $M \in \mathrm{SU}(2)$ as an element of the matrix group Q defined p. 183. Let $(\mathbb{1}, \rho_1, \rho_2, \rho_3)$ be the basis given on p. 183 for the real vector space $Q \cup \{0\}$; then

$$M = a\mathbb{1} + b\rho_1 + c\rho_2 + d\rho_3.$$

We can write

$$M^{-1}(s)M'(s)v = M^{-1}(s)\,\mathbf{d}M(v).$$

A straightforward calculation gives $\boldsymbol{\theta}_{\mathrm{MC}}(M) = M^{-1}\,\mathbf{d}M$ and $\bar{\boldsymbol{\theta}}_{\mathrm{MC}}(M) = \mathbf{d}MM^{-1}$.

$$M^{-1}\,\mathbf{d}M = \mathrm{i}\sigma_3(a\,\mathbf{d}b - b\,\mathbf{d}a - c\,\mathbf{d}d + d\,\mathbf{d}c) + \mathrm{i}\sigma_2(-a\,\mathbf{d}c - b\,\mathbf{d}d + c\,\mathbf{d}a + d\,\mathbf{d}b) + \mathrm{i}\sigma_1(a\,\mathbf{d}d - b\,\mathbf{d}c + c\,\mathbf{d}b - d\,\mathbf{d}a).$$

With $a = u^1/r$, $b = u^2/r$, $c = u^3/r$ and $d = u^4/r$, we obtain $M^{-1}\,\mathbf{d}M(\theta, \phi, \psi) = \theta^\alpha_{\mathrm{MC}}(\theta, \phi, \psi)\mathrm{i}\sigma_\alpha/2$

$$\text{where}\quad \begin{pmatrix}\theta^1_{\mathrm{MC}}\\ \theta^2_{\mathrm{MC}}\\ \theta^3_{\mathrm{MC}}\end{pmatrix} = \begin{pmatrix}\cos\psi & \sin\theta\sin\psi & 0\\ -\sin\psi & \sin\theta\cos\psi & 0\\ 0 & \cos\theta & 1\end{pmatrix}\begin{pmatrix}\mathbf{d}\theta\\ \mathbf{d}\phi\\ \mathbf{d}\psi\end{pmatrix}.$$

Similarly $\mathbf{d}M\,M^{-1} = \mathrm{i}\sigma_3(-b\,\mathbf{d}a + a\,\mathbf{d}b - d\,\mathbf{d}c + c\,\mathbf{d}d) + \mathrm{i}\sigma_2(c\,\mathbf{d}a - d\,\mathbf{d}b - a\,\mathbf{d}c + b\,\mathbf{d}d) + \mathrm{i}\sigma_1(-d\,\mathbf{d}a - c\,\mathbf{d}b + b\,\mathbf{d}c + a\,\mathbf{d}d).$

With a, b, c, d given as above, $\mathbf{d}M\,M^{-1}(\theta, \phi, \psi) = \bar{\theta}^\alpha_{\mathrm{MC}}(\theta, \phi, \psi)\mathrm{i}\sigma_\alpha/2$

$$\text{where}\quad \begin{pmatrix}\bar{\theta}^1_{\mathrm{MC}}\\ \bar{\theta}^2_{\mathrm{MC}}\\ \bar{\theta}^3_{\mathrm{MC}}\end{pmatrix} = \begin{pmatrix}\cos\phi & 0 & \sin\theta\sin\phi\\ \sin\phi & 0 & -\sin\theta\cos\phi\\ 0 & 1 & \cos\theta\end{pmatrix}\begin{pmatrix}\mathbf{d}\theta\\ \mathbf{d}\phi\\ \mathbf{d}\psi\end{pmatrix}.$$

(c) **Instanton solutions**. Let ω be a connection on the SU(2) bundle and let A_+ and A_- be the pull-backs of ω on U_+ and U_- defined by the trivializations ϕ_+ and ϕ_-. Then

$$A_-(s) = M^{-1}(s)A_+(s)M(s) + M^{-1}(s)\,\mathbf{d}M(s).$$

Thus if we choose a solution $A_+ = \alpha\,\mathbf{d}MM^{-1}$, then $A_- = (1+\alpha)M^{-1}\,\mathbf{d}M$. The given potentials are precisely of this form, hence they are related by a gauge transformation.

The field strengths are

$$\begin{aligned} F_+ &= \mathbf{d}A_+ + [A_+, A_+] \qquad \text{on } U_+ \\ &= \frac{4\mathrm{i}a^2}{\rho^2+4a^2}\left(\frac{2\rho}{\rho^2+4a^2}\,\mathbf{d}\rho \wedge \sigma_\alpha\bar{\theta}^\alpha_{\mathrm{MC}} - \sigma_\alpha\,\mathbf{d}\bar{\theta}^\alpha_{\mathrm{MC}}\right) \\ &\quad + \left(\frac{-4\alpha^2}{\rho^2+4\alpha^2}\right)^2 4\left[\frac{\mathrm{i}\sigma_\alpha}{2}\,\bar{\theta}^\alpha_{\mathrm{MC}}, \frac{\mathrm{i}\sigma_\beta}{2}\,\bar{\theta}^\beta_{\mathrm{MC}}\right]. \end{aligned}$$

Note $\mathbf{d}\bar{\theta}^\alpha_{\mathrm{MC}} = \frac{1}{2}c^\alpha_{\beta\gamma}\bar{\theta}^\beta_{\mathrm{MC}} \wedge \bar{\theta}^\gamma_{\mathrm{MC}}$ with $c^\alpha_{\beta\gamma} = \epsilon_{\alpha\beta\gamma}$, and express the commutation in terms of the exterior product (p. 374) to obtain

$$F_+ = \frac{4\mathrm{i}a^2}{(\rho^2+4a^2)^2}(2\rho\,\mathbf{d}\rho \wedge \bar{\theta}^\alpha_{\mathrm{MC}} - \rho^2 c^\alpha_{\beta\gamma}\bar{\theta}^\gamma_{\mathrm{MC}} \wedge \bar{\theta}^\beta_{\mathrm{MC}})\sigma_\alpha$$

with

$$c^\alpha_{\beta\gamma} = \epsilon_{\alpha\beta\gamma}.$$

$\boldsymbol{F}_-$ is computed similarly using $\mathbf{d}\boldsymbol{\theta}^\alpha = -\frac{1}{2}c^\alpha_{\beta\gamma}\boldsymbol{\theta}^\beta \wedge \boldsymbol{\theta}^\gamma$ (Maurer Cartan equation, p. 208)

$$\boldsymbol{F}_- = \frac{4\mathrm{i}a^2}{(\rho^2+4a^2)^2}(2\rho\, \mathbf{d}\boldsymbol{\rho} \wedge \boldsymbol{\theta}^\alpha - \rho^2 c^\alpha_{\beta\gamma}\boldsymbol{\theta}^\beta \wedge \boldsymbol{\theta}^\gamma)\boldsymbol{\sigma}_\alpha.$$

The Pauli matrices ($\boldsymbol{\sigma}_\alpha$) have been chosen (p. 183) so that ($\mathrm{i}\boldsymbol{\sigma}_\alpha/2$) form a base for $\mathcal{SU}(2)$ with $c^\alpha_{\beta\gamma} = \epsilon_{\alpha\beta\gamma}$. If one uses the matrices originally given by Pauli, one can use ($-\mathrm{i}\boldsymbol{\sigma}_\alpha/2$) as a basis with the same structure constant, but the signs of $\boldsymbol{F}_\pm$ differ from the ones obtained here. Note that, although $\boldsymbol{A}_+$ and $\boldsymbol{A}_-$ are obtained one from the other by a gauge transformation, $\boldsymbol{F}_+ \neq \boldsymbol{F}_-$.

In the absence of matter fields, the Yang–Mills equation reduces to $\mathrm{D} * \boldsymbol{F} = 0$ where $*\boldsymbol{F}$ is the dual of $\boldsymbol{F}$ (p. 295). We can check that $\boldsymbol{F}_\pm$ are self-dual, $\boldsymbol{F}_\pm = *\boldsymbol{F}_\pm$, hence $\boldsymbol{F}_\pm$ satisfy the Yang–Mills equation by virtue of the Bianchi identities $\mathrm{D}\boldsymbol{\Omega} = 0$.

We could equally well have worked with the transition functions M^{-1}, and choose $\bar{\boldsymbol{A}}_-(s) = \beta(s)\,\mathbf{d}\boldsymbol{M}^{-1}M$, then $\bar{\boldsymbol{A}}_+(s) = (1+\beta(s))M\,\mathbf{d}\boldsymbol{M}^{-1}$. If we choose $\beta(s) = -(1+\alpha(s))$, then $\bar{\boldsymbol{A}}_- = \boldsymbol{A}_-$ but if we choose $\beta(s) = -\alpha(s)$ we obtain the new gauge related potentials given in the problem. The corresponding field $\bar{\boldsymbol{F}}_\pm$ is **antiself-dual**, $\bar{\boldsymbol{F}}_\pm = -*\bar{\boldsymbol{F}}_\pm$. It satisfies also the Yang–Mills equation by virtue of the Bianchi identity.

antiself-dual

Note that when ρ tends to infinity, $\boldsymbol{A}_+$ tends to 0 and $\boldsymbol{A}_-$ tends to $\boldsymbol{\theta}_{\mathrm{MC}}$. A potential equal to or gauge related to the Maurer–Cartan form, up to a multiplicative constant, is called **pure gauge**. It can be shown that if the potential is pure gauge at infinity, the field tends to zero at infinity.

pure gauge

The solutions $\boldsymbol{F}_\pm$ and $\bar{\boldsymbol{F}}_\pm$ are called the **Yang–Mills instantons** or **Yang–Mills pseudoparticles.** It is not expected that solutions of the Yang–Mills field on S^4 rather than on spacetime have a direct classical interpretation but it has been shown that they appear in the study of the quantized Yang–Mills field.

instantons
pseudoparticles

(d) The k-th Chern class $C_k(E)$ is the cohomology class of γ_k given on p. 393. Here $\bar{\boldsymbol{\Omega}} = \boldsymbol{F}_\pm$ and $\operatorname{tr}\bar{\boldsymbol{\Omega}} = 0$ since $\operatorname{tr}\boldsymbol{\sigma}_\alpha = 0$. The only non vanishing γ_k is $\gamma_2 = (8\pi^2)^{-1}\operatorname{tr}\boldsymbol{F}\wedge\boldsymbol{F}$. The second Chern number

$$C_2 = \int_{S^4}\gamma_2 = \frac{1}{8\pi^2}\int_{U^+}\operatorname{tr}\boldsymbol{F}_+\wedge\boldsymbol{F}_+ + \frac{1}{8\pi^2}\int_{U^-}\operatorname{tr}\boldsymbol{F}_-\wedge\boldsymbol{F}_-.$$

It can be shown that $\operatorname{tr}(\boldsymbol{\Omega}\wedge\boldsymbol{\Omega}) = \mathrm{d}\operatorname{tr}(\boldsymbol{\Omega}\wedge\boldsymbol{\omega} - \frac{1}{3}(\boldsymbol{\omega})^3)$ where $(\boldsymbol{\omega})^3 \equiv \boldsymbol{\omega}\wedge\boldsymbol{\omega}\wedge\boldsymbol{\omega}$. Thus, by Stokes theorem,

$$C_2 = \frac{1}{8\pi^2}\int_{S^3} \mathrm{tr}(\boldsymbol{F}_+ \wedge \boldsymbol{A}_+ - \tfrac{1}{3}(\boldsymbol{A}_+)^3) - \mathrm{tr}(\boldsymbol{F}_- \wedge \boldsymbol{A}_- - \tfrac{1}{3}(\boldsymbol{A}_-)^3).$$

Let M^k be any transition function where M^k is the k-th power of the transition function M of paragraph b; then

$$\boldsymbol{A}_- = (M^k)^{-1}\boldsymbol{A}_+ M^k + (M^k)^{-1}\,\mathbf{d}M^k, \qquad \boldsymbol{F}_- = (M^k)^{-1}\boldsymbol{F}_+ M^k$$

and

$$C_2 = \frac{1}{24\pi^2}\int_{S^3} \mathrm{tr}((M^k)^{-1}\,\mathbf{d}M^k)^3.$$

It is easy to see that C_2 is proportional to k. For $k = 1$, with ψ, φ, θ the Euler angles,

$$C_2 = \frac{1}{24\pi^2}\int_{S^3} \mathrm{tr}\left(\boldsymbol{\theta}^\alpha_{\mathrm{MC}}\frac{\mathrm{i}\boldsymbol{\sigma}_\alpha}{2}\right)^3 = -\frac{1}{16\pi^2}\int_0^{4\pi}\mathrm{d}\psi\int_0^{2\pi}\mathrm{d}\varphi\int_0^{\pi}\sin\theta\,\mathrm{d}\theta = -1.$$

Hence $C_2 = -k$.
k is called the **instanton number**. instanton number

Atiyah, Hitchin and Singer have proved that the space of self dual SU(2) Yang–Mills fields over S^4, modulo the gauge group action, is a manifold of dimension $8k - 3$ where k is the instanton number.

References: A.A. Balavin, A.M. Polyakov, A.S. Schwartz and Yu.S. Tyupkin, "Pseudoparticle Solutions of the Yang–Mills Equations", Phys. Lett. 59B (1975) 85–87; M.F. Atiyah, N.J. Hitchin and I.M. Singer, "Deformations of Instantons", 3 Proc. Natl. Acad. Sci. USA 74 (1977) 2662–2663; R. Stora, "Yang–Mills Instantons, Geometrical Aspects", in *Invariant Wave Equations*, eds. G. Velo and A.S. Wightman, Lecture Notes in Physics 73 (Springer-Verlag, Berlin, 1978); T. Eguchi, P.B. Gilkey and A.J. Hanson, "Gravitation, Gauge Theories and Differential Geometry", Physics Reports 66 (1980) 213–393, and references therein to the uses of instantons in physics by 't Hooft, Jackiw and others.

PROBLEM 4. SPIN STRUCTURE; SPINOR FIELDS; SPIN CONNECTION* (see Problems I 1, p. 64 and III 2, p. 172)

Introduction: Spinors arise in physics from the desire to have a Lorentz invariant linear first order differential operator. The Klein–Gordon operator, $\Box + m^2$, is Lorentz invariant but of second order. It can be

*Because of its nature, this problem is not set in the usual question and answer format. It has been written in collaboration with Ted Jacobson.

written as a product of first order operators

$$\eta^{\alpha\beta}\,\partial^2/\partial x^\alpha\,\partial x^\beta + m^2 = -(\mathrm{i}\gamma^\alpha\,\partial/\partial x^\alpha + m)\,(\mathrm{i}\gamma^\beta\,\partial/\partial x^\beta - m)$$
$$\eta^{\alpha\beta} = \mathrm{diag}(1, -1, -1, -1)$$

provided that

$$\gamma^\alpha\gamma^\beta + \gamma^\beta\gamma^\alpha = 2\eta^{\alpha\beta}\,\mathbb{1}.$$

This condition cannot be satisfied by any vector since the determinant of the left hand side is then always zero. However (p. 65) the γ^α generate the Clifford algebra $C(V^4_{(1)})$ (the Dirac algebra). It is a fundamental result (see Riesz) that the Clifford algebra $C(V^{2n})$ over the complex numbers is isomorphic to the algebra M^c_n of all $2^n \times 2^n$ matrices with complex elements. It follows that the Dirac algebra is isomorphic to M^c_2. The isomorphism is not canonical. It will be used to construct a representation of the Dirac algebra on $\mathbb{C}^4$. In this representation

$$\boldsymbol{e}_\alpha \mapsto \gamma_\alpha, \quad \Lambda \mapsto \Lambda,$$

where $(\boldsymbol{e}_\alpha)$ is an orthonormal basis for $V^4_{(1)}$ and $\Lambda \in C(V^4_{(1)})$. The elements $\psi \in \mathbb{C}^4$ are said to be the components of spinors. The equation

$$\mathrm{i}\eta^{\alpha\beta}\gamma_\beta\,\partial\psi/\partial x^\alpha - m\psi = 0$$

is invariant under the Lorentz transformation (p. 66, p. 290)

$$\psi(x^\alpha) \mapsto \psi'(a^\alpha{}_\beta x^\beta) = \Lambda\psi(x^\alpha), \qquad (x^\alpha) \mapsto (a^\alpha{}_\beta x^\beta), \qquad [a^\alpha{}_\beta] \in L(4)$$

if $\Lambda\gamma_\beta\Lambda^{-1} = a^\alpha{}_\beta\gamma_\alpha$, i.e., if $\Lambda \in \mathrm{Spin}(4)$ (see p. 67).

1) Spin structure

a) The subleties involved in defining spinors arise from the fact that there is no unique choice of spinor transformation Λ corresponding to a given Lorentz transformation $[a^\alpha_\lambda]$. That is, there is no basis for choosing $\psi \mapsto \Lambda\psi$ rather than $\psi \mapsto (-\Lambda)\psi$ in defining the spinor transformation corresponding to a given Lorentz transformation. The map $\mathscr{H}$: Spin(4)→ $L(4)$ by $\Lambda \mapsto [a^\alpha{}_\beta]$ is a 2–1 homomorphism (p. 68).
To simplify the discussion we begin by considering the map

$$\mathscr{H}\colon \mathrm{Spin}_0(4) \to L_0(4)$$

where $L_0(4)$ is the proper Lorentz group and $\mathrm{Spin}_0(4) \subset \mathrm{Spin}(4)$ is its simply connected [universal] covering group, henceforth called the **proper spin group**. If $[a^\alpha{}_\beta] \in L_0(4)$ and $\mathscr{H}(\Lambda) = \mathscr{H}(-\Lambda) = [a^\alpha{}_\beta]$ then Λ and $-\Lambda$ can be thought of as corresponding to the homotopically different paths connecting $[a^\alpha{}_\beta]$ with the identity in $L_0(4)$.

proper spin group

To resolve the ambiguity in defining the transformation rule for a spinor field, we might choose a fiducial field of oriented Lorentz frames. Then given the components of a spinor relative to the fiducial frame we may uniquely define the components relative to any other frame once a path joining the two frames is given, or rather once the homotopy class of the path is given.

b) A field of oriented Lorentz frames on a space time X^4 is a section of a bundle of oriented Lorentz frames, i.e., a section of a reduction of the frame bundle $F(X^4)$ to the proper Lorentz group $L_0(4)$. Giving such a reduction is equivalent to giving a globally defined minkowskian metric tensor field (i.e., a metric of signature $+, -, -, -$), together with a choice of space and time orientation relative to this metric. Note that the reduction may not exist and if it exists, it may not have a global section.

A convenient concept for defining a spinor field is a spin structure. Given a bundle of oriented Lorentz frames $L_0(X^4)$, a (proper) **spin structure**, if it exists, is a principal bundle $\mathrm{Spin}_0(X^4)$ over X^4 of group $\mathrm{Spin}_0(4)$, together with a 2 to 1 bundle homomorphism spin structure

$$\tilde{\mathscr{H}}: \mathrm{Spin}_0(X^4) \to L_0(X^4)$$

which maps a fibre at x into a fibre at x with

$$\tilde{\mathscr{H}}(u\Lambda) = \tilde{\mathscr{H}}(u)\,\mathscr{H}(\Lambda) \qquad \text{for all } u \in \mathrm{Spin}_0(X^4) \text{ and } \Lambda \in \mathrm{Spin}_0(4).$$

c) If one wishes to define spinor fields with respect to the full Lorentz group, this definition must be generalized. There are 8 different simply connected covering groups of $L(4)$ which correspond to the various combinations of signs for P^2, T^2, $(PT)^2$ where P [respectively T] is one of the two spin transformations corresponding to spatial reflection [time reflection] (see p. 68). Hence one can consider 8 different types of spin structures. Spin(4) defined p. 67 is such that $P = \pm e_0$ and $T = \pm e_1e_2e_3$, hence such that $P^2 = 1$, $T^2 = 1$, $PTPT = -1$. Each situation would be handled differently, depending on the topology of the base X^4 and the structure desired for the vector bundle of spinors. We shall treat (p. 421) an example explicitly (spinor field on a Möbius band $\otimes \mathbb{R} \otimes \mathbb{R}$).

d) A proper spin structure can be defined for orientable ["space" and time orientable] spaces X of arbitrary dimension n. Consider an $\mathrm{SO}(n)$-bundle $[\mathrm{SO}(n-1,\ 1)$-bundle$]$ of oriented orthonormal [Lorentzian] frames. Let $\mathrm{Spin}_0(n)$ $[\mathrm{Spin}_0(n-1,\ 1)]$ denote the simply connected covering group of $\mathrm{SO}(n)$ $[\mathrm{SO}(n-1,\ 1)]$ and $\mathscr{H}$ denote the 2 to 1 covering homomorphism. A proper spin structure on X, if it exists, is a principal bundle over X of group $\mathrm{Spin}(n)$ $[\mathrm{Spin}(n-1,\ 1)]$ together with a 2 to 1 spin bundle

bundle homomorphism into the SO(n)-bundle [SO($n-1$, 1)-bundle] which is fibre preserving and group equivariant with respect to $\mathcal{H}$.

e) Existence of a spin structure. A spin structure can always be defined if the orthonormal frame bundle admits a global section. The converse is not in general true. In some cases a spin structure allows one to define spinor fields when there is no global section of the Lorentz bundle. One can show that the necessary and sufficient condition for the existence of a spin structure is the vanishing of the second Stiefel–Whitney class of X. For a general discussion see [Lichnerowicz 1964, 1967] and [Borel and Hirzebruch].
Geroch has proved that for a bundle of oriented Lorentz frames over a noncompact four-dimensional manifold X^4, a proper spin structure exists if and only if the bundle admits a global section.

f) "Generalized spin structure". In cases where no spin structure exists, one may define a generalized spin structure. In this scheme one makes topological room for a "spin structure" by mixing in with the group of a spin bundle an "internal" symmetry group, which is inextricably involved in the generalized spinor transformation rule. See for instance [Avis and Isham].

2) Spinor fields

Let $\text{Spin}(X) \to L(X)$ be a spin structure and let ρ be a representation of Spin(4) on a vector space E. One can define a **spinor field** ψ of type (ρ, E) on X^4 as a section of a vector bundle of typical fibre E associated with Spin(X) via the representation ρ. Alternatively one can define a **spinor field** ψ of type (ρ, E) on Spin(X) as an equivariant map ψ from Spin(X) to E such that

spinor field

spinor field

$$\psi(u\Lambda) = \rho(\Lambda^{-1})\psi(u) \qquad \text{for all } u \in \text{Spin}(X) \text{ and } \Lambda \in \text{Spin}(4).$$

Example 1: Contravariant spinor, covariant spinor, spin-tensors. A spinor field of type $(\rho(\Lambda) = \Lambda, \mathbb{C}^4)$ is called a **contravariant spinor field**. A spinor field ϕ of type $(\rho(\Lambda) = (\Lambda^{-1})^{\sim}, \mathbb{C}^{4\prime})$, where $\sim$ denotes transposition and $\mathbb{C}^{4\prime}$ is the dual of $\mathbb{C}^4$, is called a **covariant spinor field**. Let (e_A) and (e^A) be dual bases in $\mathbb{C}^4$ and $\mathbb{C}^{4\prime}$ respectively, the scalar

contravariant

covariant

$$\langle \phi, \psi \rangle = \sum_{A=1}^{4} \phi_A \psi^A$$

is invariant in the sense $\langle (\Lambda^{-1})^{\sim}\phi, \Lambda\psi \rangle = \langle \phi, \psi \rangle$. One can define similarly **spin tensor** fields as spinor fields of type $(\rho, \mathbb{C}^{4n} \otimes \mathbb{C}^{4p'})$ where $\rho(\Lambda)$ is a $4n \times 4n$ matrix constructed by tensor products of Λ and $\Lambda^{-1\sim}$ (see example p. 371).

spin tensor

Example 2: tensor-spinors. A **tensor-spinor** is a spinor with values in a tensor space. For example a tensor spinor at x is an element of $\otimes^p T_x X \otimes^q T_x^* X \otimes^P \mathbb{C}_x^4 \otimes^Q \mathbb{C}_x^{4}$.

3) Spin connection and covariant derivative

a) To define the covariant derivative of a spinor field one needs a spin connection, i.e., a connection on Spin(X).

Theorem. Given a spin structure $\tilde{\mathscr{H}}$: Spin(X)→$L(X)$, a connection on $L(X)$ determines a connection on Spin(X).

Proof: Although $\tilde{\mathscr{H}}$ is two to one, $\tilde{\mathscr{H}}'$: T(Spin(X))→$T(L(X))$ is an isomorphism (p. 177) on each fibre. Thus we can define the horizontal spaces of T(Spin(X)) to be the inverse under $\tilde{\mathscr{H}}'$ of those of $T(L(X))$. Since $\tilde{\mathscr{H}}$ commutes with the right group action, the horizontal spaces so defined will define a connection. ∎

Let $\boldsymbol{\omega}$: $TL(X) \to \mathscr{L}(4)$ be a connection 1-form on $L(X)$. The corresponding connection $\boldsymbol{\sigma}$ on Spin(X)

$$\boldsymbol{\sigma}: T\mathrm{Spin}(X) \to Spin(4)$$

is defined by the composition

$$\boldsymbol{\sigma} = \mathscr{H}'^{-1}(e) \circ \tilde{\mathscr{H}}^* \boldsymbol{\omega}: T(\mathrm{Spin}(X)) \to \mathscr{L}(4) \to Spin(4).$$

b) If we think of a spinor field ψ of type (ρ, E) as an equivariant map from Spin(X) to E, we can write (p. 405) the pull back of the **covariant derivative** of ψ by the section s: $X \to$ Spin(X) pull back of covariant derivative

$$s^*\nabla\psi(v) = (\psi \circ s)'(v) + \rho'(e)((s^*\boldsymbol{\sigma})(v))(\psi \circ s)(x)$$
$$s^*\boldsymbol{\sigma} = s^*(\mathscr{H}'^{-1}(e) \circ \tilde{\mathscr{H}}^*\boldsymbol{\omega}) = \mathscr{H}'^{-1}(e) \circ (\tilde{\mathscr{H}} \circ s)^*\boldsymbol{\omega}$$

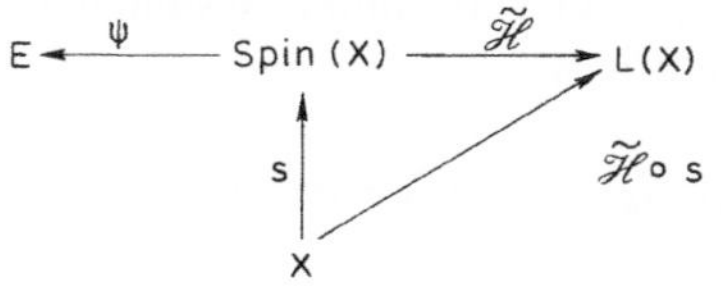

c) Components of a covariant derivative.
Assume that s has been chosen so that $\tilde{\mathscr{H}} \circ s$: $(x^\alpha) \mapsto (\boldsymbol{e}_\alpha)$ and assume that $\boldsymbol{\omega}$ is a connection on $L(X)$. Then

$$(\tilde{\mathscr{H}} \circ s)^*\boldsymbol{\omega} = \gamma^\alpha_{\mu\beta} E^\beta_\alpha \boldsymbol{\theta}^\mu$$

where, in keeping with the notations of p. 301, $\gamma^{\alpha}_{\mu\beta}$ are the connection coefficients with respect to the dual frames (e_α), (θ^α) – the indices on γ clearly distinguish it from the Dirac matrices $[\gamma^{B}_{\mu A}]$. The matrix elements of the matrix $E^{\beta}_{\alpha} \in \mathscr{L}$ are $(E^{\beta}_{\alpha})^{\mu}_{\nu} = \delta^{\mu}_{\alpha}\delta^{\beta}_{\nu}$.
$\mathscr{H}'^{-1}(e)$ has been computed p. 177. For $V \in \mathscr{L}$, i.e. $V = V^{\alpha}_{\beta}E^{\beta}_{\alpha}$

$$\mathscr{H}'^{-1}(e)(V) = -\tfrac{1}{4}V^{\alpha\beta}e_\alpha e_\beta = -\tfrac{1}{8}V^{\alpha\beta}(e_\alpha e_\beta - e_\beta e_\alpha) \text{ since } V^{\alpha\beta} = -V^{\beta\alpha}$$
$$= -\tfrac{1}{8}V^{\alpha}{}_{\beta}[e_\alpha, e^\beta]^{A}_{B}E^{B}_{A} \text{ where } E^{B}_{A} \in Spin(4) \text{ and } e^\beta = \eta^{\beta\gamma}e_\gamma.$$

Hence

$$s^*\sigma = \mathscr{H}'^{-1}(e) \circ (\tilde{\mathscr{H}} \circ s)^*\omega = -\tfrac{1}{8}\gamma^{\alpha}_{\mu\beta}[e_\alpha, e^\beta]^{A}_{B}E^{B}_{A}\theta^\mu.$$

If ψ is a spinor field of type $(\rho(\Lambda) = \Lambda, \mathbb{C}^4)$, i.e., a contravariant spinor field, then

$$\rho'(e)\colon T_e \operatorname{Spin}(4) \to T_1L(\mathbb{C}^4; \mathbb{C}^4) \text{ by } [e_\alpha, e_\beta] \mapsto [\gamma_\alpha, \gamma_\beta]$$

and finally

$$(s^*\nabla\psi(v))^A = ((\psi \circ s)'(v))^A - \tfrac{1}{8}\gamma^{\beta}_{\mu\alpha}[\gamma_\beta, \gamma^\alpha]^{A}_{B}(\psi \circ s)^B(x)v^\mu,$$

or with $\Psi = \psi \circ s$

$$\nabla_\mu\Psi^A = e_\mu(\Psi^A) - \tfrac{1}{8}\gamma^{\beta}_{\mu\alpha}[\gamma_\beta, \gamma^\alpha]^{A}_{B}\Psi^B$$
$$= e_\mu(\Psi^A) + \sigma^{A}_{\mu B}\Psi^B \quad \text{with} \quad \sigma_\mu = -\tfrac{1}{8}\gamma^{\beta}_{\mu\alpha}[\gamma_\beta, \gamma^\alpha]$$

where $e_\mu(\Psi^A)$ is the Pfaff derivative (p. 138).
For a covariant spinor field $\rho'(e)\colon \mathrm{d}\Lambda(t)/\mathrm{d}t|_{t=0} \mapsto -(\mathrm{d}\Lambda(t)/\mathrm{d}t)^{\sim}|_{t=0}$ hence the covariant derivative of a covariant spinor field ϕ is

$$\nabla_\mu\Phi_A = e_\mu(\Psi_A) - \sigma^{B}_{\mu A}\Phi_B.$$

Similarly (see detail p. 371) one obtains the covariant derivatives of any spin tensor or tensor-spinor.

Exercise: Compute the components of the covariant derivative of a tensor-spinor field $\gamma^{A}_{\alpha B}(x)$ in a coordinate system where $[\gamma^{A}_{\alpha B}(x)]$ is equal to the constant Dirac matrices.

Answer: $\nabla_\mu\gamma^{A}_{\alpha B} = e_\mu(\gamma^{A}_{\alpha B}) - \gamma^{\beta}_{\mu\alpha}\gamma^{A}_{\beta B} + \sigma^{A}_{\mu C}\gamma^{C}_{\alpha B} - \sigma^{C}_{\mu B}\gamma^{A}_{\alpha C}$.

To compute the Pfaff derivative we need to express it in terms of the coordinate derivatives, i.e. to change from the components defined by the orthonormal frames to those defined by the natural coordinates. Let $e_\mu = a^{\rho}_{\mu}\,\partial/\partial x^\rho$, then $e_\mu(\gamma^{A}_{\alpha B}) = a^{\rho}_{\mu}\,\partial(\gamma^{A}_{\alpha B})/\partial x^\rho = 0$. Recall (p. 308) that with respect to an orthonormal basis,

$$\gamma_\mu^{\beta\alpha} = -\gamma_\mu^{\alpha\beta}, \text{ since } \gamma_{\mu\alpha}^{\beta}\boldsymbol{\theta}^\mu \text{ is a 1-form with values in } \mathscr{L}(4).$$

A straightforward calculation gives $\nabla_\mu \gamma_{\alpha B}^{A} \equiv 0$. The origin of this equation is $\nabla_\mu g_{\alpha\beta} = 0$, the spin connection $\boldsymbol{\sigma}$ having been defined from a metric connection (p. 381).

d) The **laplacian** $\Box$ on a spinor is usually defined laplacian

$$\Box\psi = -\gamma^\alpha\gamma^\beta\nabla_\alpha\nabla_\beta\psi = -\tfrac{1}{2}(\gamma^\alpha\gamma^\beta + \gamma^\beta\gamma^\alpha)\nabla_\alpha\nabla_\beta\psi - \tfrac{1}{2}(\gamma^\alpha\gamma^\beta)(\nabla_\alpha\nabla_\beta - \nabla_\beta\nabla_\alpha)\psi$$
$$= (-\nabla^\alpha\nabla_\alpha + \tfrac{1}{4}R)\psi$$

where R is the Riemann curvature scalar (p. 310). Note that, if ϕ is a covariant spinor and ψ a contravariant spinor

$$\langle\Box\phi, \psi\rangle = \langle\phi, \Box\psi\rangle.$$

The Klein–Gordon equation for spinors is then

$$(-\nabla^\alpha\nabla_\alpha + \tfrac{1}{4}R + m^2)\psi = 0.$$

Example: A spinor field on a non-orientable manifold.[1]
Let $X = (\text{Möb} \times \mathbb{R}) \times \mathbb{R}$ = "space" × "time" where Möb is a Möbius band. X is not orientable, so no reduction of the frame bundle $F(X)$ to the proper Lorentz group L_0 exists. There is, however a reduction of $F(X)$ to $L_{0,s} = L_0 \cup sL_0$ where s denotes spatial reflection. Let $L(X)$ denote such a reduction.
We now define a spin structure for $L(X)$. The group of the spin bundle is a simply connected covering group $\bar{L}_{0,s}$ of $L_{0,s}$. There are two inequivalent choices for this covering in which the square of the two preimages of s equals $+1$ or -1. Let us choose for example

$$\mathscr{H}: \bar{L}_{0,s} \to L_{0,s}$$

where $\mathscr{H}$ is the restriction to $\bar{L}_{0,s}$ of the mapping defined p. 68. Then $(\mathscr{H}^{-1}(s))^2 = e_0^2 = 1$.
Let V, W be the two disjoint intersections of the two patches U_1 and U_2 covering the Möbius band (p. 126). We can take the transition functions $\eta(x) \in L_{0,s}$ to be given by

$$\eta(x) = \begin{cases} \mathbb{1} & \text{for } x \in V \times \mathbb{R} \times \mathbb{R} \\ s_2 & \text{for } x \in W \times \mathbb{R} \times \mathbb{R} \end{cases}$$

where $s_2(\boldsymbol{e}_1, \boldsymbol{e}_2, \boldsymbol{e}_3, \boldsymbol{e}_0) = (\boldsymbol{e}_1, -\boldsymbol{e}_2, \boldsymbol{e}_3, \boldsymbol{e}_0)$.
In this simple example we may define a principal bundle Spin(X) of

[1][B.S. DeWitt, Hart, Isham, p. 208].

group $L_{0,s}$ by requiring its transition functions ϕ to satisfy

$$\eta = \mathcal{H} \circ \phi.$$

Now $\mathcal{H}^{-1}(s_2) = \pm e_0 e_1 e_3$ since the equations $\Lambda e_2 \Lambda^{-1} = -e_2$ and $\Lambda e_\alpha \Lambda = e_\alpha (\alpha \neq 2)$ imply $\Lambda = \pm e_0 e_1 e_3$. Let us choose

$$\phi(x) = \begin{cases} \mathbb{1} & \text{for} \quad x \in V \times \mathbb{R} \times \mathbb{R} \quad (-\mathbb{1} \text{ is another possible choice}) \\ e_0 e_1 e_3 & \text{for} \quad x \in W \times \mathbb{R} \times \mathbb{R}. \end{cases}$$

Let $S(X)$ be the associated vector bundle of Spin(X) by the representation $(\rho(\Lambda) = \Lambda, \mathbb{C}^4)$ of $\tilde{L}_{0,s}$. Let (θ_1, U_1) and (θ_2, U_2) be a trivialization of $S(X)$. The transition functions are

$$\overset{\Delta}{\theta}_{1,x} \circ \overset{\Delta}{\theta}_{2,x} = \begin{cases} \mathbb{1} & \text{for} \quad x \in V \times \mathbb{R} \times \mathbb{R} \\ \gamma_0 \gamma_1 \gamma_3 & \text{for} \quad x \in W \times \mathbb{R} \times \mathbb{R}. \end{cases}$$

Let ψ be a section of $S(X)$. The coordinates of $\psi(x)$ are

$$\theta_{1,x} \circ \psi(x) \in \mathbb{C}^4 \text{ for } x \in U_1 \text{ and } \theta_{2,x} \circ \psi(x) \in \mathbb{C}^4 \text{ for } x \in U_2.$$

Hence, if we set $\overset{\Delta}{\theta}_{1,x} \circ \psi(x) = (\psi^A(x))$ and $\overset{\Delta}{\theta}_{2,x} \circ \psi(x) = (\bar{\psi}^A(x))$

$$\begin{aligned} \psi^A(x) &= \bar{\psi}^A(x) && \text{for} \quad x \in V \times \mathbb{R} \times \mathbb{R} \\ \psi^A(x) &= (\gamma_0 \gamma_1 \gamma_3)^A_B \bar{\psi}^B(x) && \text{for} \quad x \in W \times \mathbb{R} \times \mathbb{R}. \end{aligned}$$

References: M. Riesz, *Clifford Numbers and Spinors*, University of Maryland, Institute for Fluid Mechanic and Applied Mathematics, Lecture series, no. 38; [Choquet–Bruhat 1968]; A. Lichnerowicz, Bull. Soc. Math. France **92** (1964) 11–100; A. Lichnerowicz, article in *Battelle Rencontres 1967*, eds. C.M. DeWitt and J.A. Wheeler (Benjamin, 1968); A. Lichnerowicz, Anticommutateur du Champ Spinoriel en Relativité Générale, Comptes Rendus Acad. Sci. **9** (1961) 3742–44; A. Borel and F. Hirzebruch, Am. J. Math. **81** (1959) 315–382; R. Geroch, Spinor Structure of Space-times in General Relativity, J. Math. Phys. **9** (1968) 1739–1744; B.S. DeWitt, C.F. Hart and C.J. Isham, Topology and Quantum Field Theory, Physica **96A** (1979) 197–211; S.J. Avis and C.J. Isham, Generalized spin structures on four-dimensional space times

For an interesting presentation of spinor calculus based on two-components spinors see F.A.E. Pirani, Introduction to gravitational radiation theory, in *Lectures in General Relativity*, Vol. I, eds. A. Trautman, F.A.E. Pirani and H. Bondi (Prentice Hall, 1965) pp. 249–373; and R. Penrose, Structure of Space Time, in *Battelle Rencontres* 1967, eds. C.M. DeWitt and J.A. Wheeler (Benjamin, 1968).

T. Regge, "The group manifold approach to unified gravity" in *Relativité, Groupes et Topologie II*, eds. B.S. DeWitt and R. Stora (North-Holland, 1984), pp. 933–1006.

VI. DISTRIBUTIONS

A. TEST FUNCTIONS

1. SEMINORMS

The topology of a vector space is not always given by a norm; we shall devise a concept more general than a norm which will lead to topological vector spaces more general than normed spaces.

A **seminorm** on a vector space X over $\mathbb{K}$ is a mapping $p: X \to \mathbb{R}$ such that seminorm

$p(x+y) \leq p(x) + p(y)$ triangle inequality (subadditivity)
$p(\lambda x) = |\lambda| p(x)$ scale property.

As was the case for a norm (cf. p. 26) the definition implies that $p(0) = 0$ and $p(x) \geq 0$ for all x. Unlike a norm, a seminorm does not have the property that $p(x) = 0$ implies $x = 0$.

Examples: 1) The mapping $p: \mathbb{R}^n \to \mathbb{R}$ by $p(x) = |x^i|$ is a seminorm on $\mathbb{R}^n$. We can anticipate that, not the seminorm $p(x) = |x^i|$ but rather the family $\mathscr{P}$ of seminorms $\{p_i(x) = |x^i|;\ i = 1, \ldots n\}$ is the natural notion to introduce with a view toward a generalisation of the concept of norm. Indeed, it will be shown that the topology induced by $\mathscr{P}$ is equivalent to the Euclidean topology.

2) More generally let X be a vector space on $\mathbb{K}$. Then any linear form F on X defines a seminorm by $|\langle F, x\rangle| = |F(x)| = p(x)$.
In a separable Hilbert space $\mathscr{H}$, coordinates of a vector $f \in \mathscr{H}$ in the basis $\{g_i\}$ are defined by the scalar products $(g_i|f)$; each mapping $f \mapsto |(g_i|f)|$ is a seminorm on $\mathscr{H}$.

3) Let U be a bounded open subset of $\mathbb{R}^n$. We denote by $C^k(\bar{U})$ the space of k-times continuously differentiable functions on U which admit, together with their derivatives of order $\leq k$, continuous and bounded extensions to the closure $\bar{U}$ of U. The mapping $C^k(\bar{U}) \to \mathbb{R}$ by $f \mapsto \int_U \sum_{|j|=k} |D^j f|\, dx$ is a seminorm. It is not a norm if $k \neq 0$. $C^k(\bar{U})$

The existence, on a vector space X, of families of seminorms defined by linear forms on X raises the converse question of the existence, on a vector space X with a seminorm, of linear forms bounded by this seminorm. The Hahn–Banach theorem states the existence of such linear forms – provided they are known on one of the subspaces of X.

Hahn–Banach theorem. Let X be a vector space on $\mathbb{K}$, $p(x)$ a seminorm on X, Y a linear subspace of X and f a linear form on Y such that

$$|f(x)| \leq p(x) \qquad \forall x \in Y.$$

Then there exists a linear form F on X such that

$$F(x) = f(x) \qquad \forall x \in Y \quad \textit{and} \quad |F(x)| \leq p(x) \qquad \forall x \in X.$$

This extension is in general not unique; we shall, however, later use and prove the theorem in cases where the extension is unique (p. 425).

open ball

Topology defined by a family of seminorms. Let $\mathcal{P} = \{p_i; i \in I\}$ be a family of seminorms on the vector space X. A $\mathcal{P}$-**open ball** of center x_0 is the set of points $x \in X$ which satisfy a *finite** number of inequalities

$$p_i(x - x_0) < \epsilon_i \qquad \epsilon_i > 0 \quad i \in I.$$

$\mathcal{P}$-topology

Theorem. The $\mathcal{P}$-open balls form a base for a topology on X.

Proof: One can readily check a) that the subsets of X which consist of the unions and the finite intersections of $\mathcal{P}$-open balls are a system of subsets suitable to define a topology (p. 11); b) that the $\mathcal{P}$-open balls are a base for the topology defined by this system. This topology is called the $\mathcal{P}$**-topology.** ■

A $\mathcal{P}$-topology possesses those properties of a norm topology which are implied by the triangle inequality and the scale property, for instance:

Theorem. A $\mathcal{P}$-topology on a vector space X makes it a locally convex topological vector space. Conversely it can be proved that the topology of a locally convex vector space can always be defined by a family of seminorms.

If a $\mathcal{P}$-topology is defined by one seminorm p, which is not a norm, there exists $x \neq 0$ such that $p(x) = 0$. All such points x are in *every* neigh-

*This number has to be finite for the same reason that a base for a product topology (p. 20) contains only a finite number of proper open subsets of the constituent spaces.

borhood of the origin; hence these points are not separated from the origin and the seminorm p does not define a Hausdorff topology. A family of seminorms, on the other hand, can define a Hausdorff topology if all the seminorms do not vanish simultaneously at any point except the origin.

More precisely, a family of seminorms $\mathcal{P}$ on X is said to be **separated** if for every $x \neq 0$ in X there exists a seminorm $p \in \mathcal{P}$ such that $p(x) \neq 0$. separated family

Theorem. A separated family of seminorms defines a Hausdorff $\mathcal{P}$-topology.

Proof: Let $x, y \in X$, $x - y \neq 0$. Then there exists a seminorm p such that $p(x-y) \neq 0$, i.e. such that $p(x-y) > \epsilon$ for some $\epsilon > 0$. The neighborhoods of x and y defined respectively by $p(z-x) < \epsilon/2$ and $p(z-y) < \epsilon/2$ are disjoint because the three inequalities satisfied by the seminorms of $x-y$, $z-x$ and $z-y$ are incompatible with the triangle inequality

$$p(x-z+z-y) \leq p(z-x) + p(z-y).$$

Conversely, for the $\mathcal{P}$-topology to be Hausdorff the $\mathcal{P}$-family must be separated: if there were an $x \in X$ such that $p(x) = 0$ for every $p \in \mathcal{P}$, then every $\mathcal{P}$-ball of center 0, and hence every neighborhood of 0, would contain x; thus 0 and x would not be separated. ∎

The next theorem is a particular case of the Hahn Banach theorem.

Extension theorem. Let f be a linear form on a subspace Y of the vector space X. Let X be given a $\mathcal{P}$-topology, let Y be dense in X and f be continuous on Y for the topology induced by X. Then, there exists a unique continuous linear form F on X which extends f.

Proof: We shall now construct F. For simplicity we shall consider metrizable spaces where continuity can be stated in terms of sequences (p. 25). Because F is continuous on X, it must, if it exists, satisfy the equation $F(x) = \lim_{n=\infty} f(x_n)$ for every x in X and every sequence $\{x_n\}$ in Y which converges to x. Let $\{x_n\}$ be a sequence in Y which converges to x in the $\mathcal{P}$-topology; for each seminorm p_i of $\mathcal{P}$ and every $\epsilon_i > 0$

$$p_i(x_n - x_m) < \epsilon_i \quad \text{as soon as } n, m > N_i.$$

The continuity of f on Y for the $\mathcal{P}$-topology means that for every ϵ there exists a finite set of seminorms p_i, $i \in I$, and numbers $\eta_i > 0$, such that, if $x \in Y$,

$$p_i(x) < \eta_i,\ i \in I \quad \text{implies} \quad |f(x)| < \epsilon$$

or, in other words (cf. below an analogous deduction)

$$|f(x)| < M \sup_{i \in I} p_i(x).$$

Thus

$$|f(x_n - x_m)| = |f(x_n) - f(x_m)| < M \sup_{i \in I} p_i(x_n - x_m),$$

hence

$$|f(x_n) - f(x_m)| < \epsilon \quad \text{if} \quad n, m > N = \operatorname*{Max}_{i \in I} N_i.$$

The numbers $f(x_n)$ form a Cauchy sequence in $\mathbb{K}$, they have a unique limit which is the desired value $F(x)$. The linearity of F is an easy consequence of the linearity of f. ■

metrizable $\mathcal{P}$-topology

Theorem. Let $\mathcal{P}$ be a separated, finite or countable family of seminorms on X. The space X with the $\mathcal{P}$-topology is metrizable, that is its topology can be defined by a metric. If $\mathcal{P}$ is finite, the $\mathcal{P}$-topology is equivalent to the topology defined by the norm $\Sigma_i p_i$.

Proof: 1) If $\mathcal{P}$ is finite

$$p_i(x) < \epsilon \quad \forall i \quad \Rightarrow \quad \sum_{i=1}^{n} p_i(x) < n\epsilon,$$

conversely,

$$\sum_{i=1}^{n} p_i(x) < \epsilon \quad \Rightarrow \quad p_i(x) < \epsilon \qquad \text{for every } i.$$

Hence every open ball defined by the norm Σp_i contains a $\mathcal{P}$-open ball and conversely.

2) If $\mathcal{P}$ is countable, the $\mathcal{P}$-topology is equivalent to the topology induced by the translation invariant metric,

$$d(x, y) = \sum_{n=1}^{\infty} \frac{1}{2^n} \frac{p_n(x - y)}{1 + p_n(x - y)}.$$

We must first check that $d(x, y)$ is a metric function. The only axiom which is not immediately verified is the subadditivity of this d

function; it follows from the following inequality

$$\frac{\alpha+\beta}{1+\alpha+\beta}+\frac{\alpha}{1+\alpha+\beta}+\frac{\beta}{1+\alpha+\beta}\leq\frac{\alpha}{1+\alpha}+\frac{\beta}{1+\beta}\quad \text{for } \alpha,\beta\geq 0$$

and the fact that the function $\alpha\mapsto\alpha/(1+\alpha)$ increases monotonically. One then shows that given p_i and ϵ_i, $i\in I$ finite set (p. 424) there exists η such that $d(x,y)<\eta$ implies $p_i(x-y)<\epsilon_i$, $i\in I$ and conversely. Hence the topologies are equivalent. ■

Note that $d(x,0)$ satisfies all the properties of a norm except

$$\|\lambda x\|=|\lambda|\,\|x\|.$$

2. $\mathscr{D}$-SPACES; INDUCTIVE LIMIT TOPOLOGY

In order to define distributions, we need to give a suitable topological structure to the space $C_0^m(U)$ of functions on an open set $U\subset\mathbb{R}^n$ which are m-times continuously differentiable and which have compact support. As we shall see the matter is not trivial and to have good properties we shall need to construct a space $\mathscr{D}^m(U)$ which is not metrizable. If, however, we consider only the subset C_{0K}^m of $C_0^m(U)$ which consists of functions with support in a *fixed* compact set K, the matter is straightforward. We can give C_{0K}^m a topology which makes it a Banach space and define $\mathscr{D}_K^m$, the space of functions which are m-continuously differentiable on $\mathbb{R}^n$ and have their support in the compact set K, together with the topology induced by the finite family of seminorms $\mathscr{D}_K^m$

$$\{\sup_{x\in K}|D^j f(x)|;\ |j|\leq m\}.$$

This topology is equivalent to the topology induced by any of the following equivalent norms:

$$\sup_{|j|\leq m}\sup_{x\in K}|D^j f(x)|\quad\text{designated } p_{K,m}(f)$$ $p_{K,m}$

$$\sup_{x\in K}\sum_{|j|\leq m}|D^j f(x)|$$

$$\sum_{|j|\leq m}\sup_{x\in K}|D^j f(x)|$$

$\mathscr{D}_K^m$ is a Banach space.

Exercise: Show that

$$\mathscr{D}_K^m\hookrightarrow\mathscr{D}_{K'}^m\quad\text{if } K\subset K'.$$

Answer: If $K \subset K'$ and $\varphi \in \mathscr{D}_K^m$ then $\varphi \in \mathscr{D}_{K'}^m$ and

$$\sup_{x\in K'} |D^j\varphi(x)| = \sup_{x\in K} |D^j\varphi(x)|, \quad |j| \leq m$$

thus

$$p_{K,m}(\varphi) = p_{K',m}(\varphi), \quad \forall \varphi \in \mathscr{D}_K^m$$

and the topology induced by $\mathscr{D}_{K'}^m$ on $\mathscr{D}_K^m$ is therefore identical with the topology of $\mathscr{D}_K^m$. ∎

$\mathscr{D}_K$ The space $\mathscr{D}_K$ is the space of indefinitely differentiable functions with support on the compact set $K \subset \mathbb{R}^n$, together with the topology induced by the countable family of seminorms (in fact, norms)

$$\{p_{K,m};\ m \text{ arbitrary integer}\}.$$

$\mathscr{D}_K$ is a Fréchet space (cf. p. 26).

Consider now the set $C^m(U)$ of m-continuously differentiable functions on $U \subset \mathbb{R}^n$ which is not restricted to functions of compact support. Let $\mathscr{C}^m(U)$ be the space of functions $C^m(U)$ together with the $\mathscr{P}$-topology induced by the family of seminorms

$\mathscr{C}^m(U)$

$$\mathscr{P} = \{p_{K,m};\quad K \in \mathscr{K}\}$$

where $\mathscr{K}$ is the family of all compact subsets of U.

Theorem: *$\mathscr{C}^m(U)$ is a Fréchet space.*

Proof (cf. definitions and properties of Fréchet spaces p. 26): a) $\mathscr{C}^m(U)$ is metrizable; we shall prove that the neighborhoods of the origin admit a countable base. Let (K_i) be an increasing sequence of compact subsets of U which cover U, we shall prove that the countable collection of open balls

$$B_{i,n}(0) = \{f;\ p_{K_i,m}(f) < 1/n, \quad n \in N\}$$

form a base for the neighborhoods of the origin. Since every open set is a union of $\mathscr{P}$-open balls, it is sufficient to show that given any $\mathscr{P}$-open ball $B_{\epsilon,\mathscr{L}}(0) \equiv \{f;\ p_{K,m}(f) < \epsilon,\ K \in \mathscr{L}\}$, $\mathscr{L}$ finite subclass of $\mathscr{K}$, there exist natural numbers j, l such that $B_{j,l}(0) \subset B_{\epsilon,\mathscr{L}}(0)$. Clearly one can always choose j such that $K \subset K_j$ for all $K \in \mathscr{L}$. It then follows that $p_{K,m}(f) \leq p_{K_j,m}(f)$, and hence if $1/l < \epsilon$

$$B_{j,l}(0) \subset \{f;\ p_{K_j,m}(f) < \epsilon\} \subset B_{\epsilon,\mathscr{L}}(0).$$

b) $\mathscr{C}^m(U)$ is complete; indeed in the $\mathscr{P}$-topology the convergence of the sequence (f_n) to f is the uniform convergence of $(D^jf_n(x))$ to $D^jf(x)$ on every compact set K for all $|j| \leq m$; hence every Cauchy sequence is convergent. ■

The space $\mathscr{C}(U)$ is the space of infinitely differentiable functions on $U \subset \mathbb{R}^n$, together with the family of norms $\mathscr{C}(U)$

$$\{p_{K,m};\quad K \in \mathscr{K},\ m \text{ arbitrary integer}\}.$$

It is also denoted $\mathscr{C}^\infty(U)$. It is a Fréchet space. $\mathscr{C}^\infty(U)$

At this point it seems that the simplest topological space of functions with compact support is the subset $C_0^m(U)$ of $C^m(U)$ which consists of functions with compact support together with the topology which would make it a subspace of $\mathscr{C}^m(U)$. $C_0^m(U)$ is indeed a vector space but, alas, it is not complete in $\mathscr{C}^m(U)$ if endowed with the topology of $\mathscr{C}^m(U)$.

Proof: The sequence of functions f_n in $C_0^m(R)$ defined by

$$f_n(x) = \sum_{p=1}^{n} \frac{1}{2^p} \varphi(x-p) \qquad \varphi \in C_0^m(\mathbb{R})$$

is a Cauchy sequence in $\mathscr{C}^m(R)$ but its limit is not in $C_0^m(\mathbb{R})$ because it is not a function with compact support.

The space $C_0^m(U)$ with the topology that we shall define is called $\mathscr{D}^m(U)$. The topology of $\mathscr{D}^m(U)$ is the topology of all the seminorms (cf. p. 423) on $\mathscr{D}^m(U)$ whose restrictions to each $\mathscr{D}_K^m(U)$ are continuous: this implies that a convex subset of $\mathscr{D}^m(U)$ is open if and only if its intersection with each $\mathscr{D}_K^m(U)$, K a compact subset of U, is open. This definition may seem intricate – in effect we will not have to use much the topology of $\mathscr{D}^m(U)$, but only the following fundamental property[1].

A linear function is continuous on $\mathscr{D}^m(U)$ if and only if its restriction to each $\mathscr{D}_K^m(U)$, $K \in \mathscr{K}$, is continuous. $\mathscr{D}^m(U)$

The above definition of $\mathscr{D}^m(U)$ is indeed its definition as "locally convex inductive limit" of the Banach topologies $\mathscr{D}_K^m(U)$, due to the following inclusion properties:

1) $\mathscr{D}^m(U) = \bigcup_K \mathscr{D}_K^m(U)$.

2) $\mathscr{D}_K^m(U) \hookrightarrow \mathscr{D}_{K'}^m(U)$, if $K \subset K'$.

[1]For further study of the topology of $\mathscr{D}^m(U)$, $\mathscr{D}(U)$ and the definition of locally convex inductive limits [L. Schwartz 1966], [Trêves 1967].

3) given $\mathscr{D}^m_K(U)$ and $\mathscr{D}^m_{K'}(U)$ there exists $\mathscr{D}^m_{K''}(U)$ such that, topologically

$$\mathscr{D}^m_K(U) \hookrightarrow \mathscr{D}^m_{K''}(U) \qquad \text{and} \qquad \mathscr{D}^m_{K'}(U) \hookrightarrow \mathscr{D}^m_{K''}(U)$$

(take $K'' \supset K \cup K'$).

Exercise 1: Show that the subset:

$$p_{K,m}(\varphi) < \epsilon$$

is open in $\mathscr{D}^l(U)$ for every compact K and $m \le l$.
Conclude that the trace on $\mathscr{D}^m_K(U)$ of the $\mathscr{D}^m(U)$ topology is identical to the $\mathscr{D}^m_K(U)$ topology.

Solution: If $\varphi \in \mathscr{D}^l_{K'}(U)$ then

$$p_{K,m}(\varphi) = p_{K\cap K',m}(\varphi) \le p_{K',l}(\varphi), \text{ if } m \le l.$$

Thus the mapping $\mathscr{D}^l_{K'}(U) \to \mathbb{R}^+$ by $\varphi \mapsto p_{K,m}(\varphi)$ is continuous, the trace of $p_{K,m}(\varphi) < \epsilon$ is open in $\mathscr{D}^l_{K'}(U)$ for every K', and $m \le l$.
The trace on $\mathscr{D}^m_K(U)$ of every open set of $\mathscr{D}^m(U)$ is open by the definition of the $\mathscr{D}^m(U)$ topology, $\mathscr{D}^m_K(U) \hookrightarrow \mathscr{D}^m(U)$, and we have just shown that, conversely, every open set of $\mathscr{D}^m_K(U)$ is the trace of an open set of $\mathscr{D}^m(U)$.

Exercise 2: Show that:

$$\mathscr{D}^m(U) \hookrightarrow \mathscr{C}^m(U).$$

The proof is left to the reader.

$\mathscr{D}(U)$ The topology of $\mathscr{D}(U)$ is the inductive limit of the $\mathscr{D}_K(U)$ topologies: a convex subset of $\mathscr{D}(U)$ is open iff its intersection with each $\mathscr{D}_K(U)$ is open in $\mathscr{D}_K(U)$.
We have

$$\mathscr{D}_K(U) \hookrightarrow \mathscr{D}(U) \hookrightarrow \mathscr{C}(U).$$

$\mathscr{D}^m_K$ is a Banach space, $\mathscr{D}_K$ is a Fréchet space but $\mathscr{D}^m(U)$ and $\mathscr{D}(U)$ are not metrizable. They are sequentially complete locally convex topological vector spaces (p. 23). They are locally convex by the definition of their topology, the proof of their sequential completeness follows.

Convergence of a sequence in $\mathscr{D}^m(U)$ and in $\mathscr{D}(U)$.

Proposition: A sequence (φ_n) converges to zero in $\mathscr{D}^m(U)$ if and only if
1) all φ_n have their support in a fixed compact K_0, and
2) the sequence (φ_n) and the sequences $(D^l\varphi_n)$ of partial derivatives of order $\le m$ converge uniformly to zero on K_0.

Proof: By the definition of the topology of $\mathcal{D}^m(U)$ a sequence (φ_n) converges to zero if and only if its trace on each $\mathcal{D}^m_K(U)$ converges to zero.
If $(\varphi_n) \subset \mathcal{D}^m_{K_0}(U)$, it converges to zero in $\mathcal{D}^m_{K_0}(U)$ if and only if each sequence $(D^j\varphi_n)$, $|j| \leq m$, converges uniformly to zero on K_0.
Clearly the convergence to zero of $(\varphi_n) \subset \mathcal{D}^m_{K_0}(U)$ implies the convergence to zero of its trace in every $\mathcal{D}^m_K(U)$, hence the sufficiency part of the proposition.
To prove the necessity suppose that property 1 is not satisfied. Then there exists a sequence of integers (n_j) such that $\varphi_{n_j}(x_j) \neq 0$ where $x_j \in K_{j+1} - K_j$, (K_j) is an increasing sequence of compact sets covering U, $K_1 \neq \emptyset$.
The mapping p from $\mathcal{D}^m(U)$ into $\mathbb{R}$ defined by the semi norm

$$p(\varphi) = \sum_{j=1}^{\infty} \sup_{x \in K_{j+1} - K_j} |\varphi(x)/\varphi_{n_j}(x_j)|$$

is continuous on each $\mathcal{D}^m_K(U)$, hence on $\mathcal{D}^m(U)$, thus the set $\{\varphi \in \mathcal{D}^m(U);\ p(\varphi) < \epsilon\}$ is an open neighborhood of the origin. However, the set characterized by $p(\varphi) < 1$ contains no function φ_{n_j}, thus the sequence (φ_n) cannot converge to zero in $\mathcal{D}^m(U)$.
Property 2 results from property 1 and the definition of convergence in $\mathcal{D}^m_K(U)$. ■

Corollary. A sequence (φ_n) converges to φ in $\mathcal{D}^m(U)$ if
1. supp $\varphi_n \subset$ *fixed* K
2. $D^j\varphi_n$ *converges uniformly to* $D^j\varphi$ *for every* $|j| \leq m$.

Similar propositions hold for sequences in $\mathcal{D}(U)$: the partial derivatives of any given order have to converge uniformly.

Proposition. The spaces $\mathcal{D}^m(U)$ and $\mathcal{D}(U)$ are sequentially complete (p. 23).

Proof: If $\lim_{n,m=\infty} (f_n - f_m) = 0$ in $\mathcal{D}(U)$, we have supp $f_n \subset$ fixed K for $n > N$ and $\lim_{n,m=\infty} |D^j f_n - D^j f_m| = 0$ uniformly on K for every derivation D^j. It follows from the classical theorems of uniform convergence that there exists a function $f \in \mathcal{D}(U)$ of support K such that f_n converges to f and $D^j f_n$ converges to $D^j f$ uniformly on K for every j. ■

Examples of functions in $\mathcal{D}$. Let B_a be the open ball in $\mathbb{R}^n$ defined by

$$|x|^2 = \sum_{i=1}^{n} (x_i)^2 < a^2$$

$\theta_a(x)$

and $\bar{B}_a$ the closed ball $|x|^2 \leq a^2$. The function θ_a defined by

$$\theta_a(x) = \begin{cases} \exp(-a^2/(a^2 - |x|^2)) \text{ (for } |x| \leq a) \\ 0 \text{ for } |x| \geq a \end{cases}$$

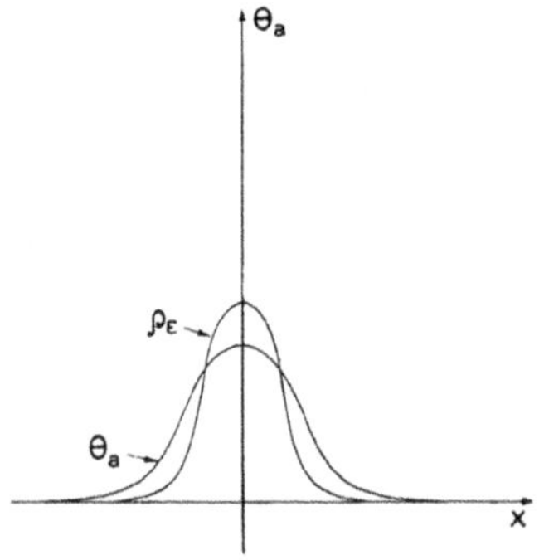

is a C^∞ function with support in $\bar{B}_a$ which vanishes together with all its derivatives at $|x| = a$. Thus $\theta_a \in \mathcal{D}(U)$ for all $U \supset \bar{B}_a$.
Note that θ_a is not real analytic for $|x| = a$ (remember there are no analytic functions with compact support, except zero).

We shall now introduce two convenient tools: regularizing sequences and truncating sequences. They will be used shortly to prove that the space $C_0^\infty(\mathbb{R}^n)$ is dense in $\mathscr{C}^m(\mathbb{R}^n)$. In other words, any m times continuously differentiable function f on $\mathbb{R}^n$ not necessarily with compact support, can be approximated by infinitely differentiable functions on $\mathbb{R}^n$ with compact support, in the sense that there is a sequence (f_n), $f_n \in C_0^\infty(\mathbb{R}^n)$, which converges to f in $\mathscr{C}^m(R^n)$.

regularizing

A family of functions ρ_ϵ on $\mathbb{R}^n$ which possesses the following properties is called **regularizing**:
1) ρ_ϵ is C^∞, $\rho_\epsilon \geq 0$
2) supp $\rho_\epsilon \subset \bar{B}_\epsilon$ the closed ball $|x|^2 \leq \epsilon^2$
3) $\int_{\mathbb{R}^n} \rho_\epsilon \, dx = 1$.
In particular set $\epsilon = 1/n$, the sequence $(\rho_{1/n})$ is called regularizing.

Example:

$$\rho_\epsilon(x) = \begin{cases} k_\epsilon \exp(-\epsilon^2/(\epsilon^2 - |x|^2)) & |x| \leq \epsilon \\ 0 & |x| \geq \epsilon \end{cases}$$

where k_ϵ is determined by the normalisation condition 3.

In the next section (p. 440) we shall need the upper bounds of ρ_ϵ and its derivatives. We have

$$\sup_{x\in\mathbb{R}^n} |\rho_\epsilon(x)| = k_\epsilon\, e^{-1}.$$

With the change of variables $x = \epsilon y$

$$k_\epsilon^{-1} = \int_{|x|<\epsilon} \exp\frac{-\epsilon^2}{\epsilon^2-|x|^2}\,dx = \epsilon^n \int_{|y|\leq 1} \exp\frac{-1}{1-|y|^2}\,dy.$$

Thus $k_\epsilon = C\epsilon^{-n}$ where C is a positive constant and

$$\sup_{x\in\mathbb{R}^n}\left|\frac{\partial}{\partial x^i}\rho_\epsilon\right| = 2k_\epsilon\epsilon^{-1}\sup_{|y|\leq 1}\frac{|y^i|}{(1-|y|^2)^2}\exp\frac{-1}{1-|y|^2}$$

$$= C_1\epsilon^{-n-1} \text{ where } C_1 \text{ is a positive constant.}$$

A similar calculation gives

$$\sup_{x\in\mathbb{R}^n}|D^\alpha\rho_\epsilon| = C_{|\alpha|}\epsilon^{-n-|\alpha|} \text{ where } C_{|\alpha|} \text{ is a positive constant.}$$

Let f be a measurable bounded function on $\mathbb{R}^n$, the operation

$$f\mapsto f_\epsilon \text{ by } f_\epsilon(x) = \int_{\mathbb{R}^n}\rho_\epsilon(x-y)f(y)\,dy$$

is called **regularization** of f. regularization

f_ϵ is a C^∞ function which does not belong to $\mathscr{D}(\mathbb{R}^n)$; indeed the domain of the integral which defines f_ϵ is compact and the theorems on derivation under an integral sign with respect to a parameter show that f_ϵ is C^∞. The regularization procedure will be used for instance to construct C^∞ truncating functions (p. 434), partition of unity (pp. 214, 215), etc. On the other hand the support of f_ϵ is arbitrarily close to the support of f which has not been assumed compact, hence supp f_ϵ is not compact and $f_\epsilon \notin \mathscr{D}(\mathbb{R}^n)$. ∎

It can be shown that if $|f(x)|\leq M$, then (Exercise 3, p. 528)

$$\sup_{\mathbb{R}^n}|D^\alpha f_\epsilon| \leq C\epsilon^{-|\alpha|}M\int_{|y|\leq 1}\left|D^\alpha\exp\frac{1}{1-|y|^2}\right|dy$$

where

$$C = \int_{|y|\leq 1}\exp\frac{-1}{1+|y|^2}\,dy.$$

truncating sequences

A sequence (τ_n) of functions on $\mathbb{R}^k$ which possesses the following properties is called **truncating**:

$$\tau_n(x) = \begin{cases} 1 & \text{for } |x| \le n \\ 0 & \text{for } |x| \ge n+1 \end{cases}$$

$$\tau_n \in C^\infty(\mathbb{R}^k).$$

Such a sequence can be obtained by regularizing a sequence of step functions which satisfies the first property.

Density theorem. The space $C_0^\infty(\mathbb{R}^n)$ is dense in $\mathscr{C}^m(\mathbb{R}^n)$.

Proof: The proof is done in three steps: first the density of $C_0^\infty(\mathbb{R}^n)$ in $\mathscr{D}^0(\mathbb{R}^n)$, then in $\mathscr{D}^m(\mathbb{R}^n)$ and finally in $\mathscr{C}^m(\mathbb{R}^n)$.

1) $f \in C_0^0(\mathbb{R}^n)$ with supp $f \subset K$; let $f_\epsilon(x) = \int_{\mathbb{R}^n} \rho_\epsilon(x-y) f(y)\,dy$.
The function f_ϵ has compact support because $\rho_\epsilon(x-y)f(y) = 0$ for every y if $x \notin K + \bar{B}_\epsilon$ and therefore supp $f_\epsilon \subset K + \bar{B}_\epsilon$.
The function f_ϵ is continuous because the integral on a compact set K of a continuous function which depends continuously on a parameter x is a continuous function of the parameter. The function f_ϵ converges uniformly to f; indeed it follows from

$$f(x) - f_\epsilon(x) = \int_{\mathbb{R}^n} \rho_\epsilon(x-y)(f(x) - f(y))\,dy$$

and from supp $\rho_\epsilon(x-y) = \{y, |x-y| \le \epsilon\}$, $\rho_\epsilon \ge 0$ that

$$|f(x) - f_\epsilon(x)| \le \sup_{|x-y|\le\epsilon} |f(x) - f(y)| \int_{\mathbb{R}^n} \rho_\epsilon(x-y)\,dy$$

$$\le \sup_{|x-y|\le\epsilon} |f(x) - f(y)|.$$

f is uniformly continuous since it is continuous and with compact support, hence f_ϵ converges uniformly to f, therefore it converges to f in $\mathscr{D}^0(R^n)$.

2) $f \in C_0^m(\mathbb{R}^n)$, $0 \le m \le \infty$. Make the change of variable $x - y = z$ in the definition of f_ϵ

$$f_\epsilon(x) = \int_{\mathbb{R}^n} \rho_\epsilon(z) f(x-z)\,dz$$

and take the derivatives of both sides with respect to x up to order m. The continuity of $D^j f_\epsilon$, $|j| \le m$, and its uniform convergence to $D^j f$ is then established in 1).

3) $f \in C^m(\mathbb{R}^n)$; let $f_n = \tau_n f$ where $\{\tau_n\}$ is a truncating sequence; $\tau_n f \in C_0^m(\mathbb{R}^n)$. Given any compact set K, $\tau_n f|_K = f|_K$ for n sufficiently large. The sequences $\{\tau_n f\}$ and $D^j(\tau_n f)$ converge uniformly to f and $D^j f$, on any compact set K. The sequence $\{\tau_n f\}$ converges to f in $\mathscr{C}^m(\mathbb{R}^n)$. ■

The same arguments show that, for any open set $U \subset \mathbb{R}^n$, $C_0^\infty(U)$ is dense in $\mathscr{C}^m(U)$. Note however, that a truncating sequence in U, relative to an increasing sequence of compact sets covering U, such that

$$\tau_n(x) = 1 \qquad \text{for } x \in K_n$$

$$\tau_n(x) = 0 \qquad \text{for } x \in K_{n+1}$$

exists, C^∞ on U, but cannot in general be chosen, if $U \neq \mathbb{R}^n$, such that $D^j \tau_n$ is uniformly bounded for $n \in N$: $C_0^\infty(U)$ is not dense in $C^m(\bar{U})$.

Remark: One says sometimes that $\mathscr{D}$ is dense in $\mathscr{C}^m$. We say C_0^∞ is dense in $\mathscr{C}^m$ because the topology of $\mathscr{D}$ does not play any role in this property.

B. DISTRIBUTIONS

1. DEFINITIONS

The space of **distributions** on $U \subset \mathbb{R}^n$ is the dual $\mathscr{D}'(U)$ of the topological vector space $\mathscr{D}(U)$ of C^∞ functions with compact support in U defined in section A. distribution

Notation: Let $T \in \mathscr{D}'(U)$ and $\varphi \in \mathscr{D}(U)$

$$T: \mathscr{D}(U) \to \mathbb{K} \text{ by } \varphi \mapsto T(\varphi) \equiv \langle T, \varphi \rangle.$$

Comments: Whereas the notation $T(\varphi)$ means only that T is a function (functional) on the space of functions $\mathscr{D}(U)$, the notation $\langle T, \varphi \rangle$ is used only when T is a continuous linear function on $\mathscr{D}(U)$ (i.e. $T \in \mathscr{D}'(U)$). Of course, if it has been explicitly stated that $T \in \mathscr{D}'(U)$ we can use either notation indifferently. Distributions appear in physics and mathematics as generalised functions; historically they were introduced as such, before the duality gave them a rigorous definition. Measures are a particular class of distributions. In Chapter I they are considered directly as set functions. We will study them in this chapter in their dual appearance.

Indeed the following inclusions (cf. p. 430)

$$\mathcal{D}(U) \hookrightarrow \mathcal{D}^m(U) \hookrightarrow \mathcal{D}^p(U), \qquad p \leq m$$

imply the converse inclusions for the duals

$$\mathcal{D}'^p(U) \hookrightarrow \mathcal{D}'^m(U) \hookrightarrow \mathcal{D}'(U), \qquad p \leq m.$$

Therefore the measures (p. 438) elements of $\mathcal{D}'^0(U)$, dual of $\mathcal{D}^0(U)$, are distributions on U.

Remark: There is no such thing as "the value of a distribution T at a point". Nevertheless we shall be able to define T uniquely by its values on open sets in U. We shall say "T is *defined* on $\mathcal{D}$", "T is *based* on U".

For brevity, when the context is clear and specially when $U = \mathbb{R}^n$ one often writes $T \in \mathcal{D}'$ and $\varphi \in \mathcal{D}$ instead of $T \in \mathcal{D}'(U)$ and $\varphi \in \mathcal{D}(U)$.

Theorem. A linear form T on $\mathcal{D}(U)$ is a distribution on U if and only if it possesses either one of the two following equivalent properties.
1. *Given a sequence $\{\varphi_n\}$ in $\mathcal{D}(U)$ which converges to zero in $\mathcal{D}(U)$ the sequence of real or complex numbers $\{T(\varphi_n)\}$ converges to zero.*
2. *For every compact set $K \subset U$, there exists an integer $M \geq 0$ and a constant $C > 0$ such that*

$$|\langle T, \varphi\rangle| \leq C p_{K,M}(\varphi) \qquad \forall \varphi \in \mathcal{D}_K(U)$$

where $p_{K,M}$ is the norm defined in section A:

$$p_{K,M}(\varphi) = \sup_{|j| \leq M} \sup_{x \in K} |D^j \varphi(x)|.$$

Proof: 1) A linear mapping on $\mathcal{D}(U)$ is continuous if and only if its restriction to any $\mathcal{D}_K$ is continuous for the $\mathcal{D}_K$ topology. The $\mathcal{D}_K$ topology is metrizable, hence property 1 follows from the theorem about continuous mappings on metrizable topological vector space (p. 25).
2) The $\mathcal{D}_K$ topology is defined by a countable family of seminorms $\{p_{K,m}$; m arbitrary$\}$; an open ball is defined by a finite number of norms equivalent to one norm $p_{K,M}$. Thus the continuity of T at the origin is expressed in the usual way: given $\epsilon > 0$ there exists η such that

$$|\langle T, \varphi\rangle| < \epsilon \qquad \text{when } p_{K,M}(\varphi) < \eta \qquad \forall \varphi \in \mathcal{D}_K.$$

On the other hand,

$$\left|\left\langle T, \frac{\mu}{\lambda}\varphi\right\rangle\right| = \frac{\mu}{\lambda}|\langle T, \varphi\rangle|.$$

Now let $\lambda = p_{K,M}(\varphi)$, $\mu < \eta$ and $C > \epsilon/\mu$, property 2 follows. ∎

Order of a distribution. A distribution is said to be of **order** q if there exists an integer q, and no smaller integer, such that

order of a distribution

$$|\langle T, \varphi\rangle| < C(K)p_{K,q}(\varphi) \qquad \forall K \subset U, \quad \varphi \in \mathscr{D}_K.$$

Example: A locally[1] integrable function on $\mathbb{R}^n$ defines a distribution on $\mathbb{R}^n$ of order 0.

Proof: Let $f \in L^1_{\text{loc}}(\mathbb{R}^m)$ (locally integrable function on $\mathbb{R}^n$). Let $\varphi \in \mathscr{D}(\mathbb{R}^n)$ with support in K. Let $\tilde{f}$ be the linear form on $\mathscr{D}$ defined by

$$\langle \tilde{f}, \varphi\rangle = \int_{\mathbb{R}^n} f\varphi \, dx;$$

we have

$$|\langle \tilde{f}, \varphi\rangle| \leq \sup_K |\varphi| \int_K |f| \, dx \leq C(K)p_{K,0}(\varphi)$$

$$\text{with } C(K) = \int_K |f| \, dx.$$ ∎

We shall say that $\tilde{f}$ is the **distribution associated to the function** f, or that the distribution $\tilde{f}$ is **equivalent to the function** f; we may also omit the tilde and say "the distribution f" or speak of "a distribution which is a continuous function" meaning "a distribution which is equivalent to a continuous function".

$\tilde{f}$ distribution associated (equivalent) to a function

Theorem. Two locally integrable functions which define the same distributions are equal almost everywhere: $f, g \in L^1_{\text{loc}}(U)$ *such that*

$$\langle f, g\rangle = \langle g, \varphi\rangle \quad \forall \varphi \in \mathscr{D}(U) \quad \textit{implies} \quad f = g \textit{ in } L^1_{\text{loc}}(U).$$

Measures.

Theorem. A distribution of order zero based on $U \subset \mathbb{R}^n$ can be extended in a unique way to a linear continuous form on $\mathscr{D}^0(U)$, hence it can be identified with a Radon measure (p. 48).

[1]A function is said to be locally integrable on $\mathbb{R}^n$ if it is a Lebesgue measurable function, defined a.e., integrable on every compact set.

Proof: μ is a linear form on $\mathscr{D}(U)$, continuous for the $\mathscr{D}^0(U)$ topology and since $\mathscr{D}$ is dense in $\mathscr{D}^0$, we can extend it uniquely to $\mathscr{D}^0(U) \supset \mathscr{D}(U)$ by the extension theorem (p. 425). ∎

regular measure

A measure, distribution of order zero, is said to be **regular** (of **finite density**) in U if it is equivalent in U to a locally integrable function.

Dirac measure

Example: A measure which is not equivalent to a locally integrable function. The **Dirac measure** at $a \in \mathbb{R}^n$, is defined by

$$\langle \delta_a, \varphi \rangle = \varphi(a);$$

δ_0 is usually abbreviated to δ. The proof that it is not a regular measure is given below.

Theorem. The Dirac measure cannot be represented by a locally integrable function.

Proof:

$$\text{Let} \quad \theta_\epsilon \in \mathscr{D} = \begin{cases} \exp(-\epsilon^2/(\epsilon^2 - x^2)) & \text{for } x < \epsilon \text{ i.e. } x \in B_\epsilon \\ 0 & \text{for } x \geq \epsilon. \end{cases}$$

Then

$$\langle \delta, \theta_\epsilon \rangle = \mathrm{e}^{-1}.$$

On the other hand let f be a locally integrable function, then

$$\langle f, \theta_\epsilon \rangle = \int_{B_\epsilon} f\theta_\epsilon \, \mathrm{d}x$$

$$|\langle f, \theta_\epsilon \rangle| \leq \mathrm{e}^{-1} \int_{B_\epsilon} |f| \, \mathrm{d}x \qquad \text{which goes to zero with } \epsilon.$$ ∎

Dirac measure on a surface

Let S be a differentiable submanifold of $\mathbb{R}^n$.
1) S is a hyperplane defined by $x^1 = 0$. Then the Dirac measure on S is defined by

$$\langle \delta_S, \varphi \rangle = \int_{\mathbb{R}^{n-1}} \varphi(0, x^2, \ldots x^n) \, \mathrm{d}x^2 \ldots \mathrm{d}x^n.$$

2) S is defined by the equation $f(x) = 0$, where $f(x)$ is a C^1 function on $\mathbb{R}^n$ without critical point. δ_S is then defined by

$$\langle \delta_S, \varphi \rangle = \int_S \varphi\omega$$

where ω is an $n-1$ exterior differential form on S (cf. Ch. IV) called the **Leray form** of S defined as follows Leray form

$$df \wedge \omega = dx^1 \wedge dx^2 \wedge \cdots \wedge dx^n.$$

Since f has no critical point, one of the following equivalent expressions for ω is valid in the neighborhood of each point of S:

$$\omega = (dx^2 \wedge \cdots \wedge dx^n)\left(\frac{\partial f}{\partial x^1}\right)^{-1}, \quad \omega = (-dx^1 \wedge dx^3 \wedge \cdots \wedge dx^n)\left(\frac{\partial f}{\partial x^2}\right)^{-1}, \ldots$$

Note: ω thus defined depends not only on the geometric definition of S, but also on the choice of its equation. If we require that ω be the volume element induced on S by the euclidean metric on $\mathbb{R}^n$, we have to replace $\partial f/\partial x^i$ by n_i in the above definition, where the n_i are the components of the unit normal to S.

Dirac measures are indeed "measures", i.e. distributions of order 0:

$$|\langle \delta_a, \varphi\rangle| < \sup |\varphi|$$

$$|\langle \delta_S, \varphi\rangle| \leq C(K) \sup |\varphi| \text{ where } C(K) = \int_{S\cap K} \omega.$$

Distributions of order p.

Theorem. A distribution is of order p if and only if it can be extended to a linear continuous form on $\mathscr{D}^p$, *that is if and only if it is in* $\mathscr{D}'^p$.

Proof: The proof is the same as the one given above for measures.

Example 1: With certain non locally integrable functions one can associate a distribution. For instance the distribution "Principal value of $1/x$" based on $\mathbb{R}$, is defined by

$$\left\langle \mathrm{Pv}\frac{1}{x}, \varphi\right\rangle = \lim_{\epsilon=0}\left(\int_{-\infty}^{-\epsilon} \frac{\varphi}{x} dx + \int_{\epsilon}^{\infty} \frac{\varphi}{x} dx\right), \qquad \varphi \in \mathscr{D}(R).$$

The limit exists if $\varphi \in \mathscr{D}^1$ and

$$\left\langle \mathrm{Pv}\frac{1}{x}, \varphi\right\rangle = \int_{-A}^{A} \frac{\varphi(x) - \varphi(0)}{x} dx, \qquad \operatorname{supp} \varphi \subset [-A, A].$$

$\mathrm{Pv}(1/x)$ is a distribution of order 1

$$\left|\left\langle \mathrm{Pv}\frac{1}{x}, \varphi\right\rangle\right| \leq 2A \sup_{|x|\leq A} |\varphi'(x)|.$$

Example 2: The distribution "Finite part of $1/x$" is defined by

$$\left\langle \mathrm{Fp}\frac{1}{x}, \varphi\right\rangle = \int_0^A \frac{\varphi(x)-\varphi(0)}{x}\,\mathrm{d}x + \varphi(0)\,\mathrm{Log}\,A, \qquad \mathrm{supp}\,\varphi \subset [-A, +A].$$

Exercise: Show that $\langle \mathrm{Fp}(1/x), \varphi\rangle$ does not depend on A; show that it is a distribution of order 1.

Example 3: Derivatives of the Dirac measure (see Example 2, p. 448)

$$\langle \delta^{(p)}, \varphi\rangle = (-1)^p (\mathrm{D}^p\varphi)(0).$$

Example 4: Distribution of infinite order on $\mathbb{R}$; for instance

$$\langle T, \varphi\rangle = \sum_1^\infty \varphi^{(m)}(m).$$

equal **Equality of distributions.** Two distributions T_1 and T_2 are said to be **equal** on $U \subset \mathbb{R}^n$ if $\langle T_1, \varphi\rangle = \langle T_2, \varphi\rangle$ for every $\varphi \in \mathscr{D}(U)$.

A distribution can be defined uniquely by its value on an open covering:

Theorem. Let $\{U_i\}$, $i \in I$ be a family of open sets in $\mathbb{R}^n$, let T_i be a distribution on U_i, for each $i \in I$. There exists a unique distribution T on $U = \bigcup_{i\in I} U_i$ such that its restriction to each U_i is T_i, if and only if $T_i = T_j$ on $U_i \cap U_j$.

Proof: The proof that the condition is necessary is trivial, the proof that it is sufficient is done by constructing a partition of unity (φ_i) relative to the open covering (U_i) of U. Let φ_i be a C^∞ function with compact support in U_i and $\{\mathrm{supp}\,\varphi_i\}$ be a locally finite covering[1] of U

$$\varphi_i \geq 0 \qquad \sum_{i\in I} \varphi_i(x) = 1 \qquad \forall x \in U.$$

[1]i.e. any compact set intersects only a finite number of the supp φ_i's. As $\mathbb{R}^n$ is locally compact, it is possible to extract a locally finite covering from any open covering.

The technique used in the construction of a truncating function (p. 434) can be used for the construction of φ_i. Now let $\psi \in \mathscr{D}(U)$, then $\psi\varphi_i \in \mathscr{D}(U_i)$. Any linear form T on $\mathscr{D}(U)$ such that $T = T_i$ in U_i necessarily satisfies the equation (for each ψ the sum is in fact finite)

$$\langle T, \psi\rangle = \left\langle T, \sum_{i\in I} \psi\varphi_i\right\rangle = \sum_{i\in I} \langle T_i, \psi\varphi_i\rangle.$$

On the other hand let us show that this equation defines a distribution T equal to T_i on each U_i. Indeed, let φ be with compact support in U_i, then

$$\langle T, \varphi\rangle = \sum_{j\in I} \langle T_j, \varphi\varphi_j\rangle$$

$$\operatorname{supp} \varphi\varphi_j \subset U_i \cap U_j \quad \text{hence} \quad T_j(\varphi\varphi_j) = T_i(\varphi\varphi_j)$$

$$\langle T, \varphi\rangle = \sum_{j\in I} \langle T_i, \varphi\varphi_j\rangle = \left\langle T_i, \sum_j \varphi\varphi_j\right\rangle = \langle T_i, \varphi\rangle \quad \forall \varphi \in \mathscr{D}(U_i). \qquad \blacksquare$$

A consequence of the theorem is that if a distribution is zero on some open sets it is zero on their union. Therefore there is a largest open set in which a distribution is zero and the following definition makes sense.

The **support of a distribution** T on U is the smallest closed set in U outside which T is zero. support

Examples

1. If a distribution is equivalent to a function, its support is the support of the function.
2. $\operatorname{supp} \delta_a = a \qquad \operatorname{supp} \delta_S = S.$
3. $\operatorname{supp} \mathrm{Pv}(1/x) = \mathbb{R}$; in the open set $x \neq 0$ this distribution is the locally integrable (C^∞) function $1/x$.
4. $\operatorname{supp} \delta'_a = a$, notice that $\langle \delta'_a, \varphi\rangle$ does not, in general vanish for functions φ which vanishes at a. However it vanishes if $\varphi'(a) = 0$.

Remark: $\varphi\delta'_a = (\varphi\delta_a)' - \varphi'\delta_a = (\varphi(a)\delta_a)' - \varphi'(a)\delta_a = \varphi(a)\delta'_a - \varphi'(a)\delta_a$. More about this later (p. 443).

Distributions with compact support.

Theorem. A distribution $T \in \mathscr{D}'(U)$ has a compact support if and only if it can be extended to a continuous linear form on $\mathscr{C}(U)$.

The space $\mathscr{C}'(U)$ of continuous linear forms on $\mathscr{C}(U)$ is also denoted $\mathscr{E}'(U)$. $\mathscr{C}'(U)$ $\mathscr{E}'(U)$

Proof: 1) $\mathscr{E}'(U) \subset \mathscr{D}'(U)$, since $\mathscr{D}(U) \hookrightarrow \mathscr{E}(U)$.
Let $T \in \mathscr{E}'(U)$. By the definition of the topology of $\mathscr{E}(U)$, where a basis at the origin consists of the open balls

$$p_{K,M}(\varphi) < \epsilon,$$

there exists a compact set K, an integer M and a constant C such that

$$|\langle T, \varphi\rangle| \leq C p_{KM}(\varphi) \qquad \forall \varphi \in \mathscr{E}(U).$$

Hence

$$\langle T, \varphi\rangle = 0 \text{ for every } \varphi \text{ with support in } U - K \equiv \complement_U K,$$

thus

$$\text{supp } T \subset K.$$

2) Let $T \in \mathscr{D}'(U)$, let supp $T = K \subset U$, let $\alpha, \beta \in \mathscr{D}(U)$ be equal to 1 in a neighborhood of K. Set

$$\langle T_\alpha, \varphi\rangle = \langle T, \alpha\varphi\rangle \qquad \forall \varphi \in \mathscr{E}(U).$$

T_α does not depend on α:

$$\langle T_\alpha, \varphi\rangle - \langle T_\beta, \varphi\rangle = \langle T, (\alpha - \beta)\varphi\rangle = 0$$

since supp $(\alpha-\beta)\varphi \cap \text{supp } T = \emptyset$; we set $\langle T, \varphi\rangle = \langle T_\alpha, \varphi\rangle$ if $\varphi \in \mathscr{E}(U)$. $\langle T, \varphi\rangle$ is continuous on $\mathscr{E}(U)$ because if supp $\alpha = K'$ is compact, there exists a constant C and an integer p such that

$$|\langle T, \varphi\rangle| = |\langle T, \alpha\varphi\rangle| \leq \sup_{|j|\leq p} \sup_{x\in K'} |\mathrm{D}^j(\alpha\varphi)|$$

$$\leq C' \sup_{|j|\leq p} \sup_{x\in K'} |\mathrm{D}^j\varphi| \qquad \forall \varphi \in \mathscr{E}(U)$$

where C' is a constant which depends on the supremum of the derivatives of α of order $\leq p$. ∎

We have also proved the following theorem.

Theorem. Distributions with compact support are of finite order.

Obviously the converse is not true: a distribution of finite order is not necessarily of compact support; for instance, a locally integrable function is a distribution of order 0.

We have remarked previously that δ'_a, whose support is $\{a\}$ does not vanish for functions ψ vanishing at a but vanishes if $\varphi(a) = \varphi'(a) = 0$. In other words supp $T \cap \{x; \varphi(x) \neq 0\} = \emptyset$ does not imply $\langle T, \varphi \rangle = 0$. However supp $T \cap$ supp $\varphi = \emptyset$ does imply $\langle T, \varphi \rangle = 0$ since then supp $\varphi \subset \complement$ supp T (see p. 11). The theorem is the following.

Theorem. Let T be a distribution of order p, then $\langle T, \varphi \rangle = 0$ *if* φ *and its derivatives of order* $\leq p$ *vanish on* supp T.

Proof: Let $S = \text{supp } T$. Let χ_ϵ be a C^∞ function equal to 1 in a neighborhood of S, vanishing outside a slightly larger neighborhood; for instance define χ_ϵ by regularizing the characteristic function of $S_{2\epsilon} = S \oplus B_{2\epsilon}$

$$\chi_\epsilon(x) = \int_{S_{2\epsilon}} \rho_\epsilon(x - y)\, dy.$$

If $x \in S_\epsilon = S \oplus B_\epsilon$ and $y \notin B_{2\epsilon}$, then $x - y \notin B_\epsilon$ and $\rho_\epsilon(x - y) = 0$. Hence

$$\chi_\epsilon(x) = \int_{\mathbb{R}^n} \rho_\epsilon(x - y)\, dy \qquad \text{if } x \in S_\epsilon$$

and $\chi_\epsilon(x) = 1 \quad$ if $x \in S_\epsilon$; furthermore sup $\chi_\epsilon \subset S_{4\epsilon}$.
Consequently, for every $\varphi \in \mathcal{D}$, $\varphi - \varphi\chi_\epsilon = 0$ in a neighborhood of S, supp $(\varphi - \varphi\chi_\epsilon) \subset \complement S$, hence

$$\langle T, \varphi \rangle = \langle T, \varphi\chi_\epsilon \rangle \qquad \forall \varphi \in \mathcal{D}.$$

For every compact set K, there exists $C(K)$ such that

$$\langle T, \varphi \rangle = \langle T, \chi_\epsilon \rangle \leq C(K) \sup_{x \in K} \sup_{|j| \leq p} |\mathrm{D}^j(\varphi\chi_e)|$$

$$\leq C(K) \sup_{x \in K \cap S_{4\epsilon}} \sup_{|j| \leq p} |\mathrm{D}^j(\varphi_\epsilon^x)|.$$

By using the upper bounds of χ_ϵ and its derivatives (p. 432) and the fact that $\mathrm{D}^j\varphi(x) = 0$ when $x \in S$ and $|j| \leq p$, it can be proved that $\langle T, \varphi \rangle = 0$ for φ satisfying the stated conditions. ■

Theorem. A distribution T whose support is the origin is a finite combination of derivatives[1] *of the Dirac measure.*

Proof: Since T is of compact support, it is of finite order; let p be the order of T; let φ be an arbitrary C^∞ function and set

$$\psi(x) = \varphi(x) - \varphi(0) - \cdots - \sum_{|j|=p} \frac{1}{j!} x^j (\mathrm{D}^j\varphi)(0).$$

[1]See p. 446, the definition of the derivatives of a distribution.

Since ψ vanishes on the support of T together with its derivatives of order $\leq p$

$$\langle T, \psi \rangle = 0$$

$$\langle T, \varphi \rangle = c_0 \varphi(0) + \cdots + \sum_{|j|=p} c_j (D^j \varphi)(0)$$

where c_j is the value of T at the C^∞ function $x \mapsto x^j / j!$. This function is not of compact support but T, being of compact support, is defined on $\mathscr{C}(U)$. Finally we get

$$T = c_0 \delta + \cdots + \sum_{|j|=p} (-1)^{|j|} c_j \delta^{(j)}. \qquad \blacksquare$$

2. OPERATIONS ON DISTRIBUTIONS

sum

1) **Sum**

$$\langle T + S, \varphi \rangle \stackrel{\text{def}}{=} \langle T, \varphi \rangle + \langle S, \varphi \rangle.$$

product by C^∞ function

2) **Product by a C^∞ function**

$$\langle fT, \varphi \rangle \stackrel{\text{def}}{=} \langle T, f\varphi \rangle.$$

This definition is meaningful only if $f\varphi$ is in the space on which T is defined. For an arbitrary distribution $\varphi \in \mathscr{D}(U)$, then f must be a C^∞ function. When such is the case the mapping $\varphi \mapsto \langle fT, \varphi \rangle$ is linear and continuous on $\mathscr{D}(U)$: the space of distribution $\mathscr{D}'(U)$ is a module over the ring of C^∞ functions.
The product of a distribution of order p by a C^p function is defined: $\varphi \in \mathscr{D}^p(U)$ and $f \in C^p(U)$ implies $f\varphi \in \mathscr{D}^p(U)$. *In general*[1] *the product of two distributions is not defined.*

division

3) If f is a C^∞ function which never vanishes, $1/f$ is also a C^∞ function and the **division** of $T \in \mathscr{D}'$ by f is uniquely defined by the product $(1/f)T$. If f vanishes at some points there may be several distributions S, or none, such that

$$fS = T.$$

Example: Consider the equation, $S \in \mathscr{D}'(\mathbb{R})$ unknown, $T \in \mathscr{D}'(\mathbb{R})$ given

$$xS = T, \quad \text{i.e.} \quad \langle T, \varphi \rangle = \langle xS, \varphi \rangle \qquad \forall \varphi \in \mathscr{D}(R).$$

[1]For further study see [Hormander 1971], [Reed-Simon II, pp. 81–108].

Every $\varphi \in \mathscr{D}$ can be written

$$\varphi(x) = \varphi(0)\theta(x) + x\psi(x)$$

where $\psi \in \mathscr{D}$, and where $\theta \in \mathscr{D}$ is fixed and satisfies $\theta(0) = 1$. Hence, S will satisfy $xS = T$ if and only if

$$\langle S, \varphi\rangle = \varphi(0)\langle S, \theta\rangle + \langle S, x\psi\rangle$$

$$\langle S, \varphi\rangle = C\varphi(0) + \langle T, \psi\rangle.$$

The general solution S is

$$S = C\delta + S_0$$

where $C\delta$ is the general solution of $xS = 0$ and S_0 the particular solution of $xS = T$ defined by

$$\langle S_0, \varphi\rangle = \left\langle T, \frac{\varphi - \varphi(0)\theta}{x}\right\rangle.$$

Exercise. 1) Show that the general solution in $\mathscr{D}'(R)$ of

$$x^p S = 0$$

is

$$S = C_1\delta + \cdots + C_p\delta^{(p)}.$$

2) Show that the general solution in $\mathscr{D}'(U)$ of

$$f(x)S = 0$$

where $f \in C^\infty(U)$, $f(a) = 0$, $f'(a) = 0, \ldots f^{(p)}(a) = 0$, $f^{(p+1)}(a) \neq 0$, and $f(x) \neq 0$ for all $x \in U$, $x \neq a$, is

$$S = c_1\delta_a + \cdots + c_p\delta_a^{(p)}.$$

3) Let f be a C^∞ function on $\mathbb{R}^n$ such that $f(x) = 0$ is a submanifold S of $\mathbb{R}^n$ with[1] $f'(x) \neq 0$ on S then

$$f(x)T = 0 \Leftrightarrow T = c\delta_S.$$

It can be proved using the definition of a distribution by its value on open sets that if f is a C^∞ function with isolated zeros of finite order, the division of any $T \in \mathscr{D}'(\mathbb{R}^n)$ by f is possible.

4) **Direct product.** Let U be an open set in $\mathbb{R}^n$ and V an open set in $\mathbb{R}^p$, then $U \times V$ is an open set in $\mathbb{R}^{n+p}$. Let $\varphi_{xy} \in \mathscr{D}(U \times V)$, $S_x \in \mathscr{D}'(U)$ and $T_y \in \mathscr{D}'(V)$. The **direct product** $S_x \times T_y$ is a distribution defined on direct product

[1] $f'(x)$ means the vector grad f (cf. Ch. II).

$\mathscr{D}(U \times V)$ by either of the following equations

$$\langle S_x \times T_y, \varphi_{xy}\rangle \overset{\text{def}}{=} \langle S_x, \langle T_y, \varphi_{xy}\rangle\rangle$$

$$\langle S_x \times T_y, \varphi_{xy}\rangle \overset{\text{def}}{=} \langle T_y, \langle S_x, \varphi_{xy}\rangle\rangle$$

where $\langle T_y, \varphi_{xy}\rangle$ is the function defined on U by

$$x \mapsto \langle T_y, \varphi_{xy}(x, .)\rangle$$

and $\langle S_x, \varphi_{xy}\rangle$ is defined on V by $y \mapsto \langle S_x, \varphi_{xy}(. , y)\rangle$.
The definitions are legitimate because these functions are respectively in $\mathscr{D}(U)$ and $\mathscr{D}(V)$; indeed $x \mapsto \langle T_y, \varphi_{xy}(x, .)\rangle$ is the composite mapping of

$$U \to \mathscr{D}(V) \quad \text{by } x \mapsto \varphi_{xy}(x, .)$$

and

$$\mathscr{D}(V) \to \mathbb{K} \quad \text{by } \varphi_{xy}(x, .) \mapsto \langle T_y, \varphi_{xy}(x, .)\rangle.$$

These two mappings are C^∞ (cf. p. 73) – the second is linear – therefore $x \mapsto \langle T_y, \varphi_{xy}(x, .)\rangle$ is C^∞. Moreover φ_{xy} having compact support K in $\mathscr{D}(U \times V)$ implies that $\langle T_y, \varphi_{xy}\rangle$ has compact support in U since $\langle T_y, \varphi_{xy}(x, .)\rangle$ vanishes, as $\varphi_{xy}(x, .)$ does, when x does not belong to the compact projection of K on U.

The equivalence of the two definitions follows from the fact that
a) if $\varphi \in \mathscr{D}(U) \times \mathscr{D}(V) \subset \mathscr{D}(U \times V)$ then $\varphi_{xy} = \xi_x \eta_y$ with $\xi_x \in \mathscr{D}(U)$ and $\eta_y \in \mathscr{D}(V)$ and

$$\langle S_x \times T_y, \xi_x \eta_y\rangle = \langle S, \xi\rangle, \langle T, \eta\rangle,$$

b) every $\varphi \in \mathscr{D}(U \times V)$ is a limit of finite sums of functions in $\mathscr{D}(U) \times \mathscr{D}(V)$.
The labels x, y which indicate on which space the functions or distributions are defined are omitted when unnecessary. The notation $\varphi(x)$ is often used instead of φ_x. Its typographical convenience will become apparent later, for example in the discussion of convolution. When it is used it must be understood not to be the complex or real value of φ at the point x but rather the mapping φ on the space where points are denoted by x. The context makes it clear when $\varphi(x) \in \mathbb{K}$ or when $\varphi(x) \equiv \varphi_x \in \mathscr{D}(U)$.

It is easy to see that

$$\operatorname{supp}(S \times T) = \operatorname{supp} S \times \operatorname{supp} T.$$

derivative of a distribution

5) **Derivation.** The **partial derivative** $\partial T/\partial x^i$ **of a distribution** is the

distribution defined by

$$\left\langle \frac{\partial T}{\partial x^i}, \varphi \right\rangle \overset{\text{def}}{=} -\left\langle T, \frac{\partial \varphi}{\partial x^i} \right\rangle \qquad \forall \varphi \in \mathscr{D}(U).$$

This definition makes sense because if $\varphi \in \mathscr{D}(U)$ then $\partial\varphi/\partial x^i \in \mathscr{D}(U)$ and $\varphi \mapsto -\langle T, \partial\varphi/\partial x^i \rangle$ is linear and continuous on $\mathscr{D}(U)$. Note that if T is of order p on a compact set then $\partial T/\partial x^i$ is of order $\leq p+1$ on the same compact set.
The definition is legitimate because if T is equivalent to a C^1 function f, then $\partial T/\partial x^i$ is equivalent to $\partial f/\partial x^i$:

$$\left\langle \frac{\partial f}{\partial x^i}, \varphi \right\rangle = \int_U \frac{\partial f}{\partial x^i} \varphi \, \mathrm{d}x = -\int_U f \frac{\partial \varphi}{\partial x^i} \mathrm{d}x = -\left\langle f, \frac{\partial \varphi}{\partial x^i} \right\rangle, \quad \forall \varphi \in \mathscr{D}(U).$$

Higher derivatives are obtained by recurrence

$$\langle \mathrm{D}^j T, \varphi \rangle = (-1)^{|j|} \langle T, \mathrm{D}^j \varphi \rangle \qquad \forall \varphi \in \mathscr{D}(U).$$

The following propositions follow immediately from the definition, they are the great virtue of distributions.

Proposition. A distribution is indefinitely differentiable.

Proposition. $\dfrac{\partial^2 T}{\partial x^i \partial x^j} = \dfrac{\partial^2 T}{\partial x^j \partial x^i}$.

Exercise: Show that the derivative of a product follows the usual rule.

Answer: With the conditions under which the product fT is defined,

$$\left\langle \frac{\partial}{\partial x^i}(fT), \varphi \right\rangle = -\left\langle fT, \frac{\partial \varphi}{\partial x^i} \right\rangle = -\left\langle T, f\frac{\partial \varphi}{\partial x^i} \right\rangle = \left\langle T, \frac{\partial}{\partial x^i}(f\varphi) - \frac{\partial f}{\partial x^i}\varphi \right\rangle$$
$$= \left\langle \frac{\partial T}{\partial x^i}, f\varphi \right\rangle + \left\langle T, \frac{\partial f}{\partial x^i}\varphi \right\rangle = \left\langle f\frac{\partial T}{\partial x^i} + \frac{\partial f}{\partial x^i} T, \varphi \right\rangle. \qquad \blacksquare$$

Examples of distributions on $\mathbb{R}$.

Example 1: The **Heaviside function** $Y(x) = \begin{cases} 1 & \text{for } x > 0 \\ 0 & \text{for } x < 0 \\ \text{undefined for } x = 0. \end{cases}$ **Heaviside function**

$$\langle Y', \varphi \rangle = -\langle Y, \varphi' \rangle = -\int_0^\infty \varphi' \mathrm{d}x = \varphi(0),$$

thus

$$Y' = \delta.$$

Example 2: Derivatives of the Dirac measure

$$\langle \delta', \varphi \rangle = -\langle \delta, \varphi' \rangle = -\varphi'(0)$$

$$\langle \delta^{(p)}, \varphi \rangle = (-1)^p \varphi^{(p)}(0).$$

Example 3: Let f be C^p in $\mathbb{R}^- \equiv (-\infty, 0]$ and in $\mathbb{R}^+ \equiv [0, +\infty)$, let $f(-0)$ and $f(+0)$ be the limits of $f(x)$ when x goes to 0 respectively in $\mathbb{R}^-$ and $\mathbb{R}^+$. Set

$$\sigma_0 = f(+0) - f(-0)$$

$$\sigma_{p-1} = f^{(p-1)}(+0) - f^{(p-1)}(-0).$$

Let $\tilde{f}$ be the distribution which is associated with f,

$$\langle \tilde{f}, \varphi \rangle = \int_U f\varphi \, \mathrm{d}x.$$

A simple computation shows that

$$\mathrm{D}^j \tilde{f} = \widetilde{\mathrm{D}^j f} + \sigma_{j-1}\delta + \cdots + \sigma_0 \delta^{(j-1)} \quad \text{valid for } j \leq p.$$

If f is C^p on $\mathbb{R}$, $\sigma_0 = \sigma_1 = \cdots \sigma_{p-1} = 0$ and $\mathrm{D}^j \tilde{f} = \widetilde{\mathrm{D}^j f}$.

Examples of distributions on $\mathbb{R}^n$. The last two examples can immediately be generalized to distributions on $\mathbb{R}^n$.

Example 1: Derivatives of the Dirac measure

$$\langle \mathrm{D}^j \delta, \varphi \rangle = (-1)^{|j|} \mathrm{D}^j \varphi(0).$$

Example 2: Let $\tilde{f}$ be a distribution equivalent to a C^p function f on U. $\mathrm{D}^j \tilde{f}$ is equivalent to $\mathrm{D}^j f$, i.e. $\tilde{\mathrm{D}}^j \tilde{f} = \widetilde{\mathrm{D}^j f}$, on U for $j \leq p$.

Example 3: Let[1] U be an open connected set in $\mathbb{R}^n$, let S be a continuously differentiable submanifold dividing U into two open disjoint sets U_+ and U_- such that both U_+ and U_- admit S as a boundary relative to U (in the sense of the Stokes formula, cf. p. 216). Let $S(x) = 0$ be an irreducible equation for S. Let f be a C^1 function in U_+ and U_- which can be extended continuously to S from U_+ [resp. U_-] by f_+ [resp. f_-], let

$$f_+ - f_- = \sigma_0 \text{ with } \sigma_0 \text{ continuous on } S.$$

We have

$$\left\langle \frac{\partial \tilde{f}}{\partial x^i}, \varphi \right\rangle = -\left\langle \tilde{f}, \frac{\partial \varphi}{\partial x^i} \right\rangle = -\int_{U^+} f \frac{\partial \varphi}{\partial x^i} \, \mathrm{d}x - \int_{U^-} f \frac{\partial \varphi}{\partial x^i} \, \mathrm{d}x, \quad \forall \varphi \in \mathcal{D}(U).$$

[1]This example is fundamental for the derivation of the shock wave equation through distribution theory. See Problem 11.

Let ω be the Leray form on S and (n_i) be the normal on S

$$dS = n_i\, dx^i, \quad n_i = \partial S/\partial x^i, \quad dS \wedge \omega = dx^1 \wedge \ldots \wedge dx^n$$

$$\omega = (dx^2 \wedge \ldots \wedge dx^n)/n_1 = -(dx^1 \wedge dx^3 \wedge \ldots \wedge dx^n)/n_2 = \cdots$$

$$\int_{U_+} f \frac{\partial \varphi}{\partial x^i}\, dx = \int_{U_+} \frac{\partial}{\partial x^i}(f\varphi)\, dx - \int_{U_+} \varphi \frac{\partial f}{\partial x^i}\, dx$$

$$= \int_{\partial U_+} f_+ \varphi n_i^+ \omega - \int_{U_+} \varphi \frac{\partial f}{\partial x^i}\, dx.$$

φ is zero on the boundary of U; on S, the i-component of the normal to ∂U_+ is opposite to the i-component of the normal to ∂U_-: $n_i^+ = -n_i^-$; hence, if n_i^- is the outward normal to U_-

$$\frac{\partial \tilde{f}}{\partial x^i} = \widetilde{\frac{\partial f}{\partial x^i}} + \sigma_0 n_i^- \delta_S.$$

The product $\sigma_0 n_i^- \delta_S$ is well defined because δ_S is a measure of support S and $\sigma_0 n_i^-$ is a continuous function on S, indeed

$$\langle \sigma_0 n_i^- \delta_S, \varphi \rangle = \int_S \sigma_0 n_i^- \varphi \omega.$$

Note that $\sigma_0 n_i^-$ is independent of the direction S is crossed since σ_0 and n_i^- both change sign if the direction is reversed.

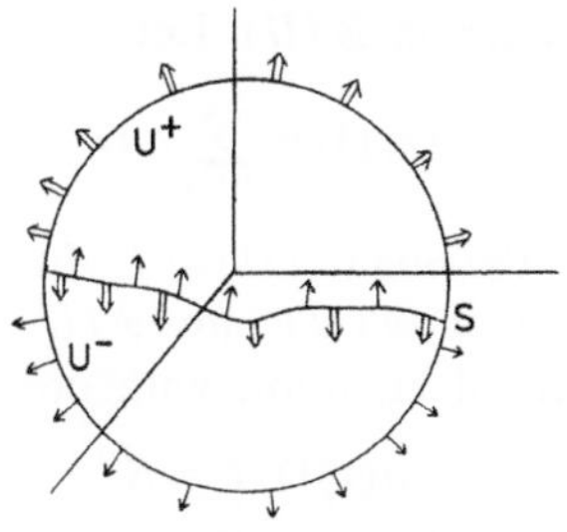

Example 4: Δr^{2-n} in $\mathbb{R}^n$ and $-\Delta \log r$ in $\mathbb{R}^2$ where $r^2 = \Sigma_{i=1}^n x_i^2$. If $n > 2$, r^{2-n} is a locally integrable function on $\mathbb{R}^n$; it is a C^∞ function on $\mathbb{R}^n - \{0\}$. Using

$$\frac{\partial^2 f(r)}{(\partial x_i)^2} = \frac{d^2 f}{dr^2}\left(\frac{x_i}{r}\right)^2 + \frac{df}{dr}\left(\frac{1}{r} - \frac{x_i^2}{r^3}\right),$$

we obtain

$$\Delta f(r) = \frac{d^2 f}{dr^2} + \frac{n-1}{r}\frac{df}{dr}$$

$$\Delta r^{2-n} = 0 \text{ in } \mathbb{R}^n - \{0\}.$$

Hence Δr^{2-n} is a distribution of order ≤ 2 whose support is the origin, it is a linear combination of $\delta^{(j)}$ for $|j| = 0, 1, 2$,

$$\langle \Delta r^{2-n}, \varphi \rangle = \langle r^{2-n}, \Delta\varphi \rangle = \sum_{|j|=0}^{2} c_j (\mathrm{D}^j \varphi)(0).$$

Let $\bar{\varphi}(r)$ be the average of φ on the $n-1$ sphere of radius r

$$\bar{\varphi}(r) = \frac{1}{S_{n-1} r^{n-1}} \int_{|x|=r} \varphi(x)\omega$$ where ω is the Leray form on the sphere.

The value S_{n-1} of the area of the $n-1$ sphere of radius 1 is

$$S_{n-1} = (2\pi)^{n/2}/\Gamma(n/2), \quad S_1 = 2\pi, \quad S_2 = 4\pi.$$

The laplacian being invariant under rotation, we have $\overline{\Delta\varphi} = \Delta\bar{\varphi}$ and

$$\langle r^{2-n}, \Delta\varphi \rangle = S_{n-1} \int_0^\infty \frac{r^{n-1}}{r^{n-2}} \Delta\bar{\varphi}(r)\,\mathrm{d}r = S_{n-1} \int_0^\infty \mathrm{d}r((r\bar{\varphi}')' + (n-2)\bar{\varphi}')$$

$$= -(n-2)S_{n-1}\bar{\varphi}(0) = -(n-2)S_{n-1}\varphi(0)$$

$$\Delta r^{2-n} = -(n-2)S_{n-1}\delta.$$

Similarly $-\Delta \log r = -2\pi\delta$ in $\mathbb{R}^2$.

Example 5: Problem 2 on the laplacian of a discontinuous function.

Linear differential operators on $\mathscr{D}'(U)$. Let

$$a(x, \mathrm{D}) \equiv \sum_{|j| \leq m} a_j \mathrm{D}^j,$$

where a_j is an element of the ring $C^\infty(U)$ of C^∞ functions on $U \subset \mathbb{R}^n$. $a(x, \mathrm{D})$ defines a linear mapping of $\mathscr{D}'(U)$ into $\mathscr{D}'(U)$, m is called its degree. A linear partial differential equation with coefficients in $C^\infty(U)$ is an equation of the form

$$a(x, \mathrm{D})T = S$$

where T is an unknown distribution and S a given distribution, both in $\mathscr{D}'(U)$.

Sometimes partial differential equations on $U \subset \mathbb{R}^n$ are just called differential equations, and ordinary differential equations when $n = 1$.

Inverse derivative of a distribution. The simplest partial differential equation is the first order differential equation on $\mathscr{D}'(\mathbb{R})$

$$\mathrm{d}T/\mathrm{d}x = S.$$

This equation is equivalent to

$$-\langle T, \varphi' \rangle = \langle S, \varphi \rangle \qquad \forall \varphi \in \mathscr{D}(\mathbb{R}).$$

It determines T only for those functions in $\mathscr{D}(\mathbb{R})$ which are derivatives of functions in $\mathscr{D}(\mathbb{R})$. We shall show that there are an infinity of inverse derivatives of the distribution S. Now let θ be an arbitrary chosen function in $\mathscr{D}(\mathbb{R})$ such that it is not the derivative of a function in $\mathscr{D}(\mathbb{R})$, i.e. $\int_{\mathbb{R}} \theta \, dx \neq 0$. *If we require the inverse derivative to take an arbitrarily chosen value at θ, then the inverse derivative is uniquely determined.*

Proof: Let us determine the number λ so that $\psi - \lambda\theta = \varphi'$ where ψ is an arbitrary function in $\mathscr{D}(\mathbb{R})$ and where φ' is the derivative of a function in $\mathscr{D}(\mathbb{R})$; the necessary and sufficient condition for φ to be in $\mathscr{D}(\mathbb{R})$ is

$$\int_{\mathbb{R}} (\psi - \lambda\theta) \, dx = 0 \text{ which implies } \lambda = \int_{\mathbb{R}} \psi \, dx \bigg/ \int_{\mathbb{R}} \theta \, dx,$$

φ is then uniquely determined by ψ. Hence for every $\psi \in \mathscr{D}(\mathbb{R})$ and the corresponding λ

$$\langle T, \psi \rangle = \lambda \langle T, \theta \rangle - \langle S, \varphi \rangle.$$

In the theory of distributions, as in the theory of functions, two inverse derivatives differ by a constant (i.e. a distribution equivalent to a constant function) indeed if we denote by C the number, independent of ψ,

$$C = \langle T, \theta \rangle \bigg/ \int_{\mathbb{R}} \theta \, dx$$

$$\langle T, \psi \rangle = C \int_{\mathbb{R}} \psi \, dx - \langle S, \varphi \rangle.$$

It follows that

$$\langle T, \psi \rangle = \langle \tilde{C}, \psi \rangle - \langle S, \varphi \rangle$$

where the constant distribution C characterises the solution T amongst the infinity of solutions. ■

If $S = \tilde{f}$ where f is a continuous function, its inverse derivatives are continuous and equivalent to the inverse derivative of f.

To what extent is the theory of distributions necessary to solve differential equations? Our first answer is to give a theorem which states when function theory is sufficient to obtain all the solutions of a system.

Theorem. The solutions in $\mathscr{D}'(\mathbb{R})$ of a linear ordinary differential equation of order m [resp. of a system of m first order differential equations with m unknowns] with coefficients in $C^\infty(\mathbb{R})$ such that the coefficient of the term of order m [resp. the determinant of the coefficients of the first order terms] does not vanish on $\mathbb{R}$ are the usual C^∞ solutions of the theory of functions.

Proof: Consider first a single first order linear differential equation

$$\frac{\mathrm{d}T}{\mathrm{d}x} + a(x)T = b(x) \qquad a, b \in C^\infty(\mathbb{R}).$$

If T is a distribution solution of this equation, then $S = e^A T$, where A is the inverse derivative of a, is a distribution solution of the equation

$$\frac{\mathrm{d}S}{\mathrm{d}x} = e^{A(x)}b(x).$$

But S is the inverse derivative of a C^∞ function, hence it is a C^∞ function and therefore so is T.
A similar demonstration holds in the case of a system of m first order equations with m unknowns and hence for one equation of order m. ∎

The theorem does not apply when the coefficient of the term of higher order vanishes on $\mathbb{R}$.

Example: $x(\mathrm{d}T/\mathrm{d}x) = 0$, then $(\mathrm{d}T/\mathrm{d}x) = C_0\delta$ and $T = C_0 Y + C_1$.

The theorem does not apply to partial differential equations.

Example: The solution of the following P.D.E. in $\mathbb{R}^2$

$$\partial T/\partial x = 0$$

is a distribution "depending only on y", more precisely $T = 1_x \times S_y$. That is for every $\varphi \in \mathscr{D}(\mathbb{R}^2)$

$$\langle T, \varphi\rangle = \langle S_y, \langle 1_x, \varphi_{xy}\rangle\rangle = \langle S, \psi\rangle$$

where S is a distribution on $\mathbb{R}$ and $\psi \in \mathscr{D}(\mathbb{R})$ is defined by

$$\psi(y) = \int_{-\infty}^{+\infty} \varphi(x, y)\,\mathrm{d}x.$$

The proof is analogous to the one previously given for the inverse derivative: write any $\varphi \in \mathscr{D}(\mathbb{R}^2)$ as $\varphi(x, y) = \theta(x)\psi(y) + \mu(x, y)$ with $\theta \in$

$\mathscr{D}(\mathbb{R})$ arbitrary, but fixed, and $\lambda \in \mathscr{D}(\mathbb{R})$ such that $\mu = \partial\psi/\partial x$, $\psi \in \mathscr{D}(\mathbb{R}^2)$:

$$\psi(y) = \int_{-\infty}^{+\infty} \varphi(x, y)\, dx \text{ (we choose } \int_{-\infty}^{+\infty} \theta(x)\, dx = 1)$$

$$\langle T, \varphi \rangle = \langle T_{xy}, \theta(x)\psi(y) \rangle = \langle \langle T_{xy}, \theta(x) \rangle, \psi(y) \rangle.$$

3. TOPOLOGY ON $\mathscr{D}'$

Several topologies can be given to the topological dual of a topological vector space of infinite dimension. We shall use here the topology called the weak topology (weak star topology, simple convergence topology). The **weak topology** on $\mathscr{D}'$ is defined by the family of seminorms $\{p_\varphi; \varphi \in \mathscr{D}\}$ with

weak star topology

$$p_\varphi(T) = |\langle T, \varphi \rangle|.$$

Proposition: The weak topology on $\mathscr{D}'$ is Hausdorff.

Proof: To say that $T_0 \in \mathscr{D}'$ is different from zero is to say that there exists $\varphi \in \mathscr{D}$ such that $\langle T_0, \varphi \rangle \neq 0$, therefore the family of seminorms $|\langle T, \varphi \rangle|$ is separated. ■

Criterion of continuity in $\mathscr{D}'$. This criterion valid for all weak topologies is stated here for the weak topology in $\mathscr{D}'$: *Let X be a topological space.*

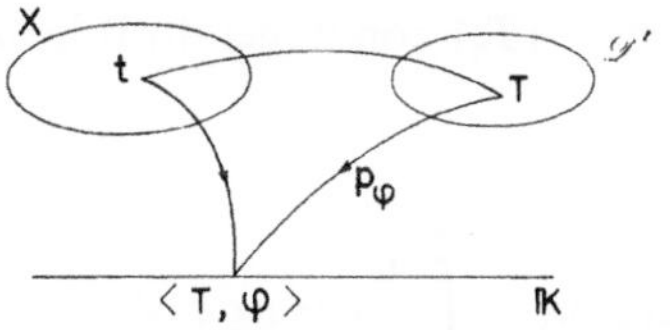

The mapping $t \in X \mapsto T \in \mathscr{D}'$ is continuous if and only if the mapping $t \mapsto \langle T, \varphi \rangle$ is continuous for every $\varphi \in \mathscr{D}$.

Proof: The mapping $T \mapsto \langle T, \varphi \rangle$ is a seminorm $p_\varphi: \mathscr{D}' \to \mathbb{K}$ and hence continuous. If $t \mapsto T$ is continuous then the mapping $t \mapsto \langle T, \varphi \rangle$ is the composite mapping of two continuous mappings, hence it is continuous. Conversely, if $t \mapsto \langle T, \varphi \rangle$ is continuous for every φ, $t \mapsto T$ is continuous since the family $|\langle T, \varphi \rangle|$ is identical with the family of semi-norms defining the topology of $\mathscr{D}'$. ■

Theorem. The derivation is a linear continuous mapping of $\mathscr{D}'$ into $\mathscr{D}'$.

Proof: The mapping $T \mapsto \partial T/\partial x^i$ is continuous if and only if the mapping: $\mathscr{D}' \to \mathbb{K}$ by $T \mapsto \langle \partial T/\partial x^i, \varphi\rangle$ is continuous for every φ.
Indeed, $\langle \partial T/\partial x^i, \varphi\rangle = -\langle T, \partial\varphi/\partial x^i\rangle = \langle T, \psi\rangle$ with $\psi = -\partial\varphi/\partial x^i$ and $|\langle T, \psi\rangle|$ is a seminorm on $\mathscr{D}'$, hence continuous. ■

Criterion of convergence in $\mathscr{D}'$. This criterion is valid for the simple convergence of linear forms on a topological vector space. It is a particular case of the criterion of continuity in the weak star topology.
A sequence (T_n) converges to T in $\mathscr{D}'$ if and only if $\langle T_n, \varphi\rangle$ converges to $\langle T, \varphi\rangle$ for every $\varphi \in \mathscr{D}$.

Proposition. Let f_n be locally integrable functions such that the sequence (f_n) converges to f in L^1 on every compact set, then the sequence (f_n) converges to f in $\mathscr{D}'$.

Proof:

$$\left|\int_{\mathbb{R}^n} (f_n - f)\varphi \,\mathrm{d}x\right| \le \sup_{x\in K} |\varphi| \int_K |f_n - f| \,\mathrm{d}x \qquad \forall \varphi \in \mathscr{D}_K. \quad ■$$

A consequence of the proposition is

$$L^1_{\mathrm{loc}} \hookrightarrow \mathscr{D}'.$$

Here are some examples of convergence in $\mathscr{D}'(\mathbb{R})$.

Example 1: $T_n = \sin nx$, (T_n) converges to 0 in $\mathscr{D}'$.

Proof:

$$\langle T_n, \varphi\rangle = \int_{-A}^{A} \sin nx\varphi \,\mathrm{d}x = \int_{-A}^{A} \frac{1}{n} \cos nx\, \varphi' \,\mathrm{d}x \qquad \forall \varphi \in \mathscr{D}([-A, A]).$$

$\int_{-A}^{A} \cos nx \,\mathrm{d}x$ is uniformly bounded, hence

$$\lim_{n=\infty} \langle T_n, \varphi\rangle = 0 \text{ in } \mathscr{D}'.$$

Example 2: $T_n = \dfrac{1}{x} \sin nx$, (T_n) converges to $\pi\delta$ in $\mathscr{D}'$.

Proof:

$$\langle T_n, \varphi\rangle = \int_{-A}^{A} \frac{1}{x} \sin nx(\varphi(x) - \varphi(0)) \,\mathrm{d}x + \varphi(0) \int_{-A}^{A} \frac{1}{x} \sin nx \,\mathrm{d}x.$$

The first integral converges to zero (cf. Example 1), the second is equal to

$$\varphi(0)\int_{-nA}^{nA}\frac{1}{y}\sin y\,\mathrm{d}y$$

and it is known that

$$\lim_{n=\infty}\int_{-nA}^{nA}\frac{1}{y}\sin y\,\mathrm{d}y=\pi.$$

Example 3: $T_n=\int_{-n}^{n}\exp(\mathrm{i}xy)\,\mathrm{d}y$, (T_n) converges to $2\pi\delta$ in $\mathscr{D}'$.

Proof:

$$T_n=\frac{1}{\mathrm{i}x}(\mathrm{e}^{\mathrm{i}nx}-\mathrm{e}^{-\mathrm{i}nx})=\frac{2}{x}\sin nx.$$

Example 4: $T_n=\int_{-n}^{n}y\exp(\mathrm{i}xy)\,\mathrm{d}y$, (T_n) converges to $-2\mathrm{i}\pi\delta'$ in $\mathscr{D}'$.

Proof:

$T_n=\dfrac{1}{\mathrm{i}}\dfrac{\mathrm{d}}{\mathrm{d}x}\displaystyle\int_{-n}^{n}\exp(\mathrm{i}xy)\,\mathrm{d}y$ and derivation is a continuous mapping in $\mathscr{D}'$.

Here are some examples of convergence in $\mathscr{D}'(\mathbb{R}^n)$.

Example 1: The previous examples can readily be generalized to $\mathbb{R}^n$

$$\lim_{n=\infty}\int_{-n}^{+n}\cdots\int_{-n}^{+n}\exp(\mathrm{i}xy)\,\mathrm{d}y=(2\pi)^n\delta(x)\text{ in }\mathscr{D}'(\mathbb{R}^n),$$

here $xy=x^1y^1+\cdots+x^ny^n$ and $\delta_x=\delta_{x^1}\times\cdots\delta_{x^n}$.

Example 2: $T_\epsilon=\rho_\epsilon$ (p. 433), (T_ϵ) converges to δ as ϵ tends to 0.

Proof:

$$\lim_{\epsilon=0}\int_{\mathbb{R}^n}\rho_\epsilon(z-x)\varphi(x)\,\mathrm{d}x=\varphi(z)$$

implies

$$\lim_{\epsilon=0}\int_{\mathbb{R}^n}\rho_\epsilon(x)\varphi(x)\,\mathrm{d}x=\varphi(0).$$

Theorem. *$\mathscr{D}'$ is sequentially complete in the weak topology.*

Proof: If (T_n) is a Cauchy sequence in $\mathscr{D}'$, $(\langle T_n, \varphi\rangle)$ is a Cauchy sequence in $\mathbb{K}$, by the definitions. Hence $(\langle T_n, \varphi\rangle)$ converges, for each φ, to a number T_φ. It is clear that T_φ depends linearly on φ. It can be proved, using the Baire theorem applied to $\mathscr{D}_K$, that it depends continuously on φ.

4. CHANGE OF VARIABLES IN $\mathbb{R}^n$

Let $x = (x^1, \ldots, x^n)$ and $y = (y^1, \ldots, y^n)$, let the mapping $\Phi: \mathbb{R}^n \to \mathbb{R}^n$ by $x \mapsto y = \Phi(x)$ define a change of coordinates, i.e. Φ is a C^∞ diffeomorphism which associates the y-coordinates $(y^1, \ldots, y^n)$ with the x-coordinates $(x^1, \ldots, x^n)$ of the same point. Let $J^i_j(x) = \partial y^i/\partial x^j$ and $\mathscr{J}(x) = \mathrm{D}(y)/\mathrm{D}(x)$.

The transformation law of a distribution T under the change of coordinates Φ is defined by

$$\langle \Phi T, \psi\rangle = \langle T, \varphi\rangle \text{ where } \varphi = |\mathscr{J}|\psi \circ \Phi, \text{ that is } \varphi(x) = \psi(y(x))|\mathscr{J}(x)|.$$

This definition makes sense because $\psi \mapsto \varphi$ is a continuous mapping from $\mathscr{D}$ to $\mathscr{D}$.

It is legitimate because, when the distribution T can be identified with a function f we have

$$\langle f, \varphi\rangle = \int_{\mathbb{R}^n} f(x)\varphi(x)\,\mathrm{d}x.$$

Thus by the usual rule for the change of variables in an integral (cf. Chapter I)

$$\langle f, \varphi\rangle = \int_{\mathbb{R}^n} (f \circ \Phi^{-1})(y) \cdot (\varphi \circ \Phi^{-1})(y) \cdot \left|\frac{\mathrm{D}(x)}{\mathrm{D}(y)}\right| \mathrm{d}y$$

which reads

$$\langle f, \varphi\rangle = \langle f \circ \Phi^{-1}, \psi\rangle, \quad \varphi = |\mathscr{J}| \cdot \psi \circ \Phi$$

and gives the usual transformation law for a function

$$\Phi f = f \circ \Phi^{-1},$$

in agreement with the following diagram.

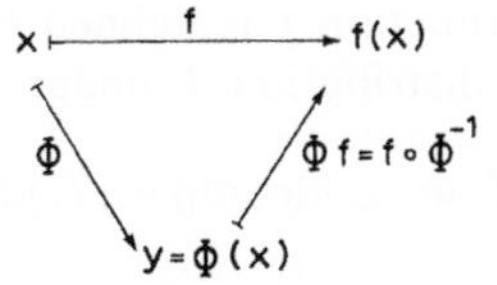

Exercise: Show that the transformation law of the Dirac measure is

$$\Phi\delta = |\mathcal{J}|_{x=0}\delta_{y(0)}.$$

The partial derivatives obey the usual rule (summation over an index which appears twice is assumed in this paragraph)

$$\frac{\partial \Phi T}{\partial y^j} = \Phi\left(\frac{\partial x^i}{\partial y^j}\frac{\partial T}{\partial x^i}\right).$$

Proof:

$$\left\langle \Phi\left(\frac{\partial x^i}{\partial y^j}\frac{\partial T}{\partial x^i}\right), \psi\right\rangle = \left\langle \frac{\partial x^i}{\partial y^j}\frac{\partial T}{\partial x^i}, \psi(y(x))|\mathcal{J}|\right\rangle$$
$$= -\left\langle T, \frac{\partial}{\partial x^i}\left[\frac{\partial x^i}{\partial y^j}\psi(y(x))|\mathcal{J}|\right]\right\rangle.$$

Using

$$\partial x^i/\partial y^j = (J^{-1})^i_j$$

we obtain

$$\frac{\partial}{\partial x^i}\frac{\partial x^i}{\partial y^j} = J^k_i\frac{\partial}{\partial y^j}(J^{-1})^i_k = -|\mathcal{J}|^{-1}\frac{\partial}{\partial y^j}|\mathcal{J}|$$

and finally

$$\left\langle \Phi\left(\frac{\partial x^i}{\partial y^j}\frac{\partial T}{\partial x^i}\right), \psi\right\rangle = -\left\langle T, |\mathcal{J}|\frac{\partial \psi}{\partial y^j}(y(x))\right\rangle.$$

On the other hand

$$\left\langle \frac{\partial \Phi T}{\partial y^j}, \psi\right\rangle = -\left\langle \Phi T, \frac{\partial \psi}{\partial y^j}\right\rangle = -\left\langle T, \frac{\partial \psi}{\partial y^j}(y(x))|\mathcal{J}|\right\rangle. \qquad \blacksquare$$

The proof could also have been made by using the fact that C^1 is dense in $\mathcal{D}'$ (cf. p. 465) and that derivation is a continuous mapping in $\mathcal{D}'$.

Transformation of a distribution under a diffeomorphism Φ. The formula is just a reinterpretation of the one given above.
Let Φ: $U \subset \mathbb{R}^n \to V \subset \mathbb{R}^n$.

The transform Φf of a function f is defined by $\Phi f = f \circ \Phi^{-1}$.
The transform ΦT of a distribution T under Φ is defined by

$$\langle \Phi T, \varphi \rangle = \langle T, \Phi^{-1}\varphi \, |\det \Phi| \rangle = \langle T, |\det \Phi| \, \varphi \circ \Phi \rangle.$$

Examples: 1) Translation operator τ_a.
τ_a is the diffeomorphism $x \mapsto y = x + a$.
Transformation of a distribution T under τ_a:

$$\langle \tau_a T, \varphi \rangle = \langle T, \tau_a^{-1}\varphi \rangle \qquad \text{since} \quad |\det \tau_a| = 1.$$

According to an incorrect but usual notation, we have, as in the case of functions

$$(\tau_a T)(x) = T(x - a).$$

2) Dilation operator ϵ_α.
ϵ_α is the diffeomorphism $x \mapsto y = \alpha x$ $(\alpha \neq 0)$ and

$$\begin{aligned}\langle \epsilon_\alpha T, \varphi \rangle &= \langle T, \epsilon_\alpha^{-1}\varphi |\det \epsilon_\alpha| \rangle \\ &= \langle T(x), \varphi(\alpha x)\alpha^n \rangle.\end{aligned}$$

Hence the notation, correct in the case of functions

$$(\epsilon_\alpha T)(x) = T(x/\alpha).$$

Remark: We have seen in Chapter I (pp. 37, 45) another definition of the image of a measure, defined as a set function. We defined there the image of a Radon measure μ by a continuous mapping (which need not be differentiable, or bijective) as

$$\int_{\Phi(U)} \varphi \, d(\Phi\mu) = \int_U \varphi \circ \Phi \, d\mu,$$

that is

$$\langle \Phi\mu, \varphi \rangle = \langle \mu, \varphi \circ \Phi \rangle.$$

The difference in the two definitions comes from the fact that a distribution of order zero on $U \subset \mathbb{R}^n$ may indeed be identified – in a given base of $\mathbb{R}^n$ – with a Radon measure, but here we want distributions that generalize functions rather than regular n-forms. The image of $f: U \to \mathbb{K}$ by a diffeomorphism $\Phi: U \to V$ is $\Phi f = f \circ \Phi^{-1}$, and the image of the measure $\mu = f \, dx$, with dx the Lebesgue measure on U, is $\Phi\mu = f \circ \Phi^{-1}|\det \Phi| \, dy$, with dy the Lebesgue measure on $\Phi(U)$.
In each coordinate system there is an isomorphism between $\mathscr{D}'^0$ and the space of Radon measures, but the natural definitions for the transformation laws are different.

Invariance. A distribution is said to be **invariant** in $\mathbb{R}^n$ under the diffeomorphism Φ if invariant

$$\Phi T = T.$$

Examples: The Dirac measure is invariant under the linear mapping $\Phi(x) = Ax$ in $\mathbb{R}^n$ when det $A = \pm 1$.
Dirac's measure δ_s on a sphere in $\mathbb{R}^n$ centered at the origin is invariant under rotation. The same is true for Dirac's multiplets which are obtained by taking the derivatives of δ_s along the normal to the sphere

$$\langle \mathrm{D}_r^{(n)}\delta_s, \varphi\rangle = (-1)^n R^{n-1} \int_{B_1} \frac{\partial^n \varphi}{\partial r^n}\, \mathrm{d}\Omega$$

where R is the radius of the sphere, B_1 the unit sphere and $\mathrm{d}\Omega$ the surface element on the unit sphere.
The distribution $\Delta\delta$, Δ the Laplace operator, is invariant under rotation.

5. CONVOLUTION

Convolution algebra $L^1(\mathbb{R}^n)$. Let f and g be two functions in $L^1(\mathbb{R}^n)$. The direct product $f \times g$ is a function on $\mathbb{R}^{2n}$ defined by

$$(f \times g)(x, y) = f(x)g(y).$$

We know (cf. Chapter I) that if f and g are in $L^1(\mathbb{R}^n)$, the direct product $f \times g$ is in $L^1(\mathbb{R}^{2n})$. Now by a change of variables $y = z - x$

$$\int_{\mathbb{R}^n}\int_{\mathbb{R}^n} (f \times g)(x, y)\, \mathrm{d}x\, \mathrm{d}y = \int_{\mathbb{R}^n}\int_{\mathbb{R}^n} f(x)g(z-x)\, \mathrm{d}x\, \mathrm{d}z$$

$$= \int_{\mathbb{R}^n} h(z)\, \mathrm{d}z$$

where

$$h(z) = \int_{\mathbb{R}^n} f(x)g(z-x)\, \mathrm{d}x = \int_{\mathbb{R}^n} g(x)f(z-x)\, \mathrm{d}x.$$

It follows from Fubini's theorem that h is defined almost everywhere and belongs to $L^1(\mathbb{R}^n)$.
h is called the **convolution product** of f and g, and is denoted convolution product

$$h = f * g = g * f.$$

The space $L^1(\mathbb{R}^n)$ together with the convolution product is a Banach algebra.

The following remarks pave the way for the definition of the convolution of distributions.
1) If f and g are in $L^1(\mathbb{R}^n)$, $h = f * g$, we have, omitting the tilde sign over distributions associated with functions,

$$\langle h, \varphi \rangle = \int_{\mathbb{R}^n} \int_{\mathbb{R}^n} f(x) g(z - x) \varphi(z) \, \mathrm{d}x \, \mathrm{d}z$$

$$= \int_{\mathbb{R}^n} \int_{\mathbb{R}^n} f(x) g(y) \varphi(x + y) \, \mathrm{d}x \, \mathrm{d}y.$$

Hence

$$\langle f * g, \varphi \rangle = \langle f \times g, \Psi \rangle \quad \text{where} \quad \Psi(x, y) = \varphi(x + y).$$

This equation is usually written, incorrectly but conveniently,

$$\langle f * g, \varphi \rangle = \langle f(x) \times g(y), \varphi(x + y) \rangle.$$

2) If f and g are only locally integrable, $\langle h, \varphi \rangle$ is not usually defined because $\varphi \in \mathscr{D}(\mathbb{R}^n)$ does not imply that φ, defined as a function on $\mathbb{R}^{2n}$ by $(x, y) \mapsto \varphi(x + y)$, is in $\mathscr{D}(\mathbb{R}^{2n})$. Indeed let the support of φ in $\mathbb{R}^n$ be $|z| \leq k$, then the support of the corresponding function in $\mathbb{R}^{2n}$ is $|x + y| \leq k$, i.e. a domain which is not compact.

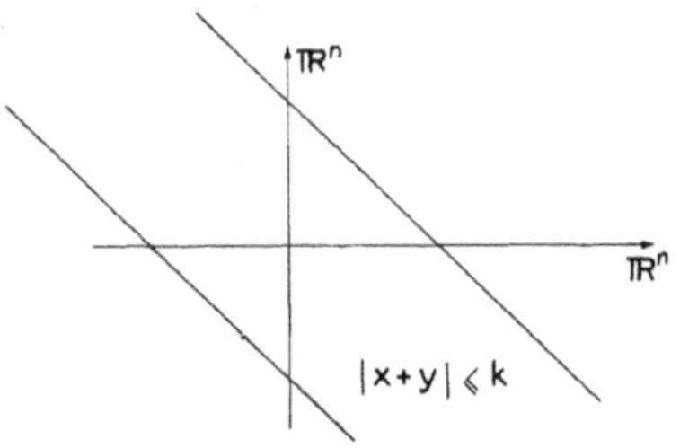

Let $I = \text{supp}\,(f \times g) \cap (\text{supp}\,\varphi \text{ in } \mathbb{R}^{2n})$, $\langle h, \varphi \rangle$ is defined if I is compact.

Let $\text{supp}\, f = A$, $\text{supp}\, g = B$, $\text{supp}\, \varphi$ in $\mathbb{R}^n = K$, the support of $(x, y) \mapsto \varphi(x + y)$ in $\mathbb{R}^{2n}$ is the set C of points $(x, y) \in \mathbb{R}^{2n}$ such that $x + y \in K$. The intersection I consists of the points $x \in A$, $y \in B$ such that $x + y \in K$, i.e. $x \in A \cap K \ominus B$, $y \in B \cap K \ominus A$.
I is compact in the following two cases.

a) Either A or B is compact.

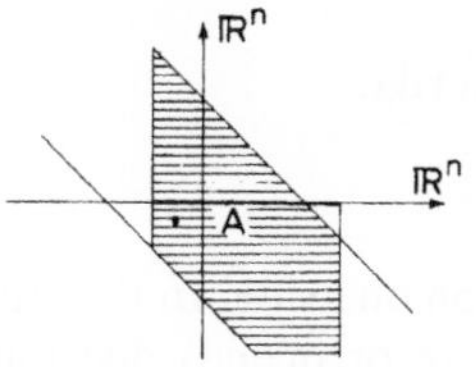

The shaded area is the set of points (x, y) where $f(x)\varphi(x+y) \neq 0$.

b) A and B are "limited on the same side"; we consider separately the cases $n = 1$ and $n > 1$.
1) $n = 1$.
A and B are limited on the left if $A = \{x; x \geq a\}$, $B = \{x; x \geq b\}$.

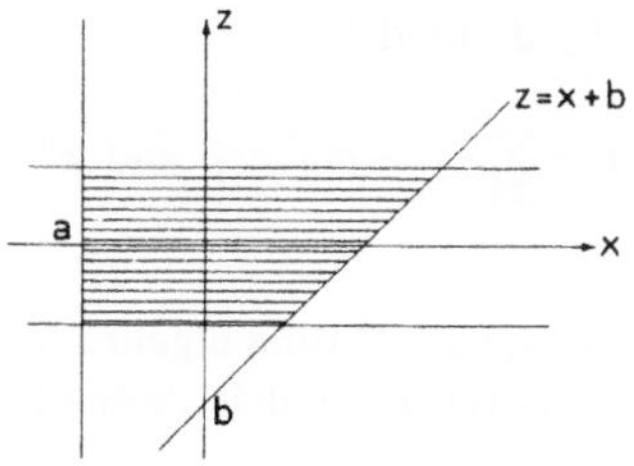

The set of points where $f(x)g(z-x)\varphi(z)$ is $\neq 0$ is $\{(x, z)\}$ such that $a \leq x \leq z - b$ together with $|z| \leq k$.
2) $n > 1$.
$A \cap K \ominus B$, $B \cap K \ominus A$ will be compact if $A \subset \Gamma_a^+$, $B \subset \Gamma_b^+$ where Γ_a^+ is a convex cone of apex a and where Γ_b^+ is the translated cone, for instance: $\Gamma_a^+ = \{x \in \mathbb{R}^n;\ x^i \geq a^i,\ i = 1, \ldots n\}$ and an analogous definition for Γ_b^+.

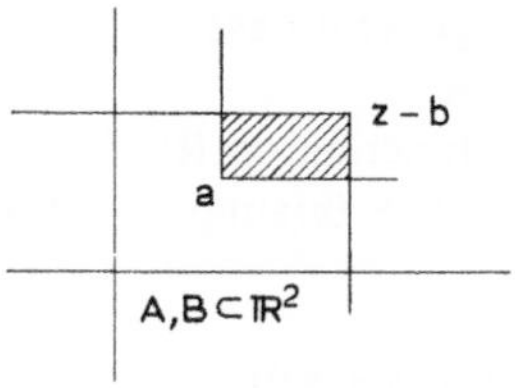

Figure in $\mathbb{R}^2$:
The dark area is the set of points $A \cap z \ominus B = \Gamma_a^+ \cap \Gamma_{z-b}^-$.

In the case $n = 1$, $\operatorname{supp} f = \{x; x \geq a\}$ and $\operatorname{supp} g = \{x; x \geq b\}$, the convolution $h = g * f$ is

$$h(z) = \int_a^{z-b} f(x)g(z-x)\,dx.$$

In the general case, if supp $f \subset \Gamma_a^+$, supp $g \subset \Gamma_b^+$

$$h(z) = \int_{\Gamma_a^+ \cap \Gamma_{z-b}^-} f(x) g(z-x)\,\mathrm{d}x.$$

Examples: Distributions with support limited on one side. In the study of time dependent processes which have a beginning (or an end), distributions on $\mathbb{R}$ with supports defined by $t \geq t_0 (t \leq t_0)$ are frequent. In relativistic physics, the support of distributions defined on space time is often the future time cone Γ_a^+ defined by

$$(x^0 - a^0)^2 - \sum_{i=1}^{3} (x^i - a^i)^2 \geq 0 \text{ and } x^0 - a^0 \geq 0$$

or the past time cone Γ_a^- defined by

$$(x^0 - a^0)^2 - \sum_{i=1}^{3} (x^i - a^i)^2 \geq 0 \text{ and } x^0 - a^0 \leq 0.$$

Convolution of distributions; convolution algebra $\mathcal{D}'^{+}, \mathcal{D}'^{-}$. The convolution $S * T$ of two distributions on $\mathbb{R}^n$, if it is defined, is a distribution on $\mathbb{R}^n$ such that

$$\langle S * T, \varphi \rangle \stackrel{\text{def}}{=} \langle S \times T, \psi \rangle \quad \text{where} \quad \psi(x, y) = \varphi(x + y) \qquad \forall \varphi \in \mathcal{D}(\mathbb{R}^n).$$

It is usually written $\langle S(x) \times T(y), \varphi(x+y) \rangle$. Note that, when defined, $S * T$ is commutative.

We have seen in the case of distributions equivalent to integrable functions that the convolution is defined when a) the support of either distribution is compact, or b) the supports of both distributions are limited on the same side. The same is true in the general case.

a) Let the support of S be a compact set K.

We have shown (p. 446) that $\langle S, \psi \rangle$ is a C^∞ function on $\mathbb{R}^n$. If, moreover, the support of S is a compact set K, then $\langle S, \psi \rangle \in \mathcal{D}(\operatorname{supp} \varphi \ominus K)$ and the convolution of $S * T$ is defined by

$$\langle S * T, \varphi \rangle = \langle T(x), \langle S(y), \varphi(x+y) \rangle \rangle.$$

It is also defined by

$$\langle S * T, \varphi \rangle = \langle S(y), \langle T(x), \varphi(x+y) \rangle \rangle$$

because $\langle T(x), \varphi(x+y) \rangle \in C^\infty(\mathbb{R}^n)$ and a distribution of compact support is defined on C^∞ functions (not necessarily with compact support). We could simply have said "because $S * T$ is commutative"; we elaborated somewhat to show the consistency of some previous theorems.

Examples:
1. $\delta * T = T \qquad \forall T \in \mathcal{D}'(\mathbb{R}^n)$.

Indeed

$$\langle \delta * T, \varphi \rangle = \langle \delta(x) \times T(y), \varphi(x+y) \rangle = \langle T(y), \langle \delta(x), \varphi(x+y) \rangle \rangle = \langle T, \varphi \rangle$$

δ is the unit element for convolution.

2. $\dfrac{\partial \delta}{\partial x^i} * T = \dfrac{\partial T}{\partial x^i}$.

3. $D^j\delta * T = D^jT$.

4. $\delta_a * T = \tau_a T$.

Proof:

$$\langle \delta_a * T, \varphi \rangle = \langle T(y), \varphi(a+y) \rangle.$$

5. Let Y be the Heaviside function

$$1 * (\delta' * Y) = 1 * \delta = 1$$
$$(1 * \delta') * Y = 0 * Y = 0.$$

This example shows that convolution is not always associative. In brief, convolution is associative if

$$\langle S(x) \times T(y) \times U(z), \varphi(x+y+z) \rangle \text{ is defined for } \quad \forall \varphi \in \mathcal{D}(\mathbb{R}^n).$$

A sufficient condition is the following.

Theorem. Convolution is associative if all distributions, with the possible exception of one, are of compact support.

Proof: Assume only one of the distributions is not of compact support; in each of the following steps, the convolution involves distributions, one of which at least, is of compact support:

$$\begin{aligned} \langle (S * T) * U, \varphi \rangle &= \langle (S * T)(x) \times U(y), \varphi(x+y) \rangle \\ &= \langle U(y), \langle (S * T)(x), \varphi(x+y) \rangle \rangle \\ &= \langle U(y), \langle S(z) \times T(u), \varphi(u+z+y) \rangle \rangle \\ &= \langle U(y) \times S(z) \times T(u), \varphi(u+z+y) \rangle. \end{aligned}$$ ■

This theorem allows for one of the distributions to be possibly of non compact support; if, however, we want to define a set of distributions which together with convolution is an algebra, we can, on the basis of this theorem, consider only the set of distributions with compact support.

b) Let the supports of both distributions be limited on the left.
Let $\mathscr{D}'(\mathbb{R}^+)$ be the space of distributions on $\mathbb{R}$ with support in a half-line $x \geq a$; more generally let $\mathscr{D}'(\Gamma^+)$ be the space of distributions on $\mathbb{R}^n$ with support in the convex cone Γ_a^+ with arbitrary apex a translated from a given convex cone with apex 0.
$\mathscr{D}'^+$ For brevity $\mathscr{D}'(\mathbb{R}^+)$ or $\mathscr{D}'(\Gamma^+)$ is often designated $\mathscr{D}'^+$; in $\mathbb{R}^n$ $\mathscr{D}'^+$ depends on the choice of Γ^+.

$\mathscr{D}^-$ Let $\mathscr{D}(\Gamma^-)$, denoted $\mathscr{D}^-$, be the space of C^∞ functions with support limited on the right relative to Γ^-. The topology on $\mathscr{D}^-$ is the inductive limit of the topologies on the space $\mathscr{D}_b^-$ of C^∞ functions with support in the cóne Γ_b^-, the topologies on $\mathscr{D}_b^-$ being the topologies induced by the topology of $\mathscr{C}^\infty$ (p. 429). The following theorem can be proved by arguments similar to the ones used in the paragraph on distributions with compact support.

Theorem. A distribution is in $\mathscr{D}'^+$ [is in $\mathscr{D}'^-$] if and only if it can be extended as a continuous linear form on $\mathscr{D}^-$ [on $\mathscr{D}^+$].

Moreover if $S \in \mathscr{D}'^+$ and $\varphi \in \mathscr{D}^-$, $\psi(y) = \langle S(x), \varphi(x+y)\rangle \in \mathscr{D}^-$. Thus if $T \in \mathscr{D}'^+$, $S * T$ is defined and belongs to $\mathscr{D}'^+$. The associativity is proved as in case a).
In summary we have the following theorem.

Theorem. $\mathscr{D}'^+$ and $\mathscr{D}'^-$ are convolution algebras.

It can be proved that the convolution algebras $\mathscr{D}'^+$ and $\mathscr{D}'^-$ have no zero divisors.[1] This is not true for arbitrary distributions; for instance

$$\frac{d\delta}{dx} * 1 = \frac{d}{dx}(1) = 0 \qquad \frac{d\delta}{dx} \in \mathscr{D}' \quad \text{and} \quad 1 \in \mathscr{C}' \text{ are different from zero.}$$

Derivation and translation of a convolution product. In the case where S or T has compact support, or S and T belong to $\mathscr{D}'^+[\mathscr{D}'^-]$, we have as a corollary of the associativity the following proposition.

[1]Cf. [L. Schwartz 1966, pp. 173 and 177].

Proposition. To take the derivative of [to translate] a convolution product, one takes the derivative of [translates] either of its factors.

Regularization of a distribution. The convolution of a distribution $T \in \mathscr{D}'(\mathbf{R}^n)$ with $\varphi \in \mathscr{D}(\mathbf{R}^n)$ is a C^∞ function $T * \varphi$ on $\mathbf{R}^n$ which is such that

$$(T * \varphi)(x) = \langle T, \psi \rangle \text{ where } \psi(y) = \varphi(x - y).$$

Proof:

$$\begin{aligned}\langle T * \varphi, \psi \rangle &= \langle T(x) \times \varphi(y), \psi(x + y) \rangle \\ &= \langle T(x), \langle \varphi(y), \psi(x + y) \rangle \rangle = \langle T(x), \int_{\mathbf{R}^n} \varphi(y) \psi(x + y)\, \mathrm{d}y \rangle \\ &= \langle T(x), \int_{\mathbf{R}^n} \varphi(z - x) \psi(z)\, \mathrm{d}z \rangle \\ &= \langle T(x), \langle \psi(z), \varphi(z - x) \rangle \rangle \\ &= \langle T(x) \times \psi(z), \varphi(z - x) \rangle \\ &= \langle \psi(z), \langle T(x), \varphi(z - x) \rangle \rangle = \int_{\mathbf{R}^n} \langle T(x), \varphi(z - x) \rangle \psi(z)\, \mathrm{d}z.\end{aligned}$$

Therefore

$$(T * \varphi)(z) = \langle T(x), \varphi(z - x) \rangle. \quad \blacksquare$$

$T * \varphi$ is called the **regularization** of T. regularization

Corollary. An arbitrary distribution T is the limit in $\mathscr{D}'$ of C^∞ functions.

Proof: Let (θ_ϵ) with $\theta_\epsilon \in \mathscr{D}$ be a sequence converging to δ in $\mathscr{D}'$ (cf. p. 455). The C^∞ function $(T * \theta_\epsilon)$ converges to $T * \delta = T$.

Theorem. $\mathscr{D}$ is dense in $\mathscr{D}'$.

The proof is done by multiplying the C^∞ functions $T * \theta_\epsilon$ by a truncating sequence (p. 434).

Support of a convolution.

Theorem. $\operatorname{supp}(S * T) \subset \overline{\operatorname{supp} S \oplus \operatorname{supp} T}$.

Proof: Set supp $S = A$, supp $T = B$. We know that $\mathrm{supp}\,(S \times T) = \overline{A \times B}$. Now let Ω be the open set which is the complement of $\overline{A \oplus B}$ in $\mathbb{R}^n$. We shall show that if $\mathrm{supp}\,\varphi \subset \Omega$, then $\langle S * T, \varphi \rangle = 0$. Indeed if, in $\mathbb{R}^n$, $\mathrm{supp}\,\varphi = K \subset \Omega$ then the support in $\mathbb{R}^{2n}$ of $(x, y) \mapsto \varphi(x + y)$ is contained in the complement of $A \times B$ because $(x, y) \in A \times B$ implies $x + y \in A \oplus B$, thus $x \times y \in \complement(A \oplus B)$ implies $(x, y) \in \complement(A \times B)$.
It follows that

$$\langle S \times T, \varphi(x + y) \rangle = 0 \qquad \forall \varphi \text{ with support in } \Omega. \quad \blacksquare$$

Remark: If either supp f or supp g is compact, then their geometrical sum is closed. If neither is compact their geometrical sum may be not closed.

Example: Let supp f be the subset of $\mathbb{R}^2$ defined by $x^1 \geq 0$, $x^1 x^2 \geq 1$; let supp g be the subset of $\mathbb{R}^2$ defined by $x^1 \geq 0$, $x^1 x^2 \leq -1$. Then supp $f \oplus$ supp g is the subset of $\mathbb{R}^2$ defined by $x^1 > 0$.

Exercise: Show the following.
a) The convolution is a mapping continuous in both its arguments separately. $* : S \to S * T$ is continuous from $\mathscr{D}'^+$ to $\mathscr{D}'^+$ for any fixed $T \in \mathscr{D}'^+$, from $\mathscr{D}'$ to $\mathscr{D}'$ for any fixed $T \in \mathscr{C}' \equiv \mathscr{E}'$.
b) The convolution product is a continuous mapping of $\mathscr{D}'^+_a \times \mathscr{D}'^+ \to \mathscr{D}'^+$, with $\mathscr{D}'^+_a$ space of distributions with support in Γ^+_a.

Remark: It can be shown[1] that the convolution product is not a continuous mapping of $\mathscr{D}'^+ \times \mathscr{D}'^+ \to \mathscr{D}'^+$.

Equations of convolution.

$$A * X = B$$

where A and B are given distributions called the "kernel" and "inhomogeneous term" respectively. If $B = 0$ the equation is said to be homogeneous.

1) $A \in \mathscr{C}'$ (the support of A is compact); $X \in \mathscr{D}'$.

Examples.
a. Partial differential equations with constant coefficients.

$$\sum_{|j| \leq m} a_j \mathrm{D}^j X \equiv \sum_{|j| \leq m} a_j \mathrm{D}^j \delta * X = B.$$

[1][L. Schwartz 1966, p. 173].

The support of the kernel is compact; indeed it is the origin.
b. Finite difference equations

$$\sum_{i=1}^{n} a_i\delta_{h_i} * X = B.$$

The support of the kernel is the n points $x = h_i$.

2) $A \in \mathscr{D}'^{+}$, $X \in \mathscr{D}'^{+}$.

Examples.
a. Volterra integral equations of the first kind.

$$f(x) + \int_0^x K(x-t)f(t)\,dt = g(x)$$

that is

$$(\delta + K) * f = g \qquad \text{supp } K, \text{supp } f, \text{supp } g \subset \mathbb{R}^{+}.$$

b. Volterra integral equations of the second kind

$$\int_0^x K(x-t)f(t)\,dt = g(x)$$

$$K * f = g \qquad \text{supp } K, \text{supp } f, \text{supp } g \subset \mathbb{R}^{+}.$$

In either case, $A \in \mathscr{E}'$ or $A \in \mathscr{D}'^{+}$, the following theorems are valid:

Theorem. The solutions of a homogeneous convolution equation $A * X = 0$ *form a closed linear subspace of* $\mathscr{D}'$ *(if* $A \in \mathscr{E}'$*), or of* $\mathscr{D}'^{+}$ *(if* $A \in \mathscr{D}'^{+}$, $X \in \mathscr{D}'^{+}$*).*

Proof: It follows immediately from the linearity and the continuity of the convolution.

Remark: Let X be a solution of the homogeneous equation $A * X = 0$, then $X * S$ is also a solution of the homogeneous equation if $A \in \mathscr{E}'$ and $S \in \mathscr{E}'$ [if A, X, S are in $\mathscr{D}'^{+}$] because the associativity then holds. In particular the derivatives of X and the regularizations of X are also solutions.

Theorem. Any solution of a homogeneous equation is a limit of solutions which are C^∞; indeed X is the limit of $(X * \theta_\epsilon)$ which are C^∞ solutions.

elementary "solutions"

An **elementary "solution"** of the equation $A * X = B$ at the origin is a distribution E such that:

$$A * E = \delta.$$

invertible

The equation $A * E = \delta$ does not necessarily have a solution; when it does, we say that A is **invertible**: $A^{-1} = E$. It can be shown, for instance, that a derivation polynomial is invertible. The elementary solution is not unique; if E is a solution, $E + E'$ is a solution whenever E' is a solution of the homogeneous equation.

Solutions X can sometimes be expressible in terms of elementary solutions by

$$X = E * B.$$

The following theorem states a case where this is so, and proves the uniqueness of A^{-1} in $\mathscr{D}'^{+}$ $[\mathscr{D}'^{-}]$ if $A \in \mathscr{D}'^{+}$ $[\mathscr{D}'^{-}]$.

Theorem. Let $\mathscr{A}$ be a convolution algebra, let A be invertible and $A^{-1} \in \mathscr{A}$, let $B \in \mathscr{A}$, then $A * X = B$ has strictly one solution in $\mathscr{A}$, $X = A^{-1} * B$.

Proof: X is solution in $\mathscr{A}$ since, by associativity

$$A * X = A * (A^{-1} * B) = (A * A^{-1}) * B = B.$$

$X = A^{-1} * B$ is the unique solution in $\mathscr{A}$; indeed let X' be such that $A * X' = B$, then

$$A^{-1} * B = A^{-1} * (A * X') = (A^{-1} * A) * X' = X'.$$

This means that the other solutions, namely the solutions obtained by adding to X an arbitrary solution of the homogeneous equation cannot be in $\mathscr{A}$. ■

Remark: If A is a distribution with compact support, A^{-1} is not in general a distribution with compact support. Existence and uniqueness have then to be formulated separately for associativity to hold.

Theorem. If $A \in \mathscr{E}'$, A^{-1} exists in $\mathscr{D}'$ and $B \in \mathscr{E}'$, then
1) $X = A^{-1} * B$ is solution of $A * X = B$
2) $A * X = B$ has at most one solution $X \in \mathscr{E}'$.

Exercise: Show that if $A = \delta_a$, $A^{-1} = \delta_{-a}$.
On $\mathbb{R}$, $A \in \mathscr{E}'$ has an inverse $A^{-1} \in \mathscr{E}'$ only if A is a multiple of a Dirac measure[1] δ_a.

[1]Cf. proof in [L. Schwartz 1969, p. 211].

Examples: Equations which do not have elementary solutions. If $A \in \mathscr{D}$, then $S * A$ is C^∞ for every $S \in \mathscr{D}$ and cannot be equal to δ.
If $A \in \mathscr{D}^+$, then $S * A$ is C^∞ for every $S \in \mathscr{D}'^+$ and cannot be equal to δ; then A is not invertible in $\mathscr{A} = \mathscr{D}'^+$. This is the case of the Volterra integral equations of the second kind with $K \in \mathscr{D}^+$.

Differential equation with constant coefficients.

Theorem. An elementary solution of a differential equation with constant coefficients

$$\frac{d^m}{dx^m}E + a_1\frac{d^{m-1}}{dx^{m-1}}E + \cdots a_m E = \delta$$

is $E = Y(x)f(x)$ *where* $f(x)$ *is the* C^∞ *function which is the solution of the homogeneous equation satisfying the following boundary conditions*

$$f(0) = f'(0) = \cdots = f^{(m-2)}(0) = 0$$
$$f^{(m-1)}(0) = 1.$$

This theorem is readily proved by using the formulae (p. 448) for the derivatives in $\mathscr{D}'$ of discontinuous functions.

Example: Find the solution X in $\mathscr{D}'(\mathbb{R}^+)$ of the equation

$$\frac{dX}{dx} + aX = B \qquad \text{where } B \in \mathscr{D}'(\mathbb{R}^+).$$

Here $A = \delta' + a\delta \in \mathscr{D}'^+$ and the elementary solution in $\mathscr{D}'^+$ is

$$E = Y(x)\,e^{-ax}.$$

The unique solution in $\mathscr{D}'^+$ of the given equation is

$$X = B * e^{-ax}Y(x).$$

If B is a locally integrable function, X is given by

$$X(x) = \int_0^x e^{-ay}B(x-y)\,dy.$$

Exercise: Show directly by solving the homogeneous equation (cf. p. 451) that E is the unique elementary solution in $\mathscr{D}'^+$. Show that it is not unique in $\mathscr{D}'$.

Example: Problem 8.

completely invertible

It is not always true even if A is invertible, that the equation $A * X = B$ has a solution for an arbitrary $B \in \mathscr{D}'$. If it does A is called **completely invertible.** There is a method to try to solve $A * X = B$ when A is invertible without B being with compact support. It is always possible to write B as a sum of distributions with compact support B_j

$$B = \sum B_j \text{ with supp } B_j \subset \text{annulus } \{x;\, b_j \leq |x| \leq b_j'\}.$$

Set $X_j = E * B_j$, the series ΣX_j is usually divergent, however it is often possible to choose solutions S_j of the homogeneous equation valid in the whole space such that $\Sigma_j(X_j - S_j)$ converges in $\mathscr{D}'$. Then

$$A * \sum_j (X_j - S_j) = \sum (A * X_j - A * S_j) = \sum B_j = B.$$

This technique can be used when A is the laplacian Δ and makes it possible to solve the Poisson equation for an arbitrary distribution. In this case the functions S_j are the harmonic polynomials.
If B is of compact support, a solution of $\Delta X = B$ in $\mathbb{R}^3$ (called the potential of B) is $X = -(1/4\pi r) * B$. If B is arbitrary in $\mathscr{D}'$, a solution is $-\Sigma_j (1/4\pi r) * (B_j - S_j)$; this solution is not unique because $\Delta X = 0$ has non-zero solutions in $\mathscr{D}'$.
Notice that the only solution of compact support of $\Delta X = 0$ is the zero solution.

Systems of convolution equations. The previous discussions can be generalized to a system of n equations in n unknowns

$$\sum_{j=1}^{n} A_i^j * X_j = B_i \qquad i = 1, \dots n.$$

We can write this equation in matrix form, with obvious notation,

$$A * X = B.$$

The convolution product is no longer commutative and we shall have both right and left elementary solutions:

$$\sum_i A_i^j * E_k^i = \delta_k^j \delta \quad \text{right elementary solution}$$

and

$$\sum_i E_i^j * A_k^i = \delta_k^j \delta \quad \text{left elementary solution.}$$

It is easy to see that, under a hypothesis on the supports analogous to those made in the case of one equation, the right [resp. left] elementary solution gives the existence [resp. uniqueness] of solutions of the equation $A * X = B$.

Exercise: Show that in a convolution algebra right and left elementary solutions are equal.

Kernels. A **kernel** on $\mathbb{R}^n$ is a distribution on $\mathbb{R}^n \times \mathbb{R}^n$, i.e. an element of the dual $\mathscr{D}(\mathbb{R}^n \times \mathbb{R}^n)$ of C^∞ functions with compact support on $\mathbb{R}^n \times \mathbb{R}^n$. kernel
$\langle K, \varphi \rangle$ with $\varphi \in \mathscr{D}(\mathbb{R}^n \times \mathbb{R}^n)$ is often written, for practical purposes as $\langle K(x, y), \varphi(x, y) \rangle$.
We define a distribution $U(y)$ on $\mathbb{R}^n$, called the left contraction of K with u

$$\langle U(y), v(y) \rangle = \langle K(x, y), u(x)v(y) \rangle, \; v \in \mathscr{D}(\mathbb{R}^n).$$

We shall denote

$$U(y) = \langle K(x, y), u(x) \rangle.$$

It is easy to show that U satisfies the conditions of linearity and continuity of distributions. The converse is Schwartz' kernel theorem whose demonstration is beyond the scope of this book; namely, any linear continuous mapping $\mathscr{D}(\mathbb{R}^n) \to \mathscr{D}'(\mathbb{R}^n)$ can be written in the form

$$U(y) = \langle K(x, y), u(x) \rangle \text{ where } K \in \mathscr{D}'(\mathbb{R}^n \times \mathbb{R}^n).$$

A kernel is said to be **semi regular** in y (right semi regular) if $U(y)$ is identifiable with a C^∞ function. A kernel which is both semi regular in y and semi regular in x is said to be regular. semi regular
If K is semi regular in y, it defines a map $\mathscr{D} \to \mathscr{C}$ by $u \mapsto U$. By the closed **graph theorem (cf. p. 64) this mapping is continuous**: it is continuous from $\mathscr{D}$ to $\mathscr{D}'$, and thus has a closed graph in $\mathscr{D} \times \mathscr{C}$, since $\mathscr{C} \subset \mathscr{D}'$ topologically.

The Volterra convolution with a right semi regular kernel $K(x, y)$. Let $U(y)$ be the left contraction of $K(x, y)$ with $u \in \mathscr{D}$, the **Volterra convolution** of $K(x, y)$ with a distribution $S(y)$, if it is defined, is the distribution $T(x)$ defined by Volterra convolution

$$\langle T(x), u(x) \rangle = \langle S(y), U(y) \rangle.$$

T is often written symbolically as

$$T(x) = \int K(x, y) S(y) \, dy.$$

T is defined principally in the following cases:

1) $S \in \mathscr{E}'$.
2) $\operatorname{supp} K \subset \mathbb{R}^n \times V$, V a compact set.
3) $S \in \mathscr{D}'^{+}$ and $U \in \mathscr{D}^{-}$.

Example: A convolution in the sense of the preceding paragraph, $A * X$, can be considered, in our present framework, as the Volterra convolution with X of a right semi regular kernel $K(x, y)$ invariant by translation: $K(x, y)$ is such that $K(x, y) = K(x - y) = A(x - y)$ that is to say, $K(x, y)$ is the kernel on $\mathbb{R}^2$

$$\langle K(x, y), \varphi(x, y)\rangle \overset{\text{def}}{=} \langle A(z), \int_{-\infty}^{+\infty} \varphi(x, x + z)\,dx\rangle.$$

Here K is semi regular in y as well as semi regular in y, hence regular.

elementary kernel

An **elementary (fundamental) kernel** E of the linear differential operator D on $\mathbb{R}^n$ with C^∞ coefficients $a_j(x)$,

$$\mathrm{D} \equiv \sum_{|j| \leq m} a_j(x)\,\mathrm{D}^j \qquad x \in R^n$$

is a kernel such that

$$\sum_{|j| \leq m} a_j(x)\mathrm{D}_x^j E(x, y) = \delta(x - y).$$

Here $\delta(x - y)$ is considered as a kernel on $\mathbb{R}^n$ defined by

$$\langle \delta(x - y), \varphi(x, y)\rangle = \int_{\mathbb{R}^n} \varphi(y, y)\,dy.$$

Notice that $\delta(x - y) = \delta(y - x)$.
$\delta(x - y)$ is a regular kernel; indeed $\langle \delta(x - y), u(x)\rangle = u(y)$ which is a C^∞ function.

In the study of the convolution equation $A * X = B$ we have stated conditions under which it has one, or at most one solution in the same subspace of $\mathscr{D}'$ as A and B. If A is the differential operator D there are similar theorems on X expressed in terms of the elementary kernels E and E^* of D and the transpose D* of D, given by

$$\mathrm{D}^* S = \sum_{|j| < m} (-1)^{|j|}\mathrm{D}^j(a_j S) \qquad S \in \mathscr{D}'.$$

Theorem. 1) *Let* $B \in \mathscr{E}'$, $\mathrm{D}X = B$ *has at least a solution in* $\mathscr{D}'$ *if* D *has an elementary kernel, right semi-regular.*
2) *Let* $B \in \mathscr{E}'$, $\mathrm{D}X = B$ *has at most one solution in* $\mathscr{E}'$ *if* D^* *has an elementary kernel, left semi-regular.*

Proof:
1) Existence. $X(x) = \int E(x, y)B(y)\,dy$ is a solution; indeed

$$\langle D_x X(x), \varphi(x)\rangle = \langle X(x), D_x^* \varphi(x)\rangle = \langle B(y), \langle E(x, y), D_x^* \varphi(x)\rangle\rangle$$
$$= \langle B(y), \langle D_x E(x, y), \varphi(x)\rangle\rangle$$
$$= \langle B(y), \varphi(y)\rangle.$$

2) Uniqueness. Let $X \in \mathscr{C}'$, we have

$$\left\langle \int E^*(x, y) D_x X(x)\,dx, \varphi(y)\right\rangle = \langle D_x X(x), \langle E^*(x, y), \varphi(y)\rangle\rangle$$
$$= \langle X(x), D_x^* \langle E^*(x, y), \varphi(y)\rangle\rangle$$
$$= \langle X(x), \langle D_x^* E(x, y), \varphi(y)\rangle\rangle$$
$$= \langle X(x), \langle \delta(x - y), \varphi(y)\rangle\rangle$$
$$= \langle X(x), \varphi(x)\rangle,$$

thus if $DX = B$, $X(y) = \int_{\mathbb{R}^n} E^*(x, y)B(x)\,dx$. ■

More compact results are obtained in a situation which generalizes the convolution algebra $\mathscr{D}'^{+}(\Gamma)$. Let us denote by Γ_y a convex cone of vertex y, translated from a given convex cone of vertex 0. Suppose that $E(x, y)$ is a regular kernel on $\mathbb{R}^{2n}$ such that, for each $y_0 \in \mathbb{R}^n$ the distribution $E(x, y_0)$ has its support in Γ_{y_0}. Then for each $u \in \mathscr{D}^{-}(\Gamma)$ the function

$$U(y) = \langle K(x, y), u(x)\rangle$$

is also in $\mathscr{D}^{-}(\Gamma)$. Thus $\int E(x, y)S(y)\,dy$ is defined for all $S \in \mathscr{D}'^{+}(\Gamma)$, and belongs to $\mathscr{D}'^{+}(\Gamma)$. It is easy to prove the following theorem.

Theorem. If the linear partial differential equation

$$\sum_{|j|=0}^{m} a_j(x) D^j X = B$$

has a fundamental kernel $E(x, y)$ satisfying the above hypothesis, then this equation has one, and only one solution for each $B \in \mathscr{D}'^{+}(\Gamma)$.

Corollary. Under the same hypothesis one has the symmetry property

$$E(x, y) = E^*(y, x).$$

We will return to these properties (p. 522) about propagators.

6. FOURIER TRANSFORMS

Fourier transform of integrable functions. Let $f \in L^1(\mathbb{R}^n)$, then its Fourier transform $\mathcal{F}f$ is a function on $\mathbb{R}^n$ – in fact on the dual of $\mathbb{R}^n$, but identified with $\mathbb{R}^n$ – defined[1] by

$$(\mathcal{F}f)(y) = \int_{\mathbb{R}^n} e^{-iy \cdot x} f(x)\, dx \qquad y \cdot x = \sum_{i=1}^{n} y_i x^i.$$

It is easy to show that $\mathcal{F}f$ is a continuous and bounded function on $\mathbb{R}^n$. The distribution associated to $\mathcal{F}f$ is such that for every $\varphi \in \mathcal{D}$

$$\langle \mathcal{F}f, \varphi \rangle = \int_{\mathbb{R}^n} (\mathcal{F}f)(y)\varphi(y)\, dy = \langle f, \mathcal{F}\varphi \rangle.$$

We have here an expression which can be generalized to distributions T; however T cannot be an arbitrary distribution, because $\varphi \in \mathcal{D}$ does not imply $\mathcal{F}\varphi \in \mathcal{D}$. We shall define a space $\mathcal{S}$ larger than $\mathcal{D}$ but smaller than L^1 such that $\mathcal{F}$ is an isomorphism of $\mathcal{S}$.

$\mathcal{S}$ $\mathcal{S}$ is the vector space of C^∞ functions on $\mathbb{R}^n$ such that for each integer $p \geq 0$ and multi-index j

$$\sup_{x \in \mathbb{R}^n} (1 + |x|^2)^p |D^j \varphi| \leq M_{pj},$$

the constant M_{pj} depends on φ, p and j. This condition implies that φ and its derivatives decrease more rapidly than any power of $1/|x|$ when $|x|$ tends to infinity. $\mathcal{S}$ is an algebra with respect to pointwise multiplication and is closed under derivation and multiplication by x^j:

$$\varphi \in \mathcal{S} \Rightarrow D^j \varphi \in \mathcal{S}$$
$$\varphi \in \mathcal{S} \Rightarrow x^j \varphi \equiv x_1^{j_1} \dots x_n^{j_n} \varphi \in \mathcal{S} \qquad |j| = j_1 + \cdots j_n.$$

Theorem. The Fourier transform is an isomorphism of $\mathcal{S}$.

Proof: a) $\varphi \in \mathcal{S} \;\Rightarrow\; \mathcal{F}\varphi \in \mathcal{S}$.
$\mathcal{F}\varphi$ has an upper bound:

$$|\mathcal{F}\varphi| = \left| \int_{\mathbb{R}^n} e^{-iy \cdot x} \varphi(x)\, dx \right| \leq \int_{\mathbb{R}^n} |\varphi(x)|\, dx \leq M_{n0} \int_{\mathbb{R}} (1 + |x|^2)^{-n}\, dx \leq N.$$

[1]The normalisation is chosen so that $\mathcal{F}\delta = 1$, 1 being the multiplicative unit element in the complex field and δ being the multiplicative unit element in convolution algebra.

$D^j\mathcal{F}\varphi$ has an upper bound:

$$(D^j\mathcal{F}\varphi)(y) = \int_{\mathbf{R}^n} (-ix)^j \, e^{-iy\cdot x}\varphi(x)\, dx = \mathcal{F}((-ix)^j\varphi)(y)$$

$$\varphi \in \mathcal{S} \;\Rightarrow\; x^j\varphi \in \mathcal{S} \;\Rightarrow\; |\mathcal{F}(-ix)^j\varphi)| = |D^j\mathcal{F}\varphi| \leq N_{0j}.$$

$(1+|y|^2)^p|D^j\mathcal{F}\varphi|$ has an upper bound. Using the fact that (integration by parts)

$$\left(\mathcal{F}\frac{\partial\varphi}{\partial x^i}\right)(y) = \int_{\mathbf{R}^n} e^{-iy\cdot x}\frac{\partial\varphi}{\partial x^i}\, dx = iy_i(\mathcal{F}\varphi)(y)$$

$$\mathcal{F}(D^p\varphi) = (iy)^p\mathcal{F}\varphi$$

we conclude that

$$\varphi \in \mathcal{S} \;\Rightarrow\; D^j\varphi \in \mathcal{S} \;\Rightarrow\; |(iy)^j\mathcal{F}\varphi| = |\mathcal{F}(D^j\varphi)| \leq N_{p0}.$$

In summary

$$\sup_{y\in\mathbf{R}^n}(1+|y|^2)^p|D^j\mathcal{F}\varphi| \leq N_{pj}.$$

b) $\mathcal{F}$ is a linear mapping of $\mathcal{S}$ into $\mathcal{S}$; we shall show that it is bijective. Let

$$\bar{\mathcal{F}}: \mathcal{S}\to\mathcal{S} \text{ by } \bar{\mathcal{F}}\varphi(u) = \frac{1}{(2\pi)^n}\int_{\mathbf{R}^n} e^{iu\cdot y}\varphi(y)\, dy.$$ $\bar{\mathcal{F}}$

We shall show that $\bar{\mathcal{F}}\mathcal{F}$ is the unit transformation on $\mathcal{S}$

$$(\bar{\mathcal{F}}\mathcal{F}\varphi)(u) = \frac{1}{(2\pi)^n}\int_{\mathbf{R}^n}\left(\int_{\mathbf{R}^n} e^{iy\cdot(u-x)}\varphi(x)\, dx\right) dy.$$

The function on $\mathbf{R}^{2n}$ defined by $(x, y)\mapsto \exp(iy(u-x))\varphi(x)$ is not integrable on $\mathbb{R}^{2n}$ but is integrable on $\mathbb{R}^n\times K$, where K is the compact set $[-A, A]^n$ and

$$(\bar{\mathcal{F}}\mathcal{F}\varphi)(u) = \lim_{A=\infty}\frac{1}{(2\pi)^n}\int_{\mathbf{R}^n\times K} e^{iy\cdot(u-x)}\varphi(x)\, dx\, dy$$

$$= \lim_{A=\infty}(f_A * \varphi)(u), \quad f_A(x) = \frac{1}{(2\pi)^n}\int_K e^{iy\cdot x}\, dy.$$

But (p. 455) in $\mathcal{D}'$

$$\lim_{A=\infty} f_A \equiv \lim_{A=\infty}\frac{1}{(2\pi)^n}\int_K e^{iyu}\, dy = \delta.$$

Hence if $\varphi \in \mathscr{S}$, by an argument analogous to one given previously

$$\bar{\mathscr{F}}\mathscr{F}\varphi = \delta * \varphi = \varphi.$$

$\mathscr{F}$ and $\mathscr{F}^{-1} \equiv \bar{\mathscr{F}}$ are both continuous in the topology of $\mathscr{S}$ defined below.

tempered distributions

Tempered distributions. $\mathscr{S}$ together with the $\mathscr{P}$-topology defined by the family of seminorms

$$s_{k,N}(\varphi) = \sup_{x \in \mathbb{R}^n} \sup_{|j| \le N} |(1 + |x|^2)^k D^j \varphi(x)|$$

is a Fréchet space. $\mathscr{D}$ is algebraically and topologically a subspace of $\mathscr{S}$. Hence the space $\mathscr{S}'$ of linear continuous forms on $\mathscr{S}$ is a linear subspace of $\mathscr{D}'$. $\mathscr{S}'$ is called the space of **tempered distributions**. A tempered distribution is a distribution which "does not increase too rapidly at infinity". Since $\mathscr{D}$ is dense in $\mathscr{S}$ a tempered distribution is defined by its values on $\mathscr{D}$. Since $\mathscr{S}$ is stable by derivation and product by x^j:

Proposition: *$\mathscr{S}'$ is a module on the ring of polynomials, it is closed under derivation.*

Examples. 1) Distributions with compact supports are tempered.
2) The polynomials are tempered distributions.
3) The function $\exp(x)$ is not a tempered distribution on $\mathbb{R}$.

Usually the topology of $\mathscr{S}'$ is the weak star dual topology of $\mathscr{S}$.

Fourier transform in $\mathscr{S}'$

Fourier transforms of tempered distributions. Having established that $\mathscr{F}$ is an isomorphism of $\mathscr{S}$, we can now define the Fourier transform of a distribution $T \in \mathscr{S}'$ by

$$\langle \mathscr{F}T, \varphi \rangle = \langle T, \mathscr{F}\varphi \rangle.$$

Note: As for the case of L^1 functions $\mathscr{F}T$ is, in fact, based on the dual of the space $\mathbb{R}^n$ in which T is based.
The following properties can easily be derived from the definition:
1) $\mathscr{F}$ is an isomorphism of $\mathscr{S}'$; the identity on $\mathscr{S}'$ is $\bar{\mathscr{F}}\mathscr{F}$
2) $\mathscr{F}(D^jT)(y) = (iy)^j \mathscr{F}T(y)$
3) $D^j(\mathscr{F}T) = \mathscr{F}((-ix)^jT)$ i.e. $D^j(\mathscr{F}T) = \mathscr{F}fT$ where $f(x) = (-ix)^j$.

If a distribution T is of compact support, $T \in \mathscr{C}' \equiv \mathscr{E}'$, then its Fourier transform can be written, using a direct product

$$\langle \mathscr{F}T, \varphi \rangle = \langle \langle T(y), e^{-iy \cdot x} \rangle, \varphi(x) \rangle.$$

Thus

$$(\mathcal{F}T)(x) = \langle T(y), e^{-iy\cdot x}\rangle, \qquad \text{if } T \in \mathscr{C}'.$$

We have already stated that if the mapping $\mathbf{R}^n \to \mathscr{D}(\mathbf{R}^n)$ by $x \mapsto f(x.)$ is continuous [differentiable] mapping and if $T \in \mathscr{D}'$, then $\langle T(y), f(x, y)\rangle$ is a continuous [differentiable] function of x. Here the mapping: $\mathbf{R}^n \to \mathscr{C}$ by $x \mapsto \exp(-iy \cdot x)$ is continuous and even differentiable, hence $\mathcal{F}T$ is a differentiable function if $T \in \mathscr{C}'$.

Examples.
1) $\mathcal{F}\delta = \langle \delta(y), e^{-iyx}\rangle = 1, \qquad \mathcal{F}1 = (2\pi)^n\delta$
2) $\mathcal{F}(D^\alpha\delta) = (iy)^\alpha$.

Theorem. The Fourier transform of a distribution with compact support is a "slowly increasing" function. More precisely, if $S \in \mathscr{C}'$, then $\mathcal{F}S \in \mathscr{C}$, and for each multi-index α there exists a positive integer p and a constant $M_{p,\alpha}$ such that

$$|(D^\alpha\mathcal{F}S)(y)| \le M_{p,\alpha}(1 + |y|^2)^p, \qquad \forall\, y \in \mathbf{R}^n.$$

Proof: If $S \in \mathscr{C}'$

$$(\mathcal{F}S)(y) = \langle S, e^{-ix\cdot y}\rangle$$

and (cf. p. 442) there exist a compact set K, and integer N such that

$$|(\mathcal{F}S)(y)| \le C \sup_{x\in K}\sup_{|j|\le N} |D_x^j e^{-ix\cdot y}|$$

and analogous inequalities resulting from

$$(D^\alpha\mathcal{F}S)(y) = \langle S, D_y^\alpha e^{-ix\cdot y}\rangle. \qquad \blacksquare$$

By extending the Fourier transform of a distribution (when it is a function) to complex values one obtains important results about the distribution, in particular about its support. The simplest of all these results is the Paley–Wiener theorem.

Paley–Wiener theorem. The Fourier transform of a distribution with compact support can be extended into an analytic function on $\mathbb{C}^n$.

Conversely if $\mathcal{F}T$ is a continuous function which can be extended into an analytic function F on $\mathbb{C}^n$ such that

$$\limsup_{|z|\to\infty} \frac{\log|F(z)|}{|z_1| + \cdots + |z_n|} \le c$$

then T has a compact support contained in the cube $|y_i| \leq c$, $i = 1, \ldots, n$.

Proof: $\mathcal{F}T$ is extended to $\mathbb{C}$ by

$$(\mathcal{F}T)(z) = \langle T(y), e^{-iy \cdot z} \rangle \qquad \forall z \in C \text{ and } T \in \mathscr{E}'.$$

$\mathcal{F}T$ satisfies the Cauchy–Riemann conditions for analyticity, because $e^{-iy \cdot z}$ satisfy them. For the converse cf. [Schwartz 1966, p. 272].

Fourier transform of a convolution. Let S and T be two distributions with compact support, then $S * T$ is a distribution with compact support, its Fourier transform is a continuous function defined by

$$\begin{aligned} \mathcal{F}(S * T) &= \langle S * T, e^{-iy \cdot x} \rangle \\ &= \langle S(u) \times T(v), e^{-i(u+v) \cdot x} \rangle \\ &= \langle S(u), e^{-iu \cdot x} \rangle \langle T(v), e^{-iv \cdot x} \rangle \\ &= \mathcal{F}S \cdot \mathcal{F}T. \end{aligned}$$

Note $\bar{\mathcal{F}}(S * T) = (2\pi)^n \bar{\mathcal{F}}S \cdot \bar{\mathcal{F}}T$.

Theorem. $\mathcal{F}(S * T) = \mathcal{F}S \cdot \mathcal{F}T$ *if one of the distributions is of compact support and the other one is a tempered distribution.*

Proof: Let $S \in \mathscr{E}'$ and $T \in \mathscr{S}'$; let $\varphi \in \mathscr{S}$; then

$$\langle \mathcal{F}(S * T), \varphi \rangle = \langle S * T, \mathcal{F}\varphi \rangle = \langle S(u) \times T(v), (\mathcal{F}\varphi)(u+v) \rangle$$

does define a distribution in $\mathscr{S}'$, indeed

$$\alpha(v) \equiv \langle S(u), (\mathcal{F}(e^{-iv \cdot u}\varphi))(u) \rangle = \langle (\mathcal{F}S)(x), e^{-iv \cdot x}\varphi(x) \rangle$$

but, since $S \in \mathscr{E}'$, $\mathcal{F}S$ is a C^∞ function, with slow increase (cf. above), and

$$\alpha(v) = \int (\mathcal{F}S)(x)\, e^{-iv \cdot x}\varphi(x)\, dx = \mathcal{F}(\mathcal{F}S \cdot \varphi)(v).$$

Thus for $\alpha \in \mathscr{S}$, the mapping: $\mathscr{S} \to \mathscr{S}$ by $\varphi \mapsto \alpha$ is continuous, hence, if $T \in \mathscr{S}'$, $\langle \mathcal{F}(S * T), \varphi \rangle = \langle T, \alpha \rangle$ is defined, depends continuously on φ, and is such that

$$\langle \mathcal{F}(S * T), \varphi \rangle = \langle T, \mathcal{F}(\mathcal{F}S \cdot \varphi) \rangle = \langle \mathcal{F}T, \mathcal{F}S \cdot \varphi \rangle = \langle \mathcal{F}S\mathcal{F}T, \varphi \rangle.$$

This property is very useful to obtain the tempered solutions of convolution equations.

Example: Elementary tempered solution of Δ in $\mathbb{R}^3$.

$$\Delta E = \delta \qquad \text{where } E \in \mathcal{S}'$$

$$\mathcal{F}\Delta\mathcal{F}E = 1$$

$$(\mathcal{F}E)(y) = -1/|y|^2 \text{ with } |y|^2 = \Sigma\,(y_i)^2$$

$$\langle \mathcal{F}E, \bar{\mathcal{F}}\varphi\rangle = -\frac{1}{(2\pi)^3}\int_{\mathbb{R}^3} \frac{1}{|y|^2}\Big(\int_{\mathbb{R}^3} e^{iy\cdot x}\varphi(x)\,dx\Big)\,dy$$

$$= \langle E, \varphi\rangle,$$

using a limit of integrals on compact sets (p. 455) and Fubini's theorem, we obtain

$$\langle E, \varphi\rangle = -\frac{1}{(2\pi)^3}\int_{\mathbb{R}^3}\Big(2\pi\int_0^\infty\int_0^\pi \frac{\exp\{i|y||x|\cos\theta\}}{|y|^2}|y|^2\sin\theta\,d\theta\,d|y|\Big)\varphi(x)\,dx$$

$$= -\frac{1}{4\pi}\int \frac{1}{|x|}\varphi(x)\,dx.$$

E is the locally integrable function defined almost everywhere by $-1/4\pi|x|$. A similar calculation in $\mathbb{R}^n$ gives

$$E = -\frac{\Gamma(n/2)}{(n-2)2\pi^{n/2}}|x|^{2-n} \qquad \text{for } n \neq 2$$

$$E = \frac{1}{2\pi}\ln|x| \qquad \text{in } \mathbb{R}^2.$$

Remark: $1/|x|$ is invariant under rotation and so is its Fourier transform $1/|y|^2$.

Theorem. The Fourier transformation and the transformation of a distribution by a rotation of $\mathbb{R}^n$ (cf. p. 458) *commute.*

Proof: Let $A: x \mapsto Ax$ be a rotation of $\mathbb{R}^n$, then

$$|\det A| = 1$$

and for any pair of vectors x, y in $\mathbb{R}^n$

$$Ax \cdot Ay = x \cdot y.$$

But, if $f \in \mathcal{S}$

$$(\mathcal{F}\varphi)(Ay) = \int_{\mathbb{R}^n} \varphi(x)\,e^{-ix\cdot Ay}\,dx$$

or, by change of variables

$$(\mathcal{F}\varphi)(Ay) = \int_{\mathbb{R}^n} \varphi(Ax)\, e^{-iAx \cdot Ay}\, dx.$$

Thus

$$(\mathcal{F}\varphi)(Ay) = \int_{\mathbb{R}^n} \varphi(Ax)\, e^{-ix \cdot y}\, dx,$$

that is

$$(\mathcal{F}\varphi) \circ A = \mathcal{F}(\varphi \circ A).$$

By the definition of a transform of a distribution by a diffeomorphism of $\mathbb{R}^n$, we have

$$\langle A\mathcal{F}T, \varphi\rangle = \langle \mathcal{F}T, \varphi \circ A\rangle$$

and, by the definition of Fourier transform and the equality just proved

$$\begin{aligned}\langle \mathcal{F}T, \varphi \circ A\rangle &= \langle T, \mathcal{F}(\varphi \circ A)\rangle = \langle T, \mathcal{F}\varphi \circ A\rangle \\ &= \langle AT, \mathcal{F}\varphi\rangle = \langle \mathcal{F}AT, \varphi\rangle.\end{aligned}$$

Finally

$$A\mathcal{F}T = \mathcal{F}AT. \qquad \blacksquare$$

An immediate consequence is the following theorem.

Theorem. The Fourier transform of a distribution invariant by rotation is itself invariant by rotation.

Exercise: Show that the Fourier transform of the Dirac measure $\delta_{r=a}$ on the sphere of radius a, centered at the origin, in $\mathbb{R}^3$ is $4\pi a \sin a\rho/\rho$.

7. DISTRIBUTIONS ON A PARACOMPACT C^∞ MANIFOLD X

Let us recall that a C^∞ manifold is a Hausdorff topological space X with a family of equivalent C^∞ atlases. A C^∞ atlas on X is a collection of charts (U_i, φ_i), $i \in I$, I is an indexing set, the domain U_i of the chart is an open subset of X, $\bigcup_{i\in I} U_i = X$, $\varphi_i\colon U_i \to \varphi_i(U_i)$ is a homeomorphism onto an open set of $\mathbb{R}^n$, such that (compatibility requirements)

$$\varphi_i \circ \varphi_j^{-1}\colon \varphi_j(U_i \cap U_j) \to \varphi_i(U_i \cap U_j)$$

is a C^∞ diffeomorphism.

We shall moreover suppose in the following that the manifold X is paracompact (p. 16) and we shall consider on it only countable, locally finite atlases.

Now let $\{(U_i, \varphi_i)\}_{i\in I}$ be an atlas on X. A distribution T on X is defined by a family of distributions $\{T_i\}_{i\in I}$ such that

$$T_i \in \mathscr{D}'(\varphi_i(U_i))$$

and, the action on T_i of the diffeomorphism $\varphi_j \circ \varphi_i^{-1}$ having been defined on p. 457,

$$T_j = (\varphi_j \circ \varphi_i^{-1})T_i \quad \text{on } \varphi_j(U_i \cap U_j).$$

Let $\{(V_i, \psi_i)\}_{i\in J}$ be another atlas on X compatible with the previous one. A family $\{S_i\}$ defines the same distribution T as the family $\{T_i\}$ if

$$S_j = (\psi_j \circ \varphi_i^{-1})T_i \quad \text{on } \psi_j(U_i \cap V_j).$$

It is not difficult to prove that this relation is an equivalence relation. The **space $\mathscr{D}'(X)$ of distributions on** X is the space of equivalence classes of families $\{T_i\}$. $\mathscr{D}'(X)$

We can analogously define a **section distribution** of a vector bundle (E, X, π, G) over X (cf. p. 125) whose typical fibre F is a finite dimensional vector space $\mathbb{R}^p$. Let $\{(U_i, \varphi_i);\ i \in I\}$ be an atlas of X; the mapping section distribution

$$\Phi_i: \pi^{-1}(U_i) \to \varphi_i(U_i) \times \mathbb{R}^p$$

by

$$t_x \mapsto (\varphi_i(x), \overset{\Delta}{\varphi}_i(t_x)), \quad \text{where } x = \pi(t_x),$$

is a C^∞ diffeomorphism. Let $\boldsymbol{T}_i = (T_i^A;\ A = 1, \ldots, p)$ be a set of p ordinary (scalar valued) distributions based on the open set $\varphi_i(U_i)$ of $\mathbb{R}^n$, the test functions being $\Phi_i(\boldsymbol{t})$, $\boldsymbol{t}$ a section of $\pi^{-1}(U_i)$.

Let $\{\boldsymbol{T}_i,\ i \in I\}$ be a collection of such sets, with the property that

$$\boldsymbol{T}_j = (\varphi_j \circ \varphi_i^{-1})(g_{ji} \circ \varphi_i^{-1} \cdot \boldsymbol{T}_i) \quad \text{on } \varphi_j(U_i \cap U_j).$$

$g_{ji} \circ \varphi_i^{-1}$ is a C^∞ mapping from $\varphi_i(U_i \cap U_j)$ into G, a subgroup[1] of the isomorphisms of $\mathbb{R}^p$. Thus $g_{ji} \circ \varphi_i^{-1} \cdot \boldsymbol{T}_i$ is a set of p ordinary distributions on $\varphi_i(U_i \cap U_j)$, $\boldsymbol{T}_j$ is the set of p ordinary distributions on $\varphi_j(U_i \cap U_j)$ defined from these by the transformation $\varphi_j \circ \varphi_i^{-1}$ (see p. 457).

A **section distribution** of the vector bundle (E, X, π, G) is an equivalence class of collections $\{\boldsymbol{T}_i\}$, the equivalence relation being defined along the same lines as in the case of an ordinary distribution.

The set $\{\boldsymbol{T}_i\}$ is called the set of **components** of the distribution $\boldsymbol{T}$ in the chart (U_i, φ_i). components of a section distribution

[1]This subgroup depends on the vector bundle over X (tangent bundle, cotangent bundle, bundle of p-tensors, of p-tensor densities, etc. . . .).

Example: E is the tangent bundle over X. A section distribution $\boldsymbol{v}$ of E has for components in a chart (U, φ), n scalar valued distributions $v^1, \ldots, v^n$. The components of $\boldsymbol{v}$ in another chart (U', φ') are $v^{i'}, \ldots, v^{n'}$ such that

$$v^i = (\varphi \circ \varphi'^{-1})\left(\frac{\partial x^i}{\partial x^{j'}} v^{j'}\right) \quad \text{on } \varphi(U \cap U').$$

The isomorphism of $\mathbb{R}^p$ (corresponding to $g_{ji} \circ \varphi_i^{-1}$) is given by the matrix with elements $\partial x^i / \partial x^{j'}$; now $(\partial x^i / \partial x^{j'}) v^{j'}$ is a distribution on $\varphi'(U \cap U')$ (linear combination of $v^{j'}$ with the C^∞ coefficients $\partial x^i / \partial x^{j'}$), and v^i is the transform of this distribution by the C^∞ mapping $\varphi \circ \varphi'^{-1}$.

8. TENSOR DISTRIBUTIONS

There is an alternative, equivalent, definition of $\mathscr{D}'(X)$ as the dual of the space $\mathscr{D}(X)$ of C^∞ functions on X, and of the section distributions $\mathscr{D}'(X, E)$ of the vector bundle E on X, as the dual of the space of C^∞ sections with compact support in X. In particular we can thus define the space of tensor [spinor] distributions of a given type. This we now do, and, to have a simpler presentation which is also useful in applications, we shall first endow the C^∞ paracompact manifold X with a C^∞ (proper) riemannian metric $\boldsymbol{h}$. The spaces $\mathscr{D}(X, E)$, $\mathscr{D}'(X, E)$ will not depend on the choice of $\boldsymbol{h}$. We denote by $\nabla \boldsymbol{t}$ the covariant derivative in the metric $\boldsymbol{h}$ of a tensor field $\boldsymbol{t}$, and by $|\boldsymbol{t}(x)|$ the length, relative to $\boldsymbol{h}$, of $\boldsymbol{t}$ at the point $x \in X$. If $\boldsymbol{t}$ is a p tensor

$$|\boldsymbol{t}(x)| = |t^{i_1 \cdots i_p}(x^j) t_{i_1 \ldots i_p}(x^j)|^{1/2}$$

where indices are elevated or lowered through the metric $\boldsymbol{h}$.

$p_{K,m}$ We define the (semi) norm $p_{K,m}$ of a C^∞ tensor $\boldsymbol{t}$ on X, with K a compact set

$$p_{K,m}(\boldsymbol{t}) = \sum_{|l| \le m} \sup_{x \in K} |(\nabla^{(l)} \boldsymbol{t})(x)|$$

where l is a non negative multi index and m is a non negative integer.

$\mathscr{E}(X, \otimes^p)$ The space of C^∞ p-tensor fields on X becomes a Fréchet space (p. 26) denoted $\mathscr{E}(X, \otimes^p)$ when endowed with the topology defined by the family of norms $p_{K,m}$.

$\mathscr{D}(X, \otimes^p)$ The space $\mathscr{D}(X, \otimes^p)$ is the space of C^∞ p-tensor fields on X of compact support, with the inductive limit topology of the Fréchet spaces $\mathscr{D}_K(X, \otimes^p)$ of C^∞ p-tensor fields with support in K and topology defined by the seminorms $p_{K,m}$.

The space of p-tensor distributions on X, $\mathscr{D}'(X, \otimes^p)$ is the dual of $\mathscr{D}(X, \otimes^p)$. As in the scalar case a linear mapping $\mathscr{D}'(X, \otimes^p)$

$$T: \mathscr{D}(X, \otimes^p) \to \mathbb{K} \text{ by } t \mapsto \langle T, t \rangle$$

is continuous iff it is continuous on each $\mathscr{D}_K$, that is if there exists $C(K)$ and $m(K)$ such that

$$|\langle T, t\rangle| \leq C(K) p_{K, m(K)}(t), \qquad \forall t \in \mathscr{D}_K(X, \otimes^p).$$

The identification of a locally integrable tensor field with a tensor distribution is defined by

$$\langle s, t \rangle = \int_X (s(x)|t(x))_h \eta_h .$$

where $(s(x)|t(x))_h$ denotes the scalar product of $s(x)$ and $t(x)$, and η_h the volume element relative to the metric h. This identification depends on the choice of h, although the space of *locally* integrable tensor fields does not.

If s is a C^1 tensor field on X, its covariant derivative in the metric h is, in the domain of a chart (U, φ), the $p+1$ tensor field ∇s such that, if t is a $p+1$ C^1 tensor field with compact support in U

$$\langle \nabla s, t \rangle = \int_{\varphi(U)} \nabla_j s_{i_1 \dots i_p}(x^l) t^{j i_1 \cdots i_p}(x^l) |\det h|^{1/2} \, dx^1 \dots dx^n;$$

thus

$$\langle \nabla s, t \rangle = - \int_{\varphi(U)} s_{i_1 \dots i_p}(x^l) \nabla_j t^{j i_1 \cdots i_p} |\det h|^{1/2} \, dx^1 \dots dx^n.$$

It is therefore legitimate to adopt the following definition.[1]

The **covariant derivative**, in the metric h, of a p-tensor distribution T is the $p+1$ tensor ∇T defined for every $t \in \mathscr{D}(X, \otimes^{p+1})$ by ∇T

$$\langle \nabla T, t \rangle = -\langle T, h \cdot \nabla t \rangle$$

where $h \cdot \nabla t$ is the contracted product of h with ∇t, that is, in local coordinates

$$(h \cdot \nabla t)^{i_1 \cdots i_p} = \nabla_j t^{j i_1 \cdots i_p}.$$

The topology of $\mathscr{D}'(X, \otimes^p)$ is defined (cf. analogous definition for $\mathscr{D}'(U)$, p. 453) by the family of seminorms

$$p_t(T) = |\langle T, t \rangle|, \qquad t \in \mathscr{D}(X, \otimes^p).$$

With this topology we have the following theorem.

[1][Lichnerowicz 1961].

Theorem. $\mathscr{D}(X, \otimes^p)$ *is dense in* $\mathscr{D}'(X, \otimes^p)$.

Proof: 1) Suppose that the support of $\boldsymbol{T} \in \mathscr{D}'(X, \otimes^p)$ is contained in the domain U of a chart. We define the components of $\boldsymbol{T}$ in the chart, $T_{i_1 \ldots i_p}$, as the scalar valued distributions defined by the restriction of the linear mapping $\boldsymbol{t} \mapsto \langle \boldsymbol{T}, \boldsymbol{t} \rangle$ to the p-tensors $\boldsymbol{t} \in \mathscr{D}(U, \otimes^p)$ which have, in the chart, only one non vanishing component $t^{i_1 \cdots i_p}$. By a previous theorem $\mathscr{D}(V)$ is dense in $\mathscr{D}'(V)$ when V is an open set of $\mathbb{R}^n$ (p. 465). Therefore there exists a sequence $(\boldsymbol{t}_\nu) \in \mathscr{D}(U, \otimes^p)$ which converges to $\boldsymbol{T}$ in $\mathscr{D}'(U, \otimes^p)$, and thus in $\mathscr{D}'(X, \otimes^p)$ since $\boldsymbol{T}$ has its support in U.
2) In the general case we take a partition of unity (p. 214) α_i relative to an atlas (U_i, φ_i) and write

$$\boldsymbol{T} = \sum_i \alpha_i \boldsymbol{T},$$

that is, for $\boldsymbol{t} \in \mathscr{D}(X, \otimes^p)$,

$$\langle \boldsymbol{T}, \boldsymbol{t} \rangle = \left\langle \boldsymbol{T}, \sum_{i=1}^{N} \alpha_i \boldsymbol{t} \right\rangle.$$

Since $\boldsymbol{t}$ has compact support the sum is finite for each $\boldsymbol{t}$. Each $\boldsymbol{T}_i = \alpha_i \boldsymbol{T}$ has its support in the domain U_i of a chart, and is a limit in $\mathscr{D}'(X, \otimes^p)$ of a sequence $(\boldsymbol{t}_\nu^i) \in \mathscr{D}(U_i, \otimes^p)$. Now set

$$\boldsymbol{t}_\nu = \sum_{i=1}^{\nu} \alpha_i \boldsymbol{t}_\nu^i \in \mathscr{D}(X, \otimes^p).$$

We have for a given $\boldsymbol{t} \in \mathscr{D}(X, \otimes^p)$

$$\begin{aligned} \lim_{\nu=\infty} \langle \boldsymbol{t}_\nu, \boldsymbol{t} \rangle &= \lim_{\nu=\infty} \left\langle \sum_{i=1}^{\nu} \alpha_i \boldsymbol{t}_\nu^i, \boldsymbol{t} \right\rangle \\ &= \lim_{\nu=\infty} \sum_{i=1}^{\nu} \langle \boldsymbol{t}_\nu^i, \alpha_i \boldsymbol{t} \rangle = \lim_{\nu=\infty} \sum_{i=1}^{N} \langle \boldsymbol{t}_\nu^i, \alpha_i \boldsymbol{t} \rangle \\ &= \sum_{i=1}^{N} \langle \boldsymbol{T}_i, \alpha_i \boldsymbol{t} \rangle = \langle \boldsymbol{T}, \boldsymbol{t} \rangle. \end{aligned}$$ ■

Currents. A particular case of tensor distributions are the totally antisymmetric tensor distributions. They have been introduced under the name of "**currents**" by De Rham[1], to generalize the notions of both chains and exterior forms on a C^∞-manifold (cf. p. 228).

Remark: The space $\mathscr{D}'(X)$ of 0-currents on X is isomorphic with the space of n-currents.

[1][De Rham].

A 0-current (generalizing a zero form) is a scalar distribution and transforms accordingly under a diffeomorphism (p. 448). An n-current (generalizing an n-form) transforms rather like the measures in Chapter I, p. 37, namely

$$\langle \Phi T, \varphi \rangle = \langle T, \Phi^* \varphi \rangle.$$

Examples: A p-form $\boldsymbol{\omega}$ defines a current of degree p (dimension $n-p$) by

$$\langle \boldsymbol{\omega}, \boldsymbol{\theta} \rangle = \int_X \boldsymbol{\omega} \wedge \boldsymbol{\theta} \qquad \text{for every } \boldsymbol{\theta} \in \mathscr{D}^{n-p}.$$

A $n-p$ chain C defines a current of degree p (dimension $n-p$) by

$$\langle C, \boldsymbol{\theta} \rangle = \int_C \boldsymbol{\theta} \qquad \text{for every } \boldsymbol{\theta} \in \hat{\mathscr{D}}^{n-p}.$$

Linear differential operator. A linear differential operator L of order m on p-tensor distributions on a manifold X with C^∞ metric $\boldsymbol{h}$ is a linear mapping which can be written:

$$L: \mathscr{D}'(X, \otimes^p) \to \mathscr{D}'(X, \otimes^q) \text{ by}$$

$$\boldsymbol{T} \mapsto \sum_{|l| \le m} a_l \nabla^{(l)} \boldsymbol{T} \qquad (l \text{ a multi index})$$

where a_l is a C^∞ section of the vector bundle over X whose fiber at x is the space of linear maps $\mathscr{L}((\otimes T_x(X))^{p+m}, (\otimes T_x(X))^q)$.

Example 1: Gradient operator

$$\mathscr{D}'(X, \otimes^p) \to \mathscr{D}'(X, \otimes^{p+1}) \text{ by } \boldsymbol{T} \mapsto \nabla \boldsymbol{T}.$$

Note that in local coordinates the components of $\nabla \boldsymbol{T}$ are given by the usual formulas:

$$\nabla_j T_{i_1 \ldots i_p} = \partial_j T_{i_1 \ldots i_p} - \Gamma^l_{ji_1} T_{li_2 \ldots i_p} - \cdots - \Gamma^l_{ji_p} T_{i_1 i_2 \ldots l}.$$

Example 2: Divergence of a vector distribution

$$\mathscr{D}'(X, \otimes^1) \to \mathscr{D}'(X) \qquad \text{by } \boldsymbol{T} \mapsto \boldsymbol{h} \cdot \nabla \boldsymbol{T}.$$

Example 3: Lie derivative

$$\mathscr{D}'(X, \otimes^1) \to \mathscr{D}'(X, \otimes^2) \qquad \text{by } \boldsymbol{V} \mapsto \mathscr{L}_V \boldsymbol{h}$$

defined in local coordinates by

$$(\mathscr{L}_V \boldsymbol{h})_{ij} = \nabla_i V_j + \nabla_j V_i.$$

Example 4: The exterior derivative d: $\boldsymbol{\omega} \mapsto \mathrm{d}\boldsymbol{\omega}$ with

$$(\mathrm{d}\boldsymbol{\omega})_{i_1 \ldots i_p} = \sum_{k=1}^{p} (-1)^{k-1} \nabla_{i_k} \omega_{i_1 \ldots \hat{i}_k \ldots i_p}$$

where $\boldsymbol{\omega}$ is an antisymmetric $p-1$ tensor-distribution (cf. p. 317).

C. SOBOLEV SPACES AND PARTIAL DIFFERENTIAL EQUATIONS

We had to restrict the space $\mathscr{D}'$ of distributions to the space $\mathscr{S}'$ of tempered distributions in order to develop a well constructed theory of the Fourier transform. We shall restrict $\mathscr{D}'$ further to develop, by the use of Banach and Hilbert space techniques, a well constructed theory of partial differential equations.

1. SOBOLEV SPACES

The theory of distributions, together with the theory of L^p spaces makes it possible to define a family of Banach spaces called Sobolev spaces which have remarkable properties and which are useful in many applications, particularly in numerical analysis.[1]

Sobolev spaces W_p^m

Let $U \subset \mathbb{R}^n$ be open; let m and p be nonnegative integers $p \geq 1$; the **Sobolev space** $W_p^m(U)$ is the space of functions in $L^p(U)$ whose partial derivatives of order $\leq m$ – in the sense of p. 446 – are also in $L^p(U)$,

$$W_p^m(U) = \{f;\, \mathrm{D}^j f \in L^p(U),\, |j| \leq m\}.$$

The norm in $W_p^m(U)$ is

$$\|f\|_{W_p^m} = \left\{ \sum_{|j| \leq m} \|\mathrm{D}^j f\|_{L^p}^p \right\}^{1/p} \text{ with } \|\mathrm{D}^j f\|_{L^p} = \left\{ \int_U |\mathrm{D}^j f|^p \,\mathrm{d}x \right\}^{1/p}.$$

H^m

The space $W_2^m(U)$ is usually called $H^m(U)$ and

$$\|f\|_{H^m} = \left\{ \sum_{|j| \leq m} \int_U |\mathrm{D}^j f|^2 \,\mathrm{d}x \right\}^{1/2}.$$

Example: The function $r^{-\alpha}$ with $\alpha > 0$ on an open bounded set $U \subset \mathbb{R}^m$ which contains the origin is in $H^1(U)$ if $\alpha < (n-2)/2$ for $n > 2$. Similarly for $n = 2$, $(\ln r)^\alpha \in H^1(U)$ if $\alpha < \frac{1}{2}$.

[1]For further properties and missing proofs see [Sobolev 1950], [Lions, Magenes 1968], [Adams 1975] and references therein.

Remark: By the Riesz representation theorem (p. 56) the topological dual of $L^p(U)$, $1 \le p < \infty$, can be identified with $L^{p'}(U)$, $1/p + 1/p' = 1$: every linear continuous function on $L^p(U)$ can be written in a *unique* way

$$T(f) = \int_U gf \, dx \qquad g \in L^{p'}(U).$$

There is a one to one correspondence between T and g and we usually write $g = \tilde{T}$ or even T; by a convenient abuse of language we often say that T is in $L^{p'}$.
Since $L^p(\Omega) \hookrightarrow L^1(\Omega)$ for each bounded open set Ω it follows that $L^p(U) \hookrightarrow \mathscr{D}'(U)$ for $p \ge 1$ (p. 18). This inclusion is topologic as well as algebraic: this means that the topology induced on L^p by the topology of $\mathscr{D}'$ is coarser than the topology of L^p. In other words, the trace on L^p of every open ball in $\mathscr{D}'$ $\{|\langle f, \varphi\rangle| < \epsilon, f \in L^p, \varphi \text{ fixed}\}$ contains an open ball in L^p $\{\|f\|_p < \epsilon/\|\varphi\|_{p'}, 1/p + 1/p' = 1\}$. The inclusion $L^p(U) \subset \mathscr{D}'(U)$ can also be obtained from the inclusion $\mathscr{D}(U) \hookrightarrow L^{p'}(U)$ because $\mathscr{D}(U)$ is dense in $L^{p'}(U)$ (cf. p. 488).
We have now introduced a new space $W_p^m(U)$ larger than $\mathscr{D}(U)$ and smaller than $L^p(U)$. Can we say that its topological dual is smaller than $\mathscr{D}'$ and larger than $L^{p'}$? Unfortunately not because, as we shall see, $\mathscr{D}(U)$ is not dense in $W_p^m(U)$. However, if we denote by $\mathring{W}_p^m$ the closure of $\mathscr{D}(U)$ in $W_p^m(U)$, we have the following inclusions $\mathring{W}_p^m$

$$\mathscr{D}(U) \subset \mathring{W}_p^m(U) \subset L^p(U)$$
$$L^{p'}(U) \subset (\mathring{W}_p^m(U))' \subset \mathscr{D}'(U).$$

We shall now prove the following fundamental theorem.

Theorem. $W_p^m(U)$ is a Banach space; $H^m(U)$ is a Hilbert space.

Proof: 1) W_p^m is a normed vector space; H^m is a prehilbert space with the sesquilinear product

$$(f|g)_{H^m} = \sum_{|j| \le m} \int_U \mathrm{D}^j f \, \overline{\mathrm{D}^j g} \, dx.$$

2) $W_p^m(U)$ is complete. Indeed, let $\{f_r\}$ be a Cauchy sequence

$$\|f_r - f_s\|_{W_p^m} < \epsilon \qquad \text{for } r, s > N.$$

It follows that $(\mathrm{D}^j f_r)$ is a Cauchy sequence in $L^p(U)$ for every $|j| \le m$ and hence converges to a limit in L^p. This limit is the limit in $\mathscr{D}'(U)$ of the sequence of distributions identified with $(\mathrm{D}^j f_r)$. Because $\mathscr{D}'$ is separated and because derivation is continuous in $\mathscr{D}'$, the limit of $(D^j f_r)$ coincides with $(D^j(\text{limit } f_r))$ hence (f_r) converges in W_p^m.

Remark: The subspace of $W_p^m(U)$ characterized by the requirement that the functions also be C^m is not complete.

Density theorems. $\mathscr{D}(\mathbb{R}^n)$ *is dense in* $W_p^m(\mathbb{R}^n)$; *in general* $\mathscr{D}(U)$ *is not dense in* $W_p^m(U)$ for $m \geq 1$; the closure of $\mathscr{D}(U)$ in $W_p^m(U)$ is designated $\mathring{W}_p^m(U)$; $\mathring{W}_p^m(\mathbb{R}^n) = W_p^m(\mathbb{R}^n)$.

The density of $\mathscr{D}(\mathbb{R}^n)$ in $W_p^m(\mathbb{R}^n)$ is proved by a method somewhat similar to the one used in proving the density of $C_0^\infty(\mathbb{R}^n)$ in $\mathscr{C}^m(\mathbb{R}^n)$ (p. 434), i.e. it is shown that every $f \in W_p^m(\mathbb{R}^n)$ is the limit, in the sense of the $W_p^m(\mathbb{R}^n)$-norm, of a sequence (f_n) of functions in $\mathscr{D}(\mathbb{R}^n)$ obtained by regularisation and truncation[1].

The density of $\mathscr{D}(U)$ in $W_p^0(U) = L^p(U)$ is proved in two steps. a) One proves the density of $C^\infty(U)$ in $L^p(U)$ by extending $f \in L^p(U)$ into $\tilde{f} \in L^p(\mathbb{R}^n)$

$$\tilde{f}|_U = f \qquad \tilde{f}|_{\mathbb{R}^n \setminus U} = 0$$

and showing that the regularized sequence $(\tilde{f}_n = \tilde{f} * \rho_n)$ of functions in $C^\infty(\mathbb{R}^n)$ converges to $\tilde{f}$ in $L^p(\mathbb{R}^n)$; then $\tilde{f}_n|_U$ converges to f in $L^p(U)$.
b) One proves the density of $\mathscr{D}(U)$ in $L^p(U)$ by truncation of the regularized sequence.

Part a of this argument fails for $W_p^m(U)$ when $m \geq 1$ because $f \in W_p^m(U)$ does not imply $\tilde{f} \in W_p^m(\mathbb{R}^n)$. For example, let $f = 1$ in the ball B, $\|x\| < 1$, $\tilde{f}$ characteristic function of B. Then $f \in W_p^m(B)$ but $\tilde{f} \notin W_p^m(\mathbb{R}^n)$, $\partial \tilde{f}/\partial x^i$ is a *measure* with support ∂B (cf. p. 448). To show that $\mathscr{D}(U)$ is not dense in $W_p^m(U)$ is to show that the closure $\mathring{W}_p^m(U)$ of $\mathscr{D}$ in W_p^m is different from $W_p^m(U)$. Indeed it can be shown that $f \in \mathring{W}_p^m(U)$ implies $\tilde{f} \in W_p^m(\mathbb{R}^n)$.

$\mathring{W}_p^m$ is a closed vector subspace of the Banach space W_p^m, hence it is a Banach space. In a loose way we can characterize this subspace by saying that $\mathring{W}_p^m(U)$ is the subspace of functions in $W_p^m(U)$ which vanish on ∂U, together with their derivatives of order $\leq m-1$.

Example: Let us consider $W_2^m(U) = H^m(U)$ and $\mathring{W}_2^m(U) = \mathring{H}^m(U)$. We shall show that $\mathring{H}^m$ has a non empty orthogonal complement $\mathscr{H}^m$ in H^m,

[1]Cf. for instance [Choquet-Bruhat 1968, p. 62].

if U is bounded.

$$h \in \mathscr{H}^m(U) \iff (h|f)_{H^m} = 0 \quad \forall f \in \mathring{H}^m(U)$$
$$\iff (h|\varphi)_{H^m} = 0 \quad \forall \varphi \in \mathscr{D}(U).$$

But

$$(h|\varphi)_{H^m} = \int_U \sum_{|j|\le m} \mathrm{D}^j h \overline{\mathrm{D}^j \varphi}\, \mathrm{d}x = \int_U \sum_{|j|\le m} (-1)^{|j|} (\mathrm{D}^{2j} h) \bar{\varphi}\, \mathrm{d}x,$$

hence

$$h \in \mathscr{H}^m(U) \iff \sum_{j \le m} (-1)^{|j|} \mathrm{D}^{2j} h = 0 \quad \text{in } \mathscr{D}',\ h \in H^m(U). \quad \blacksquare$$

For instance $h \in \mathscr{H}^1(U)$ if and only if

$$-\Delta h + h = 0,\ h \in H^1(U)$$

which has an infinity of solutions if U is bounded; it even has an infinity of C^∞ solutions. The fact that $\mathring{W}_p^m(U)$ is different from $W_p^m(U)$ is the essence of boundary value problems.

It can be proved[1] that $W_p^m(U) \neq \mathring{W}_p^m(U)$ whenever the complement of U has a positive measure and $m \ge 1$.

Friedrichs lemma. If U is a bounded open set, the scalar product

$$(f|g) = \sum_{|j|=m} \int_U \mathrm{D}^j f \overline{\mathrm{D}^j g}\, \mathrm{d}x$$

defines on $\mathring{H}^m$ a norm – and not only a semi norm – equivalent to the H^m-norm.

Indeed $(f|f) = 0$ implies $\mathrm{D}^j f = 0$ implies $\mathrm{D}^{j-1} f = $ constant; but $\mathring{H}^m$ contains no constants other than 0. $\blacksquare$

This result is important, for instance, in the solution of the Dirichlet problem for the Laplace operator.

$W_{p'}^{-m}(U)$ spaces. The dual of $\mathring{W}_p^m(U)$ is denoted $W_{p'}^{-m}(U)$. The space $W_{p'}^{-m}(U)$ is a subspace of $\mathscr{D}'(U)$. $W_{p'}^{-m}$
It can be shown that the distributions in $W_{p'}^{-m}$ are of the form

$$T = \sum_{|j|\le m} \mathrm{D}^j f_j \qquad f_j \in L^{p'}(U).$$

[1] [Lions 1967].

Fourier transform. According to the definition of $\mathcal{S}$ (p. 474), a function in $L^p(\mathbb{R}^n)$ defines a tempered distribution and hence has a Fourier transform.

Plancherel theorem. If f is in L^2, then $\mathcal{F}f$ is in L^2 and

$$\|\mathcal{F}f\|_2 = \|f\|_2 \qquad \text{with } \| \ \|_2 \equiv \| \ \|_{L^2}$$

i.e. $\mathcal{F}$ is an isometry of L^2.

Proof: 1) $\|\mathcal{F}\varphi\|_2 = \|\varphi\|_2 \quad \forall \varphi \in \mathcal{S}$. Indeed

$$\int_{\mathbb{R}^n} \mathcal{F}\varphi \cdot \psi \, \mathrm{d}x = \int_{\mathbb{R}^n} \varphi \cdot \mathcal{F}\psi \, \mathrm{d}x.$$

Set $\psi = \overline{\mathcal{F}\varphi}$, it follows that $\mathcal{F}\psi = \bar{\varphi}$.

2) $\langle \mathcal{F}f, \varphi \rangle = \langle f, \mathcal{F}\varphi \rangle = \int_{\mathbb{R}^n} f\mathcal{F}\varphi \, \mathrm{d}x \qquad f \in L^2, \ \varphi \in \mathcal{S}$

$$\left| \int_{\mathbb{R}^n} f\mathcal{F}\varphi \, \mathrm{d}x \right| \leq \|f\|_2 \ \|\mathcal{F}\varphi\|_2 = \|f\|_2 \ \|\varphi\|_2.$$

It follows that $\langle \mathcal{F}f, . \rangle$ is a linear continuous form on $\mathcal{S}$ for the L^2-norm. Hence there exists a unique $\hat{f} \in L^2$ with $\|\hat{f}\|_2 = \|f\|_2$ such that $\langle \mathcal{F}f, \varphi \rangle = \int_{\mathbb{R}^n} \hat{f}\varphi \, \mathrm{d}x \quad \forall \varphi \in \mathcal{S}$ (cf. p. 56) hence $\hat{f} = \mathcal{F}f$ in the sense of distributions.

Fourier transform of $f \in H^m(\mathbb{R}^n)$. If $f \in H^m$, then $f \in L^2$ and $\mathrm{D}^j f \in L^2$ for $|j| \leq m$. Therefore $\mathcal{F}f \in L^2$ and $\mathcal{F}\mathrm{D}^j f = \mathrm{i}^{|j|} x^j f \in L^2$ for $|j| \leq m$. It follows that

$$(1 + |x|^2)^{m/2} f \in L^2.$$

This remark leads to the following definition of a Sobolev space. The Sobolev space $H^m(\mathbb{R}^n)$ with $m \in \mathbb{R}$ is the space of tempered distributions f based on $\mathbb{R}^n$ such that their Fourier transforms *multiplied by* $(1 + |x|^2)^{m/2}$ are in $L^2(\mathbb{R}^n)$. Notice that here m is not restricted to be a positive integer. Along the lines of this definition, one defines a norm $\| \ \|'_{H^m}$ by

$\| \ \|'_{H^m}$

$$\|f\|'_{H^m} = \|(1 + |x|^2)^{m/2} \mathcal{F}f\|_2.$$

Theorem. The norm $\| \ \|'_{H^m}$ is equivalent to the original norm defined on H^m.

Proof: 1) It follows from Plancherel's theorem that

$$\|f\|^2_{H^m} = \sum_{|j| \leq m} \|\mathrm{D}^j f\|_2^2 = \sum_{|j| \leq m} \|x^j \mathcal{F}f\|_2^2 \leq c_1 \|f\|'^2_{H^m}$$

where c_1 is a positive constant.

2) It follows from the inequality

$$(1+|x|^2)^{m/2} \leq K\left(1+\sum_{i=1}^{n}|x^i|^m\right)$$

where K is a constant which depends on m, that $\|f\|'_{H^m} < c_2\|f\|_{H^m}$. ■

We have previously defined $W_{p'}^{-m}(U)$ as the dual of $\mathring{W}_p^m(U)$ with $1/p + 1/p' = 1$ which implied that $H^{-m}(U)$ is the dual of $\mathring{H}^m(U)$. Let us show that, for $U = \mathbb{R}^n$, the two definitions thus obtained for $H^{-m} \equiv H^{-m}(\mathbb{R}^n)$ coincide.

H^{-m} spaces.

$$f \in H^{-m} \quad \Leftrightarrow \quad (1+|x|^2)^{-m/2}\mathcal{F}f \in L^2 \quad \Leftrightarrow \quad \mathcal{F}f = \left(1+\sum_{i=1}^{n}|x^i|^m\right)g, \quad g \in L^2.$$

Hence f is of the form

$$f = \sum_{|j|\leq m} D^j g_j \qquad g_j \in L^2$$

which says that f is in the dual of H^m. H^{-m} is the dual of H^m.

Theorem. H^{-m} together with the norm $\|\ \|'_{H^{-m}}$ is a Hilbert space.

Theorem. The $H^m - H^{-m}$ duality can be defined by

$$\langle\langle f, g\rangle\rangle = \int_{\mathbb{R}^n} \mathcal{F}f\mathcal{F}g \, dx \qquad f \in H^{-m}, \quad g \in H^m.$$

Theorem. The strong dual topology on H^m defined by

$$\|f\|_{H^{-m}} = \sup_{g\in H^m} \frac{|\langle f, g\rangle|}{\|g\|_{H^m}}$$

is equivalent to the topology induced by the norm $\|\ \|'_{H^{-m}}$.

Sobolev inequalities (without proofs).

Theorem. If $f \in W_p^n(U)$ for every bounded open set U of $\mathbb{R}^n$, f is equivalent to a continuous function and if K is a compact set containing x

$$|f(x)| \leq c \sum_{|j|\leq n} \left\{\int_K |D^j f(y)|^p \, dy\right\}^{1/p}.$$

Corollary. If $f \in W_p^{n+k}(\mathrm{loc})$[1], f is equivalent to a C^k function. If all the

[1] i.e. if f is locally in W_p^{n+k}, $f \in W_p^{n+k}(U)$ for every bounded open set U.

derivatives of a distribution are locally in L^p this distribution is equivalent to a C^∞ function.

Theorem. $W_p^m(U) \subset C^k(U)$ *for* $k < m - n/p$. *If* $C^k(U)$ *has the topology of uniform convergence of derivatives of order* $\leq k$ *on every compact subset of* U, *the inclusion is also a topological inclusion.*

cone property

A bounded set $U \subset \mathbb{R}^n$ is said to have the **cone property** if at each point x of U one can draw an axially symmetric cone of angle α and height h which is entirely in U, α and h depending on U but not on x.
For example, a bounded open set with a differentiable or lipschitzian boundary has the cone property.

Theorem. Let U *be a bounded set which has the cone property, then the identity mapping from* $W_p^m(U)$ *into* $C^0(\bar{U})$ *is continuous if* $mp > n$. *That is* $f \in W_p^m(U)$ *is equivalent to a function in* $C^0(\bar{U})$ *and*

$$\sup_{x \in \bar{U}} |f| \leq C\|f\|_{W_p^m(U)} \qquad C \text{ depends on } n,\ p,\ U;\ mp > n.$$

Corollary. If $k < m - n/p$, *then* $W_p^m(U) \subset C^k(\bar{U})$.

Theorem. Let U *be a bounded set which has the cone property, then the identity mapping from* $W_p^{m+1}(U)$ *into* $W_p^m(U)$ *is compact.*

Theorem. If $U = \mathbb{R}^n$ *or if* U *is a bounded set of* $\mathbb{R}^n$ *having the cone property,* $H^m(U)$ *together with pointwise multiplication is an algebra provided that* $m \geq n'$, *where* n' *is the smallest integer larger than* $n/2$.

2. PARTIAL DIFFERENTIAL EQUATIONS

We shall treat linear equations with constant coefficients in paragraphs 3 to 5. Linear equations with variable coefficients are introduced in paragraph 6.

Definitions. A partial differential equation (P.D.E.) on $\mathbb{R}^n$ of order m with constant coefficients is of the form

$$\sum_{|j| \leq m} a_j \mathrm{D}^j u = f. \tag{1}$$

principal part

The **principal part** is $\sum_{|j|=m} a_j \mathrm{D}^j$.

characteristic manifold

A submanifold S of dimension $n-1$ and equation $S(x) = 0$ is said to be a **characteristic manifold** of the P.D.E. if it satisfies the following first order P.D.E.

$$\sum_{|j|=m} a_j p^j = 0 \qquad p^j = \left(\frac{\partial S}{\partial x^1}\right)^{j_1} \cdots \left(\frac{\partial S}{\partial x^n}\right)^{j_n} = (p_1)^{j_1} \dots (p_n)^{j_n}.$$

Cauchy problem. Find a C^m solution of equation (1) which, together with its derivatives of order $\leq m-1$ takes some given values on a submanifold S of dimension $n-1$, and equation $S(x)=0$.
This problem is simplified by the following change of variables:

$$x^{1\prime} = S(x)$$
$$x^{j\prime} = x^j \qquad j = 2, \dots n,$$

with a possible relabeling of variables so that $\{x^{i\prime}\}$ is a set of independent variables. Under this change of variables

$$\frac{\partial}{\partial x^i} \text{ becomes } p_i \frac{\partial}{\partial x^{1\prime}} + \sum_{j=2}^{n} \delta_i^j \frac{\partial}{\partial x^{j\prime}}, \qquad p_i = \frac{\partial S}{\partial x^i}.$$

We obtain equation (1) in the new variables which after suppression of the prime reads

$$\sum_{|j|=m} a_j p^j \left(\frac{\partial}{\partial x^1}\right)^{|j|} u + \sum_{\substack{|j|\leq m \\ j_1<m}} b_j \mathrm{D}^j u = f. \tag{2}$$

The Cauchy data for equation (2) are:

$$\left.\frac{\partial^k u}{(\partial x^1)^k}\right|_{x^1=0} = \varphi_k(x^2, \dots x^n) \qquad k = 0, \dots, m-1. \tag{3}$$

If the functions φ_k are of class C^{m-k}, they determine

$$(\mathrm{D}^j u)\ (0, x^2, \dots x^m) \quad \text{for } |j| \leq m,\ j_1 < m,$$

which in turn, by virtue of equation (2), determine

$$\frac{\partial^m u}{(\partial x^1)^m}(0, x^2, \dots x^n) \qquad \text{if } \sum_{|j|=m} a_j p^j \neq 0,$$

i.e. if S is not a characteristic manifold.

If S is a characteristic manifold, then equation (2) taken at $x^1 = 0$ gives a relationship between the Cauchy data (3). The Cauchy data cannot be independently prescribed on a characteristic manifold.

Cauchy–Kovalevski Theorem[1]. *Equation* (1) *where f and $\{a_j\}$ are analytic with analytic Cauchy data on a non-characteristic analytic manifold has strictly one analytic solution in the neighborhood of S.*

[1]For the proof see [Petrowsky].

Classifications. The properties of the solutions of a P.D.E. depend essentially on the nature of its characteristic manifolds. One is thus led to the following classification of P.D.E.'s which we shall state for equation (1) with real coefficients.

elliptic

Equation (1) is said to be **elliptic** if the equation in p^j

$$\sum_{|j|=m} a_j p^j = 0$$

has no real solution $p \neq 0$.

x^1-hyperbolic

Equation (1) is said to be x^1-**hyperbolic** if the equation in p^1

$$\sum_{|j|=m} a_j p^j = 0$$

has m real distinct roots for every system of real numbers $\{p^j;\ j = \{j_2 \ldots j_m\}\}$.

Examples: The Laplace equation $\Delta u = \sum_{i=1}^{n} \frac{\partial^2}{(\partial x^i)^2} u = 0$ is elliptic.

The wave equation $\Box u \equiv \left(\frac{\partial^2}{(\partial x^1)^2} - \sum_{i=2}^{n} \frac{\partial^2}{(\partial x^i)^2}\right) u = 0$ is x^1-hyperbolic.

An important class of equations which are neither elliptic nor hyperbolic are the parabolic equations; they are not characterized solely by their principal parts.

x^1-parabolic

Equation (1) is said to be x^1-**parabolic** if it can be written

$$\frac{\partial u}{\partial x^1} - \sum_{\substack{|j| \leq m \\ j_1 = 0}} a_j D^j u = f$$

with $\sum_{\substack{|j|=m \\ j_1=0}} a_j p^j > 0 \quad \forall p \neq 0$; thus $\sum_{\substack{|j| \leq m \\ j_1=0}} a_j D^j$ is an elliptic differential operator on $\mathbb{R}^{n-1}$.

Example: The heat equation $\partial u/\partial t - \Delta u = 0$ is t-parabolic.
The equation $\partial u/\partial t + \Delta u = 0$ is not t-parabolic, it is $(-t)$-parabolic.

Remark: On $\mathbb{R}^2$ a second order x^1-parabolic equation is necessarily of the form

$$\frac{\partial u}{\partial x^1} - a\left(\frac{\partial u}{\partial x^2}\right)^2 + b\frac{\partial u}{\partial x^2} + cu = f$$

with $a > 0$. The characteristic equation $a(p_2)^2 = 0$ has a double root.

3. ELLIPTIC EQUATIONS. LAPLACIANS

Laplace Poisson equation

Several properties proved here for the **Laplace equation** $\Delta u = 0$ and the **Poisson equation** $\Delta u = f$ are true for more general elliptic equations.[1] A distribution which satisfies Laplace's equation is said to be **harmonic**.

harmonic

Elementary solution. Regularity theorem. An elementary solution (p. 468) of the Laplacian

$$\Delta = \sum_{i=1}^{n} \frac{\partial^2}{(\partial x^i)^2}$$

is

$$E = -k_n |x|^{2-n} \qquad \text{if } n \neq 2$$

$$= \frac{1}{2\pi} \log |x| \qquad \text{if } n = 2$$

where

$$k_n = \frac{1}{(n-2)(\text{surface area of } S^{n-1})} = \frac{\Gamma(n/2)}{(n-2)2\pi^{n/2}}.$$

E is locally integrable, and C^∞ in the complement of the origin.

Regularity theorem. A distribution solution of Poisson's equation $\Delta u = f$ is C^∞ in U if f is C^∞ in U. In particular a harmonic distribution in U is C^∞ in U.

Proof: 1) If u is of compact support, which implies that f is also of compact support, then

$$u = E * f$$

E is a very regular kernel[2]. Its convolution with f is C^∞.
2) In general, consider $u\varphi$ with $\varphi \in \mathcal{D}$; then if $\Delta u = f$

$$\Delta(u\varphi) = \varphi f + 2\frac{\partial \varphi}{\partial x^i}\frac{\partial u}{\partial x^i} + u\Delta\varphi = v \qquad \text{where } v \text{ is of compact support.}$$

Set $\varphi = 1$ in U, the theorem follows because

$$v \in C^\infty(U) \;\Rightarrow\; u\varphi \in C^\infty(U) \;\Rightarrow\; u \in C^\infty(U).$$ ■

A real valued[3] distribution T is said to be **positive** if

positive distributions

$$\langle T, \varphi\rangle \geq 0 \text{ for every } \varphi \geq 0 \qquad \varphi \in \mathcal{D}.$$

[1] See for instance [Ladyzhenskaya and Uraltseva 1968] or [Morrey 1966], [Lions–Magenes 1968]. For equations on a manifold, see [Palais].
[2] See [L. Schwartz 1966].
[3] A distribution is real valued if $\langle T, \varphi\rangle$ is real for every real φ. If $\varphi \geq 0$, φ is real.

The following theorem shows that this ordering relation implies the continuity of T in the coarse topology $\mathcal{D}^0$.

Theorem. A distribution is positive if and only if it is a positive measure (distribution of order zero).

Proof: Let K be a compact set, $\varphi_0 \in \mathcal{D}$ a positive function equal to 1 in a neighborhood of K, then, if $x \in K$

$$-\varphi_0(x) \sup_{x\in K} |\varphi(x)| \leq \varphi(x) \leq \varphi_0(x) \sup_{x\in K} |\varphi(x)|.$$

Therefore

$$|\langle T, \varphi\rangle| \leq |\langle T, \varphi_0\rangle| \sup_{x\in K} |\varphi(x)| = c p_{K,0}(\varphi), \quad \forall \varphi \in \mathcal{D}_K. \qquad \blacksquare$$

subharmonic A distribution T is said to be **subharmonic** if $\Delta T \geq 0$.

Theorem. A subharmonic distribution is a locally integrable function which, inside any closed ball B_ρ of center 0 and radius ρ, admits the representation

$$T = h + E * \mu \qquad \textit{where } \mu \in \mathscr{C}'^0 \quad \textit{and } h \in C^\infty(B_\rho).$$

Proof: Set $u = E * \mu$ where μ is the measure of compact support $\Delta T|_{B_\rho}$ then

$$\Delta u = \mu.$$

Hence, inside B_ρ, $\Delta(T-u) = 0$, and $T-u$ harmonic in B_ρ implies that h is C^∞ in B_ρ.
μ having compact support together with the properties of E implies[1] $E * \mu \in L^1(\text{loc})$. $\blacksquare$

potential The **potential** u of $f \in \mathscr{C}'$ is

$$u = E * f.$$

It is a solution in the sense of distributions of the Poisson equation $\Delta u = f$.

Example 1: Volume potential. Let U be an open *bounded* set of $\mathbb{R}^n$, let f be L^∞ on U and let its support be $\bar{U}$, hence $f \in L^\infty(\mathbb{R}^n)$. The potential u of f is

$$u(y) = -k_n \int_U \frac{f(x)}{|y-x|^{n-2}} \mathrm{d}x.$$

$u(y)$ does not necessarily have second derivatives in the usual sense,

[1]See [Donoghue, p. 39].

hence Δu is not necessarily a solution of Poisson's equation $\Delta u = f$ in the usual sense. We shall give sufficient conditions on f, which are easy to prove, for u to be a usual solution of $\Delta u = f$. They are moreover easy to apply. Better conditions for mathematical purposes are formulated in Hölder spaces $C^{k,\alpha}$ (see Problem I 4) by the following theorem.

Theorem.[1] *If f is $C^{k,\alpha}$ and with compact support, then the potential is $C^{k+2,\alpha}$.*

Theorem. Let f be L^∞ in U and supp $f \subset \bar{U}$, *then the potential u of f is C^1 on $\mathbb{R}^n$, bounded and tends to zero at infinity, together with its first derivatives*:

$$\|u\|_{C^0(\mathbb{R}^n)} \leq K(U)\|f\|_{L^\infty(U)}$$

and a similar inequality holds for $\|Du\|_{C^0(\mathbb{R}^n)}$.

$K(U)$ depends only on U; more precisely we have the following inequality

$$|u(y)| \leq K(y, U)\|f\|_{L^\infty(U)}$$

where

$$K(y, U) = k_n \int_U \frac{dx}{|y-x|^{n-2}}$$

is uniformly bounded and goes to 0 when $|y|$ tends to infinity.

Proof: According to integration theory u is a continuous bounded function on $\mathbb{R}^n$. A derivative of u, in the sense of distributions, is

$$\frac{\partial u}{\partial x^i} = f * \frac{\partial E}{\partial x^i}.$$

In the complement of the origin

$$\frac{\partial E}{\partial x^i} = (n-2)k_n|x|^{1-n}\frac{x^i}{|x|}.$$

$\partial E/\partial x^i$ can be identified with a locally integrable function on $\mathbb{R}^n$, it follows that $\partial u/\partial x^i$ is a continuous bounded function on $\mathbb{R}^n$. ■

On the other hand D^jE for $|j| \geq 2$ is not locally integrable on $\mathbb{R}^n$ and the stated conditions for f are not sufficient for Δu to be defined in the usual sense. Restricting f to be a continuous function with compact support is not sufficient. The following theorem states conditions on f such that u is C^2.

[1]See for instance [Ladyzhenskaya and Uraltseva 1968], [Morrey], [Miranda].

Theorem. 1) *If* $f \in W^1_\infty(\mathbb{R}^n)$ *and* $\operatorname{supp} f \subset \bar{U}$, *then the potential* u *of* f *is* $C^2(\mathbb{R}^n)$. 2) *If* f *is* L^∞ *in* U *with* $\operatorname{supp} f \subset \bar{U}$ *then the potential* u *of* f *is* C^2 *in* U *and* C^∞ *is the complement of* $\bar{U}$, *provided that*

$\dfrac{\partial f}{\partial x^i} = g_i + h_i$ *with* g_i *satisfying the same conditions as* f *and* $\operatorname{supp} h_i \subset \partial U$.

Proof: To prove the first part of the theorem we write

$$\frac{\partial^2 u}{\partial x^i \partial x^j} = \frac{\partial f}{\partial x^i} * \frac{\partial E}{\partial x^j} \quad \text{in } \mathcal{D}' \text{ and use the previous theorem.}$$

To prove the second part we write

$$\frac{\partial^2 u}{\partial x^i \partial x^j} = g_i * \frac{\partial E}{\partial x^j} + h_i * \frac{\partial E}{\partial x^j}$$

where $\partial E/\partial x^j$ is a very regular kernel; hence its convolution with h_i is C^∞ in the complement of ∂U where h_i is zero and

$$g_i * \partial E/\partial x^j \text{ is continuous on } \mathbb{R}^n. \qquad \blacksquare$$

Remark: If U is regular in the sense of Stokes theorem and $f \in C^1(\bar{U})$ the conditions of the second part of the theorem are satisfied and $u \in C^1(\mathbb{R}^n) \cap C^2(\text{compl. of } \partial U)$.

Remark: If f is not of compact support, but is a bounded measurable function going to zero at infinity faster than $|x|^{-2}$, i.e. if $f \in L^\infty(\mathbb{R}^n)$ $\quad |f(x)| < A/|x|^{2+a}$ for $|x| > 1$, $a > 0$, then $f * E$ is C^1 and

$$u(y) = -k_n \int_{\mathbb{R}^n} \frac{f(x)}{|x-y|^{n-2}} \, dx$$

satisfies the Poisson equation.

Proof: Let $\varphi \in \mathcal{D}$, then $\Delta\varphi * E = \varphi$ and

$$\begin{aligned} \langle f, \varphi \rangle &= -k_n \int_{\mathbb{R}^n} \int_{\mathbb{R}^n} \frac{f(y)\Delta\varphi(x)}{|x-y|^{n-2}} \, dx \, dy \quad \text{and by Fubini's theorem} \\ &= \langle f * E, \Delta\varphi \rangle \\ &= \langle \Delta(f * E), \varphi \rangle. \end{aligned}$$

Moreover under the conditions stated for f, there exists a positive number M such that

$$|y|^a |u(y)| \leq M$$

and a positive number M' such that

$$|y|^{a+1}\left|\frac{\partial u}{\partial y^i}(y)\right| \leq M'.$$

It can be proved by the maximum principle (see next section) that u is the only solution of Poisson's equation which satisfies these conditions.

Example 2: **Surface potential.** Let S be a C^1 differentiable submanifold of $\mathbb{R}^n$, of dimension $n-1$, and entirely at a finite distance. Let f be a measure of support S equivalent to an integrable function f_0 on S; then, the potential u of f is

$$u(y) = -k_n \int_S \frac{f_0(x)}{|x-y|^{n-2}}\,\omega(x)$$

where $\omega(x)$ is Leray's form with respect to S taken at the point x. This potential is C^∞ in the complement of S.

Example 3: **Surface dipole potential.** With the same assumption for S, we assume that f is a dipole distribution of support S; i.e. f is a distribution of order 1 of the form

$$f = \sum_{i=1}^{n} \frac{\partial(f_0 n^i)}{\partial x^i}$$

where (n^i) are the components of the normal to S, and f_0 is a measure of support S equivalent to an integrable function f_0 on S; then, the potential u of f is

$$u(y) = f * E = \Sigma\, n^i f_0 * \frac{\partial E}{\partial x^i} = \int_S \frac{f_0(x)}{|y-x|^{n-1}}\,a\omega(x)$$

where $a = \Sigma_i\, n^i(x)(y^i - x^i)|y-x|^{-1}(n-2)k_n$.
The potential u is C^∞ in the complement of S.

Energy integral. Green's formula. Uniqueness theorem. Let U be a bounded open set in $\mathbb{R}^n$ with a C^1 differentiable boundary[1] ∂U. Let u and φ be two C^2 functions in the closure $\bar{U}$ of U; then in U

$$u\Delta\varphi = \Sigma \frac{\partial}{\partial x^i}\left(u\frac{\partial\varphi}{\partial x^i}\right) - \Sigma \frac{\partial u}{\partial x^i}\frac{\partial\varphi}{\partial x^i}$$

$$\int_U u\Delta\varphi \,\mathrm{d}x = \int_{\partial U} u\frac{\partial\varphi}{\partial x^i}\, n^i\omega - \int_U \Sigma \frac{\partial\varphi}{\partial x^i}\frac{\partial u}{\partial x^i}\,\mathrm{d}x$$

[1]In the sense of p. 218, in particular U is locally on one side of ∂U.

where (n^i) are the components of the outward normal to ∂U and ω is Leray's form with respect to ∂U. It follows that

energy integral

$$\int_U u\Delta u \,\mathrm{d}x = \int_{\partial U} u \frac{\partial u}{\partial x^i} n^i \omega - \int_U \sum \left(\frac{\partial u}{\partial x^i}\right)^2 \mathrm{d}x$$

Green formula

$$\int_U (u\Delta\varphi - \varphi\Delta u)\,\mathrm{d}x = \int_{\partial U} \sum \left(u\frac{\partial \varphi}{\partial x^i} - \varphi\frac{\partial u}{\partial x^i}\right) n^i \omega.$$

Uniqueness theorem. The equation $\Delta u = T$ has at most one solution which is in $C^2(U)\cap C^0(\bar{U})$ and takes given values on ∂U.

Proof: Let v_1 and v_2 be two solutions, we shall prove that $u \equiv v_1 - v_2$ vanishes. Let $v_1, v_2 \in C^2(U)\cap C^1(\bar{U})$. If $\Delta u = 0$ in U with $u = 0$ on ∂U, then

$$\sum \int_U \left(\frac{\partial u}{\partial x^i}\right)^2 \mathrm{d}x = 0, \quad \text{hence } \frac{\partial u}{\partial x^i} = 0 \text{ in } U.$$

Thus $u =$ constant in U and finally $u = 0$ in $\bar{U}$ since $u = 0$ on ∂U. The unicity theorem in $C^2(U)\cap C^0(\bar{U})$, and its application to unbounded domains follows from the maximum theorem given below. ∎

maximum theorem

Let f be positive [negative] in an open set U, let u be a function C^2 in U and C^0 in $\bar{U}$ which satisfies Poisson's equation $\Delta u = f$, then u cannot take in U values larger [smaller] than its largest [smallest] value on the boundary ∂U. The proof can be found in many text books.[1] It follows from the elementary remark that u cannot attain a maximum at a point where $\Delta u > 0$, since being maximum with respect to each variable separately, all second derivatives are ≤ 0 at this point.

Liouville theorem. A harmonic distribution bounded in the whole space is constant.

From Green's formula we shall obtain an expression for $u(y)$, $y \in \bar{U}$, in terms of Δu in the domain U and of u and $\partial u/\partial \nu$ in ∂U.

[1]For instance [Sobolev, lecture 9], [Mikhlin, p. 256], and for proofs and results for more general equations [Protter and Weinberger].

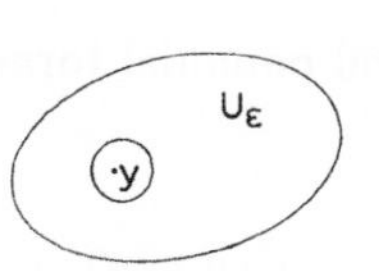

In Green's formula, set $\varphi(x) = -k_n|x-y|^{2-n}$, φ is C^∞ in the open set $x \neq y$; take for the domain of integration U_ϵ equal to U without the ball $B_\epsilon(y)$ of radius ϵ and center y. Then

$$-k_n \int_{U_\epsilon} |x-y|^{2-n} \Delta u \, \mathrm{d}x = -k_n \int_{\partial U \cup \partial B_\epsilon} \sum \left(u \frac{\partial}{\partial x^i} |x-y|^{2-n} - |x-y|^{2-n} \frac{\partial u}{\partial x^i} \right) n^i \omega.$$

The function $|x-y|^{2-n}\Delta u$ is integrable on U, its integral on U_ϵ tends to its integral on U when ϵ tends to zero. The integral on ∂B_ϵ is

$$-k_n \int_{S^{n-1}} \sum_i \left((2-n) u \epsilon^{1-n} n^i - \epsilon^{2-n} \frac{\partial u}{\partial x^i} \right) n^i \omega$$

with $\Sigma\,(n^i)^2 = 1$, $\omega = \epsilon^{n-1}\,\mathrm{d}S^{n-1}$ where S^{n-1} is the $n-1$ unit sphere. At the limit $\epsilon = 0$, the integral on ∂B_ϵ is $-k_n \int_{S^{n-1}} (2-n) u \, \mathrm{d}S^{n-1} = u(y)$. Hence

$$u(y) = -k_n \int_U |x-y|^{2-n} \Delta u(x) \, \mathrm{d}x$$
$$- k_n \int_{\partial U} \left(u \frac{\partial}{\partial \nu} |x-y|^{2-n} - |x-y|^{2-n} \frac{\partial u}{\partial \nu} \right) \omega(x).$$

This formula is called the **general potential formula**.

general potential formula

Remark: For every φ of compact support we have

$$\varphi(0) = -k_n \int |x|^{2-n} \Delta\varphi \, \mathrm{d}x, \text{ i.e.}$$

$$\langle \delta, \varphi \rangle = -\langle k_n |x|^{2-n}, \Delta\varphi \rangle = -\langle k_n \Delta |x|^{2-n}, \varphi \rangle.$$

Hence $-k_n|x|^{2-n}$ is an elementary solution of Δ, a result previously obtained by another method.

Let U be a sphere of center 0, radius R; let u satisfy $\Delta u = 0$; then for $n > 2$,

$$u(0) = -k_n \int_{\partial B_R} \left((2-n) R^{1-n} u - R^{2-n} \frac{\partial u}{\partial \nu} \right) \omega \quad \text{and using} \quad \int_{\partial U} \frac{\partial u}{\partial \nu} \omega = \int_U \Delta u \, \mathrm{d}x = 0$$

we have the **average formula** $u(0) = -k_n (2-n) R^{1-n} \int_{\partial B_R} u\omega.$

average formula

Exercise: Derive the general potential formula and the average formula for $n = 2$.

Remark: Let u be C^2 in U and C^1 in $\bar{U}$; let us extend u to $\tilde{u}$ equal to 0 in the complement of $\bar{U}$, then

$$\langle \Delta\tilde{u}, \varphi\rangle = \langle \tilde{u}, \Delta\varphi\rangle = \int_U u\Delta\varphi \, dx \qquad \varphi \in \mathscr{D}(\mathbb{R}^n).$$

Let $\widetilde{\Delta u}$ be the extension of Δu equal to 0 in the complement of U, then, by Green's formula

$$\langle \Delta\tilde{u}, \varphi\rangle = \langle \widetilde{\Delta u}, \varphi\rangle + \int_{\partial U} \sum \left(u \frac{\partial\varphi}{\partial x^i} - \varphi \frac{\partial u}{\partial x^i}\right) n^i\omega$$

$$\Delta\tilde{u} = \widetilde{\Delta u} - \frac{\partial}{\partial x^i}(u_0 n^i \delta_{\partial U}) - u_i n^i \delta_{\partial U} \equiv T$$

where u_0 is the value of u on ∂U and u_i the value of $\partial u/\partial x^i$ on ∂U. T being of compact support we have

$$\tilde{u} = E * T.$$

This formula gives back the general potential formula from which the volume potential, surface potential and dipole surface potential obtained earlier can be derived when Δu, u_0 and u_i are regular.

Boundary-value problems. Consider the problem of finding $u \in C^2(U) \cap C^1(\bar{U})$ such that

$$\Delta u = f, \qquad u = u_0 \text{ on } \partial U \text{ and } \partial u/\partial\nu = u_1 \text{ on } \partial U.$$

In general it follows from the unicity theorem that this problem has no solution: if u_0 is given there cannot be more than one solution. Thus the general potential formula cannot be the solution of the problem under consideration. The function given by the general potential formula does not satisfy the requirements when y goes to ∂U.

Dirichlet problem

Dirichlet problem: find u such that

$$\Delta u = 0 \text{ in } U \qquad u \in C^2(U) \cap C^0(\bar{U})$$

$$u(y) = u_0(y) \qquad y \in \partial U \qquad u_0 \in C^0(\partial U).$$

We have already shown that Dirichlet's problem has at most one solution if U is an open bounded set with a C^1 Stokes regular boundary. It can be shown, moreover, that there is a solution.

Neumann problem

Neumann problem: find u such that

$$\Delta u = 0 \text{ in } U \qquad u \in C^2(U) \cap C^1(\bar{U})$$

$$\frac{\partial u}{\partial \nu}(y) = u_1(y) \qquad y \in \partial U \qquad u_1 \in C^0(\partial U).$$

It can be shown that, under regularity conditions, Neumann's problem has a solution. This solution is unique, modulo an additive constant – a result which can be derived from the energy integral.

Results on the Dirichlet and Neumann problems for the Poisson equation $\Delta u = f$ can be found for instance in [Sobolev], [Petrovsky], and for more general elliptic equations in the references given on p. 497.

Green function.

Definition: Let U be an open set, a Green function is a kernel $G(x, y)$ on $U \times U$ which has the following properties.

1) $-\Delta_x G(x, y) = \delta_y$ $\quad$ G is an elementary kernel in x for $-\Delta$.

2) There exists a C^∞ function g in $U \times U$ such that

$$g(x, y) = G(x, y) + E(x, y), \quad \text{where} \quad E(x, y) = -k_n |x - y|^{2-n}.$$

3)

$$G(x, y) = 0 \qquad \text{for } x \in \partial U, \quad y \in U.$$

Remark: $\Delta_x g(x, y) = 0$ in U for $y \in U$. It follows from the unicity of the Dirichlet problem that the Green function is unique.

Theorem. The Green function is symmetric: $G(x, y) = G(y, x)$.

Proof: Here the integrals represent convolutions, in the sense of Volterra, of distributions,

$$\int_U \Delta_x G(x, y) \cdot G(x, z) \, dx = \int_U G(x, y) \Delta_x G(x, z) \, dx$$

$$\int_U \delta_y(x) G(x, z) \, dx = \int_U G(x, y) \delta_z(x) \, dx$$

$$G(y, z) = G(z, y). \qquad \blacksquare$$

Remark: $G(x, y)$ is C^∞ in U, for $x \neq y$.

Proposition. $0 \leq G(x, y) \leq k_n|x-y|^{2-n}$.

Proof: $g(x, y)$ is a harmonic function $\forall\, y \in U$ such that

$$g(x, y)|_{x\in\partial U} = E(x, y)|_{x\in\partial U} = -k_n|x-y|^{2-n}|_{x\in\partial U} \leq 0.$$

According to the maximum theorem $g(x, y) \leq 0$ in U, hence

$$G(x, y) \leq k_n|x-y|^{2-n}.$$

On the other hand $G(x, y)$ is harmonic on $U\backslash B_\epsilon$, with B_ϵ ball of center y, radius ϵ; $G(x, y)$ is zero on ∂U, positive on ∂B_ϵ for ϵ sufficiently small since $g(x, y)$ is continuous for $x = y$. According to the maximum theorem:

$$G(x, y) \geq 0 \text{ in } U. \qquad \blacksquare$$

If $G(x, y)$ is a Green function in the open set U and u a solution of the Dirichlet problem $\Delta u = -f$ with boundary value u_0 on ∂U, we see (p. 501) that

$$u(y) = \int_U G(x, y) f(x)\,\mathrm{d}x - \int_{\partial U} u_0 \frac{\partial}{\partial \nu_x} G(x, y)\omega(x).$$

Example: Green's function inside the unit ball $B = \{x; |x| \leq 1\}$ in $\mathbb{R}^3$

$$\begin{aligned} G(x, y) &= \frac{1}{4\pi|x-y|} - \frac{1}{4\pi|y||x-y/|y|^2|} \\ &= \frac{1}{4\pi}\left[\frac{1}{(|x|^2+|y|^2-2|x||y|\cos\theta)^{1/2}} - \frac{1}{(|x|^2|y|^2+1-2|x|\,|y|\cos\theta)^{1/2}}\right]. \end{aligned}$$

Proof: 1) $G(x, y) - 1/4\pi|x-y|$ is C^1 in $\bar{B} \times B$ and $\Delta_x G(x, y) = -\delta_y$ in the open set $|x| < 1$. 2) $G(x, y) = 0$ for $|x| = 1$, $y \in B$.
The derivative

$$\left.\frac{\partial G}{\partial \nu}\right|_{|x|=1} = \left.\frac{\partial G}{\partial |x|}\right|_{|x|=1} = \frac{-1}{4\pi}\frac{1-|y|^2}{(1-2|y|\cos\theta+|y|^2)^{3/2}}$$

Poisson kernel

is called **Poisson's kernel**. The solution of the Dirichlet problem $\Delta u = 0$ for the ball B is

$$u(y) = \frac{1}{4\pi}\int_{\partial B} u_0(y)\frac{1-|y|^2}{(1-2|y|\cos\theta+|y|^2)^{3/2}}\,\omega.$$

It can be shown that if y goes to y_0, $u(y)$ indeed goes to $u_0(y_0)$.

Generalized Dirichlet problem as an introduction to hilbertian methods. Let U be an open bounded set of $\mathbb{R}^n$ with differentiable boundary ∂U; let u and v be real C^1 functions in $\bar{U}$, vanishing on ∂U, let u be C^2 in U; under these conditions

$$\int_U v\Delta u \, dx = -\int_U \sum \frac{\partial u}{\partial x^i}\frac{\partial v}{\partial x^i} \, dx$$

and

$$(-\Delta u + u|v)_2 = (u|v)_{H^1}$$

where $(\ |\ )_2$ and $(\ |\ )_{H^1}$ designate the sesquilinear products in $L^2(U)$ and $H^1(U)$ respectively.

Let $f \in L^2(U)$, we shall replace the **Dirichlet problem** A L^2
(A) Find u such that $-\Delta u + u = f, \quad u|_{\partial U} = 0$
by the following problem B in functional analysis.

(B) Find $u \in \mathring{H}^1(U)$ such that $(f|v)_2 = (u|v)_{\mathring{H}^1}, \quad \forall v \in \mathring{H}^1(U)$.

The assumption that u is in $\mathring{H}^1(U)$ implies that u vanishes in a generalized sense (p. 488) on the boundary ∂U. For $u, v \in \mathring{H}^1$, $(u|v)_{H^1} = (u|v)_{\mathring{H}^1}$.

Proposition. a) *Problem* B *has strictly one solution,* b) *this solution satisfies the Poisson equation in the sense of distributions.*

Proof: a) $\|v\|_2 \leq \|v\|_{\mathring{H}^1}$, hence

$$|(f|v)_2| \leq \|f\|_2 \|v\|_2 \leq \|f\|_2 \|v\|_{\mathring{H}^1}$$

and the linear form $v \mapsto (f|v)_2$ is continuous on $\mathring{H}^1$. It follows from Riesz' theorem that there exists a unique element of $\mathring{H}^1$, say Jf such that

$$(f|v)_2 = (Jf|v)_{\mathring{H}^1} \qquad v \in \mathring{H}^1$$

and a unique solution of problem B in $\mathring{H}^1$

$$u = Jf.$$

b) let $v = \varphi \in \mathscr{D}(U)$

$$(u|\varphi)_{\mathring{H}^1} = \int_U \left(u\varphi + \sum \frac{\partial u}{\partial x^i}\frac{\partial \varphi}{\partial x^i}\right) dx \qquad \forall u \in \mathring{H}^1$$

$$= \langle -\Delta u + u, \varphi\rangle.$$

On the other hand $(f|v)_2 = \langle f, \varphi\rangle$ and equation B implies

$$\langle f, \varphi\rangle = \langle -\Delta u + u, \varphi\rangle. \qquad \blacksquare$$

Remark: If $u \in H^1(U)$ and if the boundary ∂U is lipshitzian, we can compute u on ∂U. Let $x^1 = S(x^2, \dots x^n)$ be the equation of the boundary in a neighborhood V. Let $x = (x^1, x') \in U$ when $x^1 > S(x')$; then

$$u|_{\partial U}(x') = u(x) - \int_{S(x')}^{x^1} \frac{\partial u}{\partial x^1}(\tau, x')\,\mathrm{d}\tau, \quad x' = (x^2, \dots x^n).$$

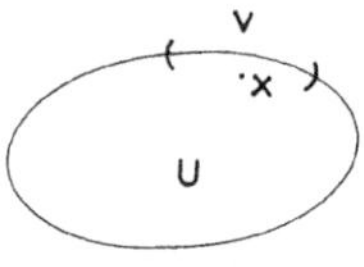

The function $u|_{\partial U}$ is defined almost everywhere on $\partial U \cap V$, it does not depend on x^1, it is square integrable in the x' variables; the value obtained for the extension of u on the boundary by this equation is equal to the value of u on the boundary when $u \in C^1(U) \cap C^0(\bar{U})$. One uses this extension of u and the Schwarz inequality (Ch. I) to prove the following theorem.

Trace theorem. Let U be a bounded open set with lipshitzian boundary, there exists a unique linear continuous mapping of $H^1(U)$ into $L^2(\partial U)$ by $u \mapsto u|_{\partial U}$ such that when $u \in C^1(\bar{U})$ the function $u|_{\partial U}$ is equal to u on the boundary.

We have

$$\|u|_{\partial U}\|_2 \le C(U)\|u\|_{H^1(U)}.$$

If $u \in \mathcal{D}(U)$ then $u|_{\partial U} = 0$; because $\mathcal{D}(U)$ is dense in $\mathring{H}^1(U)$, one can construct a sequence (u_n) with $u_n \in \mathcal{D}$ which converges to $u \in \mathring{H}^1$. Taking the limit of $u_n|_{\partial U}$ as given by the above construction we obtain

$$u|_{\partial U} = 0 \qquad \text{for } u \in \mathring{H}^1(U).$$

Remark: If U is a bounded domain with lipshitzian boundary, then the previous construction of $u|_{\partial U}$ implies

$$\|u\|_{H^1} \le \text{const}\left(\int_{\partial U} |u|^2 \omega + \int_U \sum \left(\frac{\partial u}{\partial x^i}\right)^2 \mathrm{d}x\right)^{1/2},$$

hence if $u \in \mathring{H}^1$ one obtains **Friedrichs' inequality**

Friedrichs inequality

$$\|u\|_{\mathring{H}^1} \le \text{const}\left(\int_U \sum \left(\frac{\partial u}{\partial x^i}\right)^2 \mathrm{d}x\right)^{1/2}.$$

This inequality shows that for a bounded domain with lipshitzian boundary $\int \Sigma (\partial u/\partial x^i)^2 \, dx$ defines a norm equivalent to $\mathring{H}^1$. For $u \in \mathring{H}^1$ the results obtained for $\Delta - \mathbb{1}$ can be extended to Δ. In the Neumann problem we shall see that the unknown function is not in $\mathring{H}^1$ but in H^1 and the results obtained for $\Delta - \mathbb{1}$ are no longer true for Δ.

Hilbertian methods. In all this paragraph the Hilbert spaces considered are real[1]. A bilinear form $a(u, v)$ on a Hilbert space $\mathcal{H}$ is said to be **elliptic (coercive)** if there exists a strictly positive number α such that elliptic coercive

$$a(u, u) \geq \alpha \|u\|^2_{\mathcal{H}} \qquad \forall u \in \mathcal{H}.$$

Lemma. Let $a(u, v)$ be a bilinear, continuous, elliptic form on $\mathcal{H}$, then there exists an isomorphism $\mathcal{A}$ of $\mathcal{H}$ such that

$$a(u, v) = (\mathcal{A}u|v)_{\mathcal{H}} \qquad \forall u, v \in \mathcal{H}.$$

Proof: The mapping $v \mapsto a(u, v)$ is a linear continuous form on $\mathcal{H}$ defined for each $u \in \mathcal{H}$, hence there exists a unique element on $\mathcal{H}$, say $\mathcal{A}u$ such that

$$a(u, v) = (\mathcal{A}u|v)_{\mathcal{H}}.$$

$\mathcal{A}$ is linear; $\mathcal{A}$ is an isomorphism, i.e. the equation $\mathcal{A}u = w$ has strictly one solution for each $w \in \mathcal{H}$. Indeed the ellipticity of a implies the following.

1) Unicity by minoration of $\mathcal{A}$

$$\|\mathcal{A}u\|_{\mathcal{H}} \|u\|_{\mathcal{H}} \geq (\mathcal{A}u|u)_{\mathcal{H}} \geq a(u, u) \geq \alpha \|u\|^2_{\mathcal{H}}.$$

Hence $\|\mathcal{A}u\|_{\mathcal{H}} \geq \alpha \|u\|_{\mathcal{H}}$ and $\mathcal{A}u = 0 \Rightarrow u = 0$.

2) Existence by minoration of the adjoint $\mathcal{A}^*$ of $\mathcal{A}$. By definition

$$(\mathcal{A}u|v)_{\mathcal{H}} = (u|\mathcal{A}^*v)_{\mathcal{H}} \qquad \forall u, v \in \mathcal{H},$$

hence $(u|\mathcal{A}^*u)_{\mathcal{H}} = a(u, u)$ and the same inequalities as in paragraph 1 yield

$$\|\mathcal{A}^*u\|_{\mathcal{H}} \geq \alpha \|u\|_{\mathcal{H}}.$$

u is a solution of $\mathcal{A}u = w$ if and only if

$$(\mathcal{A}u|v)_{\mathcal{H}} = (w|v)_{\mathcal{H}} \qquad \forall v \in \mathcal{H},$$

but $(\mathcal{A}u|v)_{\mathcal{H}} = (u|\mathcal{A}^*v)_{\mathcal{H}}$.

Thus the scalar product of u with the elements of the subspace of $\mathcal{H}$ of the form $\mathcal{A}^*v$ is known, it coincides, on this subspace with the linear

[1]For the case of complex Hilbert spaces, see for instance [Lions and Magenes 1970].

form defined by $\mathscr{A}^*v \mapsto (w|v)_{\mathscr{H}}$. Now

$$|(w|v)_{\mathscr{H}}| \le \|w\|_{\mathscr{H}}\|v\|_{\mathscr{H}}$$

$$|(w|v)_{\mathscr{H}}| \le \frac{1}{\alpha}\|w\|_{\mathscr{H}}\|\mathscr{A}^*v\|_{\mathscr{H}}.$$

This last inequality implies that the linear mapping $\mathscr{A}^*v \mapsto (w|v)_{\mathscr{H}}$ is continuous for the $\mathscr{H}$-topology; hence it can be extended by the Hahn–Banach theorem to the whole space $\mathscr{H}$, and defines $u \in \mathscr{H}$. ∎

Theorem. Let $a(u, v)$ be a symmetric, real, continuous, elliptic bilinear form on a Hilbert space $\mathscr{H}$. Let H be a Hilbert space such that $\mathscr{H} \subset H$ algebraically and topologically; then the equation

$$a(u, v) = (f|v)_H \qquad \forall v \in \mathscr{H}$$

has strictly one solution $u \in \mathscr{H}$ for each $f \in H$.

Proof: The linear form defined by $v \mapsto (f|v)_H$ on $\mathscr{H}$ is continuous for the $\mathscr{H}$-topology finer than the H-topology; indeed

$$\|v\|_H \le \text{const}\,\|v\|_{\mathscr{H}}$$
$$(f|v)_H \le \|f\|_H\|v\|_H \le \text{const}\,\|f\|_H\|v\|_{\mathscr{H}}.$$

Hence there exists a unique element – say Jf – of $\mathscr{H}$ such that

$$(f|v)_H = (Jf|v)_{\mathscr{H}} \qquad \forall v \in \mathscr{H}.$$

The equation $a(u, v) = (f|v)_H$ is thus equivalent to

$$(Jf|v)_{\mathscr{H}} = a(u, v) \qquad \forall v \in \mathscr{H}$$

which according to the lemma implies

$$(Jf|v)_{\mathscr{H}} = (\mathscr{A}u|v)_{\mathscr{H}} \qquad \forall v \in \mathscr{H}.$$

Since $\mathscr{A}$ is an isomorphism of $\mathscr{H}$

$$u = \mathscr{A}^{-1}Jf.$$ ∎

Theorem. Let $a(u, v)$ be a continuous, elliptic bilinear form on a Hilbert space $\mathscr{H}$. Let H be a Hilbert space such that $\mathscr{H} \subset H$ algebraically and topologically, and such that $\mathscr{H}$ is dense in H. Then the equation

$$a(u, v) = (f|v)_H \qquad \forall v \in \mathscr{H} \qquad \text{Problem A}$$

is equivalent to an equation

$$Au = f \qquad f \in H \qquad \text{Problem B}$$

where A is a linear operator, usually unbounded, of domain $N \subset \mathscr{H}$.

Proof: Let N be the set of elements u in $\mathscr{H}$ such that $v \mapsto a(u, v)$ is continuous on $\mathscr{H}$ for the topology induced by H, i.e. if $u \in N$

$$|a(u, v)| \leq C_u \|v\|_H.$$

N is clearly a linear subspace of $\mathscr{H}$, N is non empty since $|a(u, v)| \leq \|f\|_H \|v\|_H$ when u is the solution of problem A. According to the Hahn–Banach theorem, the linear form defined for $u \in N$ by $v \mapsto a(u, v)$ can be extended to a unique continuous linear form on H: such a form is given by the scalar product with an element – say Au – of H.
Let u be the solution of problem A, then

$$a(u, v) = (Au|v)_H = (f|v)_H = (f|v)_H.$$

It follows that $Au = f$. ■

Neumann problem. Let

$$a(u, v) = \int_U \left(\sum \frac{\partial u}{\partial x^i} \frac{\partial v}{\partial x^i} + \lambda uv\right) \mathrm{d}x;$$

$a(u, v)$ is a symmetric, continuous bilinear form on the Sobolev space $H^1(U)$, coercive if $\lambda > 0$.
Let us solve the problem $a(u, v) = \int_U fv \, \mathrm{d}x = (f|v)_2 \quad \forall v \in H^1$.
The problem considered has strictly one solution. A simple procedure to determine the operator A and the domain N is to take $v = \varphi \in \mathscr{D}(U)$. Then

$$a(u, \varphi) = \langle -\Delta u + \lambda u, \varphi \rangle = \langle f, \varphi \rangle,$$

hence $A = -\Delta + \lambda$.
The domain N of A is included in $H^1(U)$ and $u \in N$ implies $(-\Delta + \lambda)u \in L^2(U)$. We shall determine N by the requirement that the linear form $v \mapsto a(u, v)$ be continuous for the L^2 topology, i.e. $a(u, \varphi)$ tends to $a(u, v)$ when φ tends to v in the L^2 topology. Let v be an arbitrary function in $H^1(U)$, we can approach it in the $L^2(U)$ norm by functions φ in $\mathscr{D}$ and

$$a(u, \varphi) = \int_U (-\Delta u + \lambda u)\varphi \, \mathrm{d}x \quad \text{tends to} \quad \int_U (-\Delta u + \lambda u) v \, \mathrm{d}x.$$

Hence if $u \in N$, we have for every $v \in H^1(U)$

$$a(u, v) = \int_U \left(\sum \frac{\partial u}{\partial x^i} \frac{\partial v}{\partial x^i} + \lambda uv\right) \mathrm{d}x = \int_U (-\Delta u + \lambda u) v \, \mathrm{d}x$$

which implies that $\partial u / \partial \nu = 0$ on ∂U in a generalized sense.

More general problems are treated, for instance, in [Lions–Magenes]. For **elliptic systems** on a manifold the reference is [Palais].

4. PARABOLIC EQUATIONS, HEAT DIFFUSION

The fundamental example of parabolic equations is the heat diffusion equation in $\mathbb{R}^n$

$$\frac{\partial u}{\partial t}-\sum_{i=1}^{n}\frac{\partial^2 u}{(\partial x^i)^2}=0.$$

Elementary solution of the heat equation in $\mathbb{R}^n$. Consider first the equation on $\mathbb{R}^2$

$$\frac{\partial E}{\partial t}-\frac{\partial^2 E}{\partial x^2}=\delta(t,x) \qquad t,x\in\mathbb{R}.$$

Let $\mathscr{E}$ be the Fourier transform of E with respect to x, $\mathscr{E}=\mathscr{F}_x E$,

$$\frac{\partial \mathscr{E}}{\partial t}+y^2\mathscr{E}=\delta(t).$$

A solution in $\mathscr{D}'^+(\mathbb{R})\cap\mathscr{S}'$ is $\mathscr{E}(y,t)=Y(t)\exp(-y^2t)$.
Hence the elementary solution with support in $t\geq 0$ (Problem 6, p. 531) is

$$E(x,t)=(4\pi t)^{-1/2}\exp(-x^2/4t) \qquad \text{for } t\geq 0,$$
$$E(x,t)=0 \qquad \text{for } t<0.$$

If one uses the notation $E(x,t)=Y(t)(4\pi t)^{-1/2}\exp(-x^2/4t)\equiv Y(t)f(t)$ the time derivative of $E(x,t)$ cannot be computed by Leibniz rule because f is not differentiable at $t=0$. This elementary solution is C^∞ on $\mathbb{R}^2-\{0\}$. Similarly on $\mathbb{R}^n$,

$$E(x,t)=(4\pi t)^{-n/2}\exp(-|x|^2/4t) \qquad \text{for } t\geq 0$$
$$E(x,t)=0 \qquad \text{for } t<0.$$

Cauchy problem with given values at $t=0$. It can be stated in terms of distributions as follows. Let S be the submanifold $t=0$ in $\mathbb{R}^{n+1}$. Find a distribution $\tilde{u}$ with support in $t\geq 0$ such that

$$\frac{\partial \tilde{u}}{\partial t}-\frac{\partial^2\tilde{u}}{\partial x^2}=f\delta_S$$

where f is a given continuous function on S. A solution is

$$\tilde{u}=f\delta_S * E,$$

provided the behavior of f is such that the convolution is defined and has the good properties. Then $\tilde{u}$ can be written in the following form; for $t>0$

$$\tilde{u}(x,t)=\langle (f\delta_S)(y,z),E(x-y,t-z)\rangle$$

that is

$$u(x,t)=\int_{\mathbb{R}^n} f(y)E(x-y,t)\,dy$$
$$=\int_{\mathbb{R}^n} Y(t)(4\pi t)^{-n/2}\exp(-(x-y)^2/4t)f(y)\,dy.$$

Exercise: Show that if f is continuous and bounded on $\mathbb{R}^n$, $u(x,t)$ is a C^∞ function for $t>0$ which tends to $f(x)$ when t tends to zero.

5. HYPERBOLIC EQUATION, WAVE EQUATIONS

Wave equations are particular cases of hyperbolic equations. A fundamental property of a hyperbolic equation is that it has an elementary solution in a convolution algebra $\mathscr{D}'^{+}(\Gamma)$. This elementary solution can be used to solve Cauchy's problem, which is well defined for an initial space-like submanifold.

Elementary solution of the wave equation. Let $E(t,x)$ be an elementary solution of the wave equation in $\mathbb{R}^4$

$$\Box E \equiv \frac{\partial^2 E}{\partial t^2} - \sum_{i=1}^{3} \frac{\partial^2 E}{(\partial x^i)^2} = \delta(t,x) \qquad t \in \mathbb{R},\ x \in \mathbb{R}^3.$$

Let us consider $E(t,x)$ as a distribution based on $\mathbb{R}^3$ (variable x) with value in the space of distributions based on $\mathbb{R}$ (variable t). By definition, we set

$$\langle\langle E(t,x), \varphi(x)\rangle, \psi(t)\rangle \overset{\text{def}}{=} \langle E(t,x), \varphi(x)\psi(t)\rangle$$

for $\varphi \in \mathscr{S}(\mathbb{R}^3)$ and $\psi \in \mathscr{D}(\mathbb{R})$.
Let $\mathscr{E}(t,y)$ be its Fourier transform with respect to the x-variables,

$$\langle \mathscr{E}(t,y), \varphi(y)\rangle = \langle E(t,x), \mathscr{F}\varphi(x)\rangle.$$

$\mathscr{E}(t,y)$ is also a distribution based on $\mathbb{R}^3$ with values in the space of distributions based on $\mathbb{R}$, and hence corresponds to a distribution based on $\mathbb{R}^4$ (variables t, y).
Transforming the elementary solution equation we obtain

$$\mathscr{F}\Box E = \frac{\partial^2 \mathscr{E}}{\partial t^2} + \rho^2 \mathscr{E} = \delta(t) \qquad \rho^2 = \sum_{i=1}^{3} (y^i)^2$$

which admits the following solution with support in $t \geq 0$

$$\mathscr{E}(t,y) = Y(t)\frac{\sin \rho t}{\rho}.$$

$\mathscr{E}$ is a tempered distribution in y

$$E(t,x) = \mathscr{F}_y \mathscr{E}(t,y) = Y(t)\bar{\mathscr{F}}_y \frac{\sin \rho t}{\rho}.$$

The function $\sin \rho t/\rho$ is analytic on $\mathbb{C}^3$ for each positive value of t hence according to the Paley–Wiener theorem (p. 477) its inverse Fourier transform is a distribution with compact support in $\mathbf{R}^3$

$$\bar{\mathscr{F}}\,\frac{\sin \rho t}{\rho}=\frac{1}{4\pi}\frac{1}{r}\,\delta_{t-r} \qquad r^2=\sum_{i=1}^{3}(x^i)^2$$

where δ_{t-r} is Dirac's measure on the sphere $r=t$ in $\mathbf{R}^3$. Hence

$$\begin{aligned}\langle E(t,x),\varphi(x)\psi(t)\rangle &= \frac{1}{4\pi}\left\langle Y(t)\frac{1}{r}\,\delta_{t-r},\varphi(x)\psi(t)\right\rangle \\ &= \frac{1}{4\pi}\left\langle\frac{1}{r},\varphi(x)\psi(r)\right\rangle.\end{aligned}$$

This equation can be rewritten

$$E(t,x)=\frac{1}{4\pi}\,\delta_c$$

with $\langle \delta_c, \varphi(x)\psi(t)\rangle = \int_c \varphi(x)\psi(t)\omega_c$

$$\omega_c \wedge \tfrac{1}{2}\,\mathrm{d}(t^2-r^2)=\mathrm{d}x^1\wedge \mathrm{d}x^2\wedge \mathrm{d}x^3\wedge \mathrm{d}t \qquad \omega_c=\frac{1}{t}\,\mathrm{d}x^1\wedge \mathrm{d}x^2\wedge \mathrm{d}x^3.$$

c is the cone in $\mathbf{R}^4$ defined by $\frac{1}{2}(t^2-r^2)=0$, $t\geq 0$.
If we had considered the cone c' defined by $t^2-r^2=0$, $t\geq 0$ we would have obtained

$$E(t,x)=\frac{1}{2\pi}\,\delta_{c'}.$$

Remark 1: The elementary solution E that we have constructed is an element of a convolution algebra $\mathscr{D}'^{+}(\Gamma)$, where Γ is the convex cone $t^2-r^2\geq 0$, $t\geq 0$. The property of having an elementary solution in $\mathscr{D}'^{+}(\Gamma)$ where Γ is a convex cone is a general property of hyperbolic equations with constant coefficients.[1]

Remark 2: The support of the elementary solution E is only the surface c of the convex cone Γ. This circumstance is peculiar to the wave equation in $\mathbf{R}^{2n}$ and to a few other equations which are said to have lacunae. It is a statement of Huyghens' principle and plays an important role in electromagnetic theory (Problem 7).

Cauchy problem. The usual Cauchy problem can be formulated in the theory of distributions. We shall consider here the case of the wave equation.

[1]The generalization of this property to hyperbolic equations with variable coefficients will be discussed on pp. 516–518.

Let u be a C^2 solution of $\Box u = 0$, in the open set $t > 0$, which is C^1 for $t \geq 0$ and satisfies the Cauchy data $u = f_0$ and $\partial u/\partial t = f_1$ on the surface S defined by $t = 0$.
Let $\tilde{u}$ be the extension of u, equal to 0 when $t < 0$. Then, in $\mathscr{D}'(\mathbf{R}^4)$,

$$\Box \tilde{u} = f_1 \delta_S + f_0 \frac{\partial \delta_S}{\partial t} \equiv T.$$

T has for support the surface S whose intersection with every cone translated from Γ is compact. It follows necessarily that

$$\tilde{u} = T * E,$$

and a straightforward computation gives the **Kirchoff formula**

Kirchoff formula

$$u(x, t) = tM_t(f_1) + \frac{\partial}{\partial t}(tM_t(f_0))$$

where

$$M_t(f) = \frac{1}{4\pi} \int_0^{2\pi} \int_0^{\pi} f(x - t\alpha)\omega$$

$$\alpha^1 = \sin\theta \cos\varphi, \quad \alpha^2 = \sin\theta \sin\varphi, \quad \alpha^3 = \cos\theta$$

$$\omega = \sin\theta \, d\theta \, d\varphi.$$

Conversely if f_0 is C^3 and f_1 C^2 the Kirchoff formula solves the usual Cauchy problem: u is C^2 for $t > 0$, it tends to f_0 and $\partial u/\partial t$ tends to f_1 when t tends to zero.

Remark: The spatial character of S is fundamental for the convolution $T * E$ to be defined. If S were the surface $x^1 = 0$, the convolution $T * E$ would not be defined in general (see p. 461). Even if it were defined, the function u thus obtained for $x^1 \geq 0$ would have no reason to satisfy the Cauchy data for $x^1 = 0$ since the limit of the intersection of S and of the cone translated from Γ with vertex x does not tend to a point of S when this vertex tends to S.

Energy integral, unicity theorem. Let us consider the wave equation in $\mathbf{R}^n$

$$\Box u \equiv \frac{\partial^2 u}{\partial t^2} - \sum_{i=1}^{n-1} \frac{\partial^2 u}{(\partial x^i)^2} = f.$$

Let U be a domain in $\mathbf{R}^n$ with a regular boundary in the sense of Stokes' formula. By using

$$\frac{\partial u}{\partial t}\frac{\partial^2 u}{\partial t^2} = \frac{1}{2}\frac{\partial}{\partial t}\left(\frac{\partial u}{\partial t}\right)^2$$

$$\frac{\partial u}{\partial t}\frac{\partial^2 u}{(\partial x^i)^2} = \frac{\partial}{\partial x^i}\left(\frac{\partial u}{\partial t}\frac{\partial u}{\partial x^i}\right) - \frac{1}{2}\frac{\partial}{\partial t}\left(\frac{\partial u}{\partial x^i}\right)^2$$

we obtain

$$\int_U \frac{\partial u}{\partial t}\Box u \, dx \, dt \equiv \int_{\partial U} \left\{\tfrac{1}{2}\left[\left(\frac{\partial u}{\partial t}\right)^2 + \sum\left(\frac{\partial u}{\partial x^i}\right)^2\right] n^0 + \sum \frac{\partial u}{\partial t}\frac{\partial u}{\partial x^i} n^i\right\}\omega$$

$$= \int_U \frac{\partial u}{\partial t} f \, d^{n-1}x \, dt.$$

Let us take for $\bar{U}$ the truncated cone Ω_T of summit (t_0, x_0)

$$t_0 - t - \left(\sum_{i=1}^{n-1} (x^i - x_0^i)^2\right)^{1/2} \geq 0$$

$$0 < t \leq T < t_0.$$

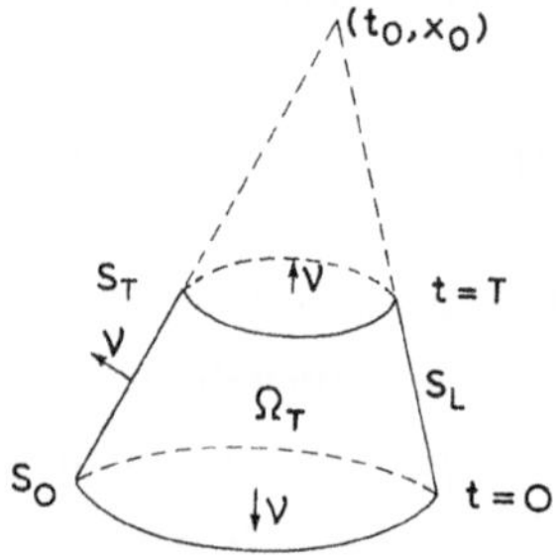

The boundary $\partial\bar{U}$ consist of the two bases S_T and S_0 and the lateral surface S_L on which

$$\nu^0 = 1, \; \nu^i = (x^i - x_0^i)/(t - t_0) \qquad \sum_{i=1}^{n-1} (\nu^i)^2 = 1.$$

Hence on S_L

$$\left\{\left(\frac{\partial u}{\partial t}\right)^2 + \sum_{i=1}^{n}\left(\frac{\partial u}{\partial x^i}\right)^2\right\} n^0 + 2\sum_{i=1}^{n-1} \frac{\partial u}{\partial t}\frac{\partial u}{\partial x^i} n^i \geq \left(\frac{\partial u}{\partial t}\right)^2 + \left(\sum_{i=1}^{n-1} \frac{\partial u}{\partial x^i} n^i\right)^2 + 2\sum \frac{\partial u}{\partial t}\frac{\partial u}{\partial x^i} n^i.$$

The last expression, being a perfect square is non negative.
The **energy inequality** follows

energy inequality

$$\frac{1}{2}\int_{S_T} \left\{\left(\frac{\partial u}{\partial t}\right)^2 + \sum_{i=1}^{n-1}\left(\frac{\partial u}{\partial x^i}\right)^2\right\} d^{n-1}x \leq \frac{1}{2}\int_{S_0} \left\{\left(\frac{\partial u}{\partial t}\right)^2 + \sum_{i=1}^{n-1}\left(\frac{\partial u}{\partial x^i}\right)^2\right\} d^{n-1}x + \int_{\bar{U}} \frac{\partial u}{\partial t} f \, d^{n-1}x \, dt.$$

An immediate consequence of this inequality is the following theorem.

Uniqueness theorem. The Cauchy problem for the wave equation $\Box u = f$ *with regular data on* S_0 *has a unique solution on the truncated cone* Ω_{t_0}.

It follows also that the value of u at the point (t_0, x_0) is determined only by the data on the ball S_0. The ball S_0 defined by $|x - x_0| \leq t_0$ is sometimes called the **domain of dependence** of (t_0, x_0). domain of dependence

Existence theorem; qualitative indication on the solution. The energy inequality, together with similar inequalities derived for more general cases, is the basis for the demonstration of existence theorems for solutions of hyperbolic equations by methods of functional analysis. We shall give an example of such a demonstration for the following problem:

$$\begin{cases} \Box u = f, & f \in L_2 \text{ in } \Omega \equiv \Omega_{t_0} \\ u = f_0, \ \dfrac{\partial u}{\partial t} = f_1 & \text{on } S_0 \quad f_0 \in H^1(S_0) \quad f_1 \in L^2(S_0). \end{cases}$$

Any regular solution of this problem satisfies the energy inequality. Set

$$\|u\|_T^1 = \left(\int_{S_T} \left((u)^2 + \left(\frac{\partial u}{\partial t}\right)^2 + \sum_{i=1}^{n-1} \left(\frac{\partial u}{\partial x^i}\right)^2 \right) \mathrm{d}^{n-1}x \right)^{1/2}.$$

We notice that

$$u(T, x) = u(0, x) + \int_0^T \frac{\partial u}{\partial t}(t, x)\, \mathrm{d}t,$$

hence $|u(T, x)| \leq |u(0, x)| + T^{1/2} \left\{ \int_0^T \left(\frac{\partial u}{\partial t}\right)^2 \mathrm{d}t \right\}^{1/2}$.

Moreover $\left| \dfrac{\partial u}{\partial t} f \right| \leq \dfrac{1}{2} \left\{ \left(\dfrac{\partial u}{\partial t}\right)^2 + f^2 \right\}$.

Thus there exist positive constants K_1, K_2, K_3 such that for every $T \leq t_0$

$$\|u\|_T^1 \leq K_1 \|u\|_0^1 + K_2 \int_0^T \|u\|_t^1\, \mathrm{d}t + K_3 \|f\|_\Omega$$

where $\|f\|_\Omega = \left(\int_\Omega f^2\, \mathrm{d}^{n-1}x\, \mathrm{d}t \right)^{1/2}$ and $\int_\Omega = \int_0^T \int_{S_T}$.

Set $y(T) = \int_0^T \|u\|_t^1\, \mathrm{d}t = \|u\|_\Omega^1$.

Then the previous inequality is of the form

$$y' \leq a(T) + by \qquad b = K_2 \geq 0, \quad a = K_1\|u\|_0^1 + K_3\|f\|_\Omega \geq 0.$$

$a(T)$ increases with T and is bounded by $a(t_0)$; $y(0) = 0$; under these conditions $y \leq y_0$ where y_0 is the solution of the differential equation, $y' = a(T) + by$ with the boundary condition $y_0(0) = 0$.
In particular

$$y(T) \leq \frac{a_0}{b}(e^{bT} - 1).$$

More precisely for $T \leq t_0$, there exist positive constants C_1 and C_2 such that

$$\|u\|_\Omega^1 \leq C_1\|u\|_0^1 + C_2\|f\|_\Omega.$$

Given these inequalities we can then proceed as follows.
Solve the problem for analytic functions f_n, and analytic Cauchy data f_{0n}, f_{1n}: according to the Cauchy–Kovalevski theorem the problem has an analytic solution of which Leray[1] has shown the existence in the whole domain Ω. The completeness of $H^1(\Omega)$, together with the above inequality are used to show the convergence in $H^1(\Omega)$ of a sequence of analytic solutions when the sequence (f_n) tends to f in $H^1(\Omega)$ the sequence (f_{0n}) tends to f_0 in $H^1(S_0)$ and the sequence (f_{1n}) tends to f_1 in $L^2(S_0)$. The limit is a solution in the sense of distributions.
It is also possible to prove directly the existence of the solution by using the $H^1 - H^{-1}$ duality – without approximating it by a sequence of analytic functions[2].

6. LERAY THEORY OF HYPERBOLIC SYSTEMS

Case of one equation

Definitions. Let $a \equiv \Sigma_{|\alpha| \leq m}\, a_\alpha(x) D^\alpha$ be a linear differential operator on an open set $U \subset \mathbb{R}^n$, a_α being assumed real.
The principal part h **of the operator (p. 493) defines at each point** $x \in U$ a cone $H_x \subset T_x^*(U)$ of vertex x and equation

$$h(x, p) \equiv \sum_{|\alpha| = m} a_\alpha(x) p^\alpha = 0, \qquad p \in T_x^*(U).$$

strictly hyperbolic at x

The operator a is said to be **strictly hyperbolic at** x if there exists a point in $T_x^*(U)$ such that each straight line through this point, that does not go through x, intersects H_x in m real distinct points. It can be shown that

[1][Leray 1953].
[2][Lax], [Gårding].

the set of such points consists of two closed, convex, opposite cones Γ_x^+ and Γ_x^- with non empty interior such that their boundaries belong to H_x. In the following "hyperbolic" is to be understood as "strictly hyperbolic".

Exercise: Show that an x^1-hyperbolic partial differential equation (p. 494) is strictly hyperbolic.

The operator a is said to be **hyperbolic on** U if hyperbolic on U
1) a is hyperbolic at each point $x \in U$.
2) $h(x, p)$ does not tend to zero when x tends to the boundary of U or when $|x|$ tends to infinity.

The operator a is said to be **globally hyperbolic on** U if it is hyperbolic and if the intersection $\Gamma^+ = \bigcap_{x \in U} \Gamma_x^+$ has a non empty interior – this intersection is understood to be the intersection of all the cones translated so as to have the same vertex. globally hyperbolic on $U \subset \mathbb{R}^n$

Let $C_x^+[C_x^-]$ be the closed convex cone dual to $\Gamma_x^+[\Gamma_x^-]$ that is the set of vectors $v \in T_x(U)$ such that

$$\langle v, p \rangle \equiv v^i p_i \geq 0, \qquad \forall p \in \Gamma_x^+[\Gamma_x^-].$$

An oriented differentiable path on U is said to be **timelike**[1] with respect to the differential operator a if its tangent at each point $x \in U$ is in C_x^+. timelike
A differentiable hypersurface S is said to be **spacelike** if its tangent vector space at x is exterior to $C_x = C_x^+ \cup C_x^-$. spacelike
The definition of global hyperbolicity implies that the cones C_x^+, **translated so as to have the same vertex, are all interior to the cone** C^+ dual of Γ^+: a timelike path has a bounded slope. It follows that, if we give to the path space the usual topology[2], and if we extend, by a limiting process, the concept of timelike path to non differentiable paths, the definition of global hyperbolicity implies that the set of timelike paths between two arbitrary points x and y of $\mathbb{R}^n$ is compact. This leads to the following alternate definition, proposed by Leray, which can be generalized to a manifold.
The operator a is said to be **globally hyperbolic on the open set** $U \subset \mathbb{R}^n$ if it is hyperbolic on U and if the set of time like paths between any two points $x, y \in U$ is compact. globally hyperbolic on $U \subset \mathbb{R}^n$
A differentiable submanifold $S \subset U$ of dimension $n-1$ is called a

[1]Time like means now "time like or null" in the sense of Chapter V.
[2]$d(\gamma_1, \gamma_2) = \inf \sup \bar{d}(\gamma_1(t), \gamma_2(t))$ where $\bar{d}$ is a metric compatible with the topology on U, and the inf is taken over all possible parametrizations.

Cauchy surface for U

Cauchy surface for U if every timelike path in U without end points in U intersects S strictly once.
If a Cauchy surface exists in U, then U is globally hyperbolic. Conversely if U is globally hyperbolic it admits a Cauchy surface[1].

quasi-linear

Quasi-linear equation. A partial differential equation of order m, with unknown u is said to be **quasi-linear** if it reads

$$h(x, \mathrm{D}^{m-1}u, \tilde{\mathrm{D}}^m)u + b(x, \mathrm{D}^{m-1}u) = 0$$

where $\tilde{\mathrm{D}}^m$ denotes the set of derivations of order m and where h is a linear differential operator of order m whose coefficients are, like b, functions of x and of the derivatives of u of order $\leq m-1$. This equation is said to be hyperbolic at x, for a set of given values, at x, of $\mathrm{D}^{m-1}u$ if the linear equation obtained by replacing $\mathrm{D}^{m-1}u$ by its values is hyperbolic[2].
The definition of hyperbolic on U [globally hyperbolic on U] is extended accordingly.

Remark: Let N be the number of partial derivatives of order $m-1$ of a function u on $\mathbb{R}^n$, the range of $\mathrm{D}^{m-1}u$ is $\mathbb{R}^N$.

Cauchy problem. To simplify the statement of the existence theorem[3] we shall assume that x does not appear explicitly in the given quasi-linear equation, as is most often the case in mathematical physics.

Leray theorem. Hypothesis:
1) *The functions b and the coefficients of the differential operator h are C^∞ functions of $\mathrm{D}^{m-1}u$, and the operator h is hyperbolic whenever $\{\mathrm{D}^{m-1}u_k\} \in Y$, an open set of $\mathbb{R}^N$.*
2) *Let S be a C^∞ submanifold of dimension $n-1$, regularly spacelike for h when $\{\mathrm{D}^{m-1}u_k\} \in Y$. Let*

$$\gamma_k^\alpha = \mathrm{D}^\alpha u_k|_S, \quad \alpha = (\alpha_0, \ldots \alpha_{n-1}), \quad |\alpha| \leq m-1$$

be the Cauchy data on S. We assume that $\{\varphi_k^\alpha\} \in Y_0$, a compact subset of Y, and that

$$\varphi^\alpha \in H^{\mathrm{loc}}_{\mu-|\alpha|}(S) \quad \text{with } \mu \geq \mu_0 = m + n'$$

where $H^{\mathrm{loc}}_\mu(S)$ is the Sobolev space with the usual notation and n' is the smallest integer $\geq n/2$.
Conclusion: The Cauchy problem has exactly one solution in a neighborhood Ω_μ of S such that

[1][Geroch].
[2]Note that the properties of b do not enter into the condition for (strict) hyperbolicity.
[3][Leray 1953].

1. $u_k \in H_\mu^{\mathrm{loc}}(\Omega_\mu)$,
2. *the operator h is globally hyperbolic on Ω_μ for the solution u_k.*

Remark: It can be shown[1] that $\Omega_\mu = \Omega_{\mu'}$ if $\mu \geq \mu' \geq \mu_0 + 1$. It follows that the solution is C^∞ in Ω_{μ_0+1} for C^∞ Cauchy data.

System of equations

Consider a linear system of N partial differential equations with N unknowns u^k

$$a^j(x, D^{B_{jk}})u^k = f^j, \qquad j, k = 1, \dots N$$

where $a^j(x, D^{B_{jk}})$ is a linear differential operator of order B_{jk}; the summation is over k, but not over j. The B_{jk} are a set of N^2 integers ≥ 0, or equal to $-\infty$, with the convention that $B_{jk} = -\infty$ if u^k does not enter the jth equation.

With each equation and each unknown of the given system, one associates a polynomial H_j^k, homogeneous in p of degree B_{jk}, by substituting p_i for $\partial/\partial x^i$ in the terms of order B_{jk} of a_j.

The **characteristic determinant** of the system is the polynomial

characteristic determinant

$$H = \det(H_j^k) \equiv \sum_\pi (\operatorname{sign} \pi) H^1_{\pi(1)} \dots H^N_{\pi(N)}$$

where π is an arbitrary permutation of the integers $1, \dots, N$.

The **degree** of the polynomial H is

degree

$$d = \sup_\pi \sum_{k=1}^N B_{\pi(k)k}.$$

The homogeneous part of order d of the polynomial H – say h – is called the **characteristic polynomial** of the system. If this polynomial is not identically zero, the system is said to be regular in the sense of Cauchy–Kovalevski.

characteristic polynomial

The following procedure simplifies the calculation of h and is useful in the hyperbolicity criteria. It can be shown easily that with the N^2 integers B_{jk} one can always associate $2N$ integers ≥ 0, m_k and n_j such that

$$B_{jk} < m_k - n_j$$

and

$$d = \sup_\pi \sum_{k=1}^N B_{\pi(k)k} = \sum_{k=1}^N m_k - \sum_{j=1}^N n_j.$$

[1][Choquet-Bruhat 1971].

This shows that one can write the given system in the form

$$a^j(x, D^{m_k - n_j})u^k = f^j$$

without modifying the characteristic polynomial h which becomes identical to the homogeneous polynomial of degree d

$$h = \det(h_j^k);$$

we have $h_j^k = 0$ if a_j does not contain any derivative of order $m_k - n_j$ of u_k.
The given system is called hyperbolic if the polynomial h is hyperbolic.

Example: The system of 2 equations with 2 unknown in $\mathbb{R}^2$

$$\frac{\partial u}{\partial x} - \frac{\partial v}{\partial y} = 0, \quad \frac{\partial^2 u}{\partial x^2} + \frac{\partial^2 v}{\partial y^2} + \frac{\partial u}{\partial x} = 0$$

$m(u) = m(v) = 2$, $n(x) = 1$, $n(y) = 0$, has the following characteristic polynomial

$$h(p) = p_1 p_2 (p_1 + p_2);$$

it is hyperbolic.

Quasi-linear systems. An arbitrary non linear system of equations reads, without summation,

$$F^j(x, D^{B_{jk}} u^k) = 0, \; k = 1, \dots N \text{ for each } j.$$

By an argument similar to the previous one and with similar definitions for m_k and n_j, it can be written

$$F^j(x, D^{m_k - n_j} u^k) = 0.$$

The system is said to be **quasilinear** if it has the following form

$$F^j \equiv h_i^j(x, D^{m_k - n_j - 1} u^k, \tilde{D}^{m_i - n_j}) u^i + b^j(x, D^{m_k - n_j - 1} u^k) = 0.$$

The characteristic polynomial $h = \det(h_i^j)$ is computed by replacing the derivation operators $\tilde{D}^{m_i - n_j}$ in h_i^j, by the corresponding components of the vector p. It depends on the unknown.

diagonal

Diagonal system. A system is said to be **diagonal** if $h_i^j \equiv 0$ for $i \neq j$, i.e., setting $h_i^i = h^i$ one has

$$F^j \equiv h^j(x, D^{m_k - n_j - 1} u^k, \tilde{D}^{m_j - n_j}) u_j + b^j(x, D^{m_k - n_j - 1} u^k) = 0.$$

hyperbolic diagonal

A diagonal system is said to be **hyperbolic**[1] at x if

[1]Definition more general than the previous one. Einstein equations in harmonic coordinates are a diagonal hyperbolic system.

1. each operator h^j is hyperbolic,
2. the intersection of the cones $\Gamma_{x,j}$ corresponding to the various h_j is non empty.

Set $\Gamma_x^+ = \cap_j \Gamma_{x,j}^+$, the timelike paths are determined by the dual C_x^+ of Γ_x^+ as above (p. 517) and the condition of global hyperbolicity is the same.

Leray theorem for a quasi-linear diagonal system. Let $h^j(\mathrm{D}^{m_k-n_j-1}u^k, \tilde{\mathrm{D}}^{m_j-n_j})u^j + b^j(\mathrm{D}^{m_k-n_j-1}u^k) = 0$. To simplify the statement of the theorem, we assume as in the case of one equation that x does not enter explicitly. Let the Cauchy data be the functions φ_j^α on a submanifold of dimension $n-1$

$$\mathrm{D}^\alpha u^j|_S = \varphi_j^\alpha \qquad |\alpha| < m_j - 1.$$

Note that if the Cauchy data were given only for $|\alpha| \leq m_j - n_j - 1$, the given system would not determine a unique solution, even in the case where coefficients and Cauchy data are analytic and S is not characteristic, because we have for each j, k

$$m_k - n_j - 1 < m_k - n_k - 1 \qquad \text{only if } n_j = 0 \text{ for every } j.$$

With the given Cauchy data, the coefficients of the given system are known on S, in fact the whole left hand side is known if $n_j > 0$. The Cauchy data must in this case satisfy some compatibility conditions. They are obtained by setting this left hand side together with its derivatives[1] of order $< n_j$ equal to zero.

Leray theorem. Hypotheses:
1) *For* $\{\mathrm{D}^{m_j-1}u^j\} \subset Y$, *an open set of* $\mathbb{R}^n$, *the system is hyperbolic and* S *regularly space like, the coefficients of* h^j *and* b^j *being* C^∞ *functions.*
2) *The Cauchy data on* S, $\mathrm{D}^\alpha u^j|_S = \varphi_j^\alpha$, $|\alpha| \leq m_j - 1$ *have their range on a compact subset* $Y_0 \subset Y$; *they satisfy the compatibility conditions and belong to the following Sobolev spaces*

$$\varphi_j^\alpha \in H^{\mathrm{loc}}_{\mu_j - |\alpha|}(S) \qquad \mu_j \geq m_j + l'.$$

Conclusion: *The Cauchy problem has a unique solution in a neighborhood of* S, $u^j \in H^{\mathrm{loc}}_{\mu_j}(U)$.

Diagonalisation. It can be shown that a quasi linear system can always to put in quasi-linear diagonal form. Let $\hat{h}_j^i$ be the minor relative to the

[1]In order to have the minimum regularity hypothesis, it is advantageous to choose the smallest positive integers for m_j and n_j which are determined modulo an additive constant.

element h_l^j of the determinant h, and also the differential operator, of order $d - m_l + n_j$, corresponding to this polynomial. Construct, by summation over j

$$\hat{h}_j^l(h_i^j(x, \mathrm{D}^{m_k - n_j - 1}u^k, \tilde{\mathrm{D}}^{m_i - n_j})u^i + b^j) = 0.$$

Attention must be paid to the following:

1) the differential operators may not commute,

2) $m_k - n_j < 0$ corresponds to the absence of the function u^k in the equation F^j,

3) the definition of quasi-linearity.

It can be shown that, if $\sup m_k - \inf n_j < d$, one obtains

$$h(x, \mathrm{D}^{d - m_l + m_k - 1}u^k, \tilde{\mathrm{D}}^d)u^l + B^l(x, \mathrm{D}^{d - m_l + m_k - 1}u^k) = 0,$$

a quasi-linear diagonal system with indices

$$\mu_l = m_l + d, \ \nu_l = m_l.$$

Remark: The following example does not satisfy the condition $d \geq \sup m_k - \inf n_j$

$$\frac{\partial u}{\partial x} = 0, \qquad \frac{\partial^3 u}{\partial x^3} - \frac{\partial v}{\partial y} = 0.$$

7. SECOND ORDER SYSTEMS. PROPAGATORS[1]

The second variation of the action of a system plays a fundamental role in quantum physics: many quantities of physical interest in a quantum system can be obtained from elementary kernels of the small disturbance equation, alias the Jacobi equation. For a system with an infinite number of degrees of freedom the Jacobi equation is of the type

$$\boldsymbol{L}\boldsymbol{T} = 0 \tag{1}$$

where L is a second order linear partial differential operator and $\boldsymbol{T}$ a tensor distribution (cf. p. 482) on a manifold.

Systems of this type occur also in many problems of classical physics.

Let $\boldsymbol{T}$ be a tensor distribution on a globally hyperbolic manifold X, with C^∞ metric g, we shall consider linear systems with C^∞ coefficients of the

[1][DeWitt, Brehme 1960], [Lichnerowicz 1961], [DeWitt 1965], [Choquet-Bruhat 1968b].

type[1]

$$LT \stackrel{\text{def}}{=} \Delta T + B^\rho \nabla_\rho T + CT = 0, \quad \Delta = \nabla_\rho \nabla^\rho.$$

The adjoint operator L^* is defined by

$$L^* U \stackrel{\text{def}}{=} \Delta U - \nabla_\rho({}^*B^\rho U) + {}^*CU$$

with the star the operation of transposition for the linear maps B^ρ, C. One checks that $\langle LT, U\rangle = \langle T, L^* U\rangle$ when U is C^∞ with compact support.

It is clear from the definitions that the operator L defines a globally hyperbolic system on the manifold X. The system

$$\Delta T + B^\rho \nabla_\rho T + CT = \varphi$$

with φ a C^∞ p-tensor with compact support on X, has exactly one solution f^+ [resp. f^-] with support in the future [resp. the past] of φ. The solution[2] is C^∞ and depends continuously on φ. Therefore it defines a regular kernel denoted $E^+(x, y)$ [resp. $E^-(x, y)$], such that

$$f^\pm(y) = \langle E^\pm(x, y), \varphi(x)\rangle.$$

In particular

$$\varphi(y) = \langle E^\pm(x, y), L\varphi(x)\rangle = \langle L_x^* E^\pm(x, y), \varphi(x)\rangle.$$

Therefore

$$L_x^* E^\pm(x, y) = \delta(x, y)$$

where $\delta(x, y)$ is the Dirac p-tensor kernel[3], defined by

$$\langle \delta(x, y), \varphi(x)\rangle = \varphi(y), \qquad \forall C^\infty p\text{-tensor } \varphi.$$

A tensor kernel is also called a **bitensor**[4]. bitensor

The elementary kernel $E^+(x, y)$ [resp. $E^-(x, y)$] of L^* has, for each y its support in the future $\mathscr{E}^+(y)$ [resp. $\mathscr{E}^-(y)$] of y. It is also called the **advanced** [resp. **retarded**] **Green function**. We shall also introduce the elementary kernels $\overset{*}{E}{}^\pm$ of the operator L. advanced retarded Green function

When X is a four dimensional manifold it has been shown[5] that, at least

[1]The local coordinates of the De Rham laplacian of a form ω, $\Delta\omega \stackrel{\text{def}}{=} (\delta\,\mathrm{d} + \mathrm{d}\delta)\omega$ computed on p. 319, have been used by [Lichnerowicz 1961] to define the De Rham–Lichnerowicz laplacian Δ_L of an arbitrary (non necessarily antisymmetric) tensor. The laplacian thus defined is a particular case of (1); like $\nabla_\rho \nabla^\rho$ it is self adjoint and commutes with contraction. If the covariant derivative of T vanishes $\Delta_L(T \otimes U) = T \otimes \Delta_L U$.

[2]cf. [Leray 1953], [Choquet-Bruhat 1968b].

[3]cf. in Problem 8 the expression for $\delta(x, y)$ in local coordinates.

[4][B. DeWitt 1965].

[5][Choquet-Bruhat 1964].

in a neighborhood of y, $\boldsymbol{E}^+(x, y)$ [resp. $\boldsymbol{E}^-(x, y)$] is for a given[1] y the sum of two terms, a tensor measure on the boundary $\partial\mathscr{C}_y^+$ [resp. $\partial\mathscr{C}_y^-$], and a C^∞ tensor which vanishes for massless fields in Minkowski space (classical wave equation, cf. p. 512). This smooth tensor is called the tail term or the diffusion term[2].

future of K past of K

Future compact and past compact. Let K be a subset of X. The **future** $\mathscr{C}^+(K)$ **of** K is the union of the futures of the points $x \in K$. The **past** $\mathscr{C}^-(K)$ is the union of their pasts.

future [past] compact

A subset K of a globally hyperbolic manifold X is said to be **future compact [past compact]** if the intersection of K with $\mathscr{C}^-(x)$ [with $\mathscr{C}^+(x)$] is either compact or empty for every $x \in X$.

Theorem. If K is past compact and K' future compact, then $\mathscr{C}^+(K) \cap \mathscr{C}^-(K')$ is compact.

Uniqueness theorem. Every tensor distribution $\boldsymbol{T}$ that is a solution of $L\boldsymbol{T} = 0$ and has past compact support [future compact support] is necessarily zero.

Proof: Let supp $\boldsymbol{T}$ be future compact. Let $\boldsymbol{V}$ be a p-tensor on X with compact support and C^∞; set

$$\boldsymbol{U}(x) = \langle \boldsymbol{E}^+(x, y), \boldsymbol{V}(y) \rangle.$$

Then supp $\boldsymbol{U} \subset \mathscr{C}^+(\text{supp } \boldsymbol{V})$ and $L^*\boldsymbol{U} = \boldsymbol{V}$.
Let K be the compact set defined by $\mathscr{C}^+(\text{supp } \boldsymbol{V}) \cap \mathscr{C}^-(\text{supp } \boldsymbol{T})$ and let $f \in \mathscr{D}$ be equal to 1 on a compact neighborhood of K,

$$0 = \langle L\boldsymbol{T}, f\boldsymbol{U} \rangle = \langle \boldsymbol{T}, L^* f\boldsymbol{U} \rangle = \langle \boldsymbol{T}, L^*\boldsymbol{U} \rangle = \langle \boldsymbol{T}, \boldsymbol{V} \rangle. \qquad \blacksquare$$

Reciprocity relations. Let $\boldsymbol{U}$ be the unique solution with past compact support of $L^*\boldsymbol{U} = \boldsymbol{V}$,

$$\langle \overset{*}{\boldsymbol{E}}{}^-(y, x), \boldsymbol{V}(y) \rangle = \langle L_y \overset{*}{\boldsymbol{E}}{}^-(y, x), \boldsymbol{U}(y) \rangle = \boldsymbol{U}(x).$$

Hence

$$\overset{*}{\boldsymbol{E}}{}^{\mp}(y, x) = \boldsymbol{E}^{\pm}(x, y).$$

When L is self-adjoint[3] $\boldsymbol{E}^{\mp}(y, x) = \boldsymbol{E}^{\pm}(x, y)$.

[1]This is meaningful because $\boldsymbol{E}^{\pm}$ are regular kernels:

$$\langle \boldsymbol{E}^{\pm}(x, y), \varphi(x) \rangle \text{ is a number for each } y.$$

[2]For an explicit calculation of a tail term see, for instance, [De Witt and DeWitt 1964].
[3]It should be noted that the small disturbance operator is always symmetric but not necessarily self-adjoint. It must be self adjoint in physical systems because the propagator $\boldsymbol{E} = \boldsymbol{E}^+ - \boldsymbol{E}^-$ is observable.

Propagators. The **propagator** $\boldsymbol{E}$ associated with L is, by definition

propagator

$$\boldsymbol{E} = \boldsymbol{E}^+ - \boldsymbol{E}^-.$$

$\boldsymbol{E}$ is also known as the **commutator function**. It stems from the analysis of the small disturbance of a system introduced by a measuring apparatus. It satisfies the homogeneous equation

commutator function

$$L_x^* \boldsymbol{E}(x, y) = 0.$$

Let $\boldsymbol{V}$ be a p-tensor distribution with future compact support, the unique solution with future compact support of

$$L\boldsymbol{T} = \boldsymbol{V}$$

is the Volterra composition (p. 471) of $\boldsymbol{E}^+$ and $\boldsymbol{V}$ which we denote

$$\boldsymbol{T}(y) = \int \boldsymbol{V}(x)\boldsymbol{E}^+(x, y)\,\mathrm{d}x.$$

One checks that $\langle L\boldsymbol{T}, \boldsymbol{U}\rangle = \langle \boldsymbol{V}, \boldsymbol{U}\rangle$ for any given C^∞ p-tensor $\boldsymbol{U}$ with compact support and that

$$\operatorname{supp} \boldsymbol{T} \subset \mathscr{E}^-(S(\boldsymbol{V})).$$

If it is defined, the Volterra composition

$$\boldsymbol{T}(y) = \int \boldsymbol{V}(x)\boldsymbol{E}(x, y)\,\mathrm{d}x$$

is a solution of $L\boldsymbol{T} = 0$.
Conversely it can be shown that if $\boldsymbol{T}$ is a solution of $L\boldsymbol{T} = 0$ it can be obtained by the Volterra composition of $\boldsymbol{E}$ with an arbitrary tensor distribution with past and future compact support.

PROBLEMS AND EXERCISES

PROBLEM 1. BOUNDED DISTRIBUTIONS

a) *We denote by* $\mathscr{D}_{L_\infty}(U)$ *the space of* C^∞ *functions on an open set* U *of* $\mathbb{R}^n$, *such that each of their derivatives is uniformly bounded on* U, *with the topology given by the family of semi norms*

$$p_j(\varphi) = \sup_{x \in U} |D^j\varphi(x)|.$$

We denote by $\mathscr{B}(U)$ *the subspace of* $\mathscr{D}_{L_\infty}(U)$ *of functions converging to*

zero at the boundary of U [at infinity if U is not bounded], together with each of their derivatives.
Show that $\mathscr{B}(U)$ is a Fréchet space.
Show that $\mathscr{D}(U)$ is not dense in $\mathscr{D}_{L_\infty}(U)$, but is dense in $\mathscr{B}(U)$.

b) *Show that the dual $\mathscr{B}'(U)$ may be identified with a subspace of $\mathscr{D}'(U)$. The distributions of $\mathscr{B}'(U)$ are called bounded distributions.*

c) *Show that a bounded positive distribution of order zero may be identified with a finite regular Borel measure on U (definition p. 36).*

Answer: a) The topology of $\mathscr{B}(U)$ is defined by a countable separated family of semi norms, and is complete.
$\mathscr{D}(U)$ is not dense in $\mathscr{D}_{L_\infty}(U)$: take $f = 1$ on U, $f \in L_\infty(U)$, but

$$\sup_{x\in U} |f(x) - \varphi(x)| \geq 1, \quad \forall \varphi \in \mathscr{D}(U).$$

The density of $\mathscr{D}(U)$ in $\mathscr{B}(U)$ is a consequence of the density of $\mathscr{D}(U)$ in $\mathscr{C}(U)$ (p. 434) and the definition of $\mathscr{B}(U)$; if $f \in \mathscr{B}(U)$ for each ϵ, there exists a compact $K \subset U$ such that

$$\sup_{x\in \complement K} |f(x)| < \epsilon.$$

b) Same arguments as the ones given on p. 439 for $\mathscr{D}'^p$ or $\mathscr{C}'$.

c) By the Riesz–Markov theorem (p. 48) a distribution of order zero, Radon measure μ, can be identified with a regular Borel measure $\hat{\mu}$ and

$$\langle \mu, \varphi \rangle = \int_U \varphi \, d\hat{\mu}.$$

$\hat{\mu}$ is finite if

$$\int_U d\hat{\mu} = \hat{\mu}(U) = \sup_{K\subset U} \hat{\mu}(K) = C < +\infty.$$

If μ, and thus $\hat{\mu}$ are positive, and $\hat{\mu}$ finite, μ is bounded since

$$\left| \int_U \varphi \, d\hat{\mu} \right| \leq \sup_{x\in U} |\varphi(x)| \hat{\mu}(U)$$

implies

$$|\langle \mu, \varphi \rangle| \leq C \sup_{x\in U} |\varphi(x)|.$$

Thus $\mu \in \mathscr{B}'(U)$.

Conversely if $\hat{\mu}$ is not finite, $\sup_{K\subset U} \hat{\mu}(K) = +\infty$. We can find a sequence of compact sets $K_n \subset K$ such that $\hat{\mu}(K_n) \geq 4^n$ and construct a sequence of functions converging to zero in $\mathscr{B}(U)$, for instance $\varphi_n = \psi_n/2^n$, with $\psi_n(x) = 1$ on K_n, such that

$$|\langle \mu, \varphi_n \rangle| \geq \frac{1}{2^n} \hat{\mu}(K_n) > 2^n.$$

Therefore $\langle \mu, \varphi_n \rangle$ does not converge to zero, μ is not continuous on $\mathscr{B}(U)$.

PROBLEM 2. LAPLACIAN OF A DISCONTINUOUS FUNCTION

Prove the Green formula

$$\int_V (f\Delta\varphi - \varphi\Delta f)\, dx + \int_{\partial V} (f\partial_\nu\varphi - \varphi\partial_\nu f)\omega = 0$$

where ν is the inward normal, by the calculus of derivatives of distributions equal to discontinuous functions, as in Example 3 (p. 449).

Answer: With the notations of Example 3 we have

$$\partial_j\partial_i\tilde{f} = \widetilde{\partial_j\partial_i f} + \sigma_i n_j^- \delta_S + \partial_j(\sigma_0 n_i^- \delta_S) \quad \text{with } \sigma_i = \partial_i f^+ - \partial_i f^-;$$
$$\langle \partial_j(\sigma_0 n_i^- \delta_S), \varphi \rangle = -\langle \sigma_0 n_i^- \delta_S, \partial_j\varphi \rangle.$$

If $i = j$ we set $\Sigma_i\, \sigma_i n_i^- \equiv \sigma_\nu$.

Note that $\sigma_\nu = \left(\frac{\partial f}{\partial \nu_+}\right)^+ + \left(\frac{\partial f}{\partial \nu_-}\right)^-$ with the convention of the drawing.

In Example 3, the orientation of n_i^- $[n_i^+]$ is opposite to the orientation of ν^- $[\nu^+]$.

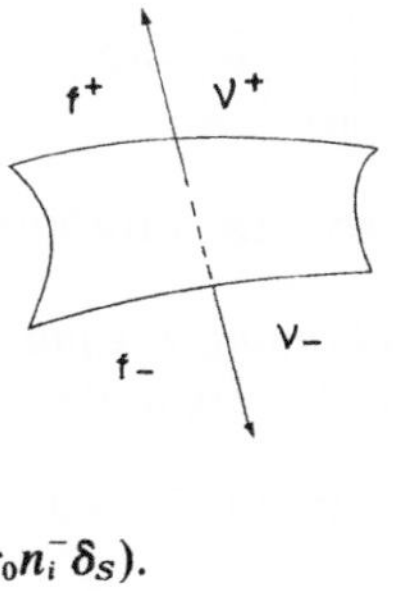

Hence $\Delta\tilde{f} = \widetilde{\Delta f} + \sigma_\nu\delta_S + \partial_i(\sigma_0 n_i^- \delta_S)$.

In particular, let f be zero outside a volume V and let ν be the inward normal, $\partial_\nu f = -\Sigma_i n_i^- \partial_i f = \sigma_\nu$, $\sigma_0 = -f$, then

$$\langle \Delta\tilde{f}, \varphi\rangle = \langle \widetilde{\Delta f}, \varphi\rangle + \langle \sigma_\nu \delta_S, \varphi\rangle + \langle \partial_i(\sigma_0 n_i^- \delta_S), \varphi\rangle$$

$$= \int_V (\Delta f)\varphi \,\mathrm{d}x + \int_{\partial V} \partial_\nu f \varphi\omega - \int_{\partial V} f\partial_\nu \varphi\omega.$$

On the other hand

$$\langle \Delta\tilde{f}, \varphi\rangle = \int f\Delta\varphi \,\mathrm{d}x.$$ ■

EXERCISE 3. REGULARIZED FUNCTIONS.

Compute the upper bounds of the regularized function f_ϵ and its derivatives (p. 433).

Answer:

$$|f_\epsilon(x)| \le M \int_{R^n} \rho_\epsilon(y-x)\,\mathrm{d}y \le M$$

$$\mathrm{D}^\alpha f_\epsilon(x) = \int_{R^n} \mathrm{D}^\alpha \rho_\epsilon(x-y) f(y)\,\mathrm{d}y$$

$$= c\epsilon^{-n} \int_{|x-y|\le\epsilon} \mathrm{D}^\alpha \exp\frac{-1}{1-|(x-y)/\epsilon|^2} f(y)\,\mathrm{d}y$$

and after the change of variables $y = \epsilon z$

$$\mathrm{D}^\alpha f_\epsilon(x) = c \int_{|x-\epsilon z|\le\epsilon} \mathrm{D}^\alpha \exp\frac{-1}{1-|x/\epsilon - z|^2} f(\epsilon z)\,\mathrm{d}z.$$

Using

$$\mathrm{D}^\alpha \exp\frac{-1}{1-|u-z|^2}\,\mathrm{d}z = \epsilon^{-|\alpha|}\left[\mathrm{D}^\alpha \exp\frac{-1}{1-|u-z|^2}\right]_{u=x/\epsilon}$$

we get

$$\sup|\mathrm{D}^\alpha f_\epsilon| \le c\epsilon^{-|\alpha|} M \int_{|u-z|\le 1} \left|\mathrm{D}^\alpha \exp\frac{-1}{1-|u-z|^2}\right| \mathrm{d}z = c_\alpha M\epsilon^{-|\alpha|}.$$

PROBLEM 4.[1] APPLICATION TO THE SCHRÖDINGER EQUATION

Derive solutions of the Schrödinger equation, given solutions of the diffusion equation, using the Lebesgue dominated convergence theorem.

[1]This example is taken from a course given by V. Bargmann. Discussions with D. Delong are gratefully acknowledged.

Answer:

$\dfrac{\partial U_\alpha}{\partial t} - (\alpha + i\beta)\Delta U_\alpha = 0,$ $\alpha > 0$, generalized diffusion equation, the physical diffusion equation corresponds to $\beta = 0$;

$\dfrac{\partial \psi}{\partial t} - i\beta \Delta \psi = 0,$ Schrödinger equation.

Let $U_\alpha(x, t)$ be the solution of the diffusion equation for $t > 0$ that satisfies the Cauchy initial condition

$$U_\alpha(x, 0) = U_{\alpha 0}(x), \quad U_{\alpha 0} \in L^1(\mathbb{R}^3) \text{ for } \alpha > 0.$$

Set

$$\bar{U}_\alpha(x, t) = U_\alpha(x, t) Y(t)$$

where Y is the Heaviside step function equal to 1 for $t > 0$ and 0 for $t < 0$

$$\partial \bar{U}_\alpha / \partial t - (\alpha + i\beta)\Delta \bar{U}_{\alpha 0}(x) = \delta(t).$$

Let $\bar{U}_{G_\alpha}$ be the elementary solution (Green function) with support in $t \geq 0$ (p. 510)

$$\frac{\partial \bar{U}_{G_\alpha}}{\partial t} - (\alpha + i\beta)\Delta \bar{U}_{G_\alpha} = \delta(x)\delta(t)$$

$$\bar{U}_{G_\alpha}(x, t) = \frac{Y(t)}{(4\pi(\alpha + i\beta)t)^{3/2}} \exp\left(-\frac{x^2}{4(\alpha + i\beta)t}\right)$$

and

$$U_\alpha(x, t) = \int_{\mathbb{R}^3} \bar{U}_{G_\alpha}(x - \xi, t) U_{\alpha 0}(\xi)\, d\xi.$$

Set

$$U(x, t) = \int_{\mathbb{R}^3} \lim_{\alpha = 0} (\bar{U}_{G_\alpha}(x - \xi, t)\, U_{\alpha 0}(\xi))\, d\xi.$$

$U(x, t)$ is defined because $U_{\alpha 0} \in L^1(\mathbb{R}^3)$ implies that the integrand is also an element of $L^1(\mathbb{R}^3)$ for $t > 0$ (p. 459). The two conditions necessary to apply the Lebesgue dominated convergence theorem are satisfied, namely:

1) the sequence $(\bar{U}_{G_\alpha}(x - \xi, t) U_{\alpha 0}(\xi))$ converges to an element of $L^1(\mathbb{R}^3)$ for $t > 0$,

2) the integrand is bounded from above and from below.

Thus

$$U(x, t) = \lim_{\alpha = 0} \int_{\mathbb{R}^3} \bar{U}_{G_\alpha}(x - \xi, t) U_{\alpha 0}(\xi)\, d\xi \quad \text{for } t > 0.$$

Moreover $U(x, 0) = U_{00}(x)$.

In summary one can obtain solutions of the Schrödinger equation by substituting $i\beta$ for α in solutions of the physical diffusion equation.

EXERCISE 5. CONVOLUTION AND LINEAR CONTINUOUS RESPONSES

Let S be a system whose response (output) $\mathscr{E}$ to an excitation (input) $i \in \mathscr{D}'^{+}(\mathbb{R})$ is the following.

1) *Linear. If $i_j \mapsto \mathscr{E}_j$, then $\Sigma\, a_j i_j \mapsto \Sigma\, a_j \mathscr{E}_j$.*

2) *Invariant under time translations. If $i(t) \mapsto \mathscr{E}(t)$, then $i(t+t_0) \mapsto \mathscr{E}(t+t_0)$.*

3) *Causal. If $i(t) = 0$ for $t < t_0$, then $\mathscr{E}(t) = 0$ for $t < t_0$.*

4) *Continuous. If $i_j \mapsto \mathscr{E}_j$ and $\langle i, \varphi\rangle = \lim \langle i_j, \varphi\rangle$, then $i \mapsto \mathscr{E}$ where $\langle \mathscr{E}, \varphi\rangle = \lim \langle \mathscr{E}_j, \varphi\rangle$.*

Prove that $i \mapsto \mathscr{E}$ is a convolution operator: $\mathscr{E} = i * Z$ where Z is the response to the unit excitation δ.

Answer: It is easy to check that if $\mathscr{E} = i * Z$, then $\mathscr{E}$ satisfies the four properties listed above. The converse – namely, given a system whose response has the four properties listed above, then its response $\mathscr{E}$ to an input i is $\mathscr{E} = i * Z$ where Z is the response to δ – can be proved as follows.

Let Z be the response of a given system to the unit impulse δ_τ.

1) Causality implies $Z(t) = 0$ for $t < \tau$.

2) Time invariance implies that the response Z_k to the unit impulse δ_{τ_k} at τ_k is such that $Z_k(t) = Z(t - \tau_k) = \delta_{\tau_k} * Z$.

3) Linearity implies that the response to $i_n = \Sigma_{k=1}^{n} q_k \delta_{\tau_k}$ is $\Sigma_{k=1}^{n} q_k Z_k$, $q_k \in \mathbb{R}$, thus $\mathscr{E}_n = i_n * Z$.

4) Continuity gives the answer for an arbitrary $i \in \mathscr{D}'_+(\mathbb{R})$, since such a distribution is a limit of finite sums i_n. Note that $\mathscr{E} \in \mathscr{D}'^{+}(\mathbb{R})$.

Example 1: Let S be an electrical circuit with resistance R, inductance L, and capacitance C; then the electromotive force $\mathscr{E} = i * Z$ where i is the electrical current and $Z = R\delta + L\delta' + C^{-1}Y^{+}$ is the impedance.

Example 2: Let S be a system whose equation of motion (field equation) is the second order (partial) differential equation with constant coefficients $\Sigma_{|j|=0}^{2} a_j \mathrm{D}^j \Psi = f$ where f is a known force (source), let E be an elementary solution of the equation of motion (field equation) $\Sigma_{|j|=0}^{2} a_j \mathrm{D}^j E = \delta$, then if the initial state of the system is such that its response to the "unit impulse" δ is E, its response to f is $\Psi = E * f$.

Reference: [L. Schwartz 1958].

PROBLEM 6. FOURIER TRANSFORM OF $\exp(-x^2)$ AND $\exp(ix^2)$

Let u be a distribution on $\mathbb{R}$ *solution of the differential equation* $\lambda\, du/dx + xu = 0$ *where* λ *is a complex number.*
Restrict λ *so that all solutions u are tempered distributions. Solve the equation when* $\lambda \neq 0$ *and* $\lambda = 0$. *Compute the Fourier transform of u.*

Answer: The general solution of the given equation is $u(x) = C\exp(-x^2/2\lambda)$; it is a tempered distribution if and only if $\operatorname{Re}\lambda \geq 0$. Set $\mathcal{F}u = v$ and take the Fourier transform of the equation satisfied by u; it becomes $\lambda y v(y) + dv(y)/dy = 0$ whose general solution is

$$v(y) = K\exp(-\lambda y^2/2).$$

When $\lambda = 0$, $u = k\delta$ and $v = k$. When $\lambda \neq 0$, we determine $K(C)$ from $v = \mathcal{F}u$. Note first that $v(y) = (K/C)u(\lambda y)$, hence

$$\begin{aligned} u(x) = \bar{\mathcal{F}}v(x) &= (2\pi)^{-1}\mathcal{F}v(-x) = (2\pi)^{-1}\mathcal{F}((K/C)u(\lambda y))(-x) \\ &= (2\pi)^{-1}(K/C)\lambda^{-1}v(-x/\lambda) = (2\pi)^{-1}(K/C)^2\lambda^{-1}u(x) \end{aligned}$$

hence

$$|K| = |C|(2\pi|\lambda|)^{1/2}.$$

Furthermore $K = (\mathcal{F}u)(0)$, hence $K/C = \int_{\mathbb{R}} \exp(-x^2/2\lambda)\, dx$. When λ is real

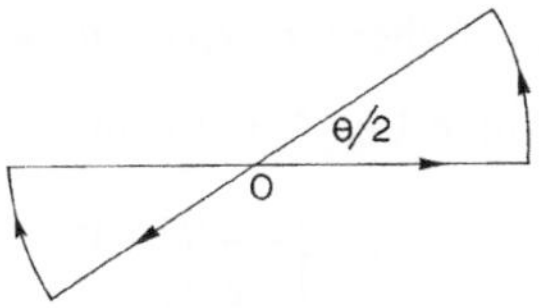

$K = C(2\pi\rho)^{1/2}$ and $\int_{\mathbb{R}} \exp(-x^2/2\rho)\, dx = (2\pi\rho)^{1/2}$. For arbitrary $\lambda = \rho\exp(i\theta)$, with $-\pi/2 < \theta < \pi/2$, use contour integration to get $\oint \exp(-x^2/2\lambda)\, dx = 0$. Hence

$$\int_{\mathbb{R}} \exp(-x^2/2\lambda)\, dx = \int_{\text{rotated line}} \exp(-x^2/2\rho\exp i\theta)\, dx$$

integrated along the line rotated from the real axis by $\theta/2$

$$\begin{aligned} &= \exp(i\theta/2)\int_{\mathbb{R}} \exp(-y^2/2\rho)\, dy \\ &= \exp(i\theta/2)(2\pi\rho)^{1/2} = (2\pi\lambda)^{1/2}. \end{aligned}$$

PROBLEM 7. FOURIER TRANSFORMS OF HEAVISIDE FUNCTIONS AND PRINCIPAL VALUE OF $1/x$

Definitions: Let Y^+ and Y^- be the Heaviside step-up and step-down functions defined by:

$$Y^+(x) = 1 \text{ for } x > 0,\ Y^+(x) = 0 \text{ for } x < 0,\ Y^+ \text{ not defined for } x = 0;$$
$$Y^-(x) = Y^+(-x).$$

Compute the Fourier transforms of $Y^+(x)\exp(-ax)$ and $Y^-(x)\exp(-ax)$ when $a > 0$; compute their limits in $\mathscr{D}'$ when a tends to zero.
Give the Fourier transform of $\mathrm{Pv}(1/x)$.

Answer: When $a > 0$, $Y^{\pm}(x)\,\mathrm{e}^{-ax} \in L^1$, their Fourier transforms are continuous

$$\mathscr{F}(Y^{\pm}(x)\,\mathrm{e}^{-ax})(\xi) = \mp \mathrm{i}/(\xi \mp \mathrm{i}a).$$

The functions $(\xi \mp \mathrm{i}a)^{-1}$ are locally integrable, they define the distributions

$$\left\langle \frac{1}{\xi \mp \mathrm{i}a}, \varphi \right\rangle = \int_{-\infty}^{+\infty} \frac{\varphi(x)}{(\xi \mp \mathrm{i}a)}\,\mathrm{d}x.$$

$\mathscr{F}$ is a continuous mapping in $\mathscr{S}'$, hence $(\mathscr{F}Y^{\pm})(\xi) = \lim_{a=0}(\mp \mathrm{i}(\xi \mp \mathrm{i}a)^{-1})$.
The distribution $(\xi \mp \mathrm{i}a)^{-1}$ is the derivative in $\mathscr{D}'$ of

$$\log(\xi \mp \mathrm{i}a) = \log r \mp \mathrm{i}\theta \quad \text{where } r\exp(\mp \mathrm{i}\theta) = \xi \mp \mathrm{i}a \quad \text{with } 0 \le \theta \le \pi.$$

Point wise, and in $\mathscr{D}'$ since $\log|\xi|$ is locally integrable we have

$$\lim_{a=0} \log(\xi \mp \mathrm{i}a) = \begin{cases} \log|\xi| \mp \mathrm{i}\pi & \text{if } \xi < 0 \\ \log|\xi| & \text{if } \xi > 0. \end{cases}$$

But the derivative is continuous in $\mathscr{D}'$, hence

$$\lim_{a=0}(\xi \mp \mathrm{i}a)^{-1} = \frac{\mathrm{d}}{\mathrm{d}\xi} \lim_{a=0} \log(\xi \mp \mathrm{i}a)$$

$$\left\langle \frac{\mathrm{d}}{\mathrm{d}\xi} \log|\xi|, \varphi \right\rangle = -\int_{-\infty}^{+\infty} \log|\xi|\, \varphi'(\xi)\,\mathrm{d}\xi = \lim_{\epsilon=0} \int_{|\xi|\ge\epsilon} -\log|\xi|\varphi'(\xi)\,\mathrm{d}\xi$$

$$= \lim_{\epsilon=0} \int_{|\xi|\ge\epsilon} \mathrm{d}\xi\varphi(\xi)/\xi = \left\langle \mathrm{Pv}\frac{1}{\xi}, \varphi \right\rangle.$$

Hence $(\mathscr{F}Y^{\pm})(\xi) = \mp \mathrm{i}\left(\mathrm{Pv}\dfrac{1}{\xi} \pm \mathrm{i}\pi\delta_\xi\right)$.

Check: $Y^+ + Y^- = 1$, $\mathcal{F}Y^+ + \mathcal{F}Y^{-1} = 2\pi\delta$,

Note:
$$\mathcal{F}\left(\mathrm{Pv}\frac{1}{x}\right)(\xi) = \mathrm{i}\pi(Y^+ - Y^-) = \mathrm{i}\pi \text{ sign } \xi.$$

PROBLEM 8.

Give an expression for the Dirac kernels (Dirac bitensors) in local coordinates in the case of
a) *p-tensors,*
b) *symmetric 2-tensor,*
c) *antisymmetric 2-tensors.*

Answer:
a) $[\boldsymbol{D}^{(p)}(x, x')]^{\alpha_1\alpha_2\ldots\alpha_p}_{\alpha'_1\alpha'_2\ldots\alpha'_p} = \delta^{\alpha_1}_{\alpha'_1}\delta^{\alpha_2}_{\alpha'_2}\ldots\delta^{\alpha_p}_{\alpha'_p}\delta(x, x')$;
$\delta^{\alpha}_{\alpha'}$: Kronecker symbol, $\delta(x, x')$: scalar Dirac kernel.

b) $[\tilde{\boldsymbol{D}}^{(2)}(x, x')]^{\alpha\beta}_{\alpha'\beta'} = \frac{1}{2}(\delta^{\alpha}_{\alpha'}\delta^{\beta}_{\beta'} + \delta^{\alpha}_{\beta'}\delta^{\beta}_{\alpha'})\delta(x, x')$.

c) $[\hat{\boldsymbol{D}}^{(2)}(x, x')]^{\alpha\beta}_{\alpha'\beta'} = \frac{1}{2}(\delta^{\alpha}_{\alpha'}\delta^{\beta}_{\beta'} - \delta^{\alpha}_{\beta'}\delta^{\beta}_{\alpha'})\delta(x, x')$.

PROBLEM 9. LEGENDRE CONDITION

Show that the following integral, where $A(x)$ and $B(x)$ are continuous functions on $[a, b]$
$$I = \int_a^b \{(A(x)(h'(x))^2 + B(x)(h(x))^2\} \,\mathrm{d}x$$
is positive (i.e. $I \geq 0$) for all C^0, piecewise C^1, functions h on $[a, b]$, with $h(a) = h(e) = 0$, only if $A(x) \geq 0$, for $a < x < b$.

Answer: Assume there exists $\bar{x}$ such that $A(\bar{x}) < 0$. Consider a series (h_n) of C^0, piecewise C^1 functions, with compact support in (a, b) such that, in $\mathcal{D}'^0$,
$$\lim_{n=\infty} (h_n)^2 = 0$$
and
$$\lim_{n=\infty} (h'_n(\bar{x}))^2 = \delta_{\bar{x}}.$$
Then
$$\lim_{n=\infty} I_n = \lim_{n=\infty} \{\langle (h'_n)^2, A\rangle + \langle (h_n)^2, B\rangle\} = A(\bar{x}) < 0.$$
therefore I_n cannot be ≥ 0 for every n.

As an example of such a series (h_n) one can take the continuous functions, linear by pieces, of fig. 1. h'_n is the function, constant by pieces,

$$\begin{aligned} h'_n(x) &= 0 \text{ for } a < x < \bar{x} - \frac{1}{2n} \\ &= \sqrt{n} \text{ for } \bar{x} - \frac{1}{2n} < x < \bar{x} \\ &= -\sqrt{n} \text{ for } \bar{x} < x < \bar{x} + \frac{1}{2n} \\ &= 0 \text{ for } \bar{x} + \frac{1}{2n} < x < b. \end{aligned}$$

Thus, if $\rho \in \mathscr{D}^0$

$$\langle (h'_n)^2, \rho \rangle = \int_a^b (h'_n(x))^2 \rho(x)\,\mathrm{d}x = \int_{\bar{x}-\frac{1}{2}n}^{\bar{x}+\frac{1}{2}n} n\rho(x)\,\mathrm{d}x$$

and

$$\lim_{n=\infty} \langle (h'_n)^2, \rho \rangle = \rho(\bar{x}).$$

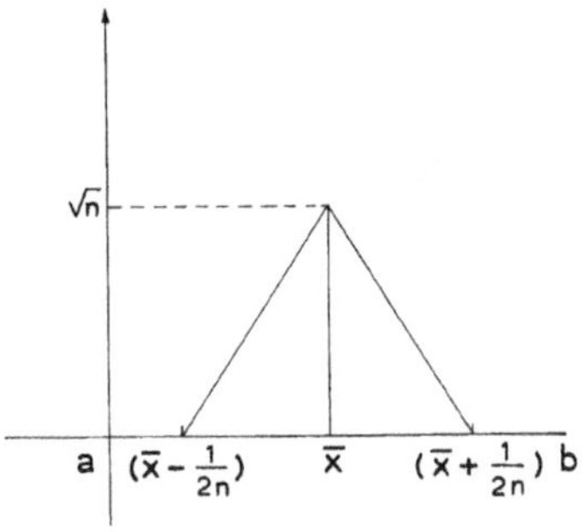

Fig. 1.

PROBLEM 10. HYPERBOLIC EQUATIONS; CHARACTERISTICS

Consider a second order partial differential equation for a function u on a manifold X^m which is linear with respect to the highest order derivatives and reads, in a local coordinate system,

$$A^{jk}(x)\frac{\partial^2 u}{\partial x^j \partial x^k} + \Phi\left(x, u, \frac{\partial u}{\partial x^1}, \dots \frac{\partial u}{\partial x^m}\right) = 0. \tag{1}$$

By definition its characteristic equation is

$$A^{jk}(x)\frac{\partial\omega}{\partial x^j}\frac{\partial\omega}{\partial x^k}=0\,(\textit{eikonal equation}). \tag{2}$$

A submanifold $\omega(x^1,\ldots,x^m)=ct$ is said to be characteristic if it satisfies the characteristic equation.
Show that the characteristic manifolds are invariant under a change of coordinates if and only if A^{ij} is a contravariant tensor.

Answer: Let $\xi^r=\xi^r(x^1,\ldots x^m)$, $r=1,\ldots m$ define a change of variables. Equation (1) becomes

$$A^{jk}\frac{\partial\xi^r}{\partial x^k}\frac{\partial\xi^s}{\partial x^j}\frac{\partial^2 u}{\partial\xi^r\partial\xi^s}+A^{jk}\frac{\partial^2\xi^r}{\partial x^j\partial x^k}\frac{\partial u}{\partial\xi^r}+\Psi\!\left(x,u,\frac{\partial u}{\partial\xi^r}\right)=0.$$

Set $\omega(x(\xi))=\tilde{\omega}(\xi)$, then the characteristic equation becomes

$$A^{jk}\frac{\partial\xi^r}{\partial x^j}\frac{\partial\xi^s}{\partial x^k}\frac{\partial\tilde{\omega}}{\partial\xi^r}\frac{\partial\tilde{\omega}}{\partial\xi^s}\equiv\tilde{A}^{rs}\frac{\partial\tilde{\omega}}{\partial\xi^r}\frac{\partial\tilde{\omega}}{\partial\xi^s}=0$$

if and only if $\tilde{A}^{rs}=A^{jk}\dfrac{\partial\xi^r}{\partial x^j}\dfrac{\partial\xi^s}{\partial x^k}$.

PROBLEM 11. ELECTROMAGNETIC SHOCK[1] WAVES: EIKONAL EQUATION: LIGHT RAYS

Consider an electromagnetic field F in a domain U of space-time. The domain U is divided by a submanifold S into two domains U^+ and U^- as on *p. 449**. Suppose that F is C^1 in $\bar{U}^+$ and $\bar{U}^-$, and denote by φ the 2-tensor defined on S by $\varphi=F^+-F^-$, where F^+ and F^- are the limits on S of F, obtained by its values in $\bar{U}^+$ and $\bar{U}^-$ respectively.*

a) *Deduce from the Maxwell equations*

$$\mathrm{d}F=0,\qquad \delta F=0,$$

written in a C^1 hyperbolic metric g on U and satisfied by F in the sense of the theory of distributions, an equation satisfied by S (called the eikonal equation).

b) *Show that S is generated by null geodesics of the metric g.*

c) *In the case where g is the Minkowski metric find solutions of the eikonal equation passing through a given 2 dimensional manifold.*

[1]The word "shock" is used here in a wider sense than usually.

Answer: a) In local coordinates the first set of Maxwell equations reads

$$\partial_\alpha F_{\beta\gamma} + \partial_\gamma F_{\alpha\beta} + \partial_\beta F_{\gamma\alpha} = 0.$$

By the calculation given p. 449 for the derivative of a discontinuous function, these equations read

$$\widetilde{\partial_\alpha F_{\beta\gamma}} + \widetilde{\partial_\gamma F_{\alpha\beta}} + \widetilde{\partial_\beta F_{\gamma\alpha}} + (l_\alpha \varphi_{\beta\gamma} + l_\gamma \varphi_{\alpha\beta} + l_\beta \varphi_{\gamma\alpha})\delta_S = 0,$$

where l_α is the normal to S, that is $l_\alpha = \partial_\alpha s$ if $s(x) = 0$ is the equation of S.

The sum of a usual function and a singular measure with support S vanishes only if each term of the sum vanishes, which means on the one hand that F satisfies $\mathrm{d}F = 0$ in U^+ and U^- and on the other hand that, on S

$$l_\alpha \varphi_{\beta\gamma} + l_\beta \varphi_{\gamma\alpha} + l_\gamma \varphi_{\alpha\beta} = 0.$$

The metric being C^1, and the connection C^0, the second set $\delta F = 0$ gives

$$\widetilde{\nabla_\alpha F^{\alpha\beta}} + l_\alpha \varphi^{\alpha\beta}\delta_S = 0.$$

F satisfies $\delta F = 0$ in U^+ and U^-, and, on S

$$l_\alpha \varphi^{\alpha\beta} = 0.$$

The vector $l_\alpha = \partial_\alpha s$ is an eigenvector of the 2-form $\varphi_{\alpha\beta}$ and of its dual $\overset{*}{\varphi}_{\alpha\beta}$, therefore it is a null (isotropic) vector (Problem V 1) and the equation satisfied by s is

$$g^{\alpha\beta}\partial_\alpha s \partial_\beta s = 0.$$

It expresses that S is a characteristic manifold for the wave operator $\nabla_l \nabla^l$ of the hyperbolic metric g.

b) The vector $l_\gamma = \partial_\gamma s$ being isotropic is both normal and tangent to S. Its trajectories are geodesics since

$$\nabla_\beta l_\gamma - \nabla_\gamma l_\beta = 0 \quad \text{implies} \quad l^\beta(\nabla_\beta l_\gamma - \nabla_\gamma l_\beta) = 0$$

and $l^\beta l_\beta = 0$ implies $l^\beta \nabla_\gamma l_\beta = 0$, thus

$$l^\beta \nabla_\beta l_\gamma = 0.$$

We can also prove that these geodesics (light rays) are solutions of the characteristic system of the eikonal equation. This system is

$$\frac{\mathrm{d}x^\alpha}{2g^{\alpha\beta}y_\beta} = \frac{-\mathrm{d}y_\alpha}{(\partial g^{\gamma\beta}/\partial x^\alpha)y_\beta y_\beta} = \frac{\mathrm{d}z}{0} = \mathrm{d}t, \quad g^{\alpha\beta}y_\alpha y_\beta = 0.$$

Thus

$$\frac{\mathrm{d}x^\alpha}{\mathrm{d}t} = 2y^\alpha, \qquad \frac{\mathrm{d}y_\alpha}{\mathrm{d}t} = -\frac{\partial g^{\gamma\beta}}{\partial x^\alpha} y_\beta y_\gamma, \qquad \frac{\mathrm{d}z}{\mathrm{d}t} = 0$$

but

$$\frac{dy^\alpha}{dt} = g^{\alpha\beta}\frac{dy_\beta}{dt} + y_\beta \frac{\partial g^{\alpha\beta}}{\partial x^\lambda}\frac{dx^\lambda}{dt}$$

which shows that the solutions of the bicharacteristic system satisfy the geodesic equation

$$\frac{d^2x^\alpha}{dt^2} + \Gamma^\alpha_{\beta\gamma}\frac{dx^\alpha}{dt}\frac{dx^\beta}{dt} = 0$$

and are isotropic

$$g_{\alpha\beta}\frac{dx^\alpha}{dt}\frac{dx^\beta}{dt} = 0.$$

c) In Minkowski space the eikonal equation is

$$\left(\frac{\partial s}{\partial x^0}\right)^2 - \sum_{i=1}^{3}\left(\frac{\partial s}{\partial x^i}\right)^2 = 0. \tag{1}$$

The characteristic system reads

$$y_0^2 - y_1^2 - y_2^2 - y_3^2 = 0 \quad \text{or} \quad 1 = p_1^2 + p_2^2 + p_3^2 \quad \text{with } p_i = y_i/y_0;$$

$$\frac{dx^0}{y_0} = \frac{dx^i}{-y_i} = \frac{dy_\alpha}{0}, \qquad dt = \frac{dx^i}{-p_i} = \frac{dp_i}{0} \quad \text{with } x^0 = t.$$

Its solutions are straight lines

$$p_i = p_i^0 \text{ and } x^i = -p^{0i}(t - t_0) + x_0^i.$$

If the initial integral manifold in $\mathbb{R}^4$ is the point $(x^i = x_0^i,\ t = t_0)$, the integral manifold of (1) is

$$s(x^\alpha) = (t - t_0)^2 - \sum_{i=1}^{3}(x^i - x_0^i)^2 = 0.$$

It is called the characteristic cone. The point (x_0^i, t_0) is a singular point of F. **Let the initial integral manifold be a surface in $\mathbb{R}^4$ given by (p. 256)**

$$(W^2, \bar{g}) \text{ where } \bar{g} \text{ is a mapping } (u^1, u^2) \mapsto (t, x^1, x^2, x^3).$$

The integral manifold of (1) defined by this initial integral manifold is a function h of x^i such that

$$h(x^i(u^1, u^2)) = t(u^1, u^2).$$

The integral manifold (W^2, g) where $g: (u_1, u_2) \mapsto (t, x^i, p_i)$ is obtained by solving the following 3 equations in the 3 unknowns p_i

$$\frac{\partial t}{\partial u^i} = p_i \frac{\partial x^i}{\partial u^i}, \qquad \sum (p_i)^2 = 1.$$

In general this system of equations has 2 solutions $p_i(u^1, u^2)$. There are 2 surfaces $S(x^\alpha) = 0$ satisfying the given Cauchy data. In the previous example the "surface" $(W^2, \bar{g})$ is a single point, $\bar{g}$ is not an immersion, the integral "manifold" is not a manifold at the point x_0, t_0.

PROBLEM 12. ELEMENTARY SOLUTION OF THE WAVE EQUATION IN $\mathbb{R}^4$ (p. 511).

The elementary solution of the wave equation in $\mathbb{R}^4$ *is* $E(t, x) = \delta_c/4\pi$ *where* δ_c *is the Dirac measure on the cone* c *defined by* $\frac{1}{2}(t^2 - r^2) = 0$. *Show that* $E(t, x)$ *is invariant under Lorentz transformations.*

Answer: Let $L: x \mapsto x'$ be a Lorentz transformation. By definition, under L, $t^2 - r^2 \mapsto t'^2 - r'^2$. The Leray form ω_c is invariant under Lorentz transformations since both the volume element and the differential $d(t^2 - r^2)$ are invariant under Lorentz transformations. The diffeomorphism $L: \mathbb{R}^4 \to \mathbb{R}^4$ induces a transformation on the distribution T based on $\mathbb{R}^4$ given by

$$\langle LT, \varphi \rangle = \langle T, \varphi \circ L|\det L| \rangle = \langle T, \varphi \circ L \rangle.$$

Hence

$$\langle L\delta_c, \varphi \rangle = \int_c (\varphi \circ L)\omega_c = \int_c (\varphi(Lx))\omega_c$$

$$= \int_c \varphi(x')L^*\omega_c = \int_c \varphi\omega_c$$

and $L\delta_c = \delta_c$. ∎

PROBLEM 13. ELEMENTARY KERNELS OF THE HARMONIC OSCILLATOR

a) *Construct the elementary kernels and the propagator for the operator* $L = -d^2/dt^2 - \rho^2$ *defined on the space of distributions on* $\mathbb{R}$. *Find their Fourier transforms.*

b) *The operator* L *can be expressed as the product of two first order operators*

$$L = L_{(+)}L_{(-)} = \left(i\frac{d}{dt} + \rho\right)\left(i\frac{d}{dt} - \rho\right).$$

Construct the elementary kernels of $L_{(+)}$ *and* $L_{(-)}$ *with support in* $\mathcal{D}'^{+}$ *and* $\mathcal{D}'^{-}$. *Compute their Fourier transforms.*

Answer: a) The elementary kernels $E(t, s)$ are solutions of

$$(-d^2/dt^2 - \rho^2)E(t, s) = \delta(t, s). \tag{4}$$

The Fourier transform of this equation is

$$\mathscr{F}((-\mathrm{d}^2/\mathrm{d}t^2-\rho^2)\delta * E)(\eta)=1,$$

$$(\eta^2-\rho^2)\mathscr{F}E=1,$$

$$(\mathscr{F}E)(\eta)=\frac{1}{2\rho}\,\mathrm{Pv}\Big(\frac{1}{\eta-\rho}-\frac{1}{\eta+\rho}\Big)+K_1\delta(\eta-\rho)+K_2\delta(\eta+\rho). \tag{5}$$

We can choose K_1 and K_2 such that the elementary kernel be in the convolution algebra $\mathscr{D}'^{+}$ or $\mathscr{D}'^{-}$. Recall (Problem VI 7) that

$$(\mathscr{F}Y^{\pm})(\eta)=\mp\mathrm{i}\Big(\mathrm{Pv}\,\frac{1}{\eta}\pm\mathrm{i}\pi\delta_\eta\Big)=\mp\mathrm{i}(\eta\mp\mathrm{i}0)^{-1}. \tag{6}$$

Hence

$E^{+}\in\mathscr{D}'^{+}$ if $K_1^{+}=-K_2^{+}=\mathrm{i}\pi/2\rho$ and

$$\mathscr{F}E^{+}=\frac{1}{2\rho}\Big(\frac{1}{\eta-\rho-\mathrm{i}0}-\frac{1}{\eta+\rho-\mathrm{i}0}\Big), \tag{7a}$$

$E^{-}\in\mathscr{D}'^{-}$ if $K_1^{-}=-K_2^{-}=-\mathrm{i}\pi/2\rho$ and

$$\mathscr{F}E^{-}=\frac{1}{2\rho}\Big(\frac{1}{\eta-\rho+\mathrm{i}0}-\frac{1}{\eta+\rho+\mathrm{i}0}\Big). \tag{7b}$$

To compute $E^{\pm}$ from (5) using (6) one can translate (see p. 458)

$$\mathrm{Pv}\Big(\frac{1}{\eta-\rho}\pm\mathrm{i}\delta(\eta-\rho)\Big)$$

by ρ and the other two terms by $-\rho$:

$$E^{\pm}(t,s)=\mp Y^{\pm}(t-s)\rho^{-1}\sin(\rho t-\rho s). \tag{8}$$

The propagator $E=E^{+}-E^{-}$ is

$$E(t,s)=-\rho^{-1}\sin(\rho t-\rho s).$$

We could, of course have obtained (8) by solving (4) according to the method developed on p. 469, which says

$$E^{\pm}(t,s)=Y^{\pm}(t-s)h^{\pm}(t-s)$$

where the C^∞ functions $h^{\pm}$ satisfy the homogeneous equation and the following boundary conditions:

$$h^{+}(0)=0,\ h^{+\prime}(0)=-1\quad\text{and}\quad h^{-}(0)=0,\ h^{-\prime}(0)=1.$$

Equation (7) suggests the following integral representation for the ele-

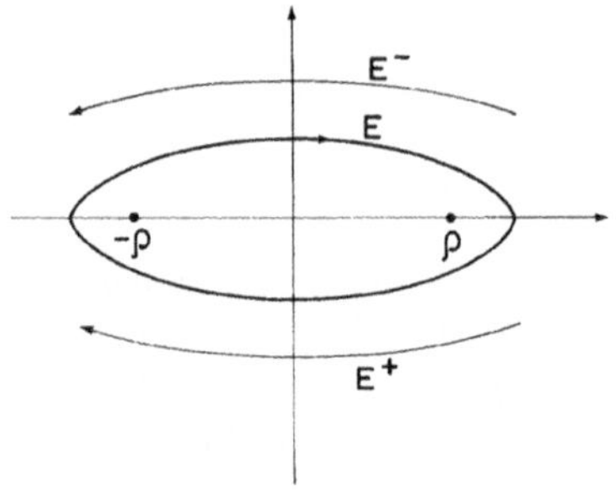

Fig. 1. The contour of integrations of $E^{\pm}$, E in the complex η plane.

mentary kernels and the propagator

$$E^{\pm}(t,s)=\frac{1}{2\pi}\int\frac{\exp \mathrm{i}\eta(t-s)}{\eta^2-\rho^2}\,\mathrm{d}\eta,$$

with the contours of integration marked on figure 1.

b) The elementary kernels of $L_{(+)}$ and $L_{(-)}$ with support in $\mathscr{D}'^{+}$ and $\mathscr{D}'^{-}$ are

$$E^{\pm}_{(+)}(t,s)=\mp\mathrm{i}Y^{\pm}(t-s)\exp \mathrm{i}\rho(t-s)$$
$$E^{\pm}_{(-)}(t-s)=\mp\mathrm{i}Y^{\pm}(t-s)\exp(-\mathrm{i}\rho(t-s))=-E^{\mp}_{(+)}(s,t)=-(E^{\pm}_{(+)}(s,t))^{*}.$$

Their Fourier transforms

$$\mathscr{E}^{\pm}_{(+)}(\eta)=-(\eta-\rho\mp \mathrm{i}0)^{-1},\ \mathscr{E}^{\pm}_{(-)}(\eta)=-(\eta+\rho\mp \mathrm{i}0)^{-1}$$

suggest the following integral representation:

$$E^{\pm}_{(+)}(t,s)=\frac{-1}{2\pi}\int\frac{\exp \mathrm{i}\eta(t-s)}{\eta-\rho}\,\mathrm{d}\eta$$
$$E^{\pm}_{(-)}(t,s)=\frac{-1}{2\pi}\int\frac{\exp \mathrm{i}\eta(t-s)}{\eta+\rho}\,\mathrm{d}\eta$$

with the contours of integration marked on figure 2.

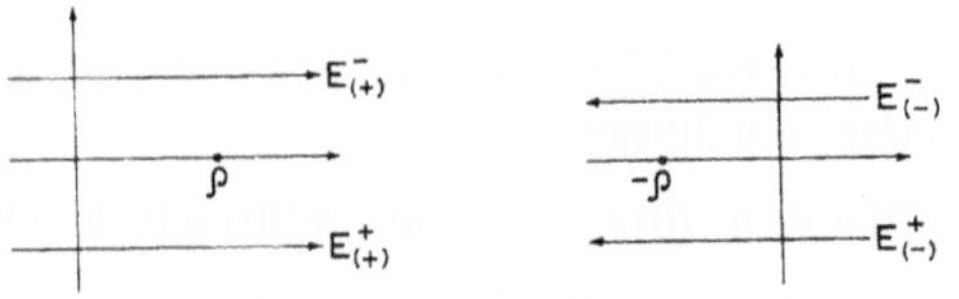

Fig. 2.

Remark: $E^+_{(+)} * E^+_{(-)} = E^+$ and $E^-_{(+)} * E^-_{(-)} = E^-$.

Proof: In the first equation the kernels are all in $\mathscr{D}'^+$, in the second they are in $\mathscr{D}'^-$. Moreover

$$(L_{(-)}L_{(+)})E^+_{(+)} * E^+_{(-)} = L_{(-)}\delta * E^+_{(-)} = L_{(-)}E^+_{(-)} = \delta = LE^+. \qquad \blacksquare$$

We could, of course, have checked that $\mathscr{F}E^+_{(+)} \cdot \mathscr{F}E^+_{(-)} = \mathscr{F}E^+$ but here again we want to relate the properties of the elementary kernels directly to their defining equation.

Remark: The labels (+) and (−) are used in physics to designate positive and negative frequencies. Here ρ is assumed positive, $E_{(+)}$ and $E_{(-)}$ are the contributions from poles on the positive and negative axis respectively.

The Feynman–Green function. The elementary kernel

$$G(t, s) = \frac{1}{2\rho}(E^-_{(-)} - E^+_{(+)}) = \frac{i}{2\rho}\exp i\rho|t - s| \tag{9}$$

is called the Feynman–Green function. Its Fourier transform

$$(\mathscr{F}G)(\eta) = (\eta^2 - (\rho + i0)^2)^{-1}$$

is used extensively and is commonly described as the Green function which "propagates negative frequencies to the past and positive frequencies to the future". The expression obtained for the Feynman–Green function in Problem II 2 differs from equation (9) because there we consider functions defined on the finite interval T whereas here we consider functions defined on $\mathbb{R}$.[1]

[1]For the analogous properties and the integral representations of the Green functions of the free scalar field in space-time see for instance D. Kastler, Introduction à l'electrodynamique quantique (Dunod, 1961) or J.D. Bjorken and S.D. Drell, Relativistic Quantum Mechanics (McGraw Hill, 1964) and Relativistic Quantum Fields (McGraw Hill, 1965).

VII. DIFFERENTIABLE MANIFOLDS, INFINITE DIMENSIONAL CASE

A. INFINITE-DIMENSIONAL MANIFOLDS

1. DEFINITIONS AND GENERAL PROPERTIES

Many definitions given in Chapter III for differentiable manifolds modelled on $\mathbb{R}^n$ can be extended to manifolds modelled on an infinite-dimensional topological vector space (p. 21). However, many useful properties are carried over only in the case of Banach manifolds (cf. below), because the implicit function theorem is not valid for arbitrary topological vector spaces.
We shall use the concepts of differential calculus defined in Chapter II.

Let X be a set and E a locally convex topological vector space. An **atlas** A of class C^k on X modelled on E is a collection of pairs (U_i, φ_i), called **local charts**, where $i \in I$ is a set of indices with the following properties. atlas local charts
1) The set $\{U_i; i \in I\}$ covers X

$$\mathcal{U} = \bigcup_{i \in I} U_i = X.$$

2) Each mapping φ_i is a bijection of U_i onto an open set of E. Let $x \in U_i \subset X$; $\varphi_i(x) \in E$ is called the **representative of** x in the chart $(U_i; \varphi_i)$. representative of $x \in X$
3) For every pair i, j the set $\varphi_i(U_i \cap U_j)$ is open in E and the mapping $\varphi_j \circ \varphi_i^{-1}$: $\varphi_i(U_i \cap U_j) \to \varphi_j(U_i \cap U_j)$ is of class C^k.
Equivalent atlases: Two C^k atlases on X, modelled on E, are said to be **compatible** if their union is another such atlas. Compatibility is an equivalence relation. compatible

A C^k **differentiable manifold** modelled on E [an E-manifold] is a set X together with an equivalence class of C^k atlases modelled on E. All the atlases within the equivalence class are said to be **admissible** on X. An admissible atlas is sufficient to define the manifold. differentiable manifold admissible
If $k = 0$, the manifold is said to be a topological manifold.

Exercise: Give a topology on X which is compatible with its manifold structure, i.e. a topology such that for all pairs (U_i, φ_i) of an atlas, and hence for all atlases, U_i is open in E and φ_i is a homeomorphism.

differentiable function

A C^p **differentiable function** on a manifold X modelled on E, $f \in C^p(X)$, is a function on X such that, if $\{(U_i, \varphi_i); i \in I\}$ is an atlas of X, then the functions $f \circ \varphi_i^{-1}$ are of class C^p on $\varphi_i(U_i) \subset E$. The definition of a C^p differentiable function on a C^k manifold, with $p \leq k$, does not depend on the choice of the atlas within the equivalence class.

germ

$G(x)$

A **germ** of differentiable functions at $x \in X$ is an equivalence class of functions differentiable in a neighborhood of x. $f \sim g$ if $f = g$ in a neighborhood of x. The algebra of these germs at x is denoted $G(x)$.

Tangent vector space. The two following definitions of $T_x(X)$ the tangent vector space to X at a point x are equivalent.

tangent vector

a) As in the finite dimensional case, a **tangent vector** at x to the manifold X is an equivalence class of curves tangent at x. Consider all curves $\gamma: I \subset \mathbb{R} \to X$ with $\gamma(0) = x$ and say that $\gamma_1 \sim \gamma_2$ if in some (and hence in every) chart (U, φ)

$$\frac{\mathrm{d}}{\mathrm{d}t}(\varphi \circ \gamma_1)\bigg|_{t=0} = \frac{\mathrm{d}}{\mathrm{d}t}(\varphi \circ \gamma_2)\bigg|_{t=0}.$$

representative of a tangent vector

Then $T_x(X)$ is defined to be the set of these equivalence classes. The vector $V \in E$, $V = \mathrm{d}/\mathrm{d}t(\varphi \circ \gamma)|_{t=0}$ is called the **representative** in the chart (U, φ) **of the vector** tangent at x to the curve γ.

Conversely, if V is a vector of E, and x a point of X, there exists a curve on X passing through x whose tangent vector at x has for its representative V in a map (U, φ) at x. Take for instance $\gamma(t) = \varphi^{-1}[\varphi(x_0) + tV]$. Thus $T_x(X)$ is isomorphic to E.

b) The transformation law for the representative of a tangent vector in a change of charts leads to the second equivalent definition. A **tangent vector** v_x, at $x \in X$, is an equivalence class of triples (U_i, φ_i, V_i) where (U_i, φ_i) is an arbitrary admissible chart of X at x, and V_i is a vector of E; two triples are equivalent if

$$V_j = (\varphi_j \circ \varphi_i^{-1})'(\varphi_i(x))V_i.$$

The representative $V_i \in E$ of the vector v_x in the map (U_i, φ_i), plays the role of the components in a local system of coordinates of a vector tangent to a finite-dimensional differentiable manifold.

Property: A tangent vector v_x defines a derivation on the algebra $G(x)$ of the germs of differentiable functions at the point x; i.e. it defines an additive operation satisfying Leibniz's rule. Let V be the representative of v_x in the map (U, φ) at x, set

$$v_x(f) = (f \circ \varphi^{-1})'(\varphi(x))V.$$

The value of $v_x(f)$ does not depend on the chart; v_x defines a derivation. The converse is not true in the case of infinite dimensions; a derivation does not define a tangent vector without additional assumptions.

Exercise: Let (U, φ) be a chart at x, let E be reflexive (cf. Ch. I), and let d be a linear mapping $G(x) \to \mathbb{R}$ such that

$$|d(f)| \leq K \|(f \circ \varphi^{-1})'(\varphi(x))\|_{\mathscr{L}(E,\mathbb{R})} \quad \text{where } K \text{ is a constant.}$$

Show that d can be identified with a tangent vector at the point x. Show that if the manifold is finite-dimensional, any derivation can be identified with a tangent vector at a point.
Proof: To each $V^* \in E^* = \mathscr{L}(E, \mathbb{R})$, we associate the element f of $G(x_0)$ defined by $f(x) = \langle V^*, \varphi(x) \rangle$, the mapping $E^* \to \mathbb{R}$ by $f \mapsto d(f)$ is continuous because

$$|d(f)| \leq K \|V^*\|_{E^*}.$$

Hence it defines an element $V \in E^{**} = E$, which is the representative of a vector tangent at the point x_0. ■

The space of vectors tangent at the point x, together with its natural vector space structure is the **tangent vector space** $T_x(X)$.

tangent vector space $T_x(X)$

It can easily be verified that the set of vector spaces tangent to a C^k manifold X, $T(X) = \cup_{x \in X} T_x(X)$ has the structure of a C^{k-1} differentiable manifold modelled on $E \times E$ by the family of charts $(\cup_{x \in U_i} T_x(X), \hat{\varphi}_i)$ where $\{(U_i, \varphi_i)\}$ is an atlas of X modelled on E, and $\hat{\varphi}_i$ the homeomorphism of $\hat{U}_i = \cup_{x \in U_i} T_x(X)$ on $U_i \times E$ defined by

tangent fibre bundle $T(X)$

$$\hat{\varphi}_i(x, v_x) = (\varphi_i(x), V_i),$$

V_i being the representative of v_x in the map (U_i, φ_i).
Moreover, the space $T(X)$ is given a fibre bundle structure by the following elements.
Base: X
Projection: $(x, v_x) \mapsto x$
Typical fibre: E
Structure group: $GL(E)$, isomorphisms of E.

A **vector field** v **on** X is a cross-section of the fibre bundle $T(X)$, i.e. a mapping $v: X \to T(X)$ by $v: x \mapsto v(x)$, such that $\pi \circ v = \text{Identity}$.

vector field

The **cotangent space** at $x \in X$, $T_x^*(X)$, is the topological dual of $T_x(X)$. It is isomorphic to E^*.

cotangent space $T_x^*(X)$

Exercise: Show that if $V_i^* \in E^*$ is the representative of a covariant vector in a map (U_i, φ_i), then its representative in the map (U_j, φ_j) is

$$V_j^* = V_i^* \circ (\varphi_i \circ \varphi_j^{-1})'(\varphi_j(x)).$$

tensor

A p **covariant tensor** at $x \in X$ is a **continuous** r linear form on $T_x(X)$. Its representative t_i in a chart (U_i, φ_i) is an element of $(\otimes E^*)^p$

$$t_i : E \times \cdots \times E \to \mathbb{R}.$$

When the space E is not reflexive, there are several kinds of tensors corresponding to the contravariant tensors defined on finite-dimensional manifolds, according to whether they are defined by tensor product or duality.

If X is a smooth Banach manifold, it results from the theorems of Chapter II, pp. 73 and 74 that if

$$\varphi_i(x) \mapsto V_i^*(\varphi_i(x))$$

is a differentiable function in the domain U_i of a chart of X, then V_j^* is differentiable in $U_i \cap U_j$. It is then possible to give to $\bigcup_{x\in X} (\otimes T_x^*)^p$ the structure of a differentiable Banach manifold and of a fiber bundle.

A **differentiable** p **covariant tensor field** is a differentiable cross section of this fiber bundle.

differential of a mapping

Let X_1 be a C^1 manifold modelled on E_1, and X_2 be a C^1 manifold modelled on E_2, let $f: X_1 \to X_2$, let (U_1, φ_1) be a chart at $x \in X_1$ and (U_2, φ_2) be a chart at $f(x)$; the mapping f is said to be **differentiable** at x if the mapping $\varphi_2 \circ f \circ \varphi_1^{-1}$ defined on a neighborhood of $\varphi_1(x)$ is differentiable at $\varphi_1(x)$. The definition is independent of the choice of charts; $(\varphi_2 \circ f \circ \varphi^{-1})'(\varphi_1(x))$ is called the **representative of the differential** $f'(x)$ in the given charts.

Property: The differential $f'(x)$ is a continuous linear mapping $T_x(X_1) \to T_{f(x)}(X_2)$ such that

$$f'(x) \cdot v_x = w_{f(x)}$$

where the representative W of $w_{f(x)}$ in the chart (U_2, φ_2) is given in terms of the representative V of v_x in the chart (U_1, φ_1) by

$$W = (\varphi_2 \circ f \circ \varphi_1^{-1})'(\varphi_1(x)) \cdot V.$$

Remark: The differential at the point x of a mapping $f: X \to \mathbb{R}$ is a covariant vector at x; its representative in the chart (U, φ) is

$$(f \circ \varphi^{-1})'(\varphi(x)) \in \mathscr{L}(E, \mathbb{R}) = E^*.$$

critical points
critical values

The **critical points** of a mapping $f: X \to \mathbb{R}$ are the points $a \in X$ such that $f'(a) = 0$. The corresponding values $f(a) = c$ are called **critical values.**

If $f: X_1 \to X_2$ is a C^p map, the property of $f'(x)$ expresses that f' defines a C^{p-1} bundle map $TX_1 \to TX_2$, that is the following diagram commutes

$$\begin{array}{ccc} TX_1 & \xrightarrow{f'} & TX_2 \\ \downarrow{\scriptstyle \pi_1} & & \downarrow{\scriptstyle \pi_2} \\ X_1 & \xrightarrow[f]{} & X_2 \end{array}$$

Let $f: X_1 \to X_2$ and $g: X_2 \to X_3$ be C^1 maps then $g \circ f$ is a C^1 map and

$$(g \circ f)' = g' \circ f'.$$

A **diffeomorphism** f of X_1 onto X_2 is a bijective mapping which, along with its inverse, is differentiable at every point. diffeomorphism

Let f be a diffeomorphism, $f'(x)$ is, at each point $x \in X_1$, an isomorphism of $T_x(X_1)$ onto $T_{f(x)}(X_2)$. The inverse function theorem (see Chapter II) provides a demonstration of the following local converse.

Theorem. Let X_1 and X_2 be two Banach manifolds, let f be a C^1 mapping from X_1 into X_2 such that, at the point $x \in X_1$ $f'(x)$ is an isomorphism of $T_x(X_1)$ onto $T_{f(x)}(X_2)$, then there exists an open neighborhood U_1 of x in X_1 and an open neighborhood U_2 of $f(x)$ in X_2 such that the restriction of f to U_1 is a diffeomorphism of U_1 onto U_2.

Proof: Let (U_1, φ_1) and (U_2, φ_2) be local charts of X_1 and X_2, respectively at x and $f(x)$ such that $f(U_1) \subset U_2$. The mapping $\varphi_2 \circ f \circ \varphi_1^{-1}$ satisfies the hypothesis of the inverse function theorem in the neighborhood of the point $\varphi_1(x)$, and therefore is a diffeomorphism of an open ball $\Omega_1 \subset \varphi_1(U_1)$ of center $\varphi_1(x)$ in E_1 onto an open ball $\Omega_2 \subset \varphi_2(U_2)$ of center $(\varphi_2 \circ f)(x)$ in E_2. According to the definitions the restriction of f to $\varphi_1^{-1}(\Omega_1)$ is a diffeomorphism on $\varphi_2^{-1}(\Omega_2)$. ■

Submanifolds. A subset Y of a C^k manifold modelled on a locally convex space E is a **submanifold** if there exists a closed subspace F of E and an atlas $\{(U_i, \varphi_i)\}$ of X such that, when $U_i \cap Y \neq \emptyset$ submanifolds

$$\varphi_i(U_i \cap Y) = \varphi_i(U_i) \cap F.$$

Exercise: Show that the set $(\tilde{U}_i, \tilde{\varphi}_i)$, where $\tilde{U}_i = U_i \cap Y$ and $\tilde{\varphi}_i$ is the restriction of φ_i to $\tilde{U}_i$, is a C^k atlas of Y modelled on F.

Answer: $\{U_i\}$ covers Y; $\tilde{\varphi}_i$ is a bijection of $\tilde{U}_i$ onto $\varphi_i(U_i) \cap F$; let (U, φ) and

(U', φ') be two charts of the atlas of X intersecting Y, $\tilde{\varphi}(\tilde{U} \cap \tilde{U}')$ is open in F because

$$\tilde{\varphi}(\tilde{U} \cap \tilde{U}') = \varphi(U \cap U') \cap \varphi(U \cap Y) = \varphi(U \cap U') \cap F;$$

the fact that $\tilde{\varphi}' \circ \tilde{\varphi}^{-1}$ is C^k on $\tilde{\varphi}(\tilde{U} \cap \tilde{U}')$ is a consequence of the C^k differentiability of $\varphi' \circ \varphi^{-1}$ on $\varphi(U \cap U')$. The C^k manifold structure of Y does not depend on the choice of atlas on X.

Example: Let X be an open set of a locally convex space E and F a closed subspace of E, then $F \cap X$ is a submanifold of X.

Proposition. Let X be a Banach manifold, let f be a differentiable mapping $f\colon X \to \mathbb{R}$, the set $Y = f^{-1}(\{c\})$ is a submanifold of X provided c is not a critical value of f.

Proof: We shall show that for every point a such that $f(a) = c$, there exists a chart (U, φ) of X such that for $x \in U$ $\quad f(x) = c + V^*(\varphi(x))$ where $V^* \in E^*$. Hence

$$\varphi(U \cap f^{-1}(\{c\}) = \varphi(U) \cap (\text{kernel of } V^*)$$

and the proposition follows.
Let (U, φ) be a chart at a such that $\varphi(a) = 0$; let F be the kernel of the linear form V^* on E defined by $(f \circ \varphi^{-1})'(0)$, let $e \in E$ be such that $V^*(e) = 1$; let Λ be the one dimensional vector space defined by $\Lambda = \{\lambda e; \lambda \in \mathbb{R}\}$, then E is the direct sum $E = F \oplus \Lambda$.
The following local diffeomorphism defines a change of charts in a neighborhood of a:

$$\Phi(V_1 + V_2) = V_1 + ((f \circ \varphi^{-1})(V_1 + V_2) - c)e, \qquad V_1 \in F,\ V_2 = \lambda e \in \Lambda.$$

According to the inverse function theorem, Φ is indeed a diffeomorphism between neighborhoods of the origin in E because

$$\Phi'(0) \cdot (h_1 + h_2) = h_1 + [V^*(h_1 + h_2)]e = h_1 + h_2$$

(note that when $h_2 = \lambda e$, $V^*(h_1 + h_2) = \lambda$). Hence $\Phi'(0)$ is the identity on E. Let $\hat{V}$ be in a neighborhood of the origin of E; according to the definition of Φ we have

$$(f \circ \varphi^{-1} \circ \Phi)(\hat{V}) = (f \circ \varphi^{-1})(V) \quad \text{with}$$
$$\hat{V} = V_1 + \hat{V}_2, \qquad V_1 \in F,\ \hat{V}_2 = \hat{\lambda} e \in \Lambda$$
$$V = \Phi^{-1}(\hat{V}) = V_1 + V_2, \qquad V_2 = \lambda e$$

where $\hat{\lambda} = (f \circ \varphi^{-1})(V_1 + V_2) - c$. However $\lambda = V^*(V)$, hence

$$f(x) = (f \circ \varphi^{-1} \circ \Phi^{-1})(\hat{V}) = c + V^*(\hat{V}).$$

The proposition is proved in the chart $\hat{\varphi} = \Phi \circ \varphi$. Note that it is a particular case of the theorem on submersions (p. 550). ■

Example: The sphere $\|x\| = \rho$ is a submanifold regularly embedded in the Hilbert space of norm $\| \quad \|$.

A submanifold is said to be **complementable** if the subspace F_1 which models it **splits**, i.e. if it can be given a topological complement[1] F_2 in E. According to the closed graph theorem, if E is a Fréchet space, then E is diffeomorphic to the direct product $F_1 \times F_2$. complementable splits

Exercise: Let X be a manifold modelled on $E = F_1 \times F_2$. Show that a subset Y of X is a complementable submanifold modelled on F_1, if and only if there exists an atlas whose charts (U, φ) meeting Y are such that

$$\varphi(U \cap Y) = U_1 \times \{0\}$$

where U_1 is an open set of F_1 and $\{0\}$ is the origin in E_2.
Solution: $\varphi(U \cap Y) \subset \varphi(U)$ is of the form $\varphi(U) \cap (F_1 \times \{0\})$.

Let X and Y be two differentiable manifolds modelled on E and F respectively; a differentiable mapping f: $Y \to X$ is said to be an **immersion** if, for every point $x \in Y$, $f'(x)$ is an isomorphism of $T_x(Y)$ onto a subspace of $T_{f(x)}(X)$ isomorphic to a fixed subspace F_1 of E, complementable in E. immersion

Theorem. Let X and Y be two Banach manifolds, let f: $Y \to X$ be an immersion, then every point $x \in Y$ has a neighborhood V, such that $f(V)$ is a submanifold of X diffeomorphic to V.

Proof: We shall prove the theorem first for the particular case where Y is an open set of a Banach space F, X is a Banach space E; let $x_0 \in Y$. $f'(x_0)$: $F \to F_1$ is an isomorphism of F onto a subspace F_1 of E; let $Y_1 = f'(x_0) Y \subset F_1$; to simplify the notations we shall take $x_0 = f(x_0) = 0$. Sct

$$g = f \circ f^{-1\prime}(0): Y_1 \to E = F_1 \times F_2.$$

[1]Note that, if E is a Hilbert space, a closed subspace F_1 always has a topological complement (the supplementary subspace $F_2 = F_1^{\perp}$).

We shall show that there exists a neighborhood U_1 of $\{0\}$ such that g is a diffeomorphism on U_1 and that $g(U_1)$ is a submanifold of E; i.e., that there exists a local diffeomorphism Φ on $F_1 \times F_2$ such that

$$\Phi(g(U_1)) = U_1 \times \{0\}.$$

Let us define a differentiable mapping Ψ on $Y_1 \times F_2$ by

$$(x_1, x_2) \mapsto (y_1, y_2) \equiv \Psi(x_1, x_2) \stackrel{\text{def}}{=} g(x_1) + x_2.$$

The differential of Ψ is the linear mapping

$$(h_1, h_2) \mapsto g'(x_1)h_1 + h_2.$$

At the origin, $\Psi'(0, 0)$ is the identity, hence an isomorphism and Ψ is a local diffeomorphism.
On the other hand, for $y = (y_1, y_2) \in g(U_1)$

$$(y_1, y_2) = g(x_1) \quad \text{for some } x_1 \in U_1,$$

hence $\Psi^{-1}(g(x_1)) = (x_1, 0)$ and in conclusion the choice $\Phi = \Psi^{-1}$. ■

The theorem is proved in the general case by using local charts.

embedding

An immersion is not always globally injective (cf. p. 241). An injective immersion is called an **embedding**.
An embedding $Y \to X$ is a homeomorphism only when the topology of the image of Y is the topology (cf. Ch. I) induced by X. An embedding which is a homeomorphism is called a **regular embedding**.

regular embedding

submersion

A C^1 mapping $f: X \to Y$ is a **submersion** on $S \subset X$ if for all $x \in S$, $f'(x)$: $T_xX \to T_{f(x)}Y$ is surjective and has a complementable kernel.

Theorem. Let X and Y be Banach manifolds, $f: X \to Y$ be a submersion on $S = f^{-1}(c)$, $c \in Y$. Then S is a (complementable) submanifold of X.

Proof: Denote by E, F the modelling spaces of X, Y, by (U, φ) and (V, ψ) charts around the points $x \in X$ and $f(x) \in Y$. Set $h = \varphi(x)$, $\bar{f} = \psi \circ f \circ \varphi^{-1}$. Denote by E_1 the kernel of the linear mapping $\bar{f}'(h)$: $E \to F$, by E_2 the complement of E_1 in E. We know that $\bar{f}'(h)$ defines an isomorphism $E_2 \to F$. The implicit function theorem shows that the mapping

$$h = (h_1, h_2) \mapsto (h_1, \bar{f}(h) - \bar{c}), \quad \bar{c} = \psi(c)$$

defines a diffeomorphism between open sets of E, hence a new chart on X where it is clear[1] that S is a submanifold, modelled on E_1.

[1]See a similar proof (p. 548).

Flow of a vector field. Let $\boldsymbol{v}$ be a vector field on the manifold X. The differential system on X defined by $\boldsymbol{v}$ is, as in the finite dimensional case (p. 144)

$$\mathrm{d}\sigma(t)/\mathrm{d}t = \boldsymbol{v}(\sigma(t)).$$

If X is a Banach manifold and $\boldsymbol{v}$ a C^1 vector field, local existence and uniqueness theorems (cf. p. 95) are valid, and the vector field $\boldsymbol{v}$ has a **local flow**, i.e. a mapping σ from an open subset $\Sigma_v \subset \mathbb{R} \times X$ into X by $(t, x) \mapsto \sigma(t, x)$, such that the curve $t \mapsto \sigma(t, x)$ is solution of the differential system and $\sigma(0, x) = x$. As in the finite dimensional case local flow

$$\sigma(t+s, x) = \sigma(t, \sigma(s, x)), \quad \text{if } (t, x), (s, x), (t+s, x) \in \Sigma_v.$$

If $\Sigma_v = I \times X$, I interval of $\mathbb{R}$, then σ may be defined for all t through the above relation, and is called the **flow** of t. For each $t \in \mathbb{R}$, $\sigma_t \colon x \mapsto \sigma(t, x)$ is a diffeomorphism of X and they have the group property flow

$$\sigma_t \circ \sigma_s = \sigma_{t+s}.$$

Differential forms (cf. analogies with the finite dimensional case, Chapter IV, Sec. A). A **differential form $\boldsymbol{\alpha}$ of degree** p (a p-**form**) on a smooth Banach manifold X is a cross section of the vector bundle $\Lambda^p(X)$ of antisymmetric covariant p-tensors on X. p-form

A differentiable function $f \colon X \to R$ is considered as a zero form, its differential f' is a 1-form.

Let $\boldsymbol{\alpha}$ be a p-form and $\boldsymbol{\beta}$ a q-form on X. The **wedge product** is defined as wedge product

$$(\boldsymbol{\alpha} \wedge \boldsymbol{\beta})_x(\boldsymbol{v}_1, \ldots, \boldsymbol{v}_{p+q}) = \frac{1}{p!q!} \sum_{\pi} (\operatorname{sign} \pi) \boldsymbol{\alpha}_x(\boldsymbol{v}_{\pi(1)}, \ldots, \boldsymbol{v}_{\pi(p)}) \times \boldsymbol{\beta}_x(\boldsymbol{v}_{\pi(p+1)}, \ldots, \boldsymbol{v}_{\pi(p+q)})$$

where the sum is over all permutations of $1, \ldots, p+q$.

Let $f \colon X \to Y$ be a C^1 mapping and $\boldsymbol{\alpha}$ be a p-form on Y, the **reciprocal image** of $\boldsymbol{\alpha}$ by f **(pull back)** is reciprocal image pull back

$$(f^*\boldsymbol{\alpha})_x(\boldsymbol{v}_1, \ldots, \boldsymbol{v}_p) = \boldsymbol{\alpha}_{f(x)}(f'(x)\boldsymbol{v}_1, \ldots, f'(x)\boldsymbol{v}_p).$$

Let $\boldsymbol{\alpha}$ be a smooth p-form on X, its **exterior differential** is the $p+1$ form defined by exterior differential

$$(\mathrm{d}\boldsymbol{\alpha})_x(\boldsymbol{v}_0, \ldots, \boldsymbol{v}_p) = \sum_{i=0}^{p} (-1)^i (\bar{\boldsymbol{\alpha}}'(x) \cdot \bar{\boldsymbol{v}}_i)(\bar{\boldsymbol{v}}_0, \ldots, \hat{\bar{\boldsymbol{v}}}_i, \ldots, \bar{\boldsymbol{v}}_p).$$

As usual $\bar{\boldsymbol{\alpha}}$ denotes the expression of $\boldsymbol{\alpha}$ in a chart (U, φ) around x, mapping from $\varphi(U)$ into $\Lambda^p(E)$, and $\bar{\boldsymbol{\alpha}}'(x) \colon E \to \Lambda^p(E)$ the derivative of this mapping at x.

One verifies that the definition is chart independent and that
1) d is linear
2) $d^2 \equiv d \circ d = 0$
3) $d(\boldsymbol{\alpha} \wedge \boldsymbol{\beta}) = d\boldsymbol{\alpha} \wedge \boldsymbol{\beta} + (-1)^p \boldsymbol{\alpha} \wedge d\boldsymbol{\beta}$, $\quad p =$ degree of $\boldsymbol{\alpha}$.

interior product

The **interior product** of a p form $\boldsymbol{\alpha}$ and a vector field $\boldsymbol{\xi}$ is the $p-1$ form defined by

$$(i_\xi \boldsymbol{\alpha})_x(v_2, \ldots, v_p) = \boldsymbol{\alpha}_x(\boldsymbol{\xi}(x), v_2, \ldots, v_p).$$

Lie derivative

The **Lie derivative** $\mathcal{L}_\xi \boldsymbol{\alpha}$ is

$$\mathcal{L}_\xi \boldsymbol{\alpha} = d i_\xi \boldsymbol{\alpha} + i_\xi \, d\boldsymbol{\alpha}.$$

The identities proved in Ch. IV §4 are still valid here.

Poincaré lemma. If $d\boldsymbol{\alpha} = 0$ *(i.e.* $\boldsymbol{\alpha}$ *is closed) there exists a neighborhood U about each point on which* $\boldsymbol{\alpha} = d\boldsymbol{\beta}$ *(i.e.* $\boldsymbol{\alpha}$ *is an exact differential).*

2. SYMPLECTIC STRUCTURES AND HAMILTONIAN SYSTEMS

Let X be a smooth manifold modelled on a Banach space E.

symplectic form

A **symplectic form** $\boldsymbol{\omega}$ on X is a 2-form such that:
1) $\boldsymbol{\omega}$ is closed: $d\boldsymbol{\omega} = 0$.
2) For each $x \in X$, $\boldsymbol{\omega}_x\colon T_xX \times T_xX \to \mathbb{R}$ is a non degenerate bilinear form (the mapping $T_xX \to T_x^*X$ by $v \mapsto \boldsymbol{\omega}_x(v, \cdot)$ is an isomorphism).

weak symplectic form

If $\boldsymbol{\omega}_x$ in 2) is weakly non degenerate (the mapping $v \mapsto \boldsymbol{\omega}_x(v, \cdot)$ is injective but not onto) $\boldsymbol{\omega}$ is a **weak symplectic form.**

In the finite dimensional case we have seen (Problem IV 6) that a skew symmetric non degenerate bilinear form exists only on vector spaces of even dimension $m = 2n$ and has the canonical form $\begin{pmatrix} 0 & I \\ -I & 0 \end{pmatrix}$ where I is the $n \times n$ identity matrix. Moreover, if X is a $2n$ dimensional manifold there

canonical coordinates

exist coordinates $(x^1, \ldots, x^n, y^1, \ldots, y^n)$ called **canonical**, about each point in which the symplectic form $\boldsymbol{\omega}$ has the expression $\boldsymbol{\omega} = \sum_{i=1}^n \mathbf{d}x^i \wedge \mathbf{d}y^i$.

These properties generalize as follows to the infinite dimensional case.

complex structure

A **complex structure** on a real vector space E is a mapping $J\colon E \to E$ such that $J^2 = -\mathbb{1}$.

Example: The matrix $\begin{pmatrix} 0 & I \\ -I & 0 \end{pmatrix}$ defines a complex structure on $\mathbb{R}^{2n}$.

By setting $iV = JV$ one gives E the structure of a complex vector space

Theorem. Let $\mathcal{H}$ be a real Hilbert space and ω a skew symmetric weakly non degenerate bilinear form on $\mathcal{H}$. Then there exists a complex structure J on $\mathcal{H}$ and a real scalar product s such that

$$\omega(v, w) = s(Jv, w).$$

Proof: Let $(\ |\ )$ be the given real (p. 30) scalar product on $\mathcal{H}$. By the Riesz theorem there exists a continuous linear operator $A: \mathcal{H} \to \mathcal{H}$ such that

$$\omega(v, w) = (Av|w).$$

Since ω is skew, $A^* = -A$, $-A^2 \geq 0$, it can be shown that the operator $-A^2$ has a symmetric non negative square root P, which is self adjoint injective, has a dense range[1] and an inverse P^{-1}. The operator $J = AP^{-1}$ defines a complex structure

$$A = JP, \qquad -A^2 = P^2 \text{ implies } J^2 = -\mathbb{1}$$

and is orthogonal

$$J^* = P^{-1*}A^* = -J = J^{-1}$$

$$\omega(v, w) = (JPv|w).$$

The scalar product s on $\mathcal{H}$ is defined by

$$s(v, w) = (Pv|w).$$

The above theorem is used in recent developments in quantum field theory to associate with a symplectic structure on a real Hilbert space a hermitian (complex prehilbert) structure defined by the sesquilinear form

$$h(v, w) = s(v, w) + \mathrm{i}\omega(v, w).$$

The Darboux theorem can be generalized as follows to the infinite dimensional case[2].

Theorem. If ω is a symplectic form on a Banach manifold there is a chart (U, φ) about each point in which ω is constant (that is $\varphi(U) \subset E \to \Lambda^2(E)$ by $\varphi(x) \mapsto \tilde{\omega}_{\varphi(x)}$ is a constant map).

This theorem is not true if ω is only weakly symplectic[3].

[1]Cf. details of the proof and further properties for instance in [Marsden 1974].
[2]See the proof in [Weinstein 1971] or [Marsden 1974]. It is a straightforward generalisation of the proof given in Problem IV 6.
[3]Cf. [Marsden 1972].

canonical 1-form on T^*X

Let X be a manifold modelled on the Banach space E, T^*X its cotangent bundle and π: $T^*X \to X$ the canonical projection. The **canonical 1-form $\boldsymbol{\theta}$** on T^*X is defined by[1]

$$\boldsymbol{\theta}_{(x,p)}(\boldsymbol{w}) = \boldsymbol{p}(\pi' \boldsymbol{w}), \qquad x \in X, \qquad \boldsymbol{p} \in T_x^*X, \qquad \boldsymbol{w} \in T_{(x,p)}(T^*X).$$

In a chart where $(\bar{x}, \bar{p}) \in E \times E^*$, $\bar{\boldsymbol{w}} = (\boldsymbol{V}, \boldsymbol{P}) \in E \times E^*$ the formula reads

$$\boldsymbol{\theta}_{(\bar{x},\bar{p})}(\boldsymbol{V}, \boldsymbol{P}) = \bar{\boldsymbol{p}}(\boldsymbol{V}).$$

If E is finite dimensional $\boldsymbol{\theta} = \Sigma\, p_i\, \mathbf{d}x^i$, cf. p. 268.

Consider the closed exact 2-form on T^*X, $\boldsymbol{\omega} = \mathbf{d}\boldsymbol{\theta}$; in a chart

$$\boldsymbol{\omega}_{(\bar{x},\bar{p})}((\boldsymbol{V}_1, \boldsymbol{P}_1), (\boldsymbol{V}_2, \boldsymbol{P}_2)) = \boldsymbol{P}_1(\boldsymbol{V}_2) - \boldsymbol{P}_2(\boldsymbol{V}_1).$$

If $E = \mathbb{R}^n$, $\boldsymbol{\omega} = \Sigma\, \mathbf{d}p_i \wedge \mathbf{d}x^i$.

canonical symplectic form on T^*X

Theorem. 1) *The form $\boldsymbol{\omega}$ is a weak symplectic form on T^*X (called* **canonical**). 2) *It is a symplectic form on T^*X if E is reflexive.*

Proof: Straightforward using the expression in a local chart[2].

symplectic transformation

A C^1 map f: $X \to X$ on a manifold X with a weak symplectic form $\boldsymbol{\omega}$ [weak symplectic manifold] is called **symplectic** when $f^*\boldsymbol{\omega} = \boldsymbol{\omega}$.

lift

Example: Let f: $X \to X$ be a diffeomorphism. Define the lift $T^*f = (f^{-1}, f^*)$

$$\text{by } (T^*f)(x, p_x) = (f^{-1}(x), (f^*p)_{f^{-1}(x)}).$$

Let $\boldsymbol{\theta}$ be the canonical 1-form on T^*X, its pull back by T^*f is such that

$$((T^*f)^*\boldsymbol{\theta})(\boldsymbol{w}) = \boldsymbol{\theta}((T^*f)'\, \boldsymbol{w}), \qquad \boldsymbol{w} \in T(T^*X).$$

Using the definitions one finds, in a few lines

$$((T^*f)^*\boldsymbol{\theta})(\boldsymbol{w}) = \boldsymbol{\theta}(\boldsymbol{w}), \qquad \boldsymbol{w} \in T(T^*X).$$

Hence the following theorem.

Theorem. The canonical 1-form, and the canonical weak symplectic 2-form on T^*X, are invariant under the lifts of diffeomorphisms of X.

locally hamiltonian

Hamiltonian vector fields. A vector field v on a symplectic manifold $(X, \boldsymbol{\omega})$ is **locally hamiltonian** if its flow leaves $\boldsymbol{\omega}$ invariant, that is if

$$\mathscr{L}_v\boldsymbol{\omega} \equiv \mathrm{d}i_v\boldsymbol{\omega} = 0.$$

[1]Varying sign conventions are in use for the canonical forms.
[2]See for instance [Marsden 1974].

When $i_v\omega$ is not only closed, but an exact differential

$$i_v\boldsymbol{\omega} = -\mathbf{d}H$$

$\boldsymbol{v}$ is said to be **globally hamiltonian**; and H, determined up to an additive constant, is called the **hamiltonian.** globally hamiltonian hamiltonian

Conversely if H is a given C^1 function, and $\boldsymbol{\omega}$ is a symplectic form on X, there exists a vector field $\boldsymbol{v}$ determined by $i_v\boldsymbol{\omega} = -\mathbf{d}H$, since the mapping $T_xX \to T_x^*X$, $\boldsymbol{v}_x \mapsto \boldsymbol{\omega}_x(\boldsymbol{v}_x, \cdot) \equiv i_{v_x}\boldsymbol{\omega}_x$, is an isomorphism.

Note that if $\boldsymbol{\omega}$ is only weakly symplectic, given H, $\boldsymbol{v}$ need not exist.

Conservation of energy theorem. The hamiltonian H of a globally hamiltonian vector field $\boldsymbol{v}$ *on the symplectic manifold* $(X, \boldsymbol{\omega})$ *is constant along the trajectories of* $\boldsymbol{v}$.

Proof: Let $f_t: X \to X$ denote the flow of $\boldsymbol{v}$

$$\frac{\mathrm{d}f_t}{\mathrm{d}t} = \boldsymbol{v}, \qquad f_0(x) = x.$$

We know, by the definition that $f_t^*\boldsymbol{\omega} = \boldsymbol{\omega}$, and

$$\frac{\mathrm{d}}{\mathrm{d}t}(H \circ f_t) = (\mathbf{d}H)(\boldsymbol{v}) = -\boldsymbol{\omega}(\boldsymbol{v}, \boldsymbol{v}) = 0. \qquad \blacksquare$$

Example: A hamiltonian differential system on an infinite dimensional manifold: Problem A, the Klein–Gordon equation.

Riemannian manifolds. Let X be a smooth Banach manifold. A **metric** on X is a section $x \mapsto \boldsymbol{g}_x$ of the bundle of symmetric 2-covariant tensors on X, such that, at each point x, the quadratic form corresponding to $\boldsymbol{g}_x$ is positive and non degenerate (positive definite cf. Chapter II) on $T_x(X)$. Note that (Chapter II) $\boldsymbol{g}_x$ then defines a scalar product, and a norm equivalent to the original norm of E, the modelling space of X. Thus E is isomorphic to a Hilbert space. metric

If the mapping $T_x(X) \to T_s^*(X)$ defined by

$$\boldsymbol{v} \mapsto \boldsymbol{g}_x(\boldsymbol{v}, \cdot)$$

is only injective we say that $\boldsymbol{g}$ is a **weak metric**. By the Banach theorem $\boldsymbol{g}_x$ is non degenerate if it is injective and onto. Weak riemannian metrics are of frequent use in mathematical physics. weak metric

A **riemannian manifold** is a pair $(X, \boldsymbol{g})$ where X is a Hilbert manifold, and $\boldsymbol{g}$ a metric on X. riemannian manifold

Examples: 1. The L_2 inner product on $C^0[0, 1]$ $g(x, y) = \int_0^1 x(t)y(t)\,dt$ is a weak metric, but not a metric.
2. The restriction to the sphere $\|x\| = 1$ of the norm $\|\ \|$ in a Hilbert space turns the sphere into a riemannian manifold.

pseudo (indefinite) metric

If the quadratic form g_x is non degenerate but not positive, g is called a **pseudo** (or **indefinite**) **metric**.

induced metric

Let S be an immersed smooth submanifold of a riemannian manifold X, with metric g. There exists on S an **induced metric** $\bar{g}$ defined by

$$\bar{g}_x(v, v) = g_{i(x)}(i'\,v, i'\,v), \qquad i: S \to X.$$

Let X be a riemannian manifold, modelled on the Hilbert space E, with metric g. The tangent bundle $T(X)$ can be canonically identified with the cotangent bundle $T^*(X)$, $v_x \in T_x(X)$ is identified with the linear form on $T_x(X)$:

$$v_x^*: w_x \mapsto g_x(v_x, w_x),\ v_x^* \in T_x^*.$$

Geodesics and riemannian connection can be defined as in the finite dimensional case, through variational calculus; they are the projection on X of the trajectories $s \mapsto (x(s), v(s))$ of a vector field on TX:

spray

$$(x, v) \mapsto (v, Q_x(v))$$

where $Q_x(v)$ is quadratic in v. Such a vector field is called a **spray**[1].

B. THEORY OF DEGREE; LERAY–SCHAUDER THEORY

The degree, at $y \in Y^n$, of a mapping f from an open bounded subset[2] D of a finite dimensional oriented manifold X^n into a finite dimensional oriented manifold Y^n is, roughly speaking, the number of times y is covered by the image $f(D)$, each covering being counted positively or negatively, according to the orientation of the mapping f in a neighborhood of the corresponding point $x \in f^{-1}(y)$.
The degree is a topological property whose precise definition will justify the formula (p. 228).

$$\int_D f^*\omega = \deg f \int_{f(D)} \omega.$$

[1]Cf. [Lang 1962, p. 110], [Marsden 1974, p. 42].
[2]i.e. with compact closure.

Numerous "fixed point" theorems arise from the fact that, by definition, the degree of f at y is zero if $y \notin f(D)$: the non-vanishing of the degree at y implies therefore $y \in f(D)$; solutions of the equation $y = f(x), x \in D$ will then exist.
The extension, by J. Leray and J. Schauder, of degree theory to compact perturbations of the identity mapping on a bounded open subset of a Banach space has many applications in the solution of functional equations, and in particular of non-linear ordinary differential equations and elliptic partial differential equations.

1. DEFINITION FOR FINITE DIMENSIONAL MANIFOLDS

Consider a mapping $f: \bar{D} \subset X^n \to Y^n$ where X^n, Y^n are oriented differentiable manifolds and D is a relatively compact subset of X^n. The definition is first given for a fairly restricted situation and is gradually refined until it applies for any continuous function f on $\bar{D}$ and any point in Y^n which is not in $f(\partial D)$.

First suppose that $f \in C^0(\bar{D}) \cap C^1(D)$. Let N denote the set of points in D at which the Jacobian J_f of f (p. 72) vanishes (the critical points of f on D). Let $y \in f(\bar{D})$ be such that $y \notin f(\partial D)$ and $f^{-1}(y) \cap N = \emptyset$. By the inverse function theorem (p. 90) the set $f^{-1}(y) \subset \bar{D}$ is discrete and therefore finite since $\bar{D}$ is compact. Moreover because the manifolds are oriented the sign of J_f is defined at each point of $f^{-1}(y)$.
The **degree** deg (f, D, y) of f at y with respect to D is thus unambiguously defined by degree

$$\deg(f, D, y) = \sum_{x \in f^{-1}(y)} \operatorname{sign} J_f(x).$$

For $y \notin f(\bar{D})$ we define

$$\deg(f, D, y) = 0.$$

Note: It follows from the definition that if $\deg(f, D, y) \neq 0$ then there exists an $x \in D$ such that $f(x) = y$.

Example 1: The identity mapping has degree 1.

Example 2: The mapping $(-a, +\infty) \subset \mathbb{R} \to \mathbb{R}$ by $x \mapsto x^2$ has degree

$$\begin{array}{ll} 0 & \text{if } y \in (-\infty, 0) \text{ or } y \in (0, a^2) \\ 1 & \text{if } y \in (a^2, +\infty). \end{array}$$

Example 3: The 2–1 mapping $S^1 \to S^1$

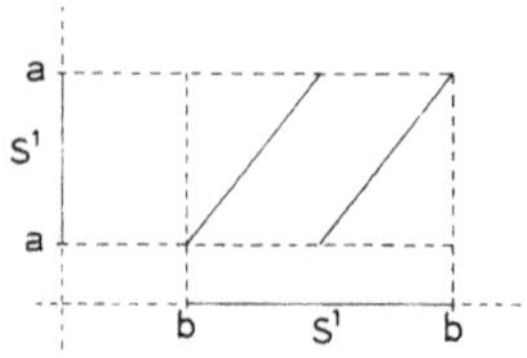

has degree 2.

Example 4: The mappings f_1, f_2: $U \subset \mathbb{R} \to \mathbb{R}$ by

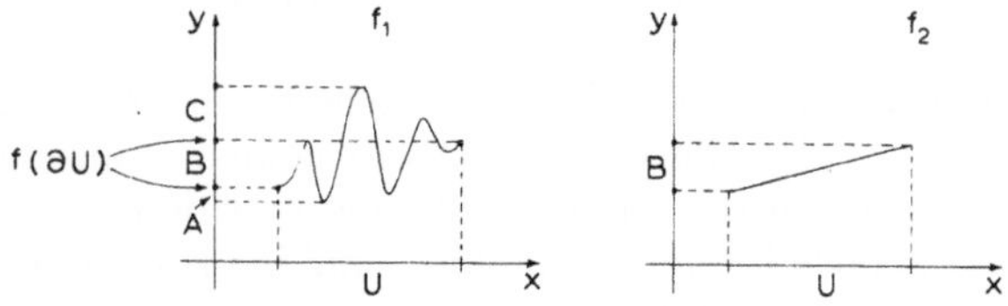

have degree

$$\begin{array}{ll} 1 & \text{for } y \in B \\ 0 & \text{for } y \in A \text{ or } y \in C. \end{array}$$

Note that f_1 and f_2 are homotopic (one can be continuously deformed into the other).

Integral formula for the degree of f. The integral formula will simplify the proofs of the properties of the degree and allow us to remove the restriction that $f^{-1}(y) \cap N = \emptyset$.

Let $\rho_{y,\epsilon}$: $Y^n \to \mathbb{R}$ be a regularizing family of functions (cf. p. 432) defined on a neighborhood of y. More precisely

$$\int_{Y^n} \rho_{y,\epsilon}\eta = 1,$$

where η is a smooth n-form on Y^n, and

$$\operatorname{supp} \rho_{y,\epsilon} \subset \bar{B}_\epsilon(y)$$

where $\bar{B}_\epsilon(y)$ is the closed ball with center y and radius ϵ defined with respect to local coordinates on a neighborhood of y.

We will show that for ϵ sufficiently small the above definition can be replaced by

$$\deg(f, D, y) = \int_{\bar{D}} (\rho_{y,\epsilon} \circ f) f^* \eta.$$

Let $f^{-1}(y) = \{x_i\}$. By the inverse function theorem, for ϵ sufficiently small each x_i has a neighborhood U_i such that $f|_{U_i}$ is a diffeomorphism onto $B_\epsilon(y)$. It follows that $\operatorname{supp}(\rho_{y,\epsilon} \circ f) \subset \cup_i U_i$ and therefore that

$$\int_{\bar{D}} (\rho_{y,\epsilon} \circ f) f^*\eta = \int_{\cup_i U_i} (\rho_{y,\epsilon} \circ f) f^*\eta$$

$$= \int_{\cup_i U_i} f^*(\rho_{y,\epsilon}\eta)$$

$$= \sum_i \operatorname{sign} J_f(x_i) \int_{B_\epsilon(y)} \rho_{y,\epsilon}\eta$$

since the orientation of the diffeomorphism $f\colon U_i \to B_\epsilon$ is determined by the sign of $J_f(x_i)$. ■

The following theorem will show that the degree is constant in every connected component of $Y^n \backslash f(\partial D)$: this constant is by definition $\deg(f, D, y)$ when $f^{-1}(y) \cap N \neq \emptyset$, $y \notin f(\partial D)$.
Sard has proved that *the set of critical values of a differentiable mapping from X^n into Y^n has measure zero.* Therefore a critical value is always the limit of non critical ones, and the definition given above coincides with the following:

$$\deg(f, D, y) = \lim \deg(f, D, z) \quad \text{when } z \text{ tends to } y,$$

where $f^{-1}(z) \cap N = \emptyset$ and $y \in f(\partial D)$.

Theorem. Let y_1 and y_2 be two points belonging to the same connected component of the open set $Y^n \backslash f(\partial D)$.
Then

$$\deg(f, D, y_1) = \deg(f, D, y_2).$$

Proof: For ϵ sufficiently small

$$\deg(f, D, y_1) - \deg(f, D, y_2) = \int_{\bar{D}} f^*[(\rho_{y_1,\epsilon} - \rho_{y_2,\epsilon})\eta].$$

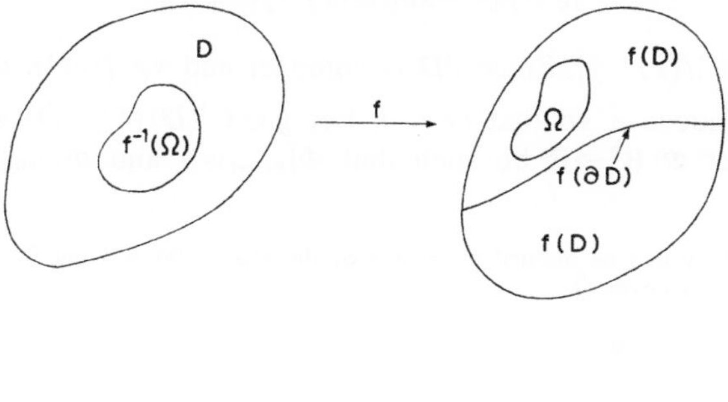

We will show that the integrand is the exterior derivative of a form with support contained in D. Then by Stoke's theorem the integral is zero. Let Ω be an open connected relatively compact subset of Y^n such that $y_1, y_2 \in \Omega$, $\bar{\Omega} \subset f(D)$ and $\bar{\Omega} \cap f(\partial D) = \emptyset$. Let $\boldsymbol{\Phi} = (\rho_{y_1,\epsilon} - \rho_{y_2,\epsilon})\boldsymbol{\eta}$ and choose ϵ so that supp $\boldsymbol{\Phi} \supset \bar{\Omega}$. Now $\boldsymbol{\Phi}$ has compact support and $\int_{Y^n} \boldsymbol{\Phi} = 0$. Therefore by de Rham's theorem (cf. Exercise p. 228) there exists an $(n-1)$-form $\boldsymbol{\omega}$ with compact support such that

$$\boldsymbol{\Phi} = \mathrm{d}\boldsymbol{\omega} \text{ and supp } \boldsymbol{\omega} \subset \bar{\Omega}.$$

Thus

$$f^*\boldsymbol{\Phi} = f^*\,\mathrm{d}\boldsymbol{\omega} = \mathrm{d}(f^*\boldsymbol{\omega})$$

and supp $f^*\boldsymbol{\omega} \subset f^{-1}(\bar{\Omega})$. But $f^{-1}(\bar{\Omega}) \subset D$ because $\bar{\Omega} \cap f(\partial D) \neq \emptyset$. ■

Suppose $y \in Y^n$ and $y \notin f(\partial D)$. Then any $z \in Y^n$ sufficiently close to y belongs to the same connected component of $Y^n \backslash f(\partial D)$.
The preceding theorem can be reformulated as follows

The degree is continuous on every connected component of $Y^n \backslash f(\partial D)$, as a mapping from Y^n into the integers, with the natural discrete topology.

Continuous mappings. We will now show that the degree can be defined for any $f \in C^0(\bar{D})$. In other words degree is a topological property. The preceding definitions are only methods of computing it in the differentiable case.

Theorem. The mapping $C^1(D) \cap C^0(\bar{D}) \to Z$ by $f \mapsto \deg(f, D, y)$, $y \notin f(\partial D)$ is continuous with respect to the C^0 topology (uniform topology (p. 423)).[1]

Proof: Since Z has the discrete topology we want to show that there exists a neighborhood N of f in the C^0 topology such that for $y \notin f(\partial D)$

$$\deg(g, D, y) = \deg(f, D, y), \qquad \forall g \in N(f).$$

Let us suppose first that $X^n = Y^n = \mathbb{R}^n$, the neighborhood N is defined by

$$\|g - f\|_{C^0} = \sup_{x \in D} |g(x) - f(x)| < \epsilon.$$

Let $d = \inf_{x \in \partial D} |f(x) - y|$. Since ∂D is compact and $y \notin f(\partial D)$, d is strictly positive. Choose ϵ so that $6\epsilon < d$. Let $g \in C^1(D) \cap C^0(\bar{D})$ satisfy $\|g - f\|_{C^0} < \epsilon$. Let $\Phi: \mathbb{R}^n \to \mathbb{R}$ be such that $\Phi|_{B_{2\epsilon}(y)} = 1$ and $\Phi_{Y^n \backslash B_{3\epsilon}(y)} = 0$, and smooth.

[1] Such a topology can be defined by means of the atlases on X^n and Y^n since a finite number of charts cover $\bar{D}$.

Consider the mapping $h: \bar{D} \to Y^n$ by

$$h(x) = [1 - (\Phi \circ f)(x)]f(x) + [(\Phi \circ f)(x)]g(x).$$

Clearly $h \in C^1(D) \cap C^0(\bar{D})$ and

$$h(x) = f(x) \quad \text{if } |f(x) - y| > 3\epsilon$$
$$h(x) = g(x) \quad \text{if } |f(x) - y| < 2\epsilon.$$

Choose $\rho_{y,5\epsilon}$ so that $\rho_{y,5\epsilon}(B_{4\epsilon}) = 0$. Then

$$\int_{\bar{D}} f^*[(\rho_{y,5\epsilon} \circ f)\boldsymbol{\eta}] = \int_{\bar{D}} h^*[(\rho_{y,5\epsilon} \circ h)\boldsymbol{\eta}].$$

Moreover

$$\int_{\bar{D}} h^*[(\rho_{y,\epsilon} \circ h)\boldsymbol{\eta}] = \int_{\bar{D}} g^*[(\rho_{y,\epsilon} \circ g)\boldsymbol{\eta}].$$

Therefore $\deg(f, D, y) = \deg(g, D, y)$. ∎

When X^n and Y^n are not $\mathbb{R}^n$ the demonstration can be made similarly, through the use of charts.

The theorem justifies the following definition of degree for a continuous mapping

$$\deg(f, D, y) = \lim \deg(f_n, D, y), \; y \notin f(\partial D)$$

where (f_n) is a sequence of C^1 mappings which converges to f in the C^0 topology.

2. PROPERTIES AND APPLICATIONS

Fundamental theorem. The mapping $(f, D, y) \to \mathbb{Z}$ *by* $(f, D, y) \mapsto \deg(f, D, y)$ *where* $f: \bar{D} \subset X^n \to Y^n$ *is continuous,* $\bar{D} \subset X^n$ *is compact and* $y \notin f(\partial D)$*, has the following properties.*

1) *It is continuous from* $C^0(\bar{D})$ *into* $\mathbb{Z}$ *if* $y \notin f(\partial D)$.

2) *It is constant on a connected component of* $Y^n \backslash f(\partial D)$. *It is zero if* $y \notin f(\bar{D})$.

3) *Homotopic invariance. If* $f_t \in C^0(\bar{D})$, $t \in [0, 1]$, *is a family of mappings which depends continuously on* t, *then*

$$\deg(f_1, D, y) = \deg(f_0, D, y),$$

when $y \notin f_t(\partial D)$ $\forall t \in [0, 1]$.

If $Y^n = \mathbb{R}^n$ *the degree depends only on boundary values. If* $f|_{\partial D} = g|_{\partial D}$ *and* $y \notin f(\partial D)$ *then* $\deg(f, D, y) = \deg(g, D, y)$.

4) *Decomposition of the domain. If D is the union of a family $\{D_i\}$ of open disjoint sets such that $\partial D_i \subset \partial D$, then for $y \notin f(\partial D)$*

$$\deg(f, D, y) = \sum_i \deg(f, D_i, y).$$

5) *Excision property. If $K \subset \bar{D}$ is closed and $y \notin f(K) \cup f(\partial D)$ then*

$$\deg(f, D, y) = \deg(f, D \backslash K, y).$$

6) *Cartesian product. If $D \subset X^n$, $D' \subset X'^n$ and $f: X^n \to Y^n$, $g: X'^n \to Y'^n$ then*

$$\deg(f \times g, D \times D', (y, z)) = \deg(f, D, y) \deg(g, D', z)$$

whenever each term makes sense.

Proof: 1) and 2) have already been proved.
3) The mapping $[0, 1] \to \mathbb{Z}$ by $t \mapsto f_t \mapsto \deg(f_t, D, y)$ is continuous since it is the composite of two continuous mappings. Since $\mathbb{Z}$ has the discrete topology and $[0, 1]$ is connected $\deg(f_t, D, y)$ is constant, if it is defined for all $t \in [0, 1]$, that is if $y \notin f_t(\partial D)$.
If $Y^n = \mathbb{R}^n$, let $\Phi_t = tf + (1-t)g$. Since f and g agree on ∂D, $y \notin f(\partial D)$ implies that $y \notin \Phi_t(\partial D)$ for any t.
4), 5) and 6) are easily demonstrated directly from the formulae for the degree. ∎

Borsuk's theorem.[1] *If D is a symmetric bounded open subset of $\mathbb{R}^n$ which contains the origin and if the mapping $f: D \to \mathbb{R}^n$ is odd*

$$f(x) = -f(-x) \quad \forall x \in D \quad \textit{then } \deg(f, D, 0) \neq 0 \textit{ if } 0 \notin f(\partial D).$$

Note: If f is even on D then $\deg(f, D, 0)$ is clearly 0 if n is odd.

Brouwer fixed point theorem. Let $f: \bar{B} \to \bar{B}$ be a continuous mapping of a closed ball $\bar{B} \subset \mathbb{R}^n$ into itself. Then there exists an $x \in \bar{B}$ such that $f(x) = x$.

Proof: We can suppose that $\bar{B}$ is centered at the origin. Let B be the open interior of $\bar{B}$. Suppose that $f(x) \neq x$ when $x \in \partial B$. Let $F_t: \bar{B} \to \mathbb{R}^n$ by $x \mapsto x - tf(x)$. Then

$$F_t(\partial B) \neq 0$$

by hypothesis for $t = 0$ and because $tf(x) \in B$ for $t < 1$. By the homo-

[1]See for instance the proof in [J. T. Schwartz 1969].

topic invariance property

$$\deg(F_1, B, 0) = \deg(F_0, B, 0) = 1.$$

Therefore, by the Note on p. 557, there exists an $x \in B$ such that $F_1(x) = 0$, i.e. $f(x) = x$. ∎

Product theorem. Let $f: \bar{D} \subset X^n \to Y^n$ and $g: Y^n \to W^n$ be continuous mappings and let Δ_i be the bounded connected components of $Y^n \backslash f(\partial D)$. Then for $y \notin (g \circ f)(\partial D)$

$$\deg(g \circ f, D, y) = \sum_i \deg(g, \Delta_i, y) \deg(f, D, \Delta_i).$$

Proof: First note that the degree of f is zero on the unbounded component of $Y^n \backslash f(\partial D)$ since $f(\bar{D})$ is compact. It is sufficient to prove the theorem for C^1 mappings. Using the fact that $J_{g \circ f} = J_g J_f$ we have

$$\begin{aligned} \deg(g \circ f, D, y) &= \sum_{x \in (g \circ f)^{-1}(y)} \operatorname{sign} J_{g \circ f}(x) \\ &= \sum_{\substack{x \in f^{-1}(z) \\ z \in g^{-1}(y)}} \operatorname{sign} J_g(z) \operatorname{sign} J_f(x) \\ &= \sum_i \Big(\deg(f, D, \Delta_i) \sum_{\substack{z \in \Delta_i \\ z \in g^{-1}(y)}} \operatorname{sign} J_g(z) \Big). \end{aligned}$$ ∎

3. LERAY–SCHAUDER THEORY

In general this theory applies to mappings, from a Banach space into itself, which are compact perturbations of the identity. However some parts of it can be extended to mappings between locally convex vector spaces. We give here only the main outlines of the theory with sketches of the proofs. For a more complete discussion see [Leray and Schauder 1934], [J. T. Schwartz 1969] or [Roseau 1970]. The last reference includes applications to physical problems.

Definitions. Let D be an open subset of the Banach space X such that $D \cap E$ is bounded for every finite dimensional vector space $E \subset X$. Let $f = \mathrm{Id} + g$ be a continuous mapping $\bar{D} \to X$. We assume that g is finite dimensional on D, that is to say that $g(D)$ is contained in some finite dimensional subspace $E_0 \subset X$. The mapping f is then called a **finite dimensional perturbation of the identity**.

finite dimensional perturbation of Id

We denote by $f_0: \bar{D} \cap E_0 \to E_0$ the restriction of f to E_0. For $y \in E_0$ and $y \notin f(\partial D)$ we define $\deg(f, D, y) = \deg(f_0, D \cap E_0, y)$.

It can be shown that the definition does not depend on the choice of E_0.

compact mapping

A continuous mapping between Banach spaces is called **compact** if the image of every bounded set is relatively compact.
We give the space of continuous mappings $\bar{D} \to X$ the C^0 topology, i.e.

$$\|f\|_{C^0(\bar{D})} = \operatorname*{Sup}_{x \in \bar{D}} \|f(x)\|.$$

Theorem. Any compact mapping $\bar{D} \to X$ is contained in the closure of the set of finite dimensional mappings $\bar{D} \to X$.

Proof: Any compact subset $K \subset X$ is totally bounded, that is, given any $\epsilon > 0$ there exists a finite number of points $x_i \in K$ such that the balls $\|x - x_i\| < \epsilon$ cover K.
To prove the theorem we construct a sequence of finite dimensional mappings $T_\epsilon \circ g$ which converge to the compact mapping g. Let E_ϵ be the finite dimensional vector subspace of X generated by the points x_i. Let $T_\epsilon: K \to E_\epsilon$ be given by

$$x \mapsto \sum_i T_i(x) x_i, \qquad T_i(x) = \frac{\mu_i(x)}{\sum_i \mu_i(x)}$$

where

$$\mu_i(x) = 2\epsilon - \|x - x_i\| \quad \text{if } \|x - x_i\| < 2\epsilon$$
$$\mu_i(x) = 0 \quad \text{if } \|x - x_i\| \geq 2\epsilon.$$

Since $\Sigma_i \mu_i(x) > 0$ for $x \in K$, the mapping T_ϵ is defined on K, moreover it is continuous and

$$\|T_\epsilon - \mathrm{Id}\|_{C^0(K)} = \sup_{x \in K} \|T_\epsilon(x) - x\| = \sup_{x \in K} \|\sum_i T_i(x)(x_i - x)\| < 2\epsilon.$$

Now with $g: D \to X$, $K = g(\bar{D})$ and $g_\epsilon = T_\epsilon \circ g$

$$\lim_{\epsilon = 0} \|g - g_\epsilon\|_{C^0(\bar{D})} = \lim_{\epsilon = 0} \|\mathrm{Id} \circ g - T_\epsilon \circ g\|_{C^0(\bar{D})}$$

$$= \lim_{\epsilon = 0} \|\mathrm{Id} - T_\epsilon\|_{C^0(K)} = 0. \qquad ■$$

degree of a compact perturbation of Id

Let $f = \mathrm{Id} + g$, $g: \bar{D} \to X$ be a compact mapping. For $y \notin g(\partial D)$ we define

$$\deg(f, D, y) = \lim_{\epsilon = 0} \deg(f_\epsilon, D \cap E_\epsilon, y)$$

where $f_\epsilon = \mathrm{Id} + g_\epsilon$ and g_ϵ the finite dimensional mappings converging to g.

It can be shown that the limit exists and is unique. More precisely it can be shown that there exists an $\eta > 0$ such that, for $\epsilon < \eta$, $\deg(f_\epsilon, D \cap E_\epsilon, y)$ is defined and independent of ϵ.

Properties. The fundamental theorem, the product theorem and Borsuk's theorem (p. 562) hold for the degree defined above. In the case of homotopic invariance it is necessary to assume that the homotopy is compact, that the mapping $[0, 1] \times \bar{D} \to X$ by $(t, x) \mapsto g_t(x)$ is compact. A fundamental application is the Leray–Schauder theorem given below.

Applications.

Schauder fixed point theorem. Every continuous mapping $f: K \to K$ on a compact convex subset K of a Banach space X has a fixed point.

Proof: It can be shown[1] that f can be extended to a continuous mapping $F: \bar{B} \to K$ where $\bar{B}$ is a closed ball centered at the origin and containing K. Consider the family of mappings $F_t: \bar{B} \to K$ where

$$F_t = \mathrm{Id} + tF.$$

The family clearly satisfies the conditions for homotopic invariance; in particular the mapping $(t, x) \mapsto tF(x)$ is compact and $F_t(\partial B) \neq 0$ (cf. proof of Brouwer fixed point theorem). Thus

$$\deg(F_1, B, 0) = \deg(F_0, B, 0) = 1.$$

It follows that there exists a point $x \in \bar{B}$ such that $F_1(x) = 0$, i.e. $x - F(x) = 0$. But since $F(\bar{B}) \subset K$, $x \in K$ and $f(x) = x$. ∎

The theorem still holds, if the conditions are changed somewhat. Let $f: A \to A$ be a compact mapping on a closed convex subset $A \subset X$. Then the closed convex hull K of $f(A)$ is compact and $K \subset A$. By the theorem $f|_K$ has a fixed point; therefore so does f.

The following theorem uses the preceding theory to determine the existence of solutions on a Banach space X of the functional equation

$$x - f(x, t) = 0 \qquad x \in X, t \in \mathbb{R}.$$

Leray–Schauder theorem. Let $\bar{\Omega}$ be the closure of a bounded open subset

[1]See for instance [J.T. Schwartz 1969, p. 120].

$\Omega \subset X$. The functional equation

$$x - f(x, t) = 0$$

has at least one solution for all $t \in [0, 1]$ if
1) *The mapping $f\colon \bar{\Omega} \times [0, 1] \to X$ is compact and the mapping $t \mapsto f(x, t)$ is uniformly continuous on $\bar{\Omega}$.*
2) *For $x \in \partial\Omega$ and $t \in [0, 1]$, $x - f(x, t) \neq 0$.*
3) *For some $t_0 \in [0, 1[$ the equation has a solution $x_0 \in \Omega$ and the index of the solution* $[\deg(\mathrm{Id} - f(\cdot, t_0), \Omega, 0)]$ *is not zero.*

Proof: We know that if the degree at 0 of the mapping $x \mapsto x - f(x, t)$ is not zero then there exists an x such that $x - f(x, t) = 0$, i.e., our equation has a solution. Now by hypothesis we know that this degree is not zero when $t = t_0$. Therefore we need only to prove that $\deg(\mathrm{Id} - f(\cdot, t), \Omega, 0)$ is constant for $t \in [0, 1]$. This follows from the homotopic invariance property.
Note that if we can find $t_0 \in [0, 1]$ such that $f(x, t_0) = 0$, then $\deg(\mathrm{Id} - f(\cdot, t_0), \Omega. 0)$ does not vanish since the degree of the identity mapping is 1. ■

If it can be shown that every solution x of the equation $x - f(x, t) = 0$ for each $t \in [0, 1]$ is bounded

$$\|x\| < M,$$

then the second hypothesis of the theorem is satisfied by the following choice of Ω

$$\Omega = \{y \in X; \|y\| < M\}.$$

The boundary of Ω is the set $\{y \in X; \|y\| = M\}$. The equation then has no solutions on $\partial\Omega$ when $t \in [0, 1]$.

The Leray–Schauder theorem still holds if $[0, 1]$ is replaced by a compact metric space T, i.e. $t \in T$.
The Leray–Schauder theory originated in the study of nonlinear elliptic equations. Problem B, p. 591 gives an example of one of the many applications to such equations[1].

[1]Particular case of a theorem given in [Choquet-Bruhat et Leray 1972].

C. MORSE THEORY[1]

1. INTRODUCTION

The properties of the critical points of a smooth function f on X are related to the topology of X. The Morse theory derives homotopy properties of X from an analysis of the second derivative of f at its critical points. It is of interest for the study of the global properties of physical systems, both classical and quantum, described by an action S: S is a smooth function on the space of paths, its critical points are the solutions of the Euler–Lagrange equations, and analysis of the second derivative of S at a critical point is based on the solutions of the small disturbance equation. For instance Morse theory has been applied to the computation of Feynman path integrals when there are conjugate points between the end points[2]. Van Hove[3] has used the theory to determine the occurrence of singularities in the elastic frequency distribution f of a crystal. The periodic structure of the crystal lattice gives to its configuration space the topological structure of a torus; the number and the types of singularities of f are derived from the homotopy properties of the torus.

2. DEFINITIONS AND THEOREMS

Smooth means C^∞, but is used in preference to C^∞ because some results do not require "infinite" smoothness.

Let X be a smooth manifold modelled on a real Hilbert space E.
Let f be a smooth real valued function on X. A point $a \in X$ is a **critical point** of f if its differential at a, $f'(a) \in \mathscr{L}(T_a(X), \mathbb{R})$ is zero. The value $f(a)$ of f at a critical point a is called a **critical value** of f.

critical point
critical value

By the theorem on p. 548 if c is not a critical value of f the subset $f^{-1}(\{c\})$ is a smooth submanifold of X, and the subset $\{x; f(x) \leq c\}$ is a smooth

[1]See further properties in [Morse 1965], [Milnor 1969], and, for the infinite dimensional case, [J.T. Schwartz 1969]. See also [Bott and Mather 1967]. For an application of the Morse theory to space time geometry see [Woodhouse 1976], [Uhlenbeck 1976].
[2][Gutzwiller 1967].
[3][Van Hove 1953].

submanifold of X with boundary – the union of $\{x; f(x) < c)$, open in X, and $\{x; f(x) = c\}$.

Palais–Smale condition

f is said to satisfy the **Palais–Smale condition** on a subset Y of X if whenever Z is a subset of Y on which f is bounded and $\|f'\|$ is not bounded away from zero[1], then the closure of Z contains a critical point of f.

This condition is always satisfied if X is finite dimensional, and Y is bounded. Indeed, in that case, Z is relatively compact. If f' is not bounded away from zero on Z, there exists a sequence $(x_n) \in Z$ such that $(f'(x_n))$ converges to zero, and from (x_n) we can extract a subsequence which converges to a point $a \in \bar{Z}$, which is a compact set. At this point $f'(a) = 0$.

Theorem (finite dimensional case). Let f be a smooth real valued function on a paracompact finite dimensional manifold X. Let $a < b$ and suppose that the set

$$f^{-1}([a, b]) \equiv Y = \{x \in X; a \leq f(x) \leq b\}$$

is compact and contains no critical point of f. Let c be prescribed in (a, b). Then there exists a diffeomorphism

$$\Gamma: (a, b) \times f^{-1}(\{c\}) \to f^{-1}((a, b)) \text{ by } (t, x) \mapsto \Gamma(t, x).$$

Corollary. The manifolds $f^{-1}(c)$, $a < c < b$ are all diffeomorphic.

Proof: The idea of the proof is to use transversal trajectories of the submanifolds $f =$ constant, namely choose a riemannian metric g on X and call v the (contravariant) vector field canonically associated with the covariant field f' on X – in local coordinates $v^i = g^{ij}\, \partial f / \partial x^j$.

Then

$$g(v, v) = g(f', f') = g^{ij} \frac{\partial f}{\partial x^i} \frac{\partial f}{\partial x^j} \neq 0$$

on Y and also on a neighborhood of Y since Y is compact. We can therefore, when X is finite dimensional and Y compact define a vector field w on X, identical to v on Y by

$$w = \rho v$$

where ρ is a C^∞ function with compact support, which is equal to

[1] $\|f'\|$ is bounded away from zero when $0 < \lambda \leq \|f'\|$.

$1/g(v, v)$ on Y. Then w (cf. Remark p. 145) generates a 1 parameter group of diffeomorphisms

$$\sigma_t: X \to Y \text{ by } x \mapsto \sigma(t, x)$$

with $t \mapsto \sigma(t, x)$ a solution of the differential equation on X

$$\frac{\mathrm{d}\sigma(t, x)}{\mathrm{d}t} = w(\sigma(t, x)), \qquad \sigma(0, x) = x.$$

$\sigma(t, x)$ is defined for all $t \in \mathbb{R}$ and $x \in X$.
By the definition of a vector field, the action of w on f is

$$wf = \frac{\mathrm{d}(f \circ \sigma_t)}{\mathrm{d}t} = \langle w, f' \rangle;$$

on the other hand

$$\langle w, f' \rangle = g(w, v) = \rho g(v, v).$$

Thus if the point $\sigma(t, x)$ lies in the set Y

$$\frac{\mathrm{d}}{\mathrm{d}t}(f \circ \sigma_t) = 1.$$

We can now prove that the mapping Γ defined on $(a, b) \times f^{-1}(\{c\})$ by $(t, x) \mapsto \sigma(t - c, x)$ is a diffeomorphism onto $f^{-1}((a, b))$:

$$\Gamma: (a, b) \times f^{-1}(\{c\}) \to f^{-1}((a, b)) \text{ by } (t, x) \mapsto \sigma(t - c, x).$$

The proof will be done in three steps, I, II, III.
I. Γ has assigned values in Y.
Let $\sigma(0, x) = x \in f^{-1}(\{c\})$. There exists h and k such that for

$$h < t < k, \qquad \sigma(t - c, x) \in f^{-1}((a, b)).$$

For $h < t < k$, $\frac{\mathrm{d}}{\mathrm{d}t} f(\sigma(t - c, x)) = 1$, thus

$$f(\sigma(t - c, x)) = t, \qquad h < t < k.$$

One proves, using the continuity of f, that if $h > a$ [if $k < b$] then $f(\sigma(h - c, x)) = h$ [then $f(\sigma(k - c, x)) = k$], thus $\sigma(t - c, x) \in f^{-1}((a, b))$ for $h \le t \le k$. It follows that $\sigma(t - c, x) \in f^{-1}((a, b))$ for $a < t < b$, and every $x \in f^{-1}(\{c\})$.
II. Γ is onto Y, and injective.
Let $y \in f^{-1}((a, b))$, $f(y) = \gamma$, $a < \gamma < b$. By I the point $\Gamma(t, x) = \sigma(t - c, x)$ coincides with y if and only if

$$t = f(\sigma(t - c, x)) \equiv f(y) = \gamma$$

$$\sigma(\gamma - c, x) = y$$

that is, by the group property of σ_t

$$x = \sigma(c - \gamma, y) \in f^{-1}(\{c\}).$$

III. The differentiability of Γ and Γ^{-1} follows from the differentiability of σ_t (Chapter III).

Example: Let X be a torus, obtained by rotation of a horizontal circle around a horizontal line, and let f measure the height above a horizontal plane. The topology of $f^{-1}(\{c\})$ changes each time c attains a value for which the torus admits a horizontal tangent plane.

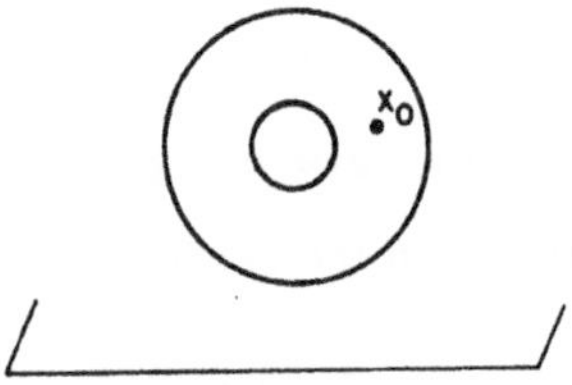

If the point x_0 is taken out of the torus, $X - \{x_0\}$ is non compact, the diffeomorphism property of $f^{-1}(\{c\})$ and $f^{-1}(\{c+\epsilon\})$ is no longer true if $c = f(x_0)$.

In the infinite dimensional case an analogous theorem is valid under a slightly stronger hypothesis.

Theorem (infinite dimensional case). Let X be a complete riemannian manifold, $f \in C^\infty(X)$, a and b real numbers; define Y as the subset

$$Y = \{x\,;\, a \le f(x) \le b\}$$

and Y_0, Y_ϵ, with $Y_0 \subset Y \subset Y_\epsilon$, as:

$$Y_0 = \{x\,;\, a < f(x) < b\}$$

$$Y_\epsilon = \{x\,;\, a - \epsilon < f(x) < b + \epsilon\}$$

where ϵ is some positive real number. We suppose that f' on Y_ϵ satisfies the Palais–Smale condition and has no critical point. The conclusions are:

1) *The manifolds $\{f^{-1}(\{c\})\}$, $a < c < b$ are all diffeomorphic. We denote one of them by Z.*

2) *Y is diffeomorphic to $(a, b) \times Z$.*

Note that Y is no longer compact, but X is now supposed complete. The proof cannot be a straightforward generalization of the one given in the

finite dimensional case since we cannot introduce a flow σ defined globally on X.
We refer the reader to the original proof[1] or to the references on p. 567.

3. INDEX OF A CRITICAL POINT

Let f be a real valued smooth function on a real Hilbert space E. Its **hessian** at a point x (p. 80) is the quadratic form $f''(x)$. If $E = \mathbb{R}^n$, then $f'' = [\partial^2 f / \partial x^i \partial x^j]$.

hessian

Now let f be a real valued smooth function on a smooth Hilbert manifold X, modelled on E. The second derivative of f at x is a chart dependent notion except when x is a critical point. Indeed the second differential of f in the chart (U, φ) at the point $\varphi(x)$ is the differential of the mapping $\varphi(x) \mapsto (f \circ \varphi^{-1})(\varphi(x))$; it belongs to $\mathscr{L}(E \times E, \mathbb{R})$. Let $(\tilde{U}, \tilde{\varphi})$ be another chart around x, we have by the composite mapping rule (cf. p. 73)

$$(f \circ \varphi^{-1})'_{\varphi(x)} = (f \circ \tilde{\varphi}^{-1})'_{\tilde{\varphi}(x)} \circ (\tilde{\varphi} \circ \varphi^{-1})'_{\varphi(x)}$$

$$(f \circ \varphi^{-1})'' = (f \circ \tilde{\varphi}^{-1})''((\tilde{\varphi} \circ \varphi^{-1})'(\tilde{\varphi} \circ \varphi^{-1})') + (f \circ \tilde{\varphi}^{-1})' \circ (\tilde{\varphi} \circ \varphi^{-1})''.$$

We see from this formula that the hessian of f *at a critical point* is a symmetric 2-covariant tensor $f'' \in \mathscr{L}(T_x(X) \times T_x(X), \mathbb{R})$.

A critical point x of f is said to be **non degenerate** if the hessian of f at x is a non degenerate quadratic form (cf. p. 83).

non degenerate

When $E = \mathbb{R}^n$ is finite dimensional it is a classical result that a quadratic form Q on E has a canonical decomposition

$$Q = -\sum_{i=1}^{p} (l_i)^2 + \sum_{i=p+1}^{p+q} (l_i)^2$$

where the l_i are independent linear forms on $\mathbb{R}^n$. The numbers p, $q - p$, and $n - (p + q)$ are called respectively the **(Morse[2]) index, signature**, and **nullity** of the quadratic form Q. These numbers are uniquely defined and coordinate independent. Therefore there is no ambiguity in defining the index and the nullity of the quadratic form f''. A critical point of f is non degenerate iff the nullity of the hessian at this point is zero ($q = n - p$). When E is a Hilbert space an analogous decomposition is valid, if Q is a non degenerate quadratic form on E: there exists $T \in \text{Isom}(E, E)$ and $P \in \mathscr{L}(E, E)$, where P is a projector ($P^2 = P$) such that

Morse
index
signature
nullity

$$Q(h, h) = \|PTh\| - \|(\text{Id} - P)Th\|, \quad \forall h \in E.$$

[1][Palais 1963, pp. 299–340], [Palais and Smale 1964, pp. 165–172].
[2]The Morse index is the negative of the index defined on p. 287.

index An **index** of Q is a pair (q, p) where q and p are the dimensions – which may be infinite cardinal numbers – of the subspaces PTE and $(\mathrm{Id} - PT)E$ appearing in the above decomposition.
The index is uniquely defined and is coordinate independent. The **index** of f, a smooth function on the Hilbert manifold X, at a critical point x is the index of $f''(x)$.
We shall see now how the index determines the change in topology of the subset of X, $\{x; f(x) \leq a\}$ when a crosses a critical level.

handle Let X and $\tilde{X}$ be two smooth manifolds, possibly with boundary. We shall say that $\tilde{X}$ has been obtained from X by attaching a **handle** of type (q, p) if the following conditions are satisfied.
1) X is a regularly embedded submanifold of $\tilde{X}$.
2) There exists a closed subset $H \subset \tilde{X}$, such that
  a) $X \cup H = \tilde{X}, X \cap H \subset \partial X$.
  b) H is homeomorphic to the product $D^q \times D^p$ where D^q[resp. D^p] is the closed unit ball in a Hilbert space with dimension q[resp. p].
If h denotes the homeomorphism

$$h: D^q \times D^p \to H,$$

then the restriction $h(\mathring{D}^q \times D^p)$ is a diffeomorphism onto $H - X$ and $h(\partial D^q \times D^p)$ is a regular embedding into ∂X. We write then

$$\tilde{X} = X \underset{h}{\cup} H(q, p).$$

Remark: In the finite n-dimensional case, $q = n - p$, the type of the handle is characterized by one number. A handle of type $(n-p, p)$ is also called a p-cell.

Example: Let X be diffeomorphic to the 2-dimensional disc $(x^1)^2 + (x^2)^2 \leq 1$ with boundary $(x^1)^2 + (x^2)^2 = 1$. If we attach to it a cell $I \times I$, where I is the unit disc (interval) in $\mathbb{R}$, we obtain a manifold with boundary homeomorphic to a half-torus.

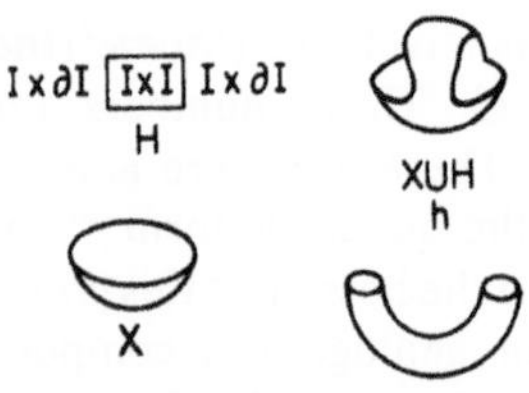

4. CRITICAL NECK THEOREM

Critical neck theorem. Let $f \in C^\infty(X)$. Assume that f has only one critical level c between the non-critical levels a and b, $a < c < b$, and that f has a *finite number of critical points $x_1, \ldots x_n$ at level c. Assume that these critical points are non degenerate. Then the following subsets are diffeomorphic*:

$$\{x \in X; f(x) \leq b\} \text{ and } \{x \in X; f(x) \leq a\} \underset{h_1}{\cup} H_1 \ldots \underset{h_n}{\cup} H_n$$

where H_i is a (k_i, l_i) handle, (k_i, l_i) is the index of x_i, and $h_i(H_i) \cap h_j(H_j) = \emptyset$.

For the proof we refer the reader to the literature[1]. It uses the following lemma:

Morse lemma: *In the neighborhood of a non degenerate critical point x there is a chart (U, φ) and a projector P such that*

$$f(y) \equiv f(x) + \|P(\varphi(y) - \varphi(x))\|^2 - \|(\mathrm{Id} - P)(\varphi(y) - \varphi(x))\|^2.$$

In the finite dimensional case this lemma means that there exist local coordinates such that in a neighborhood of x:

$$f(y) \equiv f(x) + \sum_{i=1}^{q} (y^i - x^i)^2 - \sum_{i=q+1}^{q+p} (y^i - x^i)^2.$$

Proof: Use the Taylor formula with integral remainder (p. 81) and the property of decomposition of quadratic forms recalled earlier.

Exercise: Show the consistency of the results obtained in considering the height function f on a vertical torus (cf. p. 570).

For further relations between the topology of X – for instance its Betti numbers – and the critical points of smooth functions on X, we also refer the reader to the bibliography.

D. CYLINDRICAL MEASURES, WIENER INTEGRAL

1. INTRODUCTION

In Section I.D we have defined a measure m on a space X as a countably additive set function from a σ-field $\mathscr{A}$ of subsets of X into the extended positive real numbers $\mathbb{R}^+ \cup \{+\infty\}$. We have then defined integrable func-

[1]See for instance [Bott] (finite dimensional case) or [J.T. Schwartz] (infinite dimensional case).

tions, through the definition of the integral of simple functions (step functions), and reviewed the main properties of the integral. These definitions and properties constitute what is called "abstract measure theory". They do not involve the topology of X. However, when the space X is also a topological space it is fruitful to endow X with a measure related to the topology. This has been done in section I.D with the definition of Borel sets (p. 34) and Borel measures (p. 36), in the case where X is a locally compact space. Every continuous function with compact support is then integrable, and these functions are dense in the space of integrable functions. However in many applications of interest to physics, the topological space X is not locally compact.

In most problems arising in statistical and quantum mechanics the space X to be endowed with a measure is a space of functions on another space – for instance on $\mathbb{R}^n$ or one of its subsets – and in the relevant topology X is not locally compact. For example X may be an infinite dimensional normed space, and every compact set of X has an empty interior. Then the closure of the complementary set $X \backslash K$ of an arbitrary compact set K is identical to X (Ch. I), therefore any continuous function on X with compact support is identically zero.

Much attention in recent years has been given to the fascinating subject "integration on function spaces", often with a view toward specific applications. These applications often lead to new technical difficulties. In the case where X is a locally convex topological vector (linear) space a powerful tool has been introduced in probability theory by Kolmogorov[1], and developed by a number of mathematicians and physicists under the name of "cylindrical measures" or "promeasures".

In this chapter we give some of the main ideas and results about promeasures (cylindrical measures) with a sketch of the proofs or reference to the literature. The promeasures are of particular importance in the study of the Wiener integral which plays a fundamental role in statistical physics. The formalism is also useful in computing path integrals[2] which form the backbone of the challenging formulation of quantum physics proposed by Feynman[3] in 1942.

In this chapter, X will be a locally convex, Hausdorff, topological space. We shall reserve the name "measure" on X to countably additive positive set functions on a σ-field of X (p. 33). When X is finite dimensional this σ-field will always be the Borel σ-field, and the measure a regular (positive) Borel measure, identified with a Radon measure.

[1][Kolmogorov 1936].
[2]cf. [DeWitt-Morette 1972, 1974, 1976] and [DeWitt-Morette, Maheshwari, Nelson 1979].
[3]cf. for instance [Feynman and Hibbs 1965].

2. PROMEASURES AND MEASURES ON A LOCALLY CONVEX SPACE

When X is a topological vector space, locally convex, and Hausdorff, some families of measures defined on $\mathbb{R}^n$ spaces, quotients of X by subspaces of finite codimension, play an important role in probability theory. These families, called cylindrical measures or promeasures, serve as a basis for the definition of Wiener integrals and also of some Feynman "integrals". These measures satisfy compatibility conditions given in terms of canonical "projections".

Projective system. Let $F(X)$ be the set of closed subspaces of X of finite codimension together with the partial ordering relation $\subset$; let $V \in F(X)$; the set X/V of equivalence classes $x \sim x'$ if $x - x' \in V$, is a finite dimensional vector space. Denote by P_V the canonical mapping $x \mapsto [x]$, equivalence class of x. When $W \subset V$, one can define a mapping $P_{VW}: X/W \to X/V$ through the relation $P_V = P_{VW} \circ P_W$.
The family $\{X/V, P_{VW}\}$ is called the **projective system** of finite dimensional quotients of X. projective system

It can be proved that the family $F(X)$ is determined by the topological dual X' of X; indeed: a subspace V of X is an element of $F(X)$ if and only if there exists a finite number of elements $x'_1, \dots x'_n$ of X' such that V consists of the elements $x \in X$ satisfying $\langle x'_i, x \rangle = 0$, $i = 1, \dots n$.

Example: Projective system on the space C of continuous functions on a closed interval $I = [a, b]$ of $\mathbb{R}$. The dual of C when endowed with the topology of uniform convergence is the space M of signed measures on the compact space I. An element $V \in F(C)$ is determined by n equations

$$\langle \mu_i, x \rangle = 0, \qquad \mu_i \in M, \qquad x \in C, \; i = 1, \dots n.$$

A particular such subspace V is determined by the n equations

$$\langle \delta_{t_i}, x \rangle = 0, \qquad t_i \in I, \; i = 1, \dots n.$$

$V \subset C$ is the space of functions which vanish at n given points $t_i \in I$, $i = 1, \dots n$; V is a closed subspace of finite codimension of C.
C/V is the set of equivalence classes of functions taking the same value at the points $t_1, \dots t_n$:

$$x \sim y \iff x(t_i) = y(t_i), \qquad i = 1, \dots n.$$

C/V is isomorphic to $\mathbb{R}^n$ by $[x] \mapsto \{x(t_i)\}$; $P_V: x \mapsto \{x(t_i)\}$, P_V maps x onto n of its values.
Let $W \subset C$ be another space of functions vanishing at the points $\{t'_j, j = 1, \dots p\}$; then $W \subset V$ iff $\{t'_j\} \supset \{t_i\}$.

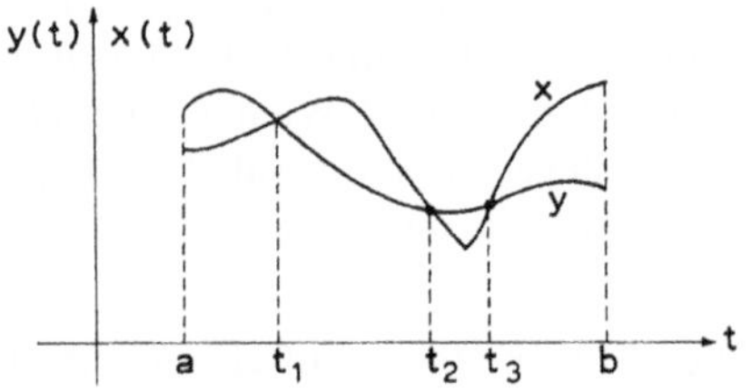

promeasure

Promeasures. A **promeasure** on X is a family $\mu = \{\mu_V; V \in F(X)\}$ such that

1) μ_V is a bounded measure on the finite dimensional space X/V.
2) When $V \supset W$, $\mu_V = P_{VW}(\mu_W)$ where P_{VW} is the canonical mapping $X/W \to X/V$.
3) The total mass $\mu_V(X/V)$ is independent of V and is called the total mass $\mu(X)$ of the promeasure μ.

We had to limit the definition to bounded measures, because the mapping P_{VW} is not proper (the reciprocal image of a compact set is a cylinder, not a compact set), therefore the compatibility condition $\mu_V = P_{VW}(\mu_W)$ would be, in general, meaningless for unbounded measures.
The expression "promeasure" comes from the fact that a promeasure on X is a *pro*jective system of measures on the *pro*jective system of finite quotient spaces of X.

cylindrical measure
cylindrical subset
cylindrical σ-field

A promeasure is sometimes called a **cylindrical measure** because the reciprocal image by P_V of a subset $A \subset X/V$ is a cylinder in X, generated by affine subspaces parallel to V. We shall reserve the name **cylindrical subset** of X for cylinders which are reciprocal images by P_V of Borel subsets of quotients X/V, $V \in F(X)$. **The cylindrical σ-field** of X will be the σ-field generated by the cylindrical subsets. The cylindrical σ-field is contained in the Borel σ-field of X, since $P_V^{-1}(A)$ is open if A is open (P_V is a continuous mapping), but it can be smaller.
The cylindrical σ-field of $C([a, b])$ is identical with its Borel σ-field generated by "quasi intervals", namely by cylinders generated by hyperplanes, defined by $\alpha_i < x(t_i) < \beta_i$, with $\{t_i\}$ a countable dense subset of $[a, b]$ (cf. Problem D1).

We shall say that a promeasure μ on X is a measure if there exists a measure m on X such that μ_V is the image of m by P_V for every

$V \in F(X)$, i.e. $\mu_V = P_V(m)$. We shall say that μ is the promeasure associated with m and write, by a convenient abuse of language, $\mu = m$.

Remark 1: If m is a measure associated with a promeasure on X, the σ-field on which m is defined contains necessarily the cylindrical σ-field, since $m_V = P_V(m)$ must be a Borel measure.
If a bounded measure m is defined on the Borel σ-field of X, there is always an associated promeasure $\{m_V = P_V(m);\ V \in F(X)\}$.

Remark 2: A promeasure μ on X defines a positive set function f on the cylindrical subsets of X. Indeed let $\Sigma = P_V^{-1}(A)$, set

$$f(\Sigma) = \mu_V(A),$$

f is finitely additive. However f cannot in general be extended to a countably additive set function on the cylindrical σ-field of X, because the cylindrical subsets do not have their base A in the same space X/V.

The following theorems give examples of spaces on which every promeasure is a measure.

Theorem 1. *If X is finite dimensional, every promeasure on X is a measure.*

Proof: $\{0\} \in F(X)$, $X/\{0\} = X$, $P_{V\{0\}} = P_V$ hence $\mu_V = P_V(\mu_{\{0\}})$ for all $V \in F(X)$; in other words the measure $m = \mu_{\{0\}}$ is such that $\tilde{m} = \mu$. ■

Now let X be the space of real functions defined on a countable set D. X is the infinite topological product $\mathbb{R}^D$, defined p. 21. Its topology is identical with the topology of point wise convergence[1]. The linear forms $\delta_t : \mathbb{R}^D \to \mathbb{R}$ by $x \mapsto x(t)$, $t \in D$ are continuous (they have been called projection mappings p. 21). It can be proved that these linear forms δ_t, $t \in D$, are a Hamel basis (p. 9) of the dual X' of $X = \mathbb{R}^D$. This leads to the following theorem which is an important step in the theory of Wiener measure.

Theorem 2. *Every promeasure on a space $\mathbb{R}^D$, with D a countable set, is a measure.*

For a proof see for instance [Bourbaki 1969, p. 71].

[1] $\mathbb{R}^D$ is the set of all mappings $x: D \to \mathbb{R}$ by $t \mapsto x(t)$. A point $x \in \mathbb{R}^D$ converges to $a \in \mathbb{R}^D$ iff its projection (defined p. 575) $P_t(x) = x(t)$ converges numerically to $a(t)$ for every $t \in D$.

Image of a promeasure. Let X and X_1 be two locally convex spaces and u be a continuous linear mapping of X into X_1. Let μ be a promeasure on X.

$$\begin{array}{ccccc} X & \xrightarrow[p_W]{} & X/W & \xrightarrow[p_{VW}]{} & X/V \\ \downarrow{\scriptstyle u} & & \downarrow{\scriptstyle u_{W_1}} & & \downarrow{\scriptstyle u_{V_1}} \\ X_1 & \xrightarrow[p_{W_1}]{} & X_1/W_1 & \xrightarrow[p_{V_1W_1}]{} & X_1/V_1 \end{array}$$

This figure is commutative.

For every $V_1 \in F(X_1)$, the subspace $V = u^{-1}(V_1)$ of X belongs to $F(X)$ and u induces a linear mapping u_{V_1}: $X/V \to X_1/V_1$. The measure ν_{V_1} on X_1/V_1, image of μ_V by μV_1, is defined by

$$\nu_{V_1} = u_{V_1}(\mu_{u^{-1}(V_1)})$$

and denoted by

$$\nu_{V_1} = u_{V_1}(\mu_V).$$

If $\mu = \{\mu_V, V \in F(X)\}$ is a promeasure on X, the family $\nu = \{\nu_{V_1}, V_1 \in F(X_1)\}$ is a promeasure on X_1, image of μ by u.
If X_1 is finite dimensional the promeasure ν is a measure which we denote by μ_u.

Integration of a cylindrical function with respect to a promeasure. A **cylindrical function**[1] on X is a couple (u, f) where u is a continuous linear mapping $X \to \mathbb{R}^n$, and f a function on $\mathbb{R}^n$; thus $F = f \circ u$ is a function on X.

cylindrical function

A cylindrical function (u, f) is said to be integrable on X with respect to the promeasure μ if f is integrable on $\mathbb{R}^n$ with respect to μ_u, image of μ by u. One then writes

$$\int_X F\, \mathrm{d}\mu = \int_{\mathbb{R}^n} f\, \mathrm{d}\mu_u.$$

If μ is a measure on X this formula is identical with the formula given in section ID for the integral with respect to an image measure.

Example: Let x' be a given element of the dual X' of X; it defines a

[1]Called a tame function by [Gross].

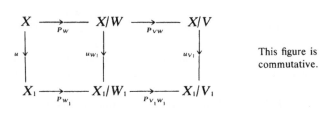

continuous linear mapping $X \to \mathbb{R}$ by $x \mapsto \langle x', x\rangle$. Consider the cylindrical function (x', f) on X with

$$f: \mathbb{R} \to \mathbb{R} \text{ by } t \mapsto |t|^n.$$

(x', f) is μ integrable on X if f is integrable on $\mathbb{R}$ with respect to $\mu_{x'}$, image of μ by x', and then

$$\int_X |\langle x', x\rangle|^n \, d\mu(x) = \int_{\mathbb{R}} |t|^n \, d\mu_{x'}(t).$$

We will see applications of this formula in the Wiener integral.

Fourier transforms.
a) Fourier transform of a bounded measure. If λ is a bounded Borel measure on X, its Fourier transform is the function on X' defined by the integral

$$(\mathcal{F}\lambda)(x') = \int_X \exp(-i\langle x', x\rangle) \, d\lambda(x)$$

which exists since $|\exp(-i\langle x', x\rangle)| = 1$ and the function $x \mapsto \exp(-i\langle x', x\rangle)$ is measurable since continuous.
Let u be a continuous map: $X \to Y$; let λ_u be the bounded Borel measure on Y image by u of the bounded Borel measure λ. The Fourier transform of λ_u is the function on Y'

$$(\mathcal{F}\lambda_u)(y') = \int_Y \exp(-i\langle y', y\rangle) \, d\lambda_u(y)$$

which is equal, by virtue of the integral equality given (p. 44) to

$$(\mathcal{F}\lambda_u)(y') = \int_X \exp(-i\langle y', u(x)\rangle) \, d\lambda(x).$$

Let $\tilde{u}: Y' \to X'$ denote the transposed map of u defined by

$$\langle \tilde{u}(y'), x\rangle = \langle y', u(x)\rangle.$$

Then

$$\mathcal{F}(\lambda_u)(y') = \int_X \exp(i\langle \tilde{u}(y'), x\rangle) \, d\lambda(x) = (\mathcal{F}\lambda \circ u)(y').$$

Let u be the linear mapping $x': X \to \mathbb{R}$ by $x \mapsto \langle x', x\rangle$, the transposed mapping $\tilde{x}': \mathbb{R} \to X'$ is such that

$$\langle \tilde{x}'(y'), x\rangle_X = \langle y', x'(x)\rangle_{\mathbb{R}}, \quad y' \in \mathbb{R}, \quad x'(x) \in \mathbb{R},$$

where the right hand side denotes the duality on $\mathbb{R}$, i.e. the ordinary product of numbers. Since here $x'(x) = \langle x', x\rangle$, we have

$$\langle \tilde{x}'(y'), x\rangle = y'\langle x', x\rangle$$

therefore, by linearity

$$x'(y') = y'x' \in X', \qquad \forall y' \in \mathbb{R}.$$

Take $u = x'$ and $y' = 1$ in $(\mathscr{F}\lambda_u)(y')$ above, we get

$$(\mathscr{F}\lambda_{x'})(1) = (\mathscr{F}\lambda)(x')$$

therefore

$$(\mathscr{F}\lambda)(x') = \int_{\mathbb{R}} e^{-it}\, d\lambda_{x'}(t).$$

b) If μ is a promeasure on X, $\mu_{x'}$ is, for every $x' \in X'$, a bounded measure on $\mathbb{R}$ image of μ by $x': X \to \mathbb{R}$. We can thus define the Fourier transform of μ by:

$$(\mathscr{F}\mu)(x') \overset{\text{def}}{=} (\mathscr{F}\mu_{x'})(1) = \int_{\mathbb{R}} e^{-it}\, d\mu_{x'}(t);$$

$\mathscr{F}\mu$ is a function on X'.

Let u be a continuous linear mapping $u: X \to Y$; let $\nu = u(\mu)$ be the image of μ by u. According to the previous definition, the Fourier transform of ν is $(\mathscr{F}\nu)(y') = \mathscr{F}\nu_{y'}(1)$ with, according to the definition of the image of a promeasure,

$$\begin{aligned}\nu_{y'} &= y'(\nu) = y'(u(\mu))\\ &= (\tilde{u}(y'))(\mu) = \mu_{\tilde{u}}(y')\end{aligned}$$

since $y' \circ u = \tilde{u}(y')$ is a linear mapping $X \to \mathbb{R}$. Therefore the formula $\mathscr{F}\nu = \mathscr{F}\mu \circ \tilde{u}$ where $\nu = u(\mu)$ remains valid for a promeasure.

In particular let u be the canonical mapping $P_V: X \to X/V$, then $\mathscr{F}\mu_V = \mathscr{F}\mu \circ \tilde{P}_V$ where $\mathscr{F}\mu_V$ is the usual Fourier transform of a bounded measure on a finite dimensional space. It follows from this formula that the coherence conditions which must be satisfied by the family $\{\mu_V\}$ imply the following condition on the family of functions $\{\mathscr{F}\mu_V\}$. When $V \supset W$,

$$\mathscr{F}\mu_V = \mathscr{F}\mu_W \circ \tilde{P}_{VW}, \text{ since } \tilde{P}_V = \tilde{P}_W \circ \tilde{P}_{VW}.$$

Conversely let a family $\{\mu_V\}$ of bounded measures on the space X/V, $V \in F(X)$, be such that their Fourier transforms satisfy the coherence conditions

$$\mathscr{F}\mu_V = \mathscr{F}\mu_W \circ \tilde{P}_{VW},$$

then it is easy to show, using the injectivity of the mapping $\mu_V \mapsto \mathscr{F}\mu_V$ that the measures define a promeasure on X, since they satisfy

$$\mu_V = P_{VW}(\mu_W)$$

It is also easy to show[1] that the mapping from the set of promeasures on X into the set of functions on X' by $\mu \mapsto \mathscr{F}\mu$ is one-one (injective).

Remark: We have for every $V \in F(X)$

$$\mu_V(X/V) = (\mathscr{F}\mu_V)(0) = (\mathscr{F}\mu \circ \tilde{P}_V)(0) = (\mathscr{F}\mu)(0).$$

3. GAUSSIAN PROMEASURES

Gaussian measures on $\mathbb{R}^n$. The **gaussian measure** $\gamma_\alpha(x)$ on $\mathbb{R}$ of covariance $\alpha > 0$ is the product of the Lebesgue measure on $\mathbb{R}$ by the rapidly decreasing C^∞ function

Gaussian measure on $\mathbb{R}$

$$\frac{1}{\sqrt{2\pi\alpha}} \exp(-x^2/2\alpha).$$

One may write

$$d\gamma_\alpha(x) = \frac{1}{\sqrt{2\pi\alpha}} \exp(-x^2/2\alpha)\, dx.$$

$\gamma_\alpha(x)$ may equivalently be defined by its Fourier transform

$$(\mathscr{F}\gamma_\alpha)(x') = \exp(-\alpha x'^2/2).$$

The **canonical gaussian measure** γ_1 **on** $\mathbb{R}$, is the gaussian of covariance $\alpha = 1$. The **canonical gaussian measure** γ **on** $\mathbb{R}^n$ is the direct product (p. 37)

Gaussian measure on $\mathbb{R}^n$

$$\gamma(x) = \gamma_1(x_1) \otimes \gamma_1(x_2) \cdots \otimes \gamma_1(x_n);$$

its Fourier transform is

$$(\mathscr{F}\gamma)(x') = \exp(-|x'|^2/2).$$

More generally, let Q be a positive quadratic form on $\mathbb{R}^n$, we shall prove that there exists a bounded measure γ_Q on $\mathbb{R}^n$ whose Fourier transform is $\exp(-Q/2)$:

$$(\mathscr{F}\gamma_Q)(x') = \exp(-Q(x')/2).$$

γ_Q is called the gaussian measure on $\mathbb{R}^n$ of variance Q. Indeed:
1) Suppose that Q is positive and non-degenerate: there is a basis in $\mathbb{R}^n$

[1][Bourbaki 1969, p. 73].

which diagonalizes Q, i.e. an isomorphism $u: \mathbb{R}^n \to \mathbb{R}^n$ such that the image Q_u of Q is of the type:

$$Q_u(x') = \sum_{j=1}^{n} \alpha_j (x'^j)^2, \qquad \alpha_j > 0.$$

The measure defined in the dual basis by the direct product

$$\gamma = \gamma_{\alpha_1}(x_1) \times \cdots \times \gamma_{\alpha_n}(x_n)$$

has Fourier transform $\exp(-Q_u(x')/2)$.
By the law of action of isomorphisms of $\mathbb{R}^n$ on Fourier transforms (p. 479), the Fourier transform of the image $\gamma_Q \overset{\text{def}}{=} \tilde{u}(\gamma)$ is $\mathcal{F}\gamma_Q = \exp(-Q/2)$. Thus γ_Q is the gaussian measure on $\mathbb{R}^n$ of variance Q.

2) Let Q be positive, but degenerate, let N be its kernel,

$$N = \{x' ; Q(x') = 0\}.$$

Let M be the supplementary subspace of N in $\mathbb{R}^n$,

$$M = \left\{ x ; \langle x', x \rangle = \sum_{i=1}^{n} x_i x'^i = 0 \qquad \forall x' \in N \right\}.$$

Let j be the canonical injection $M \to \mathbb{R}^n$, i.e. x and jx are the same point. Let $\tilde{j}$ be the transposed mapping $\mathbb{R}^n \to M$ – here a finite dimensional space and its dual are identified; $\tilde{j}$ is the orthogonal projection from $\mathbb{R}^n$ onto M, a surjective mapping whose kernel is the orthogonal complement N of M. Thus the kernel of the mapping $\mathbb{R}^n \to \mathbb{R}^n$ by $j \circ \tilde{j}$ is also N and $Q \circ j \circ \tilde{j} = Q$. The quadratic form $\bar{Q} = Q \circ j$ on M is positive non-degenerate; let $\gamma_{\bar{Q}}$ be the corresponding gaussian measure on M, its Fourier transform is $\exp(-\bar{Q}/2)$. There exists a uniquely defined measure on $\mathbb{R}^n$ such that its Fourier transform is $\exp(-Q/2)$, namely the image by j of γ_Q, indeed:

$$\mathcal{F}(j(\gamma_{\bar{Q}})) = \exp(-\bar{Q}/2) \circ \tilde{j} = \exp(-Q/2). \qquad \blacksquare$$

Gaussian promeasures. Let X be a locally convex Hausdorff space, and Q a positive quadratic form on X'.

Theorem. There exists one, and only one, promeasure on X, whose Fourier transform is $\exp(-Q/2)$.

gaussian promeasure

This promeasure is called a **gaussian promeasure of variance Q**.

Proof: The uniqueness results from the injectivity theorem (p. 581). Let us show the existence: let $V \in F(X)$, let P_V be the canonical mapping $X \to X/V$. Then $Q \circ \tilde{P}_V$ defines a positive quadratic form Q_V on $(X/V)'$. Denote by γ_V the corresponding gaussian measure on X/V. We have $Q_V = Q_W \circ \tilde{P}_{VW}$. Therefore the family (γ_V) is a promeasure on X.

Gaussian promeasures on Hilbert spaces.[1]
1) Canonical gaussian promeasure. Let X be a real Hilbert space, its dual X' can be identified with X; the mapping $I: X \to \mathbb{R}$ by $x \mapsto \|x\|^2$ is a positive quadratic form on X. The corresponding gaussian promeasure is called **canonical**. It can be shown that this promeasure is not a measure when X is infinite dimensional. canonical gaussian
2) Let u be a linear continuous operator on X, the quadratic form $Q: x' \mapsto \|\tilde{u}x'\|^2$ defines a promeasure on X. It can be shown that it is equivalent to a measure if and only if u is a Hilbert–Schmidt operator – i.e. an operator such that $\Sigma_{i \in I} \|u(e_i)\|^2 < \infty$ where $\{e_i\}$ is an orthonormal basis of X.

4. THE WIENER MEASURE

The Wiener measure was first introduced as the mathematical expression of Einstein's analysis of Brownian motion[2]: let $x(t)$ be the position of the particle at time t, the successive displacement $x(t_i) - x(t_{i-1})$ are independent random variables with a gaussian probability distribution. This approach to the Wiener measure requires a limiting procedure in which the time interval T is divided into an increasing number of intervals. This limiting procedure is delicate; another approach to Wiener measure is its definition as the image of the canonical gaussian promeasure of variance I on a real Hilbert space by the primitive mapping.

Definition: Let $\mathcal{H}$ be the Hilbert space of real square integrable functions on an interval (a, b) of $\mathbb{R}$ with scalar product $(f|g) = \int_I f(t)g(t)\,dt$ and norm $\|f\| = (f|f)^{1/2}$; the primitive mapping $P: \mathcal{H} \to \mathcal{C}$ is defined by

$$(Pf)(t) = \int_a^t f(r)\,dr = \int_a^b Y(t-r)f(r)\,dr.$$

[1][Segal 1958] has introduced this notion under the same "weak canonical distribution" and applied it to problems in quantum field theory.
[2]See for instance [Nelson 1967].

$\mathscr{C}$ denotes the space of continuous functions on $[a, b]$ vanishing at $t = a$, with the norm $|f|_{\mathscr{C}} = \sup_{t \in T} |f(t)|$.
The dual $\mathscr{C}'$ of $\mathscr{C}$ is the space of bounded measures on the semi-closed interval $T = (a, b]$. We shall denote it[1] by $\mathscr{M}^1$.

Wiener measure

Theorem. The image of the canonical gaussian promeasure on $\mathscr{H}$ by the primitive mapping P is a measure on $\mathscr{C}$. It is called the **Wiener measure.**

Before giving some indications on the proof of this theorem we shall compute the variance W of the image by P of the canonical gaussian promeasure w on $\mathscr{H}$

$$W = I \circ \tilde{P}$$

where $\tilde{P} : \mathscr{M}^1 \to \mathscr{H}'$ is the transposed mapping of P.
The dual $\mathscr{H}'$ of $\mathscr{H}$ is identified with $\mathscr{H}$ (p. 31). We have

$$\langle \tilde{P}\mu, f \rangle = \langle \mu, Pf \rangle, \qquad \forall \mu \in \mathscr{M}^1, \quad f \in \mathscr{H}.$$

Thus

$$\langle \tilde{P}\mu, f \rangle = \int_T Pf \, d\mu = \int_T \left(\int_T Y(t-r) f(r) \, dr \right) d\mu(t)$$

and by Fubini theorem

$$\langle \tilde{P}\mu, f \rangle = \int_T \int_T f(r) Y(t-r) \, d\mu(t) \, dr.$$

Hence

$$(\tilde{P}\mu)(r) = \int_T Y(t-r) \, d\mu(t).$$

The image $W = I \circ \tilde{P}$ of the canonical quadratic form $I : x \mapsto (x|x)$ on $\mathscr{H}$ by $\tilde{P}$ is then

$$W(\mu) = (\tilde{P}\mu | \tilde{P}\mu) = \int_T ((\tilde{P}\mu)(r))^2 \, dr$$
$$= \int_T \left(\int_T Y(t-r) \, d\mu(t) \int_T Y(t'-r) \, d\mu(t') \right) dr.$$

Using Fubini theorem and the fact that

$$\int_T Y(t-r) Y(t'-r) \, dr = \inf (t-a, t'-a),$$

[1] $\mathscr{M}^1$ is the space of measures on $[a, b]$ which do not charge a, cf. for instance [Bourbaki 1969, p. 81].

we obtain

$$W(\mu) = \int_T \int_T \inf (t - a, t' - a) \, \mathrm{d}\mu(t) \, \mathrm{d}\mu(t')$$
$$= \langle \mu(t) \times \mu(t'), \inf (t - a, t' - a) \rangle.$$

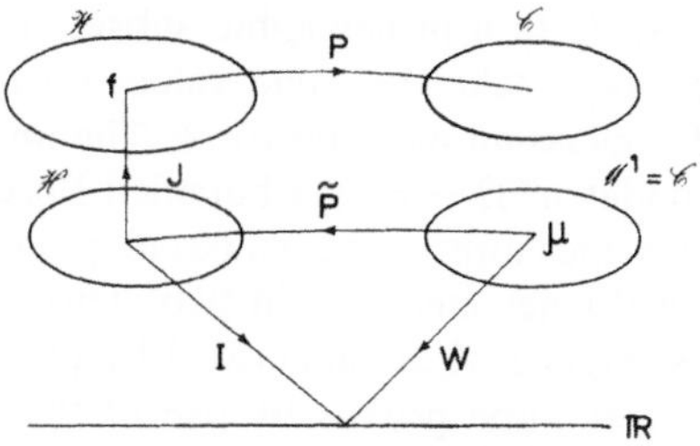

In particular let $t_1, \ldots, t_n \in T$, and $c_1, \ldots c_n$ be real numbers. Let $\mu \in \mathcal{M}^1$ be the linear combination of Dirac measures $\Sigma_{i=1}^n c_i \delta_{t_i}$, then

$$W\Big(\sum_{c=1}^{n} c_i \delta_{t_i}\Big) = \sum_{i,j=1}^{n} c_i c_j \inf (t_i - a, t_j - a).$$

This quadratic form is positive since $W(\mu)$ is positive.

We shall now outline the proof that w is a measure; from now on we set the interval $[a, b] = [0, 1]$ to simplify the notation.
1) Construction of an auxiliary space with measure m. Set $D_n = 1/2^n, 2/2^n, \ldots, 1$ and denote by D the countable set $D = \cup_{n \geq 0} D_n$; let $\Omega = \mathbb{R}^D$ be the space of real functions defined on D, together with the topology of point wise convergence. The family δ_t, $t \in D$ is a Hamel base of Ω' (p. 9). Let M be the positive quadratic form on Ω' defined by

$$M\Big(\sum_{t \in \tilde{D}} c_t \delta_t\Big) = \sum_{t, t' \in \tilde{D}} c_t c_{t'} \inf (t, t'), \quad \tilde{D} \text{ any finite subset of } D;$$

the gaussian promeasure m on Ω of covariance M is a measure since D is countable (p. 577).
2) Construction of a mapping $u: \Omega \to \mathscr{C}$. Let $g \in \Omega$, let $u_n(g)$ be the function which takes the same value as g at each point of D_n and which is affine in each interval $((k-1)/2^n, k/2^n)$.
The mapping $u_n: \Omega \to \mathscr{C}$ is Borel measurable since it is continuous.
Set $T_n(g) = u_{n+1}(g) - u_n(g)$. One shows by some precise computations that there exists a set $\Omega_0 \subset \Omega$ such that $\Omega \backslash \Omega_0$ is m-negligible[1] and such

[1]Contained in a subset of m-measure zero.

that $\Sigma_{n=0}^{\infty} T_n(g)$ is absolutely convergent in $\mathscr{C}$ for every $g \in \Omega_0$. Set

$$u(g) = \begin{cases} \sum_{n=0}^{\infty} T_n(g) = \lim_{n=\infty} u_n(g) & \text{for } g \in \Omega_0 \\ 0 & \text{for } g \in \Omega/\Omega_0. \end{cases}$$

u is an m-measurable mapping from Ω to $\mathscr{C}$: the reciprocal image of a Borel subset in $\mathscr{C}$ is, up to a m-negligible subset, a Borel subset of Ω. The functions $u(g)$ and g take the same values on D for every $g \in \Omega_0$.
3) Construction of a gaussian measure on $\mathscr{C}$. The image of the bounded measure m on Ω under $u: \Omega \to \mathscr{C}$, is a bounded Borel measure $\hat{w}$ on $\mathscr{C}$. One shows that this measure is the gaussian promeasure w on $\mathscr{C}$ of variance W, i.e. the Wiener measure, in two steps.
a) Let $\mathscr{M}_D$ be the subspace of $\mathscr{M}^1$ generated by $\{\delta_t$ for $t \in D\}$, then for every measure μ in $\mathscr{M}_D$, one proves by use of the relevant definitions that

$$\int \exp(-\mathrm{i}\langle\mu, f\rangle)\,\mathrm{d}\hat{w}(f) = \exp(-\tfrac{1}{2}W(\mu)).$$

b) Let $\mu \in \mathscr{M}^1$; it can be proved that there exists a sequence $(\mu_n \in \mathscr{M}_D)$ such that $\lim_{n=\infty} \langle\mu_n, f\rangle = \langle\mu, f\rangle$ for every $f \in \mathscr{C}$ and $\lim_{n=\infty} W(\mu_n) = W(\mu)$.

Wiener integral. The theory of integration given in section I.D applies to integration with respect to the Wiener measure. The Wiener measure is a countably additive set function on the Borel σ-field of $\mathscr{C}$, which is identical with its cylindrical σ-field (cf. Problem D1). In particular a function F on $\mathscr{C}$ is w-integrable if it is w-measurable and $F \leq G$ with G w-integrable. The general properties given (Ch. ID) for measurable functions are valid for Wiener measurability.
The continuous mappings u from $\mathscr{C}$ into a finite dimensional space Y are w-measurable (the reciprocal image of the open set in Y is an open cylindrical set in $\mathscr{C}$).

Example of the calculation of a Wiener integral: The Wiener integral of the quadratic function $f \mapsto |\langle\mu, f\rangle|^2$ defined on $\mathscr{C}$ by a given bounded measure μ is

$$\int_{\mathscr{C}} |\langle\mu, f\rangle|^2\,\mathrm{d}w(f) = \int_{\mathbb{R}} |t|^2\,\mathrm{d}w_\mu(t)$$

where w_μ is the image of the Wiener measure on $\mathscr{C}$ by the linear continuous mapping $\mathscr{C} \to \mathbb{R}$ by $f \mapsto \langle\mu, f\rangle$. It is therefore the gaussian measure of covariance $W \circ \tilde{\mu}$. Since $\tilde{\mu}: R \to \mathscr{M}^1$ is given by $t' \mapsto t'\mu$ and

W is quadratic we have

$$(W \circ \tilde{\mu})(t') = W(t'\mu) = t'^2 W(\mu).$$

w_μ is the gaussian measure on $\mathbb{R}$ with covariance $W(\mu)$ (cf. p. 581)

$$\mathrm{d}w_\mu = \frac{1}{\sqrt{2\pi W(\mu)}} \exp\left(\frac{-t^2}{2W(\mu)}\right) \mathrm{d}t,$$

thus

$$\int_{\mathscr{C}} |\langle \mu, f\rangle|^2 \,\mathrm{d}w(f) = \frac{1}{\sqrt{2\pi W(\mu)}} \int_{\mathbb{R}} t^2 \exp\left(-\frac{t^2}{2W(\mu)}\right) \mathrm{d}t$$

$$= (2\pi)^{-1/2} W(\mu) \int_{\mathbb{R}} t^2 \,\mathrm{e}^{-t^2/2} \,\mathrm{d}t = W(\mu).$$

Application: Let us take $\mu = \delta_t$, then

$$\int_{\mathscr{C}} |\langle \delta_t, f\rangle|^2 \,\mathrm{d}w(f) = W(\delta_t) = t.$$

The physical interpretation of this equality is that the square of the distance a particle diffuses in the Brownian motion is proportional to the time it has been diffusing.

Sequential Wiener integral. It follows from the definition of w as the image of $m = w_{\mathbb{R}^D}$ by u that the function $F: \mathscr{C} \to \mathbb{R}$ is integrable with respect to w iff $F \circ u$ is integrable with respect to m on $\mathbb{R}^D$; then

$$\int_{\mathscr{C}} F(f) \,\mathrm{d}w(f) = \int_{\mathbb{R}^D} (F \circ u)(g) \,\mathrm{d}m(g).$$

u has been defined as a limit of mappings $u_n: \mathbb{R}^D \to \mathscr{C}$, therefore

$$\int_{\mathbb{R}^D} (F \circ u)(g) \,\mathrm{d}m(g) = \int_{\mathbb{R}^D} (F \circ \lim_{n=\infty} u_n)(g) \,\mathrm{d}m(g).$$

On the other hand we know that the range of u_n is a finite dimensional subspace $E_n \subset \mathscr{C}$, where E_n is the space of functions which are affine in each interval $((k-1)/2^n, k/2^n)$.

If the function $F \circ u_n$ is m integrable on $\mathbb{R}^D$ we have

$$\int_{\mathbb{R}^D} (F \circ u_n)(g) \,\mathrm{d}m(g) = \int_{\mathbb{R}^D} (F|_{E_n} \circ u_n)(g) \,\mathrm{d}m(g)$$

$$= \int_{E_n} F|_{E_n}(x) \,\mathrm{d}w_n(x)$$

where $F|_{E_n}$ stands for the restriction of F to E_n, and where w_n is the image by u_n of the gaussian measure $m = w_\Omega$, i.e. w_n is the gaussian measure on E_n of variance $W \circ \tilde{u}_n$.
The space E_n is isomorphic to $\mathbb{R}^N$ with $N = 2^n$ by the mapping

$$i_N : x \mapsto q = \{q^i = x(t_i) - x(t_{i-1})\} \qquad i = 1, \dots N$$

where $x : t \mapsto x(t)$ is a function of E_n and

$$t_0 = 0, t_1 = 1/2^n, t_2 = 2/2^n, \dots, t_N = 1.$$

The reason for choosing this isomorphism will become clear later. Hence

$$\int_{E_n} (f|_{E_n})(x)\, \mathrm{d}w_n(x) = \int_{\mathbb{R}^n} (f|_{E_n} \circ i_N^{-1})(q)\, \mathrm{d}w_N(q)$$

where w_N is the gaussian measure on $\mathbb{R}^n$ of variance

$$W_N = W \circ \tilde{u}_n \circ \tilde{i}_N = W \circ (\widetilde{i_N \circ u_n}).$$

The mapping $v_N = i_N \circ u_n : \Omega \to \mathbb{R}^N$ is defined by

$$v_N : g \mapsto v_N(g) = \{g(t_i) - g(t_{i-1})\}.$$

The transposed mapping from $\mathbb{R}^n$ into Ω' is such that

$$\langle \tilde{v}_n(q), g \rangle = \langle q, v_N(g) \rangle$$

where the left hand side bracket is the duality of Ω' and Ω and where the right hand side one is the duality of $\mathbb{R}^N$ and $\mathbb{R}^N$, i.e.

$$\langle q, v_N(g) \rangle = \sum_{i=1}^{n} q_i(g(t_i) - g(t_{i-1})).$$

Hence

$$\langle \tilde{v}_N(q), g \rangle = \left\langle \sum_{i=1}^{n} q_i(\delta_{t_i} - \delta_{t_{i-1}}), g \right\rangle$$

where δ_{t_i} is the Dirac measure at t_i.
Thus $\tilde{v}_N(q)$ is the bounded measure, element of Ω', equal to

$$\tilde{v}_N(q) = \sum_{i=1}^{n} q_i(\delta_{t_i} - \delta_{t_{i-1}}).$$

W being the infimum variance previously defined, the variance W_N is equal to

$$W_N(q) = W\Big(\sum_{i=1}^{N} q_i(\delta_{t_i} - \delta_{t_{i-1}})\Big) = \sum_{i,j=1}^{N} q_i q_j C^{ij}.$$

Let W designate both the quadratic form W and the corresponding bilinear form, then

$$C^{ij} = W(\delta_{t_i} - \delta_{t_{i-1}}, \delta_{t_j} - \delta_{t_{j-1}})$$
$$= \delta_{ij}(t_i - t_{i-1}) \text{ where } \delta_{ij} \text{ is the Kronecker symbol.}$$

Hence, thanks to the choice made for i_N we obtain for W_N the diagonalized expression

$$W_N(q) = \exp\left(-\sum_{i=1}^{N} \alpha_i (q_i)^2\right), \qquad \alpha_i = t_i - t_{i-1}.$$

The measure w_N, in the dual basis of $\mathbb{R}^N$ is the gaussian measure

$$w_N = \gamma_{\alpha_1} \otimes \cdots \otimes \gamma_{\alpha_N}.$$

By definition the sequential Wiener integral is the limit, if it exists

$$\int_{\mathscr{C}}^{\text{sw}} F(f)\, \mathrm{d}w(f) = \lim_{N=\infty} \int_{\mathbb{R}^N} (F|_{E_n} \circ v_N)(q) \times \prod_{i=1}^{N} \frac{1}{\sqrt{2\pi\alpha_i}} \exp(-q_i^2/2\alpha_i)\, \mathrm{d}q_i.$$

This is indeed the original definition of the Wiener integral, mathematical foundation of the Brownian motion where successive displacements q_i are treated as independent random variables with a gaussian probability distribution. This definition coincides with the previous one for those functions F for which it is legitimate to interchange the limits as n tends to infinity: it is not always the case, but it is so for a fairly wide class of smooth functions[1] F on $\mathscr{C}$.

PROBLEMS AND EXERCISES

PROBLEM A. THE KLEIN–GORDON EQUATION

Consider the Klein–Gordon equation on $\mathbb{R}^n \times \mathbb{R}$

$$\frac{\partial^2 u}{\partial t^2} = \Delta u - m^2 u, \quad \Delta = \sum_{i=1}^{n} \frac{\partial^2}{(\partial x^i)^2}.$$

Suppose $u(t, \cdot) \in H^1(\mathbb{R}^n)$, $\dfrac{\partial u}{\partial t}(t, \cdot) \equiv \dot{u}(t, \cdot) \in L^2(\mathbb{R}^n)$.

[1]cf. [Cameron 1960].

Denote by ω the 2 form on $E = H^1(\mathbb{R}^n) \times L^2(\mathbb{R}^n)$ defined by the Lebesgue measure on $\mathbb{R}^n$ by

$$\omega((U_0, V_0), (U_1, V_1)) = \int_{\mathbb{R}^n} (V_0 U_1 - V_1 U_0)\,\mathrm{d}x.$$

Show that ω is a weak symplectic form on E, and that the Klein–Gordon equation can be written as a hamiltonian system.

Answer: Consider the 1-form on E:

$$\theta_{(u,v)}((U, V)) = \int_{\mathbb{R}^n} vU\,\mathrm{d}x, \qquad (u, v) \in E, \qquad (U, V) \in E.$$

Its differential is (cf. p. 551)

$$\begin{aligned}(\mathrm{d}\theta)_{(u,v)}((U_0, V_0), (U_1, V_1)) &= (\theta'_{(u,v)} \cdot (U_0, V_0))(U_1, V_1) \\ &\quad - (\theta'_{(u,v)} \cdot (U_1, V_1))(U_0, V_0) \\ &= \int_{\mathbb{R}^n} (V_0 U_1 - V_1 U_0)\,\mathrm{d}x\end{aligned}$$

therefore ω (which is a constant form on E) is the differential of θ, and hence a weak symplectic form on E.

Let us show that the vector field defined on the subspace $F = H^2 \times H^1$ of E by

$$X_{(u,v)} = v \in H^1, \qquad Y_{(u,v)} = \Delta u - m^2 u \in L^2$$

is hamiltonian for ω. Indeed

$$(i_{(X,Y)}\omega)(U, V) = \omega((X, Y), (U, V)) = \int_{\mathbb{R}^n} ((\Delta u - m^2 u)U - Vv)\,\mathrm{d}x$$

$$= -\mathrm{d}\Big(\int_{\mathbb{R}^n} (\tfrac{1}{2}v^2 + \tfrac{1}{2}\textstyle\sum (\partial u/\partial x^i)^2 + \tfrac{1}{2}m^2u^2)\,\mathrm{d}x\Big)(U, V).$$

The hamiltonian H is the "energy integral" (cf. p. 514)

$$H(u, v) = \tfrac{1}{2}\int_{\mathbb{R}^n} (v^2 + \textstyle\sum (\partial u/\partial x^i)^2 + m^2u^2)\,\mathrm{d}x.$$

The associated differential system is

$$\mathrm{d}u/\mathrm{d}t = v, \qquad \mathrm{d}v/\mathrm{d}t = \Delta u - m^2 u,$$

it is not a lipshitzian differential system on E (it is not even defined on E, but only on the subspace F of E); it is not either a differential system on F since $(v, \Delta u - m^2 u)$ is not in F. The fact that this differential system has a flow: $(u_t, v_t) \in F$ if $(u_0, v_0) \in F$ is a consequence of the existence properties proved for the solutions of the wave equation.

Reference: [Lichnerowicz 1961]; I. Segal, J. Math. Pures et Appl. XLIV (1965) 71–113.

PROBLEM B. APPLICATION OF THE LERAY–SCHAUDER THEOREM

We consider the second order partial differential equation on the bounded open set $U \subset \mathbf{R}^n$ *with* ∂U *regular*

$$\Delta u + f(x, u) = 0, \qquad x \in U, \qquad \Delta = \sum_{i=1}^{n} \frac{\partial^2}{(\partial x^i)^2}$$

where $f: \bar{U} \times I \to \mathbf{R}$ *is* C^1, I *interval* $(a, b) \subset \mathbf{R}$. *We assume that there exist constants* l *and* m *such that*

$$a < l < m < b$$

and

$$f(x, l) > 0, \qquad f(x, m) < 0, \qquad \forall x \in \bar{U}.$$

Prove that the partial differential equation has at least one C^2 *solution which satisfies the boundary condition*

$u|_{\partial U} = \varphi$, $\varphi \in C^1(\partial U)$, $l < \varphi < m$ *and* $l < u(x) < m$, $\forall x \in \bar{U}$.

Answer: On U the Laplace equation has a unique C^2 solution u_0 satisfying the boundary condition (cf. VIC, §3). By the maximum theorem (p. 500)

$$l < u_0 < m.$$

Consider the mapping $F: \Omega \times [0, 1] \to C^0(\bar{U})$ by $(u, t) \mapsto F(u, t)$, $\Omega = \{u \in C^0(\bar{U});\ l < u < m\}$, where the function $v = F(u, t)$ is given by the formula

$$v(x) = -t \int_U G(x, y) f(y, u(y))\, dy + u_0(x);$$

with $G(x, y)$ the Green function of Δ relative to U.
v is in $C^1(U)$ and its first derivatives are uniformly bounded for $u \in \bar{\Omega}$ and $t \in [0, 1]$. Therefore F is a compact mapping on the bounded subset $\Omega \subset C^0(\bar{U})$.
We are now dealing with the functional equation

$$u - F(u, t) = 0$$

or

$$u=-t\int_U G(\cdot,y)f(y,u(y))\,dy+u_0.$$

When $t=0$ the equation $u-F(u,t)=0$ has an unique solution $u=u_0$ and $\deg(\mathrm{Id}-F(\cdot,0),\Omega,0)=1$.
If we can show that the equation $u-F(u,t)=0$ has no solution on the boundary of Ω, then all the conditions of the Leray–Schauder theorem are satisfied. On the other hand since $u\in C^1(\bar U)$ it follows that (Ch. VI §3) if $u-F(u,t)=0$ then $u\in C^2(U)$ and u satisfies

$$\Delta u+tf(x,u)=0.$$

In other words the functional equation and our original equation are equivalent when $t=1$.
To prove that $\Delta u+tf(x,u)=0$, $u|_{\partial U}=\varphi$ has no solution on $\partial\Omega$ we show that if a solution satisfies $l\leq u\leq m$ [i.e. $u\in\bar\Omega$] it satisfies $l<u<m$ [i.e. $u\in\Omega$]. The property is already known for $t=0$. Now if $t\in[0,1]$

$$\Delta(u-l)+t[f(x,u)-f(x,l)]=-tf(x,l)<0$$

$$\Delta(u-m)+t[f(x,u)-f(x,m)]=-tf(x,m)>0.$$

A point x where $u(x)=l$ [resp. $u(x)=m$] is a minimum [resp. a maximum] of u, and interior to U: this is incompatible with the above inequalities.

PROBLEM C1. THE REEB THEOREM

Show that if X is a closed n-dimensional manifold and f smooth on X with only two critical points which are non degenerate, then X is homeomorphic to a sphere.

Answer: X being closed (compact without boundary) f must attain on X its minimum a at a point x_1 and its maximum b at x_2. Thus x_1 and x_2 are the two critical points, and their indices are respectively zero and n since $f''(x_1)$ is positive definite and $f''(x_2)$ negative definite (cf. p. 84). Therefore, if $c<b-a$, the set

$$\{x\in X,\ f(x)\leq a+c\}$$

is diffeomorphic to a closed n ball and the whole of X is diffeomorphic to the manifold obtained by attaching a n-cell to this n ball. It is easy to see that X is therefore homeomorphic to S^n. X is not necessarily diffeomorphic to S^n with its usual differentiable structure if $n \geq 7$ (cf. [Milnor]).

PROBLEM C2. THE METHOD OF STATIONARY PHASE

This method gives asymptotic evaluations, for large real k, of integrals of the form

$$I(k) = \int_X a(x) \exp(\mathrm{i}kf(x))\, \mathrm{d}\mu(x)$$

where a and f are real valued C^∞ functions on a C^∞, n-dimensional manifold X, and $\mathrm{d}\mu(x)$ is the volume element of some given riemannian metric on X.

a) *Let h be a C^2 function on* $\mathbb{R}$ *which, together with its first two derivatives, is bounded by a number M. Show that the improper* (p. 47) *riemann integral* $\int_{-\infty}^{+\infty} \exp(\mathrm{i}kt^2/2)h(t)\,\mathrm{d}t$ *is defined for every $k \neq 0$ and bounded if $|k| \geq 1$ by a number depending only on M.*

b) *Show that if a has compact support and f has no critical point* (p. 546) *on* Supp a, *then*

$$I(k) = \mathrm{O}(k^{-N}) \qquad \textit{for any } N.$$

c) *Show that if a has compact support and f has one, non degenerate, critical point y on* Supp a, *then*

$$I(k) = (2\pi/k)^{n/2} \exp(\mathrm{i}kf(y)) \exp(\mathrm{i}\pi \operatorname{sign} Hf(y)/4) a(y)/|Hf(y)|^{1/2} + \mathrm{O}(k^{-n/2-1})$$

where sign $Hf = q - p = n - 2p$ *where p is the index of the hessian* (p. 571) *of f at the point y and $|Hf|$ its scalar determinant relative to g, that is in coordinates*

$$|Hf| = |\det \partial^2 f/\partial x^i\, \partial x^j|\, |\det g_{ij}|.$$

Answer: a) Let $0 < A < B$. By partial integrations we get for non zero $\lambda \in \mathbb{C}$

$$\int_A^B \exp(-\lambda t^2/2)h(t)\,\mathrm{d}t = \frac{-1}{\lambda t}\left(h(t) - \frac{h(t)}{\lambda t^2} + \frac{h'(t)}{\lambda t}\right)\exp(-\lambda t^2/2)\Big|_A^B$$

$$+ \frac{1}{\lambda^2 t^2}\int_A^B \exp(-\lambda t^2/2)\left(3\frac{h(t)}{t^2} - 3\frac{h'(t)}{t} + h''(t)\right)\mathrm{d}t.$$

b) Let $\boldsymbol{v}$ denote the contravariant vector field, associated by the metric $\boldsymbol{g}$ with the covariant vector field $\boldsymbol{f}'$ (p. 568), $v^i = g^{ij}\,\partial f/\partial x^j$. The action of $\boldsymbol{v}$ on the function exp (ikf) (derivation p. 544) is

$$v(\exp(\mathrm{i}kf)) = \mathrm{i}kg(v, v)\exp(\mathrm{i}kf) = \mathrm{i}kg^{ij}\frac{\partial f}{\partial x^i}\frac{\partial f}{\partial x^j}\exp(\mathrm{i}kf).$$

Thus if we denote $w = -(\mathrm{i}/k)v/g(v, v)$ we have $w(\exp \mathrm{i}kf) = \exp(\mathrm{i}kf)$. Hence

$$\int_X a(x)\exp(\mathrm{i}kf)\,\mathrm{d}\mu = \int_X a(x)w(\exp \mathrm{i}kf(x))\,\mathrm{d}\mu$$

$$= -\int_X \exp(\mathrm{i}kf)\operatorname{div}(aw)\,\mathrm{d}\mu = (\mathrm{i}/k)\int_X b\exp(\mathrm{i}kf)\,\mathrm{d}\mu$$

where $b = \operatorname{div} av/g(v, v)$ is a C^∞ function with compact support, independent of k. By repeating this operation we see that $I_k = O(k^{-N})$.

c) By the Morse lemma there exists a neighborhood U of y and coordinates $(x^1, \ldots, x^n)$ such that if we also choose the coordinates such that $y^i = 0$, $i = 1, \ldots, n$

$$f(x) = f(y) + Q(x^1, \ldots, x^n) \qquad \text{in } U$$

where Q is the quadratic form with signature

$$Q = \frac{1}{2}\sum_{i=1}^{q}(x^i)^2 - \frac{1}{2}\sum_{i=q+1}^{p+q}(x^i)^2.$$

In these coordinates

$$|\det \partial^2 f/\partial x^i\,\partial x^j| = 1$$

and the integral $I(k)$ reads:

$$I(k) = \int_{\mathbb{R}^n} \bar{b}(x^1, \ldots, x^n)\exp(\mathrm{i}kQ(x^1, \ldots, x^n))\,\mathrm{d}x^1 \ldots \mathrm{d}x^n$$

with

$$\bar{b}(x^1, \ldots, x^n) = \exp(\mathrm{i}kf(y))\bar{a}(x^1, \ldots, x^n)|\overline{Hf}(x^1, \ldots, x^n)|^{-1/2}.$$

This smooth function can always be written

$$\bar{b}(x^1, \ldots, x^n) = b(y) + \sum_i x^i b_i(x^1, \ldots, x^n)$$

where the b_i are C^∞ functions and

$$b(y) = \bar{b}(0, \ldots, 0) = \exp(\mathrm{i}kf(y))a(y)|Hf(y)|^{-1/2}.$$

The integral $I(k)$ is defined (absolutely convergent) since b has compact support. It can, in particular, be computed by the limiting procedure:

$$I(k) = \lim_{A=+\infty} \int_{-A}^{A} \cdots \int_{-A}^{A} \bar{b}(x^1, \ldots, x^n) \exp(ikQ(x^1, \ldots, x^n))\, dx^1 \ldots dx^n$$

we know the limit

$$\begin{aligned} I_0(k) &= \lim_{A=+\infty} \int_{-A}^{A} \cdots \int_{-A}^{A} b(y) \exp(ikQ(x^1, \ldots, x^n))\, dx^1 \ldots dx^n \\ &= b(y)\Big(\int_{\mathbf{R}} \exp(ikt^2/2)\, dt\Big)^q \Big(\int_{\mathbf{R}} \exp(-ikt^2/2)\, dt\Big)^p \\ &= b(y)(2\pi/k)^{n/2} \exp(i\pi \operatorname{sign} Hf/4). \end{aligned}$$

The remainder $I_1(k) = I(k) - I_0(k)$ is estimated as follows

$$I_1(k) = \lim_{A=+\infty} \int_{-A}^{A} \cdots \int_{-A}^{A} \exp(ikQ(x^1, \ldots, x^n) x^j b_j(x^1, \ldots, x^n)\, dx^1 \ldots dx^n.$$

We note that if $h(A) = h(-A)$

$$\int_{-A}^{A} \exp(ikt^2/2) t h(t)\, dt = -\frac{1}{ik} \int_{-A}^{A} \exp(ikt^2/2) h'(t)\, dt;$$

the integral on the left hand side is therefore the product by k^{-1} of a bounded integral (when A tends to infinity), if h is a C^3 function with bounded first three derivatives.

The functions b_j can be chosen, since b has compact support, such that $b_j(x^1, \ldots, A, \ldots, x^n) = -b_j(x^1, \ldots, -A, \ldots, x^n)$ (take $b_1(x^1, \ldots, x^n) = (b(x^1, \ldots, x^n))/x_1, \ldots, b_n(x^1, \ldots, x^n) = (b(0, \ldots, x^n) - b(0, \ldots, 0))/x^n$. We thus obtain for $I_1(k)$ the product of k^{-1} by an integral of the form of $I(k)$ itself, thus the indicated estimate.

Remark: If f has a finite number N of critical points on X and if these points are not degenerate, $I(k)$ is obviously the sum of N terms, each given by an expression of the type derived above.

PROBLEM D1. A METRIC ON THE SPACE OF PATHS WITH FIXED END POINTS

Let X be a smooth finite dimensional proper riemannian manifold. The points of $\Omega(X, x, y)$ are the piecewise differentiable maps $C: [0, 1] \to X$ with $c(0) = x$, $c(1) = y$, which are parametrized proportionally to arc length. The distance between two points c and c' in $\Omega(X, x, y)$ is given by

$$d_1(c, c') = \max_{t \in [0,1]} \{d(c(t), c'(t))\} + |J(c) - J(c')|$$

where d is the distance and J the length function in the metric of X.
1) *Prove that d_1 is a distance on $\Omega(X, x, y)$.*
2) *Prove that $\Omega(X, x, y)$ is not complete (p. 25) in this metric topology.*

Proof: 1) Straightforward.
2) Consider the sequence (c_n) of paths in $\mathbb{R}^2$, where the path c_n is composed of n triangles of height $1/n$ along the unit interval I of the

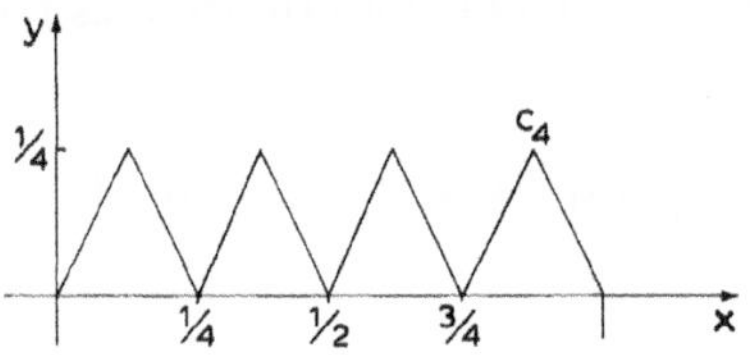

x-axis. The length of each path is $\sqrt{5}$, so that

$$d_1(c_n, c_k) = \max_{t \in [0,1]} \{d(c_n(t), c_k(t))\} \leq \max\left(\frac{1}{n}, \frac{1}{k}\right).$$

Thus (c_n) is a Cauchy sequence. If the sequence converges to a path f, it follows that

$$\lim_{n \to \infty} \left[\max_{t \in [0,1]} \{d(c_n(t), f(t))\} \right] = 0$$

and that $J(f) = \sqrt{5}$. The first condition implies that $f = I$ which is impossible since $J(I) \neq \sqrt{5}$.
The advantage of this definition is that in this topology of $\Omega(X, x, y)$ $J(c)$, the length of c, is a continuous function.

Reference: R. Bott, The stable homotopy of the classical groups, Annals of Mathematics 70 (1959) 313.

PROBLEM D2. MEASURES INVARIANT UNDER TRANSLATION

a) *Show that there is no measure invariant by translation on a Hilbert space E, such that the measure of every bounded open ball*

$$\|x\| < \rho$$

is finite.
b) *Show that there is no Borel measure on $L^2(0, 1)$ invariant by translation.*

Answer: a) Let a be the measure of balls of radius ϵ, $\epsilon < 1$. Consider the family of balls B_n

$$\|x - x_n\| < \epsilon$$

with $x_n = \lambda e_n$, where $\{e_n\}$ is an orthonormal, infinite subset:

$$\|x_n - x_m\| = \lambda \|e_n - e_m\| = \lambda\sqrt{2}.$$

The family is included in the unit ball if $\epsilon + \lambda < 1$, and disjoint if $\lambda\sqrt{2} > 2\epsilon$, since if B_n and B_m have a common point x:

$$\|x_n - x_m\| < \|x_n - x\| + \|x_m - x\| < 2\epsilon.$$

The measure of the unit ball should therefore by additivity, be infinite.
b) Use the above arguments with the compact subsets of $L^2(0, 1)$ (cf. p. 492)

$$\|f - f_n\|_{H_1(0,1)} < \epsilon.$$

PROBLEM D3. CYLINDRICAL σ-FIELD OF $C([a, b])$

a) *Show that the cylindrical σ-field of $C([a, b])$ is identical with its Borel σ-field, and also with the σ-field generated by the quasi intervals (p. 576) $\alpha_i < \langle \delta_{t_i}, x\rangle < \beta_i$, $x \in C([a, b])$, $\alpha_i \in \mathbb{R}$ if $\{t_i\}$ is a countable dense subset of $[a, b]$.*
b) *Same question for $\mathscr{C}([a, b])$, subspace of $C([a, b])$ of functions vanishing for $t = a$.*

Answer: a) The cylinder $\alpha_i < \langle \delta_{t_i}, x\rangle \equiv x(t_i) < \beta_i$ is a cylinder set based on an open interval of $\mathbb{R}$. The σ-field $\mathscr{H}$ generated by such cylinders is therefore contained in the cylindrical σ-field of $C([a, b])$, which is itself contained in the Borel σ-field $\mathscr{B}$. We shall show that if $\{t_i\}$ is countable and dense in $[a, b]$, the σ-field $\mathscr{H}$ generated by the quasi intervals is identical with the Borel σ-field $\mathscr{B}$. To prove this, it is enough to prove that any open set is an element of $\mathcal{H}$, or that any open ball is an element of $\mathcal{H}$ (since an open set in $C([a, b])$ is a countable union of open balls). An open ball $\|x\|_{\mathscr{C}([a,b])} \equiv \sup_{t\in[a,b]}|x(t)| < \epsilon$ is in $\mathcal{H}$ if the mapping $x \mapsto \|x\|_{C([a,b])}$ is $\mathcal{H}$-measurable. But, if $\{t_i\}$ is a dense subset of $[a, b]$

$$\|x\|_{C[(a,b])} = \sup_i |x(t_i)|$$

each mapping $x \mapsto x(t_i)$ is $\mathcal{H}$-measurable by the very definition of $\mathcal{H}$; therefore the same is true for $x \mapsto |x(t_i)|$ and $\sup_i |x(t_i)|$ if t_i is a countable subset, by the general properties of measurable mappings.

b) $\mathscr{C}([a, b])$ is a closed subspace of $C([a, b])$ and an element of $\mathcal{H}$, its Borel σ-field and its $\mathcal{H}$ σ-field are the traces of the Borel σ-field and $\mathcal{H}$ σ-fields of $C([a, b])$.

PROBLEM D4. GENERALIZED WIENER INTEGRAL OF A CYLINDRICAL FUNCTION FOR PATHS f WITH VALUES IN $\mathbb{R}^n$; $f: T \to \mathbb{R}^n$

Let the variance W of the generalized Wiener measure w on $\mathscr{C}(T)$ be defined by the bilinear form W on $\mathcal{M}^1(T)$

$$W(\mu, \nu) = \int_T d\mu_\alpha(r) \int_T d\nu_\beta(s) G^{\alpha\beta}(r, s)$$

where the covariance G is a vector valued continuous two point function on T.

Let the cylindrical function $F = u \circ P$ where $P : \mathscr{C}(T) \to \mathbb{R}^{pn}$ by $f \mapsto x$, x is

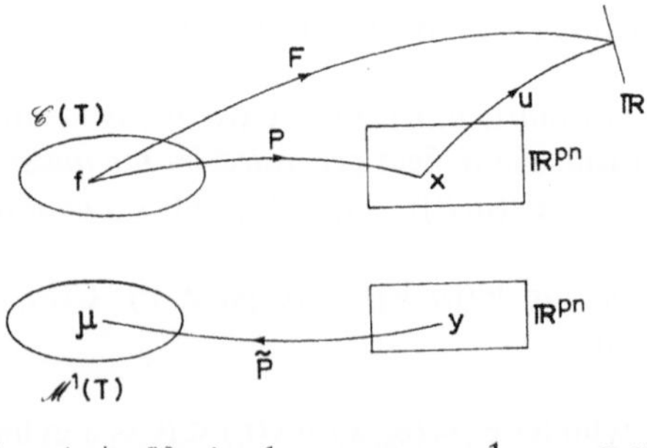

the pn tuple $\{x^{i\alpha} = \langle \mu^{i\alpha}, f \rangle\}$, $i = 1, \ldots, p$; $\alpha = 1, \ldots, n$ with $\mu^{i\alpha}$ bounded measures on $\mathbb{R}$. For instance $\mu^{i\alpha} = \delta^\alpha_{t_i}$ is the Dirac vector valued measure at t_i,

$$\langle \delta^\alpha_{t_i}, f \rangle = f^\alpha(t_i).$$

a) *Compute the image, w_P under P of the Wiener measure w on $\mathscr{C}(T)$.*

b) *Compute* $I = \int_{\mathscr{C}(T)}\langle\mu^1, V(f)\rangle\langle\mu^2, V(f)\rangle \cdots \langle\mu^p, V(f)\rangle\, dw(f)$ *where* V *is a linear continuous mapping from* $\mathscr{C}(T)$ *into* $\mathscr{C}(T)$: $V(f)(t) = v(t)f(t)$.

c) *Change of order of integration. Set*

$$I(\mu^1, \dots \mu^p) = \int_{\mathscr{C}(T)} dw(f) \int_T d\mu^1(t_1)f(t_1) \dots \int_T d\mu^p(t_p)f(t_p)$$

$$J(\mu^1, \dots \mu^p) = \int_T d\mu^1(t_1) \dots \int_T d\mu^p(t_p)K$$

where

$$K = \int f(t_1) \dots f(t_p)\, dw(f) = I(\delta_{t_1}, \dots \delta_{t_p})$$

Show that $I = J$, *i.e. show that it is possible to change the order of integration over* T *and over* $\mathscr{C}(T)$.

d) *Sketch the Feynman diagram technique.*

Answer: a) The image w_P under P of the Wiener measure on $\mathscr{C}(T)$ is the normalized gaussian measure on $\mathbb{R}^{pn}$ of variance

$$\mathscr{W} = W \circ \tilde{P}$$

$\tilde{P} : \mathbb{R}^{pn} \to \mathscr{M}^1(T)$. Since $\langle\tilde{P}y, f\rangle = \langle y, Pf\rangle = \Sigma_{i,\alpha}\, y_{i\alpha}\langle\mu^{i\alpha}, f\rangle$, $\tilde{P}y = \Sigma\, y_{i\alpha}\mu^{i\alpha}$ and

$$\mathscr{W}(y) = W\left(\sum y_{i\alpha}\mu^{i\alpha}\right) = \sum y_{i\alpha}y_{j\beta}W(\mu^{i\alpha}, \mu^{j\beta}).$$

Set $\mathscr{W}^{i\alpha j\beta} = W(\mu^{i\alpha}, \mu^{j\beta})$ the covariance of w_P

$$dw_P(x) = dx^{11} \dots dx^{pn}(2\pi)^{-pn/2}(\det \mathscr{W}^{-1}_{i\alpha j\beta})^{1/2} \exp(-\tfrac{1}{2}x^{i\alpha}x^{j\beta}\mathscr{W}^{-1}_{i\alpha j\beta})$$

and the generalized Wiener integral of a cylindrical function is

$$\int_{\mathscr{C}} F(f)\, dw(f) = \int_{\mathbb{R}^{pn}} u(x)\, dw_P(x).$$

b) One dimensional case, $f: T \to \mathbb{R}$

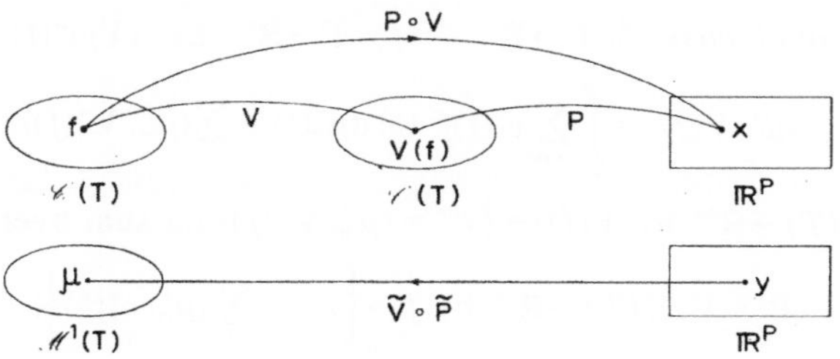

Let $P: \mathscr{C}(T) \to \mathbb{R}^p$ by $V(f) \mapsto x$ where x is the p-tuple $(x^i = \langle \mu^i, V(f) \rangle)$. $P \circ V: \mathscr{C}(T) \to \mathbb{R}^p$ by $f \mapsto x$.

$$I = \int_{\mathbb{R}^p} x^1 \dots x^p \, dw_{P \circ V}(x).$$

$w_{P \circ V}$ is the normalized gaussian on $\mathbb{R}^p$ of variance $W \circ \tilde{V} \circ \tilde{P}$ and covariance $\mathscr{W}^{ij} = (W \circ \tilde{V})(\mu^i, \mu^j)$. For instance if λ is the Lebesgue measure

$$\langle \lambda, V(f) \rangle = \int_T v(t) f(t) \, dt \text{ and}$$

$$(W \circ \tilde{V})(\lambda, \lambda) = \int_T dr \int_T ds \, v(r) v(s) G(r, s).$$

We can compute I directly. It is simpler, however, to compute it as follows:
Reexpress the integrand as follows.

$$\langle \mu^1, V(f) \rangle \langle \mu^2, V(f) \rangle \cdots \langle \mu^p, V(f) \rangle$$
$$= \left\langle \sum_i \mu^i, V(f) \right\rangle^p - \sum_k \langle \mu^1 + \cdots + \hat{\mu}^k + \cdots \mu^n, V(f) \rangle^p + \sum_{k,l} \cdots - \sum_i \langle \mu^i, V(f) \rangle^p.$$

Use

$$J = \int_{\mathscr{C}(T)} \langle \nu, g \rangle^p \, dw(g) = \frac{p!}{2^{p/2}(p/2)!} (W(\nu))^{p/2} \quad \text{if } p \text{ is even}$$
$$= 0 \quad \text{if } p \text{ is odd.}$$

One obtains

$$I = \sum_{\text{part}} \mathscr{W}^{i_1 i_2} \mathscr{W}^{i_3 i_4} \dots \mathscr{W}^{i_{p-1} i_p} \quad \text{if } p \text{ is even}$$
$$= 0 \quad \text{if } p \text{ is odd,}$$

the sum is taken over all partitions of $\{1, \dots p\}$. There are $p!/2^{p/2}(p/2)! = 1 \cdot 3 \cdot 5 \dots (p-1)$ terms in this sum, as can be seen readily by setting $\mu^1 = \mu^2 = \cdots = \mu^p = \nu$ in I and comparing the result with J.

n dimensional case, $f: T \to \mathbb{R}^n$; $V(f): T \to \mathbb{R}^m$, *i.e.* $(Vf)^\alpha(t) = v^\alpha_\beta(t) f^\beta(t)$

$$\langle \mu^i, V(f) \rangle = \int_T \sum_{\alpha, \beta} v^\alpha_\beta(t) f^\beta(t) \, d\mu^i_\alpha(t) = \sum_\alpha \langle \mu^i_\alpha, V^\alpha(f) \rangle.$$

Let $P: \mathscr{C}(T) \to \mathbb{R}^{pn}$ by $V(f) \mapsto \{x^{i\alpha} = \langle \mu^i_\alpha, V^\alpha(f) \rangle$ no sum over $\alpha\}$

$$P \circ V: \mathscr{C}(T) \to \mathbb{R}^{pn} \text{ by } f \to \left\{ x^{i\alpha} = \sum_\beta \langle \mu^i_\alpha, v^\alpha_\beta f^\beta \rangle \right\}.$$

By an argument similar to the previous one, one obtains

$$J = \sum_{\text{part}} \mathcal{W}^{i_1\alpha_1 i_2\alpha_2} \dots \mathcal{W}^{i_{p-1}\alpha_{p-1} i_p\alpha_p} \quad \text{if } p \text{ is even}$$

$$= 0 \qquad\qquad \text{if } p \text{ is odd}$$

where

$$\mathcal{W}^{i\alpha j\beta} = (W \circ \tilde{V})(\mu^{\,i}_{\alpha}, \mu^{\,j}_{\beta}).$$

c) Change of order of integration: the result is a consequence of the Fubini theorem. We shall, for illustration, compute it when $V(f) = f$ and $f: T \to \mathbb{R}$. We have then:

$$I = \sum_{\text{part}} W(\mu^{i_1}, \mu^{i_2}) \dots W(\mu^{i_{p-1}}, \mu^{i_p})$$

on the other hand

$$J = \int_T d\mu^1(t_1) \dots \int_T d\mu^p(t_p) \sum_{\text{part}} W(\delta(t_{i_1}), \delta(t_{i_2})) \dots W(\delta(t_{i_{p-1}}), \delta(t_{i_p}))$$

where for typographical convenience $\delta(t_i) = \delta_{t_i}$, hence $J = I$.

d) *Feynman diagram technique.* The diagram technique was invented to compute expressions of the type

$$K = \int_{\mathscr{C}(T)} P_{\alpha_1 \dots \alpha_p}(t_1, \dots t_p) \langle \delta^{\alpha_1}_{t_1}, f \rangle \dots \langle \delta^{\alpha_p}_{t_p}, f \rangle \, dw(f)$$

and is used extensively in physics. Using the previous results we obtain:

$$K = \sum_{\text{part}} P_{\alpha_1 \dots \alpha_p}(t_1, \dots t_p) G^{\alpha_{i_1}\alpha_{i_2}}(t_{i_1}, t_{i_2}) \dots G^{\alpha_{i_{p-1}}\alpha_{i_p}}(t_{i_{p-1}}, t_{i_p}).$$

When $P_{\alpha_1 \dots \alpha_p}(t_1, \dots t_p) = P_{\alpha_1}(t_1) \dots P_{\alpha_2}(t_p)$ one speaks of vertex functions $P_\alpha(t)$ and propagator lines $G^{\alpha\beta}(r, s)$; the propagator lines are hooked up to the vertices in all possible ways, and the Σ_{part} is represented by $1 \cdot 3 \cdots (p-1)$ different diagrams. The diagram technique was first invented for the case $\langle \lambda, V(f) \rangle = \int_T v_\alpha(t) f^\alpha(t)\, dt$. The integral I was computed by performing the change of order of integration justified in paragraph 3; and finally K was computed as outlined above. Note that one does not need to change the order of integration to compute I.

REFERENCES

The references listed here do not necessarily refer to the original work, nor to the most thorough presentation. They usually indicate books which can easily be consulted by readers familiar with the approach and notation used in this book.

This list includes the books and articles which are referenced in the text, together with a few additional basic references and suggestions for further study. Other references can be found at the end of some of the problems.

ABRAHAM, R. and J. MARSDEN, *Foundations of Mechanics*, The Mathematical Physics Monograph series (W.A. Benjamin, New York, 1967; Revised Edition 1979).

ADAMS, R.A., *Sobolev Spaces* (Academic Press, New York, 1975).

ARNOL'D, V.I., *Ordinary Differential Equations* (translated by R.A. Silverman, M.I.T. Press, Cambridge, 1973).

ARNOL'D, V.I. *Mathematical Methods of Classical Mechanics* (translated by K. Vogtmann and A. Weinstein, Springer-Verlag, New York, 1978).

ARNOL'D, V.I. and A. AVEZ, *Problèmes ergodiques de la Mécanique classique* (Gauthier-Villars, Paris, 1967).

BARTLE, R.G., *Elements of Integration* (John Wiley & Sons, New York, 1966).

BERGER, M., P. GAUDUCHON and E. MAZET, *Le spectre d'une variété riemanienne*, Lectures Notes in Mathematics 194 (Springer Verlag, 1971).

BISHOP, R.L. and R.J. CRITTENDEN, *Geometry of Manifolds* (Academic Press, New York, London, 1964).

BOLZA, O., *Lectures on the calculus of variations* (Chelsea Publishing Co., New York, 1973).

BOTT, R. and J. MATHER, "Topics in Topology and Differential Geometry", in *Battelle Rencontres*, eds. C.M. DeWitt and J.A. Wheeler, p. 460–515.

BOURBAKI, N., *Eléments de Mathématique*, Livre VI, Ch. IX, "Intégration sur les espaces topologiques séparés", Actualités scientifiques et Industrielles 1343 (Hermann, Paris, 1969). See in particular "Note historique" at the end of this volume.

BOURBAKI, N., *Lie Groups and Lie Algebras* (Hermann, Paris, 1975).

BROWN, A.L. and A. PAGE, *Elements of Functional Analysis* (Van Nostrand Reinhold Co., New York, 1971).

BUCK, R.C., *Advanced Calculus* (McGraw-Hill, New York, 1956).

CAMERON, R.H., "A Family of Integrals Serving to Connect the Wiener and Feynman Integrals", J. Math. & Phys. 39 (1960) 126–140.

CARTAN, E., *Leçons sur la géométrie des espaces de Riemann* (Gauthier-Villars, Paris, 1952–1955).

CARTAN, H., *Formes différentielles*, Collection Méthodes (Hermann, Paris, 1967a).

CARTAN, H., *Calcul différentiel*, Collection Méthodes (Hermann, Paris 1967b).

CARTER, B., "Underlying Mathematical Structures of Classical Gravitation Theory", in *Recent Developments in Gravitation, Cargèse 1978*, eds. M. Levy and S. Deser (Plenum Press, New York, 1979).

CHERN, S.S., *Selected Papers* (Springer-Verlag, New York, 1978).

CHERNOFF, P.R. and J.E. MARSDEN, *Properties of Infinite Dimensional Hamiltonian Systems*, Lecture Notes in Mathematics 425 (Springer-Verlag, Berlin, New York, 1974).

CHEVALLEY, Cl., *Theory of Lie Groups* (Princeton University Press, Princeton, N.J., 1946).

CHOQUET-BRUHAT, Y., *The Cauchy Problem*, ch. IV of "Gravitation, an introduction to current research", ed. L. Witten (J. Wiley, New York, 1962) p. 130–168.

CHOQUET-BRUHAT, Y., "Sur la théorie des Propagateurs", Ann. di Mat., série 4, t. 64 (1964).

CHOQUET-BRUHAT, Y., "Hyperbolic Partial Differential Equations on a Manifold", in *Battelle Rencontres*, eds. C.M. DeWitt and J.A. Wheeler (1967) p. 84–106.

CHOQUET-BRUHAT, Y., "Diagonalisation des systèmes quasi linéaires et hyperbolicité non stricte", J. Maths. Pures et Appl. 45 (1966) 371–386.

CHOQUET-BRUHAT, Y., *Géométrie différentielle et systèmes extérieurs*, Monographies universitaires de Math. No. 28 (Dunod, Paris, 1968).

CHOQUET-BRUHAT, Y., *Distributions, théories et problèmes* (Masson, Paris, 1972).

CHOQUET-BRUHAT, Y., "C^∞ solutions of hyperbolic non linear equations", J. of General Relativity and Gravitation 2, No. 4 (1972) 359–362.

CHOQUET-BRUHAT, Y. and J. LERAY, "Sur le problème de Dirichlet quasi-linéaire d'ordre 2", C.R. Acad. Sciences, Paris, t. 274 (1972) 81–85.

CHOQUET, G., *Lectures on Analysis*, Mathematics Lecture Note series (W.A. Benjamin, New York, 1969).

COURANT, R. and D. HILBERT, *Methods of Mathematical Physics*, Vol. II (Interscience, New York, 1966) ch. 2.

CRONIN, J., *Fixed Points and Topological Degree in Nonlinear Analysis*, Mathematical Surveys 11 (Am. Math. Soc., 1964).

DESCOMBES, R., *Intégration* (Hermann, Paris, 1973).

DEWITT, B.S., *Dynamical Theory of Groups and Fields* (Gordon and Breach Science Publishers, New York, 1965).

DEWITT, B.S. and R.W. BREHME, "Radiation Damping in a Gravitational Field", Ann. of Phys. 9 (1960) 220–259.

DEWITT, Cecile MORETTE and B.S. DEWITT, "Falling charges", Physics 1 (1964) 3–20.

DEWITT-MORETTE, C., "Feynman's Path Integral, Definition without Limiting Procedure", Comm. Math. Phys. 28 (1972) 47–67.

DEWITT-MORETTE, C., "Feynman Path Integrals: I. Linear and Affine Techniques, II. The Feynman-Green Function", Comm. Math. Phys. 37 (1974) 63–81.

DEWITT-MORETTE, C., "The Semiclassical Expansion", Ann. Phys. 97 (1976) 367–399, erratum 101 (1976) 682–683.

DEWITT-MORETTE C., A. MAHESHWARI and B. NELSON, "Path integration in non-relativistic quantum mechanics", Phys. Reports 50 (1979) 255–372.

DEWITT, C.M. and J.A. WHEELER, *Battelle Rencontres: 1967 Lectures in Mathematics and Physics* (W.A. Benjamin, New York, 1968).

DIEUDONNE, J., Elements d'Analyse Tome IV chap. XVIII á XX (Gauthier-Villars, Paris, 1971).

DINCULEANU, N., *Vector Measures*, International Series of Monographs in Pure and Applied Mathematics, Vol. 95 (Pergamon Press Ltd., Oxford, New York, 1967).

DONOGHUE Jr., W.F., *Distributions and Fourier transforms* (Academic Press, 1969).

DRECHSLER, W. and M.E. MAYER, *Fibre Bundle Techniques in Gauge Theories*, Lecture Notes in Physics 67 (Springer-Verlag, Berlin, 1977).

DUNFORD, N. and J.T. SCHWARTZ, *Linear Operators*, Pure and Appl. Mathematics, Series of Texts and Monographs (Interscience, New York, 1958).

EISENHART, L.P., *Riemannian Geometry* (Princeton University Press, 1926).

ESTABROOK, F.B., "Some Old and New Techniques for the Practical Use of Exterior Differential Forms", in Backlund Transformation, the Inverse Scattering Method. Solitons and their Application, ed. Robert N. Miura, Lecture Notes in Mathematics 515 (Springer-Verlag, Berlin, New York, 1976).

FEYNMAN, R.P. and A.R. HIBBS, *Quantum mechanics and path integrals* (McGraw-Hill, New York, 1965).

FLANDERS, H., *Differential Forms with Applications for the Physical Sciences* (Academic Press, New York, 1963).

FREIFELD, C., "One-Parameter Subgroups do not Fill a Neighborhood of the Identity in an Infinite-Dimensional Lie (Pseudo-) Group", in *Battelle Rencontres*, eds. C.M. DeWitt and J.A. Wheeler, p. 538–543.

FRIEDLANDER, F.G., *The wave equation on a curved space-time* (Cambridge University Press, 1975).

FRIEDMAN, A., *Partial Differential Equations of Parabolic Type* (Prentice-Hall, Englewood Cliffs, NJ, 1964).

FRIEDRICHS, K.O., "Symmetric hyperbolic linear differential equations", Comms. Pure Appl. Maths. 7 (1954) 345–392.

GARDING, L., *Solution directe du problème de Cauchy pour les équations hyperboliques*, Coll. Int. C.N.R.S. Nancy (1956) 71–90.

GEL'FAND, I.M., *Generalized Functions*, 5 volumes (Academic Press, New York, 1964–1968).

GEL'FAND, I.M. and G.E. SHILOV, *Les Distributions* (Dunod, 1964).

GEL'FAND, I.M. and S.V. FOMIN, *Calculus of variations* (Prentice-Hall, Englewood Cliffs, NJ, 1963).

GEROCH, R.P., "The domain of dependence", J. Math. Phys. 11 (1970) 437–449.

GILMORE, R., *Lie Groups, Lie Algebras, and Some of Their Applications* (John Wiley and Sons, New York, 1974).

GODBILLON, C., *Géométrie différentielle et Mécanique Analytique*, Collection Méthodes (Hermann, Paris, 1969).

GOFMANN, C. and G. PERDRICK, *First course in functional analysis*, Prentice-Hall series in modern analysis (Prentice-Hall, Englewood Cliffs, NJ, 1965).

GOLDSTEIN, H., *Classical Mechanics* (2nd ed., Addison-Wesley Pub. Co., Reading, 1980).

GROSS, L., *Analysis in Function Space*, eds. W. Martin and I. Segal (MIT Press, Cambridge, Mass., 1964).

GUILLEMIN, V. and A. POLLACK, *Differential Topology* (Prentice-Hall, Englewood Cliffs, NJ, 1974).

GUTZWILLER, M.C., "Phase-Integral Approximation in Momentum Space and the Bound States of an Atom", J. Math. Phys. 8 (1967) 1979–2000.

HALMOS, P.R., *Measure Theory*, The University series in higher mathematics (D.Van Nostrand, Princeton, NJ, 1950).

HALMOS, P.R., *Naive set theory* (D. Van Nostrand, Princeton, NJ, 1967).

HALMOS, P.R., *Finite-dimensional Vector Spaces*, Appendix on Hilbert Spaces (D. Van Nostrand, Princeton, NJ, 1958).

HARRISON, B.K. and F.B. ESTABROOK, "Geometric Approach to Invariance Groups and Solution of Partial Differential Systems", J. Math. Phys. 12 (1971) 653.

HAWKING, S.W. and G.F.R. ELLIS, *The large scale structure of space time*, Cambridge monographs on mathematical physics (The University Press, Cambridge, England, 1973).

HELGASON, S., *Differential Geometry and Symmetric Spaces* (Academic Press, New York, London, 1962).

HICKS, N.J., *Notes on Differential Geometry*, Van Nostrand mathematical studies 3 (D. Van Nostrand, Princeton, NJ, 1965).

HIRSCH, M.W. and S. SMALE, *Differential Equations, Dynamical Systems and Linear Algebra* (Academic Press, New York, 1974).

HORMANDER, L., *The Analysis of Linear Partial Differential Operators.* Volume I, *Distribution Theory and Fourier Analysis;* Volume II, *Differential Operators with Constant Coefficient;* Volume III, *Pseudo-differential Operators;* Volume IV, *Fourier Integral Operators* (Springer Verlag, New York, Tokyo, 1983).

HORMANDER, L., "Fourier Integral Operators", Acta Matem. 127 (1971) 79–183.

ITZYKSON, C. and J.B. ZUBER, *Quantum Field Theory* (McGraw-Hill, New York, 1980).

KELLEY, J.L., *General Topology* (D. Van Nostrand, Princeton, NJ, 1963).

KELLEY, J.L., *Linear Topological Spaces* (Springer-Verlag, Berlin, New York, 1976).

KOBAYASHI, S. and K. NOMIZU, *Foundations of differential geometry*, vols. 1 and 2 (Interscience, New York, 1969).

KOLMOGOROV, A.N., *Principes fondamentaux de la théorie des probabilités* (Moscou, 1936); Foundations of the theory of probability (Chelsea Publishing Co., New York, 1956).

KOLMOGOROV, A.N. and S.V. FOMIN, *Elements of the Theory of Functions and Functional Analysis.* Volume I, *Metric and Normed Spaces*; Volume II, *Measure, the Lebesgue Integral, Hilbert Spaces*; Translated from the 1954 and 1960 Russian editions respectively (Graylock Press, Rochester, NY, 1957).

KREE, P., "Introduction aux théories des distributions en dimension finie", Bull. Soc. Math., France, Mém. 46 (1976) 143–162.

LADYZHENSKAYA, O.A. and N.N. URAL'TSEVA, *Linear and quasilinear equations of elliptic type* (Academic Press, New York, 1968).

LANG, S., *Introduction to Differentiable Manifolds* (John Wiley and Sons, Inc., New York, 1962).

LAX, P.D., "On Cauchy's problem for hyperbolic equations and the differentiability of solutions of elliptic equations", Comm. Pure App. Maths. 8 (1955) 615–633.

LERAY, J., *Hyperbolic Differential Equations* (The Institute for Advanced Study, Princeton, NJ, 1953).

LERAY, J. and J. SCHAUDER, "Topologie et Equations Fonctionnelles", Annales Ecole Normale Supérieure (1934) 45–78.

LICHNEROWICZ, A., Géométrie des groupes de transformations (Dunod, Paris, 1958).

LICHNEROWICZ, A., *Propagateurs, Commutateurs et Anticommutateurs en Relativité Générale,* Publication I.H.E.S. No. 10 (1961).

LICHNEROWICZ, A., "Commutativité de l'algèbre des opérateurs différentiels invariants sur un espace symétrique", in *Battelle Rencontres 1967,* Lecture Notes in Math. and physics, eds. C.M. DeWitt and J.A. Wheeler (Benjamin, 1968) p. 73.

LICHNEROWICZ, A., "Cohomologie 1-differentiable des algèbres de Lie attachées à une variété symplectique ou de contact", J. Maths. Pures et Appl. 53 (1974) 459.

LICHNEROWICZ, A., "Variétés symplectiques, variétés canoniques et systèmes dynamiques", E.T. Davies memorial volume (Rund. Forbes Ed., 1975).

LICHNEROWICZ, A., *Global theory of connections and holonomy groups* (Noordhoff International Publishing, 1976). French edition published in 1955.

LIONS, J.L., *Cours d'Analyse Numérique,* polycopié, Ecole Polytechnique.

LIONS, J.L., *Equations aux dérivées partielles et calcul des variations. Application à la théorie du contrôle,* Cours à l'I.H.P. (Faculté des Sciences à Paris, 1967).

LIONS, J.L., *Quelques Méthodes de Résolution des Problèmes aux Limites Nonlinéaires* (Dunod, Paris, 1969).

LIONS, J.L. and E. MAGENES, *Problèmes aux limites non holomorphes et applications*, Volumes 1 et 2 (Dunod, Paris, 1970).

MACKEY, G.W., *Mathematical Foundations of Quantum Mechanics*, The Mathematical Physics Monograph series (W.A. Benjamin, New York, 1963).

MacLANE, S., *Geometrical Mechanics*, 2 volumes, Mathematical Lecture Notes (University of Chicago, 1968).

MARKUS, L., "Line Element Fields and Lorentz Structures on Differentiable Manifolds", Annals of Math. serie 2, 62 (1955) p. 411.

MARLE, C., *Mesures et Probabilités* (Hermann, Paris, 1974).

MARSDEN, J., "Darboux's Theorem Fails for Weak Symplectic Forms", Proc. Amer. Math. Soc. 32 (1972) 590–592.

MARSDEN, J., *Applications of global analysis on Mathematical Physics* (Publish or Perish, Inc., Boston, 1974).

MIKHLIN, S.G., *Mathematical Physics, An Advanced Course*, North-Holland Series in Applied Mathematics and Mechanics (North-Holland, Amsterdam, The Netherlands, 1970).

MILNOR, J.W., *Topology from the differentiable viewpoint* (Univ. Press of Virginia, Charlottesville, 1965).

MILNOR, J.W., *Morse Theory*, Annals of Mathematics, Studies No. 51 (Princeton University Press, 1969).

MIRANDA, C., *Partial Differential Equations of Elliptic Type* (Springer-Verlag, Heidelberg, New York, 1970).

MISNER, Ch.W., Kip S. THORNE and J.A. WHEELER, *Gravitation* (Freeman and Company, San Francisco, 1973).

MORREY, C.B., *Multiple integrals in the calculus of variations* (Springer-Verlag, Berlin, New York, 1966).

MORSE, M., "Calculus of variations in the large", Ann. Math. Soc. Coll. (4th ed., 1965).

NAMIOKA, I. and J.L. KELLEY, *Linear Topological Spaces*, The University series in higher mathematics (D. Van Nostrand, Princeton, NJ, 1963).

NELSON, B. and B. SHEEKS, "Fredholm determinants associated with Wiener integrals", J. Math. Phys. 1981.

NELSON, E., *Dynamical theories in Brownian motion*, Mathematics Notes (Princeton, University Press, 1967).

ODEN, J.T. and J.N. REDDY, *An Introduction to the Mathematical Theory of Finite Elements* (John Wiley and Sons, New York, 1976).

O'RAIFEARTAIGH, L., "Hidden Gauge Symmetry", Rep. Prog. Phys. 42 (1979) 159–223.

PALAIS, R.S., *Morse Theory on Hilbert Manifolds*, Topology 2, p. 299–340.

PALAIS, R.S., *Seminar on the Atiyah-Singer index theorems*, Annals of Math. Studies (Princeton University Press, 1965).

PALAIS, R.S., *Foundations of global non linear analysis* (Benjamin, New York, 1968).

PALAIS, R.S. and S. SMALE, "A Generalized Morse Theory", Bull. AMS 70 (1964) 165–172.

PATTERSON, E.M., *Topology* (Oliver and Boyd, Edinburgh, Scotland, also Interscience, New York, 1959).

PETROVSKY, I.G., *Lectures on Partial Differential Equations* (Interscience, New York, 1954).

PHAM, F., *Introduction à l'étude topologique des singularités de Landau*, Memorial des Sciences Mathematiques, fascicule CLXIV (Gauthier-Villars, Paris, 1967).

PHAM MAU QUAN, *Introduction à la géométrie des variétés différentiables*, Monographies Universitaires de Mathématiques 29 (Dunod, Paris, 1969).

PONTRYAGIN, L.S., *Topological Groups* (second ed., Gordon and Breach, New York, 1966).

PROTTER, M.H. and H.F. WEINBERGER, *Maximum Principle in differential equations* (Prentice-Hall, 1967).

REED, M. and B. SIMON, *Methods of modern mathematical physics*, I Functional Analysis (1972), II Fourier Analysis, Self Adjointness (1975), III Scattering Theory (1979), IV Analysis of Operators (1978) (Acad. Press, New York).

de RHAM, G., *Variétés différentiables*, Actualités Scientifiques et Industrielles 1222 (Hermann, Paris, 1960).

ROSEAU, M., *Solutions périodiques ou presque périodiques des systèmes différentiels de la mécanique non linéaire* (International Centre for Mechanics Sciences, Udine, France, 1970).

ROYDEN, H.L., *Real Analysis* (Macmillan, New York, 1963).

RUDIN, W., *Real and Complex Analysis* (McGraw-Hill, New York, 1966).

RUDIN, W., *Functional Analysis* (McGraw-Hill, New York, 1973).

RUDIN, W., *Fourier Analysis on Groups* (Second Printing, Interscience, Publishers, New York, 1967).

SAMELSON, H., "A Theorem on Differentiable Manifolds", Portugaliae Math. 10 (1951) 129.

SCHAEFFER, H.H., *Topological Vector Spaces*, Macmillan series in Advanced Mathematics and Theoretical Physics (Macmillan, New York, 1966).

SCHUTZ, B., *Geometrical methods of mathematical physics* (Cambridge University Press, 1980).

SCHWARTZ, L., *Lectures on Complex Analytic Manifolds* (Tata Institute of Fundamental Research, Bombay, India, 1955).

SCHWARTZ, L., *Méthodes mathématiques de la physique* (Hermann, Paris, 1965).

SCHWARTZ, L., *Théorie des Distributions* (2e ed., Hermann, Paris, 1966).

SCHWARTZ, J.T., *Nonlinear functional analysis* (Gordon and Breach Science Publishers, New York, 1969).

SEGAL, I.E., "Differential Operators in the Manifold of Solutions of a Non-Linear Differential Equation", J. Maths. Pure and Appl. 49 (1965) 71–132.

SEGAL, I.E., "Distributions in Hilbert Space and Canonical Systems of Operators", Trans. Amer. Math. Soc. 88 (1958) 12–42.

SEGAL, I.E., *Mathematical cosmology and extragalactic astronomy* (Academic Press, 1976).

SIMMONS, G.F., *Introduction to Topology and Modern Analysis* (McGraw-Hill, New York, 1963).

SIMMS, D.J. and N.M.J. WOODHOUSE, *Lectures on Geometric Quantization*, Lecture Notes in Physics series (Springer-Verlag, Berlin, New York, 1976).

SINGER, I.M. and J.A. THORPE, *Lecture notes on Elementary Topology and Geometry* (Scott, Foresman and Company, Dallas, TX, 1967).

SOBOLEV, S.L., *Application of Functional Analysis in Mathematical Physics*, Translations of Mathematical Monographs, Vol. 7 (American Mathematical Society, Providence, RI, 1963).

SOBOLEV, S.L., *Partial Differential Equations of Mathematical Physics*, Translated from the third Russian edition by E.R. Dawson (Pergamon Press, Oxford, England, New York, 1964).

SOURIAU, J.M., *Structure des Systèmes Dynamiques* (Dunod, Paris, 1970).

SOURIAU, J.M., "Thermodynamique des Fluides" (Preprint, 1976).

SPIVAK, M., *Calculus on Manifolds*, Mathematical Monograph Series (W.A. Benjamin, New York, 1965).

SPIVAK, M., *A Comprehensive Introduction to Differential Geometry*, 5 volumes (2nd ed., Publish or Perish, Inc., Boston, 1979).

STEENROD, N., *The Topology of Fibre Bundles* (Princeton Univ. Press, 1951).

STERNBERG, S., *Lectures on differential geometry* (Prentice-Hall, Englewood Cliffs, NJ, 1964).

TAYLOR, A.E., *General Theory of Functions and Integration* (Blaisdell, 1965).

THIRRING, W., *Classical Theory of Fields* (Springer-Verlag, Berlin, 1979).

TRAUTMAN, A., "Fiber Bundles Associated with Space Time", Reports on Math. Phys. 1 (1970) 1–34.

TREVES, F., *Topological vector spaces, distribution and kernels* (Academic Press, 1969).

UHLENBECK, K., "A Morse Theory for Geodesics on a Lorentz Manifold", Topology 14, pp. 69–90 (Pergamon Press, 1975).

Van HOVE, L., "The Occurrance of Singularities on the Elastic Frequency Distribution of a Crystal", Phys. Rev. 89 (1953) 1189–1193.

VICK, J.W., *Homology Theory. An Introduction to Algebraic Topology* (Academic Press, 1973).

WARNER, F.W., *Foundations of Differentiable Manifolds and Lie Groups* (Scott-Foresman, Dallas, TX, 1971).

WEINSTEIN, M.A., Advances in Math. 6 (1971) 329–346.

WOODHOUSE, N.M.J., "An Application of Morse Theory to Space-Time Geometry", Comm. Math. Phys. 46 (1976) 135–152.

YANG, C.N., "Magnetic Monopoles, Fiber Bundles and Gauge Fields", Ann. N.Y. Acad. Sciences 294 (1977) 86–97.

YANG, C.N., "Einstein and his impact on the physics of the second half of the twentieth century", Ref. TH 2710-CERN (1979).

YOSIDA, K., *Functional Analysis* (second ed., Springer-Verlag, Berlin, New York, 1965).

SYMBOLS

This list is not exhaustive. Besides the mathematical symbols listed below, we have used brackets to avoid writing the same sentence twice when only a few words had to be interchanged. Parentheses are in general used only to give synonymous expressions.

1. General

i.e.:	id est
e.g.:	exempli gratia
$\forall x$:	for every x
$\exists$:	there exists
$\in$:	included in
$\ni$:	which contains
$\subset$, $\subsetneq$:	subset, proper subset
iff:	if and only if
$\Rightarrow$:	implies
$\Leftrightarrow$:	is logically equivalent to
$\{\}$, $\{0\}$:	set, set consisting of the zero element
$a_1 \dots \hat{a}_j \dots a_n$:	exclude a_j
a.e.:	almost everywhere
$[a, b[$ or $[a, b)$:	semi closed, semi open interval
$\complement_X Y$, $X \setminus Y$:	complement of Y relative to X
$X \times Y$:	direct product (usually called cartesian product if X and Y are sets without any other structure)
$\otimes$:	tensor product
$\oplus$:	geometrical sum
$\ominus$:	geometrical difference
$\mathbb{R}$:	real line, field of reals
$\mathbb{R}^n$:	$\mathbb{R} \times \mathbb{R} \cdots \times \mathbb{R}$, n times
$\mathbb{C}$:	complex plane, field of complex; $\bar{\alpha}$ complex conjugate of α
$\mathbb{K}$:	$\mathbb{R}$ or $\mathbb{C}$
$\mathbb{R}^+$:	non negative real numbers
$\mathbb{Z}$:	sėt of integers
$\rightsquigarrow$:	converges to (in reference to sequences and nets)
$\hookrightarrow$:	algebraic and topological inclusion

$\mathbb{N}$:	natural numbers: positive integers
$\mathbb{T}$:	$\mathbb{R}/\mathbb{Z}$, 1-torus
X^*:	algebraic dual of X
X':	topological dual of x
$\langle x', x\rangle$:	$x'(x)$ for $x \in X$, $x' \in X'$, duality
$f(x^i)$:	often $f(x^1, \ldots, x^n)$
$f \circ g$:	mapping composition
∂_i:	in general $\partial/\partial x^i$; however on p. 138 ∂_i is reserved for a Pfaff derivative
$f', \mathrm{D}f$:	derivative, differential of f
$f^{(p)}, \mathrm{D}^p f$:	p-th derivative (see multi index)
$[a^i_{\ j}]$:	matrix
$\det [a^i_{\ j}]$:	determinant of the matrix $[a^i_{\ j}]$
g:	determinant of the metric tensor $[g_{ij}]$
$\mathcal{J}, \mathrm{D}(x^i)/\mathrm{D}(y^j)$:	determinant of the matrix $[\partial x^i/\partial y^j]$
$(a\|b)$:	scalar product
Supp:	support
sup, inf:	supremum, infimum
$\\| \ \\|_{L^p}, \\| \ \\|_p$:	L^p-norm
$\epsilon^{k_1 \ldots k_p}_{i_1 \ldots i_p}$:	Kronecker tensor p. 142

A multi index j is a set $\{j_1, \ldots j_n\}$. One defines its order $|j| = j_1 + j_2 + \cdots j_n$, its rank $J = \sup\{j_1, \ldots j_n\}$, its factorial $j! = j_1! j_2! \ldots j_n!$. It is used for instance

in the binomial coefficient $C^j_k = C^{j_1}_{k_1} C^{j_2}_{k_2} \ldots C^{j_n}_{k_n}$ with $C^{j_i}_{k_i} = k_i!/j_i!(k_i - j_i)!$;

as an exponent $x^j = x_1^{j_1} x_2^{j_2} \ldots x_n^{j_n}$;

for partial derivatives $\mathrm{D}^j = \partial^{|j|}/\partial x_1^{j_1} \partial x_2^{j_2} \ldots \partial x_n^{j_n}$.

Example: The Taylor expansion of a function on $\mathbb{R}^n$ can be written

$$f(x) = \sum_{|j|=0}^{\infty} \mathrm{D}^j f(0) x^j / j!$$

where the sum over $|j|$ means the sum over all values of $j_1 \ldots j_n$ such that $|j| = 0, 1, \ldots \infty$.

2. Symbols introduced in Chapter III

(U, φ):	chart	p. 111
$\bar{f}$:	$f \circ \varphi^{-1}$	113

3. Symbols introduced in Chapter IV

4. Symbols introduced in Chapter V

5. Symbols introduced in Chapter VI

including some function spaces introduced in other chapters. X is a topological space, U an open set in $\mathbb{R}^n$, K a compact set in U.

INDEX

Printed and bound by CPI Group (UK) Ltd, Croydon, CR0 4YY

Printed and bound by CPI Group (UK) Ltd, Croydon, CR0 4YY
03/10/2024
01040428-0014